LAKE BIWA

MONOGRAPHIAE BIOLOGICAE

VOLUME 54

Series Editor

H.J. Dumont

1984 **DR W. JUNK PUBLISHERS**
a member of the KLUWER ACADEMIC PUBLISHERS GROUP
DORDRECHT / BOSTON / LANCASTER

LAKE BIWA

Edited by

SHOJI HORIE

1984 **DR W. JUNK PUBLISHERS**
a member of the KLUWER ACADEMIC PUBLISHERS GROUP
DORDRECHT / BOSTON / LANCASTER

Distributors

for the United States and Canada: Kluwer Academic Publishers, 190 Old Derby Street, Hingham, MA 02043, USA
for the UK and Ireland: Kluwer Academic Publishers, MTP Press Limited, Falcon House, Queen Square, Lancaster LA1 1RN, England
for all other countries: Kluwer Academic Publishers Group, Distribution Center, P.O. Box 322, 3300 AH Dordrecht, The Netherlands

Library of Congress Cataloging in Publication Data

Main entry under title:

Lake Biwa.

(Monographiae biologicae ; v. 54)
Includes bibliographical references and index.
1. Limnology--Japan--Biwa, Lake. 2. Paleolimnology--Japan--Biwa, Lake. 3. Biwa, Lake (Japan) I. Horie, Shoji, 1926- . II. Series.
QP1.P37 vol. 54 [QH188] 574s [551.48'2'095218] 83-9888
ISBN 90-6193-095-2

ISBN 90-6193-095-2 (this volume)

Copyright

PRINTED IN THE NETHERLANDS

Contents

List of Contributors

Norio Fuji
Institute of Earth Science,
Kanazawa University

Keiichiro Fuwa
Department of Chemistry,
Faculty of Science,
Tokyo University

Nobuhiko Handa
Water Research Institute,
Nagoya University

Eiji Harada
Seto Marine Biological Laboratory,
Faculty of Science,
Kyoto University

Fujio Hirao
Shiga Prefectural Educational Center,
Yasu-chō, Shiga-ken

Shoji Horie
Institute of Paleolimnology and
Paleoenvironment on Lake Biwa,
Faculty of Science,
Kyoto University

Isao Ikusima
Department of Biology,
Faculty of Science,
Chiba University

Norihisa Imasato
Geophysical Institute,
Faculty of Science,
Kyoto University

Ryoshi Ishiwatari
Department of Chemistry,
Faculty of Science,
Tokyo Metropolitan University

Tatsuya Iwashima
Laboratory for Climatic,
Change Research,
Faculty of Science,
Kyoto University

Hajime Kadota
Department of Fishery,
Faculty of Agriculture,
Kyoto University

Sadami Kadota
College of Agriculture and
Veterinary Medicine,
Nihon University

Tadao Kamei
Department of Geology and
Mineralogy,
Faculty of Science,
Kyoto University

Seiichi Kanari
Department of Geophysics,
Faculty of Science,
Hokkaido University

Naoto Kawai (late professor)
High Pressure Research Laboratory,
Osaka University

Hiroya Kawanabe
Department of Zoology,
Faculty of Science,
Kyoto University

Munetsugu Kawashima
Department of Chemistry,
Shiga University

Kazuo Kotoda
Institute of Geoscience,
The University of Tsukuba

Mutsuo Koyama
Research Reactor Institute,
Kyoto University

Takayuki Mizuyama
Institute of Geography,
Kyoto University of Education

Shinobu Mori
Department of Earth Sciences,
Faculty of Science,
Nagoya University

Syuiti Mori
Shiga University

Isamu Murai
Earthquake Research Institute,
University of Tokyo

Nobuyuki Nakai
Department of Earth Sciences,
Faculty of Science,
Nagoya University

Masami Nakanishi
Otsu Hydrobiological Station,
Faculty of Science,
Kyoto University

Tetsuya Narita
Otsu Hydrobiological Station,
Faculty of Science,
Kyoto University

Yo Naruse
Osaka University of Economics

Susumu Nishimura
Department of Geology and
Mineralogy,
Faculty of Science,
Kyoto University

Kazuko Ogura
Department of Chemistry,
Faculty of Science,
Tokyo Metropolitan University

Iwao Okamoto
Institute of Earth Science,
Shiga University

Kunihiro Okamoto
Otsu Hydrobiological Station,
Kyoto University

Setsuo Okuda
Disaster Prevention Research
Institute,
Kyoto University

Yoko Ota
Department of Geography,
Yokohama National University

Yatsuka Saijo
Water Research Institute,
Nagoya University

Kazuo Shigesawa
Institute of Earth Science,
Department of Liberal Arts,
Kyoto University

Mitsuru Skamoto
Water Research Institute,
Nagoya University

Takejiro Takamatsu
National Institute for Environmental
Studies

Nobuhiko Tanaka
National Research Institute
of Aquaculture,
Mie-ken

Toshinobu Tokui
Tokui Institute for Freshwater
Fishery Biology,
Ise-shi, Mie-ken

Yoshimasa Toyoda
Department of Oceanography,
Tōkai University

Masuzo Ueno
Kōnan Women's University

Atsuyuki Yamamoto
Osaka Electro-Communication
University

Ryōzaburo Yamamoto
Laboratory for Climatic
Change Research,
Faculty of Science,
Kyoto University

Takuo Yokoyama
Laboratory of Earth Science,
Faculty of Technology,
Doshisha University

Introduction

Lake Biwa which is located in the center of the Honshu Island exhibits the following morphometry:

Altitude	85 m
Length	68.0 km
Maximum breadth	22.6 km
Length of shoreline	188.0 km
Area	674.4 km^2
Shore development	2.04
Maximum depth	104.0 m
Mean depth	41.2 m
Volume	27.8 km^3

It is apparent from these parameters that a cryptodepression denotes 19 m. Although this lake is remarkable as the biggest lake in Japan, its oldness is of more interest to our limnological study. The fission-track method for absolute dating applied to volcanic ash layers obtained at an outcrop of the Kobiwako (Paleo-Biwa) Group indicates that the lake is five million years old. This is therefore one of the most ancient lakes on the globe.

Limnological study of Lake Biwa poses many kinds of questions. Apart from those which are treated in common with other large lakes in the world, questions most attractive not only to limnologists and hydrobiologists but also to scientists in fields interdisciplinary with limnology are ancient lake problems such as geological history of the surrounding areas, crustal movements, volcanic eruptions, drainage pattern changes, biological characteristics such as speciation, migration of species during each age of the geological time, and also combined problems, for instance, migration of the lake basin and its effect on lake organisms. Besides those, there are many other paleolimnological features that have already been dealt with in the series of publications, *Paleolimnology of Lake Biwa and the Japanese Pleistocene,* Vols. 1 to 9 (1972 – 1981).

The objective of this book is to elucidate all limnological features in the

Horie, S. Lake Biwa

ISBN 90 6193 095 2. Printed in The Netherlands

ancient lake of Biwa. Particularly, some of the writers have devoted their efforts to the sections of 'Geoscientific features on an ancient lake' and 'Paleolimnology', which are significant areas of research in the study of an ancient lake; nevertheless, it is still in the young stage as compared with other subjects which have a long history of study.

This book is a compendium of papers by many scholars who are active and leading Japanese limnologists. They appear in alphabetical order in the List of contributors. Shoji Horie is responsible for the editorial work.

We express our thanks to Drs. D. G. Frey (Indiana University; Animal Microfossil), Fujino, Y. (Toshi Chosa-kai; Public Works), Furukawa, M. (Shiga Fishery Experiment Station; Fishes), Kawakatsu, M. (Fuji Women's College; Fishes), Nakamura, M. (National Science Museum; Fishes), Nakazawa, S. (Yamagata University; Fishes).

Finally we all sincerely hope that this book will be of service to all scholars of the world, limnologists in particular, as a milestone in the long road to full understanding of limnological features of ancient lakes which, though few, hold many unsolved academic problems on our globe.

Shoji Horie
Editor
Takashima, Japan

General features of the ancient Lake Biwa as the basis for reconsideration of limnology

Shoji Horie

Limnology, having started in European countries such as Switzerland and Sweden as well as in the northern region of that continent, was of course initially closely connected with the existence of naturally impounded waters.

Generally speaking, water-occupied areas on the earth have various kinds of origin. The best classification is given in the book of Hutchinson (1957). Simply stated, those lakes treated by early limnologists in Europe have glacial-concerned, volcanic lakes. Lakes and ponds in the Fenno-Scandinavia are the typical examples of the Pleistocene glaciation. Glaciers covered that area at least several times with a considerable thickness of ice counted in the km unit. Needless to say, placed under such an enormity of thick ice, the earth's surface has its relief flattened, not only by the weight and pressure of the huge ice which caused the isostatic downwarping and afterwards upwarping but by slow movements of the flow action of the ice.

Therefore, most of the earth's surface is stripped and the gentle-slope relief was left only after the continental ice sheet retreated. Lakes located in that area after the glacial recession have also many kinds of origin. The simplest of these origins is the impoundment of water at lowlands between hillsides; however, other kinds of origin such as esker, (osar), kame, moraine, and glacial dammed ponds are also usually found. In other words, glacial lakes are complex in origin. In fact, when we look over these lakes connected with the glacial action, almost all of them are not very deep in morphometry of the lake basin, indicating that these lakes are not closely constructed for the morphometric oligotrophy in lake typology. Actually, these lakes are comparatively shallow, and they sometimes disappear when the outlet discharge increases or dislocation takes action through tilting of the area by a rebound of the earth's crust. Burial is also combined with that tilting.

These lakes are simple in morphometry. For instance, the shoreline development is small when we compare them with those in volcanic regions interfered by the volcanic lava flow. Similarly simple are other elements such as length, maximum breadth, length of shoreline, maximum depth, and mean depth. Accordingly, these lakes have a condition favorable for early development into a eutrophic lake type.

Horie, S. Lake Biwa

ISBN 90 6193 095 2. Printed in The Netherlands

During the last half century, most limnological textbooks have stated that the transition of lake trophy is mostly unidirectional, that is a lake that has started from the oligotrophic condition moves simply towards the eutrophic condition.

It is certain that during several decades covering one generation of our life, we occasionally meet the birth of a lake. For instance, volcanic eruptions generated lakes. One that is well known is found at the northern foot of Volcano Bandai where a big eruption took place in 1888 (Horie 1964). As a matter of fact, many lakes of various trophic types were formed (Yoshimura 1932, 1936) as a result of landslides with the collapse of volcanic body caused by sudden explosions (Horie 1964). Naturally, these lakes are closely related to an acidotrophic tendency, but at any rate they were in the oligotrophic stage in general. Neither vegetation nor organisms prevailed. At present, vegetations and freshwater organisms are alive since the conditions have become favorable to these organisms. Naturally, limnological conditions of lake trophy have been changed, especially in the recent state of human activity, i.e. pollution. A similar case is also found in a crater lake in which a lake or pond is found after the volcanic activity has ceased. Although these cases are common in the Japanese Islands, they are usually formed in an inland area. However, even in the middle of an ocean, volcanic activity has formed new cones on new islands (Schwabe 1974). In such a case, a lake may advance slowly from an entirely oligotrophic state to a eutrophic state.

The point with which the writer is concerned is that in the above-mentioned example, the trophic advance *may* take a unidirectional step if the natural condition remains almost constant in the surrounding area. But it is a geologically well-known fact that the natural condition tends to oscillate through the geological time. Most remarkable events that took place are climatic changes which may accompany ocean-level fluctuations. Besides, tectonic displacements such as folding, warping, faulting, and volcanic activity have been alternating. Those changes of climate cause not only temperature and precipitation fluctuations, but variations of vegetation, of the process of erosion, of the rate of sedimentation and so on. Consequently, all kinds of conditions of limnology of the lake are controlled directly or indirectly. More concretely, a rise or fall of water temperature, under the influence of air temperature, determines the pattern of the thermal condition; oscillations of solar radiation may intervene and control the lake productivity. Below the surface water level, the decomposition process is also controlled by it. Actually, it is not easy to judge whether or not high temperature is beneficial to lake community's production and decomposition. This problem also is directly concerned with the rate of sedimentation. Precipitation, the main source of supply of water in the drainage area and, of course, the lake level itself, is also subject to the changing pattern of the climate.

Daly (1934) proposed the change of the ocean level which is universally evident along the sea shore of the world, but at present, this change is problematical (Sugimura 1977). Even so, the instability of the ocean level cannot be doubted as there are many kinds of evidence on land and below the sea level as noted from many points of view. In any case, water, which may be supplied from vapor in the atmosphere evaporated from the world ocean, forms moist, rain, snow and finally ice in high latitude districts and in high places of mountains. Accordingly, the pluvial climate prevailed in some parts of the globe which shifted according to the changing pattern of the global climate. These two features, namely, the glacial advance in high latitude regions and high mountain cirques, and lake-level oscillations not only in that glacial area but even in middle-latitude low plains, imply a shift of the climatic zone on a global scale; some parts suffer from the pluvial climate and other parts suffer from the drought as the lake level declines in altitude and finally disappears. Such a case is well known in the western United States (Hubs & Miller 1948). However, the problem is more complex. A rise or fall of the lake level is directly concerned with lake trophy. Theinemann (1921) took note of the lake trophy transition and Naumann (1921, 1927, 1932) had already pointed out the classification of lake trophy or lake types. Lindeman (1942) also discussed the nature of lake metabolism and succession from the trophic dynamic aspect. In those discussions conducted by many scholars, it is true that lake trophy is considered to be rapidly changeable toward the eutrophic level if sufficient nutrients of allochthonous or autochthonous origin are supplied into the metabolic cycle of the lake together with the discharge of lake water accompanying the burial of the lake basin by many kinds of seston. On the other hand, an increase or a decrease of precipitation with a similar process of evaporation must cause a level change of the lake basin leading to a change in the ratio between trophogenic and tropholytic layers. It should be the controlling factor of the lake trophy. In the case of the above-mentioned 'eutrophication', this mechanism may accelerate 'eutrophication' when the lake level sinks and vice versa. Therefore, the writer proposes here in the global sense the term, *climate-controlled-multitrophication* (Oligotrophication, Mesotrophication or Eutrophication), which is reversible and alternatable. This designation is particularly important in a closed lake basin in which the level of lake water is virtually controlled by the climatic condition. A good example is found in Lake Yogo (Horie 1967a, b, 1968a). It is a naturally closed lake and the best indicator of its former lake level is the buried forest *in situ* (Horie et al. 1975). The ages of three strata of buried forest, intervened by lacustrine deposits, are 3,100, 6,500 and 8,000 years BP., showing beautiful rhythmic alternations of the lacustrine stage and the swamp or terrestrial stage on the deposits composition. Horie and others (1975) argued that the recurrence surface nature in Japan is comparable with that in western Europe. Von Post's discussion (1946) of Blytt-Sernander's scheme is a standard one in the world and the uniqueness of the above-stated oriental example is that this lakeshore evidence is combined with a core sample proof obtained at the center of the same lake. The writer and his collaborators have already tried Livingstone borer operations two or three times and the results have been published elsewhere (Horie 1966, 1968a, b, 1969a, b, 1971). Another core sample was obtained by a number of workers in the autumn of 1975 when the International Symposium on 'Global-scale Paleolimnology and Paleoclimate' was held in Kyōto. Some of the analytical work by an international team has also been reported (Horie 1976, 1977) and all the other works in progress. Especially, absolute dating of the core is important for further study of the mechanism of the lake-level change discussed above. Oscillations in the amount of nitrogen in the core, grain size and pollen grains density variation is composer for integration on paleolimnological succession of lake type. These data should be examined with respect to the absolute ages of various horizons in order to determine their correlations with the glacial and interglacial evidence on Japanese high mountains. At present, the writer merely places an emphasis on the importance of *climate-controlled-multitrophication*.

What cannot be neglected in Paleolimnology in conjunction with the precipitation and evaporation change in the lake basin is the temperature variation. So far as we are concerned with the temperature variation in lake water itself, the most conspicuous mechanism is its effect on the circulation pattern of lake water itself. Forel (1892) proposed a classification of the circulation pattern into three types, namely, the Tropical type, the Temperate type, and the Polar type. A more detailed discussion was given by Hutchinson (1957); the Japanese example in question here is a dimictic lake type and warm monomictic lake type oscillation in the ancient lake of Biwa. Tanaka citing Forel's classification, tried to draw a demarcation between the Tropical and Temperate types on the line connecting Lake Yogo and Lake Motosu. In the sense of Hutchinson, this boundary is actually a warm monomictic — dimictic lake type. It is, however, notable that such a borderline is determined under the present climate, as shown in Horie's previous paper. (Horie 1975). The borderline is controlled by the present pattern of isotherm on the globe. The isothermic distribution at present and the abovementioned borderline coincides with the isotherm of an annual temperature of 14°C. However, the paleoclimatic isotherm during the glacial age, as deduced by other workers, is of course, different. Naturally, during the glacial age, the borderline must have been shifted southward as compared with the present one; it would be several degrees south in latitude and both Lakes Yogo and Motosu were included in the zone of dimictic lake type. Lake Biwa also should take the dimictic circulation pattern, though in the present climate it shows the warm monomictic circulation pattern as Okamoto, Kanari and others explain in a later chapter. Needless to say, the amount of dissolved oxygen is important for the metabolic process of a lake; a change of the circulation pattern accompanied by the glacial and interglacial climate is not negligible to the lake metabolism as a whole.

In contrast, the present temperature as Interglacial is almost high though Hypsithermal Interval had shown 2–3°C higher value as shown in both the Caribbean Sea core and the Lake Biwa core (Emiliani 1955; Deevey & Flint 1957; Fuji & Horie 1977; Sasajima 1977). In any way, the interglacial circulation pattern is analogous with the present pattern.

The above-stated physical limnological transition affecting all limnological features may be called the 'Shift of the lacustrine climatic zonation'. *The writer now stresses the importance of the geological change of the natural condition for all features of limnology.* Without considering it, an ancient lake history gives rise to a misleading interpretation.

Another fact which must be noted by us on an ancient lake is the nature of earth's crust structure in that region. If water is filled simply in a lowland area, it may not last for an immense length of time, that is, in the geological sense. Although cases of an ancient origin are scanty when we look over the world's distribution of lakes, the few extant examples should be given some comments. Why could they continue their existence? For instance, Lake Baikal has its life over at least 10,000,000 years. According to the fission-track dating of volcanic ash layer, Lake Biwa had its appearance 5,000,000 years BP. The Caspian and Aral Seas began their lake type after they were separated from the Tethys Sea by the Alps — Himalayan orogenic movements. Such orogenic movements had lasted since the Oligocene (30 million years BP.) towards the Miocene (20 million years BP.) (Hutchinson 1957; Zenkevitch 1957; Dunbar Rodgers 1957). Other examples of ancient lakes are Tanganyika, Ohrid, Prespa, Titicaca, Tahoe, and Koso-Gol (Horie 1981), all of which have peculiar earth's crust structures such as severe seismic areas with or without active volcanic activities. Except for the existence of the Pleistocene volcanic body, these feature are shared also by Lake Biwa. Particularly, a recent investigation on the negative gravity anomaly (Bouguer Anomaly) has clarified a striking negative nature that suggests the existence of extremely thick low density material, probably sand, silt, or pebble which is much lower in density than the bed rock; they are probably mostly lacustrine sediments. If so, the sedimentation process that continued for a tremendous length of years had formed such extremely thick lacustrine, partly marine or brackish sediments, around 1,400 m (Horie & Sasajima 1978) in accordance with the continuous downwarping of the Biwa Basin. In keeping the oligotrophic state of Biwa, those crustal movements must have played an important role. Do similar movements continue? Before the writer began his work on the Paleolimnology of Lake Biwa, he had been engaged in a geological and geomorphological field work around the lake. One noteworthy datum is concerned with the former altitude of the lake level at the paleo-lake stage. The shoreline with the same age as that of the lake must have been formed at the same altitude. Nevertheless, the present height of the shoreline varies from spot to spot. After the writer finished his mapping of altitudes measured by a barometer, he found that the basin-making movement had evidently been continuing in this tectonic basin (Horie 1961). Strikingly, the isoaltitude line exhibits a shape that resembles the present outline of Lake Biwa. This feature is similar to the movement of the isostasy at Lake Bonneville Basin (Crittenden 1963). It is because of the continuation of the basin-making movements that Lake Biwa has been able to continue its existence since the late Pliocene. The downwarping is also complex in its mode. The distribution of lacustrine terraces indicates that the highest terrace surface is the oldest one and its construction as a terrace may have appeared 200,000 – 300,000 years BP., the time of Rokko Movements' one Climax. Morphology that shows an entirely flat surface over a wide area during the Paleo-Biwa age suggests a considerably long calm interval in the lake-level variation, after which climax of Rokko Movements occurred. Before that age, the lake level may have been kept at the same altitude for a considerable length of time though a short gap was found in the core record. That gap is the faunistic and other evidence of shallow-water stage of approximately 85 m depth of the 200 m core (Kanari 1978; Kawanabe 1978). This relationship is shown in Fig. 1 but it is too early to describe a detailed scheme of lake-level fluctuations in view of the inadequacy of the time scale and level-movement data. It is also difficult to judge whether the basin-making

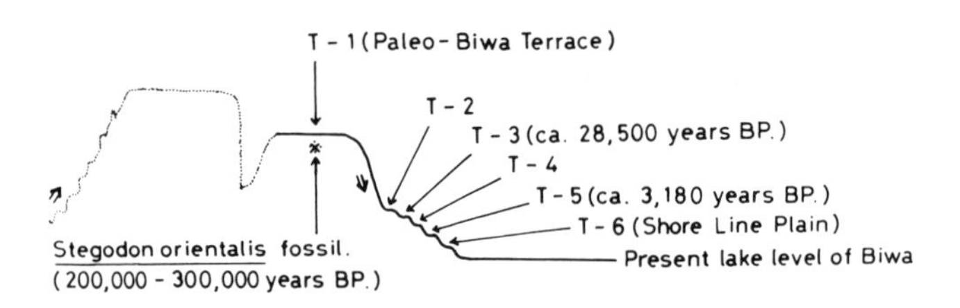

Fig. 1. Lake level rise, repose, intermittent movements and drop. (It may be interpreted as the result of crustal movements; in that case, the scheme must be dealt with in *Reverse* to deduce the mode of the crustal movements. However, the combination with the climatic effect is not negligible. More detailed information will be obtained by an analytical work on that core and a new extremely long core obtained in 1982.)

movement happened radically or continued slowly with other factors affecting the lake-level drop. It is not possible to determine the subsequent altitude of the shoreline because the T-2 – T-5 terraces are formed only inside the T-1 terrace; in other words they are only preserved as river terraces morphology corresponding to each lake level drop. They do not show any extended coastal plain remnance because time between each level drop was short and thereby it was insufficient to form coastal plain by accumulation through inflowing rivers. One reason is the short time over which each terrace age lasted; as a matter of fact, five (exactly speaking four) terraces appeared during the last 28,500 years (Horie 1967a). However, as the T-1 terrace passes to a steep scarp fronting the present Lake Biwa, at least one sudden drop of the lake level must have taken place between T-1 and T-2, then followed possibly by periodic uplifts like pulse beats (Fig. 1).

Because of those basin-making crustal movements, Lake Biwa could keep a deep morphometric situation, i.e., oligotrophy. However, in the natural condition of lake trophy during the last million years that has been demonstrated by various analyses of the 200 m core, lake trophy was not constant oligotrophy as the writer mentioned it above with respect to the 85 m core depth. The core was obtained at the water depth of 65 m at which the amount of dissolved oxygen was several cc/1 throughout the summer stagnation period (Horie 1968c) before the present pollution came to interfere. That is based entirely on oligotrophic evidence, but even at this spot, the age of 85 m depth shows a shallow littoral feature not only in terms of fauna but of grain size. If we neglect other factors, the time might have been a *more*-eutrophic stage. It is still unknown how the lake level dropped and then was restored. Climatic changes and the sudden discharge from an outlet may combine together with the crustal movements. The solution of the problem must await a future study. In addition, the effect of turbulence in the lake must also be taken into consideration.

If a drop of the lake level is assumed, tha littoral region might be shallower. In the case of Lake Yogo, buried forest and the amount of nitrogen indicate an alternation of lake trophy and, during the forest age, a peat layer was formed in the littoral region; then, the peat layer might show a dystrophic-like feature in Lake Yogo even though the lake itself was a meso- or eutrophic type. In conjunction with this evidence, it may be that the littoral region of Lake Biwa was considerably shallow during this shallow-water stage and a peat layer was formed. This is only a possibility, but the existence of a peat layer may support this interpretation.

Thienemann (1921) had proposed a terminology, dystrophic lake type to Naumann's classification (Naumann 1921); later, Järnefelt (1925, 1956) added his own opinion to it (Hutchinson 1969; Rodhe 1969). If seston derived from the shore peat layer is full in lake surface, it may look like the so-called dystrophic lake type but is not a dystrophic type by Naumann in terms of morphometric data. Accordingly, previous classification of lake trophy must be revised; under the natural condition, if the lake surface is relatively small and shallow to the drainage area, a dystrophic feature must have appeared in any lake. On this occasion, the writer wishes to note that lake typology is a complex subject; even at the same time and age, oligotrophic dystrophy might have happened, but its proof is poor in the sediments since the organic material has advanced decomposition in such deep morphometry. This deep morphometry is called oligotrophy; however, when one of the complex elements prevails, one type of lake is not beneficial. Although the poverty in nutrients and the low production are common in an oligotrophic lake type with a large mean depth, deep morphometry does not necessarily imply oligotrophy. The writer, therefore, proposes here new terms of Oligotrophoform (oligotrophic deep morphological form), Mesotrophoform (mesotrophic medium), and Eutrophoform (eutrophic shallow). Oligotrophoform refers to a deep concave lake basin where allochthonous nutrients might be plentiful and also high productivity might be shown. This is not the purpose of this paper; the writer's real intention is to pay attention to a long-ranged ancient lake, whose lake typology is not simple. It suffers from influences of various kinds of natural environmental succession and Oligotrophoform may pass on to Mesotrophoform and/or to Eutrophoform and *vice versa,* i.e., the alternation of trophoform (Horie 1969a; 1975).

Acidotrophic, Siderotrophic, Alkalitrophic, Gypsotrophic, and Argillotrophic elements are not the same as Dystrophic elements, because they are not concerned with depth oscillation except the case in which changes of drainage pattern of surface rock by uncovering process or new volcanic activity have taken place. When we consider the floristic and faunistic composition in ancient lakes, we must be careful with historical events which affected the biology of such a lake.

Biogeography, particularly speciation problems, have already been discussed in detail by Brooks (1950). It is a quite important problem in ancient lakes. All problems of Lake Biwa are to be discussed in later chapters; here, however, the writer wishes to point out a few questions from the viewpoint of paleolimnology.

Firstly, we must pay our attention to the already mentioned mechanism of the alternation of lake trophoform. Every species which has survived is resistant to many kinds of environmental changes, implying that environmental barriers have not been uniform. Therefore, changes of species, composition and so on must be carefully dealt with.

Secondly, as the writer mentioned above, fluctuations of the lake level during at least 500,000 years have not been simple. However, even in the shallow water-depth stage, many species, possibly either stenothermal, eurythermal/stenotopic, or eurytopic species, could continue to live, suggesting that quite a deep region possibly existed in other parts of Lake Biwa. Otherwise, these species might have perished under the influence of the changing environment. They might have moved to environmentally better districts in the lake when a sudden change had happened. Therefore, removal of these species inside the lake itself occurred in accordance with the environmental transition.

Thirdly, these species were not carried into the lake all at once. Actually, they came into the lake over a number of events, particularly upon the change of climate. As will be mentioned in a later chapter, the core sample of 200 m indicates that a number of climatic events happened in the Lake Biwa area. For instance, boreal elements might have invaded the lake when the climate was extremely cold, but later the climate improved; these cold and warm climates repeated cyclically; probably boreal elements were pushed into deep cold hypolimnion as an escape when the climate was improved. Species which had spread northward during the interglacial climate just like today might not have been able to continue to live or to escape from the Biwa region during the glacial climate. Besides, marine species can now be found as endemic species to Biwa. Marine species might have migrated from the Japan Sea. It is another problem. As the writer has already discussed (Horie 1961), the entry of these marine species into the lake might have been from the northernmost area where a channel or some other connection might have existed. This is evidenced in the continuation of Kobiwako (Paleo-Biwa) Group which could be traced near the pass separating Lake Biwa and the Japan Sea. Biological evidence will be explained in a later chapter. However, another possibility concerning the living of marine species in Lake Biwa is that remnant marine organisms were able to continue to live. Their age may be traced to the beginning of Lake Biwa. In the writer's opinion, in the very early stage of its history, Lake Biwa may have been a lagoon-like area at several spots along the Japan Sea coast. That is the peneplain stage of approximately 5,000,000-10,000,000 years BP. when peneplanation took place on the Japanese Islands. The remnant feature of that peneplain is well preserved in the Kinki and Chūgoku Districts, particularly Mino Plateau and Tamba Plateau, both of which surround the Lake Biwa Basin. As Okayama (1956) stressed, the present drainage pattern which began closely on that peneplain was not disturbed by a big fault scarp named the 'Ōmi-Iga Great Fault' (Nakamura 1934); in other words, the fault movement which demarcates the eastern limit of the Biwa Tectonic Basin took place before the drainage pattern was established. If so, the subsidence of the Biwa Basin must have occurred at a very early stage of the peneplain, probably antedating its elevation as we can see at present at the Chugoku Elevated Peneplain including the Tamba Plateau or the Mino-Hida Elevated Peneplain. Because Peneplain was approximately at the same level as the ocean, except for monadnocks, at this early stage of Biwa subsidence, ocean water may have been intercalated with freshwater in spots of that basin as entirely salt water or brackish water. The present existence of *Corbicula sandai* in both Lake Biwa and Lake Yogo (Pleistocene inner bay of Biwa) may have been derived from such brackish water condition in one age. This interpretation is also supported by the existence of other organisms of marine origin in Lake Biwa such as *Kamaka biwae, Anisogammarus annandalei* and *Plecoglossus altivelis.* Anyhow, such a geomorphological feature that has a close connection with the Japan Sea is an important basis of Biwa history. More detailed features will be revealed by *1982's* deep boring operations reaching to the bed rock of this basin. In Lake Baikal and the Caspian Sea, it is well known that seal is living. Kozhov (1963) pointed out, on the basis of data given by L. Berg (1935) and Verescagin (1940) that, as regards Lake Baikal, these were forced to come to Lake Baikal by ice sheets. But he did not discuss when the migration took place through routes like the Lena River. Certainly, Baikal is a special case in zoogeography as L. Berg (1935) pointed out but the number of ice-sheet expansion from the North Polar Sea and pushing effect to these animals must be multiple as the writer mentioned above. Our study of an ancient lake is still in the young stage. However, zoogeographical features in those areas must not be overlooked. In the case of Biwa, Miyadi (1933) already stressed that Lake Biwa is an exceptional case as far as benthic fauna is concerned. Such a special feature in zoogeography must be mainly derived from its oldness and complex composition through many changes of the natural environment.

References

Berg, L. S., 1935. Ueber die vermeintlichen marinen Elementen in der Fauna und Flora des Baikalsees. Zoogeographica 2: 455-483.
Brooks, J. L., 1950. Speciation in ancient lakes. Quart. Rev. Biol. 25: 30-60, 131-176.
Crittenden, M. D., Jr., 1963. New data on the isostatic deformation of Lake Bonneville. U.S. Geol. Surv. Prof. Pap. 454-E, 31 pp.
Daly, R. A., 1934. The changing world of the ice age. New Haven, Yale Univ. Press, 271 pp.
Deevey, E. S. and Flint, R. F., 1957. Postglacial Hypsithermal Interval. Science 125: 182-184.
Dunbar, C. O. and Rodgers, J., 1957. Principles of Stratigraphy. New York, John Wiley and Sons, Inc., 356 pp.
Emiliani, C., 1955. Pleistocene temperatures. Jour. Geol. 63: 538-578.
Forel, F. A., 1892. Thermique des lacs l'eau douce. Verh. Schweiz. Naturf. Gesell. 75: 5-8.
Fuji, N. and Horie, S., 1977. Palynological study of a 200-meter core sample from Lake Biwa, Central Japan. I Palaeoclimate during the last 600,000 years. Proc. Japan Acad. 53: 139-142.
Horie, S., 1961. Paleolimnological Problems of Lake Biwa-ko. Mem. Coll. Sci. Univ. Kyoto, Ser. B, 28 (Biology): 53-71.
Horie, S., 1964. Nihon no Mizuumi – Sono Shizen to Kagaku. Nikkei Shinsho 11, Tokyo, Nihon Keizai Shinbunsha, 226 pp.
Horie, S., 1966. Paleolimnological study on ancient lake sediments in Japan. Verh. Internat. Verein. Limnol. 16: 274-281.
Horie, S., 1967a. Limnological studies of Lake Yogo-ko (I). Bull. Disaster Prev. Res. Inst., Kyoto Univ. 17. 1-8.
Horie, S., 1967b. Limnological studies of Lake Yogo-ko (II). Bull. Disaster Prev. Res. Inst., Kyoto Univ. 17. 31-46.
Horie, S., 1968a. Limnological studies of Lake Yogo-ko (III). Bull. Disaster Prev. Res. Inst., Kyoto Univ. 17. 21-28.
Horie, S., 1968b. Late-Pleistocene climatic changes inferred from the stratigraphic sequences of Japanese lake sediments. VIIth Internat. Quaternary Congr., (Boulder, 1965), Rept., 311-324.
Horie, S., 1968c. Second Report of the regular limnological survey of Lake Biwa (1967) IV. Abiotic environment. Mem. Fac. Sci., Kyoto Univ. Ser. Biol. 2: 125-136.
Horie, S., 1969a. Asian Lakes. In Eutrophication; Causes, Consequences, Correctives, Washington, D.C., Nat. Acad. Sci., 98-123.
Horie, S., 1969b. Late Pleistocene limnetic history of Japanese ancient lakes Biwa, Yogo, Suwa, and Kizaki. Mitt. Internat. Verein. Limnol. 17: 436-445.
Horie, S. and Miyake, H., 1971. The developmental history of Lake Yogo-ko inferred from the granulometric analyses. Disaster Prev. Res. Inst. Ann., no. 14, B, 763-769.
Horie, S., 1975. Lake trophy and its instability. Ann. Rept. Co-ope. Res. (Environment & Human Survival), Tokyo, Minist. Educ., 174-182.
Horie, S., Kanari, S. and Nakao, K., 1975. Buried forest in Lake Yogo-ko and its significance for the study of past bio-environments. Proc. Japan Acad. 51: 669-674.
Horie, S., 1976. Achievement by international analytical team on Lake Yogo core sample. Paleolimnology of Lake Biwa and the Japanese Pleistocene, 4: ed. S. Horie, 743-809.

Horie, S., 1977. Achievement by international analytical team on Lake Yogo core sample. Paleolimnology of Lake Biwa and the Japanese Pleistocene, 5: ed. S. Horie, 347-353.
Horie, S. and Sasajima, S., 1978. On the significance of study for ancient lakes Paleolimnology and Pleistocene climatic succession, notably on Lake Biwa core samples of 200 m and 1,000 m. Abstract. Verh. Internat. Verein. Limnol. 20: 2656-2657.
Horie, S., 1981. On the significance of Paleolimnological studies of ancient lakes – Lake Biwa and other relict lakes. Verh. Internat. Verein. Limnol. 21: 13-44.
Hubbs, C. L. and Miller, R. R., 1948. The Zoological Evidence. In: The Great Basin with Emphasis on Glacial and Postglacial Times. Bull. Univ. Utah, 38: No. 20, 17-166.
Hutchinson, G. E., 1957. A Treatise on Limnology. 1. New York, John Wiley and Sons, 1015 pp.
Hutchinson, G. E., 1969. Eutrophication, Past and Present. In: Eutrophication; Causes, Consequences, Correctives, Washington, D.C., Nat. Acad. Sci., 17-26.
Järnefelt, H., 1925. Zur Limnologie einiger Gewässer Finnlands. Ann. Soc. Zool. Bot. Fennicae Vanamo 2: 185-352.
Järnefelt, H., 1956. Zur Limnologie einiger Gewässer Finnlands. XVI Mit besonderer Berücksichtigung des Planktons. Suomal. Eläin-ja Kasvit seuran. vanamon Eläin-t. Julk, 17: N:o 1, 201.
Kanari, S., 1978. A consolidation model of lake sediments. Verh. Internat. Verein. Limnol. 20: 2658-2662.
Kawanabe, H., 1978. Some biological problems. Verh. Internat. Verein. Limnol. 20: 2674-2677.
Kozhov, M., 1963. Lake Baikal and Its Life. Monographiae Biologicae 11, The Hague, Dr. W. Junk, 344 pp.
Lindeman, R. L., 1942. The trophic dynamic aspect of Ecology. Ecol. 23: 399-418.
Miyadi, D., 1933. Studies on the bottom fauna of Japanese lakes. X. Regional characteristics and a system of Japanese lakes based on the bottom fauna. Jap. Jour. Zool. 4: 417-437.
Nakamura, S., 1934. Tectonic Lines in Central Kinki. Chikyu 22: 155-163, 328-337.
Naumann, E., 1921. Einige Grundlinien der regionalen Limnologie. Lunds Univ. Årsskr. N.F. Avd. 2, 17: Nr. 8, 1-22.
Naumann, E., 1927. Ziel und Hauptprobleme der regionalen Limnologie. Bot. Notiser För År 1927, Ht. 2, 81-103.
Naumann, E., 1932. Grundzüge der regionalen Limnologie. Binnengewässer 11: 176ss.
Okayama, T., 1956. Yanagase Fault and Tsuruga Bay – Ise Bay Line. Sundai Shigaku, no. 7, 75-101.
Rohde, W., 1969. Crystallization of Eutrophication Concepts in Northern Europe. In: Eutrophication; Causes, Consequences, Correctives, Washington, D.C., Nat. Acad. Sci., 50-64.
Sasajima, S., 1977. Paleoenvironmental changes deduced from a 200 m core sample in Lake Biwa, central Japan: A review. Paleolimnology of Lake Biwa and the Japanese Pleistocene, 5: ed. S. Horie, 19-35.
Schwabe, G. H., 1974. "Landnahme" auf Surtsey. In: Island, ed. F. K. von Linden und H. Weyer, München, Bern, Wien, Georg Verlag. Bern, 178-208.
Sugimura, A., 1977. Ice, Land, and Ocean. Kagaku 47: 749-755.
Thienemann, A., 1921. Seetypen. Naturwissenschaften Neunter Jahrgang, 343-346.
Verescagin, G. J., 1940. Origine et histoire du Baikal, de sa faune et de sa flore. Akad. Nauk S.S.S.R. Trav. Stat. Limnol. Baikal 10: 73-227.
Von Post, L., 1946. The prospect for pollen analysis in the study of the Earth's climatic history. New Phytol. 45: 193-217.
Yoshimura, S., 1932. Reconnaissance of the Regional Limnology of the Lakes surrounding Volcano Bandai, Hukusima, Japan. Geogr. Rev. Japan 8: 782-802, 860-880, 933-976.
Yoshimura, S., Negoro, K. and Yamamoto, S., 1936. Limnological Reconnaissance of Gosikinuma Lake Group of Volcano Bandai, Hukusima Prefecture. Geogr. Rev. Japan 12: 1-17, 126-153.
Zenkevitch, L., 1963. Biology of the Seas of the U.S.S.R. London, New York, Interscience Publishers, 955 pp.

Geomorphology around Lake Biwa

Isamu Murai

Introduction

The origin of Lake Biwa is inferred to be very old. Since the birth of the lake, depression has proceeded through each period of the later Cenozoic era, and the lake has survived up to the present in spite of a thick pile of accumulated sediments on the lake floor. The formation and the development of the lake were controlled by fault movements that occurred through the late Cenozoic era.

The area around Lake Biwa and its surroundings is one of the important tectonic regions in the Japanese Islands. Distinct geological structures and morphological features of a large depression basin are recognized in this area, and also intensive negative Bouguer anomalies of gravity are found (Figs. 2, 3). The distribution pattern of the anomalies conforms entirely to the shape of the lake, and it was inferred that thick sediments of more than 1,000 m might have been deposited beneath the lake bottom (Abe & Sasajima 1974). The Old Lake Biwa (Paleo-Biwa or Kobiwako) in the earlier stages is considered to have been three times larger than the present Lake Biwa. Parts of the sediments deposited on the bottom of this old lake are distributed on hills and terraces in the area around Lake Biwa and the Iga Basin. They are called the Kobiwako (Paleo-Biwa) Group, strata of Pliocene to earlier Pleistocene of 1,000 to 1,500 m thickness (Ikebe 1933; Takaya 1963; Yokoyama 1969; Hayashi 1974).

Lake Biwa fills the vast depression of the Ōmi Basin which is situated near the northern apex of the 'Kinki Triangle', which is proposed by Huzita in 1962 as one of the neotectonic provinces in Southwest Japan, characterized by the alternating arrangement of small basins and short mountain ranges (Figs. 4, 5). The basin of Lake Biwa is held between the Hira and Tamba Mountains to the west and the Mino Mountains to the east. The characteristic geomorphological features of the Kinki Triangle result from severe crustal movements, named the 'Rokkō Movements', which occurred mainly during the Quaternary period (Ikebe 1956; Ikebe & Huzita 1966). The large depression of Lake Biwa was formed and developed through the process of fault movements and foldings of the 'Rokkō Movements'. Distinct fault structures are developed in the area

Horie, S. Lake Biwa

ISBN 90 6193 095 2. Printed in The Netherlands

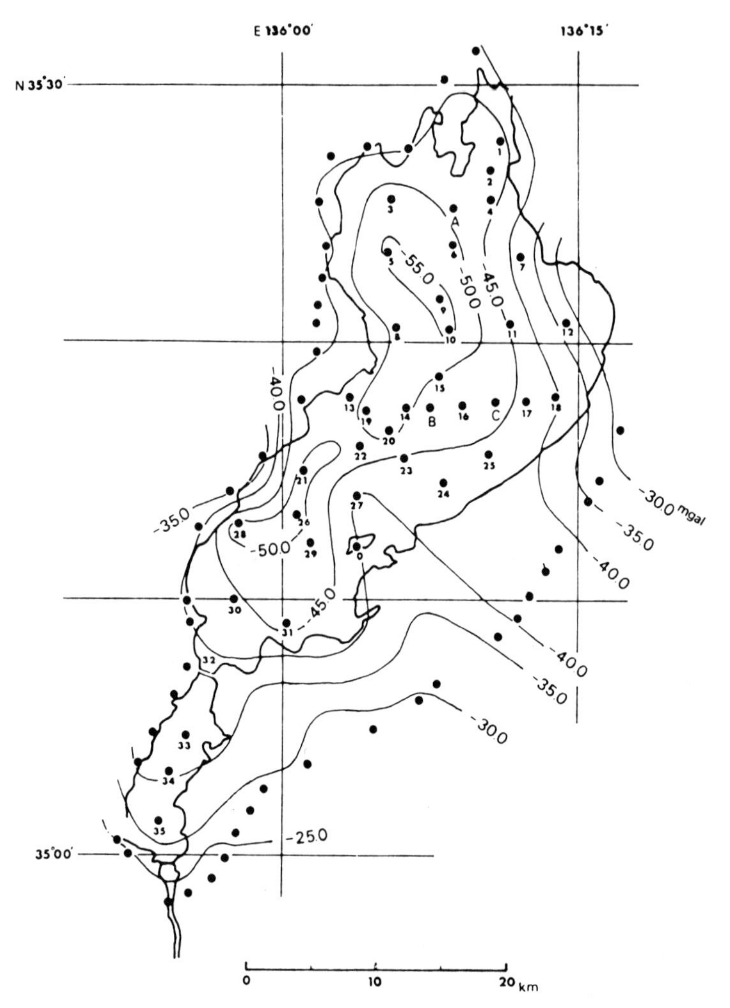

Fig. 2. Distribution of Bouguer anomaly by means of 'Under-water Gravimeter' in Lake Biwa (Abe et al. unpublished)

around Lake Biwa and its surroundings. Some of them are active faults which show evidence of recent movements.

Yamasaki and Tada made a basic study of active faults in the area around Lake Biwa in 1927. Okayama made a preliminary work about the fault topography of the Mino-Hida Mountains, the adjacent areas of Lake Biwa, in 1931. The writer suggested in 1955 the left lateral displacement of the Yanagase Fault which borders the eastern margin of the area around Lake Biwa, and Okayama gave in 1956 some interpretation of the character of such displacement; Sugimura also made a survey of lateral displacement of the Yanagase Fault in 1963. Recently, many studies of the Quaternary geology of the Kinki

Fig. 3. Terrain corrected Bouguer anomaly distribution in the Kinki and Chūbu District. Unit: mgal. (Hagiwara 1967).

Triangle have been carried out and several pieces of information on the neotectonics of this area have been successively accumulated. Huzita et al. (1973), and Huzita (1974) drew distribution maps of active faults of the Kinki District, which show the outline of the structural framework of recent crustal movements in this region. Togo in 1974 analyzed the geomorphology on the Nosaka Mountains to the north of Lake Biwa, and showed the development of many active faults of lateral displacement. The present writer studied and

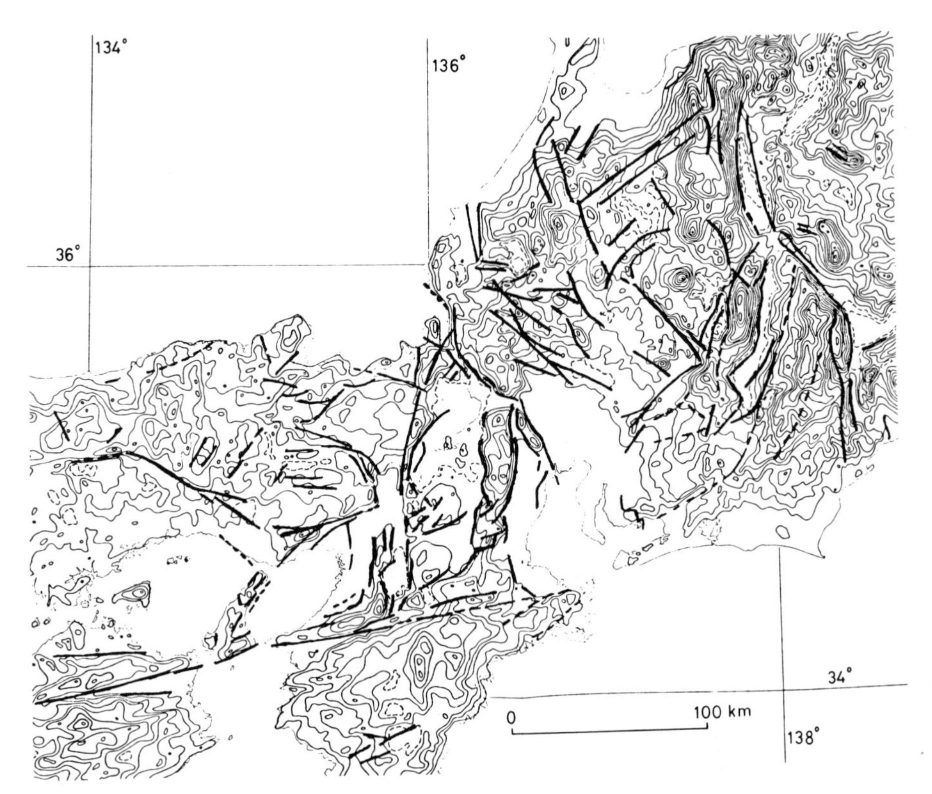

Fig. 4. Active faults in Kinki and western Chūbu. (Murai & Kaneko 1975, partly modified)

reported the tectonic structures and morphogenesis of the area around Lake Biwa (Murai & Kaneko 1975).

The tectonic landforms and the structural framework of the area around Lake Biwa

Complicated fault systems and tectonic landforms of various kinds were developed remarkably in the area around Lake Biwa.

The outline of distribution of active faults in this area is shown by Huzita (1974) and Huzita et al. (1973). Huzita (1974) divided the fault systems developed in the Kinki Triangle and the surrounding areas into four systems according to their trends and characteristics. The first is a left-lateral fault system of a NW–SE trend, the second is a right-lateral fault system with a dip-slip component of a NE–SW trend, the third is a thrust system of a N–S trend, and the fourth is a system of fracture zones developed along the boundaries of geological units. The Yamazaki, Mitoke and other faults belong to the first one. Faults developed along the southern border of the Rokkō Mountains, the Iga

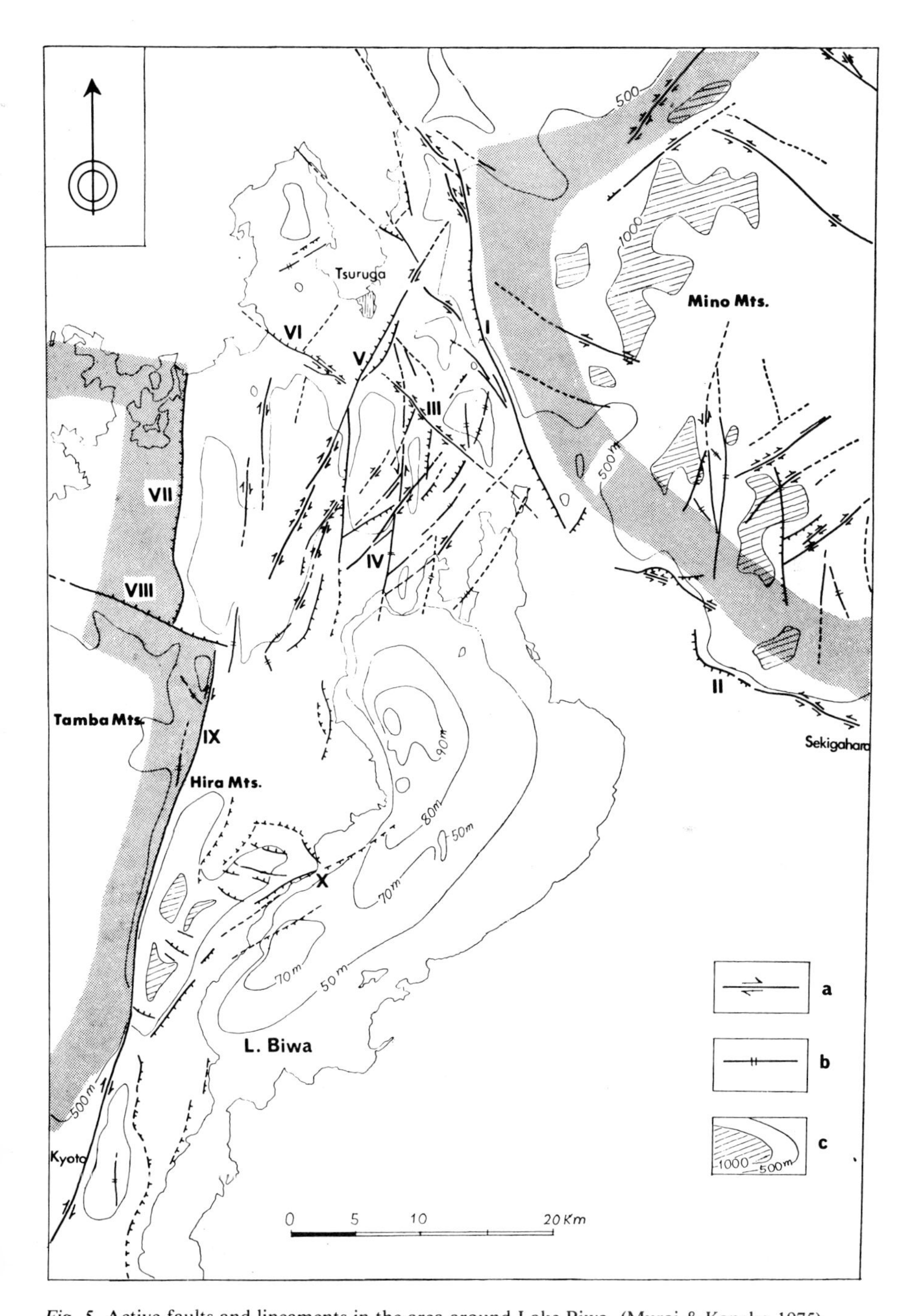

Fig. 5. Active faults and lineaments in the area around Lake Biwa. (Murai & Kaneko 1975)
a: lateral displacement, b: lineament, c: contour line.
I: Yanagase Fault, II: Sekigahara Fault, III: Shufukuji Fault, IV: Chihara Fault, V: Tsuruga Fault, VI: Sekitoge Fault, VII: Mikata Fault, VIII: Obama Fault, IX: Hanaori Fault, X: East-Hira Fault.

Fault and others belong to the second one. The Ikoma, Tongu and other faults belong to the third. The Arima-Takatsuki Tectonic-line, the Median Tectonic Line, the Hanaori, Yanagase, and other faults belong to the fourth system. Among them, the first and the second are in a conjugate relation and are considered to have been formed under the tectonic compression of an E–W direction.

The area around Lake Biwa is situated near the apex of the Kinki Triangle, held between two cratonic masses of the Mino Mountains and the Tamba Mountains, and is inferred to have been compressed and depressed under a lateral tectonic compression in a E–W direction. Many faults were developed in a complicated manner by the Quaternary tectonic movements along the boundaries of the tectonic units as well as within each of the units themselves. The Hanaori and Yanagase Faults are developed along the borders between the Tamba and the Mino Mountains and the Kinki Triangle respectively and are associated with wide fracture zones. On the Tamba and Mino Mountains the basement of Paleozoic formations shows wide distribution. On the contrary, the basement of the area around Lake Biwa consists of granitic rocks and acidic effusives of Cretaceous to early Tertiary. The Tamba Mountains do not show the development of young active faults, and are inferred to be a cratonic mass. On the other hand, the Mino Mountains show many features of recent movements, being cut by many active faults, and are regarded as a resurgent cratonic mass. Throughout the Mino Mountains, active faults of left-lateral displacement in NNW–SSE to NW–SE directions show distinct development. Active faults of right-lateral displacement in NE–SW to ENE–WSW directions are also developed and considered to be in a conjugate relationship with the former ones. Left-lateral active faults in a N--S direction and right-lateral active faults in a E–W direction are developed subordinately. To the north of Sekigahara, the developments of lineations in a N–S direction are recognized. These are considered to represent linear valleys formed by selective erosion along old sheared zones, and do not show any features of recent movements.

The Hanaori Fault, which borders the eastern margin of the Tamba Mountains, forms a distinct fault valley and is accompanied with a shear zone. Minor geomorphological features which indicate right-lateral displacement in recent ages are found in a few places along this fault. The valley which is developed along the Hanaori Fault shows the features of a rejuvenated fault-line valley. To the east of the fault, the Hiei-Hira Mountains rise higher than the Tamba Mountains, and then to the east of the Hiei-Hira Mountains the large depression of Lake Biwa is located. The eastern margin of the depression basin of Lake Biwa is bordered by the Yanagase Fault which is accompanied with a wide shear zone. On the northern parts of the fault, geomorphological features indicating recent left-lateral displacements are recognized. To the east of this fault, the Mino Mountains rise remarkably higher than the lake basin. The area around Lake Biwa is bordered by these two large faults, and the depression basin was formed and downwarped by the tectonic compression of the

Quaternary tectonic movements, being held between two cratonic masses of Paleozoic basements. The active faults and lineaments can be classified into three systems. The first is a system of dip-slip faults and lineaments in a N–S direction. This system may represent old crushed structures and scarcely shows any recent displacements. The second is a system of left-lateral-slip faults in NW–SE to WNW–ESE directions, and the third is of right-lateral-slip faults in NNE–SSW to NE–SW directions. The latter two make a conjugate set of active fault systems. The dip-slip faults of the first fault system border small mountain ranges lying in a N–S direction to the north of Lake Biwa. these dip-slip faults have not been active recently, but are inferred to have been active after the formation of the erosion surface on the summits of the mountain ranges, judging from the vertical displacement of these summit flat surfaces. The lateral-slip fault systems show fresh tectonic landforms and diagonally cut the mountain ranges lying in a N—S trend. Lateral offsets up to 1 km are observed along these lateral-slip faults. These three systems of faults and lineaments show intimate interrelationships to the surface relief and topographical features. Morphogenesis of the area around Lake Biwa may be controlled by the fault movements which proceeded under a lateral compression in a E—W direction during the Quaternary tectonic movements.

Active faults in the cratonic masses of the Tamba and Mino Mountains are generally developed along old faults. In these mountains, many veins of metalic minerals are distributed in association with fracture systems in Paleozoic-Mesozoic strata and Cretaceous-early Tertiary granitic masses as well as in Tertiary formations and igneous rocks. These veins show distinct preferred orientation. These veins are considered to have been formed by igneous activities in the ages of Cretaceous to early Tertiary and the Neogene, but show fairly good accordance with the development of active faults systems. Consequently, it is inferred that the active fault systems in these masses may have been formed by the rejuvenation of the fault movements along old fracture systems. Major structures of fractures might have become active repeatedly throughout the long periods of the geological past. For instance, in the southern margin of the Rokkō Mountains, many porphyrite dykes and aplites in directions of NE–SW and NW–SE intruded in the granite masses, and distinct fracture systems in the same directions are developed. Active faults such as the Ōtsuki, Gosukebashi, and Ashiya Faults are formed along these fractures. In the Mino Mountains, the Neodani Fault, along which the great destructive earthquake occurred in 1891, is formed along the old fault structure. The development of minor fracture systems in the Paleozoic formations distributed in the areas along the Neodani Fault show an intimate relationship with the active fault (Murai 1970). The Hanaori and Yanagase Faults are major faults which border both sides of the area around Lake Biwa. These faults are also formed by the rejuvenation of fault movements along the old fracture zones.

The major active faults

The structure of the area around Lake Biwa is highly complicated with the development of many active faults as mapped in Figs. 4 and 5. The present writer will present descriptions of several major faults.

Yanagase Fault (Plates 1, 2)

The Yanagase Fault (Okayama 1956) extends from Itadori to Kinomoto along River Yogo with 25 km length, passing through Tochinoki-Pass, Nakanokōchi, Tsubakizaka-Pass, Yanagase, and Nakanogō. River Yogo flows in a long linear valley which many geologists assumed to be a fault-line valley formed by erosion along a N–S-striking fault line. However, traces of recent fault movements are recognized in several places in the valley, and the valley of River Yogo may be considered to be a fault-valley. On the summits of the mountains on both sides of the Yanagase Fault, low-relief surfaces are distributed in many places. Such low-relief surfaces are considered to be the remnants of the flat surface which were formed by erosion in a certain geological age, and are useful for identifying the fault differential movements among the mountain blocks. The summit surfaces on the east side of the fault stand 200–300 m higher than those on the west side of the fault. This indicates the upthrow movement of the mountains' block on the east side against the block on the west side. Besides this fact, on the east side of the southern half of the fault, west-facing fault scarps and wind-gaps are developed. Along the base of the fault scarps, taluses and cones are developed although no sign of dislocation is observed on the surfaces of taluses and cone deposits. At the northern end of the fault, an offset of topography which indicates the left-slip displacement is recognized to the west of Itadori.

The most important information about the Yanagase Fault was obtained during the excavation of the Hokuriku Tunnel crossing the northern part of the fault (Ikeda 1958; Takahashi & Shirai 1956). A main fault plane with a strike of N9°W and the dip of 80°W is exposed on the wall of the tunnel, being associated with a crushed zone of about 200 m width. The central part of this crushed zone with 30–50 m thickness is completely crushed, and many subordinate faults are developed around the main fault. Sugimura (1963) proved geologically the actual displacements of left-slip from the offsets of dykes on both sides of a fault in the crushed zone. The displacements amount to 75 m and 100 m respectively for two dykes. He also mentioned that the minor faults with strikes of nearly N–S directions and vertical dips have horizontal or low angle striations on the fault plane.

Plate 1. Yanagase Fault, extending along a linear valley of River Yogo. (Air-photo published by Geographical Survey Institute, Japan.)

Sekigahara Fault (Plate 3)

On the north of Sekigahara, the writer discovered sinistrally offsetted shutter ridges and a fresh low scarplet which indicate recent fault movements. Plio-

Plate 2. Yanagase Fault, bordering the western edge of Mino Mountains, to the east of Lake Yogo. (Air-photo published by Geographical Survey Institute, Japan.)

Pleistocene sedimentary rocks are distributed on the south side of this fault, being in contact with Paleozoic formations on the north side. Limestone strata are distributed on Mt. Ibuki on the north side of the fault, whose extension on the south side seems to have been shifted left-laterally for 3–4 kms by the fault

Plate 3. Sekigahara Fault, trending in a WNW—ESE direction at the north of Sekigahara. (Airphoto published by Geographical Survey Institute, Japan.)

displacements. To the northwest of this fault trace, at the north of Asai-machi, some pieces of evidence which suggest recent left-lateral displacements are identified at three places. The recent displacements are considered to have occurred along the fault system developed in the Permian formations (Isomi 1956).

A destructive earthquake occurred in 1909 near Sekigahara. But it is not certain that this earthquake (Gōnō Earthquake) with the magnitude of 6.9 occurred along the Sekigahara Fault. Some vertical displacements of land amounting to 15 cm or more were revealed by the resurvey of levelling on the route from Hikone to Ōgaki (Imamura 1928). From the pattern of the vertical displacements, the upheaving movement of the northeast side block was assumed. However, the site of the presumed fault lies apart westward from the Sekigahara Fault.

There is a controversial problem regarding the interrelationship between the Sekigahara Fault and the Yanagase Fault. Sugimura (1963) considered that the Yanagase Fault extends toward the Asai Fault (Koto 1910) running from Asai-machi to Sekigahara, which is the equivalent of the Sekigahara Fault. In the map of active faults in the Kinki District by Huzita (1974), the southeastern extension of the Yanagase Fault is not drawn clearly. As to the fault movements of the Yanagase Fault, the vertical displacements exceed greatly the left-lateral displacements, and its activities seem to date mainly in the earlier stages of Quaternary. On the contrary, the Sekigahara Fault shows fresh tectonic features indicating recent left-lateral displacements. It may be reasonable to consider that the Yanagase Fault belongs to the older fault system with a N–S trend developed in the basements of the cratonic mass, while the Sekigahara Fault belongs to the younger fault system of left-lateral with a NW–SE trend. The Sekigahara Fault lies over the southeastern extension of the Shūfukuji Fault. These two faults are both left-lateral, and it may be possible to infer that they are segments of the same fault.

Shūfukuji Fault

Recently a very active fault with a NW—SE direction that developed in the mountains to the north of Lake Biwa was studied and named the Shūfukuji Fault (Ito & Huzita 1971; Huzita et al. (1971). Huzita et al. (1973) and Huzita (1974) plotted the position of this left-lateral active fault in the distribution map of active faults in the Kinki and Chūbu Districts. Togo (1974) pointed out the fact that the Pleistocene terrace deposits distributed near Kutsukake are cut by this fault, with a vertical displacement of 30 m. He also indicated recent sinistral horizontal movements up to about 100 m as inferred from the offsets of valleys.

The Shūfukuji Fault runs northwestward from Shūfukuji, extending through Kutsukake and Hukasaka-goe. Offsets of spurs and gullies are developed along the whole of the fault trace. Fukasaka-goe on the northern segment of the fault is a tectonic saddle formed on a small mountain block with a N–S trend, where the mountain block is offsetted left-laterally for about 1 km by the fault. In the southern segment of the fault, many spurs and gullies are offset left-laterally for about 100 m, and in the terrace deposits on the foot of hills the traces of recent fault displacements are recognized clearly. A distinct shear zone is developed along the fault trace. It is very noteworthy that the Shūfukuji Fault traverses both the north-south-striking faults and the northeast-southwest-striking faults and causes remarkable lateral displacements for them. Judging from the tectonic features of the structures and landforms of the Shūfukuji Fault as mentioned above, this fault is a first-class active fault in the area around Lake Biwa. It is inferred that the southern end of the fault extends to Nishikuroda to the southeast of Lake Yogo. The whole length of the fault reaches to about 17 kms. It is possible that the southeastern extension of the fault connects with the Sekigahara Fault. Similar sinistral active faults with a NW—SE direction are developed near Ikenokōchi to the east of Tsuruga and to the west of Tsubakizaka-Pass. They belong to the same fault system with the same direction developed in the Mino Mountains.

Chihara Fault

Togo (1974) studied a right-lateral active fault developed to the north of Lake Biwa and named it the Chihara Fault. The present writer also surveyed this fault independently and reached a conclusion that this fault is one of the most interesting active faults in the area around Lake Biwa. The Chihara Fault is developed in the mountains of granites and Paleozoic rocks to the north of Lake Biwa with a direction of NE–SW, passing the north of Chihara in the upper course of River Chinai. To the north of Chihara, a distinct crushed zone is developed along the fault, where the southward course of River Chinai and the block mountain range with a N–S trend are dislocated dextrally up to about 600 m along the fault. Valleys in the upper courses of River Chinai and River Goi may have been formed under the influence of selective erosion along an older crushed zone with a N–S direction. However, there are some differences in the altitudes of the remnants of erosional surfaces on the summits of the block mountains on both sides of the valleys. Considering this fact, it is inferred that the fault displacements along the old structure of a N–S direction may have occurred in the earlier stages of the Quaternary tectonics, although movements along this old crushed zone seem to be not so active recently. The Chihara Fault traverses this crushed zone and caused a right-lateral displacement for about 600 m. New minor offsets of spurs and gullies are developed along the Chihara Fault. Undoubtedly, the horizontal fault movements which resulted in the characteristic features of diverted streams and offsets of topography have occurred repeatedly and the amount of displacement may have been accumulated gradually.

Tsuruga Fault (Plate 4)

About 2 kms southeast of Tsuruga, a remarkable fault in a NE–SW direction, associated with many subordinate faults and a conspicuous crushed zone, is observed and called the Kinome-gawa Fault. The northern part of this fault crosses the mountains consisting of intensely disturbed Paleozoic formations. Near Koshisaka at the northern end of the fault, stream offsets by dextral horizontal displacements are recognized. On the other hand, the middle section of the fault is characterized by a high standing fault scarp facing northwestward on an alluvial plain at the southeast of Tsuruga City. The distinct fault scarp fringed by many alluvial cones may indicate considerable amounts of vertical displacements along the fault. Besides the upheaving of the mountain block on the southeast side of the fault, the downthrow movements of the land mass on the northwest side may have occurred, judging from the flattened shapes of the alluvial aprons at the foot of the scarp. The component of the vertical displacement in the fault movements seems to decrease rapidly toward the northeastern and southwestern segments of the fault, where the recent move-

Plate 4. Tsuruga Fault, trending in a NE—SW direction, to the southeast of Tsuruga. (Air-photo published by Geographical Survey Institute, Japan.)

ments are dominantly strike-slip. At the southern segment of the fault extending into the mountains, numerous scarplets, trenches and offsets are developed along the fault trace. Horizontal displacements are entirely dextral, and the amount of the lateral-slips reach to about 300 m or more, although in many cases they are not so large.

To the southeast of the Tsuruga Fault, other active faults of right-lateral slips with a NE–SW direction are distributed. The Chihara Fault mentioned above is one of the most prominent. In each case of these faults, recent fault traces traverse older geologic structures in the basement complex. Fault movements may have progressed and rejuvenated through the later stages of the Quaternary Period. Besides those active faults of a NE—SW direction, faults of a N—S direction are developed in the mountains on the southeastern side of the Tsuruga Fault. Some of them seem to show features suggesting recent lateral displacements.

Sekitōge Fault (Plate 5)

A conspicuous fault of a NW–SE direction, named here the Sekitōge Fault, is recognized to the southwest of Tsuruga. This fault can be traced along a linear front of a mountain block, tectonic trenches, saddles and scarplets. In the southeast segment on the foot of the mountains to the southeast of Sekitōge, fault trenches with left-lateral stream-offsets up to 50–60 m and a fresh reverse scarplet with the height of several meters are formed on alluvial cone deposits. The northeast side of the fault is relatively downthrown. This fault is equivalent to the northern half of the Kurokawagawa Fault Scarp noted by Yamasaki and Tada (1927). Togo (1974) named this fault the Nosaka Fault.

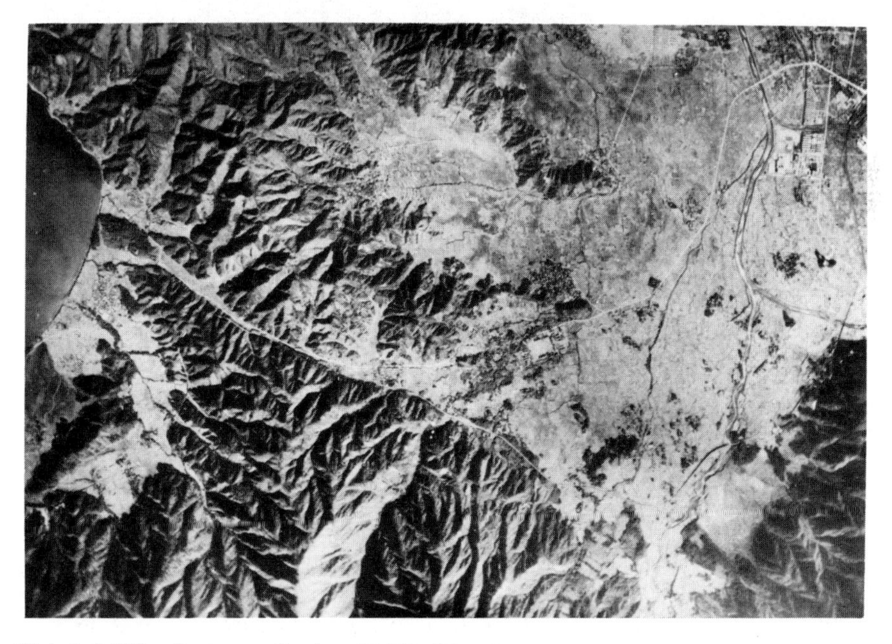

Plate 5. Sekitōge Fault, trending in a WNW—ESE direction along a linear front of a mountain block, to the southwest of Tsuruga. (Air-photo published by Geographical Survey Institute, Japan.)

Mikata Fault (Plate 6)

The structure of the Nosaka Mountains to the north of Lake Biwa is geologically highly complicated with the development of numerous faults. The landscape of the mountains is characterized by the arrangement of smallscale mountain blocks trending in a N–S direction. Along the western border of the

Plate 6. Mikata Fault, trending in a N—S direction, to the east of the subsiding block of Mikata Five Lakes (Mikata-goko). (Air-photo published by Geographical Survey Institute, Japan.)

mountains, the Mikata Fault is developed in a N–S direction, passing through Mikata-machi (Miura et al. 1969). Undoubtedly, the block on the western side of the fault seems to be downthrown relatively, but the amount of its displacement is not clearly known. At the northern extremity of the fault, a Holocene coastal terrace is traversed and dislocated by the fault, showing a downthrow movement on the western-side block. The evidence of strike-slip movements is not recognized along this fault.

Obama Fault

The fault trending in a WNW–ESE direction along the valley of River Kita, from Kumagawa to Obama, is called the Kumagawa Fault (Yamasaki & Tada 1927). Many investigators have mapped this fault with no descriptions. It may probably be presumed that the block on the northern side of the fault has been subsided recently, judging from the development of the rias coast along the northern shoreline of the mountain block. Traces of horizontal displacements may be not clearly recognized. The writer's information about this fault is not enough to let him determine the characteristics of the recent movements along the fault. The fault traverses the Paleozoic terrace, and the rocks along the fault are intensely disturbed and sheared by faulting. It may be inferred that an important structural line is located along this fault. The eastern end of the fault seems to be traversed by a fault with a N–S direction. However, it is inferred from the gross physiographical features of the lake basin that its extension may possibly reach into the basements of Lake Biwa. The deepest part of the lake is located over its eastern extension, where the basin axis, which trends in a NNE–SSW direction in the southern half of the lake, bends sharply toward a northwest direction. This may positively suggest a conjecture on the eastern extension of the fault, although no precise knowledge of the geological structures in the basements of the lake basin has yet been obtained. The western extension of the Obama Fault may reach to the south of Asakurahana across the Obama Bay.

Hanaori (*Hanaore*) *Fault* (Plate 7)

To the southwest of Lake Biwa, there is an upheaved landmass of the Hiei-Hira Mountains with a N–S trend. The western margin of the mountains is demarcated by long continuous valleys of Rivers Ado and Takano which are formed along a remarkable fault called the Hanaori Fault. The Hanaori Fault (Nakamura 1928) extends in a NNE–SSW direction along these valleys passing Hanaori-Pass and Tochū-goe, and reaches to the Kyoto Basin. The Tamba Plateau on the western side of the fault is a cratonic landmass. Low-relief remnants of the erosional surface on the summits of the plateau are clearly 200–300 m lower than those on the summits of the Hiei-Hira Mountains. The fault is

Plate 7. Hanaori Fault, trending in a NNE—SSW direction along the western margin of the Hiei-Hira Mountains. (Air-photo published by Geographical Survey Institute, Japan.)

exposed as a normal fault with a steep-dip to the west at Sakashita, north of Hanaori-Pass, being associated with a layer of fault clay with 2 m thickness and a crushed zone of 2 m thickness. Although the valleys of River Ado and River Takano are probably formed under the influence of selective erosion along the crushed zone on the fault line, they also show many tectonic features which indicate recent movements such as tectonic saddles, scarplets, shutter ridges and offsets of minute topography. The Tamba Plateau consisting of Paleozoic formations and granitic rocks is downthrown against the Hiei-Hira Mountains to the east, judging from the difference in the altitudes of the summit plains.

The strike-slip movements suggested by the tectonic features such as shutter ridges and offsets are dextral, but the amounts of their displacement are usually slight and difficult to measure. Near Ginkakuji, Kyoto, horizontal displacements along one of subordinate faults, diverging from the main fault in the southern segment of the fault, have caused a dextral offset of 250 m on the geological boundary between Paleozoic formations and granitic rocks on both sides (Matsushita 1961). The other signs of recent fault movements are also observed in the area around Kyoto, where the faults generally located at the western foot of the mountains to the west of the Kyoto Basin. Near Shisendō, the granitic mass of the mountains is upthrown on the Osaka Group of Pleistocene with a low-angle thrust. In the area around the Kyoto Basin, the amount of the recent vertical displacements along the Hanaori Fault is estimated at about 300 m or more on the basis of an inference from the difference between the height of Pleistocene terrace deposits on the summits of the mountains and the depth of the basement of the Kyoto Basin (Nakazawa 1961).

East-Hira Fault (Plate 8)

The Hiei-Hira Mountains is an up-arched horst, bordered by faults on both the eastern and western margins. To the east of the mountains, a steep scarp of 500–600 m height rises on the hilly area along the west coast of the lake. The fault developed along the foot of this precipitous scarp is named here the East-Hira Fault. This is an equivalent of what is called the Hiei Fault by Takaya in 1963 and the Hira Fault by Huzita in 1974. The East-Hira Fault shows the fresh feature of a fault scarp, a part of which seems to extend into the bottom of the lake. The base of the fault scarp is fringed by confluent alluvial aprons. However, the present writer could not identify any traces of strike-slip faulting along the fault line on these alluvial deposits. Immediately to the south of the fault scarp, faults with a NE–SW direction are developed in parallel to the East-Hira Fault. Tectonic trenches and scarplets are preserved on the lake floor and can be traced over about 25 km in length.

Many geological and geomorphological studies had been carried out since 1933 on the southern part of the hilly area, known as the Katata Hills, in front of the Hiei-Hira Mountains (Ikebe 1933; Takaya 1963; Okumura et al. 1972; Hayashi 1974). The Katata Hills consist of lacustrine deposits of the Kobiwako (Paleo-Biwa) Group from late Pliocene to middle Pleistocene. At the southwest of Kamiryūge, a low-angle thrust with a strike of N5°E and a dip of 20–30°W is

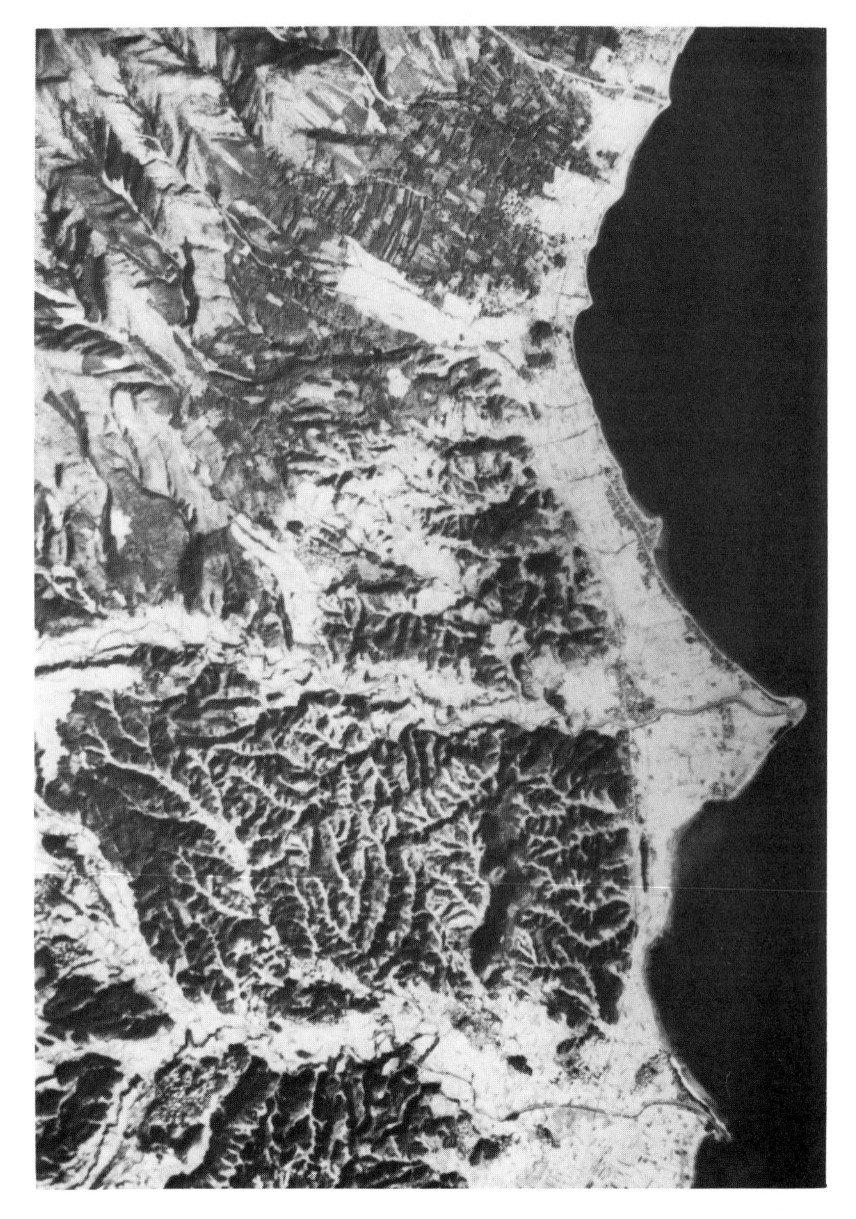

Plate 8. Southern end of the East-Hira Fault and the Katata Fault, bordering the western and eastern edges of the Katata Hills. (Air-photo published by Geographical Survey Institute, Japan.)

observed. Along this thrust, granitic rocks of the west side are upthrown against terrace deposits on the hills (Okumura et al. 1972). At Namatsu, high-angle thrusts are developed in steeply dipping strata of the Kobiwako Group (Takaya 1963). Traces of landslides and rockfalls are distributed along the fault line (Shibasaki 1956). This may indicate the development of crushed zone along the fault. Many dykes are developed in granitic rocks of the Hiei-Hira Mountains in parallel to the East-Hira Fault. The writer infers that the recent fault movements have occurred along the old crushed structures. Small-scale transverse faults are developed in the Hiei-Hira Mountains. They are considered to be formed secondarily with the up-arched movement of the mountains. Lacustrine deposits on the Katata Hills have been deformed to form asymmetrical dome structures in a N–S direction. Along the eastern edge of the Katata Hills, parallel rows of distinct flexures or asymmetrical dome structures in a N—S direction are developed in the Kobiwako Group. On the east wing of these structures, lacustrine formations are inclined steeply to the east or sometimes overturn. It is possible to infer that formations on the eastern edge of the hills are affected by the recent movements of flexure or fault in the basement. A fault, called the Katata Fault, is presumed to be located beneath alluvial deposits along these structures. Fault dislocations with NNE–SSW or N–S directions have been discovered near the hill edge, both in the Kobiwako Group and in terrace deposits covering the surface of the Katata Hills.

Several terrace plains are developed on the surface of the Katata Hills. These plains have been deformed to form gentle anticlinal structures in the middle and eastern parts of the hills. The grades of deformation are stronger in the older terrace plains than in the younger ones. This may indicate that movements such as the fault displacements along the East-Hira Fault and the Katata Fault have occurred repeatedly through the periods of terrace formation, and the effects of the movements have accumulated gradually. Strata of the Kobiwako Group composing the Katata Hills are inclined westward, and indicate the tilt block structure of the hilly area. Such movements of tilting to the west have dominated in the area around Lake Biwa. The basin of Lake Biwa itself is an asymmetrical tilt block of depression, being inclined toward the fault on the western border.

Consideration on the tectonic relief of the area around Lake Biwa

As mentioned in the preceding sections, distinct fault systems are developed in the area around Lake Biwa. Almost all of the faults developed in this area are considered to be active faults. Fault systems in directions of NE–SW and NW–SE are characterized by remarkable tectonic landforms which indicate recent lateral displacements. Faults of a N–S direction bordered both sides of parallel rows of small mountain ranges, and are associated with wide crushed zones. Linear valleys which seem to be fault line valleys formed by selective erosions on crushed zones are developed along these faults, and apparently there may be no direct evidence of recent activities of fault movements. However, low relief surfaces which are considered as erosion plains in the geological ages lie on the summits of every row of the mountain ranges, and there are some variations in the altitude of these summit surfaces. The ages of formation of such erosion surfaces in the Kinki District have not yet been determined precisely, but are inferred to be in the latest Miocene or the early Pliocene, althouth in some cases may be in the more older stages of the Tertiary period. (Nakazawa 1961; Itihara 1966). As Huzita and Kishimoto mentioned, it is thought that the climax of the Rokkō Movements was in the middle Pleistocene. In the area around Lake Biwa and in the Rokkō Mountains, terrace surfaces which were formed in the middle Pleistocene rise at the altitudes of 200—300 m or more, and the summit surfaces of the mountain ranges also upheaved in concordance with such displacement of terrace surfaces. Consequently, it may be inferred that the upheaving and the fault movements which caused the variation in the altitude of the summit surfaces may not be so old. Although the time of the activity and the rate of displacement of each fault system have not yet been precisely clarified, it is considered that many faults of the area around Lake Biwa were developed by the Quaternary tectonic movements. The surface relief and other morphological features of the area around Lake Biwa are considered to have resulted from the block movements which proceeded mainly in the Quaternary period. The large depression of Lake Biwa itself is considered to have been resulted from the fault movements, judging from the structural framework of the lake basin and the adjacent mountain ranges, although the process of formation and development of the basin may have been complicated. Roughly speaking, the structure of the basin may be an asymmetrical tilt block of depression, which was bordered by fault structures along the western margin. Each row of small mountain ranges of the Nosaka Mountains shows a tendency of tilting to the west. The Tamba Mountains also show a tendency of tilting to the west. Such westward tilt blocks are generally recognized as the characteristic structures in the Kinki and Chūbu Districts (Matsuzawa 1968). The outline of Lake Biwa may be controlled by the faults which are inferred to be developed in the basement of the basin.

The process of development and the mechanism of formation of complicated structures and tectonic relief in the area around Lake Biwa have not yet been clarified in detail. As mentioned repeatedly in the preceding sections, it is considered that the basin of Lake Biwa was held between two huge cratonic masses of the Tamba and Mino Mountains and downwarped under the tectonic compression in an E—W direction. The development of conjugate sets of lateral-slip fault systems in the Nosaka and Mino Mountains clearly shows the influence of such tectonic compression in an E—W direction. Fundamental structures of the Kinki and the western Chūbu Districts, the area around Lake Biwa and its surroundings, show some variations in types of tectonic framework. The structures of the Kinki Triangle are characterized by the alternating arrangement of small basins and narrow mountain blocks and the development of thrust faults of a N–S trend, whereas the structures of cratonic masses of the Tamba and Mino Mountains are characterized by fault blocks cut by conjugate shear fractures with lateral displacements. Along the boundaries of these different tectonic provinces, another type of fault with components of lateral displacement and thrusting was developed. Huzita et al. (1973) considered that such differences in tectonics between the Kinki Triangle and the surrounding areas may be due to the physical properties of rocks, although in every province the structures were formed under the tectonic compression in an E—W direction.

However, actual conditions of the crustal movements, and the mechanism of processes of fault movements in the area around Lake Biwa and its surroundings may not be unresolved by such a simple interpretation. In the Nosaka Mountains, both the small mountain ranges bordered by the thrust system in a N–S direction and the lateral-slip fault systems cutting the mountain ranges diagonally are developed together. As mentioned in the preceding sections, active faults are accompanied with distinct crushed zones, and recent fault movements occurred along the old fractures. Fault displacements along the thrusts in a N–S direction are dominantly dip-slip, whereas those along the faults in NE–SW and NW–SE directions are strike-slip. Apparently the dip-slip movements along the former fault system preceded the strike-slip movements along the latter fault systems. This difference in the character of fault movement along each fault system may be partly due to variations in the orientation of fault plane against the direction of the tectonic compression in an E—W direction. Actually, the strike-slip component of displacements is recognized when the trend of the fault plane of a dip-slip fault of the N—S direction shifts from N—S. The writer considers, that the transition of the direction of fault movements, from vertical to lateral, may be caused by the change of stress condition that occurred during the process of tectonic movements. Similar examples of the transition of movement direction of faults, from vertical to lateral, are recognized in the Tanzawa Mountains where the intense crustal movements proceeded since Neogene (Kaneko 1964). We can deal with the stress condition of fracturing in the upper layers of the crust under the lateral tectonic compression by following the idea of Price who, in 1959, attempted to give a detailed discussion on the stress condition of fracturing on the basis of an elastic model. It can be considered that the stress distribution which enables the development of the strike-slip faults may exist in the intermediate depth and the stress distribution for the dip-slip faults may lie in the upper layers near the surface and the deeper parts beneath the intermediate depth, although we have to discuss carefully the problem of conditions of fracturing. Consequently, the writer crudely assumes that in the earlier stages of the tectonic movements foldings of a N–S trend and associated thrusts were formed mainly in the upper layers, and in the later stages, the tectonic compression increased and caused fracturing in the depth of the terrains, and block movements proceeded to the accompaniment of lateral displacement along the faults in NE–SW and NW–SE directions.

Relation between the seismic activities and the active faults in the area around Lake Biwa

The active faults developed in the area around Lake Biwa are inferred to have been inactive through every stage of the Quaternary tectonic movements. The displacements of the active fault reach to several hundred meters or more. They are considered to have been resulted by the accumulation of fault movements which occurred intermittently with the proceeding of the tectonic movements. Seismic activities in the upper layers of the crust are inferred to be associated intimately with fault movements. So it assumed that severe earthquakes may have occurred together with distinct movements along major active faults. In some cases fault displacements may have occurred by creep without any seismic activities as in the case of the present movements of the Egesan Fault on the southern side of the Rokkō Mountains (Tsuda 1972). However, severe seismic activities with fault displacements are sure to have occurred through the Quaternary tectonic movements, as the destructive earthquakes (Fig. 6) occurred frequently in the area around Lake Biwa and the adjacent areas in the historical ages.

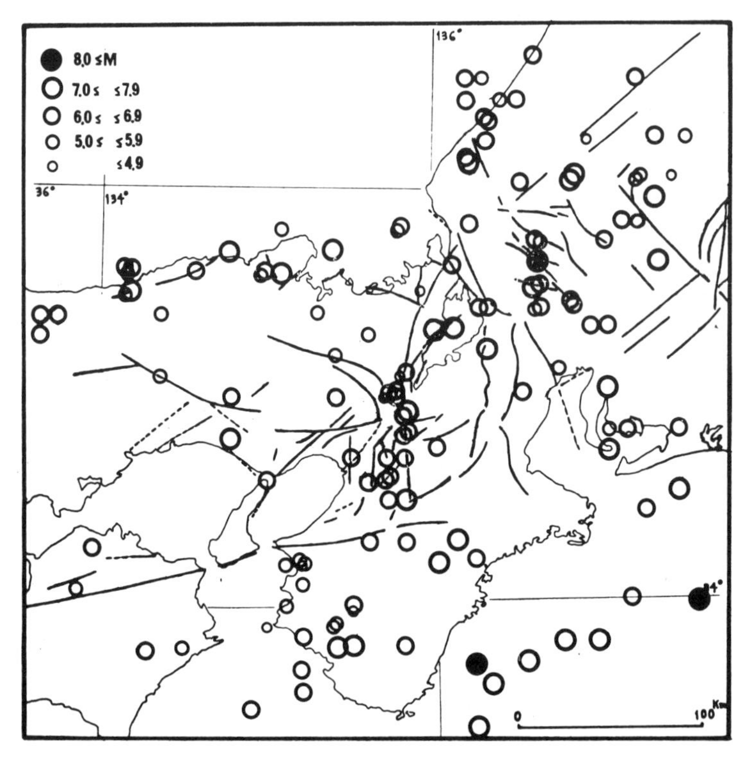

Fig. 6. Map showing the distribution of destructive earthquakes in Kinki and Western Chūbu (Usami 1966, 1974).

The Rokkō Movements which caused the formation of the characteristic structures and tectonic relief of the Kinki Triangle are considered to have not yet ceased. The present seismic activities in the Kinki Triangle and its surroundings may represent the continuance of the Quaternary tectonic movements. Micro-earthquakes in the Kinki District (Fig. 7) and the western parts of the Chūbu District show the intimate relationships to the geologic structures and the active fault systems. It has been already clarified that the distribution of the epicenters of these micro-earthquakes concentrated along the major active faults developed in the margins of the Kinki Triangle and the lateral-slip active faults in the cratonic masses such as the Yamazaki, Mitoke and Neodani Faults (Huzita et al. 1973). As already mentioned, the active fault systems in the area around Lake Biwa and the adjacent areas have been formed under the tectonic compression in an E—W direction. Ichikawa (1971) showed (Fig. 8) that the directions of the main compression of the focal mechanism on the distinct earthquakes in the inner belts of southwest Japan, in the Chūbu and Kinki Districts, lie generally in a direction of E–W. Kishimoto and Nishida (1973) also

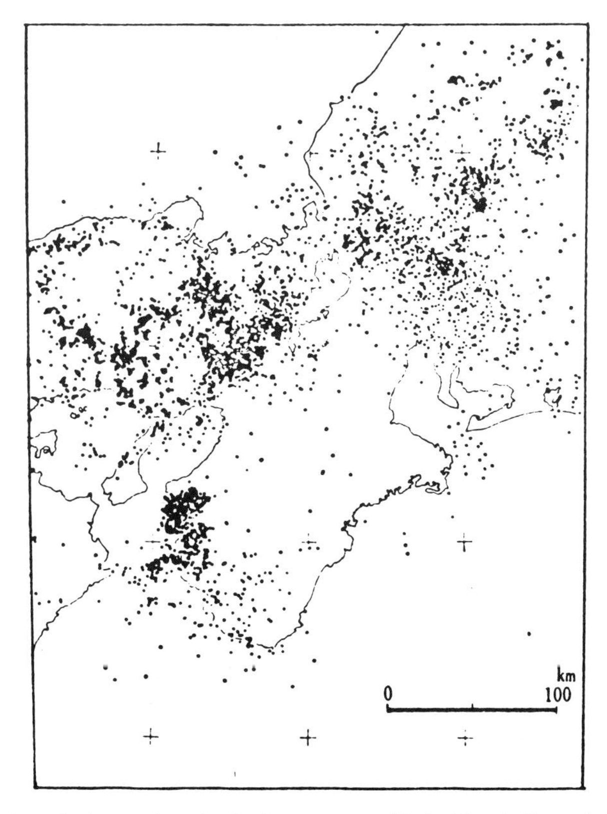

Fig. 7. Distribution of microearthquakes in the area around Lake Biwa in the period from 1963 to 1972 (Matsumura & Oike 1973).

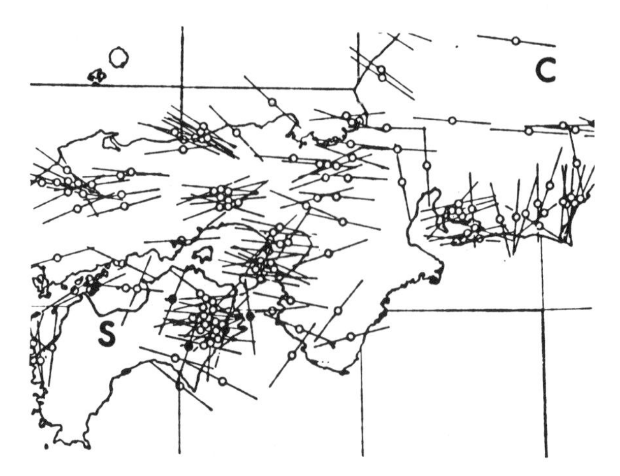

Fig. 8. Directions of pressure axes of very shallow earthquake in central Japan (Ichikawa 1971).

showed the fact that the directions of the main compression of the focal mechanism on the micro-earthquakes along the Yamazaki Fault lie mostly in a direction of E–W. Ito and Watanabe (1977) showed that the focal mechanisms of very shallow earthquakes along the west and the east coasts of Lake Biwa are reverse faulting type with the maximum pressure axes in the direction of nearly east-west or southeast-northwest, and those of the earthquakes in the other regions of north Kinki and west Chūbu Districts are strike-slip faulting type with the same pressure direction. So the stress condition of faulting from earthquake focal mechanism solutions is consistent with that of active fault systems developed in the area around Lake Biwa. Microseismic activities along the west and east coasts of Lake Biwa are very active, while those under the Lake Biwa are very low.

The Nōbi Earthquake with the magnitude of 8.4 occurred along the Neodani Active Fault, being accompanied with the appearance of the remarkable earthquake fault. In the area around Lake Biwa, such historical earthquakes being accompanied with the fault displacements as the Nōbi Earthquake have not been recognized. However, old documents said that the lands on the western coast of Lake Biwa subsided beneath the lake water at the times of the earthquakes with the magnitude of 7.4 in 1185 and with the magnitude of 7.6 in 1662, whose epicenters are inferred to have been located in the Hira Mountains on the west coast of Lake Biwa. It may be inferred that some fault movements may possibly have occurred with the earthquakes. Epicenters of the destructive earthquakes in 1317 ($M=6.7$) and in 1325 ($M=6.7$) are located near the Hanaori Fault and the Tsuruga Fault respectively. Epicenters of the earthquakes in 1449 ($M=6.4$) and in 1838 ($M=6.4$), which caused severe damage to the towns of Kyoto, are also possibly located on the extension of the Hanaori Fault. As already mentioned, the Gōnō Earthquake in 1909 is inferred to have some relation with the Sekigahara Fault or the Yanagase Fault (Figs. 9, 10).

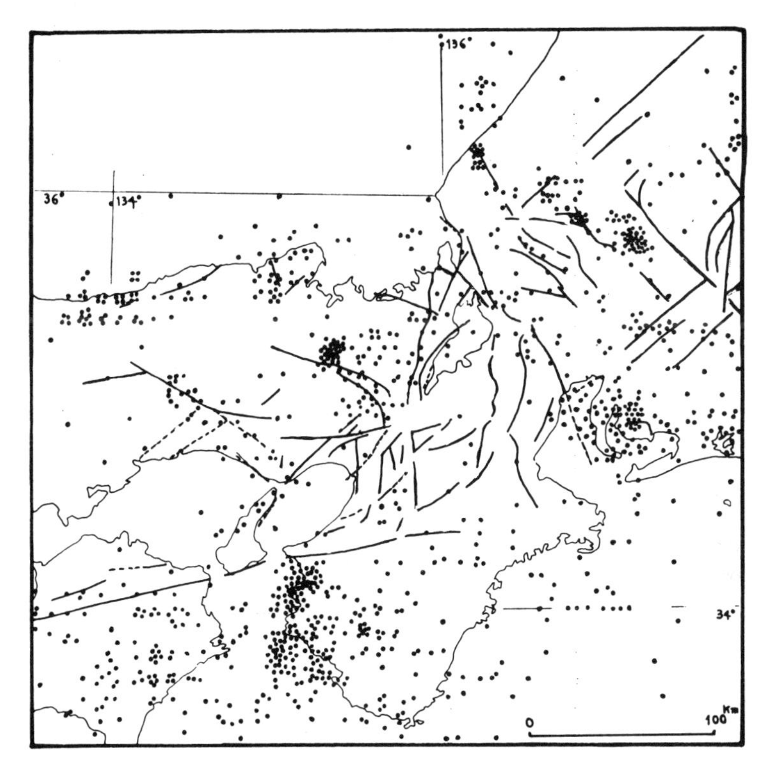

Fig. 9. Map showing the distribution of epicenters in the area around Lake Biwa during 1925-1971 (Murai & Kaneko 1975).

Conclusion

The writer has outlined the active fault systems developed in the area around Lake Biwa and its surroundings. They are characterized by remarkable photogeologic lineaments and tectonic landforms. The faults are classified into three systems: dip-slip faults in a N–S direction, left lateral-slip faults in NW–SE to WNW–ESE directions, and right-lateral-slip faults in NNE–SSW to NE–SW directions. They were developed through the Quaternary tectonic movements under the lateral compression of an E-W direction. The dip-slip faults bordered the small-scale mountain ranges lying in a N-S direction to the north of Lake Biwa, and are considered to be developed along the old crushed structures. The vertical displacement along these faults may have occurred after the formation of the flat summit surfaces on the mountain ranges, although the recent movements seem to be not so active. The conjugate set of lateral-slip faults shows fresh tectonic landforms, cutting diagonally the mountain ranges lying in a N–S trend. The lateral offsets of topography by the fault displacements reach to 1 km. These faults are considered to be the first-class active faults. The surface

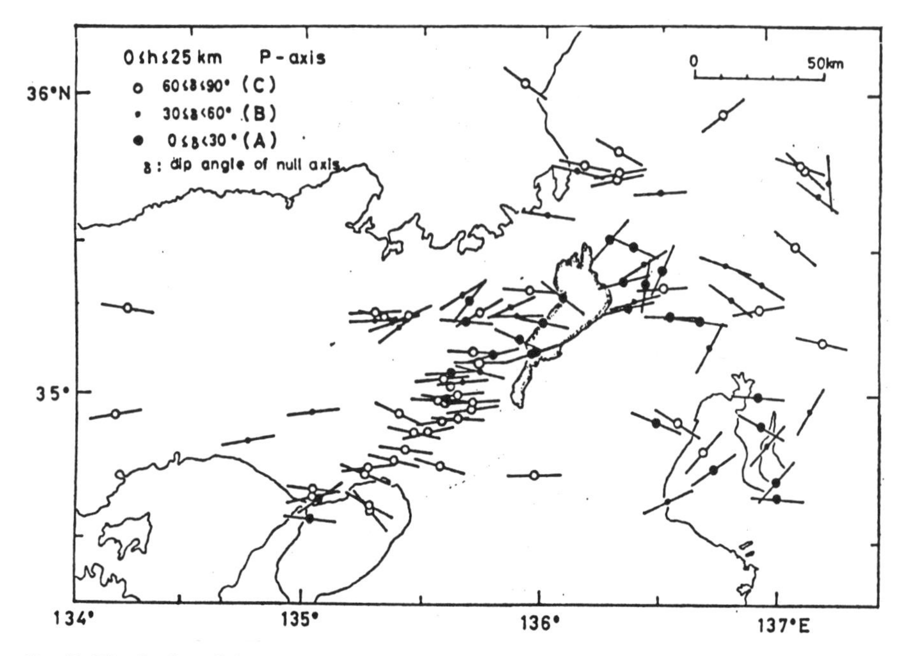

Fig. 10. Distribution of 'fault type' and horizontal projection of the axis of maximum compression for the earthquakes ($M \geqq 2.0$) occurred in the north Kinki District.'fault-types', A, B and C, are based on the dip angle of null axis (θ). A; $0° \leq \theta < 30°$ (large solid circle), B; $30° \leq \theta < 60°$ (small solid circle), C; $60° \leq \theta \leq 90°$ (open circle). Because the maximum pressure is nearly horizontal, A and C roughly correspond to reverse faulting and strike-slip faulting, respectively (by Ito & Watanabe 1977).

relief as well as the geologic structures of the area around Lake Biwa are considered to have resulted from the block movements which proceeded along these active faults during the Quaternary tectonic movements. In order to decipher details of the mechanism and processes of the tectonic movements, careful studies on the topography and the geological structures must further be carried out.

Stratigraphy of the Quaternary system around Lake Biwa and geohistory of the ancient Lake Biwa

Takuo Yokoyama

Introduction

Lake Biwa is the biggest lake in Japan. Not only is it the oldest lake in Japan, but also one of the oldest freshwater lakes in the world. This fact had been proved by evidence in both geological and biological investigations. It is estimated from fission-track dating and paleomagnetism that this lake appeared about 5.0—4.5 million years (m.y.) ago. It is an outstanding piece of information about the biological field that such endemic species live in Lake Biwa and in the surrounding water areas as *Gnathopogon elongatus carulescens, Carassius auratus grandculis, C. auratus cuvieri, Parasilurus lithophilus, P. biwaensis* and *Chaenogobius isaza*. In the geological field, on the other hand, there are many sediments around Lake Biwa, called the Kobiwako (Paleo-Biwa) Group and others, which are about 1,800—2,000 m thick. Some fossils, yielded in the sediments of the ancient lake, belong to the endemic species of the lake, such as *Lanceolaria oxyrhyncha, Unio biwae, Cristaria plicata clessini* and *Corbicula sandai*. The sedimentary basin of the ancient Lake Biwa is about 1,500 km^2 wide. Its diameter along a N—S direction is about 100 km long and one along an E—W direction is about 40 km long (Fig. 11). In these sedimentary basins, many Quaternary sediments were accumulated. Now they are found in the hills around Lake Biwa, the height of which ranges from about 20 to 200 meters above the water level of Lake Biwa. The summary of the Quaternary system around Lake Biwa is given in Fig. 12. The Quaternary system around Lake Biwa is divided into the following three categories in Yokoyama (1969):

1. Alluvium.
2. Terrace deposits.
3. Plio-Pleistocene Kobiwako Group.

In view of the results of the two deep drillings, carried out in or by Lake Biwa in 1971 and 1975–1976, we must recognize the Quaternary system around Lake Biwa as one unit. In these two boring examinations all sediments were continuous without any disconformities from the ancient lake sediments to the actual bottom surface of Lake Biwa. Consequently, the mutual relations of

Horie, S. Lake Biwa

ISBN 90 6193 095 2. Printed in The Netherlands

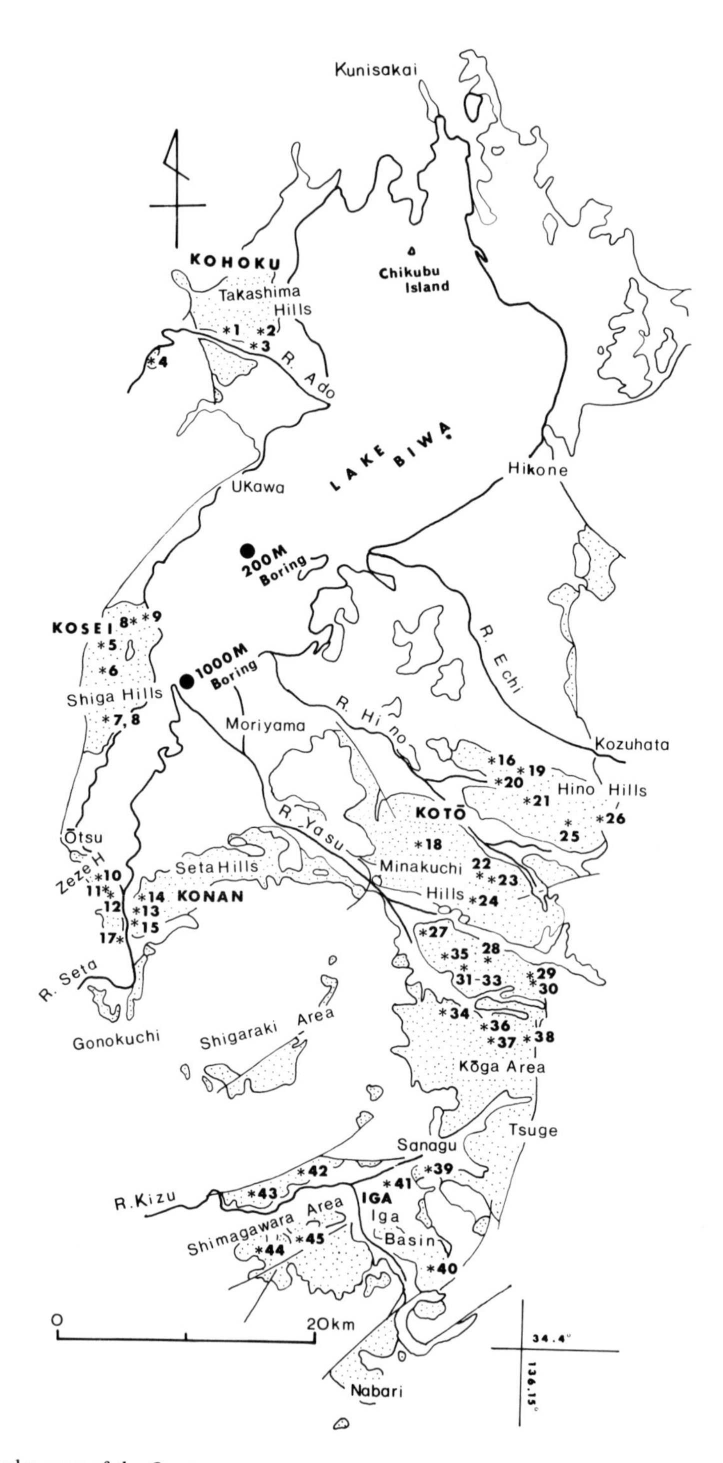

Fig. 11. Index map of the Quaternary systems around Lake Biwa.
Dotted area; Hills composed of the Kobiwako Group. *1-45; Type locality of the formations (see Table 5).

various Quaternary sediments around Lake Biwa have been recognized as shown in Fig. 12.

It is certain that the investigations of the Kobiwako Group and the sediments of Lake Biwa are very important to Quaternary geology and limnology of Lake Biwa, because these sediments have the following convenient features in the

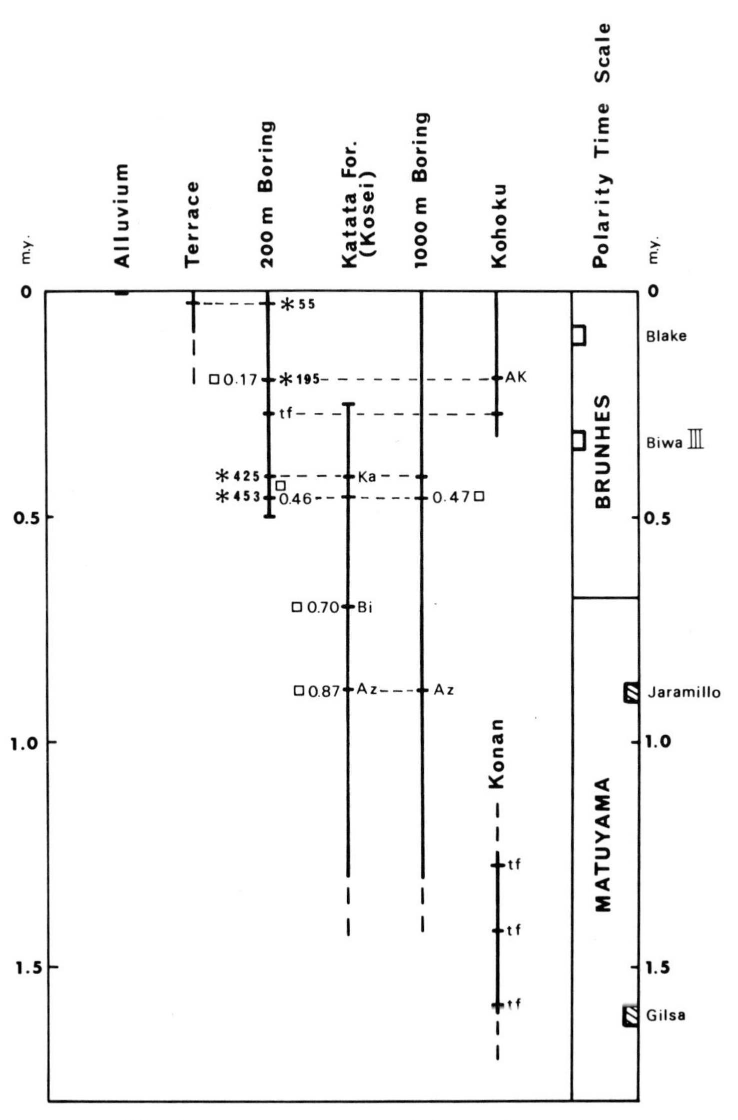

Fig. 12. Time range of the Quaternary systems around the actual Lake Biwa. □0.17: fission-track age (m.y.) (see Table 3).
Volcanic ashes: *55 ~ 453; BB 55 ~ 453, Ak: Akatsuki, Az: Azuki, Bi: Biotite, Ka: Kamiōgi.

studies of the Quaternary systems. There are few sedimentary bodies in the world which have such various useful points.

1. Existence of the continuous sediments from 5.0 m.y. ago to recent, as similar to the core samples from deep sea bottom.
2. As it is larger in sedimentation rate than the deep sea core samples, we can obtain much and precise information in geology and natural history.
3. Homogeneous sedimentological environments because of no influence of the sea.
4. Fossiliferous such as elephant, plant, mollusc, pollen, diatom etc.
5. The recent biological studies such as the endemic species, can be carried out.
6. Established tephro- and chronostratigraphy, by means of the many volcanic ash layers.
7. Tission-track ages and precise magnetostratigraphy.
8. Because of the correct correlation by some volcanic ash layers to the Osaka Group, which distributes in the adjacent sedimentary basin and is intercalated with some marine facies, we can know the relations to the sea level change in Quaternary period.

In the present paper, the writer will present the litho-, chrono-, tephra-, and magneto-stratigraphy of the Quaternary systems around Lake Biwa, and then will propose the image of the geohistory of Lake Biwa in tectonic and sedimentologic senses.

Historical review of geologic study around Lake Biwa

The first report of the ancient lake deposits of this lake was published in 1930 (Nakamura 1930). Nakamura called the sediments the Kobiwako Group. It contained comments on them at the Shiga Hills and in the Seta-Gonokuchi regions. Ikebe (1933, 1934) reported on the detailed stratigraphy of the Kobiwako Group of the hills in the Katata area in Kosei, which was the west coast of Lake Biwa and at the Kōga area in Kotō, southeastern region of the lake. He divided the Kobiwako (Paleo-Biwa) Group at Katata into two members in an ascending order, Minamishō Clays and Ryūge Sands, and at the Kōga area in the same order, the Aburahi and Sayama Stages. Then, the Minamishō Clays were correlated to the Sayama Stages. This work was the first division of the Kobiwako (Paleo-Biwa) Group from the lithological viewpoint. Afterwards, because of the Second World War, investigation of the Kobiwako (Paleo-Biwa) Group was discontinued for about 30 years until 1960.

In 1960, an andesitic volcanic ash layer, which contained two pyroxene crystals, was found in the Katata region and it was correlated to the Azuki Volcanic Ash Layer of the Osaka Group (Araki 1960). Takaya (1963) carried out a field survey of almost distributed areas of the Kobiwako (Paleo-Biwa) Group and, by volcanic ash, divided this group into three formations. The total image of the development of the ancient Lake Biwa was proposed by Takaya's work. The origin of the endemic species of the lake was discussed in his paper.

The reexamination of Takaya's ideas were carried out by many investigators from 1968 to now, for example, Yokoyama et al. (1968), Kondo (1968), Yokoyama (1969), Kaigake Research Group (1972), Hayashi (1974), Yokoyama (1975b), Ishida et al. (1976a), Tamura et al. (1977), Kobiwako Research Group (1977), Yokoyama et al. (1978) in stratigraphical field (Fig. 13), Yokoyama (1968b, 1969, 1975a) in paleocurrent and paleogeography, Ishida & Yokoyama (1969), Yokoyama (1969, 1972, 1973b, 1975b) in tephrochronology, Ishida et al. (1969), Maenaka et al. (1977), Hayashida et al. (1976), Tamura et al. (1977), Kobiwako Reasearch Group (1977) in magnetostratigraphy and Nishimura & Sasajima (1970), Nishimura & Yokoyama (1974, 1975), Yokoyama et al. (1978) in the fission-track dating and so on. In Yokoyama (1969, 1971, 1973a, 1974a), Ishida (1976), Ishida et al. (1969), Maenaka et al. (1977), a new total image of the development of Lake Biwa is described and discussed. On the other hand, the two deep drilling examinations of the Lake Biwa sediments, the chief director of which was S. Horie, Kyoto University, were done in 1971 and 1975—1976. Many remarkable results in various fields were obtained and were summarized in 'Paleolimnology of Lake Biwa and the Japanese Pleistocene' Vol. 1—9 (ed. S. Horie 1972, 1974, 1975, 1976, 1977, 1978, 1979, 1980, 1981).

Review of the Plio-Pleistocene series in Kinki and Tōkai Districts in Southwest Japan

Southwest Japan of the Neogene Period has been divided into the following three zonal geologic provinces from the north to the south:

1. Hokuriku-Sanin Province,
2. Setouchi Province,
3. Nankai Province.

The Hokuriku-Sanin Province belongs to the 'green tuff region' characterized by voluminous intermediate or basic volcanic materials, while almost all of the Neogene sediments in the Nankai Province are terrigenous clastics of the Miocene in age.

The Cenozoic strata in the Setouchi Province are divided as follows;

1. Alluvial deposits,
2. Terrace deposits,
3. Plio-Pleistocene Series (Second Setouchi Series),
4. Setouchi Volcanic Series,
5. Miocene Series (First Setouchi Series).

The basements contain various kinds of pre-Cenozoic rock units, namely, Upper Paleozoic rocks, Triassic strata, Ryoke granitic rocks, welded tuffs and so on.

The Plio-Pleistocene series in the Setouchi Province are exposed in the

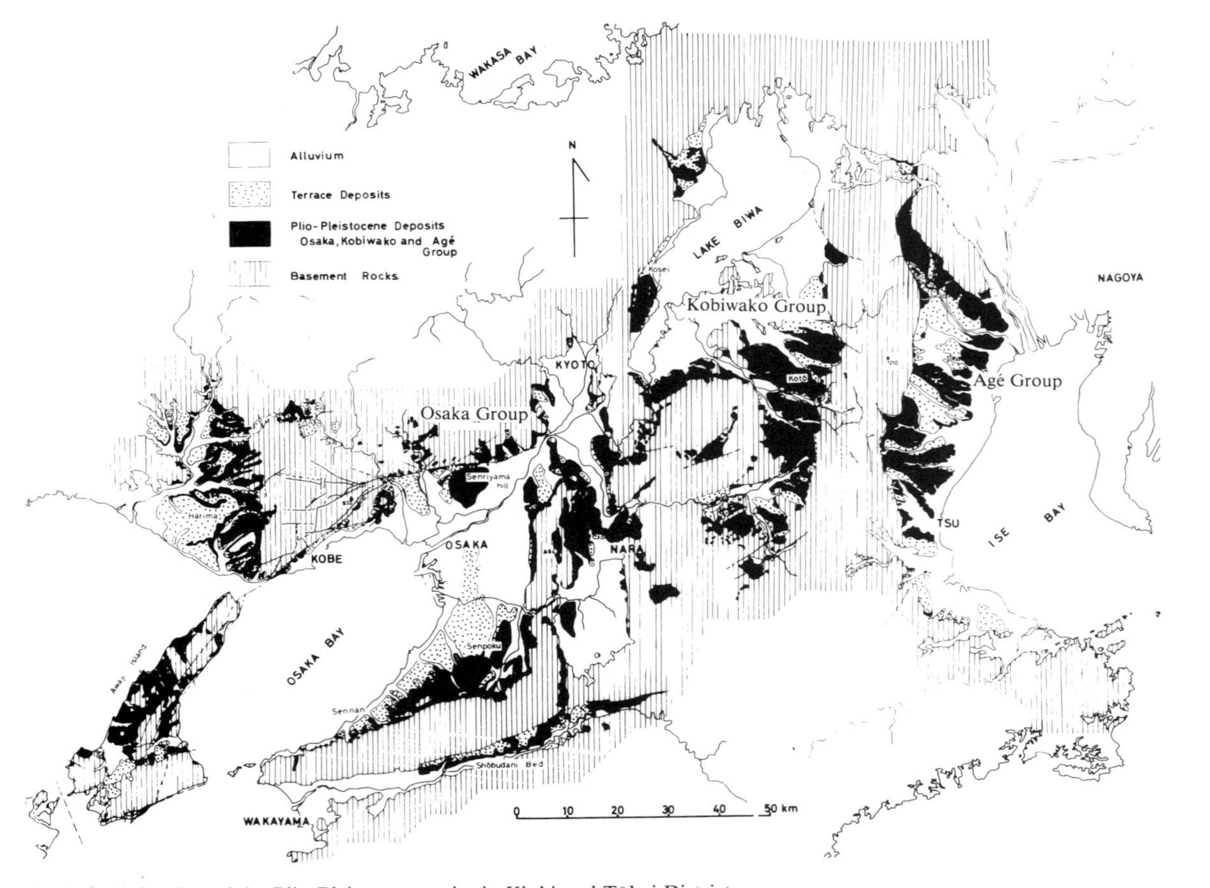

Fig. 13. Distribution of the Plio-Pleistocene series in Kinki and Tōkai Districts
(Agé Group is a subgroup of the Tōkai Group)

foothills of the mountainlands and are called Kuchinotsu, Oita, Mitoyo, Osaka, Kobiwako and Tōkai Groups respectively. The Osaka, Kobiwako and Tōkai Groups are widely distributed in the separate basins of Kinki and Tōkai Districts (Figs. 13, 14). They yield rich fossils: elephants, plants, diatoms, pollens and molluscs. These groups have been studied by many investigators during the last 20 years: Huzita et al. (1951), Itihara (1960), Takaya (1963), Takehara (1961), Yokoyama (1969), Ishida & Yokoyama (1969), Ishida et al. (1969), Nishimura

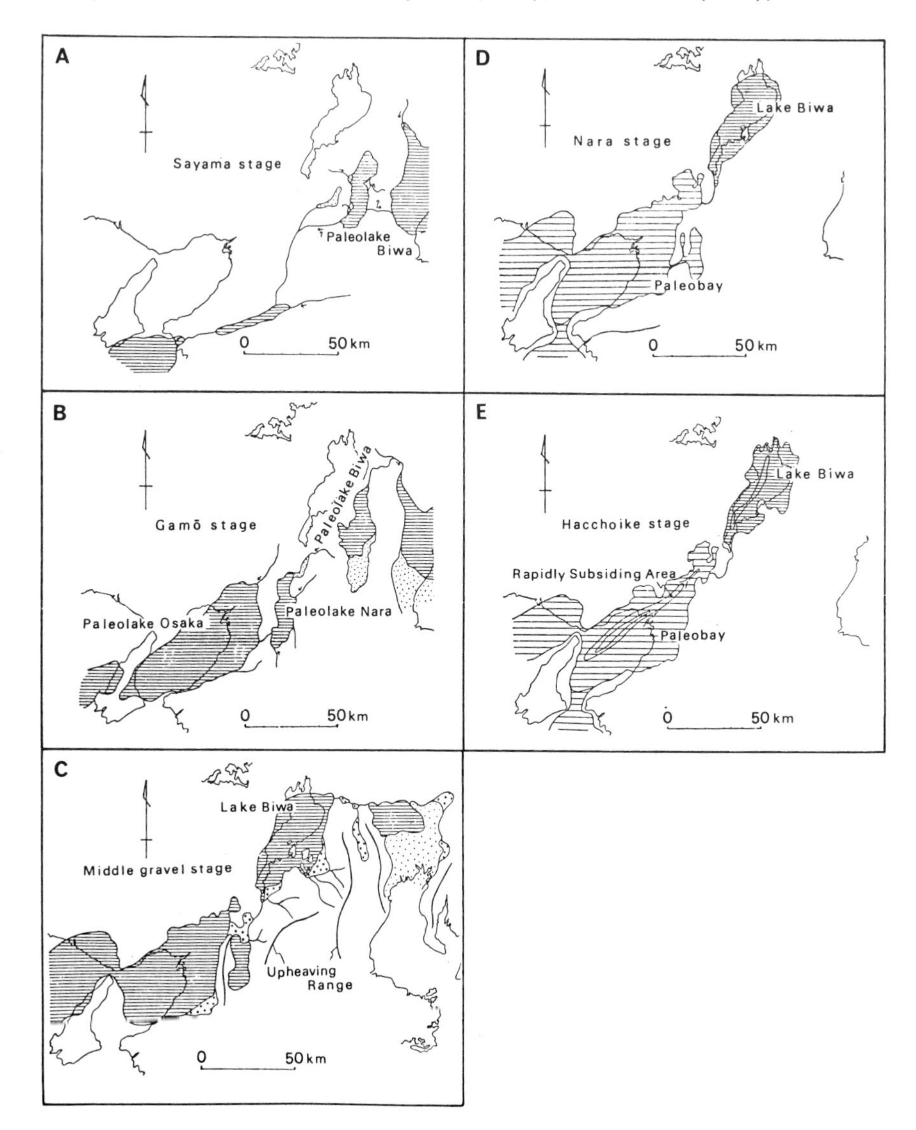

Fig. 14. Paleogeographic map of the Plio-Pleistocene series in Kinki Area (Yokoyama 1969, 1973a). A: 3.2—2.5 m.y., B: 2.5—1.5 m.y., C: 1.5—1.3 m.y., D: 1.3—1.0 m.y., E: 0.9-0.5 m.y. Sedimentary basins migrated from south to north. Lake Biwa basins had been lacustrine areas since their appearance, while those of Osaka were marine environments since 1.3 m.y. ago.

& Sasajima (1970), Itihara et al. (1975), Maenaka et al. (1977), Yokoyama et al. (1977a), Yokoyama (1977).

Sedimentary Basins of the Plio-Pleistocene in Kinki and Tōkai Districts

Roughly speaking, sedimentary basins are arranged latitudinally in Southwest Japan. They are separated by meridional highlands, for example, Kiso, Suzuka, Shigaraki and Ikoma Mountain Ranges. The Awaji-shima Island also belongs to these highlands. There are several independent basins in Kinki and Tōkai Districts, that is, Harima, Osaka, Kyoto-Nara, Ōmi-Iga (Biwako) and Nōbi. Hence, we have three group names for the Plio-Pleistocene, namely, Osaka, Kobiwako and Tōkai.

Generally, the strata are inclined northward in these sedimentary basins. Thus, older sediments are seen in the southern part of these basins. Yokoyama (1969) suggested the migration of the sedimentary basin in the Ōmi-Iga Basin from south to north (Fig. 14). The subsided center was moved northward at intervals of 20–30 km distance. This fact may indicate that the Ōmi-Iga Basin, in which Lake Biwa exists, is not a simple basin. The basement complex under the Kobiwako (Paleo-Biwa) Group is separated into many blocks. Figs. 15 and 16 indicate the features of faults and crush zones in the Ōmi-Iga Basin. They correspond to the boundary parts of the basement blocks. Generally speaking, these blocks range about 10 to 20 km^2 in width. The shape and situation of the sedimentary basins may be determined by characters of the mutual block movements of the basements.

Moreover, the Ōmi-Iga Basin is made of the assemblage of several small sedimentary basins as shown in Fig. 17. From south to north these small basins become younger. Other basins, such as Kyoto-Nara and Nōbi, are also the assemblage of small sedimentary basins. Just as in the Ōmi-Iga Basin, the sedimentary fills become older to southward. The lithostratigraphy of the Plio-Pleistocene Series in these basins are given in Table 1.

Outline of the Osaka Group

General Review. The Plio-Pleistocene strata in Kinki and Tōkai Districts are mainly composed of clastic sediments such as gravels, sands, and clays with thin seams of peat and volcanic ash layer. The Osaka Group occupies Harima, Osaka, Kyoto and Nara Basins. The lower part is composed of lacustrine and fluviatile sediments, while the upper part is composed of alternating lacustrine and marine deposits. Most of marine facies are represented by continuous clay beds well traceable as horizon markers with characteristic pyroclatic layers. Eleven marine clays recognized in the peripheral hill sides are represented by symbols; Ma 0, Ma 1, Ma 2, ..., Ma 10 in an ascending order. Additional clays,

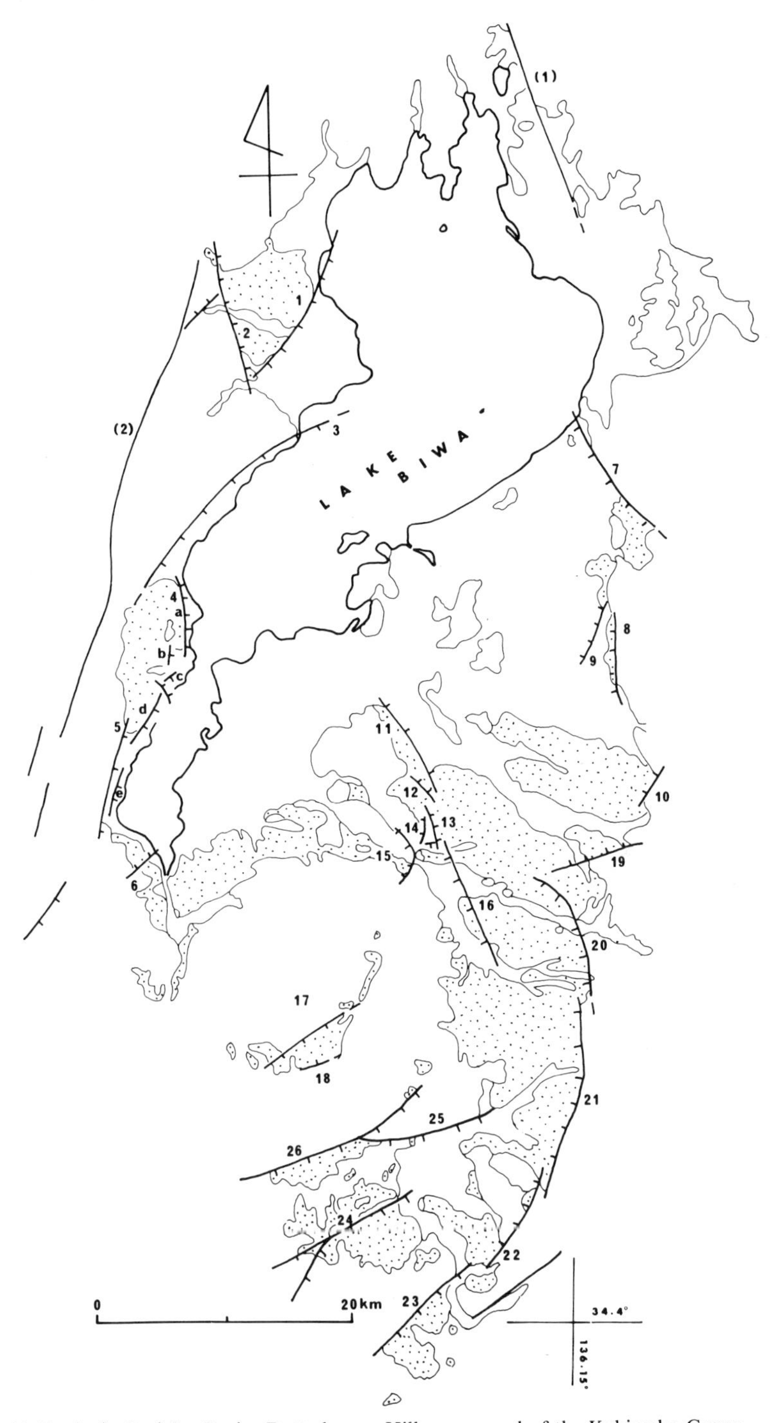

Fig. 15. Faults in Omi-Iga Basin. Dotted area: Hills composed of the Kobiwako Group. Fault Name; (1) Yanagase (2) Hanaore (Hanaori).
1. Takashima; 2. Taisanji; 3. Ukawa; 4. Katata; 5. Ōtsu; 6. Chausuyama; 7. Taga; 8. Kozuhata; 9. Usogawa; 10. Kono; 11—13. Shimoda; 14. Sasagatani; 15. Mikumo; 16. Kazuraki; 17—18. Shigaraki; 19. Kaigake; 20. Tongu; 21—22. Tsuge; 23. Nabari; 24. Hananoki; 25. Iga; 26. Shimagawara (after Yokoyama, 1979a).

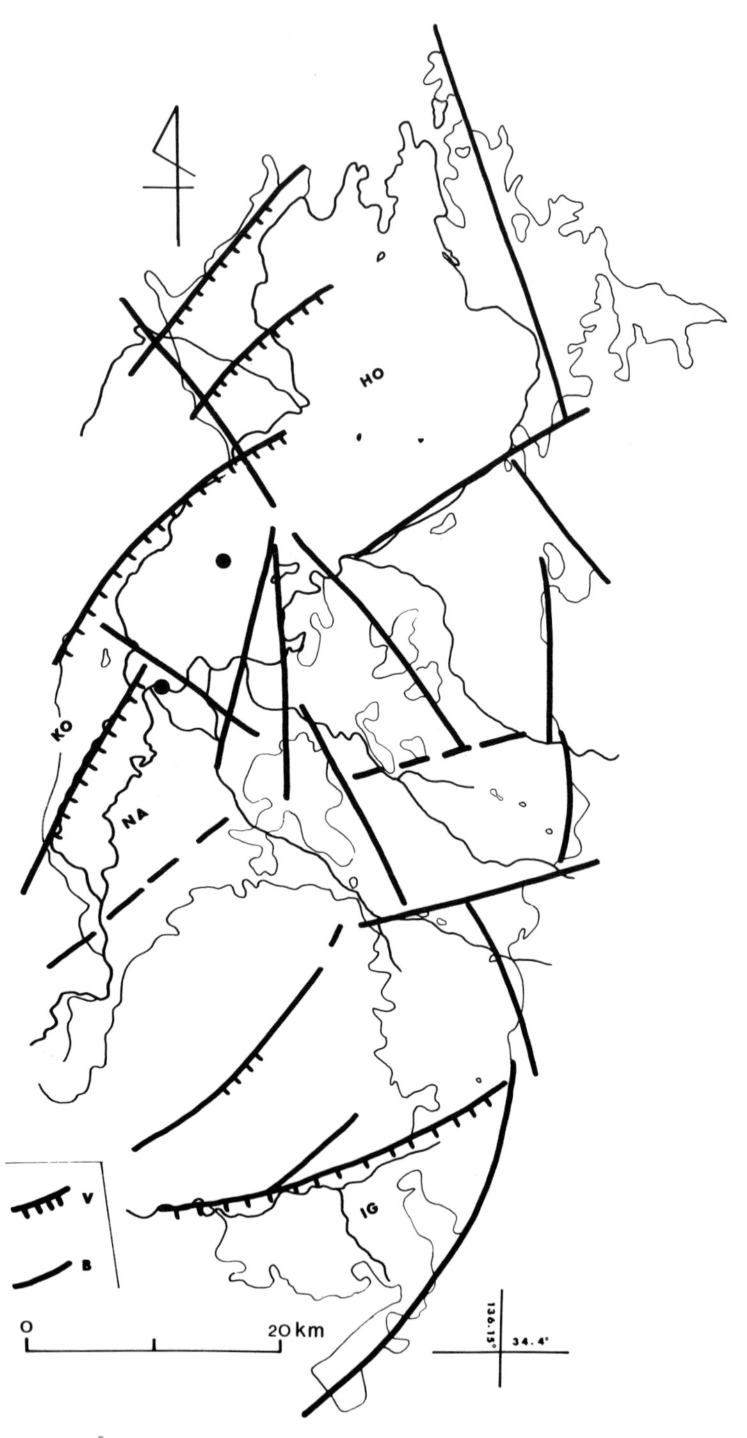

Fig. 16. Blocks in Ōmi-Iga Basin.
HO: Hokko (North lake), KO: Kosei, NA: Nanko (South lake), IG: Iga Blocks, V: Fault having vertical slip, B: Boundary of blocks.

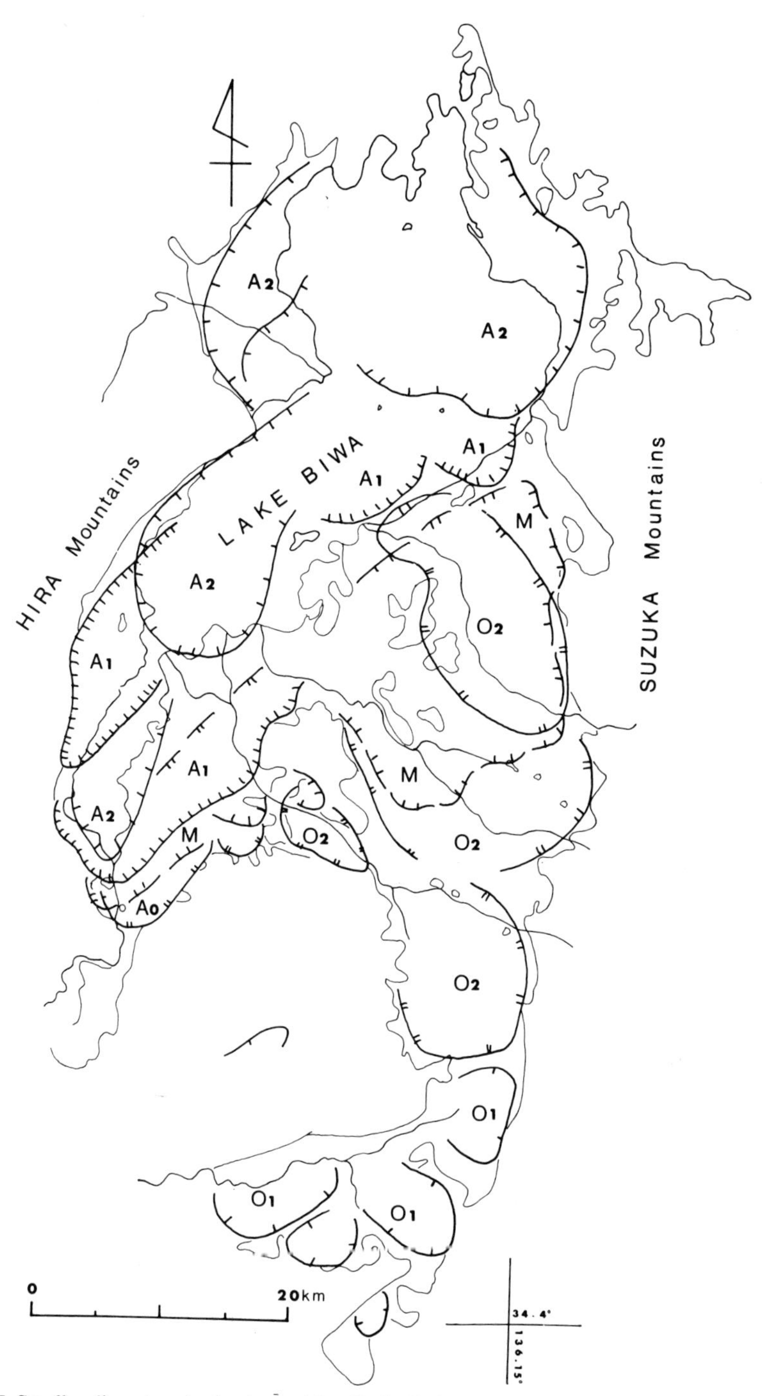

Fig. 17. Small sedimentary basins in Ōmi-Iga Basin (Yokoyama 1979a).
O_1: Old Stage I (From west, Shimagawara, Iga and Tsuge Sub-basins)
O_2: Old Stage II (From South, Sayama, Gamō and Echi Sub-basins)
M: Middle Gravel Stage
A_0: Actual sedimentary basin in Pre-A, Stage.
A_1: Actual Stage I
A_2: Actual Stage II

Table 1. Lithostratigraphy of the Plio-Pleistocene Series in Kinki and Tōkai Districts.

Osaka Group	
Manchidani Facies (20 ~ 50 m)	Mainly gravels with sands and marine clay beds (Ma7 ~ Ma10)
	Partial unconformity
Hacchoike Facies (30 ~ 80 m)	Alternations of marine clays (Ma3–Ma6) and lacustrine sands and muds
Nara Facies (50 ~ 80 m)	Mainly sands and gravels with some marine clay beds (Ma0 ~ Ma2)
	Unconformity
Senriyama Facies (70 m +)	dominant sand beds with some lacustrine muddy layers
Middle gravel Facies (0 ~ 200 m)	Pebble ~ cobble gravels
	Partial unconformity
Sennan Facies (200 m +)	Alternations of muds, sands and gravels in lacustrine origin

Kobiwako Group	
Takashima Formation (130–150 m)	Old Lake Biwa Basin
Katata Formation (470–600 m)	
Zeze Formation (155–165 m)	
Seta Gravel I (65 m)	Yōkaichi Formation (90 m)
Nangō Alternations (70 m)	Gamō Formation (484–504 m)
Actual Lake Biwa Basin	Sayama Formation (86–139 m)
	Iga-Aburahi Formation (500 m)
	Shimagawara Formation (100m)

Tōkai Group	
Komeno Formation (300 m)	Pebble — cobble gravels
Ōizumi Formation (360 m)	Alternations of lacustrine muds and sands
Kuragari Formation (0–140 m)	Mainly pebble gravels
Ichinohara Formation (220–470 m)	Alternations of lacustrine muds and gravels
Kono Formation (25–110 m)	Alternations of lacustrine muds and sands with lignites
Biroku Formation (0–25 m)	Cobble — boulder gravels (marginal gravel)

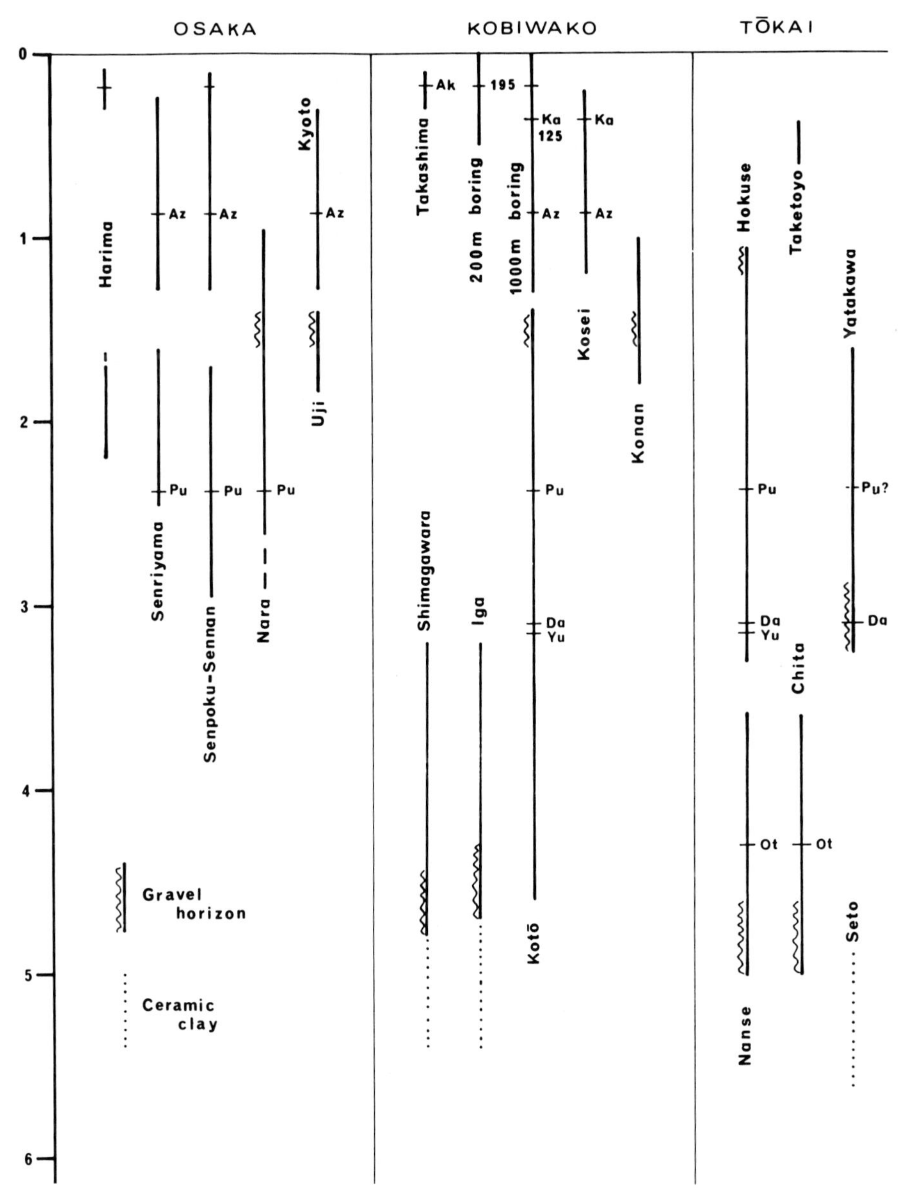

Fig. 18. Chronostratigraphy of the Osaka, Kobiwako (Paleo-Biwa) and Tōkai Groups.
Name of volcanic ash: Az: Azuki, Pu: Pumice, Ak: Akatsuki, 195; BB 195, Ka: Kamiōgi, 125; db 125, Da: Dacite, Yu: Yubunc, Ot: Ohta.

Ma 11 to Ma 13, have been found by core examinations of testing wells at several places in the Osaka Plain.

Two marine transgressions are recognized in Ma 0 from fossil diatoms (Nakanishi et al. 1969). The lower marine part, which contains the Grey Ash, yields *Coscinodiscus* cf. *commutatus, Cos. Rothii* var. *Normani, Cyclotella striata,* and *Cyc. stylorum.* The middle lacustrine part, containing the Yellow Ash, yields *Gommphonema augur,* and *Stephanodiscus niagarae,* etc., and the upper marine part contains the same diatoms as the lower one.

The lowermost portion of the Osaka Group yields plant remains such as *Ginkgo, Ketereelia, Pseudolarix, Metasequoia, Sequoia, Liquidambar,* etc., of the late Tertiary Age (Miki 1948). The assemblage of these fossil plants is called the *Metasequoia* flora (Itihara 1960). Characteristic elements of *Metasequoia* flora, with a few exceptions, disappeared during the early Pleistocene from Kinki District. Itihara (1960) distinguished between the flourishing age and the extinction age of *Metasequoia* flora. Peat layers of the upper part of the group yielded some forms suggesting a cold climate, for instance, *Pinus koraiensis, Menyanthes trifoliata, Larix gmelinii,* etc. The lowest horizon of all the beds which contain cold plant remains is subjacent to the Kamimura Volcanic Ash Layer as reported by Ibaraki Research Group (1966). The upper part of this group contains a layer with *Syzygium* and others which existed under a warm climatic condition. Some fossil elephants such as *Stegodon orientalis, Elephas shigensis* and *Stegodon akashiensis* were found in this group. The occurrences of the fossils and magnetostratigraphy in the Kobiwako Group were summarized in Table 2 and Fig. 19.

Table 2. Fossil list of the Kobiwako Group ('P' means 'Prefecture').

A. Mammals

Stegodon cf. *elephantoides*

1) Madei Igaueno city, Iga. (personal communciation with Dr. T. Kamei) – about 300 m below the Yubune Volcanic Ash. (Iga-Aburahi Formation).

Stegodon insignis sugiyamai

1) Middle stream of the River Sakura, Hino-cho, Kotō. (personal communication with Dr. T. Kamei & Mr. C. Matsuoka) — horizon of the Naka Volcanic Ash (Hino Alternations of Gamō Formation).

Elephas shigensis

1) Ono, Shiga-cho, Kosei; Matsumoto and Ozaki (1957) – about 7—8 m below the Ono Volcanic Ash (Ogoto Sands, Minamishō Clays, Katata Formation).
2) Mano, Kosei; Ikebe et al. (1966) — a few meters below the Ono Volcanic Ash.
3) Ōgi, Ōtsu City, Kosei; Ikebe et al. (1966) – lowermost part of the Kamiōgi Clays, Minamishō Clays, Katata Formation.

Stegodon orientalis

1) Minamishō, Katata-cho, Kosei; Ikebe et al. (1966) – about 6 m upper of the Kamiōgi Volcanic Ash (upper part of Kamiōgi Clays, Minamishō Clays, Katata Formation).

Cervus sp.

1) Minamishō, Ōtsu City, Kosei (personal communication with Mr. M. Tamura & Mr. Y. Okazaki) – upper part of the Kamiōgi Clays, Minamishō Clay, Katata Formation.

B. Reptiles

Tomistoma sp.

1) Kibōgaoka, Kōnan-cho, Kotō (personal communication with Mr. C. Matsuoka & Mr. Y. Okazaki) — horizon of the Iwamuro Volcanic Ash (Iwamuro Clays, Sayama Formation).

C. Molluscs

1) Shindō, Ayama-mura, Mie P.; Ikebe and Nakagawa cited in Matsushita (1953) *Anodonta* sp. *Viviparus* longispira – lower part of the Sayama Formation.

Table 2. continued

2) River Iwamuro, Kōga-cho, Shiga P.; *Cristaria* sp., *Unio* sp. – lower part of the Sayama Formation.
3) Jimpo, Kōnan-cho, Shiga P.; Ikebe and Nakagawa cited in Matsushita (1953) *Lanceolaria oxyrhyncha, Unio* sp., *Anodonta* sp., *Corbicula* sp., *Viviparus* sp. – middle part of the Sayama Formation.
4) Kita aki, Hino-cho, Shiga P.; Ikebe and Nakagawa cited in Matsushita (1953) *Cristaria* sp., *Anodonta* sp. – upper part of the Hino Alternations, Gamō Formation.
5) River Mano, south of Ono, Ōtsu City, Kosei (Ikebe 1933 and personal communication with Dr. N. Ikebe) *Corbicula sandai* – Just below the Ono Volcanic Ash (Kamiōgi Clays, Minamishō Clays, Katata Formation).
6) River Ashiarai, south of Imasen-no, Ōtsu City, Kosei (Ikebe 1933 and personal communication with Dr. N. Ikebe). *Viviparus longispira* – lower part of the Kamiōgi Clays, Minamishō Clays, Katata Formation.
7) River Ogoto, west of Ogoto, Ōtsu City, Kosei (Ikebe 1933 and personal communication with Dr. N. Ikebe). *Viviparus longispira, Semisulcospira libertina Phragmites* sp. – just above the Ogoto Volcanic Ash (Ogoto Sands, Minamishō Clays, Katata Formation).
8) River Ogoto, northwest of Sen-no, Ōtsu City, Kosei (Ikebe 1933 and personal communication with Dr. N. Ikebe), *Viviparus longispira* – lowest part of the Kamiōgi Clays, Minamishō Clays, Katata Formation.
9) River Tenjin, west of Katata, Ōtsu City, Kosei (Ikebe 1933 and personal communication with Dr. N. Ikebe), *Unio douglasiae, U. biwae, Inversidens brandti, I. hirasei, I. japanensis, I. reiniana, Cristaria plicata spatiosa, Viviparus japonicus.* – about 10 m below the White Volcanic Ash (Kamiōgi Clays, Minamishō, Clays, Katata Formation).
10) River Mano, north of Minamishō, Ōtsu City, Kosei (Ikebe 1933 and personal communication with Dr. N. Ikebe), *Viviparus longispira* – between the White and Kinukawa Volcanic Ashes (middle part of the Kamiōgi Clays, Minamishō Clays, Katata Formation).

D. Plants

1) Shidadani, Shimagahara mura, Miè P.; Miki (1948) *Glyptostrobus pencilis, Metasequoia disticha, Alnus japonica, Corylus heterophylla, Fagus hayatae, Nuphar akashiensis, Cocculus trilobus, Stephania japonica, Magnolia stellata, Schizandra nigra, Hamamelis parrotioides, Spondias axillaris* var. *polymeris, Sabia japonica, Berchemia racemosa, Paliurus nipponicus, Vitis Thunbergii, Nyssa sylvatica, Trapa incisa, T. mammillifera, Cornus controversa, Pieris* sp., *Pterostyrax corymbosum, Styrax japonica, S. rugosa, Brasenia purpurea, Sparganium protojaponicum* – Shimagawara Formation.
2) Kami-iso, Kōnan-cho, Shiga P.; Takaya (1963) *Metasequoia disticha, Alnus japonica, Styrax japonica* — Nojiri Clays, Sayama Formation.
3) Takano, Kōga-cho, Shiga P.; Maenaka et al. (1977), *Pinus thunbergii* — Nojiri Clays, Sayama Formation.
4. River Sōma, Kibukawa, Minakuchi-cho, Shiga P.; Miki (1948) *Glyptostrobus pencilis, Metasequoia disticha, Sequoia couttisie* – Nunobikiyama Alternations, Gamō Formation.
5) Bessho, Hino-cho, Shiga P.; Miki (1956) *Metasequoia districha, Liquidambar formosana, Sapium sebiferum, Styrax japonica* – Yohkigaoka Clays, Hino Clays, Gamō Formation.
6. Naka, Yokkaichi city, Shiga P.; *Metasequoia disticha* — horizon of the Naka Volcanic Ash (Hino Alternations, Gamō Formation).
7) Wani, Ōtsu City; PO2 of Hayashi (1974) *Juglans* sp. (*Juglans mandshurica* type) – Minamishō Clays, Katata Formation.
8) Shimoōgi, Ōtsu City; PO6 of Hayashi (1974) *Paliurus nipponicus, Trapa macropoda* – upper part of the Minamishō Clays, Katata Formation.
9) Kamiōgi, Ōtsu City; PO9 of Hayashi (1974) *Alnus japonica, Styrax japonica* – upper part of the Minamishō Clays, Katata Formation.
10) Akatsuki kaidō, Shin-Asahi-cho, Takashima, Kohoku. Yokoyama et al.(1977c), *Carex* sp., *Sparganium* sp., *Scirpus?, Menyanthes trifoliata, Ceratophyllum demersum, Monochoria* sp., *Styrax* sp., *Sapium* sp., *Xanthoxylum* cf. *simulans, X.* cf. *piperitum, X.* sp., *Trichosanthes* sp.?, *Prunus?, Polygonum?, Trapa* sp., *Nuphar* sp., *Sambucus sieboldiana, Vitis* sp., – Akatsukikaidō Sandy Bed, Takashima Formation (See Fig. 36 and Table 11).

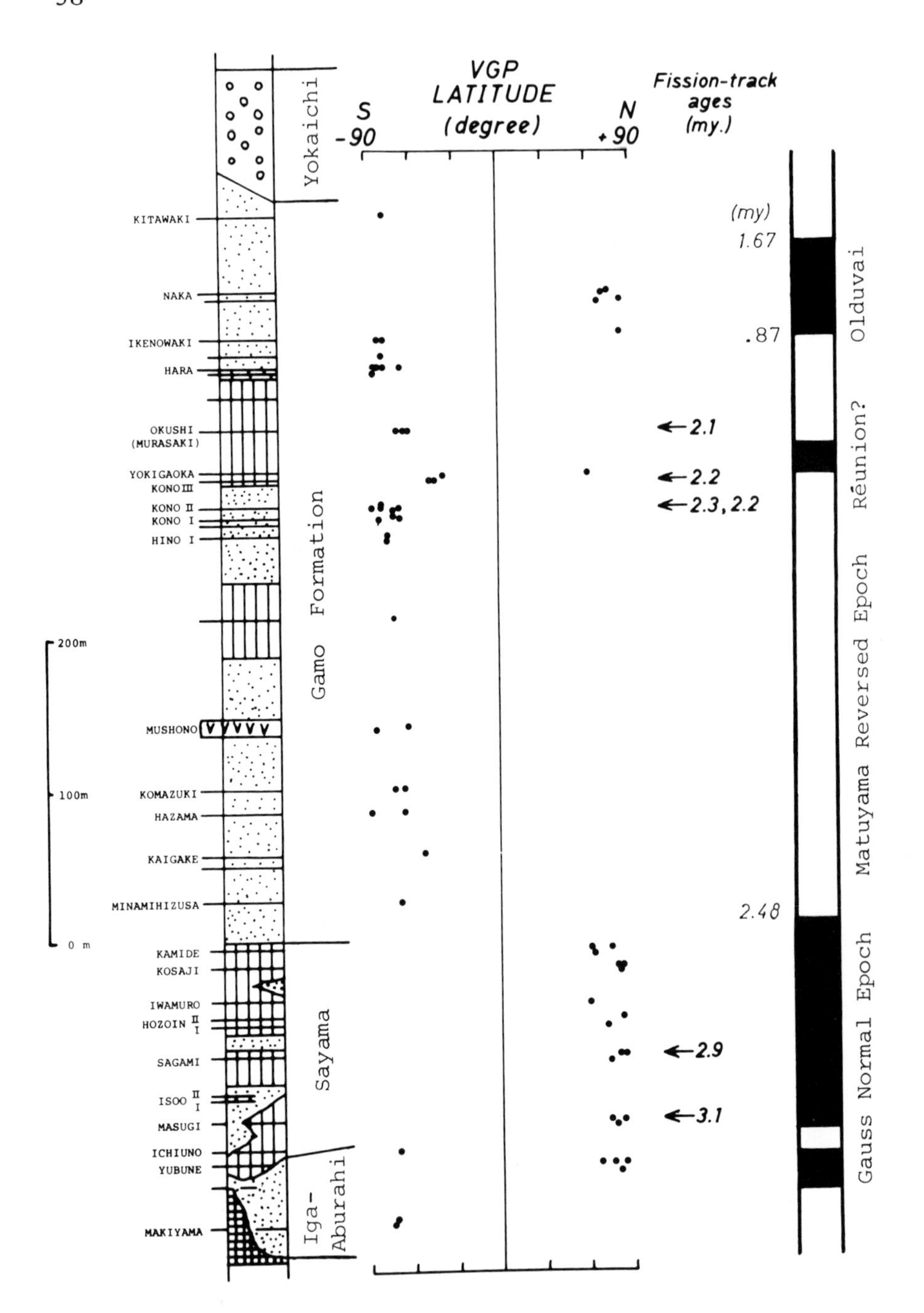

Fig. 19. Plot of VGP latitudes vs. stratigraphic positions of sampling sites, and correlation of the magnetozones to paleomagnetic time scale (Mankinen & Dalrymple 1979). Fission-track ages of volcanic ashes are also shown after Nishimura and Sasajima (1970), and Yokoyama et al. (1977b). Geologic columnar section (left) was compiled from Yokoyama et al. (1968), and Tamura et al. (1977) (after Hayashida & Yokoyama, 1979).

Table 3. Fission-track ages of volcanic ashes in the Plio-Pleistocene and Holocene sediments in Kinki and Tōkai Districts.

Sample number	Sampling site	Name of volcanic ash	Horizon (Depth)	Age (m.y.)
1. Kobiwako Group				
BB.85			(37.1 m)	0.08 ± 0.02[a]
BB.153			(62.4 m)	0.11 ± 0.02[a]
BB.195	200 m boring core		(82.8 m)	0.17 ± 0.03[b]
BB.239			(99.2 m)	0.18 ± 0.04[a]
BB.267			(110.3 m)	0.27 ± 0.05[b]
BB.453			(182.2 m)	0.46 ± 0.09[b]
db.161	1,000 m boring core	Kinukawa	(378.7 m)	0.47 ± 0.09[c]
BW.2	River Kisen	Biotite	Ogoto Clays of Katata F.	0.70 ± 0.14[b]
BE.1001	Okushi	Murasaki	Hino Clays of Gamō F.	2.1 ± 0.4[e]
BE.1002	Okushi	Kono III		2.2 ± 0.4[e]
BE.1004	Okushi	Kono II		2.3 ± 0.5[e]
BE.246	Nishiōji	Kono II	Hino Clays of Gamō F.	2.2 ± 0.3[d]
BE.49	Sagami	Sagami		2.9 ± 0.4[d]
BE.202	Kamimasugi	Masugi	Sayama F.	3.1 ± 0.5[d]
2. Osaka Group				
Os.45	Osaka Univ.	Kasuri	Just below Ma 8	0.37 ± 0.04[d] 0.38 ± 0.03[d]
Os.48	River Senri	Azuki	Ma 3	0.87 ± 0.07
Os.32	Kōmyōike	Kōmyōike	between Ma 1 and Ma 2	1.1 ± 0.1[d]
Os.39	Shiba	Grey	Ma 0	1.2 ± 0.2[i]
Os.58	Toyonaka	Shima-kuma yama		2.3 ± 0.2[d] 2.4 ± 0.3[d]
Ky.108	Fukenji	Fukenji	First Clay of Tanabe	1.6 ± 0.3[f]
Ky.94	Higashihata	Higashihata	under Middle Gravel	2.2 ± 0.4[f]
3. Tōkai Group				
A.7772	Rokkoku	Komeno I	Upper part of Ōizumi F.	1.5 ± 0.3[g]
A.888	Kono	Dacite	Kono F.	3.00 ± 0.6[g]
A.786	River Tashida	Dacite	Kono F.	2.8 ± 0.6[g]
A.268	Ichinohara	Dacite	Kono F.	3.05 ± 0.6[h]

[a] Nishimura & Yokoyama (1973)
[b] Nishimura & Yokoyama (1975)
[c] Takemura et al. (1976)
[d] Nishimura & Sasajima (1970)
[e] Yokoyama et al. (1977b)
[f] Yokoyama et al. (1978)
[g] Yokoyama et al. (1980)
[h] Miyamura et al. (1976)
[i] Yokoyama (1979b)

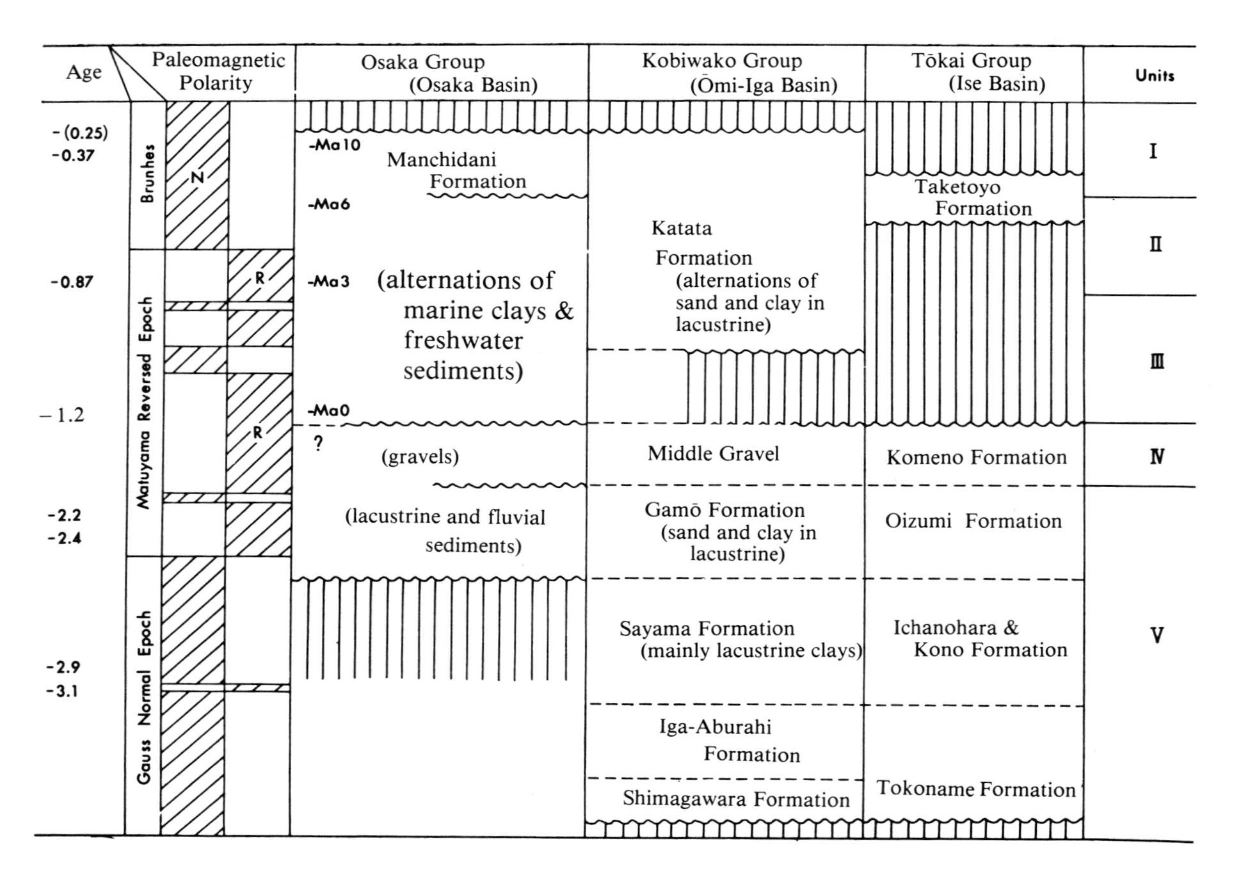

Fig. 20. Outline of lithostratigraphy of the Plio-Pleistocene in Kinki and Tōkai Districts (Yokoyama et al. 1977a).

Fission-track age determinations on phenocrysts, mostly zircon grains, which were separated from some volcanic ash layers were carried out (Nishimura & Sasajima 1970, Nishimura & Yokoyama 1973, 1975, Yokoyama et al. 1976, Yokoyama et al. 1978). All fission-track ages in regard to the Quaternary and Plio-Pleistocene sediments are summarized in Table 3.

Lithostratigraphic Review. Recently, three unconformities were drawn in this group. Yokoyama et al. (1977a) stated: 'The outline of lithostratigraphy and chronostratigraphy of the Plio-Pleistocene sediments in three isolated basins and the radiometric ages of intercalated volcanic ashes are shown in Fig. 20 and Table 3. From this information, we can observe the following points as regards the Plio-Pleistocene sediments of Southwest Japan:

1. The range of radiometric age is from 4.0 to 0.26 m.y. It corresponds to Gauss to Brunhes Epoch in paleomegnetism.
2. Since 1.7 m.y. ago, several regressions and transgressions have taken place in this area.

In the Osaka Group, we can find unconformities at three horizons, that is, the basal horizon of the Manchidani Formation, a few meters below Ma 0, and the basal horizon of the Middle Gravels. Their radiometric ages are about 0.45 m.y., 1.2 – 1.3 m.y. and 2.0 m.y., respectively. The unconformity at the base of the Manchidani Formation is always found in marginal areas of the sedimentary basin. In the center of the basin, sediments are conformably accumulated at this horizon. Also the unconformity at the basal part of the Middle Gravels is found only in marginal regions. We therefore think that these two were caused by the tectonic movement.

On the other hand, the unconformity at the horizon a few meters below Ma 0, is thought to have been caused by a transgression. We call this the Shiba Unconformity and found one outcrop in the Senriyama Hills, a type locality of the Osaka Group, at the end of last year. Up to that time, we had thought that the Plio-Pleistocene series was a continuous sequence in the center of the sedimentary basin. Now, we do not deny the possibility of such unconformity.

In the outcrop, the paleomagnetic polarities are as follows:

a. The upper part above this unconformity has reversed polarities, which correspond to the reversed time between the Oluduvai and Jaramillo events. This correlation is reasonable, judging from the fission-track age of the Grey Volcanic Ash Layer in the Ma 0 marine clay. The age is 1.2 ± 0.2 m.y.

b. The part under the Shiba Unconformity has normal polarities, and the part about 20 m below this horizon has reversed polarities. This normal polarity therefore belongs to the Oluduvai event. This is substantiated by the fission-track age of the Shimakumayama Volcanic Ash, which is intercalated in the horizon about 20 to 30 m below this horizon. The age of this volcanic ash is 2.4 ± 0.3 m.y.

c. The thickness of sediments, which lack by this unconformity, may attain to 75 meters or more.

From these pieces of information, we can speculate on a transgression in this area in the earliest Quaternary Period, just after the Oluduvai event. If this could be the same as the Calabrian transgression, the phenomenon could be a global one.

The Kobiwako (Paleo-Biwa) Group is divided into two essential units. The older one is distributed in the southern area. The Middle Gravel Formation represents the regressive gravels of an ancient basin which disappeared about 1.3 m.y. ago. The new one, actual basin of Lake Biwa, appeared 1.7 to 2.0 m.y. ago, and accumulated very thick sediments about 1,000 m thick.'

The lower part of the Osaka Group, that is Unit V in Fig. 20, is mainly composed of lacustrine, grey-greenish clay, sand and gravel beds. It is widely distributed in the Sennan-Senhoku Area, a southern part of Osaka Prefecture, the Nara Basin and the Awaji-shima Island. And it is also somewhat seen in the Harima Basin, Senriyama Hills and foothills of Tamba Mountain Land.

Their lithology is given in the columnar sections of Table 1. There are outstanding gravel facies in the middle horizon of this group. We call this 'the Middle Gravel Facies'. We can see at the north of Nara City and Tanabe District, Kyoto Prefecture that the Middle Gravel Facies washed out or cut off the sedimentary bodies of the lower horizon. This gravel facies is well distributed at the Uji Hills and the Tanabe Hills in southern Kyoto. But it is not distributed in the Senriyama Hills. It may be changed to coarse sand facies or washed out by the Shiba Uncomformity under Ma 0. The upper part of the Osaka Group is composed of alternations of marine clay beds and fresh water sediments.

Outline of the Tōkai Group

General Review. The Tōkai Group is distributed widely and exposed in the hills surrounding the Nōbi Plain and Mie Prefecture. It is composed largely of terrestrial sediments deposited in lacustrine and fluvial environments. The *Pinus trifolia* flora is found in the Seto-ceramic Clay Bed, the lowermost part of the group. In Mie Prefecture, *Stegodon elephantoides* occurs in the lower part of this group and *Stegodon akashiensis* is found in the middle part.

Recently, detailed lithostratigraphical, magnetostratigraphical and tephro-stratigraphical surveys were carried out (Yokoyama 1971, Itoigawa 1971, Mori 1971a, b, Makinouchi 1975a, b, Takemura 1978 and so on). Then fission-track age determinations of some volcanic ash layers were conducted(Yokoyama et al. 1980).

Lithostratigraphy of the Kobiwako Group

The Plio-Pleistocene Kobiwako Group occupies the Ōmi-Iga Basin of the Lake Biwa area. It is mainly composed of gravels, sands and clays with some peats and volcanic ash layers, which are wholly lacustrine or fluvial in origin. The total thickness of this group is estimated at 1,800–2,000 meters. They yield many lacustrine molluscs akin to the living forms in the present Lake Biwa.

Elephas shigensis and *Stegodon orientalis* are found on the west side of Lake Biwa (Kosei). *Metasequoia* flora also occurs. The radiogenic ages of minerals contained in some volcanic ash layers of this group were determined by the fission-track method. These ages range from 3.0 to 0.3 million years. The absolute age of the basal part of it was estimated to be 4.5-5.0 million years ago from now by the radiogenic age determination and the thickness of sediments (Fig. 39).

The Kobiwako (Paleo Biwa) Group is divided into six formations in an ascending order: Shimagahara, Iga-Aburahi, Sayama, Gamō, Yōkaichi and Katata Formations (Yokoyama 1969). In the present paper, the writer recognizes a few additional formations, that is, Zeze and Takashima Formations. Stratigraphical relations among them are summarized in Table 4, and Figs. 21 and 22. The type localities of individual beds are given in Table 5 and Fig. 11.

Table 4. Lithostratigraphy of the Kobiwako Group.

1. Actual Lake Biwa Basin		
Takashima Formation	Aibano Gravel Bed (50–70 m)	Cobble — pebble gravels
	Akatsukikaidō Sandy Bed (50 m +)	Coarse sands with coaly clays and gravels
	Inokuchi Gravel Bed (10 m+)	Boulder — cobble gravel
	Shiratsuchidani Bed (20 m)	Coaly muds with gravels
Katata Formation	Ryūge Sands and Gravels (100 m)	Cobble — pebble gravels
	Minamishō Clays	
	a) Kamiōgi Clays (110–130 m)	Massive muds
	b) Ogoto Sands (50–70 m)	Grey fine and brown coarse sands
	c) Ogoto Clays (110–200 m)	Alternations of predominant clay and slightly sand
	Wani Sands (100 m+)	Coarse sands with thin clays and silts
Zeze Formation	Chausuyama Bed (40 m)	Sands and gravels
	Fujimidai Bed (50–60 m)	Alternations of sand and mud
	Akibadai Bed (30 m+)	Medium sands
	Seta Gravel II (15 m)	Pebble — cobble gravels
	Zinryō Sands (20 m)	Medium — coarse sands with gravels
Seta Gravel I (Yōkaichi Formation)	(65 m)	Pebble — cobble gravels

Table 4. continued

Nangō Alternations (Gamō Formations)	(70 m+)	Alternations of sand and mud
2. Old Lake Biwa Basin		
Yōkaichi Formation	(90 m)	Cobble — granule gravels
Gamō Formation	Hino Alternations	
	a) Uriuzutōge Sands and Gravels (50–70 m)	Sands and gravels with some mud and silt layers
	b) Rengeji Fossil Wood Bed (4 m)	Silty and sandy muds bearing many fossil woods
	c) Nakazaiji Sands and Clays (60 m)	Alternations of mud and dominant sand
	Hino Clay (140–190 m)	Massive clays and sandy silts with some sand layers
	Nunobikiyama Alternations (180 m)	Medium brown sands with clay and silt
Sayama Formation	Kosaji Clays (25 m)	Bluish grey massive clays
	Iwamuro Sands (0–4 m)	Brown medium sands
	Iwamuro Clays (12 m)	Bluish grey massive clays
	Sunazaka Sands (6 m)	Brown medium — coarse sands
	Nojiri Clays (27–42 m)	Bluish grey clays with slightly fine — medium sands
	Kazuraki Sands (0–50 m)	Brown coarse — medium sands with gravels and muds
	Wata-Ichiuno Clays (6–20 m)	Bluish grey clays with some sandy layers
Iga-Aburahi Formation	Aburahi Sands (350 m)	Brown sands with slightly silts and muds
	Iga Clay (50 m?)	Greenish grey massive muds
	Iga Sands and Gravels (50 m)	Very coarse sands and dominant pebble — cobble gravels
Shimagahara Formation*	(100 m)	Ceramic clay facies coarse sand and gravel with lignites and coaly clays

* Both Shimagawara and Shimagahara are at the same location but have a different pronunciation.

Kotō and Iga Areas

Shimagahara Formation (100 m). This is a ceramic clay facies in the Iga Basin and its western highlands (Shigaraki and Shimagahara Area). It is mainly composed of arkosic coarse sands and gravels containing some lignites and coaly clays called 'Kibushi Clay'. The sediments are about 100 meters thick in the center of the Iga Basin, while they are about 30 meters thick in the western highlands. These sediments were deposited in some hollows, the diameter of which is a few kilometers. Two typical columns of this formation are illustrated in Fig. 23.

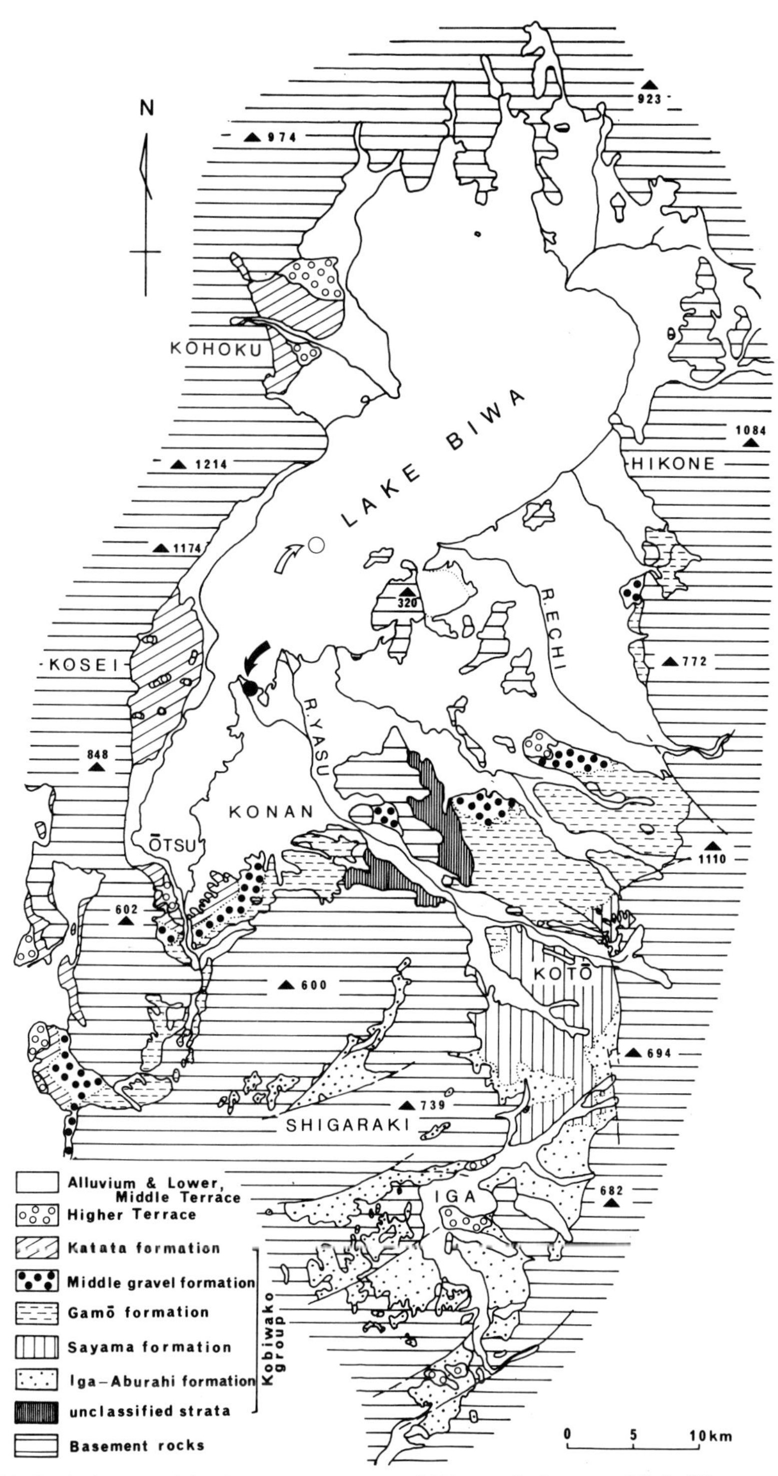

Fig. 21. Geologic map of the Quaternary system and Pliocene Series around Lake Biwa. ○: 200 m drilling point, ●: 1,000 m drilling point.

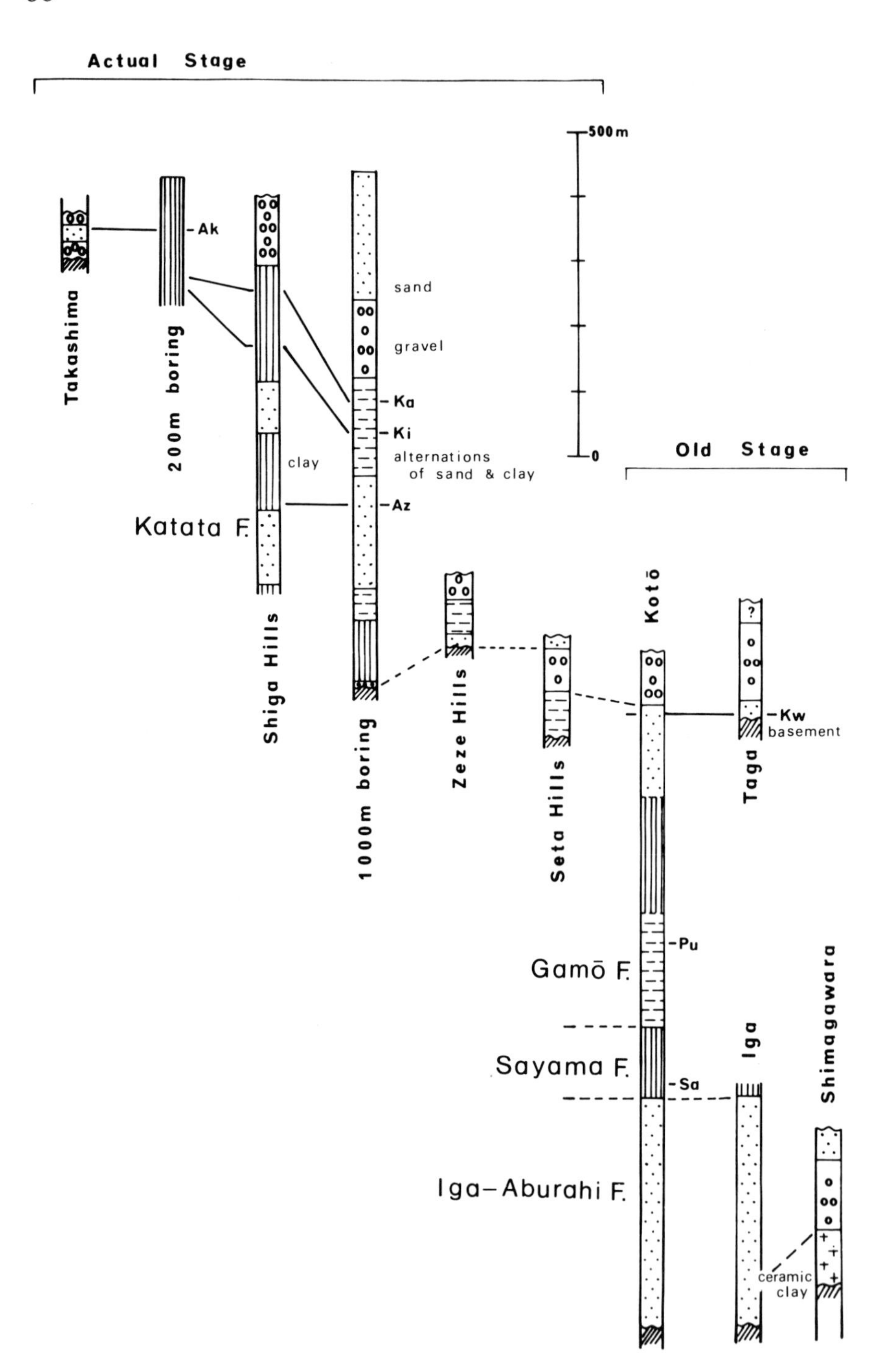

Fig. 22. Columnar sections of the Kobiwako Group in Ōmi-Iga Basin.
Name of volcanic ash: Ak: Akatsuki, Ka: Kamiōgi, Ki: Kinukawa, Az: Azuki, Kw: Kitawaki, Pu: Mushono, Sa: Sagami.

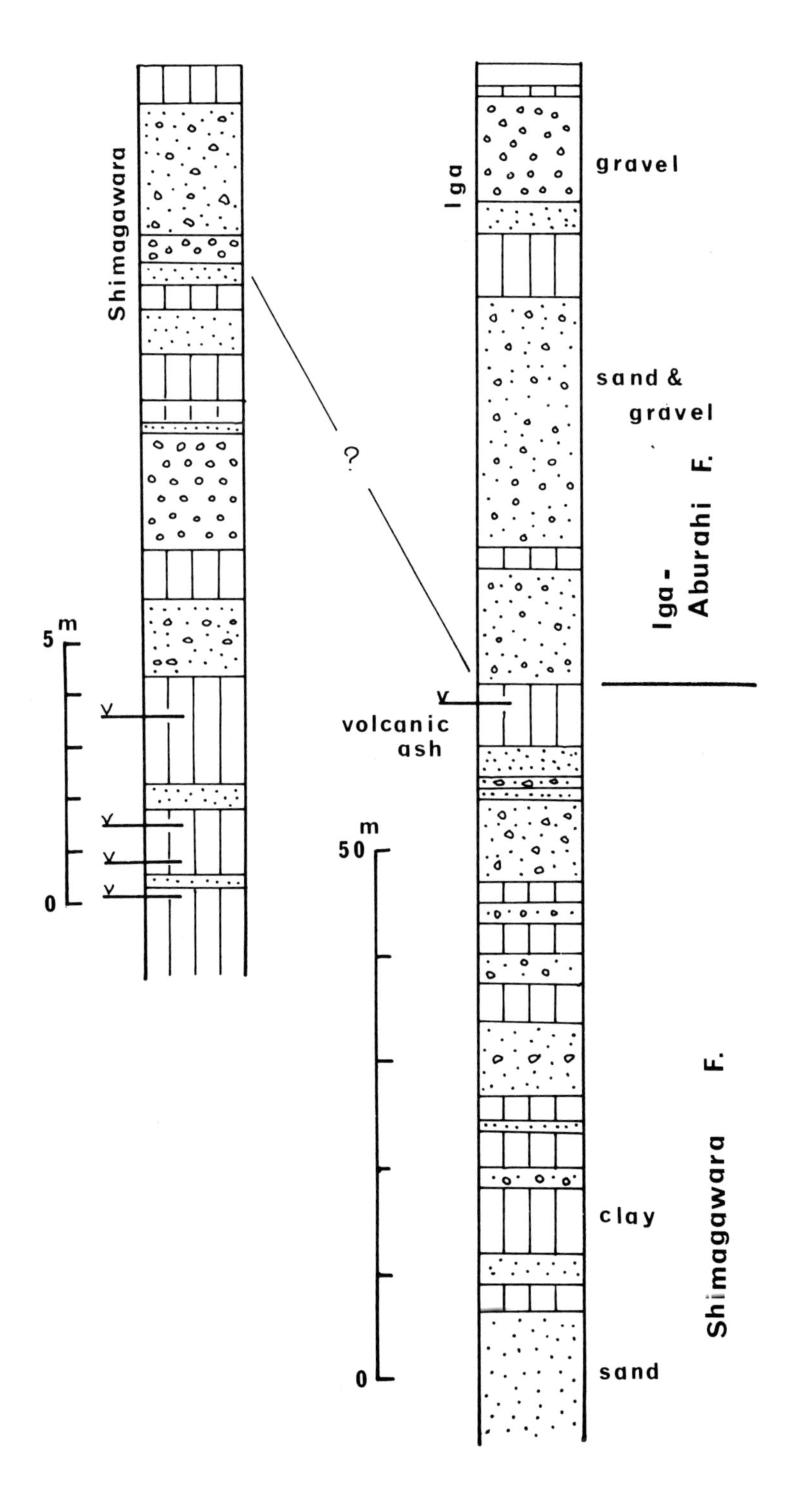

Fig. 23. Columnar sections of the Iga-Aburahi and Shimagawara Formations (after Yokoyama et al. 1979a).

Table 5. Type locality of Formation and Bed of the Kobiwako Group.

Name of Formation	Name of Bed	Type Locality (P. is Prefecture)	Area
Takashima Formation	**1** Aibano Gravel Bed	Northeast of Shimokoga, Adogawa-cho, Takashima-gun, Shiga Pref.	Takashima Hills
	2 Akatsukikaidō Sandy Bed	Roadside of Road Akatsuki kaidō, Shin-asahi-cho, Takashima-gun, Shiga Pref.	
	3 Inokuchi Gravel Bed	North of Inokuchi, Shin-asahi-cho, Takashima-gun, Shiga Pref.	
	4 Shiratsuchidani Bed	Valley Shiratsuchidani, Miyamae, Kutsuki-mura, Takashima-gun, Shiga Pref.	Middle stream of the River Ado
Katata Formation	**5** Ryūge Sands and Gravels	South of Ryūge, cliffs of the River Yowatari, Ōtsu City, Shiga Pref.	Shiga Hills
	Minamishō Clays		
	6 Kamiōgi Clays	Minamishō and Kamiōgi, Ōtsu City, Shiga Pref.	
	7 Ogoto Sands	West of Ogoto, Ōtsu City, Shiga Pref.	
	8 Ogoto Clays	West of Ogoto, Ōtsu City and cliffs of the River Kisen, Shiga-cho, Shiga-gun, Shiga Pref.	
	9 Wani Sands	Cliffs along the down-stream of the River Kisen, Shiga-cho, Shiga-gun, Shiga Pref.	
Zeze Formation	**10** Chausuyama Bed	Chausuyama, Zeze, Ōtsu City, Shiga Pref.	Zeze Hills
	11 Fujimidai Bed	Fujimidai, Zeze, Ōtsu City, Shiga Pref.	
	12 Akibadai Bed	Akibadai, Zeze, Ōtsu City, Shiga Pref.	
	13 Seta Gravel II	Top of the hills at southeastern part of Seta, Seta-cho, Ōtsu City, Shiga Pref.	Seta Hills
	14 Zinryō Sands	Zinryo, Seta-cho, Ōtsu City, Shiga Pref.	
Yōkaichi Formation	**15** Seta Gravel I	Southern part of Seta Hills, Seta-cho, Ōtsu City, Shiga Pref.	Seta Hills
	16 Yōkaichi Gravels	Nunobikiyama Hills, Yōkaichi City, Shiga Pref.	Hino Hills
Gamō Formation	**17** Nangō Alternations	Nangō area, Ōtsu City, Shiga Pref.	Seta Hills
	18 Kasuga Alternations	Kasuga, Minakuchi-cho, Kōga-gun Shiga Pref.	Minakuchi Hills

	Hino Alternations		
	19 Uriuzutōge Sands and Gravels	Uriuzutōge, Nunobikiyama Hills. Hino-cho, Gamō-gun, Shiga Pref.	Hino Hills
	20 Rengeji Fossil Wood Bed	Cliffs of the River Sakura, South of Rengeji, Hino-cho, Gamō-gun, Shiga Pref.	
	21 Nakazaiji Sands and Clays	Cliffs of the River Sakura, South of Nakazaiji, Hino-cho, Gamō-gun, Shiga Pref.	
	Hino Clays		
	22 Yohkigaoka Clays	Yohkigaoka, Hino-cho, Gamō-gun, Shiga Pref.	Minakuchi Hills
	23 Bessho Sands	South of Bessho, Hino-cho, Gamō-gun, Shiga Pref.	
	24 Nakahata Clays	North of Nakahata, Minakuchi-cho, Kōga-gun, Shiga Pref.	
	25 Hino Clays	Cliffs of the middle stream of the River Sakura, Hino-cho, Gamō-gun, Shiga Pref.	Hino Hills
	26 Sakura Alternations	Cliffs of the upper stream of the River Sakura, Hino-cho, Gamō-gun, Shiga Pref.	
	27 Nunobikiyama Alternations	Northeast of Mushono, Kibukawa-cho, Kōga-gun, Shiga Pref.	Kōga Area
Sayama Formation	**28** Kosaji Clays	North of Kosaji, Kōga-cho, Kōga-gun, Shiga Pref.	Kōga Area
	29 Iwamuro Sands	South of Iwamuro, Kōga-cho, Kōga-gun, Shiga Pref.	
	30 Iwamuro Clays	South of Iwamuro, Kōga-cho, Kōga-gun, Shiga Pref.	
	31 Sunazaka Sands	Sunazaka, Oki, Kōga-cho, Kōga-gun, Shiga Pref.	
	32 Sunazaka Clays	Sunazaka, Oki, Kōga-cho, Kōga-gun, Shiga Pref.	
	33 Sunazaka Sands I	Sunazaka, Oki, Kōga-cho, Kōga-gun, Shiga Pref.	
	34 Nojiri Clays	Between Nojiri and Koji, Kōnan-cho, Kōga-gun, Shiga Pref.	
	35 Kazuraki Sands	Ōmitobashi, Kōnan-cho, Kōga-gun, Shiga Pref.	
	36 Ichiuno Clays	Cliffs of the River Aburahi, Mobira, Kōga-cho, Kōga-gun, Shiga Pref.	
	37 Wata Clays	Wata, Kōga-cho, Kōga-gun, Shiga Pref.	
Iga-Aburahi Formation	**38** Aburahi Sands	Aburahi, Kōga-cho, Kōga-gun, Shiga Pref.	Kōga Area
	39 Sanagu Sands	South of Sanagu Station, Ueno City, Mie Pref.	Iga Basin
	40 Iga Clays	Shoda, Nabari City, Mie Pref.	
	41 Iga Gravels	Ueno City, Mie Pref.	
	42 Nakaya Sands	Nakaya, Shimagahara-mura, Ayama-gun, Mie Pref.	Shimagahara and Tayama Area
	43 Okuda Alternations	Okuda, Shimagahara-mura, Ayama-gun, Mie Pref.	
	44 Tayama Gravels	Tayama, Sōraku-gun, Kyoto Pref.	
Shimagahara Formation	**45**	Kawamoto Kōzan, Shimagahara-mura, Ayama-gun, Mie Pref.	

Iga-Aburahi Formation (350 *m*). This formation is distributed in the Iga Basin and its surrounding hills. The main exposed area are divided into the following three; i.e., Kōga-Tsuge area, Iga Basin and its western highland.

a. Kōga-Tsuge area. Aburahi Sands are mainly exposed in the western foot-area of the Suzuka Mountain Land, chiefly composed of medium to coarse sand layers with some silt and sandy clay beds. The thickness is about 350 m at main exposed area, but about 100 m in the center and western margin of the sedimentary basin. In the western margin, Makiyama Area, a volcanic ash is seen. This is white, very fine volcanic ash, and about 5 cm in thickness. This is called 'Makiyama Volcanic Ash.' Recently, *Stegodon elephantoides* was found in this bed. The horizon of this fossil elephant is about 300 m below from the Yubune Volcanic Ash.

b. Iga area. Iga Clays are mainly composed of silty clay and silt layers bearing the sandy layers. This is exposed at the eastern area of the Iga Basin. Many fossil molluscs were found in this bed; for example, *Viviparus* sp., *Sinotaia* sp., *Anodonta* sp., and *Lanceolaria* sp. The thickness is about 50 m (Table 6).

Iga Gravels are composed of pebble gravels, and are distributed in the underground of Iga Basin. The thickness is about 60 m.

c. Western highland (Shimagahara and Tayama area). Nakaya Sands are white medium sand layers and are exposed at southern foothills of the Shigaraki Hills. The thickness is 10 m.

Okuda Alternations are composed of the alternations of mud, sand and gravelly sand. A volcanic ash layer, called 'Okuda Volcanic Ash', was found in a mud layer. This may be the same bed to the Makiyama Volcanic Ash in the Aburahi Sands.

Tayama Gravels are composed of cobble to pebble gravels. This is exposed on the western highlands of Iga Basin, that is Shimagahara and Tayama Area. The thickness is about 50 m.

Table 6. Lithostratigraphy of the Iga-Aburahi Formation.

	Western Highland	Iga Basin	Kōga — Tsuge Area
	Nakaya Sands (10 m) Medium — coarse sands and pebble gravels	Sanagu Sands (50 m) Medium — coarse sands with some silt layers	Aburahi Sands (350 m) Alternations of clay and predominant medium sand
Iga-Aburahi Formation	Okuda Alternations (30 m)	Iga Clays (50 m?) Greenish silty muds with sands	
	Tayama Gravels (50 m) Cobble — pebble gravels	Iga Gravels (60 m) Pebble gravels	

Sayama Formation (100 *m*). Yokoyama et al. (1968) described the stratigraphy of this formation in detail. It is composed of alternations of thick clay and thin medium-grained sand, except that Kazuraki Sands contain some pebble gravels. Because outstanding volcanic ash layers are intercalated in these clay facies, the stratigraphical horizons can be determined in detail. The volcanic ashes are called in an ascending order: Tōge, Yubune, Ichiuno, Masugi, Iso I, II, Sagami, Sunazaka, Hōzōin I, II, Iwamuro, Kosaji and Kamide Volcanic Ash Layers (Yokoyama et al. 1968 and Yokoyama et al. 1979a). This formation is divided into ten members: Wata Clays, Kazuraki Sands, Kosaji Clays and so on, as shown in Table 7 and Figs. 24, 25.

Gamō Formation (500 *m*). Most recently, the stratigraphy of this formation has been clarified in detail by Tamura et al. (1977) and Kobiwako Research Group (1977). It is divided by Tamura et al. (1977) into three members in an ascending order: Nunobikiyama Alternations, Hino Clays and Kasuga Alternations. Now, we divide it into the following three members (Table 8):

a. Nunobikiyama Alternations. In this member, medium grained, brown sand beds are dominant and greenish grey colored clay and silt layers are common. At least five remarkable volcanic ashes are seen in this member. They are called in an ascending order: Minamihizusa, Kaigake, Hazama, Komazuki and Mushono Volcanic Ashes (Figs. 26 ~ 28).

b. Hino Clays. Hino Clays are mostly composed of massive clays and sandy silts which make many high cliffs along the river. In the western half of the sedimentary basin, Minakuchi Hills, there are some sandy parts called Bessho

Table 7. Lithostratigraphy of the Sayama Formation.

Name of bed	Lithofacies	Volcanic Ash
Kosaji Clays (25 m)	Bluish grey massive clays	Kamide Kosaji
Iwamuro Sands (0–4 m)	Brown medium sands, distributed only eastern margin of sedimentary basin	
Iwamuro Clays (12 m)	Bluish grey massive clays with some fine — medium sands	Iwamuro Hōzōin I, II
Sunazaka Sands II (6 m)	Brown medium — coarse sands	
Sunazaka Clays (13 m)	Bluish grey massive clays with some fine — medium sands	Sunazaka
Sunazaka Sands I (4 m)	Brown medium — coarse sands	
Nojiri Clays (10–25 m)	Bluish grey massive clays	Sagami
Kazuraki Sands (0–50 m)	Coarse sands with pebble — granule gravels Lower half interfingers with the Ichiuno Clays	Iso I, II
Ichiuno & Wata Clays (6–20 m)	Alternations of sand and predominant bluish grey massive clay	Masugi Yubune Tōge

Fig. 24. Geologic map of the Kobiwako Group, mainly Sayama Formations in Kōga Area (Partly modified after Yokoyama et al. 1968). V: Kobiwako Group (symbols of volcanic ashes are same as in Fig. 25), W: Miocene Ayukawa Group, X: Paleozoic, Y: Granite, Z: Dip and strike.

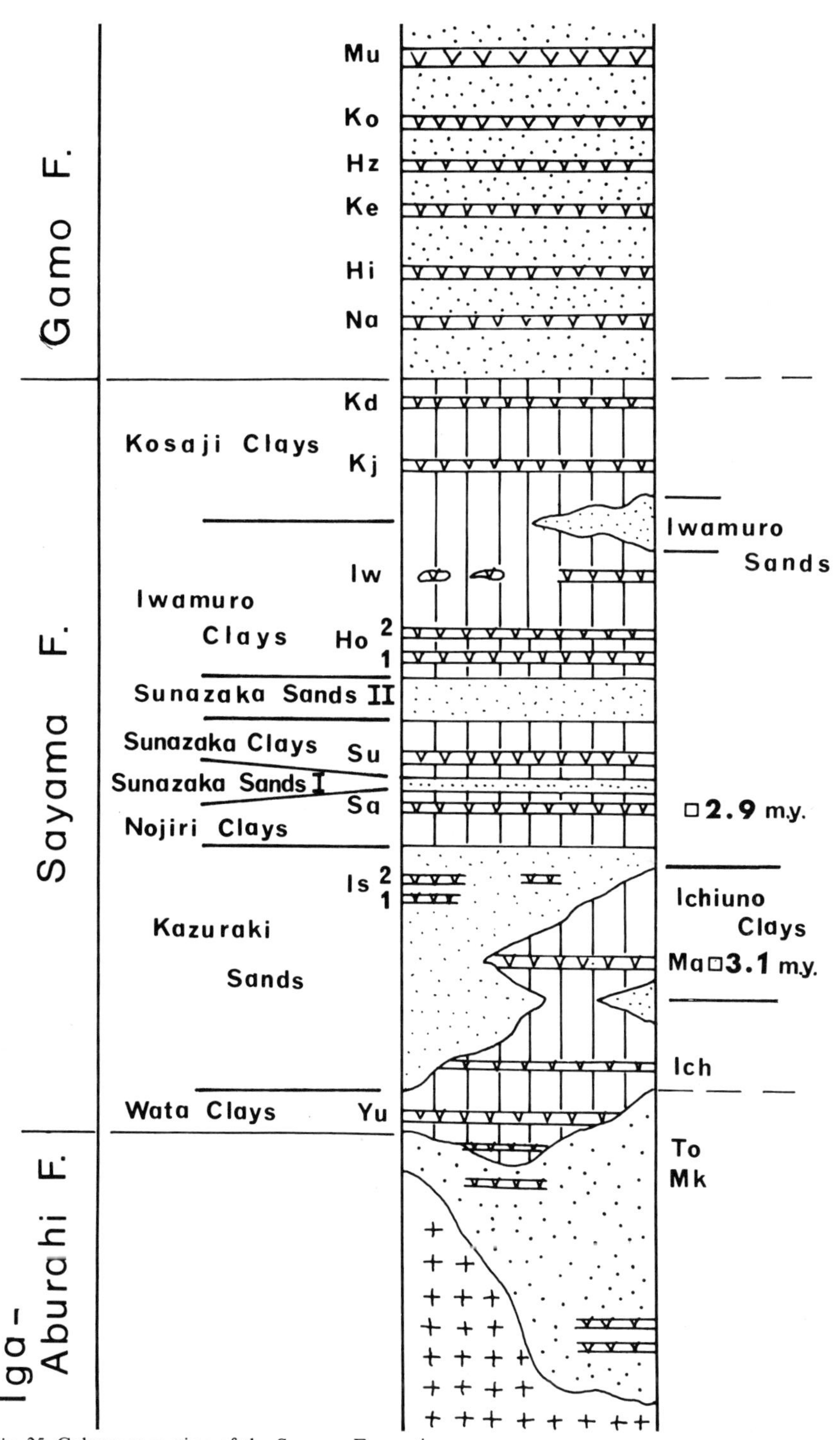

Fig. 25. Columnar section of the Sayama Formation.
Name of volcanic ash — Mu: Mushono, Ko: Komazuki, Hz: Hazama, Ke: Kaigake, Hi: Minamihizasa, Na: Naiki, Kd: Kamide, Kj: Kosaji, Iw: Iwamuro, Ho: Hōzōin, Su: Sunazaka, Sa: Sagami, Is: Iso, Ma: Masugi, Ich: Ishiuno, Yu: Yubune, To: Tōge, Mk: Makiyama.

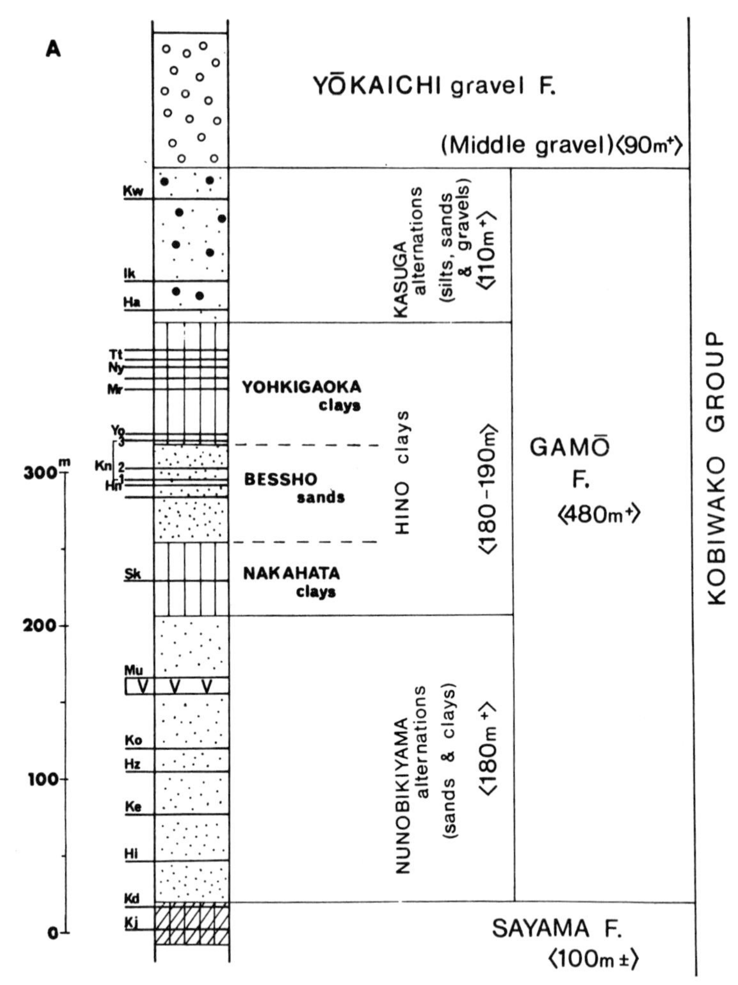

Fig. 26. Columnar section of the Gamō Formation.
A. Minakuchi Hills (Tamura et al. 1977); B. Hino Hills (Yokohama et al. 1979a).
Name of volcanic ash – Kw: Kitawaki, Nk: Naka, Ik: Ikenowaki, Ha: Hara, Tt: Toyota, Ny: Nakayama, Mr: Murasaki, Yo: Yohkigaoka, Kn: Kono, Hn: Hino, Sk: Sakuradani.

Sands in the middle horizon of this member. At least ten outstanding volcanic ash layers are found in this member. They are called in an ascending order: Sakuradani, Hino I, II, Kono I, II, III, Yohkigaoka, Murasaki, Nakayama and Toyota Volcanic Ashes.

In 1976, a stratigraphic and tephrostratigraphic study of Hino Clays and the Hino Alternations at the eastern margin of the sedimentary basin, the northern part of Hino Hills, was carried out in detail by K. Amemori (Yokoyama et al. 1979a). The former is mud dominant alternations of sand and clay in this area. It is about 140 m thick. The lithology is mostly mud facies, which is composed of massive, clay silty clay and silt beds of about 1—4 m in thickness. Some medium

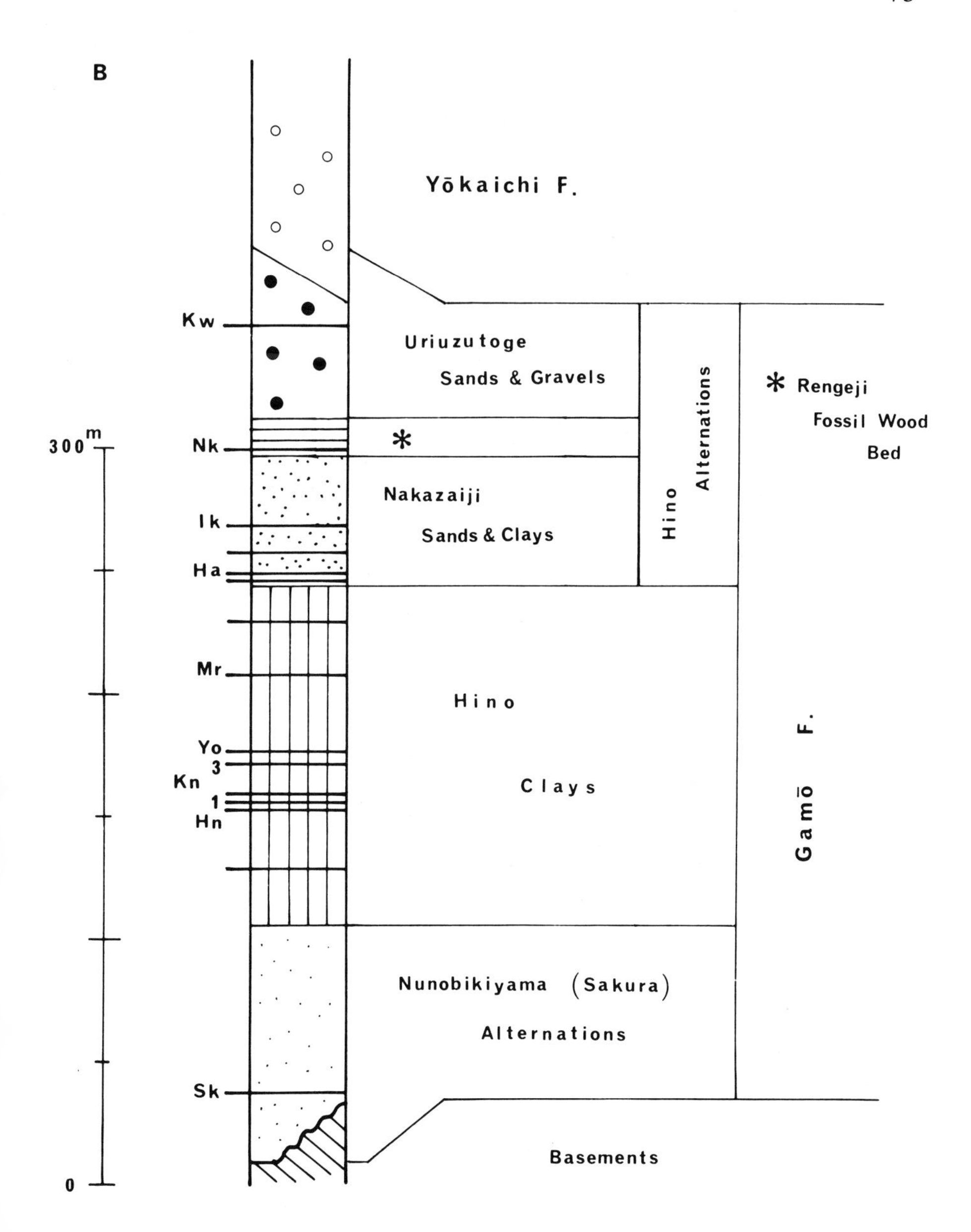

sand layers, less than 1 m thick, are often intercalated. Nine outstanding volcanic ash beds are found in it: Hino I, II, Kono I, II, III, Yohkigaoka, Murasaki and Toyota Volcanic Ashes in an ascending order (Table 8, Fig. 26A, Fig. 28).

c. Hino Alternations. They are composed of sands, gravels and clays and are about 120 m thick. They differ from Hino Clays in the existence of gravel layers, much change of lithofacies and many plant remains. The Hino Alternations are

Table 8. Lithostratigraphy of the Gamō Formation.

Minakuchi Hills				Hino Hills			
Name of stratum (thickness)		Lithofacies	Ash	Name of stratum (thickness)		Lithofacies	Ash
Yōkaichi Formation							
Kasuga Alternation (110 m)		Sands and gravels with some greenish silt and sandy mud layers	Kitawaki Naka Ikenowaki	Hino Alternations	Uriuzutōge Sands and Gravels (70 m)	Sands and gravels with some mud and silt layers	Kitawaki
Hino Clays	Yohkigaoka Clay (80 m)	Alternations of medium–fine sands and predominant bluish grey clays	Hara Toyota Nakayama Murasaki Yohkigaoka Kono III		Rengeji fossil Wood Bed (4 m)	Silty and sandy muds bearing many fossil woods	Naka
					Nakazaiji Sands and Clays (60 m)	Alternations of mud and dominant sand	Ikenowaki
	Bessho Sand (50 ~ 60 m)	Coarse–medium sands with greenish grey silts and muds	Kono I, II Hino I, II	Hino Clay (140 m)		Massive clays and sandy silts with some sand layers	Hara Murasaki Yohkigaoka Kono III Kono II Kono I Sakura II, III
	Nakahata Clay (50 m)	Greenish grey massive clays with brown medium-fine sands	Sakuradani				
Nunobikiyama Alternations (180 m)		Medium brown sands with clay and silt	Mushono Komazuki Hazama Kaigake Minamihizusa Naiki	Sakura Alternations (80 m)		Alternations of sand and gravels with some thin muddy layers and lignites	Sakura I Sakuradani
Sayama Formation (Kosaji Clays)							
after (Tamura et al. 1977)				after Yokoyama et al. 1979a			

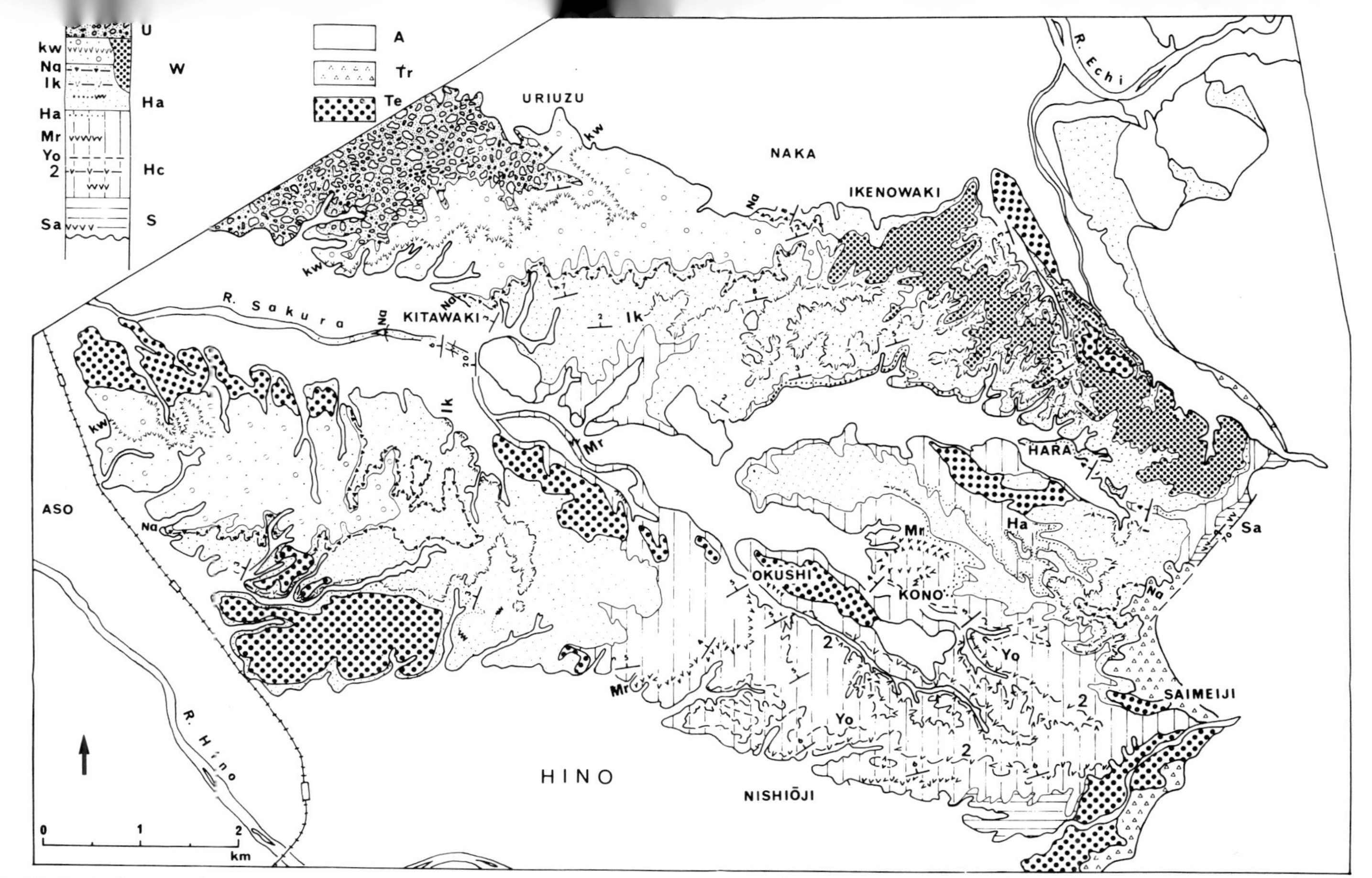

Fig. 27. Geologic map of the Kobiwako Group, mainly Gamō Formation in Hino Hills(after Yokoyama et al. 1979a).
A: Alluvium, Tr: Fan deposit, Te: Terrace, W: Kobiwako Group, U: Yokaichi Formation, Ha: Hino Alternations, Hc: Hino Clays, S: Sakura Alternations (symbols of volcanic ashes are same as in Fig. 26).

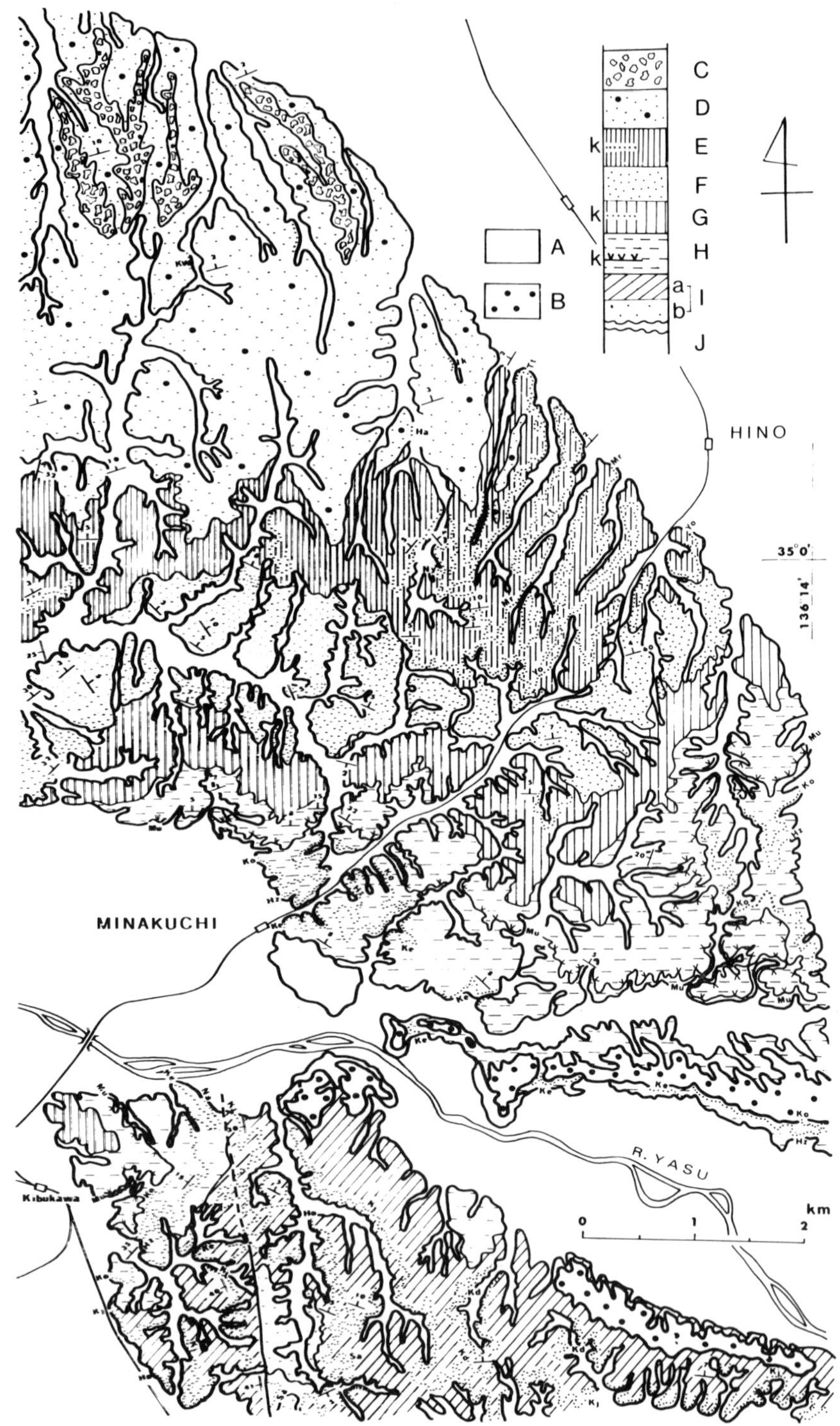

Fig. 28. Geologic map of the Kobiwako Group, mainly Gamō and Sayama Formations, at Minakuchi Hills (Tamura et al. 1977).
A: Alluvium, B: Terrace, C: Yōkaichi Formation, D: Kasuga Alternations, E ~ G: Hino Clays — E: Yohkigaoka Clays, F: Bessho Sands, G: Nakahata Clays, H: Nunobikiyama Alternations, I: Sayama Formation, J: Basements.

divided into three parts in an ascending order: Nakazaiji Sands and Clays, Rengeji Fossil Wood Beds and Uriuzutōge Sands and Gravels. Moreover, a marginal gravel facies, called 'Wanami Gravels', is seen at the eastern margin of the sedimentary basin. The lithofacies and thickness are given in Table 8 and Fig. 26B.

Yōkaichi Formation (90 *m*). This is an outstanding gravel horizon in the Kobiwako Group, in which cobble to granule gravels of chert, sandstone and other Paleozic rocks are abundant. This is probably the top gravel of the older sedimentary basin of the ancient Lake Biwa and the source of this gravel was the upheaving of the Suzuka Mountain Ranges.

Konan Area (Figs. 29 and 30)

There are a few reports on the Kobiwako Group in this area. Nakazawa and Ishida (1959) reported two volcanic ashes and some plant fossils: *Styrax* sp. cf. *microcarpa, Pterocarya paliurus, Stewartia monadelpha* and *Sapium sebiferum* var. *pleistoceaca* from the foundation of the dam, called Nangō-Araizeki, when it was under construction. The stratigraphy of the Kobiwako (Paleo-Biwa) Group in the Seta-Nangō area was proposed first by Yokoyama and Nishiyama Research Group (1969). The columnar sections of this group at the Zeze and Seta-Nangō areas were contained in the above report with the description of eight ash layers and some plant remains: *Juglans* cf. *megacinerea, J. mandshurica* and *Picea maximowiczii.*

Yokoyama (1969) also divided the group of this area into the following five members in an ascending order: Nangō Alternations, Seta Gravels I, Zinryō Sands, Seta Gravels II and Zeze Alternations. Sekinotsu Gravels are added below the above-mentioned five members, and the Zeze Alternations are divided into three parts, the Sonoyama, Zeze and Chausuyama Formations (Ishida et al. 1976a). At that time, Seta Gravels I are called the Ishiyama Formation. Now, the lithology and thickness of these strata are shown in Table 9. Sekinotsu Gravels may be correlated to Seta Gravels II (Ishiyama Formation).

Nangō Alternations (70 m^+) = *Gamō Formation.* This is mainly composed of alternations of sand and mud, and is divided into three parts: sand-rich lower and upper parts and a mud-rich middle part.

The lower part is mainly composed of fine to medium grained bedded sand and is about 20 meters in thickness. It is well exposed along the northern side of the River Daido. Good outcrops of it are seen to west of Ishisue. At the top of the lower part, there is a gravel bed which is about 3 meters thick. This bed is also exposed at the west of Ishisue. The middle part cropping out west of Nangō is about 10–16 meters in thickness. Two mud layers are a good marker bed in this area. The upper part at Akao is composed of coarse sand with some pebble to

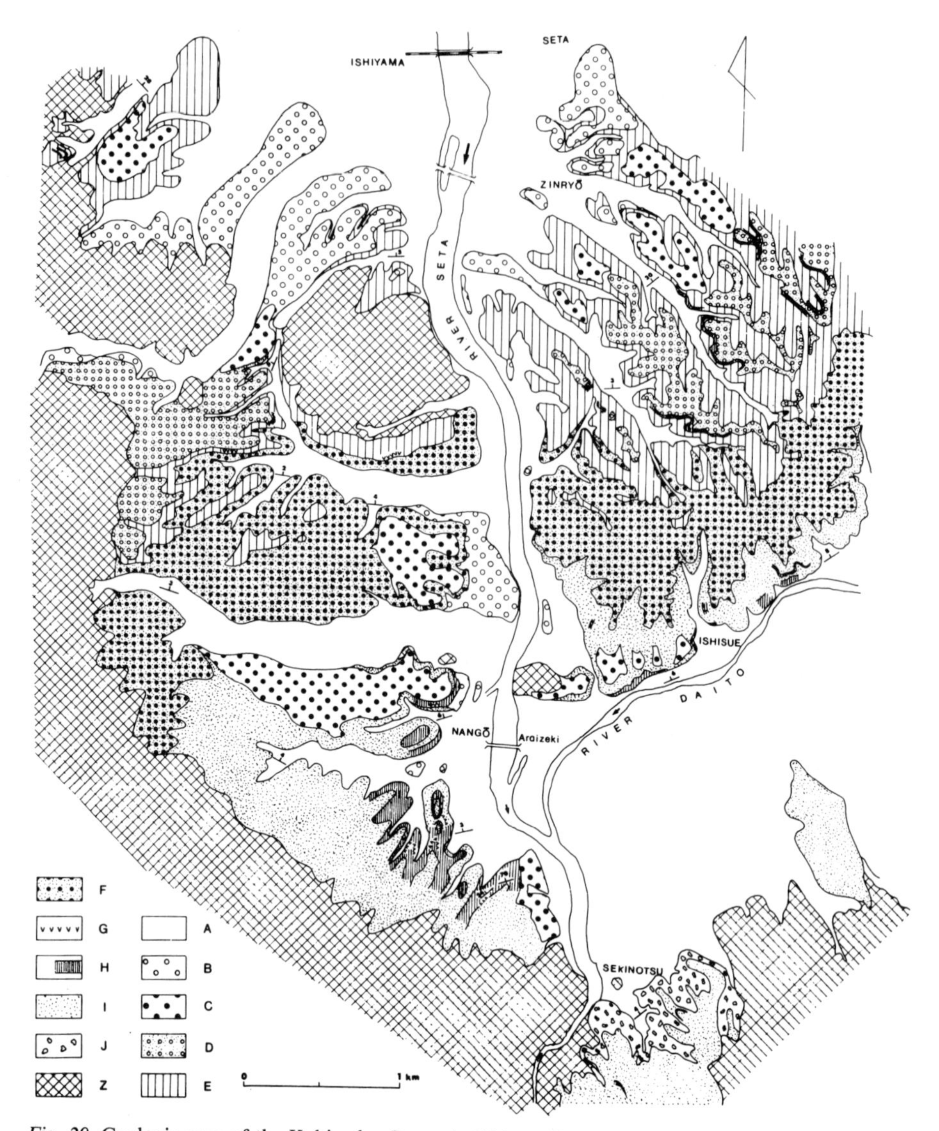

Fig. 29. Geologic map of the Kobiwako Group in Ishiyata-Seta region (Ishida et al. 1976a). A: Alluvium, B: Lower Terrace, C: Higher Terrace, D: Seta Gravel I, E: Zinryō Sands, F: Seta Gravel II, G: volanic ash, H: mud, I: Nangō Alternations, J: Sekinotsu Gravel, Z: Basements.

granule gravels and laminated silt-fine sands. It is about 20 meters thick.

Six volcanic ash layers are intercalated with the Nangō Alternations. They are called Nangō I–VI Volcanic Ashes by Yokoyama (1969). The lowest two ashes were discovered at the bottom of the River Seta (Nakazawa and Ishida 1959), but now they are invisible. Nakazawa and Ishida (1959) reported that these ashes contained orthorhombic pyroxene and hornblende crystals.

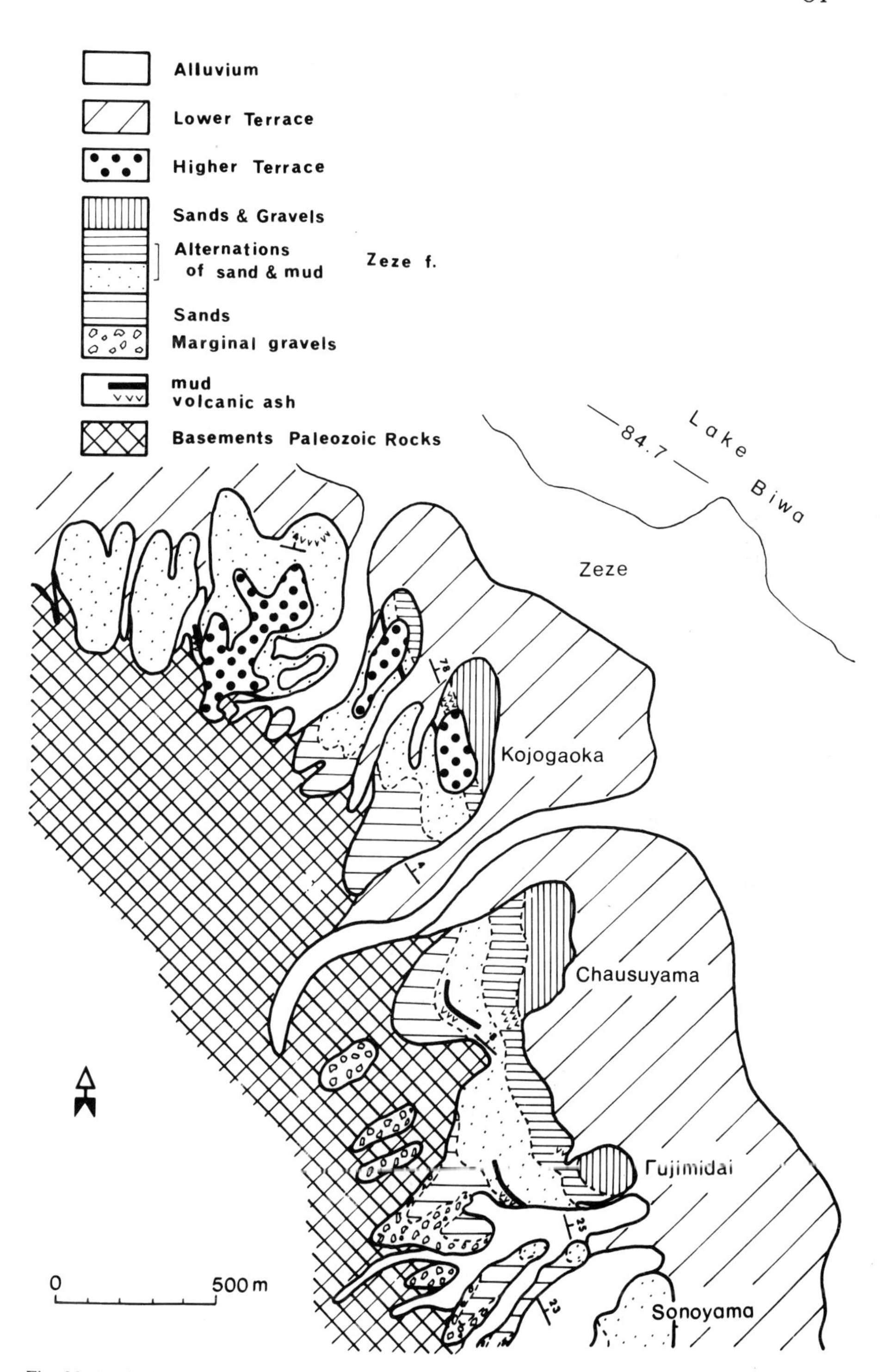

Fig. 30. Geologic map of the Kobiwako Group in Zeze-Ishiyama area (Ishida et al. 1976a).

Table 9. Stratigraphy of the Kobiwako Group in Konan Region.

	Actual stage of Lake Biwa (Zeze Area)			Old stage of Lake Biwa (Seta Hills)		
	Name of stratum	Lithofacies	Volcanic ash	Name of stratum	Lithofacies	Volcanic ash
Zeze Formation	Chausuyama Bed (40 m)	Coarse sands and pebble–cobble gravels	Zeze II			
	Fujimidai Bed (50–60 m)	Alternations of sand and mud	Zeze I			
	Akibadai Bed (30 m)	Mainly cross bedded medium–coarse sands	Toyo			
	(marginal basal gravels) (10–20 m)	Angular cobble–boulder gravels (matrix is muddy sand)		Seta Gravel II (15 m)	Pebble–cobble gravels (very weathered)	
Yōkaichi Formation				Zinryō Sands (20 m)	Medium–coarse sands with gravels and coaly muds	Ishiyama I, II Terabe
				Seta Gravel I (65 m)	Upper half: pebble–cobble gravels Lower half: sands and gravels	
Gamō Formation				Nangō Alternations (70 m)	Alternations of sand and mud	Nangō I ~ VI

Seta Gravels I (*about* 65 *m*) = *Yōkaichi Formation.* This is composed of pebble and cobble gravels. The gravels are mostly from Paleozoic. Chert is very rich and sandstone is as common as in the Sekinotsu Gravels. This is the Ishiyama Formation of Ishida et al. (1976a).

Zeze Formation

a. Zinryō Sand (*about* 20 *m*). Zinryō Sand is mainly composed of medium to coarse sand with granule and pebble gravels. There is a muddy upper part about 3 meters thick. In this muddy bed, there is a coaly part, containing fruit remains of *Trapa* sp.

A volcanic ash seam, Terabe Ash, was found in this formation. This ash contains green hornblendes. Takaya (1963) correlated this to Ikenowaki Ash in the Kotō Region. In this formation, another two volcanic ashes were discovered at the west of the Ishiyama Prison. These two are called Ishiyama I and II Volcanic Ashes (Yokoyama 1969). In these ashes, hornblende is rich and pyroxenes are few. Plant remains such as *Juglans* cf. *megacinerea, J. mandshurica, Picea maximowiczii, Styrax japonica* and *Alnus japonica* were obtained from peaty sands and clays intercalating in gravels 1 meter below Ishiyama I Ash.

b. Seta Gravels II (15 m^+). This is distributed on the top parts of the southern Seta Hills. These gravels were involved as the old terrace deposits by Takaya (1963). But this is divided into two formations. The lower belongs to the Kobiwako (Paleo-Biwa) Group and the upper is the terrace gravels. This is the Seta Formation of Ishida et al. (1976a).

c. Zeze Alternations. This is divided into three parts in an ascending order: Sonoyama, Zeze and Chausuyama Formations by Ishida et al. (1976a), and Akibadai, Fujimidai and Chausuyama Beds by Yokoyama et al. (1979a).

*c*1. *Akibadai Bed* (30 m^+). This is mainly composed of medium sands. Its lower half abuts on the basements, which are composed of the Paleozoic rocks. The marginal gravels are less than 10 meters in thickness. This may be above or lateral to Seta Gravels II.

*c*2. *Fujimidai Bed* (50–60 *m*). This is composed of rhythmic alternations of sand and mud. Five cycles are found. Generally speaking, the mud layers are about 3–6 meters in thickness and the sand layers are 3–10 meters in thickness. A volcanic ash seam, 0.5–3 cm thick, is intercalated with the mud of the lowermost cycle. The mud of the third from the base of this bed at Tōyō Seiki yields lacustrine molluscs: *Inversidens* cf. *hirasei, I.* cf. *reiniana, Cristaria plicata spatiosa, Anodonta lauta* and *Lanceolaria* sp. by S. Ishida. In the upper part of this bed, a pinkish white colored volcanic ash was found at Chausuyama to the south of Zeze. This volcanic ash is about 5 cm thick and heavily weathered; this was called Zeze I Ash by Yokoyama and Maenaka (1971).

*c*3. *Chausuyama Bed* (*about* 40 *m*). This is composed of sands and gravels, and is divided into three parts: lower gravels, middle sands and upper gravels. In the lower gravels, a white ash, about 20–25 cm thick, is present. This was weathered and called Zeze II Ash by Yokoyama and Maenaka (1971). The lower gravels are mainly composed of pebbles, which came from the Paleozoic. The matrix sands are generally coarse. The middle sands are mainly composed of coarse sands with small pebbles and granules. The upper gravels are mainly composed of cobble to pebble gravels which came from the Paleozoic rocks.

Kosei Area

The Kobiwako Group of this area is called 'Katata Formation'. The detailed stratigraphy of the Katata Formation was presented first by Ikebe (1933). Subsequent discussions were made by a few other investigators (Araki 1960, Takaya 1963, Hayashi 1974, Yokoyama 1975b, Hayashida et al. 1976).

Some fossil elephants such as *Stegodon orientalis* and *Elephas shigensis* were found in the Katata Formation. The horizon of these fossils is important also for the correlations to the Plio-Pleistocene series in other areas, especially the far eastern part of Asia.

There is a hilly country between the Hira Mountains and the west coast of Lake Biwa. Its height ranges from 10 to 200 meters above the water surface of the lake, which is about 84 m above the sea level. This is called the 'Shiga Hills' and is composed of Plio-Pleistocene loose sediments which are the Katata Formation.

The writer divided the Katata Formation into three members in a previous paper (Yokoyama 1969); Wani Sands, Minamisho Clays and Ryūge Sands and Gravels in an ascending order. It was divided into eight by Hayashi (1974): Nijigaoka Clays, Kitahama Sands, Kisen Clays, Takajo Alternations, Hiraen Clays, Kurihara Alternations, Sakawa Clays and Ryūge Sands and Gravels in the same order. The relation of the two divisions and lithofacies of each member is shown in Table 10. The columnar section and the geologic map of the Katata Formation are given in Figs. 31 ~ 35.

Wani Sands (100 m^+). The outcrops of Wani Sands are seen in the northeastern part of the Shiga Hills. They are well exposed along the River Kisen and at the Wani Area, and are mainly composed of coarse-grained sand layers with thin bluish grey clays and silts. Two white volcanic ash beds, called Nijigaoka I, II, are seen in the lowest part of this member.

Minamishō Clays (270–400 *m*). In the Ogoto Area, Minamishō Clays can be subdivided into three parts: lower clays, middle sands and upper clays. The writer (1975b) calls them Kamiōgi Clays, Ogoto Sands and Ogoto Clays respectively. But in the northern River Kisen Area, there is no such clear difference among the three as shown in Figs. 34, 35.

a. Ogoto Clays (110–200 *m*). In the Ogoto Area, they are mainly composed of alternations of predominant clay beds and slightly sand beds. The thickness of both clay and silt layers is generally about 2 meters and the maximum value of

Table 10. Lithostratigraphy of the Katata Formation (after Yokoyama 1975b).

Yokoyama						Hayashi (1974)
Fission-track age ×10^5 Yrs	Volcanic ash	Member			Lithofacies	
3.0		Ryūge Sands and Gravels (100 m)			pebble–cobble gravels	Ryūge Sands and Gravels (90 m)
5.0	Kamiōgi Kinukawa White Ono	Minamishō Clays (270 m)	Upper clays	Kamiōgo Clays (110 m)	predominant clays with a little sandy layers	Sakawa Clays (40—60 m)
7.0	Ogoto Biotite		Middle sands	Ogoto Sands (50 m)	alternations of clays and predominant fine–coarse sands	Kurihara Alternations (70 m + −) Hiraen Clays (35—40 m)
9.0 10.0	Azuki		Lower clays	Kisengawa Clays (110 m)	alternations of medium sands and dominant clays	Takajo Alternations (35 m+ −)
		Wani Sands (110 m)			alternations of a little clays and predominant sands	Kitahama Sands (45—50 m)
12.0					clay dominant	Nijigaoka Clays (15 m)

clay facies is 10 meters in thickness. On the other hand, the sand beds are 50–20 cm thick and the maximum value is about 3 meters thick. Ogoto Clays are intercalated with two outstanding volcanic ash layers: Azuki and Biotite Volcanic Ash Layers. They are well exposed at cliffs along the Rivers Kisen and Ten at the northern end of the Shiga Hills. Azuki Volcanic Ash is also seen at the Ogoto Area. It is also the most useful key bed in the Osaka Group.

b. Ogoto Sands (50–70 *m*). Ogoto Sands are mainly composed of sandy facies, grey fine sand and brown coarse sand. There is a horizon of muddy facies about 10 m thick in the middle part, in which two white volcanic ashes are found. They are called Ogoto and Kurihara Volcanic Ashes. All fossils of *Elephas shigensis* in

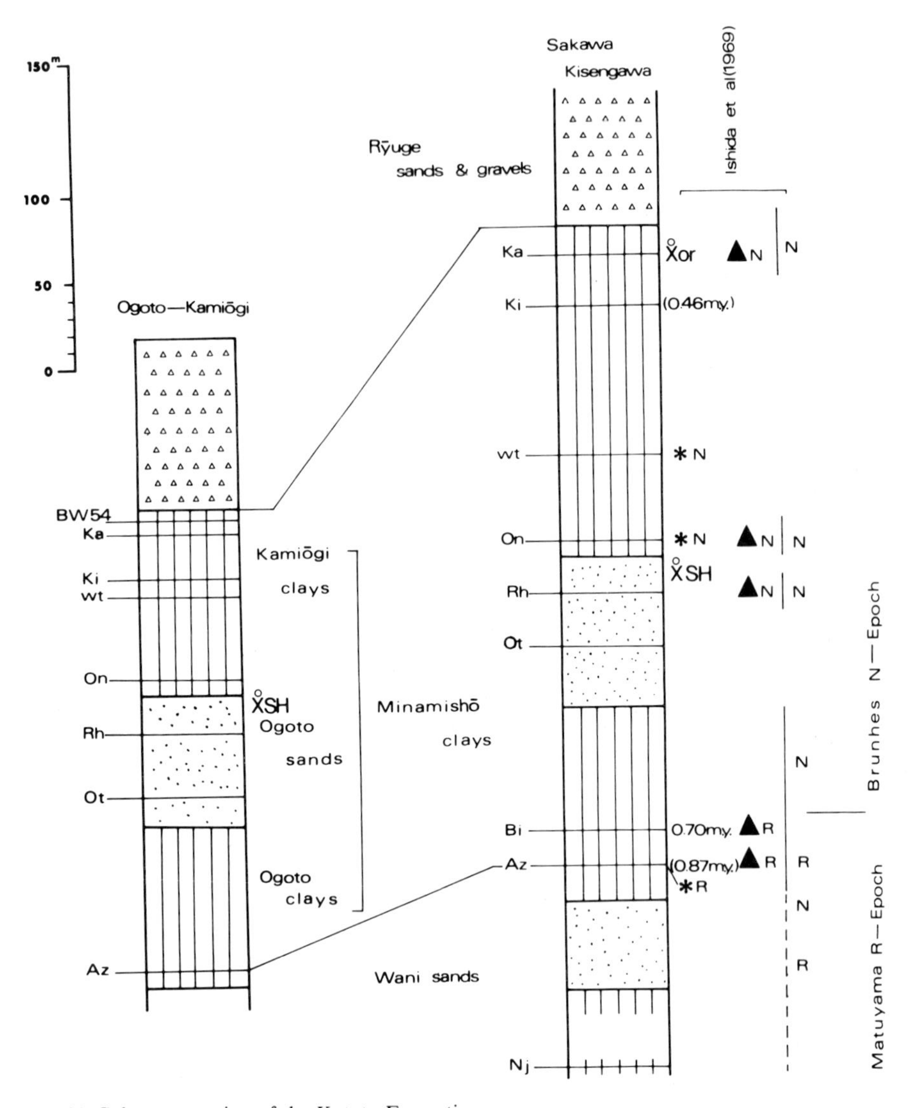

Fig. 31. Columnar section of the Katata Formation.
Name of volcanic ash — Ka: Kamiōgi, Ki: Kinukawa, Wt: White, On: Ono, Rh: Kurihara, Ot: Ogoto, Bi: Biotite, Az: Azuki, Nj: Nijigaoka, ẊOr: horizon of *Stegodon orientalis,* ẊSH: horizon of *Elephas shigensis,* N: normal polarity, R: reversed polarity.

the Kobiwako (Paleo-Biwa) Group were found in the uppermost part of Ogoto Sands, that is, just below the Ono Volcanic Ash Bed.

c. Kamiōgi Clays (110–130 *m*). Kamiōgi Clays have facies similar to Ogoto Clays. Generally speaking, clays and silts are dominant in the western half of the hills and sands are dominant in the eastern half. Kamiōgi Clays are intercalated with four remarkable volcanic ashes: Ono, White, Kinukawa and Kamiōgi Volcanic Ashes in an ascending order.

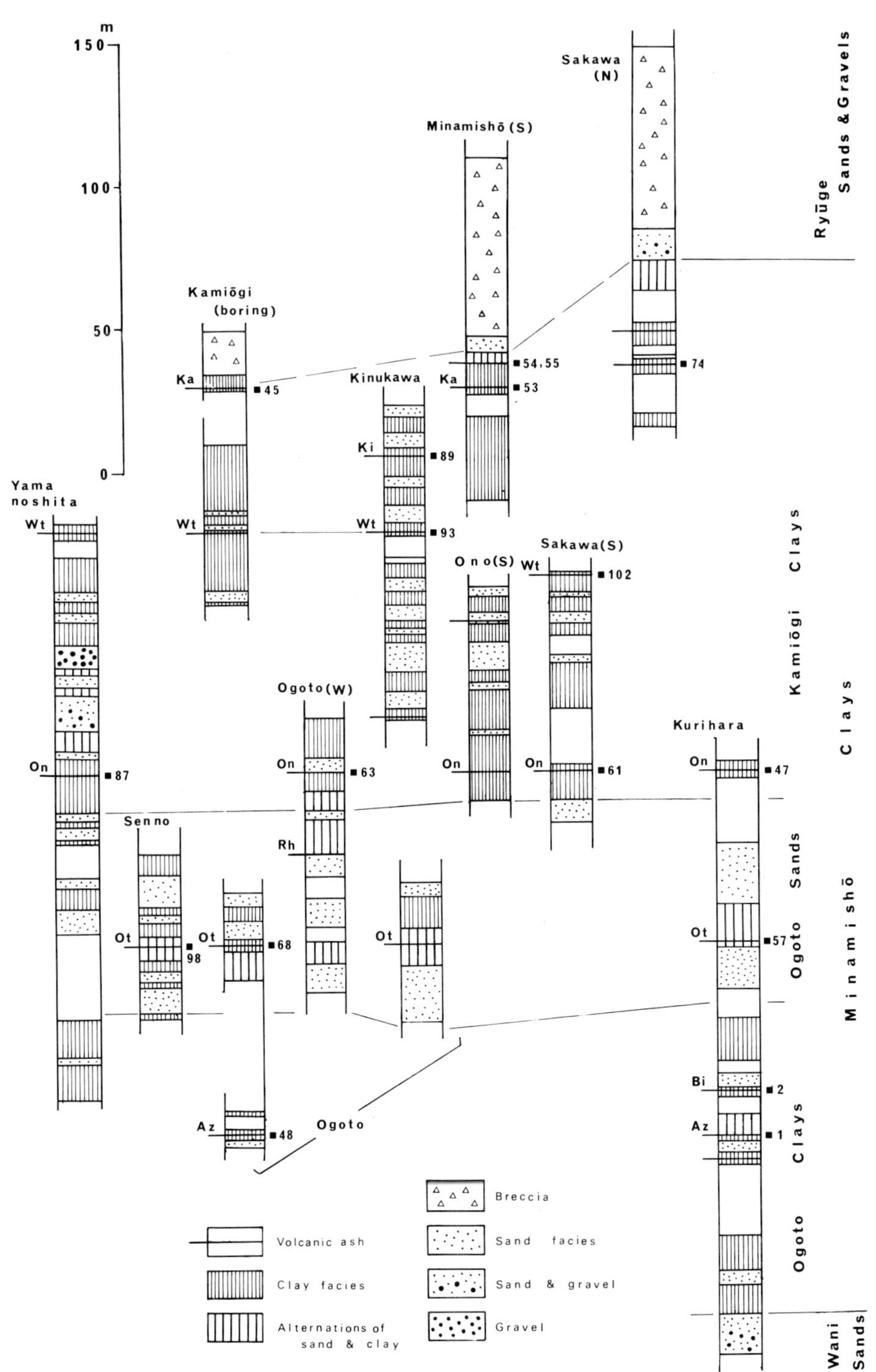

Fig. 32. Columns of the Kobiwako Group on the west coast of Lake Biwa, Kosei (after Yokoyama 1975b).
Name of volcanic ash — Az: Azuki, Bi: Biotite, Ot: Ogoto, Rh: Kurihara, On: Ono, Wt: White, Ki: Kinukawa, Ka: Kamiōgi, ■86: Sampling site of BW 86 (Yokoyama 1975b).

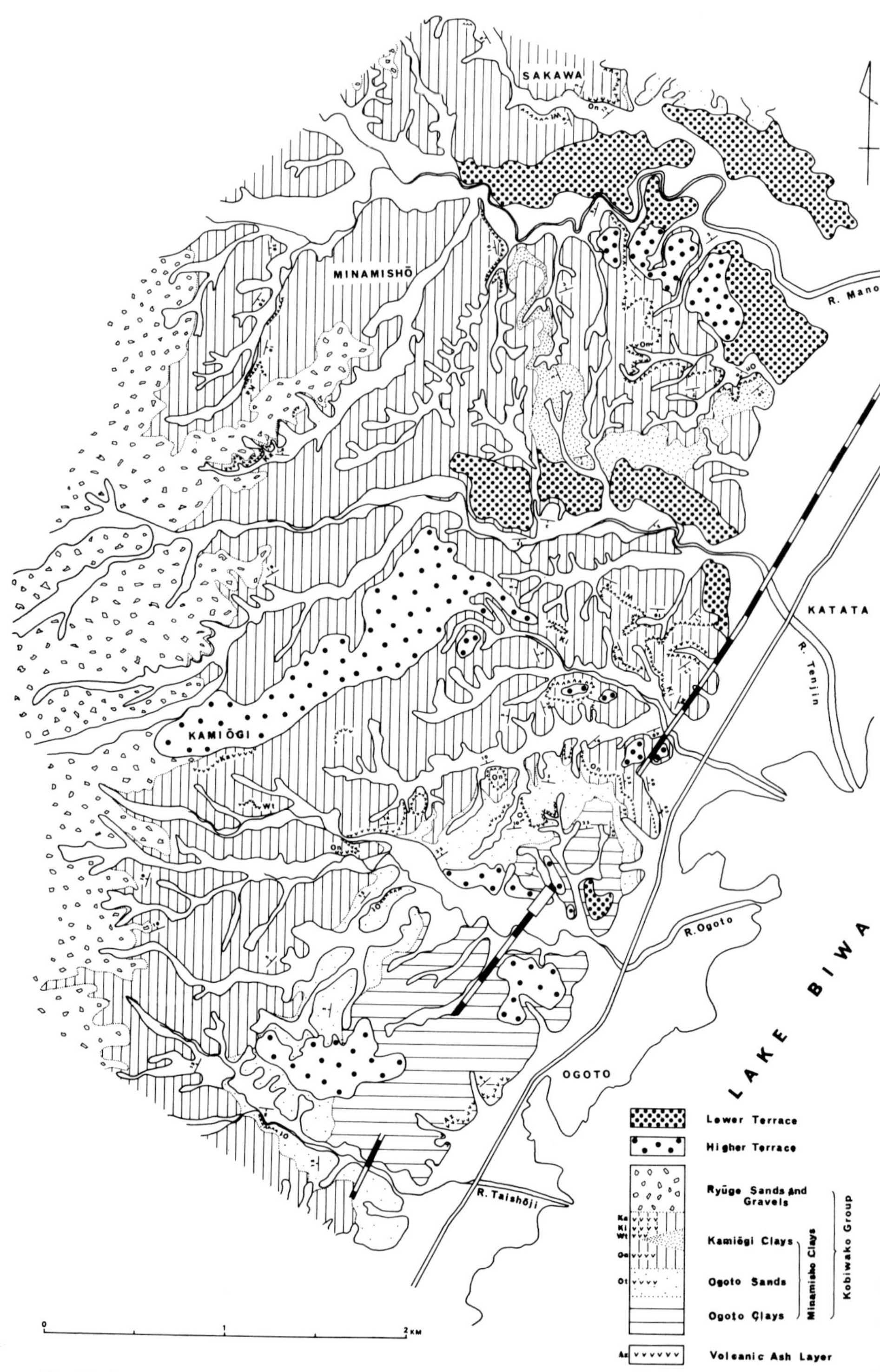

Fig. 33. Geologic map of the Katata Formation at Ogoto Area, Shiga Hills (symbols of volcanic ashes are the same as in Fig. 31) (Yokoyama 1975b).

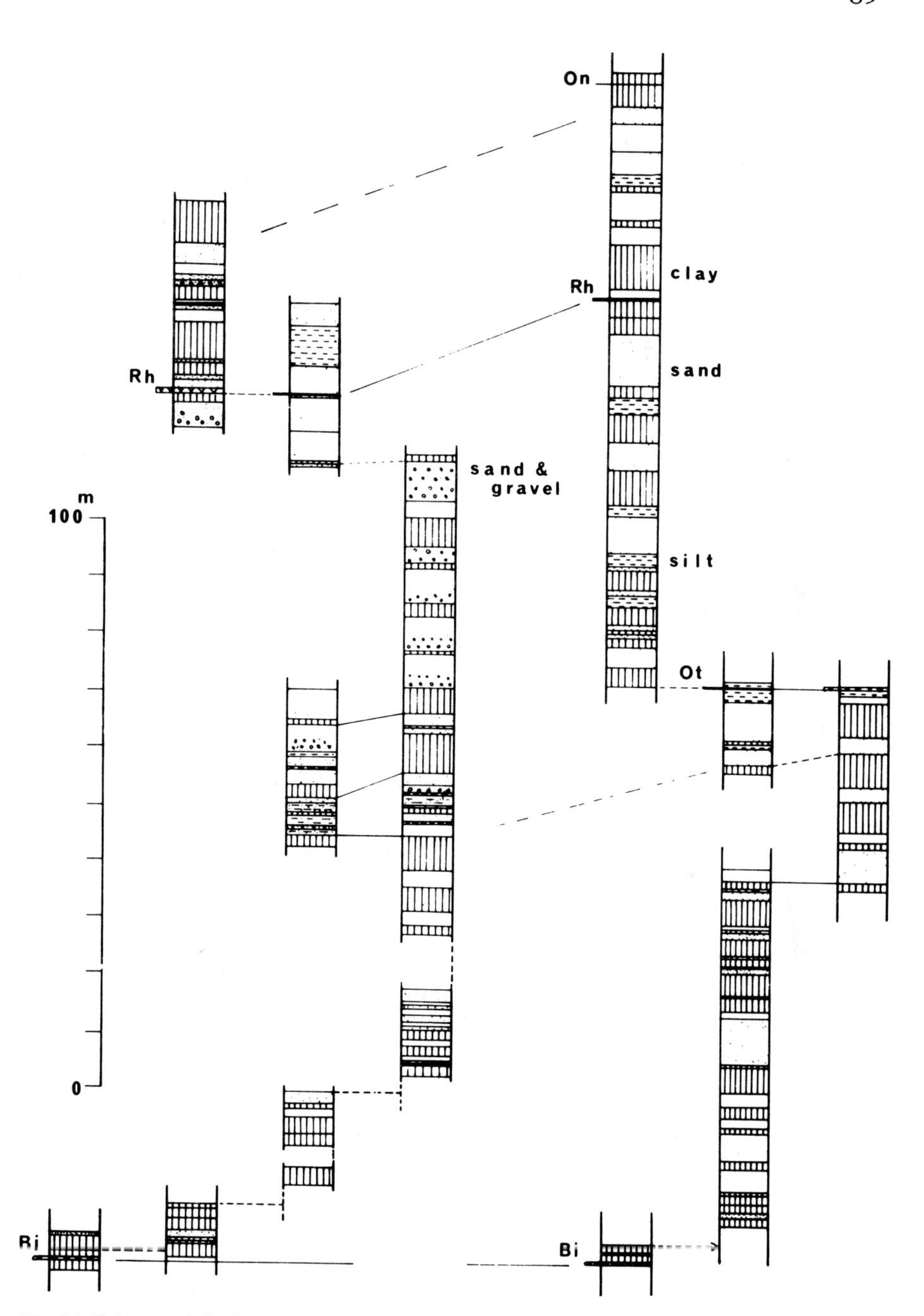

Fig. 34. Columns of the Katata Formation at the River Kisen in Shiga Hills (symbols of volcanic ashes are the same in Fig. 31).

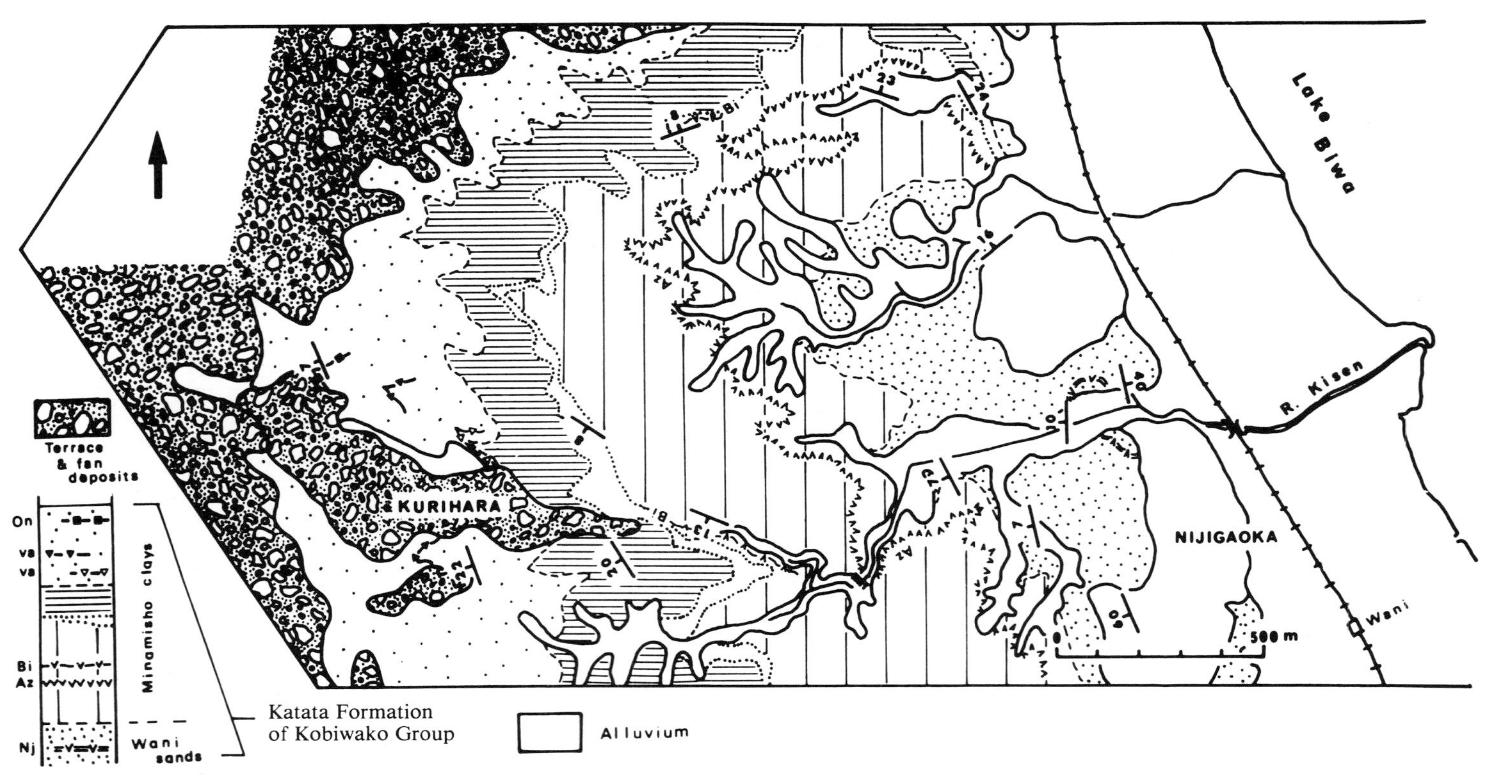

Fig. 35. Geological map of the Katata Formation along the River Kisen (Hayashida et al. 1976). Name of volcanic ash; On; Ono, Bi; Biotite, Az; Azuki, Nj; Nijigaoka. va; volcanic ash.

Stegodon orientalis was found in the upper part of Kamiōgi Clays, just under the Kamiōgi Volcanic Ash Layer.

Ryūge Sands and Gravels (100 m^+). These are the top gravels of the Katata Formation. They consist exclusively of dominant cobble-pebble gravels with slightly coarse grained sands and silts.

Kohoku Area

Yokoyama et al. (1977c) called the Kobiwako Group of this area the 'Takashima Formation'. They divided this into three lithologic facies; lower muddy, middle gravelly and upper sandy facies, and reported the occurrence of some plant remains, containing the cool temperate flora such as *Menyanthes trifoliata*, etc. A stratigraphical survey has been conducted since 1978 by the present writer and his co-workers. We can divide the Takashima Formation into four stratigraphic units in an ascending order: Lowest Shiratsuchidani Bed, Lower Inokuchi Gravel Bed, Middle Akatsukikaidō Sandy Bed and Upper Shimokoga Gravel Bed (Yokoyama et al. 1977c).

Shiratsuchidani Bed (20 *m*). This is mainly composed of coaly mud beds accompanied by gravel beds in the basal part, which is a marginal basal gravel of the Kobiwako (Paleo-Biwa) Group. This bed is found at Miyamae, Kutsuki-mura, especially around the Valley Shiratsuchidani and the Nishibiwako Country Club. This is intercalated with one thick, glassy white volcanic ash, about 2–3 m thick, which is called 'Shiratsuchidani Volcanic Ash' (Yokoyama 1969).

Inokuchi Gravel Bed (10 m^+). This is composed of cobble-boulder gravels. Most gravel grains are rounded to subrounded and come from the Paleozoic area. Thus, sandstone, chert and shale are dominant though some gravels of welded tuff are contained.

Akatsukikaidō Sandy Bed (50 m^+). This is mainly composed of coarse sand beds, intercalated with some coaly clay and gravel layers. A characteristic volcanic ash bed, called the Akatsuki Volcanic Ash by Yokoyama (1969), are found in this facies. In a field survey conducted by Yokoyama, two other volcanic ash beds were found in this bed. One is a yellowish medium grained pumiceous ash and another ash is a pinkish, glassy one, the paleomagnetic polarity of which is reversed. This reversed polarity may be correlated to one of the Biwa I—III events by Kawai et al. (1972). Yokoyama et al. (1977c, 1979b) discovered some plant remains in the middle and upper horizon of the AkatsukikaidōBed. The lithofacies of this horizon consists mainly of the coarse sand and gravel layers, as shown in the columns in Fig.36.

Some fossil plants, as shown in Table 11, were yielded from these lignites and coaly beds. Their sites are also given in Fig. 36.

Shimokoga Gravel Bed (50–70 *m*). This is mainly composed of cobble to pebble gravels which came from the Paleozoic and Triassic areas. Gravels become larger from east to west. In the foothill area of the western mountains, we can see large cobble to boulder gravels.

Terrace Fan and Alluvial deposits

Though many terrace deposits are found in the Ōmi-Iga Basin, the stratigraphic correlation among them has not yet been established. The writer has proposed the initial image of the terrace sediments of this basin by comparing them in some partial districts. The correlation among them is given in Table 12.

Along the Rivers Yasu and Hino

The terrace surface can be divided by height into four geomorphological units. The terrace deposits are called Saimyōji Bed, Nagaike Bed, Tongy Bed and Kitahata Bed.

Most of these sediments are gravel layers as shown in Table 13. A glassy, pinkish volcanic ash seam was found at the cliff of the River Saimyōji to the northeast of Kitahata. This is intercalated in the Kitahata Bed and is called 'Hacchono Volcanic Ash'. It may be correlated to the BB 55 volcanic ash in the 200 m core samples at the bottom of Lake Biwa, because of much resemblance of mineral assemblage and physical appearances.

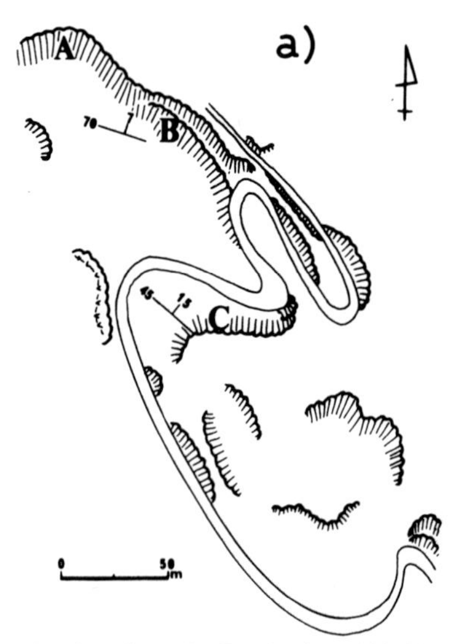

Fig. 36. a) Route map of the sites along the Road Akatsukikaidō where many plant remains were discoverd (Yokoyama et al. 1977c).

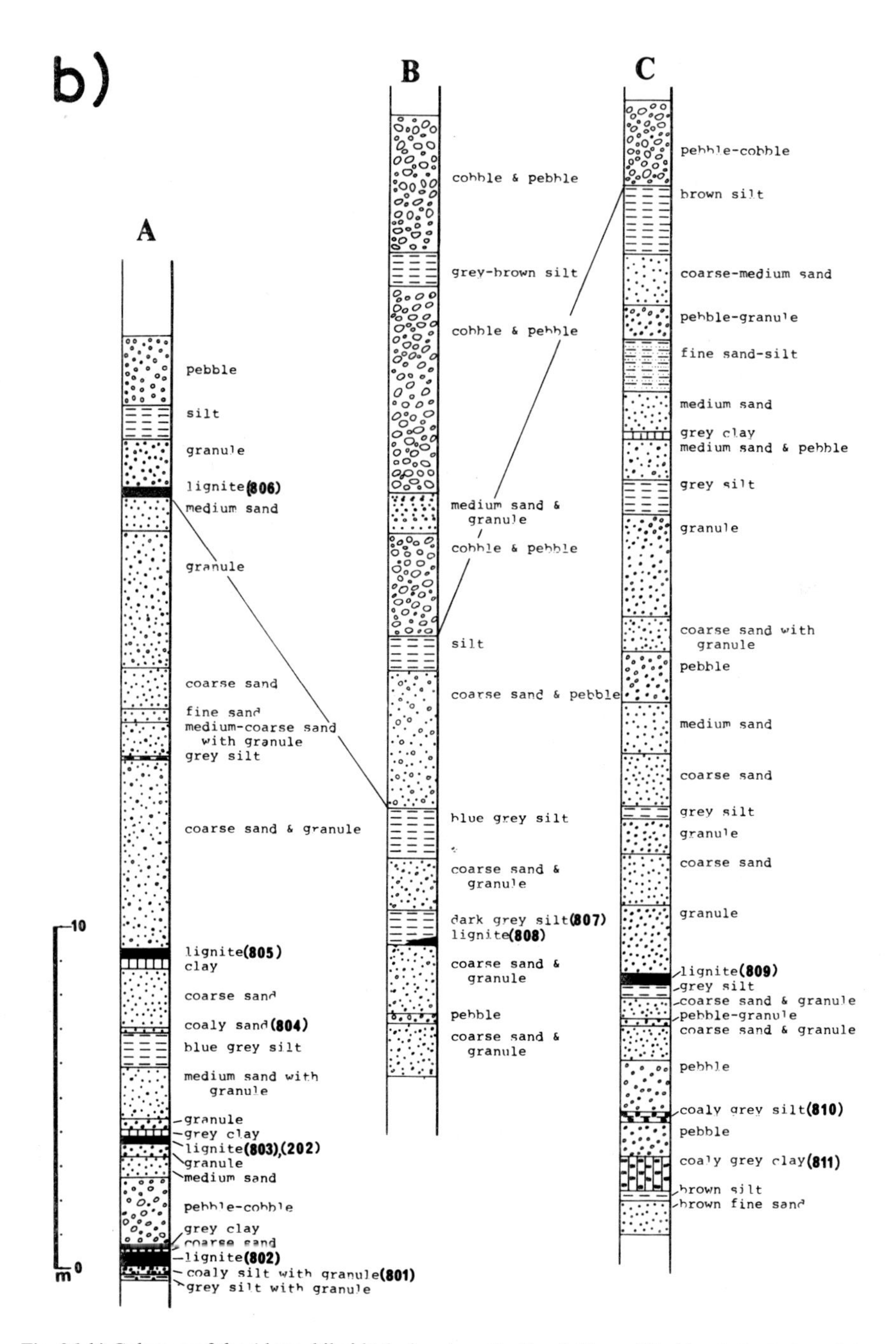

Fig. 36. b) Columns of the Akatsukikaidō facies along the Road Akatsukikaidō. A, B, C: Localities shown in Fig. 36 a), (801): Horizon number of Table 11.

Along the River Echi

In a geomorphologic sense, wide terrace surfaces are seen along the River Echi. All of the sediments are gravel layers only, except small lenses of sand and silt deposits. Miki (1956) reported some plant remains such as *Pinus koraiensis, Tsuga diversifolia* and so on in terrace deposits near Kozuhata. The values of C^{14} are 30,000 ± 1,700 years B.P. and 33,200 + 4,100/2,700 years B.P. (Itihara & Takaya 1965). Terrace deposits in this area are shown in Table 14.

Kosei an Kohoku Areas

There are two outstanding geomorphic planes. The higher one, made of the top of the hills, has cobble-boulder gravelly sediments about 3–10 m thick. The lower one is widely distributed along the river side. Hayashi (1974) discovered a volcanic ash seam in this sediment. The thickness of this volcanic ash is 10 cm. This may be the same as the BB 55 ash in the 200 m core sample as the two contain two pyroxene crystals. Terrace deposits of this area are summarized in Table 15.

A large quantity of terrace deposits are seen at the foothill area in the Ōmi Basin. But the age and mutual relation cannot be clarified because of the absence of geologic key beds. The writer found a volcanic ash seam in the terrace deposit, distributed at Hino-cho, but this seam is not yet found in other sediments. In this paper, the writer reports on some terrace deposits as follows:

Tenkawa Bed (20 m^+). In the foothills of Kosei, namely, at the upper stream of the River Ten in the northern part of the Shiga Hills, there is a loose boulder bed which contains big boulder gravels with the maximum diameter of 1–2 m. The gravel is almost granite and the matrix of this layer is mostly arkosic coarse sand. This bed overlies unconformably the Katata Formation, the horizon of which is Ogoto Sands at the upper stream of the River Ten and is Kamiōgi Clays at the upper stream of the River Kisen. A similar bed is also exposed at the Ukawa Area. It also overlies unconformably the Kobiwako Group, perhaps the Akatsukikaidō Bed of the Takashima Formation.

Kunisakai Bed (30 *m*). This bed was reported first by Yokoyama (1968a). The Kunisakai Bed can be divided into three parts by the lithologic viewpoint. The lowest part, about 3 m thick, is composed of subrounded boulder gravels. This part overlies directly on the basement rocks of granite.

The middle part, about 20 m thick, is composed mainly of sands and gravels with some lenticular shaped coaly clay beds containing some plant remains, such as *Pinus koraiensis, Tsuga diversifolia, Fagus crenata* and *Chamaecyparis obtusa.* In this bed, there are four volcanic ashes, named Kunisakai Volcanic Ash I—IV. One of the Kunisakai Volcanic Ashes is perhaps the same as the BB 55 ash in the 200 m core samples from the lake. The upper part is composed of breccia bed, which is cobble-boulder in size. The thickness is thicker than 3 m.

Table 11. Plant Remains from the Takashima Formation (after Yokoyama et al. 1977c).

Site-number		Fossil plant	Others
801	a	*Scirpus?, Ceratophyllum demersum, Monochoria* sp.	many fragments
	b	*Carex* sp. *Sparganium* sp. *Scirpus?, Menyanthes trifoliata, Ceratophyllum demersum, Monochoria* sp.	
802		*Menyanthes trifoliata*	wood fragments
803		*Styrax* sp., *Sapium* sp., *Xanthoxylum* cf. *simulans, Prunus?*	many wood fragments
804	a	*Styrax* sp.	wood fragments
	b	*Carex?, Scirpus?*	many fragments
805		*Sparganium* sp., *Carex* sp., *Polygonum?*	woods, seats
806		*Menyanthes trifoliata, Carex* sp.	many wood fragments
807		*Trapa* sp. *Nuphar* sp.	
808		*Menyanthes trifoliata, Polygonum* sp.	many wood fragments
809			woods
810		*Sambucus sieboldiana*	
811		*Carex??, Polygonum??*	
201		*Menyanthes trifoliata*	
202		*Sapium* sp., *Styrax* sp., *Vitis* sp., *Xanthoxylum* cf. *piperitum, X.* sp., *Trichosanthes* sp.	

Alluvium

Alluvial deposits are distributed only at the shore area of the lake. In the hilly area, there is little distribution of the Alluvium, and the flat plain in the valley is made of the youngest terrace deposits, the age of which may be less than 0.01 m.y., thus indicating that it may be Holocene Series.

At the mouth of the River Yasu, alluvial deposits are mostly sand layers with a few coaly silt-clay beds. Cross-beddings are commonly seen in the sandy part. The thickness of alluvium may be 11–12 m. In Moriyama City, two muddy beds were seen in the Alluvium.

We have the following three C^{14} ages from the alluvial deposits around Lake Biwa (Yokoyama et al. 1979a).

A. $14{,}480 \pm 550$ years BP.

This age was obtained from the coaly material which was gotten from 4 m

Table 12. Terrace and Terrace Deposits in the Ōmi-Iga Basin.

Kunisakai	200 m Boring	Kosei	R. Seta	R. Yasu	R. Echi	others
	BB.15	Tenkawa Bed		Saimyōji volcanic ash		Terrace deposits
Kunisakai Bed Kunisakai volcanic ash	BB.55	Lower Terrace volcanic ash	Lower Terrace	Kitahata Bed Hacchono volcanic ash	Lower Terrace	Terrace surface is very flat
			Middle Terrace	Tongu Bed	Kozuhata Bed	Fan deposits a little distribution
		Upper Terrace		Nagaike Bed	Ikenowaki Bed	Terrace surface is flat
				Saimyōji Bed		

BB.15, and BB.55 are Volcanic ash in the 200 m boring core sample.

Table 13. Lithology of Terrace Deposits along the Rivers Yasu and Hino.

Name of bed	River Yasu	River Hino
Lower terrace	Minakuchi Bed *2–5 m thick *150–320 m in height *Gravels	Kitahata Bed *2–5 m thick *150–250 m in height *Gravels with silt and sand lense
Middle terrace	Tongu Bed *10 m thick *240–300 m in height *Gravels	
High terrace	Nagaike Bed *5–15 m thick *230–300 m in height *cobble gravels	Yamamotoshinden Bed *10 m thick *140–300 m in height *Pebble–cobble gravels
Highest terrace		Saimyōji Bed *20 m thick *390–420 m in height *Boulder gravels

Table 14. Lithology of Terrace Deposits along the River Echi.

Name of bed	Height	Lithofacies and others
Lower Terrace (2–5 m)	120–290 m	Pebble — cobble gravels
Kozuhata Bed (Middle terrace) (7 m)	150–300 m	Sands, silts, gravelly sands and gravels C14 age; 30,000 ± 1,700 years BP./33,200 + 4,100, – 2,700 years BP.
Ikenowaki Bed (5–7 m)	230–270 m	Cobble — pebble gravels (subrounded — rounded)

Table 15. Lithology of the Terrace Deposits in Kosei Area.

Name of bed	Height	Lithofacies and others (thickness)
Lower terrace	110–200 m	Mainly pebble — cobble gravels (1–3 m), and partly some silt and peat layers. A andesitic volcanic ash seam was found by Hayashi (1974). This is same to the BB.55 volcanic ash in the 200 m boring at the bottom of Lake Biwa
Upper terrace	125–300 m	Boulder — cobble gravels (3–10 m), and the matrix is coarse sand

below the ground surface, at No. 8 bridge of the River Shiratori in Ōmihachiman City. The altitude of this point is 87.8 m.
B. 2,900 ± 1,700 years BP.
This age was obtained from the coaly mud at Niihama, Kusatsu City. It was gotten from 5.5—5.7 m below the ground surface by drilling.
C. 33,600 years BP.
This age was obtained from the wood by the same drilling to above B. It was gotten from 16.7—17.0 m below the ground surface.

The 200 m and 1,000 m deep drilled core samples and their stratigraphical horizon

200 *m core sample*

The present writer and his co-workers succeeded in obtaining sediments by core drilling from the bottom of Lake Biwa in 1971, directed by Shoji Horie of Kyoto University. The sediment core is about 200 meters in length and composed of loose homogeneous clay. Up until now, many investigations, such as paleomagnetic, paleontological, paleoclimatic, mineralogical and geochemical studies, have been carried out (*Paleolimnology of Lake Biwa and the Japanese Pleistocene,* 1—9, ed. S. Horie). As shown in Fig. 11, the drilling was carried out the center of Lake Biwa, at a spot on the bottom about 65.2 m deep. The top of the core sample obtained corresponds to the depth of 3 m below the bottom surface of Lake Biwa. The bottom of Lake Biwa is roughly divided into two geographical categories: submerged terraces and the bottom surface. The latter spreads at the depth of about 30 meters below the water surface of the lake where homogeneous lacustrine clay sediments are distributed all over. On the other hand, the former is constructed by coarser sediments, sand and gravel. Clay sediments which are a few meters thick have been deposited on the terrace surface.

Roughly speaking, the lithofacies of these core sediments are mostly clay facies containing some fossil diatoms and plants except some aerial materials, such as volcanic ash and aerolite. Some crystals of vivianite are often seen in the clay, but coarser grains, that is sands and gravels, are not found. Clay's appearance is bluish in color on its fresh surface. It is very homogeneous in every part. Sedimentary structures, such as graded bedding and lamination etc., cannot be seen by unaided eyes. The sandy grains that originated in hills and mountains cannot be found after size analysis in almost any part. All grains coarser than 0.062 mm in diameter contained in many horizons, originated from volcanic ash. This core sample is intercalated with many thin volcanic ash layers, at least 30 seams in this 200 m core sample, similar to the Kobiwako (Paleo-Biwa)

Group. They are chiefly composed of volcanic glass flakes and pumice grains. They are roughly divided into two types from the petrological viewpoint. Some kinds of volcanic ashes are andesitic and others are more acidic, perhaps rhyolitic. The former contains two pyroxenes, hornblende and iron ore as the main heavy minerals in volcanic glasses. The latter generally contains orthorombic pyroxene, hornblende, biotite and quartz as the crystalline components in volcanic glasses and pumice grains. In addition to these minerals, they contain other crystalline minerals, such as apatite, zircon and feldspar (Yokoyama, 1973b).

1,000 *m core sample*

In 1976, 1,000 m core boring on the east coast of Lake Biwa was carried out by our research group, directed by late Professor Naoto Kawai of Osaka University. The drilling was undertaken to examine changes in geomagnetic polarity since 2.0 million years ago, especially the paleomagnetism of the boundary age of Brunhes — Matuyama Paleomagnetic Epoch.

In addition to the paleomagnetic works, various other analyses are being conducted at present, for example, geological, sedimentological, geo-chemical and biological investigations. We can obtain a considerable amount of data of global changes of the natural environments as Lake Biwa is a very old lake.
Biwa is a very old lake.

In synthesizing the data, it is necessary to settle two problems. Firstly, the well was not straight. Secondly, the strata were not horizontal. Fortunately, these two difficulties were solved. The inclination of the well was measured (Ishida et al. 1976b). The dips and strikes of the strata were estimated by means of the paleomagnetic method (Yokoyama 1976). Ishida et al. (1976b) proposed that the inclined direction of the well was generally N10°W and that the inclination of the well was less than 10 degrees from 0 to 200 m in depth and about 20 to 30 degrees in the part deeper than 200 m.

On the other hand, the writer (1976) estimated that the general strikes of the underground strata were N29°E and that the dips were 20–25 degrees toward the west.

Judging from the two above-mentioned works, we need to shorten the drilled length by about 16% to get at the thickness of sediments. Of the 16%, 6% is the effect of the inclination of the well and 10% is that of dips of strata. After calculating the above 16%, the columnar section of the core boring was arranged and the sampling sites were given their points as a result of the calculations.

Division of the sediments:

The core samples were mainly composed of fine to very fine sand layers with some clays, silts, medium to coarse sands, gravels, peats and volcanic ashes. But this fact does not indicate that all stratum is mainly composed of the fine to very fine sand beds, because the loose sediment coarser than fine sand were washed away by the water moving through the drilling works. Therefore, from the drillers we obtained the data for the loose sediments washed away through the drillings. The columnar section was made by summarizing all this information. We can divide the columnar section into 13 parts from the lithofacies as given in Fig. 37. We call them Beds A to M in a descending order.

Bed A. This is composed of very coarse, loose sand layers, which are deposits at the mouth of the River Yasu. The cross-beddings of sand are common in this bed and seen in a hall excavated on the ground surface. There is a silty layer about 2 m thick, and 8 to 10 m deep.

Bed B. This is composed of dark grey silts and clays, intercalated with some loose sands.

Bed C. This is mainly composed of loose, grey-colored coarse sand layers with some gravels. There are four gravel rich horizons, that is, 72–74, 84–85, 100.5–101.6 and 124–129 meters in depth.

Bed D (61.7 *m*). This is composed of alternations of grey loose sands and bluish muddy layers, in which sand is rich in volume.

Bed E (27.4 *m*). This bed is composed of gravel layers with some sands and clays. Gravels obtained by the boring machine are 4.5–2.0 cm in diameter, and are made of chert, sandstone and granite.

Bed F. This is mainly composed of loose, coarse to very coarse sands with some gravels. Some mud layers, which are bluish grey or greenish grey colored, intercalate with this bed. This volcanic ash is located at 266.37 m in depth, namely, 259.77 m below the ground surface in 'horizon'. Here, what is meant by the term 'horizon' is the thickness of strata from the ground surface.

Bed G. This bed is mainly composed of medium to fine sand layers. Obtained core samples were mostly fine to very fine grained sands with some thin silts and clays, the additional thickness of which is about 1.5 m. So, the silt and clay component in this bed may be about 6.6%. These core samples contain a little peat and plant remains. There is a volcanic ash bed (db. 49), 279.8 m in depth and 271.8 m in 'horizon'. The thickness is about 3 mm.

Bed H (35.8 *m*). This bed, like Bed G, is mainly composed of medium to fine sands but there are more silt-clay parts than in Bed G. Their component is about 29.6%. Obtained core samples were mainly bluish grey colored silts and clays, in which peats and coaly clays, with the maximum thickness of 12 cm, are very rich,

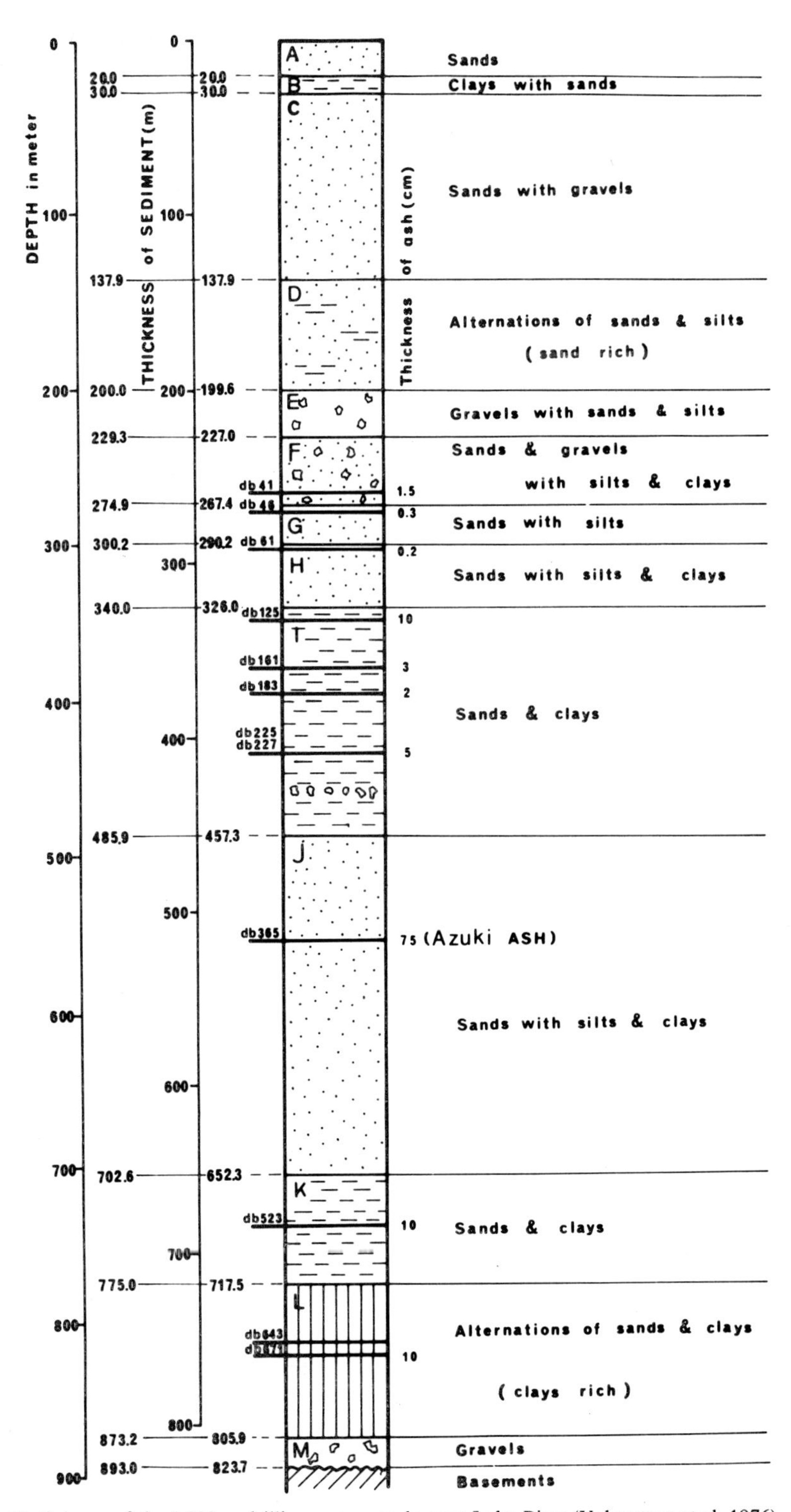

Fig. 37. Column of the 1,000 m drilling core sample near Lake Biwa (Yokoyama et al. 1976).

especially in the core samples of No. 67, and 83–91. Vivianite was also found in the clay beds (No. 79, 83, etc).

This bed also contains a volcanic ash seam (db. 61), 300.7 m in depth and 290.6 m in 'horizon'.

Bed I (131.3 *m*). This is mainly composed of alternations of sands and clays. The content of silts and clays is about 36.8% in total volume. More than 24% of sands are fine to very fine grained. Obtained core samples, 62.2% of them are mostly bluish grey colored clays and silts. Lignites and coaly clays with plant remains in them are common.

Four volcanic ash layers were found in this bed; db. 125, 161, 183 and 227. They are located 348.7–348.9 m, 378.7 m, 394.4 m and 434.1 m in depth, namely, 333.8–334.0 m, 360.8 m, 375.0 m and 410.7 m in 'horizon', respectively.

Bed J (195.0 *m*). This bed is mainly composed of coarse to very coarse sand layers with some silts, clays and fine to very fine sands. Almost all silts, clays, and very fine sand layers irregularly contain granule or very coarse sand grains. These beds are generally greenish grey colored. There are a few pure clay beds. Though it is difficult to calculate the silt-clay content, it may be less than 15%. Plant remains are found in almost all samples. An outstanding volcanic ash (db. 365) is found in this bed, 553.1–554.0 m in depth and 517.8–518.6 m in 'horizon'.

Bed K (65.2 *m*). This is composed of clays, silts with coarse sand grains, very fine to granule sands. Both sandy layers and silt clay beds are common, but the sandy layer occur more frequently. Silts and very fine sands contain commonly grains of coarse sand and granule. These beds are greenish colored like those in Bed J. On the other hand, coarse to very coarse sand layers contain many plant remains. These beds are grey or bluish grey colored.

There is a volcanic ash (db. 523) in this bed, 735.4–735.5 m in depth and 682.8–682.9 m in 'horizon'.

Bed L (88.4 *m*). This is mainly composed of alternations of sands and clays. Silt-clay beds are dominant and sand layers are common. Clay is dark greenish grey — dark grey colored. In the uppermost part of this bed, plant remains are very rich and are also commonly found in other parts.

There are two volcanic seams (db. 643.671) in this bed, 806.9 m and 819.4—819.5 m in depth and 746.8 m and 757.4—757.5 m in 'horizon'.

Bed M (17.2 m^+). This may be composed of gravel beds. Gravels obtained by boring are generally small pebble to pebble in size, and the matrix of them may be coarse sand. Some coaly silt layers are intercalated with these gravel layers.

Correlation of the Kobiwako (Paleo-Biwa) Group, the 200 m and 1,000 m core samples

As already mentioned, many tephra were found in these three sequences. All features of these tephra were compared with each other in detail for the purpose of correlation (Table 16).

It is considered that the Kamiōgi Volcanic Ash Layer may be the same as the volcanic seam intercalated in the lower part of the 200 m core sample. This is called BB. 425 in a previous paper (Yokoyama 1975b). Both volcanic ash layers are pinkish in color and are composed almost entirely of volcanic glass flakes. Both contain two pyroxenes, hornblende and apatite as the main heavy minerals. The paleomagnetic polarity of these two beds is also identical with normal polarity.

Table 16. Similarities of properties in the Volcanic Ash layers of the Kobiwako Group, 200 m and 1,000 m Core Samples (after Takemura et al. 1976).

Name of volcanic ash layer	Curie point (°C)	Type of thermo-magnetic curve[a]	Type of glass[b]	Heavy mineral composition[c] (%)						No. of references
				oPx	cPx	Am	Bi	Ap	Zr	
Kamiōgi	510	A	AB	27.0	20.0	43.0	6.0	3.0	1.0	1)
BB425	500	A	AB	40.5	15.5	16.0	0.0	28.0	0.5	2)
db125	510	A	AB	36.0	14.0	45.0	3.5	1.0	0.5	3)
Kinukawa	460	A	C	5.0	0.0	91.5	3.5	0.0	0.0	1)
BB453	430	A	AC	46.0	0.0	47.0	6.5	0.0	0.5	2)
db161	485	A	C	5.0	0.0	94.0	0.5	0.0	0.5	3)
White			A	29.0	4.0	62.0	0.0	5.0	0.0	1)
db183	180	B	A	+		+++				3)
Ono	90, 540	AB or B	AB	14.2	0.3	80.2	2.0	0.3	3.0	1)
db225	75, 400+	AB	AB	29.5	5.0	48.0	11.5	5.0	1.0	3)
db227	70, 500	AB	AB	53.0	19.5	10.5	14.5	2.0	0.5	3)
Azuki	60, 285, 500	A_2B	AB	33.5	26.0	31.5	1.5	7.5	0.0	1)
db365	25, 320, 500	A_2B	AB	44.0	34.0	16.0	0.5	5.0	0.5	3)

1) Yokoyama 1975b; 2) Yokoyama 1973b, 1975b; 3) Takemura et al. 1976.
[a]see the report by Yokoyama 1975b.
[b]see the report by Yokoyama 1969, 1975b.
[c]see Table 1 of Takemura et al. 1976.

The Kamiōgi Volcanic Ash Layer is contained in the upper part of the Kobiwako Group, namely, the upper part of the Katata Formation. *Stegodon orientalis* was yielded from the horizon of this volcanic ash layer.

The BB 425 volcanic ash layer is intercalated at the level of about 169 m below the present bottom of Lake Biwa.

On the other hand, the BB 453 volcanic ash layer is very similar to the Kinukawa Volcanic Ash Layer in the Katata Formation as shown in Table 16.

These two volcanic ashes have the same petrological features as follows:

1. They are rich in crystal grains;
2. They commonly contain feldspar crystals and are wrapped up in glass;
3. They contain many isomorphic crystals such as orthorombic pyroxene, apatite and hornblende, and they are also wrapped up in glass.

These two pairs, Kamiōgi — BB 425 and Kinukawa — BB 453, are good keys to the correlation between the Kobiwako Group and the 200 m core samples. This correlation is considered reasonable in view of the fission-track age of the BB 453 volcanic ash, 4.6×10^5 years obtained by Nishimura and Yokoyama (1975), because the absolute age of the Kamiōgi Volcanic Ash Layer had been estimated to be about 0.4–0.45 million years ago by paleomagnetic age determination .

Takemura et al. (1976) examined minerals in volcanic ashes of the 1,000 m core samples for the correlation to the Kobiwako (Paleo-Biwa) Group on land. They asserted that the db. 227, 161 and 125 ashes were correlated to the Ono, Kinukawa and Kamiōgi Volcanic Ashes in the Katata Formation (Yokoyama 1975b). The fission-track age was measured by S. Nishimura. The age obtained from db. 161 is 0.47 m.y. which coincides very much with 0.46 m.y. of the BB 453 Ash in the 200 m core samples.

Moreover, the db. 183 and 365 ashes in the 1,000 m core samples are considered to be the same as the White and Azuki Volcanic Ashes in the Katata Formation because of similarities in the petrologic properties of the volcanic material (Table 16).

Consequently, the correlation among the 200 m, 1,000 m cores and the Katata Formation was established by the key of these tephra as summarized in Fig. 38.

Biostratigraphy of Quaternary systems around Lake Biwa

Biostratigraphy of the Kobiwako Group was recently summarized in a paper by the writer and his co-workers in conjunction with paleomagnetic polarity changes and climatic oscillations of the Osaka and Tōkai Groups (Maenaka et al. 1977). But some fossils were discovered after this summary was published. The writer now reports on fossils unreported in that summary, mainly in regard to molluscs and plants of the Kobiwako (Paleo-Biwa) Group and plant fossils in Holocene sediments. The data are summarized in Table 2 along with the previously reported data. The new discovery of plant remains is made mostly by Yokoyama et al. (1977c) from the Takashima Formation as shown in Table 11. Another datum on fossil plant is from the Kunisakai Bed: *Pinus koraiensis, Tsuga diversifolia*, etc. (Yokoyama 1968a), and Kozuhata Bed; *Pinus koraiensis, Tsuga diversifolia*, etc. (Miki 1956).

These flora, in all cases, contain plant fossils indicating the cool temperate

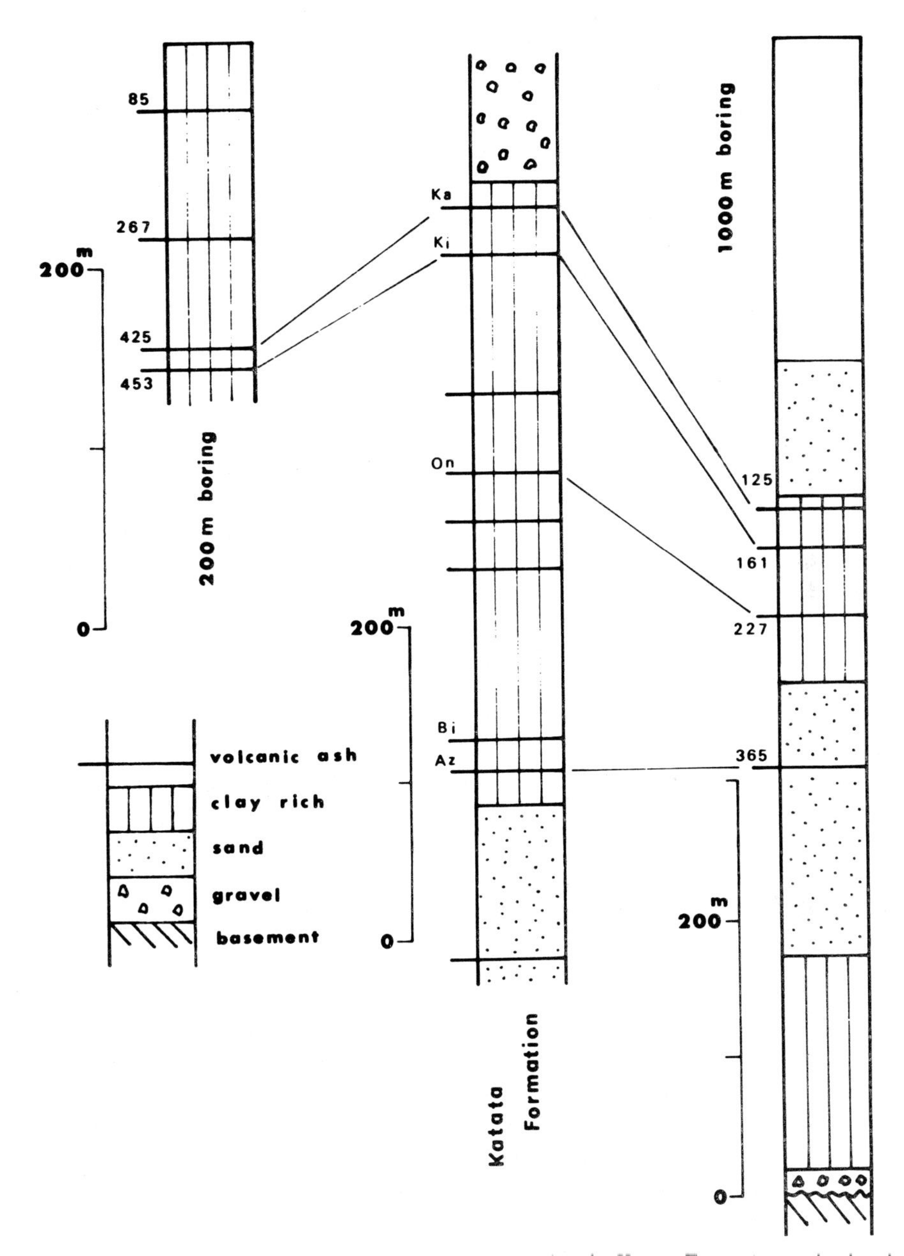

Fig. 38. Correlation by tephra among the 200 m core samples, the Katata Formation on land and the 1,000 m core samples (after Takemura et al. 1976).
Ka: Kamiōgi, Ki: Kinukawa, On: Ono, Bi: Biotite, Az: Azuki volcanic ashes.

climate. Then, the ages of these flora are involved in time of the 200 m core samples, but we cannot exactly correlate the cool temperate ages to the Co–1–24 by Fuji and Horie (1972) of the 200 m core sample, except the Kunisakai Bed may be the same as the Co–1 because the Kunisakai Volcanic Ash III may coincide with the BB 15 volcanic ash.

Ikebe (1933) reported many molluscan fossils from the Katata Formation. Now, their horizon can be exactly determined as follows:

Locality number (by Ikebe)	Horizon	Fossil name
Ik. 0.	Just below the Ono Volcanic Ash	*Corbicula sandai.*
Ik. 3.	Lower part of Kamiōgi Clays	*Viviparus longispira.*
Ik. 20.	Just above the Ogoto Volcanic Ash	*Viviparus longispira, Semisulcospira libertina.*
Ik. 21.	Same as above	*Viviparus longispira, Phragmite* sp.
Ik. 28.	Lowest part of Kamiōgi Clays	*Viviparus longispira.*
Ik. 72.	Just below the Kinukawa Volcanic Ash	*Corbicula* sp.
Ik. 127 & 128.	About 10 m below the White Volcanic Ash	*Unio douglasiae, U. biwae, Inversidens brandti, I. hirasei, I. japanensis, I. reiniana, Cristaria plicata spatiosa, Viviparus japonicus, Trapa macropoda.*
Ik. 63.	Below the White Volcanic Ash	*Trapa macropoda.*
Ik. 163 & 164	Between the White and Kinukawa Volcanic Ashes	*Viviparus longispira.*
Ik. 165	Just above the Ik. 163 and 164	*Trapa macropoda.*

(by personal communications from Dr. N. Ikebe)

Tephrostratigraphy, magnetostratigraphy and radiometric ages of the Quaternary system around Lake Biwa

Tephrostratigraphy and magnetostratigraphy of the Quaternary system in the Ōmi-Iga Basin are to be presented in other reports.

Up to now, 15 fission-track ages of volcanic seams have been obtained around Lake Biwa (Nishimura & Sasajima 1970, Nishimura & Yokoyama 1973, 1974, 1975, Takamura et al. 1976, Yokoyama et al. 1977b, 1978, 1980). They are listed in Table 3. The values range from 3.1 to 0.08 million years. Fig. 39 represents the relation between the fission-track ages and the thickness of sediments. This figure indicates that the thickness of sediment is concordant with the fission-track ages and paleomagnetic polarity changes. We can read the ages of each horizon of sediment from Fig. 39.

Geologic structure of Lake Biwa Basin

Fig. 17 shows the tectonic structure of the Ōmi-Iga Basin and Fig. 21 is a geologic map of the Kobiwako Group. Some characteristic features of the

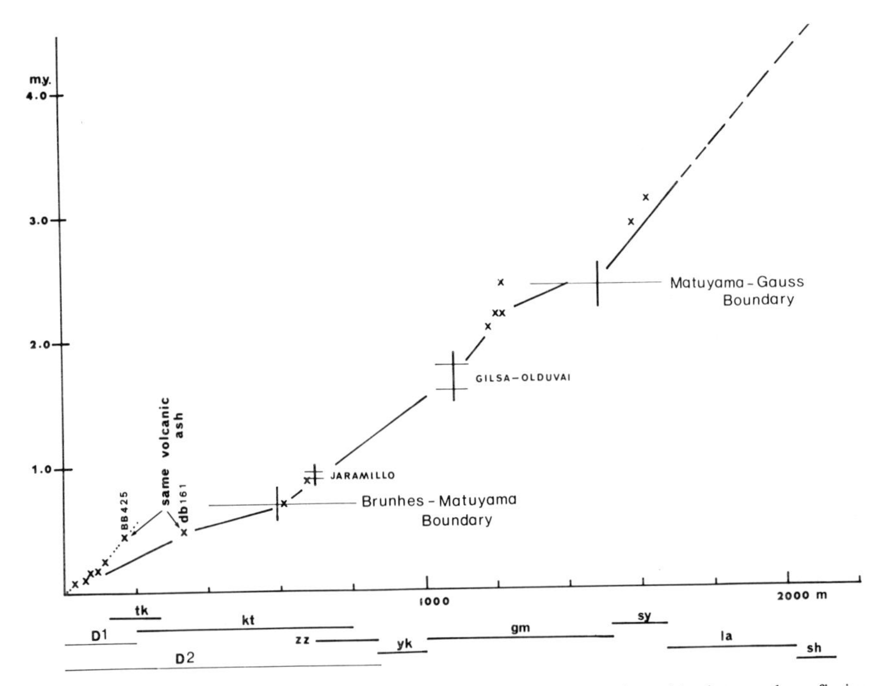

Fig. 39. Relation between the thickness of sediments and chronostratigraphic data such as fission-track dating and paleomagnetism.
D_1: 200 m drilling, D_2: 1,000 m drilling, tk: Takashima Formation, Kt: Katata Formation, zz: Zeze Formation, yk: Yōkaichi Formation, gm: Gamō Formation, sy: Sayama Formation, Ia: Iga-Aburahi Formation, sh: Shimagawara Formation.

tectonic movement and migration of the Lake Biwa Basin are clearly represented in Fig. 12, 14, 15, 16, 17 and 20 in Table 4. They may be summarized as follows:

Direction of faults

The faults running along the west foothill area of the Suzuka Mountains, such as Tsuge and Tongu Faults, are the most essential ones in the Ōmi-Iga Basin. Faults other than them in this basin can be divided into two categories by their directions. One is characterized by a NW–SE trend, and another by a NE–SW to ENE—WSW trend as shown in Fig. 15. In both categories, faults located at the southern area are older than those in the northern area of the basin. It is thought that the essential ones along the Suzuka Ranges belong to the former category, for both of them are left lateral faults in the geomorphic feature, while the faults having the NE–SW trend involve right lateral faults. Generally speaking, the former is similar to the Yamazaki Fault, one of the famous active faults of Southwest Japan, and the latter is contained in the category of Median Tectonic Fault Zone which is also made of many right lateral faults.

Forming of fault blocks in Basement Rock

As the above-mentioned faults grew up, the Basement Rock was separated into some massifs. In their boundary, faults and crush zones were formed. Most massifs were inclined to a north or northwest direction, and faults accompanied with active vertical movements were made at the northwestern edges of some massifs, for example, the Katata and Takashima Faults. The subsiding speed of the Nanko Block can be estimated by the 1,000 m deep drilling. It is about 0.6 mm/year and equal to the vertical separation of the Katata Fault. This is the fastest rate even in the Japanese Islands. Perhaps the Katata Fault began its active movement about 0.4 million years ago, when the Ryūge Sands and Gravels began to deposit. But this fault is not so active in comparison with the Ukawa and Takashima Faults (Yokoyama, 1979b).

Geohistory of Lake Biwa

Paleocurrent direction

Paleocurrent directions provide good information on the movement of fault blocks, restoration of the ancient geography, and migration of sedimentary basins. In the Kobiwako (Paleo-Biwa) Group, the paleocurrent directions were deduced from the maximum inclination of tabular type cross-beddings. Cross beds which are inclined to the principal surface of accumulation are frequently observed in sandy facies of the Plio-Pleistocene in Kinki and Tōkai Districts. Tabular and trough types are dominant and others are absent.

It has been shown that the directions of the maximum dip are highly concentrated in the Plio-Pleistocene sediments, especially in the limnetic deposits (Yokoyama 1968b). The paleocurrent directions are thought to be parallel to the mean directions of the maximum dip of cross beddings in the Plio-Pleistocene of Kinki and Tōkai areas as discussed in many other reports (Matsui 1966, Yokoyama 1968b, 1969, 1970, 1974b, 1975a).

In these investigations, angles of inclinations are azimuths of directions. The maximum dips of cross beddings are measured in some areas as presented in Fig. 40. The method of measurement has been reported by Yokoyama (1968b). Moreover, the writer has some unpublished data, which are also shown in Fig. 40 and Table 17.

Kōga area. The paleocurrent directions were measured in the Kazuraki and Sunazaka Sands, the middle part of the Sayama Formation. The results were reported by the writer (1968b). The paleocurrent directions are shown in Fig. 41. They have the azimuth as the mean value of vectors from north to south in the Kazuraki Sands while the directions turn north in the Sunazawa Sands.

The thickness and maximum grain-size in Kazuraki Sand are shown in Fig. 42. Variations in maximum grain sizes parallel to the paleocurrent are shown in Fig. 43. Kazuraki Sand increases in thickness and decreases in maximum grain size downstream as shown by cross beddings. The paleocurrent

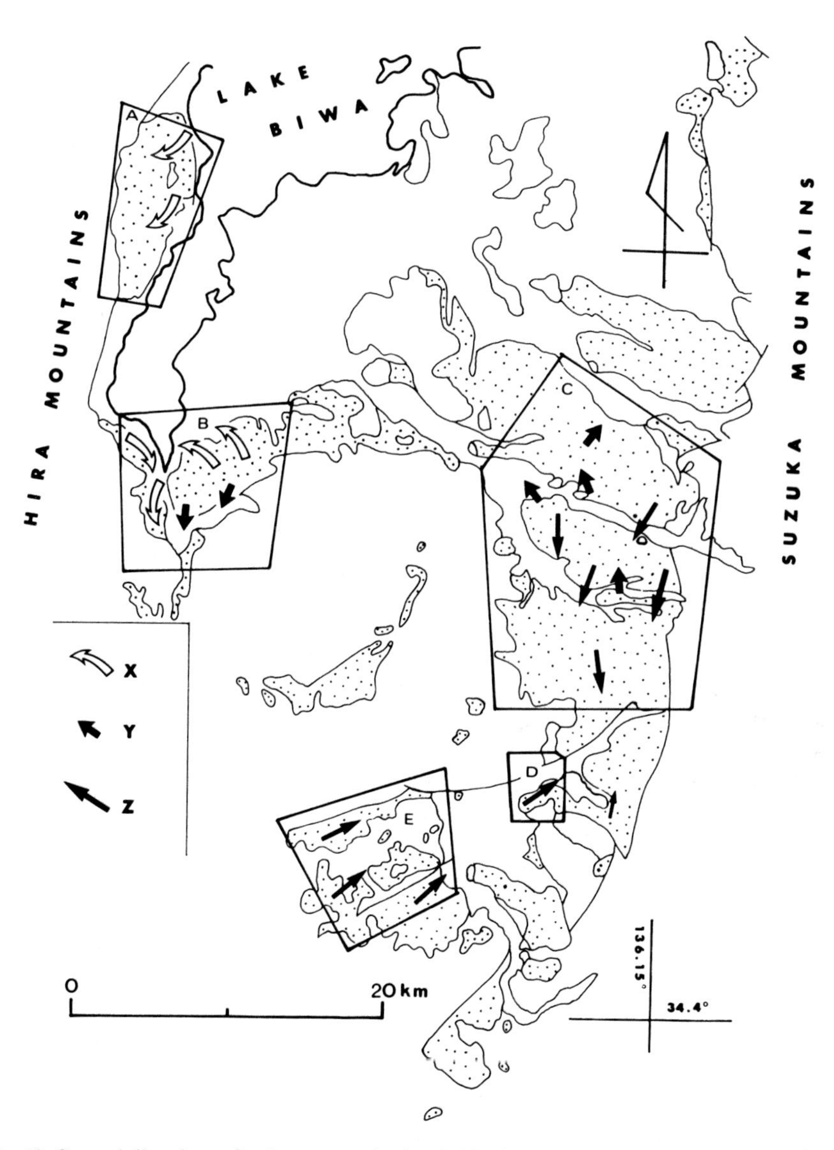

Fig. 40. General directions of paleocurrents in the Kobiwako Group.
A: Kosei, B: Seta, C: Kōga-Gamō, D: Sanagu, E: Shimagawara Areas, X: Upper horizon (Zeze and Katata Formation), Y: Middle horizon (upper part of Sayama Formation and Gamō Formation), Z: Lower horizon (Iga-Aburahi Formation and lower part of Sayama Formation).

Table 17. Paleocurrent Directions of the Kobiwako Group.

Formation	Member	Horizon	Site number	Locality	Direction of vector mean	Consistency ratio	Measured number	Average dip angle
		Ryūge Sands and Gravels	BW.101	South of Minamisho	69.0	98.8	5	23.4
Katata Formation	Minamishō Clays	Kamiōgi Clays	BW.82	Yamanoshita	179.9	99.3	16	36.3
			BW.90	East of Shimoōgi	273.6	99.4	15	34.3
			BW.97	North of Shimoōgi	219.1	97.3	7	13.6
			BW.108	East of Minamisho	217.7	99.2	6	20.7
			BW.111	North of Ono	208.0	99.6	10	22.4
		Ogoto Sands	*BW.7a	South of Kurihara	177.0	99.5	10	29.7
			*BW.7b	South of Kurihara	183.0	99.8	6	33.8
			BW.95	South of Ogoto	185.2	99.3	12	33.5
			BW.96	Ogoto	234.0	99.2	5	14.0
		Ogoto Clays	*BW.2	East of Kurihara	105.0	95.3	6	25.0
			*BW.5	West of Nijigaoka	255.0	94.5	15	31.0
Zeze Formation		Fujimidai Bed	Su.29	Fujimidai	129.0	99.7	5	29.6
			Su.44	South of Chausuyama	110.9	99.4	8	26.9
			Su.45	South of Chausuyama	146.0	98.7	15	19.5
		Zinryō Sands	*Su.26	South of Ishiyama	189.0	97.3	15	15.0
			*Su.27	South of Ishiyama	183.0	95.3	15	15.9
			*Su.28	South of Ishiyama	170.0	95.5	20	14.3
			Su.32	Zinryō	290.2	98.0	15	31.6
			Su.33	Zinryō	259.9	99.4	18	15.9
			Su.36a	South of Zinryō	334.1	93.6	7	27.7
			Su.36b	South of Zinryō	24.0	98.2	3	20.0
			Su.24	Shiga Univ.	220.0	95.9	15	11.6
			Su.25	Shiga Univ.	235.0	99.4	5	15.2
			Su.37	North of Dō	256.0	98.7	12	25.3
			(Su.46)	Sekinotsu	247.0	99.9	5	15.0

Table 17 (continued)

Formation	Horizon	Site number	Locality	Direction of vector mean	Consistency ratio	Measured number	Average dip angle
Gamō Formation	Nangō Alternations	Su.34a	Inatsu-cho	209.0	99.2	5	25.4
		Su.34b	Inatsu-cho	240.0	96.0	5	20.2
		Su.34c	Inatsu-cho	257.6	99.6	6	18.3
		Su.38a	North of Imamura	221.9	99.5	7	26.2
		Su.38b	North of Imamura	232.5	90.4	11	20.1
		Su.39	West of Ishisue	180.6	91.7	14	17.7
		Su.41	Dō	208.5	93.0	17	19.1
		Su.42	Shibahara	201.0	99.7	10	22.2
		Su.43	Shibahara	210.6	94.2	16	20.6
	Hino Clays (Bessho sands)	BE.515a	South of Nishi-ichiba	55.2	99.3	6	42.7
		BE.515b	South of Nishi-ichiba	77.4	99.4	3	35.3
		BE.515c	South of Nishi-ichiba	64.0	99.8	3	35.7
		BE.561c	South of Bessho	54.0	99.9	5	24.0
		BE.561d	South of Bessho	67.4	99.7	3	26.7
		BE.561a	South of Bessho	14.0	99.8	5	22.2
		BE.561b	South of Bessho	4.7	98.7	7	31.0
	Nunobikiyama Alternations	*BE.195	Minakuchibashi	285.0	99.2	13	16.9
		*BE.11	Shinjō	298.0	95.2	10	14.5
Sayama Formation	Sunazaka Sands	*BE.7	Oki	358.0	98.0	20	27.1
		*BE.13	Uchikoshi	317.0	97.7	15	24.8
		*BE.37	Fukawa	42.0	98.0	10	18.2
		*BE.53	Oharaueda	358.0	95.3	15	13.2
	Kazuraki Sands	*BE.28	Mobira	170.0	96.7	7	26.4
		*BE.16	Mobira	217.0	94.6	15	17.2
		*BE.46a	Takano	208.0	98.3	12	24.8
		*BE.46b	Takano	204.0	99.7	10	26.6
		*BE.57	Kami	205.0	98.5	8	9.5

Table 17. (continued)

Horizon		Site number	Locality	Direction of vector mean	Consistency ratio	Measured number	Average dip angle
Sayama Formation		*BE.165	Dodoike	35.0	97.7	13	15.2
		*BE.185	Iwamuro	245.0	98.7	8	18.0
		*BE.211	Terasho	160.0	96.6	12	20.2
		*BE.3a	Momoyama	223.0	96.7	11	24.3
		*BE.3b	Momoyama	202.0	96.9	13	22.9
		*BE.3c	Momoyama	190.0	90.0	15	21.6
		*BE.3d	Momoyama	245.0	99.5	20	29.2
		*BE.210	Kazuraki	168.0	91.3	15	20.2
		*BE.33a	Ōmitobashi	248.0	96.0	15	22.7
		*BE.33b	Ōmitobashi	260.0	99.5	10	18.5
		*BE.175	East of Mushono	214.0	98.0	20	22.7
		BE.263	North of Komazuki	216.0	99.3	5	22.0
Iga–Aburahi Formation	Aburahi Sands	*BE.204	Higashide	190.0	98.4	10	20.0
		*BE.155	Ichiuno	283.0	96.7	9	11.6
		*BE.59	East of Ichiuno	240.0	98.0	15	12.7
	Sanagu Sands	**Iu.1	South of Sanagu	68.4	99.1	12	21.6
		**Iu.1a		43.4	99.2	5	22.4
		**Iu.1b		42.1	99.4	13	22.0
		**Iu.1c		29.1	99.8	10	6.1
		**Iu.2a		52.0	99.4	3	10.3
		**Iu.2b		54.2	98.1	9	19.1
		**Iu.3		43.4	99.9	60	20.5
		**Iu.4	Ichinomiya	52.0	99.2	12	25.8
		**Iu.7	Northwest of Kashiki	36.1	99.0	4	10.0
		**Iu.8	South of Kanaya	337.3	99.2	13	18.0
		**Iu.9		51.4	97.8	17	15.8
		**Iu.10	East of Senzai	20.1	96.4	6	21.8

Table 17 (continued)

Horizon		Site number	Locality	Direction of vector mean	Consistency ratio	Measured number	Average dip angle
Iga–Aburahi Formation	Nakaya Sands	Nu.5a	North of Okuda	87.2	98.5	11	17.7
		Nu.5b		90.3	99.4	4	17.5
		Nu.9	East of Okuda	93.9	99.5	5	15.0
		Nu.12	Sakaide	93.0	92.4	10	24.9
		Nu.13	Okumura	101.6	97.4	77	18.6
	Okuda Alternations	Nu.8	Southeast of Okuda	25.2	99.8	8	19.4
		Nu.10	North of Shimagawara	130.0	99.8	5	21.0
		Nu.111	Shimagawara	90.9	99.7	5	18.0
		Nu.16a	Shirakashi	64.0	98.8	10	24.4
		Nu.16b		52.0	99.1	10	22.3
		Nu.17	Ishiuchi	80.7	97.8	6	14.1
		Nu.18		86.2	95.2	7	26.4
	Tayama Gravels	Nu.2	South of Oshihara	102.5	99.6	13	23.3
		Nu.3		131.1	98.0	10	15.1
		Nu.4	Kitamata	89.9	99.5	5	12.4
		Nu.6	South of Okuda	84.0	99.9	5	24.0
		Nu.7	East of Tsukigase St.	32.0	99.8	5	23.0

*Yokoyama 1969; ** Yokoyama 1975b.

directions accord with other sedimentological criteria in general. It must be noticed that the mean directions of the paleocurrent were turned to the reversed direction, that is, from south to north, in the middle horizon of the Sayama Formation. It is certain that the center of the sedimentary basin changed the location to the north at this time, as shown in Fig. 44.

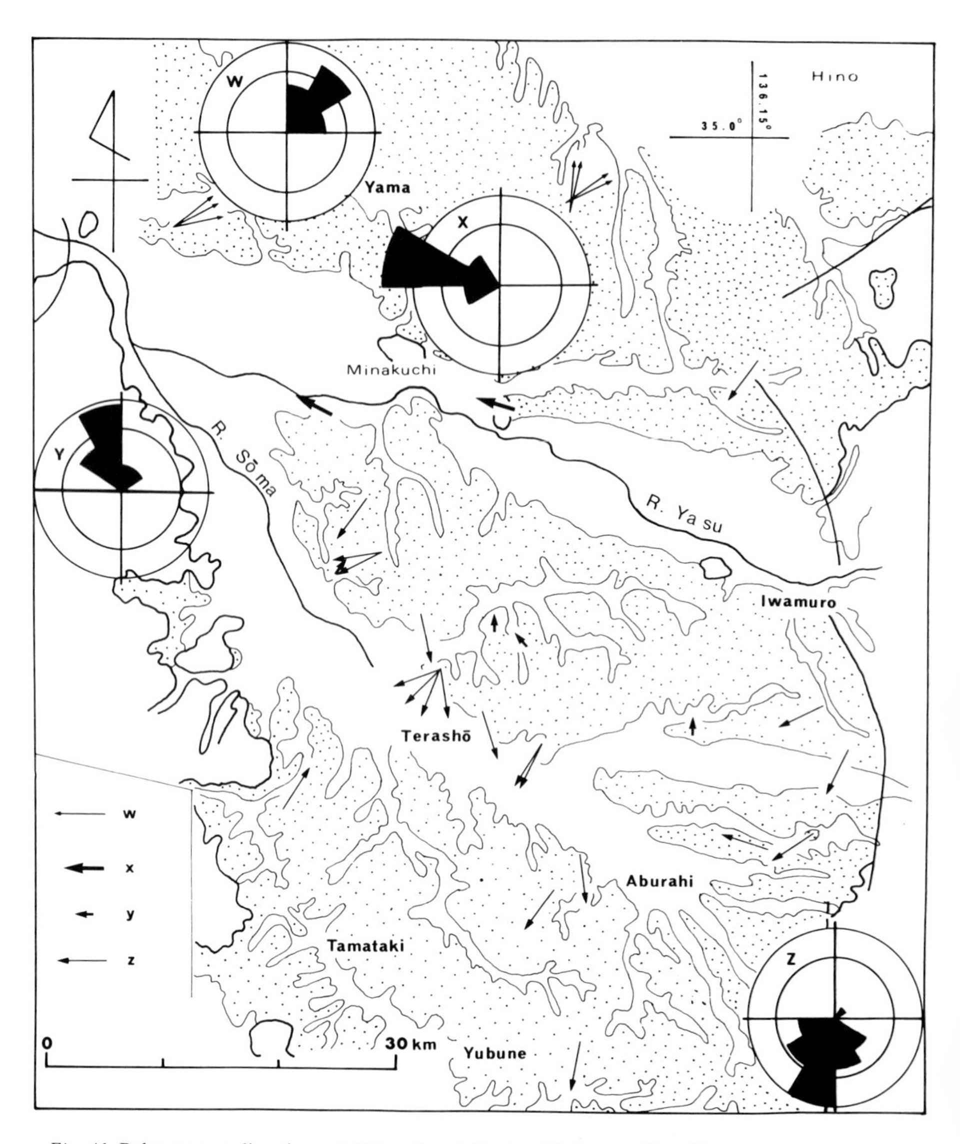

Fig. 41. Paleocurrent directions at Kōga-Gamō Region (C-Area at Fig. 40).
w: Bessho Sands of Hino Clays in Gamō Formation, x: Nunobikiyama Alternations in Gamō Formation, y: Sunazaka Sands in Sayama Formation, z: Kazuraki Sands in Sayama Formation and lower horizon.

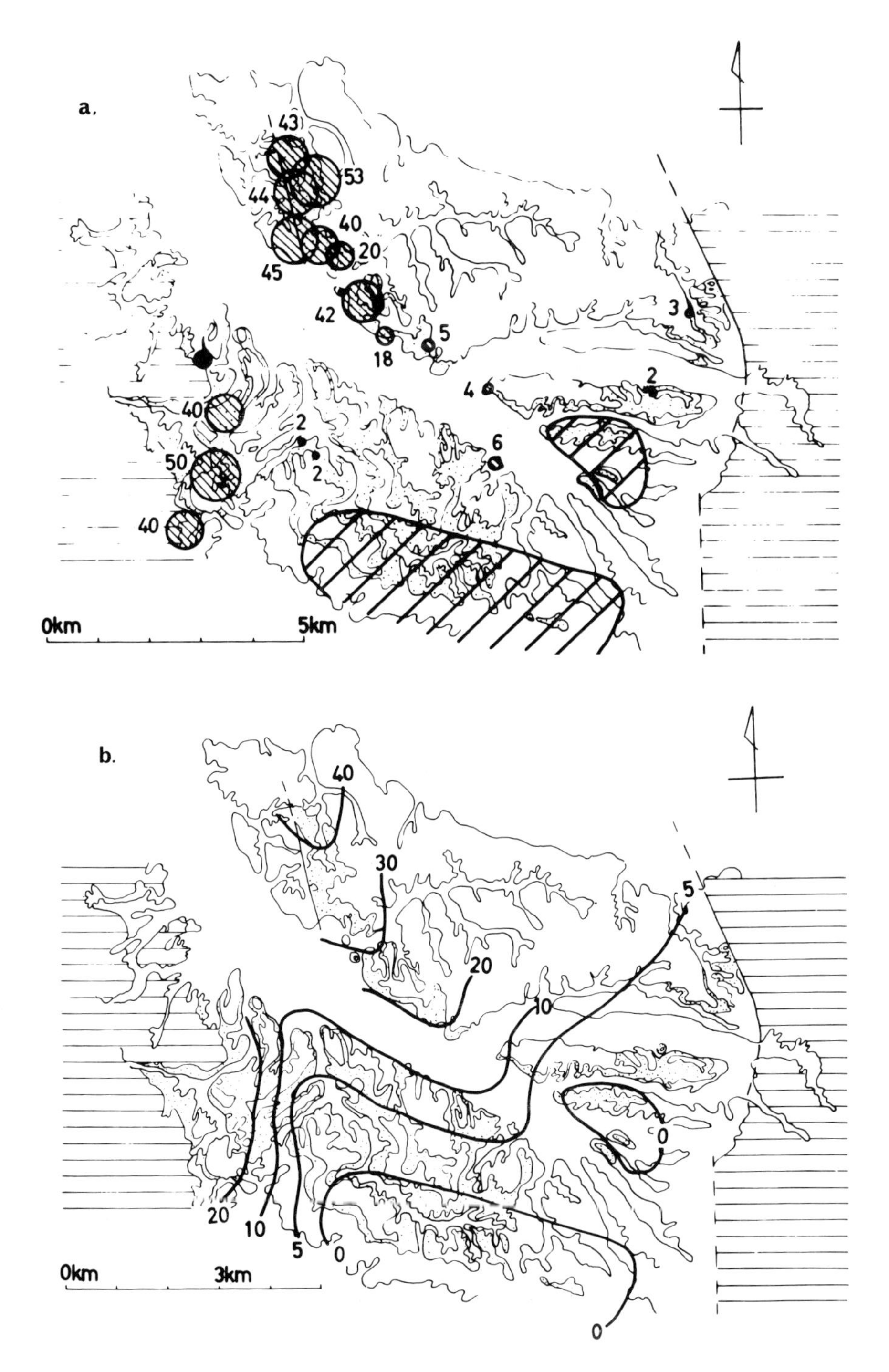

Fig. 42. a) Maximum grain size of the Kazuraki Sands, Sayama Formation. The area of inclined line: only clay beds in the horizon of the Kazuraki Sands. 40: 40 mm in diameter of the maximum grain. • ; 200 cm in diameter of the maximum grain (Yokoyama 1968b).
b) The isopach map of the Kazuraki Sands, Sayama Formation. 40: 40 m in thickness.

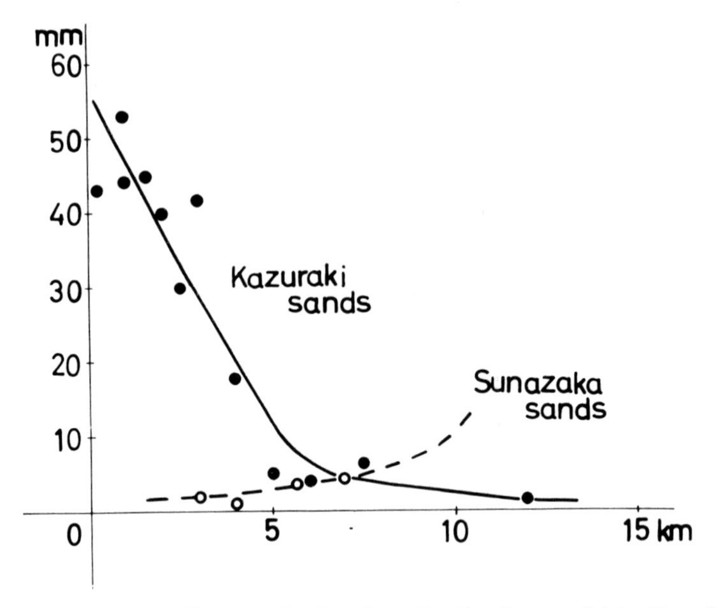

Fig. 43. Change of the maximum grain size along the direction parallel to the paleocurrent (Yokoyama 1968b).

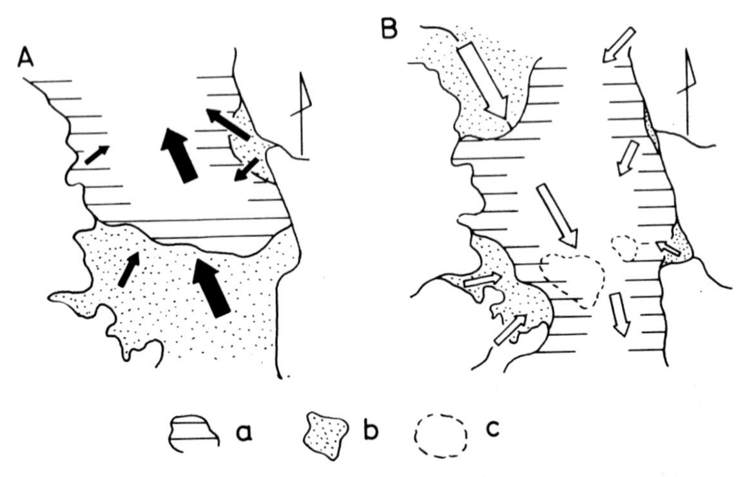

Fig. 44. Paleogeography in the lower and upper horizons of the Sayama Formation. (Yokoyama 1968b)
A: Upper horizon than Sunazaka Sands.
B: Lower horizon than Kazuraki Sands.
a: lake, b: plain, c: area where clay sediment only deposited.

Shimagahara area. In this area, the paleocurrent directions were obtained in Nakaya Sands and Tayama Gravels on the western highlands of the Iga Basin. By means of the distribution and stratigraphy of the Kobiwako Group, Takaya (1963) proposed that the water of the ancient Lake Biwa flowed out westward

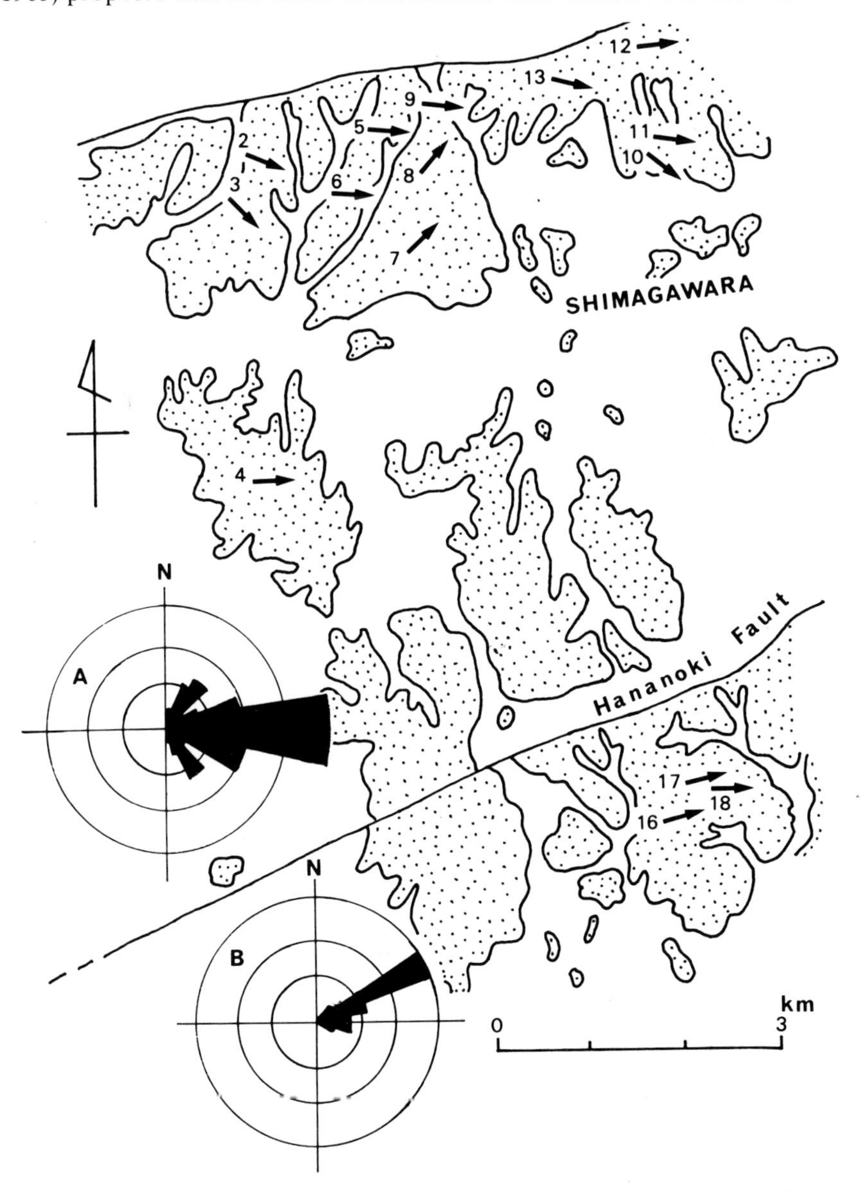

Fig. 45. Paleocurrent directions of the Iga-Aburahi Formation at Shimagawara Area (E-Area at Fig. 40).
A: Northern part of the Hananoki Fault.
B: Southern part of the Hananoki Fault.

from the Iga Basin in the oldest stage and reached to the Nara Basin. If this image were true, the paleocurrent directions must have been westward. But, as shown in Fig. 45, all of them have an east direction.

Sanagu area. As shown in Fig. 17, the older sedimentary basin of the ancient Lake Biwa is roughly divided into several small basins, namely, Gamō-Kōga, Tsuge, Iga Basins and so on. As mentioned above, the writer (1968b) verified that the paleocurrent directions of the lower part of the Kobiwako (Paleo-Biwa) Group, namely, the Iga-Aburahi Formation and the lower half of the Sayama Formation in Kōga to Tsuge, supported Takaya's opinion of 1963. On the other hand, the paleocurrent directions of the Iga-Aburahi Formation in the Shimagahara Area, the western part of the Iga Basin, which connects the Iga and Nara Basins, were eastward. The Sanagu Region is the junction point of the Tsuge and Iga Basins. It is for this reason that the paleocurrent directions of this area are very important for the paleogeography of the ancient Lake Biwa. The results are presented in Figs. 46 and 47. Every paleocurrent was in an eastward direction in this area. This fact indicates that the water of this ancient lake might have flowed out eastward across the Suzuka Mountain Ranges. At that time, the ranges may have been not so high with the height of about 200–300 meters (Fig. 48).

Seta and Zeze area. The River Seta, which is the only one water outlet of Lake Biwa, starts southward from this area. The birth of this river must correspond to the beginning of the actual sedimentary basin of the lake. So, the paleocurrent directions in this area are also very important for the paleogeography of Lake Biwa. The aspect of them at the Seta Area is shown in Fig. 40; the striking fact is that they have a west or southwest direction at the nearest region of the southern mountains. This fact indicates that the River Seta existed here in the earliest stage of the actual sedimentary basin of Lake Biwa, and that the center of that basin was located near and along the southern mountain land. Then, the paleocurrent directions changed to northward when the center of the sedimentary basin moved to north in the age of Seta Gravels I. At the Zeze Area, all of the paleocurrent direction is from NW to SE.

Kosei area. This area is involved in the actual sedimentary basin of Lake Biwa. The paleocurrent directions are represented in Fig. 49 and Table 17. Their outstanding feature is that the paleocurrent directions at the eastern margin of hills are SW, namely, they direct from the coast of the lake to the mountains. This fact may indicate that the center of the sedimentary basin located in the west half of the Shiga Hills and that the area where the southern part of Lake Biwa now exists was a coastal plain at that age when the Katata Formation was depositing.

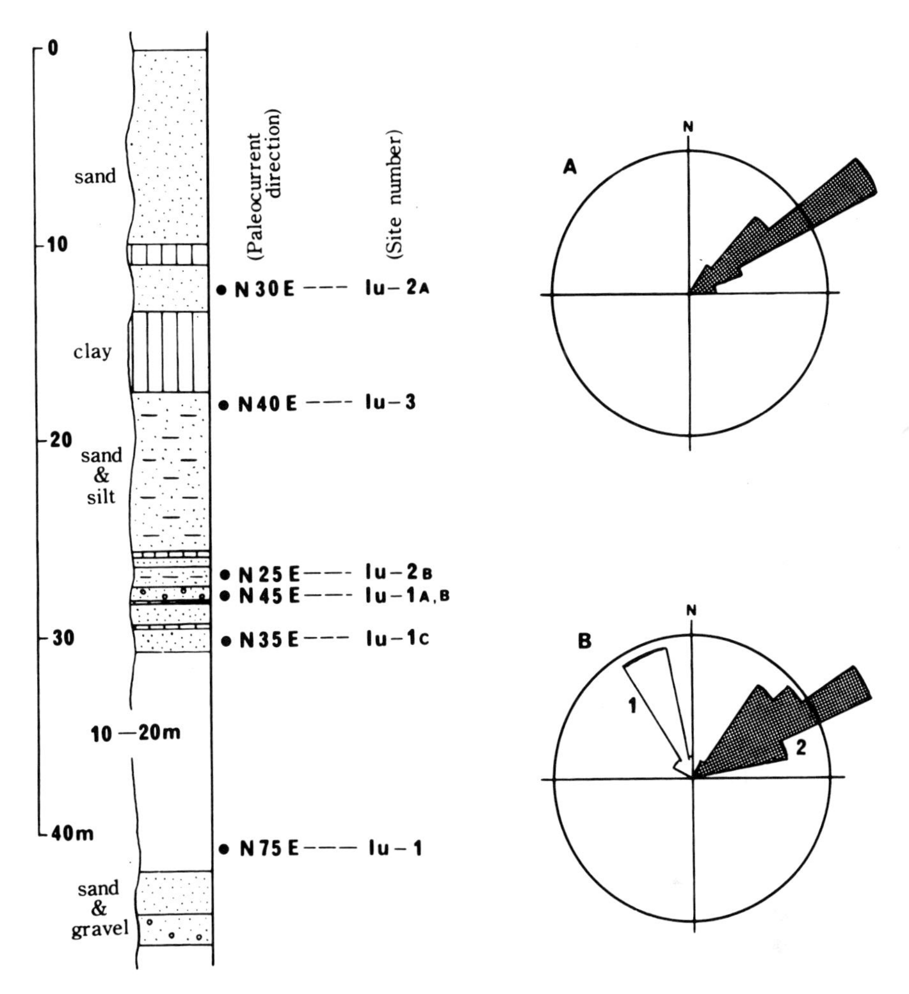

Fig. 46. Paleocurrent directions of the Kobiwako Group at the cliff near Sanagu Interchange of Meihan High Way (Yokoyama 1975a).
A: Current rose diagram of the maximum dip directions of cross beddings at the cliff near Sanagu Interchange (Iu-1, 2, 3).
B: Current rose diagram of the maximum dip directions of cross beddings at other locations (Iu-4, 7, 8, 9, 10).
1: Tsuge area, 2: Iga-Ueno area.

Tectonic development of Lake Biwa Basin

The tectonic history of the sedimentary basin of Lake Biwa is now to be reconstructed on the basis of all the data that have been presented in this report. At the time the lake was born, the Suzuka Mountain Ranges were not so high as the present one. Their height was perhaps about 200–300 m from the water level of the ancient lake. The Kobiwako Group can now be found at the mountain area with the height of about 500–600 m. Those sediments are not so coarse. They are mostly composed of sand layers, intercalated with lignites and

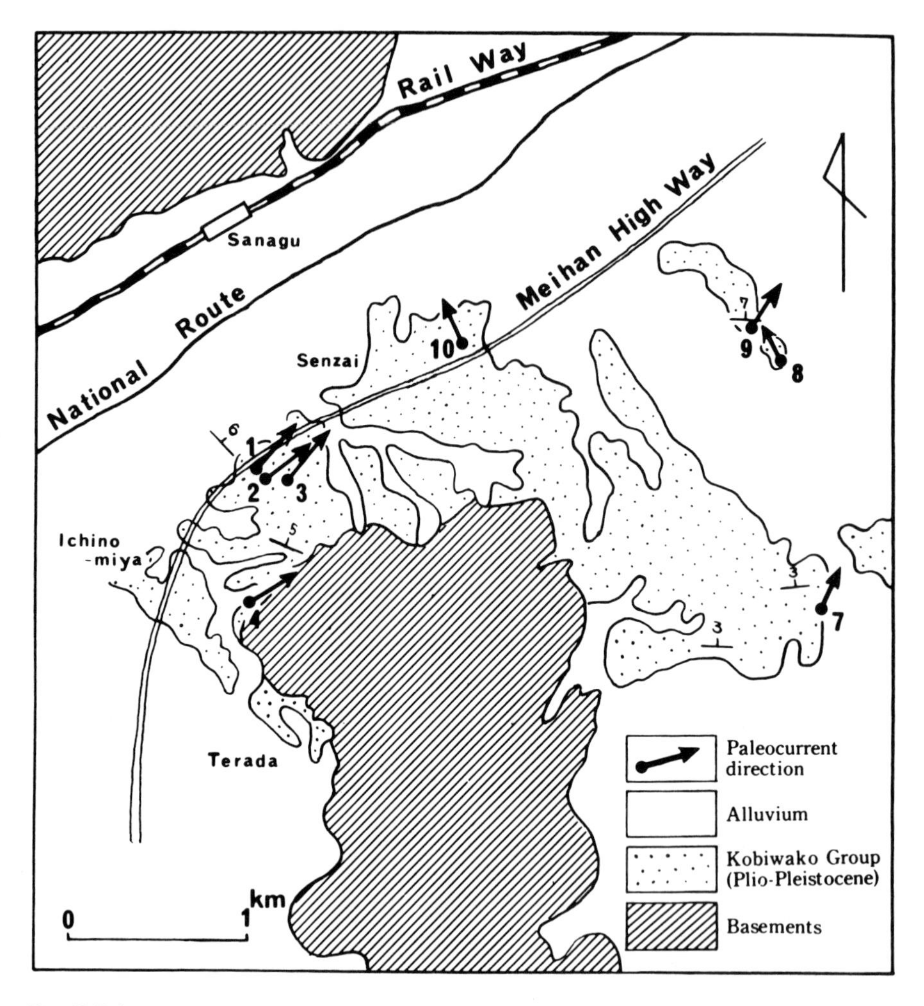

Fig. 47. Paleocurrent directions of the Sanagu Sands in the Iga-Aburahi Formation at Sanagu Area (D-Area in Fig. 40) (Yokoyama 1975a).

coaly silts. Further, one can imagine from the paleocurrent directions in the age of Aburahi Sands, Iga Aburahi Formation (Figs. 40 ~ 47) that the water outlet of the lake at that time was a river across the Suzuka Ranges and that the water reached to a big ancient eastern lake called 'Tōkaiko' (Yokoyama 1975a), while the mountain lands, located in the west and northwest region, might have been higher than the Suzuka Ranges because Iga Gravels were produced in the western mountains. In Kazuraki Sand, large cobble gravels, even boulders, were seen at the western margin of the sedimentary basin. According to these pieces of geologic information, the writer made the image of tectonic developments and paleogeographic maps of this sedimentary basin as Fig. 50.

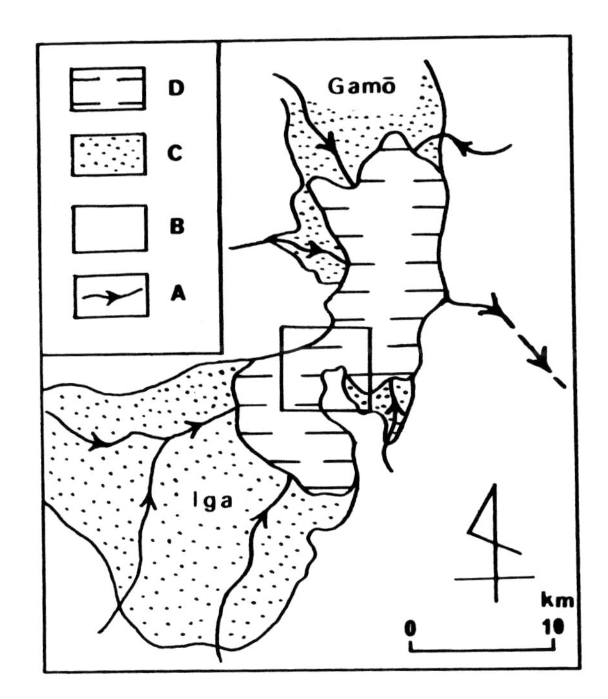

Fig. 48. Paleogeographical map of ancient Lake Biwa in the oldest stage (from middle horizon of Iga-Aburahi Formation to lower horizon of Sayama Formation, about 3.2–2.7 million years ago). A: River course, B: Hills, C: Alluvial plain, D: Lake. (Yokoyama 1975a)

Origin of Corbicula sandai

Corbicula sandai is a famous endemic species in Lake Biwa which migrated from the sea. This shell has been believed to be an exact piece of evidence that Lake Biwa changed from the sea in origin. But this idea is geologically very doubtful. The Kobiwako (Paleo Biwa) Group never contains marine facies in all its horizons. The writer would like to point out a possibility that this species came from the Osaka and Kyoto Areas. Fossils of *Corbicula sandai* were found only in the Katata Formation in the Kobiwako (Paleo-Biwa) Group. Its lowest horizon is just below the Ono Volcanic Ash Layer which is correlated to the Ma 5 horizon of the Osaka Group. The first appearance of this endemic species in the Osaka Group is slightly earlier than the Kobiwako Group, At that time, alternations of the lacustrine and internal bay environment happened several times: It is thought that this environmental change was the cause for the birth of *Corbicula sandai.*

Migration of the sedimentary basins

It has been noticed by many investigators that the sedimentary basins are

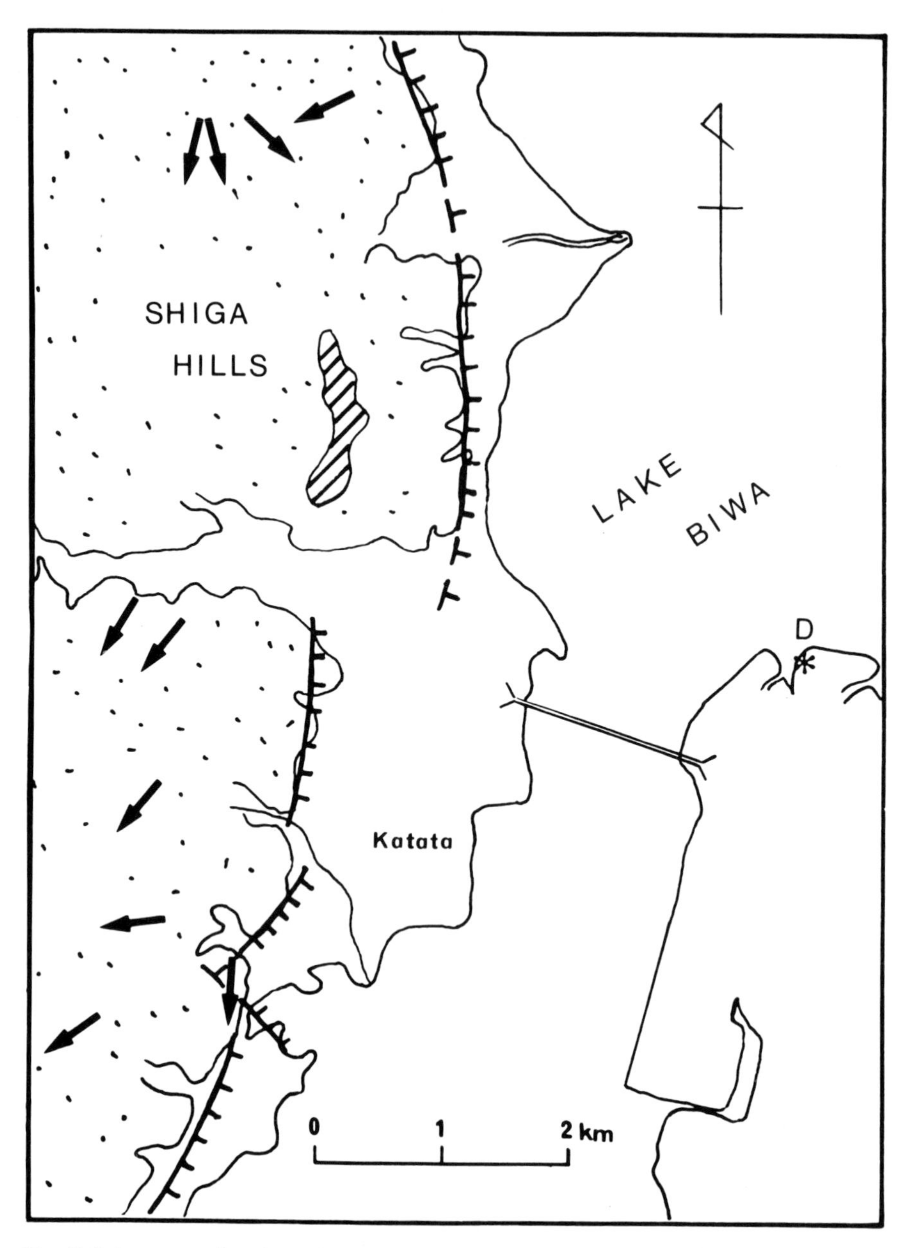

Fig. 49. Paleocurrent directions of the Katata Formation (D: 1,000 m drilling point).

considered to have migrated from south to north in the Ōmi-Iga Basin as well as in other sedimentary basins of the Setouchi Geologic Province. The writer (1968b) mentioned that the mountain ranges having the meridional trend, for example, the Suzuka Ranges, began to upheave before the sedimentation of the

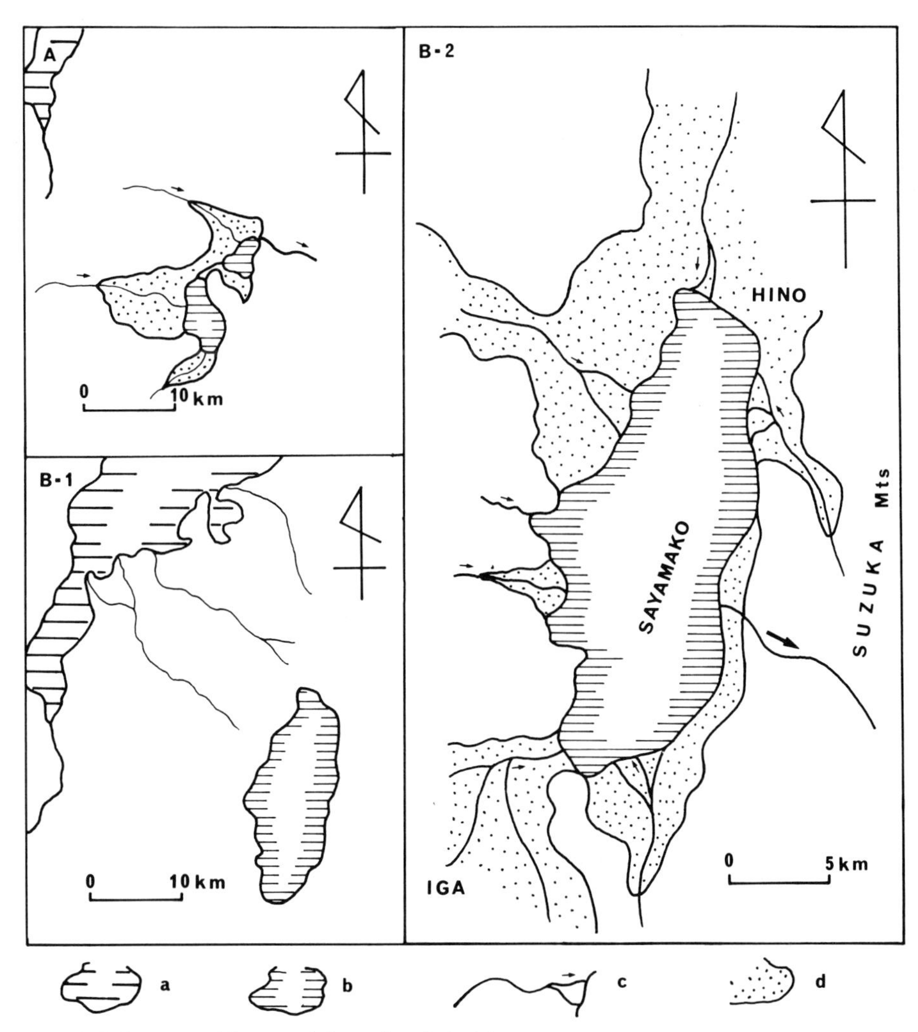

Fig. 50. Paleogeographic map of the ancient Lake Biwa.
a: actual lake, b: ancient lake, c: ancient river, d: ancient alluvial plain.
A: Old stage (Late Iga-Aburahi Age, older than 3.2 m.y.).
B: Old stage II (Sayama Age, 3.2 ~ 2.5 m.y.).

Plio-Pleistocene series so that the fundamental form of the present topography of the Kinki Area was originated before the sedimentation of the Kobiwako and Osaka Groups and that the present geomorphological features are not different essentially from those of the early Kobiwako Age, though the mountains were not so high as now. The migration of the sedimentary basins where the Plio-Pleistocene series deposited is considered to be as shown in Figs. 17 and 50. These figures were made of the sedimentological informations such as paleocurrent directions are in the same horizon that is determined from many tephras.

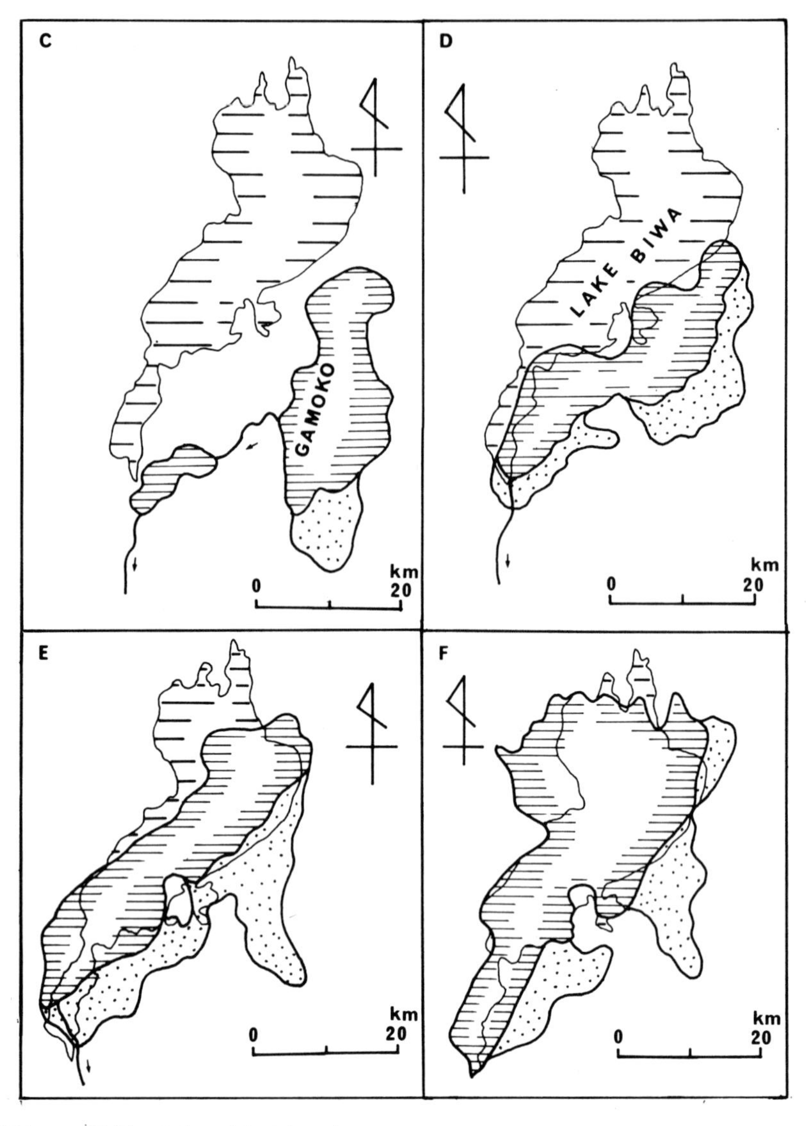

C: Old stage II (Gamō Age, 2.5 ~ 1.5 m.y.).
D: Middle Gravel stage (1.3 ~ 1.2 m.y.).
E: Actual stage I (Minamishō Age, 1.0 ~ 0.4 m.y.).
F: Actual stage II (after 0.4 m.y.).

Old and actual lake stages in the geohistory of Lake Biwa

When the writer proposed the geohistorical image of Lake Biwa in 1973 and 1974 (Yokoyama 1973a, 1974a), two categories of the sedimentary basin were pointed out in the Ōmi-Iga Basin. One is the 'Old Stage of Lake Biwa'. The Iga,

Tsuge, Kōga and Gamō Sub-basins belong to this stage. Ikebe and Yokoyama (1976) called this the 'First Kobiwako'. Another is the 'Actual Stage of Lake Biwa'. The Nanko, Kosei, Hokko, and Takashima Sub-basins belong to this new stage. Ikebe & Yokoyama (1976) called this the 'Second Kobiwako'. Both of the Old and Actual Stages can also be divided into two substages. In addition to the above two, the writer must recognize a different kind of a sedimentary basin in the boundary area of the two. This is the sedimentary basin of the Yōkaichi Gravels and Seta Gravels I Formations. As they are called 'Middle Gravel' (Ishida & Yokoyama 1969), this is a drastic one in the geohistory of Lake Biwa (Middle Gravel Stage of Yokoyama 1969). The ages of these stages may be as follows:

a. Old Stage I 5.0–3.0 m.y.
b. Old Stage II 3.0–1.5 m.y.
c. Middle Gravel Stage 1.5 ± 2.0 m.y.
d. Actual Stage I 1.3–Present
e. Actual Stage II 0.45–Present

Changes in these sedimentary basins are to be discussed in the next chapter.

Stages in Lake Biwa

Pre-Kobiwako stage. No sedimentation took place throughout the 4–5 million years before the Plio-Pleistocene Kobiwako Group began to deposit about 4.5–5.0 million years ago. Though the topography in this stage cannot be clarified in detail because there is no sediment, it is believed that the topographical features had a slightly undulating form like a peneplain throughout this stage.

Old stage I

a. Shimagahara age. This is characterized by ceramic clays which deposited in small low grounds at the Shimagahara and Iga Regions. The basement of these low grounds was composed mainly of granitic rocks. The diameter of these depressions ranges from a few hundred meters to a few kilometers, scattered in the wide and evenly undulated plain as the crustal movement was not so active.

b. Iga-Aburahi age. This age is represented by basal gravels of the Kobiwako Group. Coarse sediments such as cobble or pebble gravels and very coarse sands are dominant. This fact indicates the beginning not only of the subsidence of the southern zone, but also of the upheaval of the west mountainlands. In the latest period of this stage, a few small ancient-lakes appeared in the Kōga, Tsuge and Aburahi regions. The water flowed out to the eastern ancient lake 'Tōkaiko' across the Suzuka Ranges (Fig. 50). This speculation is supported also by the fact that the molluscan fauna of the 'Tōkaiko' is very similar to that of the ancient Lake Biwa.

Old stage II

a. Sayama age. In this age, an ancient lake was formed in the Sayama region.

Fine-grained materials such as clay and fine or medium sand of the Sayama Formation were deposited in this ancient lake. Thus, this age is characterized by the predominance of lacustrine clays. The subsidence of the southern zone such as the Iga and Tsuge sub-basins, ended in the middle period of this stage. Then, the northern zone began to subside continuously, as deduced from the 180° turn of the paleocurrent directions. Judging from the sedimentologic information and the occurrence of fossil plants as only one of *Pinus thunbergii* is seen in the thick homogeneous clay, the paleolake Sayama was the largest and deepest lake in the history of Lake Biwa.

b. Gamō age. The subsiding center of the basin had been migrating to the north since the late Sayama Age. It reached to the Hino and Yōkaichi Area, which corresponds to the Gamō and Echi sub-basins. This age is characterized by alternations of sand and clay with some marginal gravels. The River Seta had already appeared at the early time of this age and the water flowed out through this outlet, as indicated by the paleocurrent directions in the Nangō Alternations.

Middle gravel stage. In this stage, coarser sediments such as cobble or pebble gravels became very dominant in the Plio-Pleistocene series, for example, the Yōkaichi Formation (Table 1). Nakagawa and Yokoyama (1975) pointed out that the sedimentary basins of the 'Middle Gravel Facies' have an elliptical shape, the longitudinal axis of which is parallel to that of Lake Biwa. They are located in the boundary area between the sedimentary basin-chains of Old and Actual Stage. We cannot find out this gravel layer at the base of the well of the 1,000 m drilling at the coast of Lake Biwa. So, it is supposed that in the area including neighbouring basins the width of the sedimentary basins may be about 5-6 km and that the length reaches to about 50-70 km from Tanabe to Inabe. Then, it was recently proposed by Yokoyama et al. (1978) that these gravels, especially in the Osaka Group, were removed from older sediments, because of its abnormal high content of chert gravels, about 70–80% in number. The source is not the Paleozoic area. Perhaps it is basal gravels of the Kobiwako Group. These coarse sediments were overlaid on the southern mountainland and the meridional mountain ranges that are separating the basins. It may be certain not only that these mountainlands rapidly upheaved in this stage but also that the northern zone (the Biwako-Osaka Subsidence Zone) began to subside successively. This tectonic movement is named the 'Suzuka phase in the Rokkō crustal movement' (Yokoyama 1969).

Actual stage I

a. Wani age. In the 1,000 m deep well, the mud dominant facies are at 702.6–873.2 m in depth. This is correlated to the Zeze Alternations distributed in the southern margin of the Nanko Sub-basin. If this correlation is definite, the horizon of the Zeze Alternations corresponds to about 135 m below the Azuki Volcanic Ash, that is, Wani Sands. This stage can be correlated to the Nara Stage of Yokoyama (1969). The center of the sedimentary basin existed at the present Nanko region. This stage is characterized by the downwarping of the massif center. So, the coarser sediments were deposited at the margin and fine-grained materials were settled at the center of the sedimentary basin. At about 1.3 million years ago, the actual Lake Biwa area began to subside. At that time, the Kosei and Nanko Area was connected as one massif and had been inclining to the northwest. Then, this massif separated to two blocks, namely, the Nanko and Kosei Blocks, when the Katata Fault began to form at about 0.45 million years ago. The writer calls the age after the separation of the above two the Actual Stage II (Fig. 51).

b. Minamishō age. This age is characterized by the rapid subsidence of the zonal area along the northwest margin of the sedimentary basin. At the Nanko Sub-basin, the sandy facies were mainly deposited, and the muddy layers such as Minamishō Clays were made in the Kosei and Hokko Basins. In this age, a few chain-lakes were recognized in Kinki and Tokai districts. The writer (1969) called the age 'Hacchoike Stage'. The writer stated: 'This stage is a period when the Biwako-Osaka subsidence zone was continuously subsiding. It is characterized by alternations of marine and lacustrine environments in Osaka and Kyoto Districts. The Hacchoike Alternations are composed of alternating marine clays, Ma 3—Ma 6, and lacustrine sands in the Senriyama hill-lands and other hill-lands in the Osaka area. These marine clays often yield some molluska which lived in an internal bay environment, viz., *Anadara granosa, Dosinia angulosa, D. japonica, Theora lubrica* and *Raeta pulchella.* The lacustrine clays yield some fresh-water molluscan species living in the present Lake Biwa: *Anodonta, Lanceolaria, Viviparus* etc.' In this period, the internal bay came within the shortest distance from Lake Biwa in the geohistory of Lake Biwa, and *Corbicula sandai* migrated from this internal bay to live in the lake at the middle of this age. On the other hand, *Metasequoia disticha* and members of the *Metasequoia* flora were extinct at this time, exactly at the horizon just underneath Ma 3.

Actual stage II. This stage is characterized by the upheaval of the Kosei Block and the rapid subsidence of the northwestern margin of the Nanko Block and the Hokko Basin. In the Kosei Area, gravel layers such as the Ryūge Sands and Gravels correspond to the top gravels made by the upheaval of the fault block. The Western Mountains, Hira Mountain Lands, upheaved with the Kosei Massif at this stage, when the Takashima Formation had been depositing in the Takashima district. Then, the Takashima Fault began to move actively about 0.2—0.25 million years ago. Consequently, the Takashima district upheaved rapidly to deposit Aibano Gravels, while the Hokko Block began to subside. The movement continues up to now.

Concluding remarks

At present, many geoscientific investigations of Lake Biwa are being carried out and a mass of data is being accumulated. It is therefore expected that the geohistorical image of Lake Biwa will be subject to dramatic change even immediately after this book has been published.

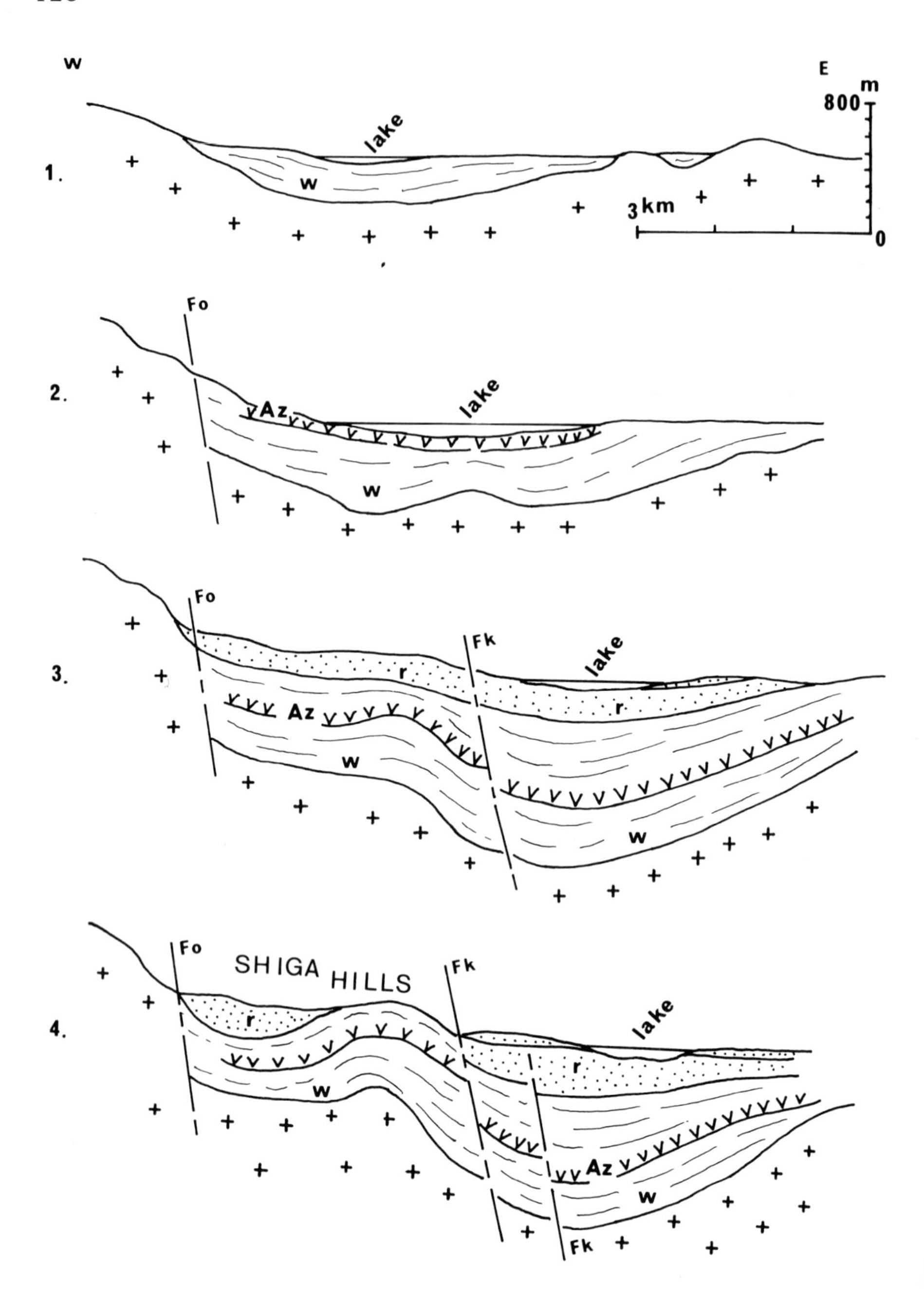

Fig. 51. Tectonic development of Lake Biwa in the Nanko Basin.
W: West, E: East, Fo: Ōtsu Fault, Fk: Katata Fault, Az: Azuki Volcanic Ash, Ẇ: Wani Sands, r: Ryūge Sands and Gravels.
1. Age of Wani Sands (1.0 ~ 1.3 m.y.).
2. Age of Ogoto Clays, lower part of Minamishō Clays (0.8 ~ 0.9 m.y.).
3. Age of Ryūge Sands and Gravels (0.3 ~ 0.45 m.y.).
4. Recent.

The center of the lake migrates to eastward. The Katata Fault was appeared in Age 3.

Geotectonic setting of Lake Biwa and its surroundings

Isamu Murai

In the central part of the Japanese Islands, several major tectonic depressions are developed, such as Osaka Bay bordered with a coastal plain and the Ōmi Basin, where the largest lake in Japan lies, in the Kinki District and Ise Bay fringed with vast alluvial plains in the Tōkai District. They are developed in the Inner zone of the Southwest Japan separated with mountain ranges which run in the north-south direction.

The Japanese Islands are divided into two tectonic provinces, Northeast Japan and Southwest Japan by the Itoigawa-Shizuoka Line or Fossa Magna which is considered to be a great depression belt. A new active arc structure has been developed along the Itoigawa-Shizuoka Line or Fossa Magna since the Neogene. It runs along the inner side of Northeast Japan, across the old arc structure of the Honshū Arc at a right angle through Fossa Magna, and extends to the southern islands of the Izu-Mariana Arc. This active belt is considered to be a late Cenozoic orogenic belt. Distinct crustal movements have proceeded in this belt since the Miocene, and a large geosyncline had been formed with severe volcanic activities, while Southwest Japan, especially the Inner Zone, has shown the history of crustal movement as rather cratonical masses during the Cenozoic times. Inner Zone of Southwest Japan is chiefly composed of Paleozoic rocks and Cretaceous intrusives, and has obtained a cratonic character.

During the Neogene Periods, some gentle deformation of the basement occurred and only narrow parts of the land in the Inner Zone of Southwest Japan were invaded with sea water. Some depressions have been formed by the down-warping of the basement, and Neogene to Quaternary sediments are scattered on basement rocks. Three series of the Cenozoic sediments lie on sedimentary basins developed in the Inner Zone of Southwest Japan. They are the First Setouchi Series which belongs to the middle Miocene, the Setouchi Volcanic Series extruded in the periods from the latest Miocene to the earliest Pliocene, and the Second Setouchi Series of the Plio-Pleistocene ages. The First Setouchi Series filled up depressional zones which had extended in the east-west direction in the central part of the Kinki District. In the Inner Zone, especially in the Kinki District, crustal movements that have resulted to form sedimentary

Horie, S. Lake Biwa

ISBN 90 6193 095 2. Printed in The Netherlands

basins occurred after the volcanic activity of an acidic magma of the Setouchi Volcanic Series. These crustal movements were designated as the Rokkō Movements by Ikebe (1956). The Second Setouchi Series were piled up in the basins, being associated with the proceeding of subsidence of the basins and the uplifting of the hinterlands. In the basin of Osaka Bay and its surroundings, in the area of the ancient lake on the Ōmi and Iga Basins, and in the basin of Ise Bay and its bordering areas, thick piles of Plio-Pleistocene sediments are deposited. They are called the Osaka, Kobiwako and Tōkai Groups respectively.

The Rokkō Movements reached the climax phases in the latest early Pleistocene to the earliest middle Pleistocene. In the early stages of the movements, the structural trend is similar to that in the Miocene. Depressional zones with the east-west direction and the thrust system with the same direction controlled the deposition of the lower parts of the Second Setouchi Series. Then a change in the structural trend occurred in the middle of the Pleistocene, although it caused no remarkable disconformity in the piles of the Second Setouchi Series. The change of stress direction which resulted in the pattern of distribution of faults and folding may have occurred. The direction of the compressive stress which has acted on the Inner Zone of Southwest Japan may have shifted from north-south to west-northwest – east-southeast. In the later stages of the Rokkō Movements, east-west trending depressions began to be separated by the north-south trending foundation folding. The depression of the Plio-Pleistocene age in Kinki and Tōkai Districs has the northeast-southwest axis and the depression areas shifted to the west. Then, strike-slip fault systems appeared in the northwest-southeast and northeast-southwest directions. Some of the older dip-slip faults have been rejuvenated as such strike-slip faults. The later movements have produced various types of structures, such as weak warping associated with strike-slip faults or foundation folding with north-south trending thrustings. Such later crustal movements, the Rokkō Movements, are clearly expressed in the Kinki Triangle.

The Kinki Triangle proposed by Huzita (1962) is a neotectonic province of Southwest Japan, which occupies the central part of the Kinki District. It is considered to have been a relatively stable cratonic mass during the Neogene, but became a tectonic site in the late Pliocene. It is characterized by the alternating arrangement of many elliptical basins and bordering narrow mountain ranges running almost in the north-south direction. In this area, several faultings and foldings proceeded mainly in the culminating stages during the Quaternary. Many fault blocks have been elevated quickly, and also the Osaka and Ōmi Basins have subsided continuously. At the same time, the general uplift of the Japanese Islands have gone on. Such rapid movement of the uplift of bordering mountain ranges and the subsidence of sedimentary basins resulted in the development of the characteristic morphological features in the Kinki Triangle. The Rokkō Mountains on the western fringe of the Kinki Triangle have been uplifted as high as 900 m, and the sedimentary basin of

Ōsaka Bay and its surroundings piled up sediments to more than several hundred meters. Lake Biwa is located in the central part of this Kinki Triangle.

Lake Biwa lies in a large fault basin which is developed in the basement of the Inner Zone of Southwest Japan. The basement rocks are mainly composed of Paleozoic formations and acidic intrusives and extrusives of the Cretaceous Period. On the east and west sides of the lake, Ibuki, Suzuka, and Hiei-Hira Mountains rise up at the altitude of more than 1,000 m, while on the norht and south sides relatively low mountains extend. These bordering mountains are composed of Paleozoic formations and granitic intrusives. Paleozoic formations consist of chert, shale, sandstone, schalstein, clayslate and limestone. They suffered intense contact metamorphism along the boundary with granites and altered to compact hornfels. The high peaks of Mt. Ibuki and Mt. Ryōsen to the east of the lake consist of limestone of the Permian Period. Many hill such as Kōjin-yama, Kannonji-yama, Azuchi-yama, etc. on the east coastal plain, and small islands in the lake such as Oki-no-shima are composed of acidic welded tuffs which effused in the Cretaceous or in the early Tertiary ages. The western extension of these welded tuffs may reach to the western side of the lake floor, and the eastern extension attains on the western foots of the Suzuka Mountains. On the upper stream of River Yasu to the north of Suzuka-Pass, a small distribution of the Miocene sediments called the Ayukawa Formation is recognized on the Paleozoic formations.

Thus the foundation of the basin in which Lake Biwa lies consists of these pre-Pliocene rocks, such as Paleozoic rocks, Cretaceous granites, Cretaceous to early Tertiary welded tuffs, and Miocene formation. On this basement complex, lacustrine deposits accumulated in the ancient lake, and the terrace deposits and alluvial deposits on the areas around the lake are distributed widely. The Kobiwako Group which consists of lacustrine sediments in the Plio-Pleistocene ages now develops forming mainly hilly lands of 200–300 m in height on the west and south sides of the lake. It is composed of loose sediments of gravel, sand, silt and clay. The stratigraphy of this group is clarified by tracing the volcanic ash layers. From the paleoecological studies of this group it is clarified that the ancient Lake Biwa has three stages, old closed lake stage, open lake stage and young closed lake stage.

The terrace deposits composed predominantly of gravels and sands form a blanket on the margin of the hills of the Kobiwako Group. They are classified into the Old Terrace and the Young Terrace; and the Old Terrace is divided into the Upper Old Terrace and the Lower Old Terrace. The Upper Old Terrace is distributed on the tops of hilly lands along the coast of the lake and along stream courses which pour into the lake at the altitude of 125 to 340 m. The Lower Old Terrace is situated lower than the Upper Old Terrace, lying over the gently sloped hills on the lake sides and along the stream courses. The Young Terrace is distributed only on the upper course of the large rivers. Its highest depositional surface comes up to over 15 m, higher than the present stream bed, but it lowers in the down stream and disappears beneath the alluvial plain along the mid-stream.

References

Abe, E., Katsura, K., Nishimura, S., Kanari, S. and Hashimoto, S., (unpublished). Distribution of gravity anomaly in Lake Biwa, Japan.

Abe, E. and Sasajima, S., 1974. Probable sedimentary thickness in Lake Biwa inferred from the gravity measurement. IPPCCE Newsletter, no. 1, ed. S. Horie, 2-3.

Araki, N., 1960. Kobiwako Group at Kosei Area. Mc. Thesis, Dept. Geol. and Mineral., Fac. Sci., Kyoto Univ.

Fuji, N. and Horie, S., 1972. Palynological study on 200 meters core sample of Lake Biwa in Japan. Proc. Japan Acad. 48: 500-504.

Hagiwara, Y., 1967. Analyses of gravity values in Japan. Bull. Earthq. Res. Inst. Univ. Tokyo, 45: 1091-1228.

Hayashi, T., 1974. The Kobiwako Group in the Katata Hills, Shiga Prefecture, Japan. Jour. Geol. Soc. Japan. 80: 261-276.

Hayashida, A., Yokoyama, T., Takemura, K., Danhara, T. and Sasajima, S., 1976. Preliminary report on magnetostratigraphy of the Kobiwako Group on the west coast of Lake Biwa, Central Japan. Paleolimnology of Lake Biwa and the Japanese Pleistocene, 4: ed. S. Horie, 96-108.

Hayashida, A. and Yokoyama, T., 1979. Paleomagnetic chronology of the Plio-Pleistocene Kobiwako Group on the east coast of Lake Biwa, central Japan. Rock Magnetism and Paleogeophysics 6: 48-51.

Horie, S. (ed.), 1972. Paleolimnology of Lake Biwa and the Japanese Pleistocene, 1, 93 pp.

Horie, S. (ed.), 1974. Paleolimnology of Lake Biwa and the Japanese Pleistocene, 2, 288 pp.

Horie, S. (ed.), 1975. Paleolimnology of Lake Biwa and the Japanese Pleistocene, 3, 577 pp.

Horie, S. (ed.), 1976. Paleolimnology of Lake Biwa and the Japanese Pleistocene, 4, 836 pp.

Horie, S. (ed.), 1977. Paleolimnology of Lake Biwa and the Japanese Pleistocene, 5, 372 pp.

Horie, S. (ed.), 1978. Paleolimnology of Lake Biwa and the Japanese Pleistocene, 6, 366 pp. & (Supplement One), 20 pp.

Horie, S. (ed.), 1979. Paleolimnology of Lake Biwa and the Japanese Pleistocene, 7, 467 pp.

Horie, S. (ed.), 1980. Paleolimnology of Lake Biwa and the Japanese Pleistocene, 8, 333 pp.

Horie, S. (ed.), 1981. Paleolimnology of Lake Biwa and the Japanese Pleistocene, 9, 224 pp.

Huzita, K., Ikebè, N., Itihara, M., Kobatake, N., Morishima, M., Morishita, A., Nakagawa, C. and Nakaseko, K., 1951. The Osaka Group and the related Cainozoic Formations. Earth Sci. (Chikyu Kagaku), no. 6, 49-60.

Huzita, K., 1962. Tectonic development of the Median Zone (Setouchi) of southwest Japan, since Miocene. Jour. Geosci., Osaka City Univ. 6: 103-144.

Huzita, K., Kasama, T., Hirano, M., Shinoda, T. and Tanaka-Yamashita, M., 1971. Geology and geomorphology of the Rokko area, Kinki district, Japan with special reference to Quaternary tectonics. Jour. Geosci., Osaka City Univ. 14: 71-124.

Huzita, K., Kishimoto, Y. and Shiono, K., 1973. Neotectonics and seismicity in the Kinki Area, south west Japan. Jour. Geosci., Osaka City Univ. 16: 93-124.

Huzita, K. (ed.), 1974. Quaternary tectonic map, Kinki. Geol. Surv., Japan.

Ibaragi Research Group, 1966. The Osaka Group of Fukui Area, north of Ibaragi (Ibaraki), Osaka Prefecture and occurrence of *Elephas shigensis*-The research of younger Cenozoic strata in Kinki Province, Part 6-Mem. Vol. Prof. Susumu Matsushita, Kyoto, Kyoto Univ., 117-130.

Ichikawa, M., 1971. Reanalyses of mechanism of earthquakes which occurred in and near Japan, and statistical studies on the nodal plane solutions obtained, 1926-1968. Geophys. Mag. 35: 207-274.
Ikebe, No., 1933, Paleo-Biwa series, a Pleistocene deposits in the west side of Lake Biwa. Chikyu 20: 241-260.
Ikebe, N., 1934. Cainozoic stratigraphy of the eastern part of Koga-gori, Shiga-ken. Thesis, Dept. Geol. and Mineral., Fac. Sci., Kyoto Univ.
Ikebe, N., 1956. Cenozoic geohistory of Japan. Proc. 8th Pacific Sci. Congr. 2: 446-456.
Ikebe, N., Chiji, M. and Ishida, S., 1966a. Catalogue of the late Cenozoic Proboscidea in the Kinki District, Japan. Jour. Geosci., Osaka City Univ. 9: 47-87.
Ikebe, N., and Huzita, K., 1966b. The Rokko movements, the Pliocene-Pleistocene crustal movements in Japan. Quaternaria 8: 277-287.
Ikebe, N. and Yokoyama, T., 1976. General explanation of the Kobiwako Group–Ancient lake deposits of Lake Biwa – Paleolimnology of Lake Biwa and the Japanese Pleistocene, 4: ed. S. Horie, 31-51.
Ikeda, K., 1958. Geological investigations on the area around the Itadori Fault, Hokuriku Tunnel. Tetsudō Gijitsu Kenkyusho Sokuhō, No. 58-93. 5pp.
Imamura, A., 1928. On the topographical changes preceding and following the Anegawa earthquake of 1909. Proc. Imp. Acad. 4: 371-373.
Ishida, S., Maenaka, K. and Yokoyama, T., 1969. Paleomagnetic chronology of volcanic ash of the Plio-Pleistocene series in Kinki district, Japan. – The research of younger Cenozoic strata in Kinki district, Part 12 – Jour. Geol. Soc. Japan 75: 183-197.
Ishida, S. and Yokoyama, T., 1969. Tephrochronology, paleogeography and tectonic development of Plio-Pleistocene in Kinki and Tokai districts, Japan. – The research of younger Cenozoic strata in Kinki province, Part 10 – Quaternary Res. (Daiyonki Kenkyu) 8: 31-43.
Ishida, S., 1976. Significance of paleoecologic study in Lake Biwa. In: Land Palaeoecology – Monograph on Palaeoecology I – ed. Geol. Soc. Japan, Palaeontol. Soc. Japan, Tokyo, Kyoritsu Shuppan k.k., 42–62.
Ishida, S., Nakagawa, Y., Nasu, T. and Nishiyama Research Group, 1976a. Stratigraphy of the Kobiwako Group in Konan Area, south of Lake Biwa, Central Japan. Paleolimnology of Lake Biwa and the Japanese Pleistocene, 4: ed. S. Horie, 109-124.
Ishida, S., Horie, S. and Yokoyama, T., 1976b. Comments on the inclination of the well of the 1,000 M core boring on east coast of Lake Biwa. Paleolimnology of Lake Biwa and the Japanese Pleistocene, 4: ed. S. Horie, 74-78.
Isomi, H., 1956. Explanation text of the geological map of Japan, scale 1:50,000, 'Ominagahama'. Geol. Surv. Japan.
Itihara, M., 1960. Some problems of the Quaternary sedimentaries, Osaka and Akasi Areas. Earth Sci. (Chikyu Kagaku), No. 49, 15-25.
Itihara, M. and Takaya, Y., 1965. The absolute age of the Kōzuhata plant bearing bed. – ^{14}C-Age of the Quarternay deposits in Japan XVII –, Earth Sci. (Chikyu Kagaku), no. 76, 41.
Itihara, M., 1966. The Osaka Group and the Rokko movements. Earth Sci. (Chikyu Kagaku), No. 85-86, 12-18.
Itihara, M., Yoshikawa, S., Inoue, K., Hayashi, T., Tateishi, M. and Nakajima, K., 1975. Stratigraphy of the Plio-Pleistocene Osaka Group in Sennan-Senpoku area, south of Osaka, Japan. Jour. Geosci., Osaka City Univ. 19: 1-29.
Ito, H. and Huzita, K., 1971. The flow of the earth's crust considered from the Quaternay crustal movements in southwest Japan. Jour. Soc. Materials Sci. Japan 20: 190–196.
Ito, K. and Watanabe, K., 1977. Focal mechanisms of very shallow earthquakes in the vicinity of Lake Biwa. Zisin, (Jour. Seismol. Soc. Japan), second Ser., 30: 43-54.
Itoigawa, J., 1971. The stratigraphy of the Tokoname Formation in the Environs of the Chita Peninsula, Japan. Research on the Seto Group, Part 2. – Mem. Vol. Prof. Heiichi Takehara, Nagoya, Nagoya Univ., 83-98.

Kaigake Research Group, 1972. Stratigraphy of the Kobiwako Group in the Kaigake-Komazuki area, Shiga Prefecture, Japan. Jour. Geol. Soc. Japan 78: 601-609.

Kaneko, S., 1964. Tectonic relief in South Kanto, Japan. Trans. Roy. Soc. N. Z. 2: 187-204.

Kawai, N., Yaskawa, K., Nakajima, T., Torii, M. and Horie, S., 1972. Oscillating geomagnetic field with a recurring reversal discovered from Lake Biwa. Proc. Japan Acad. 48: 186-190.

Kishimoto, Y. and Nishida, R., 1973. Mechanisms of microearthquakes and their relation to geological structures. Bull. Disaster Prev. Res. Inst., Kyoto Univ. 23: 1-25.

Kobiwako Research Group, 1977. The Kobiwako Group in the western part of Minakuchi Hills, Shiga Prefecture, Japan. Earth Sci. (Chikyu Kagaku), No. 31: 115-129.

Kondo, Y., 1968. Studies on structural geology of the Iga tectonic basin. Rept. Geol. Surv. Japan, No. 231, 1-30.

Koto, B., 1910. The Gono earthquake from the geological viewpoint. Bull. Imp. Earthq. Invest. Comm., No. 69, 1-15.

Maenaka, K., Yokoyama, T. and Ishida, S., 1977. Paleomagnetic stratigraphy and biostratigraphy of the Plio-Pleistocene in the Kinki district, Japan. Quaternary Res. 7: 341-362.

Makinouchi, T., 1975a. The Tokoname Group in the southern part of the Chita Peninsula, Central Japan. Jour. Geol. Soc. Japan 81: 67-80.

Makinouchi, T., 1975b. The Taketoyo Formation in the southern part of the Chita Peninsula, Central Japan. Jour. Geol. Soc. Japan 81: 185-196.

Mankinen, E. A. and Dalrymple, G. B., 1979. Revised geomagnetic polarity time scale for the interval 0-5 m.y. B. P. Jour. Geophys. Res. 84: No. B2, 615-626.

Matsui, H., 1966. Transporting direction of sediments of the Kuragari Formation. Mem. Vol. Prof. Susumu Matsushita, Kyoto, Kyoto Univ., 89-95.

Matsumura, K. and Oike, K., 1973. The microseismicities in and around Japan. Disaster Prev. Res. Inst. Ann., Kyoto Univ., No. 16, B, 77–87.

Matsushita, S., 1953. Nihon Chihō Chishitsu-shi 'Kinki Chihō'. Tokyo, Asakura Shoten, 293 pp.

Matsushita, S., 1961. Geology of Mt. Hiei. In: Mt. Hiei – its nature and culture, ed. S. Kitamura et al., Kyoto, Kyoto Shinbun-sha, 3-18.

Matsuzawa, I., 1968. On the tilt movements in the Chubu District of Honshu. Jour. Geol. Soc. Japan 74: 61-71.

Miki, S., 1948. Floral remains in Kinki and adjacent districts since the Pliocene with description 8 new species. Mineral. and Geol. No. 9, 105-144.

Miki, S., 1956. Remains of *Pinus koraiensis* S. et Z. and associated remains in Japan. Bot. Mag., 69: 447-454.

Miura, S. & Wakasa Quaternay Research Group, 1969. The Pleistocene deposits in the Wakasa District, Fukui Prefecture, Central Japan. Mem. Fac. Educ., Fukui Univ. Ser. II (Nat. Sci.), No. 19, Pt. 3, 57-70.

Miyamura, M., Mimura, K. and Yokoyama, T., 1976. Geology of the Hikone Tobu District. Geol. Surv. Japan, 1-49.

Mori, Sh., 1971a. Some Volcanic Ash Layers in the Seto and Agé Groups. Researches on the Seto Group, Part 3. Mem. Vol. Prof. Heiichi Takehara, Nagoya, Nagoya Univ., 99-111.

Mori, Sh., 1971b. The Yadagawa formation of the Seto Group in the east of Nagoya City, Aichi Prefecture. Jour. Geol. Soc. Japan 77: 635-644.

Murai, I., 1955. Tectonic analysis of the district surrounding Fukui Plain. Bull. Earthq. Res. Inst., Tokyo Univ. 33: 121-151.

Murai, I., 1970. Geologic structure of the epicentral area and its surroundings of the earthquake of the central part of Gifu Prefecture, September 9, 1969. Bull. Earthq. Res. Inst. Univ. Tokyo 48: 1251-1266.

Murai, I. and Kaneko, S., 1975. Active fault systems developed in the area around Lake Biwa. Paleolimnology of Lake Biwa and the Japanese Pleistocene, 3: ed. S. Horie, 30-59.

Nakagawa, Y. and Yokoyama, T., 1975. New data on the ancient River Seta. Pre-print, Geol. Soc., Japan, 397.

Nakamura, S., 1928. Preliminary notes on the Hanaori Fault. Chikyu 10: 327-334.

Nakamura, S., 1930. Division of the Pleistocene strata in Japan. Nihon Gakujitsu Kyokai Hōkoku 5: 115-117.

Nakanishi, A., Nakagawa, Y. and Yokoyama, T., 1969. Marine facies and diatom flora in the horizon of yellow volcanic ash layer of the Osaka Group, Plio-Pleistocene in Central Japan. Quaternary Res. (Daiyonki Kenkyu) 8: 131-137.

Nakazawa, K., 1961. Geological history of Mt. Hiei. In: Mt. Hiei – its nature and culture, ed. S. Kitamura et al., Kyoto, Kyoto Shinbun-sha, 19-26.

Nakazawa, K. and Ishida, S., 1959. The fossil-woods at the bottom of the River Seta. Chigaku-Kenkyu 11: 138-143.

Nishimura, S. and Sasajima, S., 1970. Fission-track age of volcanic ash-layers of the Plio-Pleistocene series in Kinki district, Japan. Earth Sci. (Chikyu Kagaku), 24: 222-224.

Nishimura, S. and Yokoyama, T., 1973. Fission-track ages of volcanic ashes in 200 m core sample of Lake Biwa, Japan. Proc. Japan Acad. 49: 615-618.

Nishimura, S. and Yokoyama, T., 1974. Fission-track ages of volcanic ashes of core samples of Lake Biwa and Kobiwako Group. Paleolimnology of Lake Biwa and the Japanese Pleistocene, 2: ed. S. Horie, 38-46.

Nishimura, S. and Yokoyama, T., 1975. Fission-track ages of volcanic ashes of core samples of Lake Biwa and the Kobiwako Group (2). Paleolimnology of Lake Biwa and the Japanese Pleistocene, 3: ed. S. Horie, 138-142.

Okayama, T., 1931. Fault topography of the interior part of the Mino-Echizen Mountainland. Geogr. Rev. Japan 7: 920-942, 1035-1062.

Okayama, T., 1956. Yanagase Fault and Tsuruga Bay – Ise Bay Line. Sundai-shigaku, No. 7, 75-101.

Okumura, Y., Nakagawa, N. and Togo, M., 1972. Some considerations on the deformation of the Shiga Hills, Shiga Prefecture. Geogr. Rept. Hōsei Univ., No. 1, 29-40.

Price, N. J., 1959. Mechanics of jointing in rocks. Geol. Mag. 96: 149-167.

Shibasaki, T., 1956. On the Tsuruga – Otsu sheared zone. Earth Sci. (Chikyu Kagaku), No. 29, 1-8.

Sugimura, A., 1963. Notes on the Yanagase Fault., Japan. Quaternary Res. (Daiyonki Kenkyu), 2: 220-231.

Takahashi, H. and Shirai, K., 1959. Engineering geology of Hokuriku Tunnel. Railway Tech. Res. Rept., No. 74, 61 pp.

Takaya, Y., 1963. Stratigraphy of the Paleo-Biwa Group and the paleogeography of Lake Biwa with special reference to the origin of the endemic species in Lake Biwa. Mem. Coll. Sci., Univ. Kyoto, Ser. B, 30: 81-119.

Takehara, H., 1961. Stratigraphy of the Agé Group, Northern Mie Pref. Japan. Mem. Vol. Prof. Jiro Makiyama, Kyoto, Kyoto Univ., 45-50.

Takemura, K., Nishimura, S., Danhara, T. and Yokoyama, T., 1976. Properties and fission track age of volcanic ashes in the 1000 M core sample of Lake Biwa with special reference to correlation by tephra among the 1000 M, 200 M boring core samples of Lake Biwa and the Kobiwako Group. Paleolimnology of Lake Biwa and the Japanese Pleistocene, 4: ed. S. Horie, 79–95.

Takemura, K., 1978. Stratigraphy of the Plio-Pleistocene Ag'e Group in the northern part of Mie Prefecture. MC. Thesis, Dept. Geol and Mineral., Fac. Sci., Kyoto Univ.

Tamura, M., Matsuoka, C. and Yokoyama, T., 1977. Stratigraphy of the Gamo Formation of Plio-Pleistocene Kobiwako Group in the northern hills of Minakuchi-cho, south of Lake Biwa, Central Japan. Jour. Geol. Soc. Japan 83: 749-762.

Togo, M., 1974. Tectonic morphogenesis in the Nosaka Mountains, north of Lake Biwa, Central Japan. Geogr. Rev. Japan 47: 669-683.

Tsuda, K., 1972. Effects of the active faults on the constructions in the Rokko Area. Eng. Geol. 13: 101-111.

Usami, T., 1966. Descriptive table of major earthquakes in and near Japan which were accompanied by damages. Bull. Earthq. Res. Inst. 44: 1571-1622.

Usami, T., 1974. Error estimation for epicenters of Japanese historical earthquakes. Special Rept. Earthq. Res. Inst., Tokyo Univ., No. 12, 1-29.

Yamasaki, N. and Tada, F., 1927. On the morphology and tectonics of the districts near the Lake Biwa. Bull. Earthq. Res. Inst. Tokyo Imp. Univ., 2: 85-108.

Yokoyama, T., Matsuoka, T., Nasu, T. and Tamura, M., 1968. Stratigraphy and structure of the Sayama Formation, the lower part of the Kobiwako Group, southeastern part of Lake Biwa, Central Japan. – Researches on younger Cenozoic strata in Kinki Province, Part 9, Jour. Geol. Soc. Japan 74: 327-341.

Yokoyama, T., 1968a. Note on the Quaternary systems around Lake Biwa. Daiyonki, No. 13, 75-80.

Yokoyama, T., 1968b. Transition of paleo-lake Biwa during the Pliocene deduced from paleocurrent directions and other sedimentological features of sandy facies in the Kobiwako Group, Central Japan – The research of younger Cenozoic strata in Kinki province, Part 11 –. Jour. Geol. Soc. Japan 74: 623-632.

Yokoyama, T., 1969. Tephrochronology and paleogeography of the Plio-Pleistocene in the Eastern Setouchi Geologic Province, southwest Japan. Mem. Fac. Sci., Kyoto Univ., Ser. Geol. Mineral. 36: 19-85.

Yokoyama, T. and Nishiyama Research Group, 1969. Horizon of the Kobiwako Group at the westside of the Seta River in Otsu City. Osakasogun Soken-Renrakushi, No. 4, 35-38.

Yokoyama, T., 1970. Migration of the Seta and Kizu Rivers in the early Pleistocene. Pre-print, Geol. Soc. Japan, 139.

Yokoyama, T., 1971. Upheaval of the Suzuka Mountain Range (1), with special reference to the stratigraphy of the Plio-Pleistocene series in Northern Mie Prefecture, Central Japan. The research of younger Cenozoic strata in Kinki District, Part 19. Mem. Vol. Prof. Heiichi Takehara, Nagoya, Nagoya Univ., 55-67.

Yokoyama, T. and Maenaka, K., 1971. Ferromagnetic minerals of the volcanic ash layers in Plio-Pleistocene Osaka and Kobiwako Groups. Daiyonki, No. 16, 44-52.

Yokoyama, T., 1972. Discrimination of the volcanic ash layers by means of the features of volcanic glass, with special reference to the difference of the titanium contents in the volcanic glass of the Plio-Pleistocene Osaka Group, Japan. Quaternary Res. (Daiyonki Kenkyu) 11: 247-253.

Yokoyama, T., 1973a. Introduction to geohistory of Lake Biwa. Nature Study 19: 2-8.

Yokoyama, T., 1973b. An outline of Quaternary system around Lake Biwa-ko and geologic observations of 200 m core sample obtained from that lake, with special reference to mineralogic analysis of volcanic ash layers found in the core sample. Jap. Jour. Limnol. 34: 111-118.

Yokoyama, T., 1974a. Natural history of River Yodo and Lake Biwa. In: Yodogawa Hyakunenshi, Osaka, Kensetsu-shō, Kinki Chihō Kensetsu-kyoku, 3-28.

Yokoyama, T., 1974b. Stratigraphy of the Kobiwako Group and paleocurrent directions at west part of the Iga Basin. Pre-print, Geol. Soc. Japan, 81.

Yokoyama, T., 1975a. Paleogeography of ancient Lake Biwa in late Pliocene, deduced from paleocurrent directions of Kobiwako Group at Sanagu area in Mie Prefecture, Japan. Sci. Eng. Rev. Doshisha Univ. 15: 37-44.

Yokoyama, T., 1975b. Plio-Pleistocene Kobiwako Group on the west coast of Lake Biwa with special reference to correlation to the 200 M core sample of Lake Biwa by tephra. Paleolimnology of Lake Biwa and the Japanese Pleistocene, 3: ed. S. Horie, 114-137.

Yokoyama, T., 1976. Paleomagnetic estimation of the dip and strike of stratum in 1000 M deep core boring near Lake Biwa, Japan. Paleolimnology of Lake Biwa and the Japanese Pleistocene, 4: ed. S. Horie, 67-73.

Yokoyama, T., Ishida, S., Danhara, T., Hashimoto, S., Hayashi, T., Hayashida, A., Nakagawa, Y., Nakajima, T., Natsuhara, N., Nishida, J., Otofuji, Y., Sakamoto, M., Takemura, K., Tanaka, N., Torii, M., Yamada, K., Yoshikawa, S. and Horie, S., 1976. Lithofacies of the 1000 M core samples on the east coast of Lake Biwa, Japan. Paleolimnology of Lake Biwa and the Japanese Pleistocene, 4: ed. S. Horie, 52-66.

Yokoyama, T., 1977. Magnetostratigraphy of the Plio-Pleistocene Osaka and Kobiwako Groups, Kinki District, Japan. Quaternary Res. (Daiyonki Kenkyu) 16: 139-148.

Yokoyama, T., Nakagawa, Y., Makinouchi, T. and Ishida, S., 1977a. Subdivision of Plio-Pleistocene series in Kinki and Tokai District, Japan. Proc. 1st Internat. Congr. Pacific Neogene Stratigr. 1976, CPNS, 408-412.

Yokoyama, T., Danhara, T., Kobata, Y. and Nishimura, S., 1977b. Fission-track ages of volcanic ashes of core samples of Lake Biwa and the Kobiwako Group (3). Paleolimnology of Lake Biwa and the Japanese Pleistocene, 5: ed. S. Horie, 44-53.

Yokoyama, T., Takemura, K. and Matsuoka, K., 1977c. Preliminary report on the Takashima Formation, uppermost part of the Kobiwako Group, Plio-Pleistocene sediments around Lake Biwa, Japan. Paleolimnology of Lake Biwa and the Japanese Pleistocene, 5: ed. S. Horie, 54-64.

Yokoyama, T., Nakagawa, Y., Makinouchi, T., Matsuda, T., Takemura, K. Hayashida, A., Danhara, T. and Kobata, Y., 1978. Geology of the Dōshisha-Tanabe-Kōchi. – Geohistory of the south Yamashiro District, Kyoto. – Kyoto, Dōshisha Univ., Dōshisha Daigaku Kōchi Gakujutsu Chōsa Iinkai Chōsa Shiryō, No. 13, 68 pp.

Yokoyama, T., 1979a. Tectonic development of the Nanko Basin, southern part of Lake Biwa, Central Japan. Paleolimnology of Lake Biwa and the Japanese Pleistocene, 7: ed. S. Horie, 115-134.

Yokoyama, T., 1979b. Re-measurement of fission track age on the grey volcanic ash in the MaO, Osaka Group, Japan. Tsukumo Earth Sci., No. 14, 32-37.

Yokoyama, T., Matsuoka, C., Tamura, M. and Amemori, K., 1979a. On the Plio-Pleistocene Kobiwako Group. In: Land and Life in Shiga – Scientific Studies of Shiga Prefecture, Japan, 309-389.

Yokoyama, T., Nakagawa, Y., Takemura, K., Mori, Sh., Makinouchi, T., Hayashida, A. Iida, Y. and Matsuoka, K., 1979b. Stratigraphy of the Takashima Formation of the Plio-Pleistocene Kobiwako Group, Japan. Paleolimnology of Lake Biwa and the Japanese Pleistocene, 7: ed. S. Horie, 100-114.

Yokoyama, T., Matsuda, T. and Takemura, K., 1980. Fission-track ages of volcanic ashes in the Tokai Group, Central Japan (I). Quaternary Res. (Daiyonki Kenkyu) 19: 301-309.

Meteorology

Fujio Hirao, Tatsuya Iwashima and Ryozaburo Yamamoto

Introduction

Outline of weather conditions over Japan

Lake Biwa is located in the nearly central part of the Honshū Island. General characteristics of mid-latitude weather conditions in the east coast of the continent prevail over the island. The meteorological conditions over the Honshū Island show clear seasonal variations, which can be demonstrated with the synoptic weather maps (Figs. 52a, b, c, d and e). In the winter season, the polar continental air mass from the Siberian High dominates over the Honshū Island. On the way from Siberia to Japan, this air mass is so modified by the warm sea surface of the Japan Sea as to be heated below and to contain more moisture, and brings much snow to the coastal region on the Japan Sea side of the Island. In spring and fall, travelling cyclones and anticyclones come from the East China Sea one after another, and we have a fine and rainy weather quasi-periodically. In summer, tropical maritime air mass from the Pacific High prevails over Japan, and the conditions are warm and moist, with thunder storms sometimes.

In early summer, a stationary front which is called the *Baiu* front is generated across the Honshū Island. Then, Japan has come into the rainy season which is called Baiu (Fig. 52c) and it continuously rains and occasionally heavily somewhere. In August and September, typhoons sometimes come near or strike the Honshū Island, and the frequency is, on the average, three or four a year.

General pictures of meteorological and climatological characteristics around Lake Biwa

Lake Biwa and the surrounding region have local specific features of weather conditions and climate. The network of meteorological and/or climatological observational stations is given in Fig. 53 and Table 18. Of those, the Hikone

Horie, S. Lake Biwa

ISBN 90 6193 095 2. Printed in The Netherlands

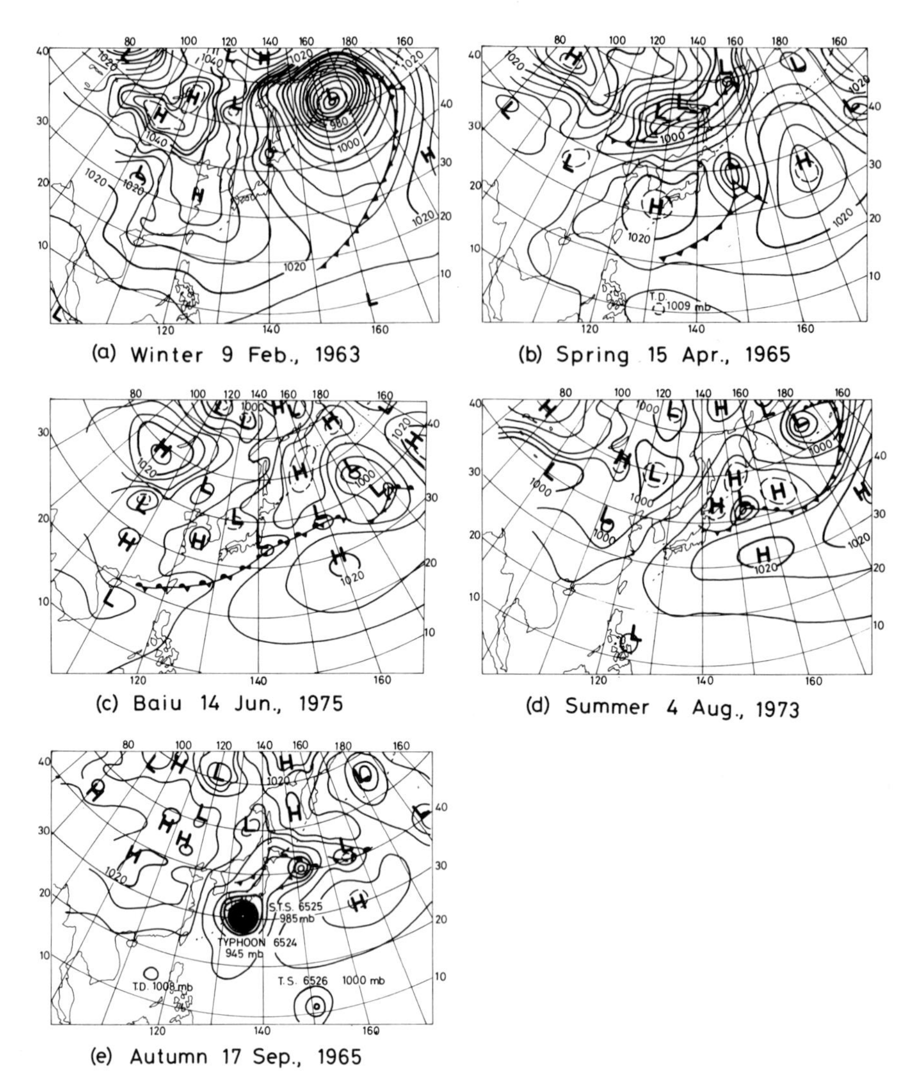

Fig. 52. Typical pressure pattern near Japan. (Japan Meteorological Agency 1963, 1965, 1973, 1975).

Local Meteorological Observatory of the Japan Meteorological Agency furnishes most reliable data of many kinds.

The climatic elements at Hikone are listed on monthly basis in Table 19. The annual mean surface air temperature and the annual precipitation at Hikone are 13.9°C and 1,696.4 mm, respectively. They are approximately equal to the values averaged over the western part of the Honshū Island. Within the Ōmi Basin, there is a considerable climatic contrast between the northern and southern parts: the former has climatic characteristics of the Japan Sea side, while the latter has those of the Pacific Ocean side. Such a contrast is shown in

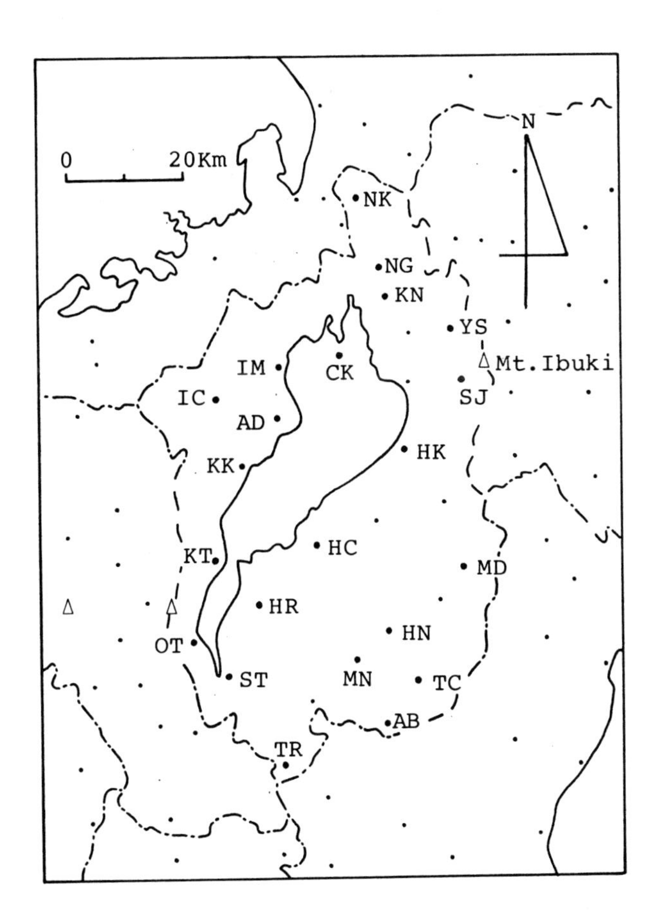

Fig. 53. Location of meteorological observation station.

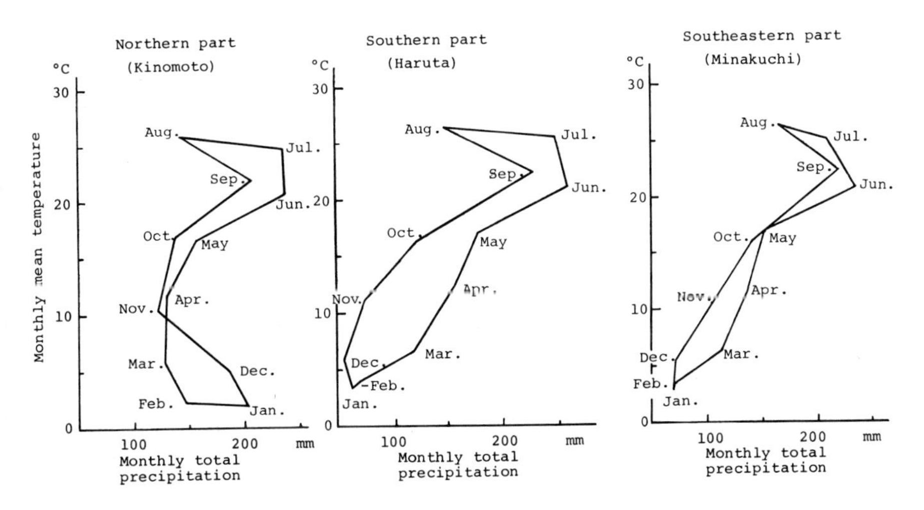

Fig. 54. Climograph around Lake Biwa (data period: 1941–1970).
(a) Northern part (Kinomoto), (b) Southern part (Haruta), (c) Southeastern part (Minakuchi).

Table 18. Location of meteorological observation stations around Lake Biwa; that of Hikone is one of the Local Meteorological Observatory of Japan Meteorological Agency. (Japan Meteorological Agency 1972).

Code	Station	Latitude	Longitude	Altitude above sea level
HK	Hikone	35°16′ N	136°15′ E	87 m
NK	Nakanokawachi	35°40′ N	136°10′ E	425 m
NG	Nakanogō	35°33′ N	136°12′ E	167 m
KN	Kinomoto	35°30′ N	136°14′ E	107 m
YS	Yoshitsuki	35°27′ N	136°22′ E	270 m
CK	Chikubushima	35°25′ N	136°09′ E	101 m
IM	Imazu	35°24′ N	136°02′ E	88 m
SJ	Shunjō	35°23′ N	136°23′ E	161 m
IC	Ichiba	35°21′ N	135°55′ E	185 m
AD	Adogawa	35°20′ N	136°00′ E	95 m
KK	Kitakomatsu	35°15′ N	135°58′ E	87 m
HC	Hachiman	35°08′ N	136°05′ E	88 m
KT	Katata	35°07′ N	135°55′ E	87 m
MD	Mandokoro	35°06′ N	136°22′ E	300 m
HR	Haruta	35°02′ N	135°58′ E	95 m
HN	Hino	35°01′ N	136°14′ E	163 m
OT	Ōtsu	35°00′ N	135°52′ E	140 m
MN	Minakuchi	34°58′ N	136°10′ E	174 m
TC	Tsuchiyama	34°56′ N	136°17′ E	263 m
ST	Setagawa	34°56′ N	135°55′ E	89 m
AB	Aburahi	34°52′ N	136°15′ E	263 m
TR	Tarao	34°48′ N	136°02′ E	485 m

climographs of three stations (Fig. 54). No appreciable difference is found in the warm season (April to October), but much more precipitation in the cold season appears in the northern part than in the southern part. This implies that the snowfall associated with northwest monsoons does not extend to the southern part of the Ōmi Basin. In the rainy season (June and July), a somewhat smaller amount of precipitation is noticed in the southeastern part than in the southern part. This may be explained by the fact that most of the precipitation associated with the prevailing southwest wind is confined to the southwestern part of the Ōmi Basin.

Thus, appreciable differences are seen among the northern part near Wakasa Bay, the southern part near Osaka Bay and the southeastern part near Ise Bay. These climatic patterns are sometimes called the Japan Sea pattern, the Setouchi pattern and the Tōkai pattern, respectively.

The mean surface air temperature shows a systematic distribution associated with the topography around Lake Biwa. The predominant wind direction around Lake Biwa is of the southeast in summer, and of the north-northwest in other seasons. The surface winds within the basin are remarkably modified by the surrounding mountains. Land and lake breeze can be observed under the conditions of calm or weakness of predominant winds. A severe and short-duration local wind which is called 'Hira Hakkō' should be noted. More detailed description of climatic and meteorological elements around Lake Biwa will be given in the following sections.

Incoming solar radiation and sunshine duration

The observation of solar radiation has been made only in Hikone among the above-mentioned stations. Tabel 19 gives the average values of global radiation for the past 10 years (1961—1970). The global solar radiation is defined as the sum of direct and diffuse solar radiation. The maximum of the monthly average of solar radiation is 470 cal/cm²day in May, and the minimum 204 cal/cm²day in December. The annual mean is 352 cal/cm²/day.

The observed values may reflect a few factors such as the altitude angle of the sun, the transmissivity of the air and the cloud amount. Although the solar altitude angle is high in June, the global solar radiation is less than that in May because of the large amount of cloud in the rainy month of June.

Fig. 55 shows the annual variation of the sunshine duration in four observational stations: Kinomoto in the north of the basin, Hikone in the central place, Haruta in the southwest and Aburahi in the south. The maxima can be seen in May and August. The sunshine duration is shortened, because of more cloudy days, in June and July. The maximum sunshine duration has been recorded in August for such reasons that the sun is in the highest altitude and that the fine weather, comparatively speaking, continues.

In general, the sunshine duration becomes shorter in proportion to the latitude: The difference between those at the north and south ends of the latitude band is 1.2 hours in January and 0.8 hours in August, respectively. The sun-

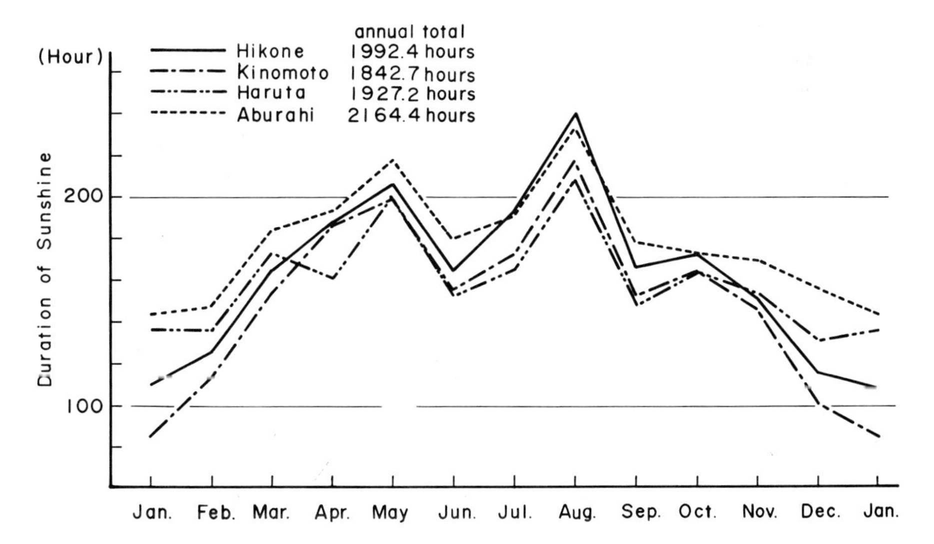

Fig. 55. Annual variation of sunshine duration (data period: 1941–1970).

Table 19. Climatic table of Hikone; in 'Climatic Table of Japan' (Japan Meteorological Agency 1971).

	JAN.	FEB.	MAR.	APR.	MAY	JUN.	JUL.	AUG.	SEP.	OCT.	NOV.	DEC.	ANN.	NN
PRESSURE (0.1 MB)														
STATION LEVEL	10071	10081	10067	10063	10021	9990	9987	9994	10021	10068	10095	10089	10045	20
SEA LEVEL	10183	10189	10179	10166	10126	10093	10090	10096	10124	10174	10201	10196	10151	30
TEMPERATURE (0.1 DEG. C)														
MEAN	30	32	60	114	164	207	251	264	223	161	107	57	139	30
MEAN MAXIMUM	65	69	105	167	218	252	297	316	270	209	153	95	185	30
MEAN MINIMUM	−03	00	22	69	119	171	217	226	187	120	65	22	101	30
RELATIVE HUMIDITY (PERCENT)	78	77	75	76	77	80	81	78	79	78	78	78	78	30
PRECIPITATION (0.1 MM)	1116	1001	1196	1362	1456	2250	2199	1327	2008	1289	833	929	16964	30
GLOBAL RADIATION (CAL/CM²/DAY)	215	284	374	412	470	427	433	453	372	326	256	204	352	10
SUNSHINE														
DURATION (0.1 H)	1088	1243	1639	1870	2063	1639	1928	2401	1660	1723	1517	1153	19924	30
PERCENTAGE (PERCENT)	35	41	44	48	48	38	44	58	45	48	49	38	45	30
WIND SPEED (0.1 M/S)	38	37	34	28	25	22	20	23	25	29	30	32	29	10
PREVAILING WIND														
1ST DIRECTION	NW	NNW	NNW	NNW	NNW	NW	S	S	NW	NNW	NNW	SSE		20
FREQUENCY (0.1 PERCENT)	400	700	550	500	500	450	350	350	350	500	400	350		
2ND DIRECTION	S	NW	NW	NW	NW	NNW	NW	NW	NNW	NW	NW	S		
FREQUENCY (0.1 PERCENT)	300	300	350	400	350	400	300	250	250	200	200	200		
3RD DIRECTION	NNW	·	N	N	N	N	N	SSE	S	N	N	NNW		
FREQUENCY (0.1 PERCENT)	100	·	100	100	150	100	150	100	150	150	200	150		
NUMBER OF DAYS (0.1)														
TEMPERATURE (DEG. C)														
MIN. 25.0 OR ABOVE	·	·	·	·	·	·	07	21	02	·	·	·	29	30
MAX. 30.0 OR ABOVE	·	·	·	·	01	20	163	250	64	00	·	·	498	30
MAX. 25.0 OR ABOVE	·	·	·	06	62	165	276	307	214	21	·	·	1052	30
MIN. BELOW 0.0	170	154	86	07	·	·	·	·	·	·	04	66	488	30
MAX. BELOW 0.0	01	02	·	·	·	·	·	·	·	·	·	·	02	30

Table 19 (continued)

	JAN.	FEB.	MAR.	APR.	MAY	JUN.	JUL.	AUG.	SEP.	OCT.	NOV.	DEC.	ANN.	NN
PRECIPITATION														
1 MM OR MORE	141	119	132	109	108	128	125	84	117	96	87	114	1359	20
10 MM OR MORE	43	38	43	56	56	65	73	34	55	36	28	31	557	20
30 MM OR MORE	03	03	05	12	12	27	29	15	19	07	04	02	137	20
SNOWCOVER DEPTH														
LESS THAN 10 CM	78	78	30	01	.	.	.	.	.	.	.	26	213	30
1 MM OR MORE	53	41	09	.	.	.	.	.	.	.	.	10	114	30
10 MM OR MORE	29	19	04	.	.	.	.	.	.	.	.	05	57	30
30 MM OR MORE	02	03	00	.	.	.	.	.	.	.	.	00	06	30
100 CM OR MORE	.	.	.	.	.	.	.	.	.	.	.	.	.	
MAXIMUM WIND SPEED														
10 M/S OR MORE	100	89	77	47	22	18	09	19	28	40	63	73	585	10
15 M/S OR MORE	02	05	01	.	.	.	02	02	05	01	03	02	23	10
29 M/S OR MORE	.	.	.	.	.	.	.	.	.	.	.	.	.	
METEORS														
SUNLESS DAY	38	34	45	50	43	59	43	15	40	45	33	35	482	30
SNOW	127	128	66	03	.	.	.	.	.	.	05	55	384	30
FOG	07	05	.09	10	07	03	01	00	00	02	06	07	59	30
THUNDERSTORM	02	01	02	06	07	20	42	40	27	04	02	00	153	30

DATE OF FIRST AND LAST SNOWFALL	DATE	YEAR	PERIOD
FIRST SNOWFALL			
NORMAL	7 DEC.		1941–1970
EARLIEST	6 NOV.	1904	1896–1970
LAST SNOWFALL			
NORMAL	30 MAR.		1941–1970
LATEST	22 APR.	1947	1896–1970

DATE OF FIRST AND LAST HOAR-FROST	DATE	YEAR	PERIOD
FIRST HOAR-FROST			
NORMAL	13 NOV.		1941–1970
EARLIEST	21 OCT.	1926	1896–1970
LAST HOAR-FROST			
NORMAL	18 APR.		1941–1970
LATEST	16 MAY	1908	1896–1970

shine duration in Kinomoto is 85.1 hours in January and 217.3 hours in August; the respective quantities in Aburahi are 143.6 hours and 233.1 hours. The difference between the values observed in Kinomoto and Aburahi is larger, at least by one order of magnitude, than that due to the difference of the latitude, especially in January: in winter, it is the shortest in Kinomoto because fine weather continues there not so long as in other places. The above difference of the sunshine duration is mostly attributed to meteorological and/or topographical factors.

Air temperature and humidity

Distribution of air temperature

Fig. 56 gives the distribution of the monthly mean air temperature in each season. We can find that a high temperature area in the basin slightly varies from season to season.

Figs. 57 and 58 show the geographical distributions of the maximum and minimum air temperatures respectively. The two figures contrast with one another: the highest temperature zone in Fig. 57 appears in the plain of the basin, but it appears above the lake in Fig. 58. These facts indicate that the temperature in the district of the plain rises more easily than above the lake.

The geographical distribution of the air temperature is influenced by Lake Biwa. Reduction of the temperature data observed on the ground surface to mean sea level is not made. We may clarify the effect of Lake Biwa on the temperature distribution, by using the procedure employed by the Hikone Local Meteorological Observatory (1969) to remove the effect due to the difference of the altitude and the latitude.

The air temperature (T) on the ground level is assumed to depend on four factors — the macro-climatic condition, the height above the sea level (h), the latitude (ϕ) and the topographical condition (G):

$$T = T_0 + \frac{\partial T}{\partial h}\Delta h + \frac{\partial T}{\partial \phi}\Delta\phi + \frac{\partial T}{\partial G}\Delta G.$$

Here, the first term T_0 on the right-hand side depends on the macroclimatic condition and is assumed to be the air temperature at the standard station; the second one, height-difference from the level of the standard station; the third one, distance from the station; and the fourth one, difference of the topographical condition. The temperature distribution for the fourth term $\frac{\partial T}{\partial G}\Delta G$ may represent the actual topographical feature very well.

Figs. 59a and b show the geographical distributions of the term $\frac{\partial T}{\partial G}\Delta G$ for the minimum air temperature in January and the maximum in August, respectively. The standard station used is the Hikone Local Meteorological Observatory

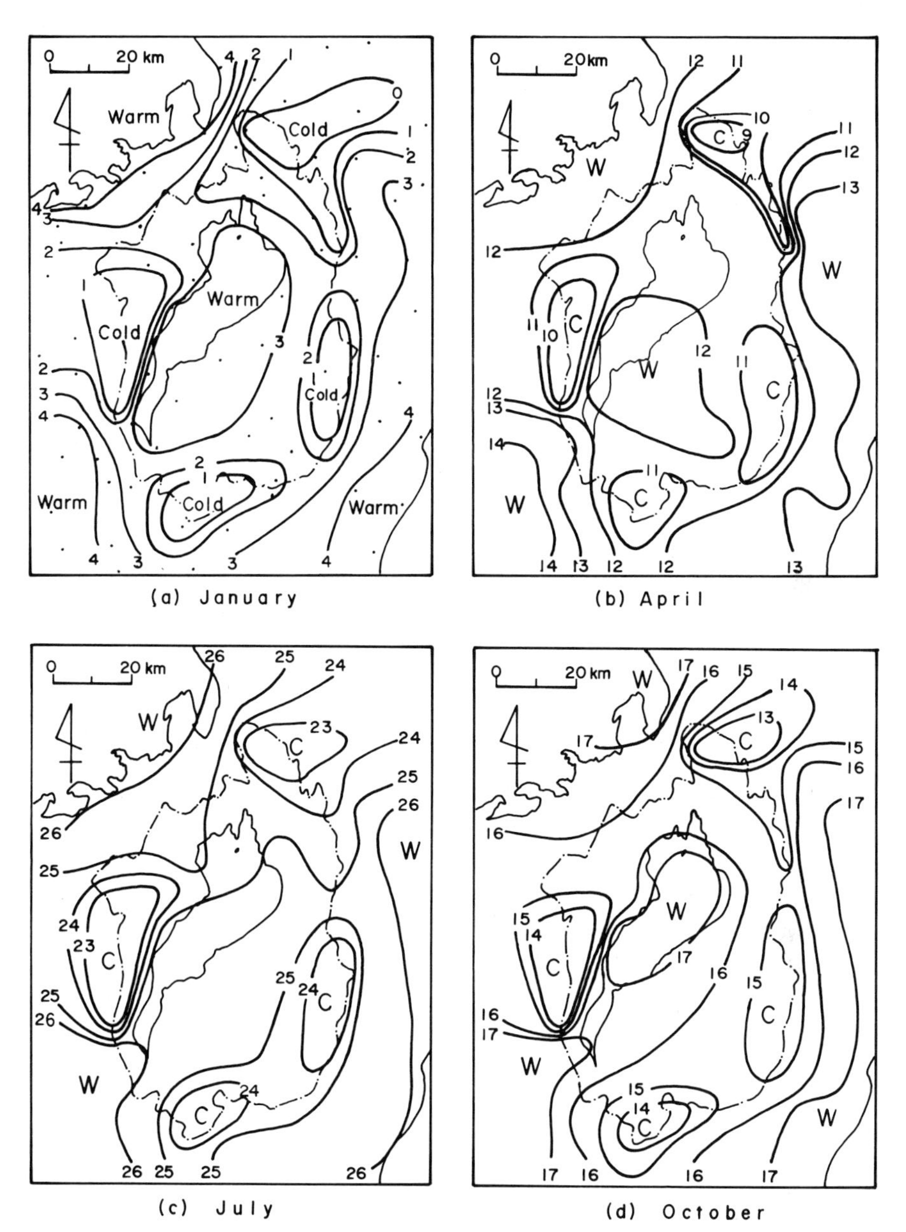

Fig. 56. Monthly mean surface air temperature (data period: 1941–1970). Unit is °C. (a) January, (b) April, (c) July, (d) October.

marked by the sign ⊗ in the figure: the positive (negative) sign indicates that the temperature is higher (lower) than the standard value of Hikone.

The temperature on the lake is evidently different from that in the plain. In the distribution of the maximum air temperature, the positive zone is in the plain

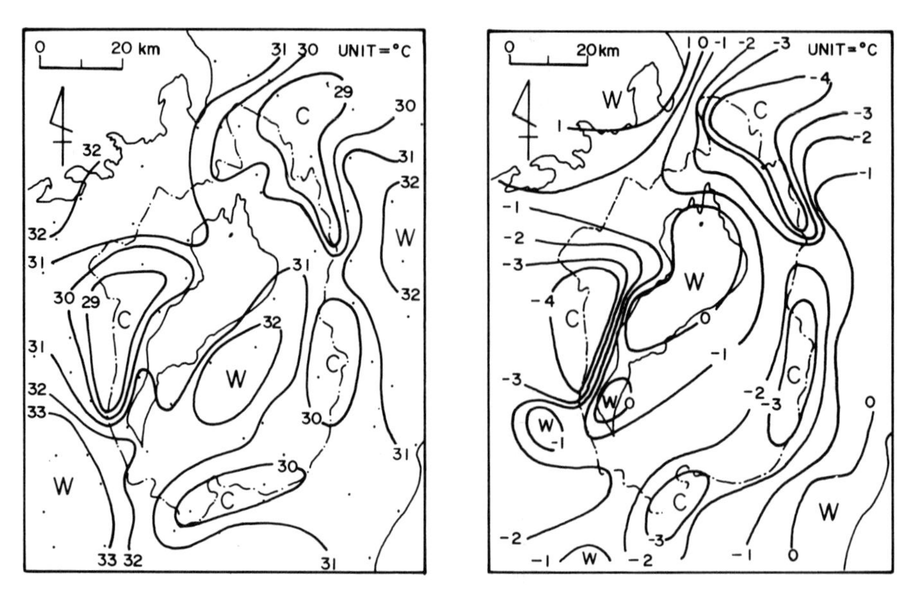

Fig. 57. Mean maximum surface air temperature in August (°C) (data period: 1941–1970).
Fig. 58. Mean minimum surface air temperature in January (°C) (data period: 1941–1970).

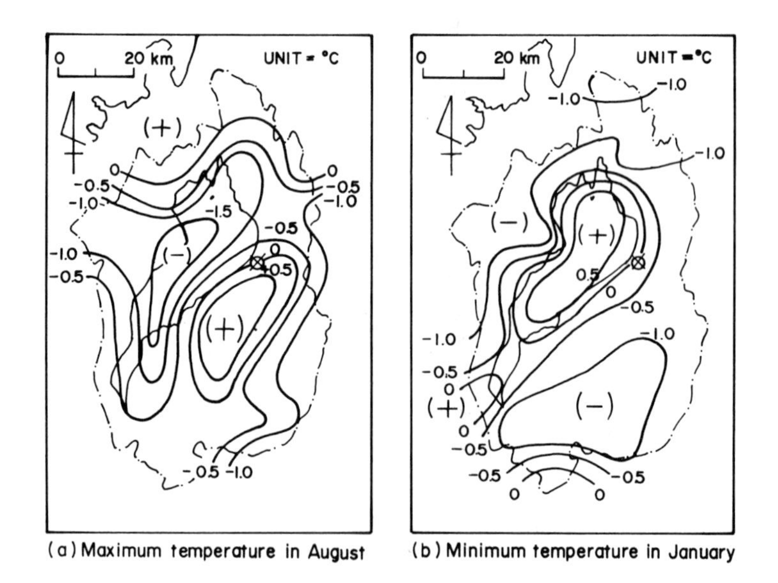

Fig. 59. Corrected air temperature deviation from that in Hikone due to the effect of topographical condition (data period: 1941—1970).
(a) Maximum temperature in August.
(b) Minimum temperature in January.

and the negative zone on the lake, while the positive and negative areas are reversed in the distribution of the minimum air temperature. In other words, Lake Biwa operates to weaken the maximum and the minimum of the air temperature.

Diurnal range of air temperature

The difference between the maximum and minimum daily air temperature is one of the important characteristic indices. It is small on the sea, the coast, or mountainous districts and large in inland parts of the plains or at high latitudes. Diurnal variation of the air temperature is mainly controlled by radiation; the range of variation is larger in a fine day than in a cloudy or rainy day.

The diurnal range of air temperature near the Chikubu-shima Island is small all year round. It gets larger, the farther away from Lake Biwa. This is made clear by comparing the data of Hikone with those of Minakuchi. The maximum value from April to May and that from October to November depends on the duration of the fine weather under the travelling anticyclone. The diurnal range of the air temperature becomes smaller in the rainy season from June to July in Japan. It somewhat decreases in September as the bad weather continues with the front, and decreases in winter in every place. The decrease of the temperature range in the northern area is partly caused by rain and snowfall. We should note that the temperature range decreases more in December in the northern observation stations Hikone and Kinomoto than in any other southern parts. This causes a difference between the two districts.

During winter when the northwest seasonal wind prevails, the diurnal range of air temperature in Hikone is the smallest in the data of all observed points except that in Chikubu-shima Island. It may be influenced by Lake Biwa which spreads in the northwest portion of Hikone.

Fig. 60 shows the distribution of the diurnal range of air temperature on and around Lake Biwa. Within the Ōmi Basin, its lowest zone appears on the lake. Comparing the distribution in January with that in August, we notice that the zone with a smaller range spreads over the main lake basin and the west coastal part of Lake Biwa in the direction from southwest to northeast. In addition, it extends in the direction from southeast to northwest in January. As mentioned above, it may be due to the difference of direction of the prevailing seasonal wind; a southeast wind in summer and a northwest wind in winter.

Hoarfrost

It often freezes in the Ōmi Basin from November to April, when the temperature inversion causes radiative cooling under the travelling anticyclone.

The temperature inversion frequently appears in such a place to which the

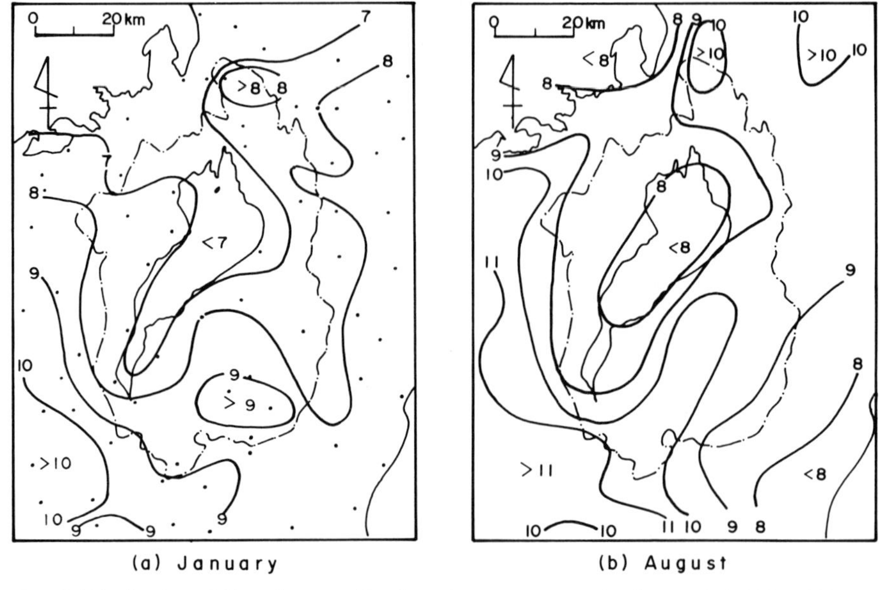

Fig. 60. Distribution of diurnal range of surface air temperature (°C) (data period: 1941–1970). (a) January, (b) August.

cold air mass is easily confined. The land and lake breeze along Lake Biwa seems to prevent the inversion layer from growing. The frosting period in the Ōmi Basin seems to be shorter than in other surrounding places: for example, it is five days shorter in Hikone than in Kyoto.

Table 20 gives the mean of the frosting period measured around Lake Biwa for the last twenty years (1956–1975). The first date of frosting varies considerably from place to place, though the last hoarfrost occurs simultaneously in the middle of April. The frosting period increases in proportion to the distance from the lake coast. It depends on the altitude above the sea level. In the southern part of the plain, the dense fog readily appears and lasts long. This kind of phenomenon is frequently seen in a district where the air current is easily stagnant.

The results of forecasting the minimum air temperature on the ground surface

Table 20. 20-year mean data of first and last hoarfrost (data period: 1956–1976).

Station	First hoarfrost	Last hoarfrost
Kinomoto	30 Oct.	18 Apr.
Chikubu-shima Island	25 Nov.	3 Apr.
Hikone	15 Nov.	13 Apr.
Adogawa	15 Nov.	14 Apr.
Haruta	6 Nov.	17 Apr.
Minakuchi	3 Nov.	15 Apr.

and at the anemometer level around Lake Biwa have been reported by the Hikone Local Meteorological Observatory (1969). It gives the following formula for the minimum ground temperature (Tgm) related to the wet bulb temperature at 1500 hours on the previous day (Tw).

1. When it is fine all night long,
   Tgm = 1.3 Tw — 14.7 (standard deviation, 1.4°C).
   If the wind velocity at night is above 5 m/s, about 6°C is added to the above value.
2. When a thin or high cloud is in the night sky,
   Tgm = 1.5 Tw — 13.7 (standard deviation, 2.2°C).
3. When it is cloudy all night long,
   Tgm = 1.5 Tw — 8.5 (standard deviation, 1.9°C).

The minimum air temperature (Tm) in the last hoarfrost period is related to the ground temperature (Tgm) as is shown in the following formula:

Tgm = 1.1 Tm — 3.6 (standard deviation, 0.9°C).

Though these data in Hikone cannot necessarily be applied to those at other stations around Lake Biwa, they may be assumed to indicate a mean value of Tm and Tgm in the last hoarfrost.

Surface air and water temperatures

Fig. 61 shows the annual variation of the monthly mean surface air temperature in Hikone and surface water temperature of Lake Biwa over the last ten years (1966–1975). Since daily data of water temperature have not been available up to this time, the writers tentatively use the data observed in the daytime of every 15th day. Therefore, they cannot precisely represent the monthly mean value. Several observations show that the water temperature from the surface to a depth near bottom equally reaches about 7°C during the period from January to March, owing to convection. During the period from winter to summer, the air temperature is a little higher than the water temperature, and vice versa during the other seasons. Particularly, its difference gets larger as the winter closes: the difference is 5.2°C in December.

Studies on the heat balance condition over Lake Biwa were made by several investigators. Ito and Okamoto (1974) estimated the heat transfer from the surface of the lake to the air by the bulk aerodynamic method as given in Table 21. The heat is transferred from the lake to the air from August to March as is clearly suggested in Fig. 61.

Humidity

The annual mean relative humidity in Hikone is 78%. This belongs to the higher class of humidity in Japan. One of the reasons may be the influence of Lake Biwa: the dominant southwest wind across Lake Biwa brings the moist air to Hikone.

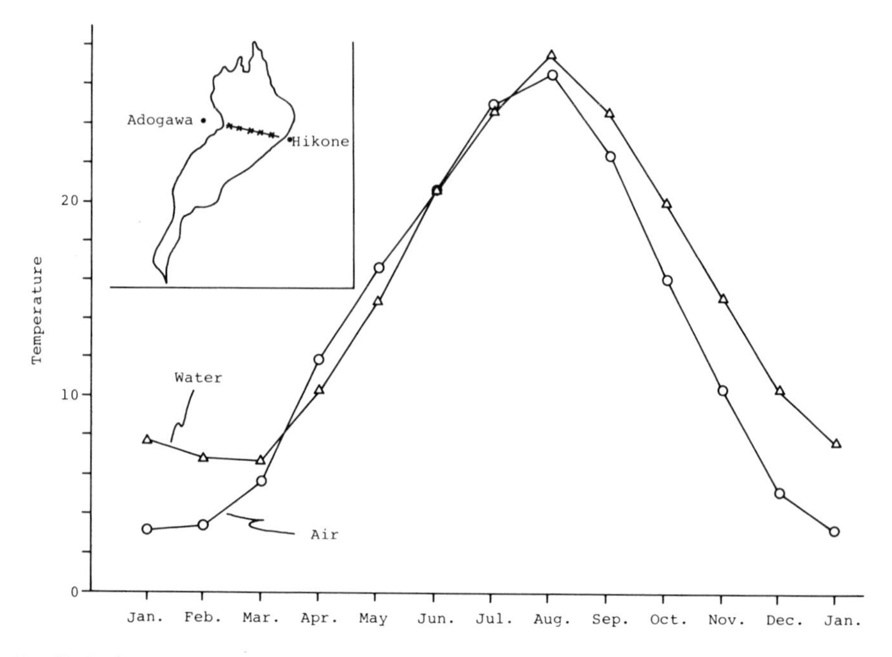

Fig. 61. Surface water temperature of Lake Biwa and monthly mean surface air temperature at Hikone. The former values are estimated from the data of five stations shown by x mark in the upper figure (data period: 1966—1975).

There are many uncertain points in the estimation of evaporation. The annual mean value of evaporation is estimated by the method of the heat balance. It is approximately 770 mm. The lake, which will never dry up, may influence humidity. The fog sometimes appears on the lake in winter when the water temperature rises above the air temperature.

Fig. 62 gives the annual variation of relative humidity and water vapour in Hikone. The maximum and minimum of the relative humidity are respectively 81% in July and 75% in March, differing only by 6% between them. The maximum and minimum of the water vapour pressure are 26.8 mb in August and 5.9 mb in January and February, respectively.

The absolute humidity in Japan is usually higher in summer than in winter. In Hikone, the annual variation of relative humidity is small because the water vapour changes in proportion to air temperature.

Winds

Wind system

In Hikone, N–NW and S–SE winds are predominant as shown in the wind rose of Fig. 63. The frequency of the N–NW wind amounts to 37.8% of the annual

Table 21. Annual variation of sensible and latent heat transfer and the evaporation from the northern part of Lake Biwa to the atmosphere (1968–1970) (Ito & Okamoto 1974).

	JAN.	FEB.	MAR.	APR.	MAY	JUN.	JUL.	AUG.	SEP.	OCT.	NOV.	DEC.	TOTAL
Heat transfer ($\times 10^{16}$cal/month)	5.0	3.6	2.0	0.3	0.2	0.1	1.8	3.2	4.8	5.0	4.4	4.7	35.1
Evaporation (mm/month)	83.5	61.3	50.6	10.3	3.0	9.1	70.0	81.4	119.0	108.6	85.1	84.2	766.1

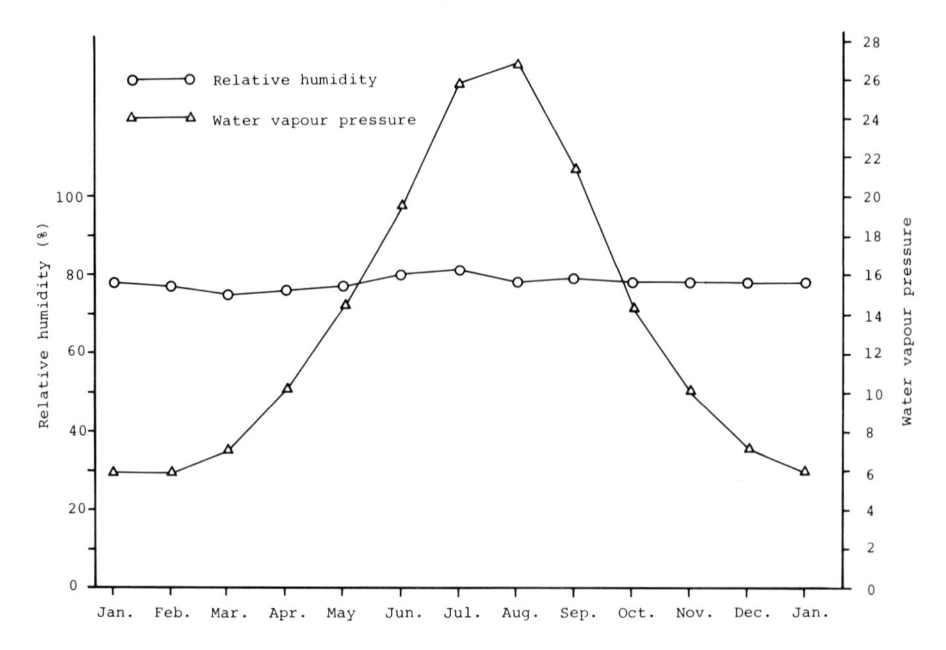

Fig. 62. Annual variation of relative humidity and water vapour pressure in Hikone (data period: 1941–1970).

average (Fig. 63a), and that of the S–SE wind to 26.6%; the frequency of the wind from these two directions accounts for about two-thirds of the whole.

The pattern of the wind rose in each season is similar to that of the annual average wind. In Fig. 63b for January, we note that the S–SSE wind and the northwest seasonal wind blow at almost the same rate. This may be due to a land breeze. Since the surface water temperature of the lake is higher than the surface air temperature in winter as shown in Fig. 61, the land breeze may prevail.

A broken line in the figure of wind rose represents the frequency of the calm day. In winter, the average wind speed is large and the frequency of the calm is small, but they are reversed in summer. The maximum and minimum of monthly average wind speeds are 3.8 m/s in January and 2.0 m/s in July respectively, as shown in Table 19. The maximum instantaneous wind speed in Hikone is 42.5 m/s registered by the typhoon on 3 September, 1950.

A feature of wind types in the surroundings of Lake Biwa is that the air current flows into the basin over the lowland in the mountains around Lake Biwa and flows out of the other one. Several gate-ways are the gorges of the Nosaka Mountains in the north, the Sekigahara Lowlands between the Ibuki Mountains and the Suzuka Range in the east, and the lowland along the River Yodo and the southern-most part of the Suzuka Range in the south.

The Suzuka Range and the Hira Mountains act as a large obstacle to the air current. When the southeast wind with a low speed is blowing, the wind

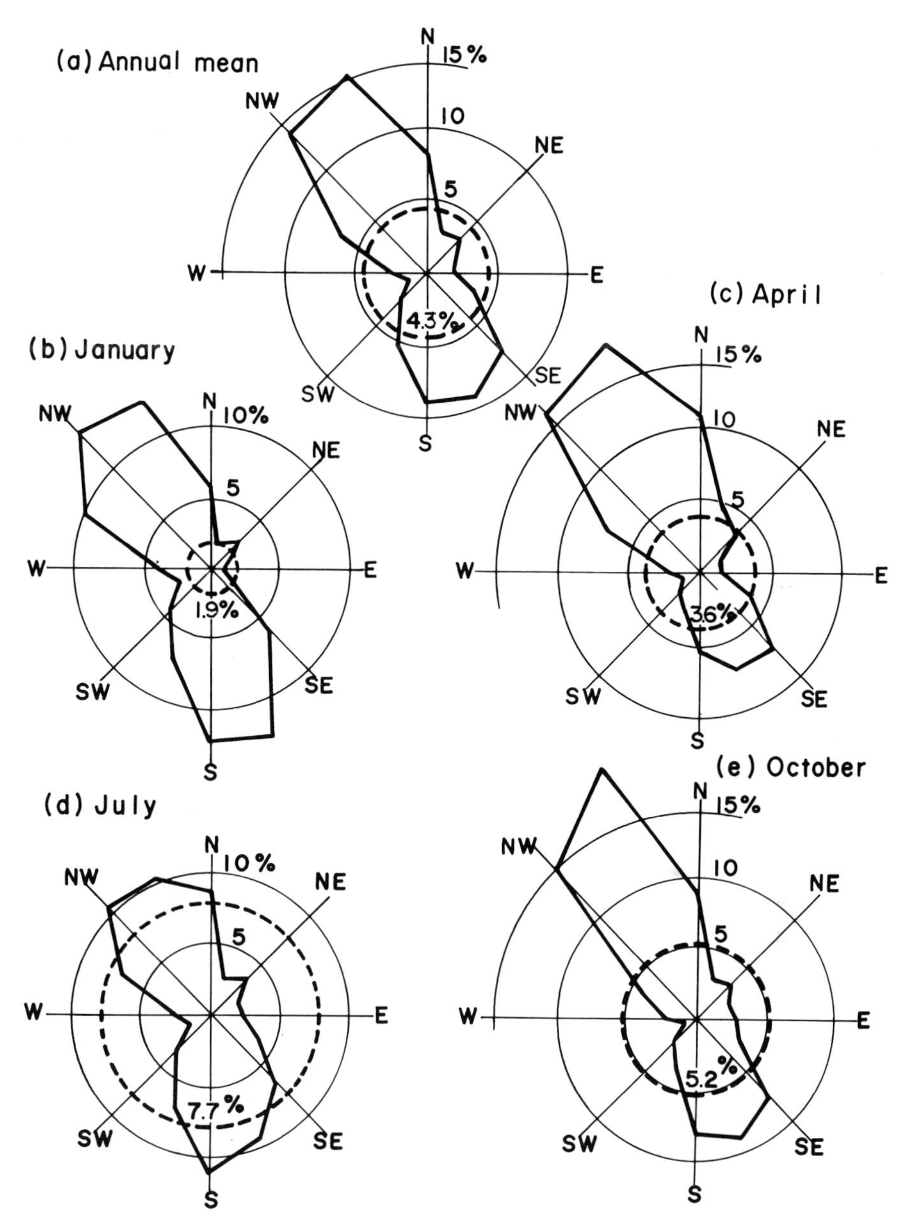

Fig. 63. Wind-rose in Hikone (data period: 1959-1968). Broken circles show a percentage of calm days.
(a) Annual mean, (b) January, (c) April, (d) July, (e) October.

direction on the lee side of the Suzuka Range in the basin is sometimes reversed to that of the windward side. When the wind blows with an inclination toward the W–NW wind direction, two air currents which pass around both north and south ends of the Hira Mountains meet in the basin (Kodama 1966).

Land and lake breeze

A land and lake breeze has frequently been observed in the surroundings of a large lake. Since the air temperature on the lake is lower than on the ground in the daytime, the wind blows landward owing to this atmospheric condition. This is called a lake wind. As the reverse condition occurs at night, a land wind subsequently blows lakeward. The vertical variation of the land and lake breeze on Lake Biwa was studied by the Kobe Marine Observatory in 1928. Kodama (1973) reported about the geographical distribution of the land and lake breeze.

Fig. 64 shows the wind system on a slightly windy day. Kodama (1973) noted that this figure included the mountain and valley wind in addition to the land and lake breeze, since both wind directions agree, and that the influence of a land and lake breeze was limited to within a distance of 8 km off the coast of the lake.

Local wind

A blast of wind, which is called 'Hira Hakkō' or 'Miidera Oroshi', blows at times in the Hira Mountains and the southern area of Lake Biwa.

The Hira Mountains sweep off in the direction of NNE–SSW. It is about 12 km long and 4–5 km wide. It consists of a series of mountains which are almost 1,000–1,200 m high above the sea level. The east and west sides of the mountains are both sheer precipices. Lake Biwa is situated at a distance of about 1 km from the eastern foots. According to a report of the Hikone Local Meteorological Observatory (1969), the phenomenon 'Oroshi' occurs when a

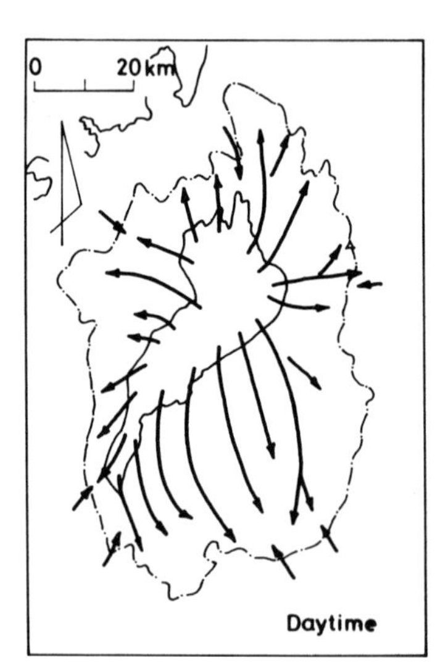

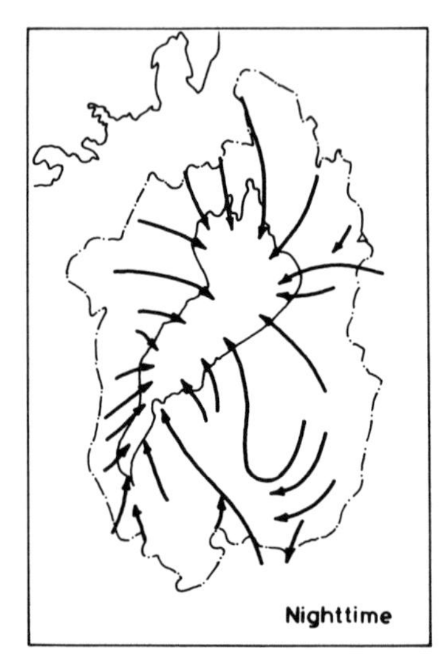

Fig. 64. Land and lake breeze around Lake Biwa (Kodama 1973).

northwest wind blows. A blast of wind above 15 m/s occasionally comes down from the tops of mountains. The influence of this local wind, however, is limited to within a very small area from the east piedmont to a part of Lake Biwa which is usually 4–5 km or about 8 km even in the strongest case. Generally speaking, it continues for 4-5 hours in the afternoon (2-6 o'clock).

Precipitation

Precipitation and frequency of rainy days

Figs. 65 and 66 show the geographical distribution of precipitation around Lake Biwa. The amount of precipitation in mountains is larger than that in plains.

Fig. 65 shows the annual amount of precipitation around Lake Biwa; 2,600–3,000 mm around the Ibuki Mountains (in the northern district), 2,000–2,400 mm around the Suzuka Range (in the eastern district), 1,800–2,600 mm around the Hira and Nosaka Mountains (western district), and 1,600–1,800 mm on Lake Biwa and its surrounding plains. On the whole, the annual amount of precipitation is much in the north and little in the south: the enormous difference in the annual precipitation between the north and the south is due to the difference in winter.

Fig. 66 gives the geographical distribution of precipitation in each season. Precipitation in the north in January is mostly due to snowfall. In January, precipitation in the north is more than three times as much as that in the south.

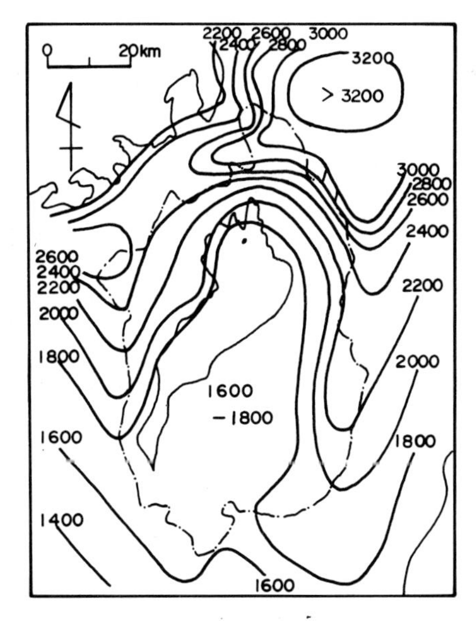

Fig. 65. Annual amount of precipitation (mm) (data period: 1941-1970).

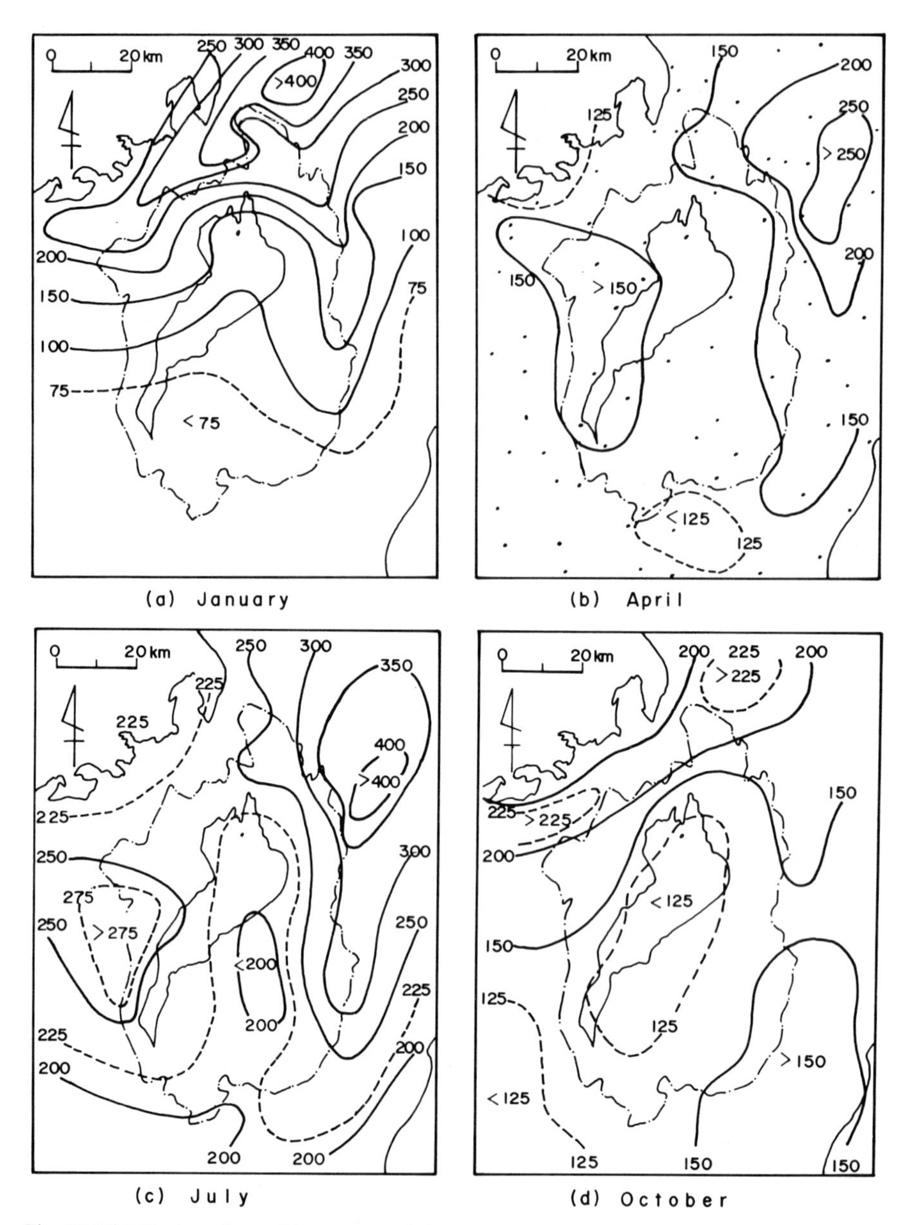

Fig. 66. Distribution of monthly mean precipitation (mm) (data period: 1941-1970).

In April and July, there is a small amount of precipitation on the Nosaka Mountains in the north, and much in the east and west sides of the basin. Around the Nosaka Mountains, the total amount of precipitation increases in October (due to snowfall in winter).

The geographical distribution of precipitation differs from one season to another, mainly due to the difference in the direction of prevailing wind. In autumn and winter, the northwest seasonal wind prevails, and the ascending current crossing the northern mountains brings rain in. The geographical distribution of precipitation in spring and summer shows that the southeast wind brings rain in. The center of the heavy precipitation area is situated closely to the east mountains around the basin. Thunderstorms increase precipitation in summer. The frequency of occurrence or strike of thunderstorms varies from district to district. From June to September, the maximum frequency is 24 times in some districts, and the minimum is 9 times in other districts. It is not so frequent in the north and northwest districts. The average number of rainy days in each month with more than 1 mm precipitation falls in the range of 8—15 days at Hikone. The maximum is 14.1 days in January and the minimum 8.4 days in August.

Through all seasons, the difference in the frequency of rainy days between the plain and the mountains is not so large as the difference in the amount of precipitation. This implies that one rainy occasion brings more precipitation on the mountains than on the plain. For example, the mean daily precipitation on the plain is 17.3 mm/day in July, as estimated from the data of 13 rainy days and 225 mm of total precipitation, and on the mountains the precipitation is 21.4 mm/day, from the data of 14 rainy days and 300 mm of total precipitation.

It often showers in the north in the late autumn. A local shower with gusts of wind, which is called 'Takashima-shigure', is observed at the foot of the Nosaka Mountains. This shower changes gradually to snow, as the season goes from autumn to winter.

Annual variation of precipitation

The pattern of annual variation of precipitation in the northern district is different from that in the southern one. Both variations in the north and south districts have a peak of precipitation in June. In the north district a peak appears in winter. The south district has a small amount of precipitation in winter. In the rainy season of June and July, there is heavy precipitation at Adogawa or Haruta (in the western part of the basin). This may be caused by the prevailing southeast wind on a rainy occasion. The amount of precipitation in August, when the weather is stable, is comparatively little. The precipitation in September is prominent, because of continuous rain due to the front and to some typhoons: a great quantity of precipitation is observed in the area of the Suzuka Range and Tsuchiyama.

These patterns of annual variation of precipitation in various districts reflect respective characteristics of Wakasa Bay, Osaka Bay, and Ise Bay, close to the Ōmi Basin; the climatic characteristics belong to the Japan Sea pattern in the north basin close to Wakasa Bay, to the Setouchi pattern in the southern area close to Osaka Bay, and to the Tōkai pattern in the southeast area close to Ise Bay.

Snowfall

Most of the northern part of the Ōmi Basin can be regarded as a heavy-snowfall area. At the top of Mt. Ibuki (1,376 m) which is the southern extremity of the Ibuki Mountains, the deepest snowfall amounting to 1,182 cm was recorded on February 14, 1927. This is a world-wide record. The record of the deepest snowfall in dwelling areas is 565 cm in Naka-no-kawachi (the northeast part of the basin) on March 2, 1936. At the lake side, the amount of 290 cm was recorded in Imazu on February 5 and 6, 1945. In Hikone, it begins to snow on December 7 and ends on March 30 in the normal year. The first snowfall is seen on November 6 and the last one on April 22 (Table 19).

Fig. 67 gives the geographical distribution of the frequency of snowy days. A heavy-snowfall area extends from the Ibuki Mountains to the Nosaka Mountains and expands in the southwest direction. The heavy snowfall is due to the northwest monsoon, which comes across the Japan Sea, and runs against mountains in the northern part of the Ōmi Basin. One of the reasons why the maximum snowfall occurs in the Ibuki Mountains (1,000–1,300 m) is that it is higher than the Nosaka Mountains (500–900 m).

The number of days of snowfall, even a little, is less than ten in a year in the southern area, and 60 to 100 days in the northern area. The snow-covered days above 50 cm in depth are hardly observed in the south, but the northern area has such deep snow for 30—80 days in a year.

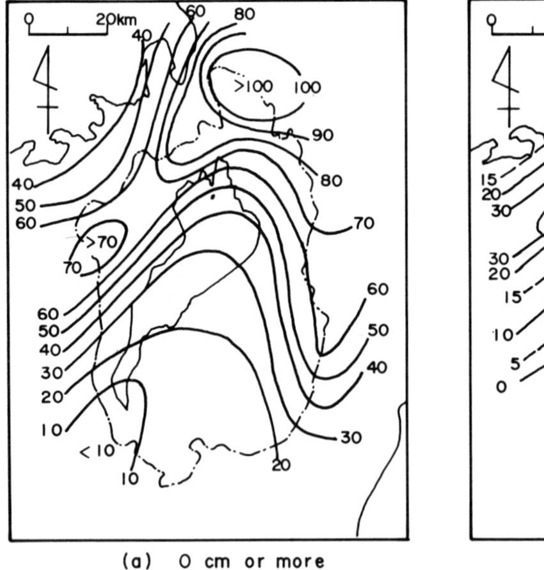

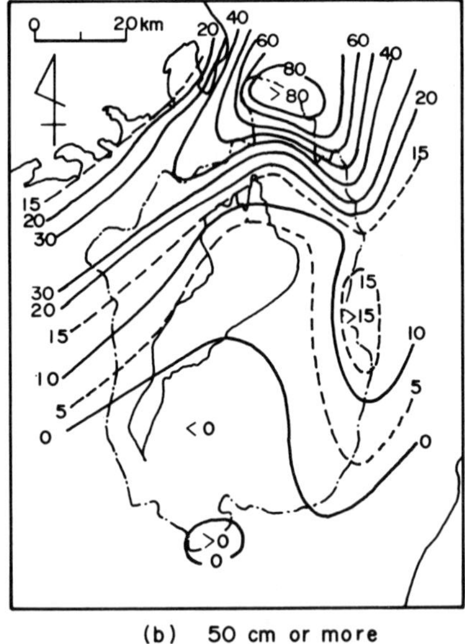

Fig. 67. Frequency of snowy days (data period: 1931—1960).
(a) 0 cm or more, (b) 50 cm or more. (Hikone Local Meteorological Observatory 1969).

In a heavily snowing year, a flood is occasionally caused at rivers in the north when snow melts: when a growing cyclone passes over the Japan Sea, the air temperature rapidly rises, and thaws snow. The rip of the embankment of a river in the north was recorded twice in February in 1922 and 1923.

Several types are observed in the geographical distribution of snowfall around Lake Biwa.

North type: It snows heavily in the northern area, mainly in the Ibuki Mountains.

Central type: It snows much on the belt area which lies from the Hira Mountains to the Suzuka Range.

Mixed type: It snows much on the Ibuki Mountains and the Hira Mountains.

South type: It snows much in the southern area.

The difference of direction of the monsoon wind causes the above four types in the distribution of snowfall as it is related to topographical features. The westerly monsoon which runs across the Ibuki Mountains brings snowfall to the northern area. On the other hand, it seldom snows in the southern part of the Ōmi Basin, because most snow falls on the western side of the Hira Mountains.

The northerly monsoon blows over the Nosaka Mountains and the basin. It sometimes snows much over the Hira Mountains and the Suzuka Range and, then, brings snowfall to the central part of the basin. This is the Central type.

The Mixed type has a combined feature of the above two types.

The South type is observed when the center of the low pressure system passes along the Pacific coast. This type seldom appears.

Precipitation due to heavy rainfall

The geographical distribution of precipitation by a heavy rain is very characteristic. Areas where we have a frequent rainfall by the cyclone and the front are distributed in a zone that crosses the Ōmi Basin in the direction from northeast to southwest and extends to Osaka Bay. Therefore the moist tongue (extended from Osaka Bay), i.e., the wet southwest air current, may bring a heavy rainfall to the zone. However, the geographical distribution of heavy rainfall due to the typhoon has a character different from the above description. Concerning the rain due to a typhoon, the Suzuka Range and the Hira Mountains are placed in the center of frequent occurrences of heavy rainfall, but heavy rainfall seldom occurs on Lake Biwa and the plain. This markedly contrasts with heavy rainfall duc to the cyclone or the front. The southeast wind, which blows from Ise Bay and crosses the Suzuka Range and the Hira Mountains until the typhoon passes over, brings heavy rainfall to these mountains. The northwest wind also brings much precipitation to these mountains after passing of the center of a typhoon.

It is very much of interest that the geographical distribution of heavy rainfall and snowfall is explained by a relation between topographical features and the direction of wind.

The heavy rainfall is concentrated in a few months from mid-June to early October; the heavy rainfall in August or September often brings damage to us. The heavy rainfall above 100 mm/day is observed about one and a half times in the plain and about twice in mountains in a year; precipitation above 200 mm/day is observed once every four years in the plain and once every three years in mountains.

Long-term variation of precipitation and water level of the lake

Precipitation varies widely in time and place.

There are considerable variations in the amount of precipitation in every *Baiu* rainy season: for example, the maximum (482 mm in 1896) precipitation in Hikone in June is fourteen times as much as the minimum (34 mm in 1924). The maximum annual amount of precipitation (3,069 mm in 1896) is three times as much as the minimum (1,101 mm in 1939).

These variations in the amount of precipitation directly affect changes in the water level of Lake Biwa. Solid lines in Fig. 68 show variations of the annual amount of precipitation in Hikone, of the effluence from Lake Biwa, and of the water level of the lake. Dotted lines denote the five-year moving averages.

Periodic changes in various climatic elements have been observed: for example, 3 to 4-year cycle, 11-year cycle, and 35-year cycle. Roughly speaking, the annual amount of precipitation in Hikone seems to have periodic changes with 2 to 4- and 35-year cycles during the period from 1894 to 1975, although the data record is statistically not long enough.

The effluence from Lake Biwa contains the controlled discharge from a dam of the River Seta, water-power generation and service water for neighboring cities. The water from Lake Biwa flows finally into Osaka Bay. The water level of Lake Biwa has been lowered by an increase in the water demand since 1933 when the discharge in winter was begun in addition to that in the other three seasons.

Although the quantity of effluence is controlled artificially, year-to-year variation of the annual mean discharge and that of the annual amount of precipitation are approximately in phase (Fig. 68).

Evaporation from the lake causes a loss of water before its effluence. The quantity of the loss is estimated by two methods: one method estimates the evaporation from the difference between the annual mean of precipitation on the Ōmi Basin and the annual mean effluence from the lake. The other method estimates the evaporation from the lake on a certain assumption of the heat balance. The value estimated by the former method is about 500–650 mm in a year (Matsuoka 1970). This value is small as compared with the amount of evaporation from the lake surface 770 mm in a year as estimated by the second

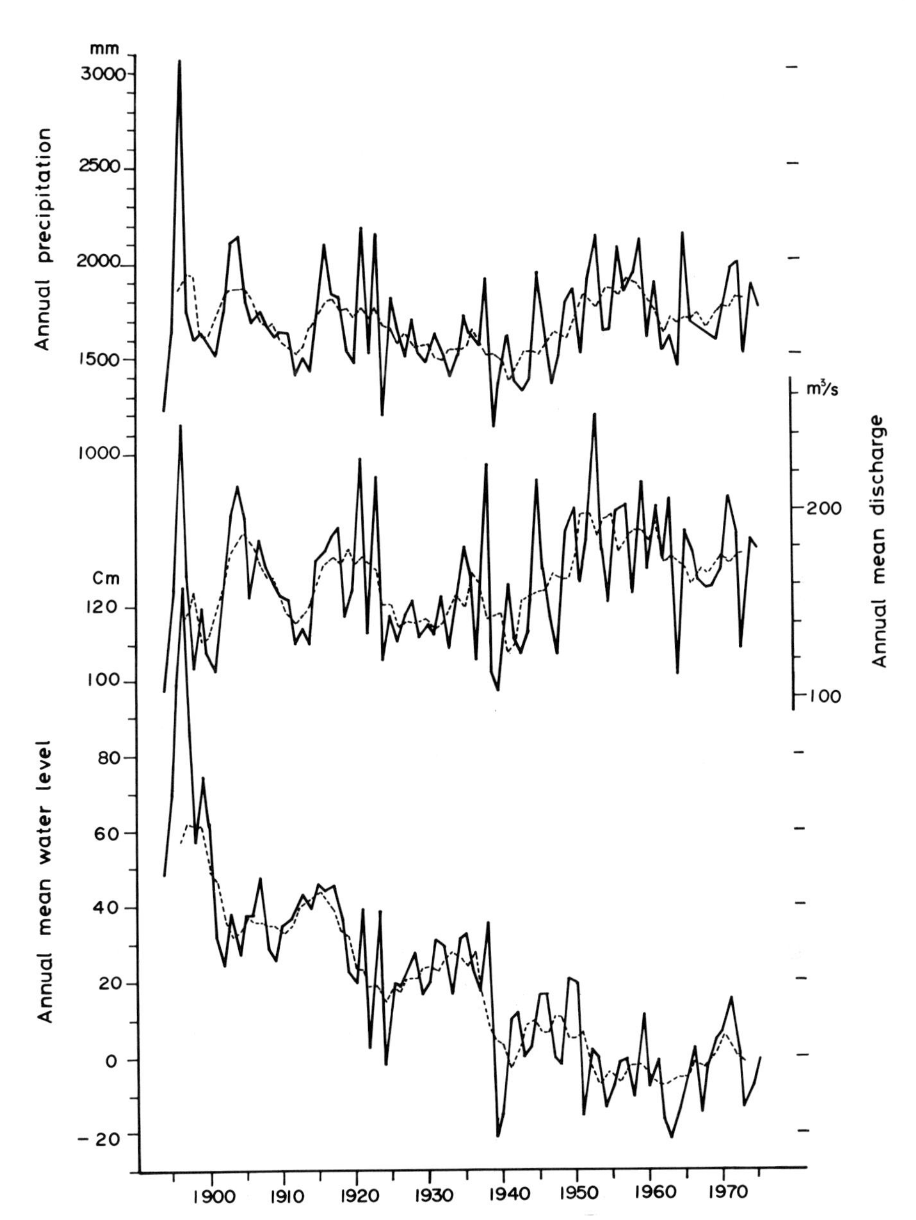

Fig. 68. Variation of annual amount of precipitation, annual mean discharge from Lake Biwa and annual mean water level of the lake: The solid and dotted lines denote the value of these elements and the five-year moving average, respectively.

method. We should note that both estimates have the following questions to be solved: the former method, in estimation of the annual mean precipitation averaged over the basin; the latter, in using the values observed at one station (water temperature, the air temperature above the lake and the amount of cloud) as a mean averaged over the lake surface. Taking into consideration the fact that the evaporation from the lake surface is about 1 m deep in a year, we may say that the estimation of the amount of evaporation from the lake is one of the significant subjects.

Water balance

Kazuo Kotoda and Takayuki Mizuyama

Introduction

The water balance of a lake is closely related not only to climatic conditions such as precipitation and evaporation but to the inflow, the catchment area and the outflow from the lake. Among them, the inflow water gives much influence upon the water balance of the lake. It is important to clarify the relationship of the water balance between the lake and its catchment area.

Lake Biwa covers an area of 681 km^2* water surface (in 1970) located at an altitude of 85 m, with a volume of about 276×10^8 m^3* and a maximum depth of 104 m. The lake consists of two basins, the main basin (northern) has an area of 623 km^2* and the mean depth of 41 m*, while the sub-basin (southern) has 58 km^2* and 3.6 m*, respectively.

The lake has two outlets at the southern coast of the sub-basin, one is the River Seta and the other is the Kyoto Canal which was constructed in 1890 for the purpose of irrigation and power generation.

The catchment area is about 3,131 km^2 and the altitude of the mountain ranges surrounding the catchment is from 800 m to 1,000 m. About 73% of the catchment area is a mountainous region and the plain is for the most part cultivated land with an area of about 845 km^2 which includes 680 km^2 of rice (paddy) fields.

The heat balance

The Birgean heat budget

Table 22 shows the monthly temperature of different layers in Lake Biwa calculated for the period 1936/41 and 1947/55 (Takahashi 1959). The thermal type of Lake Biwa is classified as a subtropical lake as defined by Yoshimura (1935—1936), with the surface temperature never below 4°C; a large annual amplitude of temperature variation; sharp summer stratification with a large vertical

* In the present paper, our calculation is based on the above-mentioned morphometric figure on lake level altitude, area and volume.

Horie, S. Lake Biwa

ISBN 90 6193 095 2. Printed in The Netherlands

Table 22. Mean monthly temperatures in Lake Biwa, period 1936/41 and 1947/55 (after Takahashi, 1959).

Depth in m	I	II	III	IV	V	VI	VII	VIII	IX	X	XI	XII	Annual mean
0	8.37	7.13	7.69	9.82	15.45	20.90	24.89	28.35	24.84	19.87	15.50	11.70	16.21
5	8.35	7.14	7.07	8.80	13.69	19.16	23.61	27.06	24.40	19.51	15.45	11.65	15.49
10	8.31	7.09	7.02	8.55	12.22	17.25	19.92	22.65	24.06	19.53	15.36	11.64	14.47
15	8.34	7.10	7.00	8.36	11.03	13.70	16.02	17.32	19.20	19.22	15.39	11.69	12.86
20	8.34	7.07	6.94	8.12	9.72	11.28	12.34	13.03	14.28	15.71	15.32	11.65	11.15
30	8.32	7.09	6.93	7.78	8.38	9.27	9.62	9.88	9.65	10.00	10.84	11.56	9.11
40	8.31	7.03	6.90	7.55	7.82	8.34	8.50	8.31	8.51	8.35	8.53	9.46	8.13
50	8.08	7.03	6.89	7.31	7.40	7.87	7.95	7.89	7.84	7.75	7.94	8.26	7.68
60	7.92	7.07	6.88	7.23	7.20	7.41	7.56	7.44	7.54	7.43	7.60	7.93	7.43
70	7.66	7.06	6.83	6.99	7.05	7.32	7.39	7.21	7.29	7.22	7.40	7.71	7.26

thermal gradient; and one circulation period in winter. Lake Biwa is one of the subtropical lakes situated in the most northern boundary region between the subtropical and the temperate lake region in Japan.

Table 23 shows the total amount of heat that enters the lake between the time of its lowest and highest heat content known as the Birgean heat budget (Birge 1915).

The analytical heat balance and lake evaporation

The heat balance equation for the lake may be given by

$$Q_B = R_N - LE_W - H \quad [1]$$

where

Q_B = the change of heat storage
R_N = the net radiation flux received at the lake surface
LE_W = the latent heat flux (L being the latent heat of evaporation)
H = the sensible heat flux

Assuming the identity of the transfer coefficients and constancy variation with the height of the fluxes H and E_W from the water surface to the upper measuring point, and using the Bowen ratio, the evaporation is given as follows.

$$E_W = \frac{R_N - Q_B}{L}\left(\frac{1}{1+R}\right) \quad [2]$$

where R is the Bowen ratio given by

$$R = 0.66\frac{T_s - T_a}{e_s - e_a} \quad [3]$$

where e_s is the saturated vapor pressure at the water surface temperature T_s, e_a is the vapor pressure in the air and T_a is the air temperature.

Table 23. Annual heat balance (Birgean heat budget) of Lake Biwa (in 10^3 cal/cm^2, after Takahashi, 1959).

Layer in m	Time of q max	Time of q min	Q =q max −q min
0— 5	IX	III	39.0
5— 10	IX	III	30.3
10— 15	XI	III	21.8
15— 20	XII	III	15.6
20— 30	XII	III	12.2
30— 40	XII	III	7.5
40— 50	XII	III	3.9
50— 60	XII	III	2.2
60—104 (bottom)	XII	III	1.0

If we know the net short wave radiation R_S and the net long wave radiation R_L, the net radiation R_N is obtained as $R_N = R_S + R_L$.

The annual march of the heat balance terms for Lake Biwa (north basin) is shown in Fig. 69 (Ito & Okamoto 1974). The maximum of the monthly mean net short wave radiation is about 12 Kcal/cm^2 month in May, and the minimum is about 4 Kcal/cm^2 month. While, the maximum of the monthly mean net long radiation is around -4 Kcal/cm^2 month in winter, and the minimum is -2 Kcal/cm^2 month in early summer. The change of heat storage Q_B increases rapidly during the period of January to May until it reaches to the maximum value (about 9 to 10 Kcal/cm^2 month) and, then, it decreases slowly during the period of June to December.

The amount of annual evaporation from Lake Biwa may be 700 mm/year to 800 mm/year. Using the heat balance method, Ito and Okamoto (1974) estimated 770 mm/year of annual evaporation from the Lake Biwa. Yamamoto et al. (1972) computed 754 mm/year by a method consisting of heat conduction in the lake together with the heat balance equation. Kotoda and Mizuyama (1975) obtained 811 mm/year by use of the aerodynamic method determined for Lake Biwa.

These annual variations of lake evaporation are shown in Fig. 70, with the values of pan evaporation (20 cm in diameters) measured at the Hikone Local Meteorological Observatory, and potential evaporation for the shallow depth

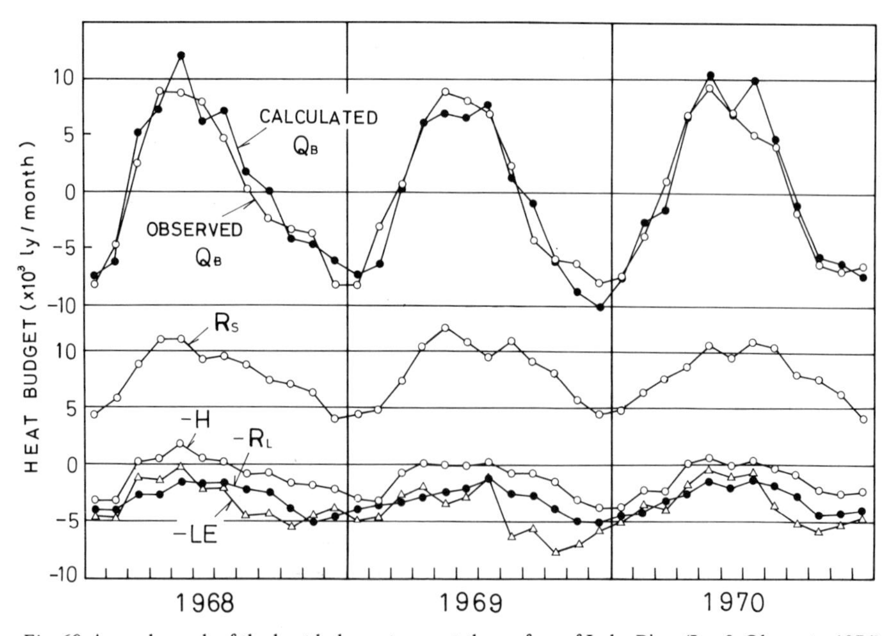

Fig. 69. Annual march of the heat balance terms at the surface of Lake Biwa (Ito & Okamoto 1974). Q_B: the change of heat storage, R_S: the net short wave radiation, R_L: the net long wave radiation, H: the sensible heat, LE: the latent heat (ly = cal/cm²).

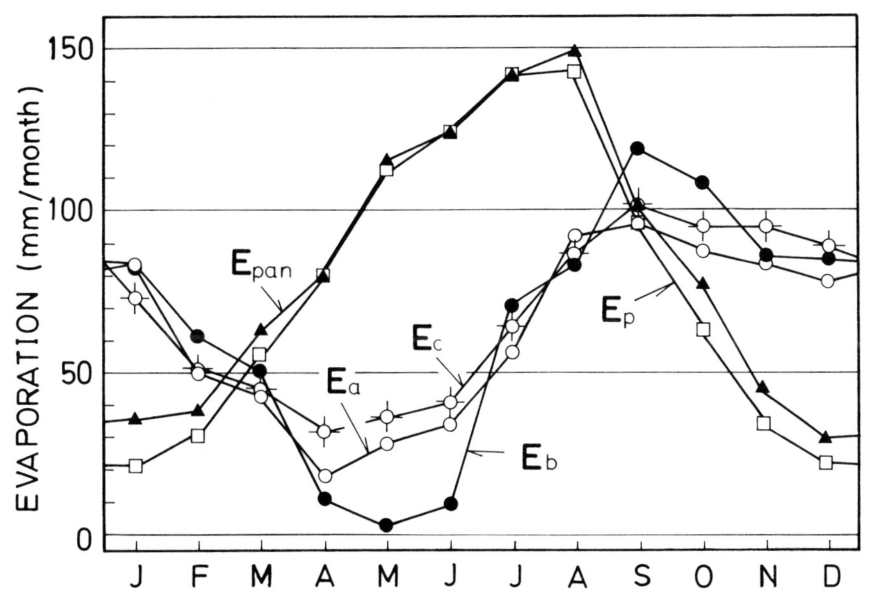

Fig. 70. Evaporation from Lake Biwa, pan evaporation and Penman's potential evaporation. E_a: Yamamoto et al. (1972), E_b: Ito & Okamoto (1974), E_c: Kotoda & Mizuyama (1975), E_{pan}: pan evaporation (Hikone, 1926/35), E_p: Penman's potential evaporation (open water, 1926/35).

openwater calculated by means of Penman's method (Penman 1963). As seen in the figure, it is worthy of notice that the evaporation from a deep lake like Lake Biwa during the period of autumn to winter exceeds that of spring to summer. It suggests that the characteristics of heat storage capacity of water mass is an important determinant of the behavior of lake evaporation.

The evaporation formula

Many studies of aerodynamic methods have been done to estimate the evaporation from the lake or sea surface. Some of the results applied to Lake Biwa are shown in Fig. 71. Jacobs (1942) obtained the wind factor f(u) in the Dalton type evaporation formula given as $E = f(u)(e_s - e_a)$ on the sea surface. Jacobs' value of the wind factor $f(u) = 0.134u$ (where evaporation E is in mm/day, the vapor pressures e_s and e_a are in mb, and the wind velocity u is in m/sec.) agrees fairly well with the value of 0.136 obtained by Kotoda and Mizuyama (1975) who used the data of the Hikone Local Meteorological Observatory. Ito and Okamoto (1974) also proposed the value of 0.17 as the wind factor of the evaporation formula, which, however, seems to be a little overestimated. Recently, Fujita (1977) proposed the two height bulk method using measurements at two levels (1 m and 4 m) in order to take an account of the effects of

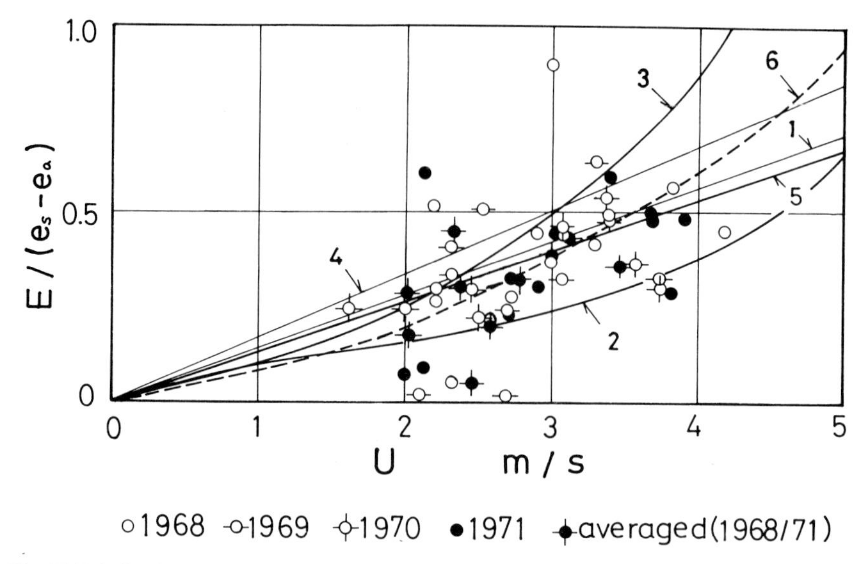

Fig. 71. Relation between $E/(e_s-e_a)$ and wind velocity (u).
1: Jacobs (1942) ... $E=0.143u(e_s-e_a)$, uz = 6m,ez = 6m
2: Yamamoto & Kondo (1964).
3: Kondo & Watabe (1969) ... $E=4.6l^{-0.1}u^{1.8}(e_s-e_a)/L$.
4: Ito & Okamoto (1974) ... $E=0.17u(e_s-e_a)$.
5: Kotoda & Mizuyama (1975) ... $E=0.136u(e_s-e_a)$, uz = 17m,ez = 1.5m.
6: Kotoda (1977) ... $E=0.087l^{-0.1}u^{1.7}(e_s-e_a)$,uz = 17m,ez = 1.5m.
E: mm/day, u: m/sec, e: mb, $l=2\sqrt{A/\pi}$, A: surface area of lake (km^2), uz,ez: the measurement height of u and e, respectively, L: the latent heat of evaporation.

thermal stability on fluxes over the sea. Kotoda (1977) presented the following simple empirical formula:

$$E=\frac{0.137u(e_s-e_a)}{1+0.375\exp(-10.5\zeta)},\ \zeta=(T_s-T_a)/u^2$$

where E is evaporation in mm/day, u is the wind velocity in m/sec at the measurement height of 17 m, e_s and e_a are the water vapour pressures of air at the lake surface and at the height of 1.5 m, respectively. T_s and T_a are the water temperature at the lake surface and the air temperature at the height of 1.5 m, respectively.

The water balance

The water balance of Lake Biwa

The water balance equation for the lake may be written as follows:

$$P_W=(D_{WO}-D_{Wi})+E_W+\Delta B_W \qquad [4]$$

where the notations are:

P_W = the precipitation falling on the lake
D_{WO} = the outflow discharge from the lake
D_{Wi} = the inflow discharge from the catchment area into the lake
E_W = the evaporation from the lake surface
ΔB_W = the change of water storage.

In Eq. [4], it is very difficult to estimate E_W and D_{Wi} from direct observation data. However E_W can be estimated by the heat balance method using hydroclimatological data and then, unknown term D_{Wi} can be computed from Eq. [4].

The water balance of the catchment area

The water balance equation for the catchment area may be given as

$$P_L = D_L + E_L + \Delta B_L \qquad [5]$$

where P_L is the precipitation fall on the catchment area, D_L is the outflow discharge from the catchment area, E_L is the evapotranspiration and ΔB_L is the change of water storage.

As the amount of outflow from the catchment is equal to that of inflow into the lake, the term D_L in Eq. [5] is given as follows:

$$D_L = \frac{A_W}{A_L} D_{Wi} \qquad [6]$$

where A_W and A_L are the areas of the lake and the catchment, respectively.

It is expected that ΔB_L and ΔB_W have marked variations in a short time scale of water-balance computation. However, if the long time scale such as the annual water balance is considered, ΔB_L and ΔB_W could be neglected without an undue loss of accuracy, because the both terms are generally small in comparison to the other quantities in the water balance equation.

To estimate the monthly evapotranspiration from the catchment area, the following assumptions are made:

1. Actual monthly evapotranspiration is proportional to the potential evaporation.
2. The proportional constant does not change during the year.

Then, the following relation is obtained:

$$E_L = fE_O \qquad [7]$$

where E_O is the potential evaporation and f is the proportional constant.

Using the values of annual evapotranspiration E_L and E_O (open water) obtained by the Penman's method, f is estimated from Eq. [7]. In the case of the catchment area of Lake Biwa, f=0.73 was obtained (Kotoda & Mizuyama 1975).

From the above equations, the monthly values of ΔB_L may be given by

$$\Delta B_L = P_L - fE_O - \frac{A_W}{A_L}(D_{WO} + E_W + \Delta B_W - P_W) \qquad [8]$$

The annual water balance

The annual water balance of Lake Biwa and its catchment area is given in Fig. 72. Lake Biwa annually receives 12×10^8 m^3 (1,637 mm) of precipitation, while the catchment receives 61×10^8 m^3 (1,941 mm) of precipitation. The difference of areal precipitation between the two areas is approximately 300 mm/year. About $6 \times 10^8 m^3$ (811 mm) of water is lost by evaporation from the lake surface and $21 \times 10^8 m^3$ (676 mm) is lost by evapotranspiration from the land surface. Therefore, about 27×10^8 m^3 (701 mm in height) of water is transferred

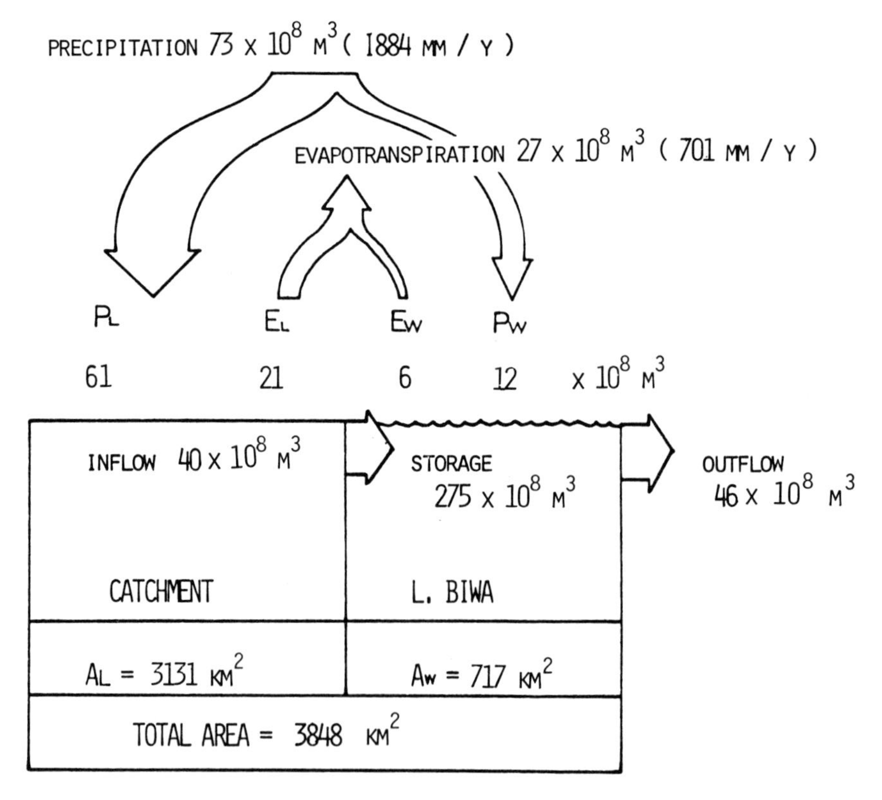

Fig. 72. Annual water balance of Lake Biwa (1926/1935). Including former (now dried up) inner lakes (shore lagoon) (Kotada 1975).

into the atmosphere by evapotranspiration from the lake and its catchment as a whole.

The total amount of affluents carried by numerous rivers, irrigation canals and ground water flow into the lake is 40×10^8 m^3, and 46×10^8 m^3 of the lake water is carried out by the River Seta and the Kyoto Canal.

The annual runoff coefficient of the total area is approximately 62%. It means that the nearly two-thirds of precipitation is lost by runoff from the lake.

Annual variation of water balance components

Table 24 shows the annual variation of the water balance components for Lake Biwa and its catchment area. As seen in the table, the affluents (D_L) from the catchment into the lake are at the minimum in June, although the precipitation is good enough to maintain the affluents. This can be explained as follows. In this region, as well as in the southwest Japan, June is the beginning of the irrigation period of rice paddy fields, so that it needs a large amount of water for irrigation. It seems that this result is reflected in the decrement of the affluents. In contrast to the affluents, the change of water storage (ΔB_L) is positive (31 mm/month) in June. It indicates that about 97×10^8 m^3 of the volume of water was stored in the catchment in the month of June. As the total area of rice paddy fields in this catchment is about 680 km^2 of that time (1926–1935), if it is assumed that the above water is stored in those rice fields, the height of the water-column will reach to 13.4 cm. This value is not intolerant, because each field has a capacity to hold this much water.

After the period of transplanting in the rice fields, not so much water is needed for irrigation. The change of water storage in the catchment land is complicated depending on the customary water-use of each farmer and on a temporary heavy storm which is accompanied by a typhoon. Fig. 73 shows the monthly variation of the runoff coefficients of the catchment.

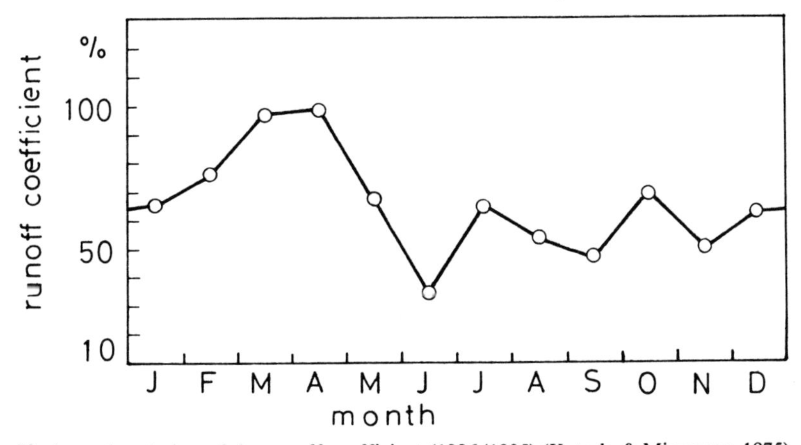

Fig. 73. Annual variation of the runoff coefficient (1926/1935) (Kotoda & Mizuyama 1975).

Table 24. Water balance of Lake Biwa and of its catchment, period 1926/35 (in mm water column, after Kotoda & Mizuyama, 1975).

			1	2	3	4	5	6	7	8	9	10	11	12	Annual
Catchment	Precipitation	P_L	163	133	145	145	130	186	209	176	218	137	145	154	1941
	Evapotranspiration	E_L	16	22	40	56	81	92	106	107	71	45	23	16	676
	Outflow	D_L	108	104	144	147	90	63	140	94	106	96	74	98	1265
	Change of storage	$\triangle B_L$	39	7	−39	−58	−41	31	−37	−25	41	−4	48	40	0
Lake Biwa	Precipitation	P_W	114	103	120	129	127	191	188	134	183	117	120	112	1637
	Inflow	D_{Wi}	473	454	630	641	395	276	612	409	463	419	324	427	5525
	Evaporation	E_{WO}	74	51	45	32	37	41	65	86	101	95	95	89	811
	Outflow	D_W	476	497	663	749	569	407	705	520	553	447	367	398	6351
	Change of storage	$\triangle B_W$	37	9	43	−11	−84	19	30	−63	−8	−6	−18	52	0

1 mm of the catchment = $3{,}131 \times 10^3$ m^3, 1 mm of Lake Biwa = 717×10^3 m^3.

Water currents

Iwao Okamoto

Introduction

Since water currents in a lake are usually very weak and unstable, it is not an easy problem to perform their observations with high accuracy reliable enough to yield reasonable interpretations. It is absolutely necessary, therefore, that systematic and continuous observations are carefully repeated before obtaining knowledge of even a single aspect of the lake currents. It is for this reason that there are very few lakes in Japan for which such systematic investigations have been repeatedly conducted up to the present: Lake Biwa could be the only exception.

In 1926, the first systematic observations of water movements were carried out for Lake Biwa by the Kōbe Marine Observatory which discovered a gyre system in the surface layer of the north basin. Since then, a number of projects, such as BST (Biwa-ko Seibutsu Shigen Tyōsa-dan [Lake Biwa Biological Resources Research Group]), IBP (International Biological Program), IHD (International Hydrological Decade) or LBI (Lake Biwa Institution), have been conducted one after another, especially in the recent 15 years.

One of the important objectives of these projects was to clarify behaviors and mechanisms of the lake currents because it should be considered most essential both for the study of water-mass-distributions and for the analysis of biological productions. For pursuing this objective, not only direct measurements of water movements were obtained but also an indirect method was employed, that is, at least the main current in the north basin can be inferred from the temperature distributions easily obtainable by means of a BT cast or a thermistor thermometer (Okamoto & Morikawa 1961).

In addition to these observational studies, numerical experiments have also been attempted in order to develop simulation models of water movements (Imasato et al. 1971).

Horie, S. Lake Biwa

ISBN 90 6193 095 2. Printed in The Netherlands

Observations and general characters

As is the case with the ocean, various types of currents due to variety of mechanisms have been found also for Lake Biwa. It is to be noted, however, that the dominant current can be variable not only from place to place but also from time to time. For example, in the north basin which is deep and extensive, both gyratory movements and internal waves are the dominant currents during the season of thermal stratification, while the internal waves cannot exist during the winter (time Okamoto & Yamashita 1973). In the south basin which is shallow and channel-shaped, oscillatory currents caused by developed seiches are frequently superimposed onto the southward current steadily flowing out into the River Seta. It has been observed that the water movements in bays and caves located inside the north basin are greatly influenced by the internal seiches originated in the basin.

Formerly, an Ekman-Merz current-meter used to be employed on board for measuring lake currents (Morikawa & Okamoto 1962). The greatest difficulty with this method, however, lies in the virtual impossibility of suppressing irregular motions of a vessel even anchored onto the lake bottom. Another method of the 'Eulerian type' has therefore been adopted recently. In this method, we use an anchored buoy for the deep water and a tower platform system for the shallow, both with a set of current-meters attached. There are still several other methods now being used. For instance, in the 'Lagrangian-type' method as many as 10,000 drift-bottles or cards are released onto the lake surface (Morikawa & Okamoto 1960). In order to observe the water diffusion or dispersion along with detecting the lake current itself, current drogues are often used; and the method of radio tracking has recently been adopted (Okumura & Yamamoto 1976). A remote sensing method such as aerophotography or satellite-photography can also be utilized, particularly after a heavy rainfall because loci of turbid river water-masses can distinctly be traced under this condition (Okuda et al. 1970).

Besides these direct methods mentioned above, an indirect method consisting of synoptic sounding and continuous recording of temperatures is frequently used in the north basin to study the main current (including oscillatory movements occasionally developed for internal waves) for the period of thermal stratifications. This is indispensable for dynamical analysis of the lake currents as is to be shown below.

In Lake Biwa, the water temperature is usually in the state of stratification except in the winter time, and the thermocline is developed during summer and autumn. Since distributions of water-masses may depend upon those of water-densities which are strong functions of the temperature, the main pattern of the lake currents should be inferred from the pressure gradients due to the mass distributions. 'Dynamical calculations' for the water currents must therefore be very important and useful also for Lake Biwa as well as for the ocean (Okamoto & Morikawa 1961; Kunishi et al. 1967).

Geostrophic flows

Geostrophic flows are formed under the condition of geostrophic balance, by which is meant such a state that the pressure gradient is balanced with the Coriolis force caused by the earth's rotation. It is often said that the geostrophic flow is visible only in Lake Biwa of all inland waters in this country. This must be due to the fact that the north basin of this lake is deep and extensive enough to establish the condition of geostrophic balance.

Fig. 74 shows an example of the topography of the thermocline represented by an isothermal surface of 16°C. As seen in this figure, the thermocline is inclined from the south to the north under the influence of internal waves. According to the conception of geostrophic balance, the rise in the north corresponds to a counter-clockwise gyre. As shown by the arrows in Fig. 74, the geostrophic currents in the northern hemisphere on the earth flow in such a manner as to keep deeper contours at their righthand side, that is, clockwise or counter-clockwise along contours. The counter-clockwise gyre shown in Fig. 74 is the biggest and the most stable, while several other gyres are occasionally visible around the middle or the southern part of the basin (Okamoto & Yamashita 1973).

In Fig. 75, a typical pattern of temperature distributions in a vertical section is given. As seen in this figure, the thermocline shows a rise at the middle of the section corresponding to the center of the gyre given in Fig. 74. This rise seems to exist consistently as far as we can judge from the observations that are frequently made up to the present. The horizontally circulating flows seem to be confined to the layer above the thermocline below which no gyratory currents seem to appear in the hypolimnion, presumably due to the disappearance of the horizontal pressure gradient. The mean value of the flow velocity in the gyre is usually around 10 cm s^{-1} although it may exceptionally reach to the speed of 30 cm s^{-1}. The rate of volume transport is estimated at some 10^3 m^3 s^{-1}, and the kinetic energy calculated for the whole circulating mass is in the order of 10^{17} erg. It is not clear yet if the gyre system remains also during winter when the thermal stratifications have disappeared.

In order to clarify the mechanism both for the origin and the maintenance of the gyre system, simulation trials have been made for flow patterns using a two-layer model (Endoh et al. 1976). The results show that no stable gyre can exist without dispersing into internal waves for a uniformly blowing wind, while in the case of a non-uniform wind, that is, when the curl of wind stress is not zero, a stable gyre can be developed and maintained for a long time. It may be necessary therefore that the observations of overall wind-velocity distributions are carried out in the future.

It is found that periodically oscillating flows accompanied with long internal waves are affected by the Coriolis force in such a manner that the velocity hodograph of water particles takes an elliptic shape, circulating counter-clockwise in a period (Kanari 1975b). Prevailing flows seen in the hypolimnion may be such oscillatory movements.

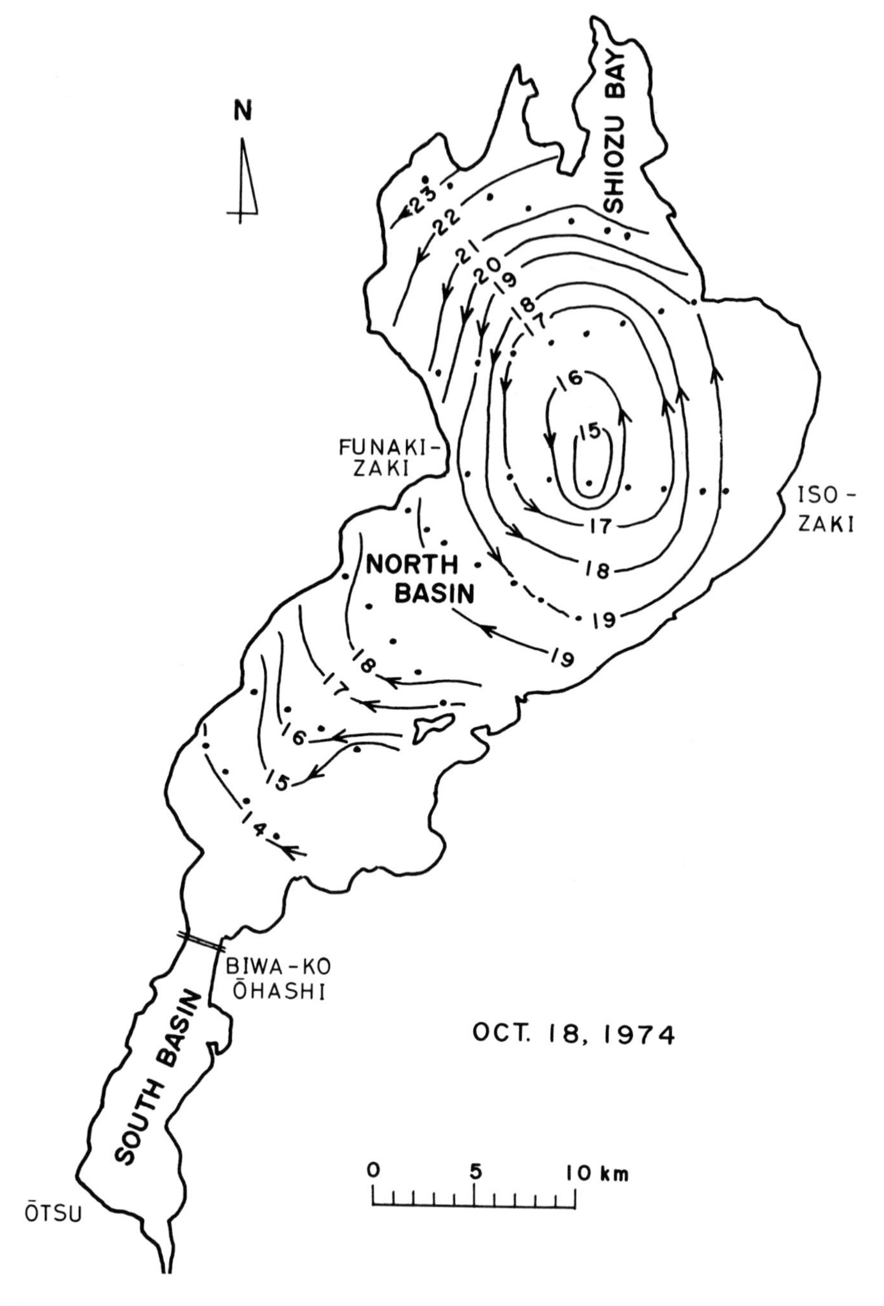

Fig. 74. Topography of the 16°C-isothermal surface representing the thermocline core. Contours are given in meters. The rise of the thermocline in the north area corresponds to a counter-clockwise gyre (BIWA-KO ŌHASHI means Lake Biwa Big Bridge) (Okuda 1978).

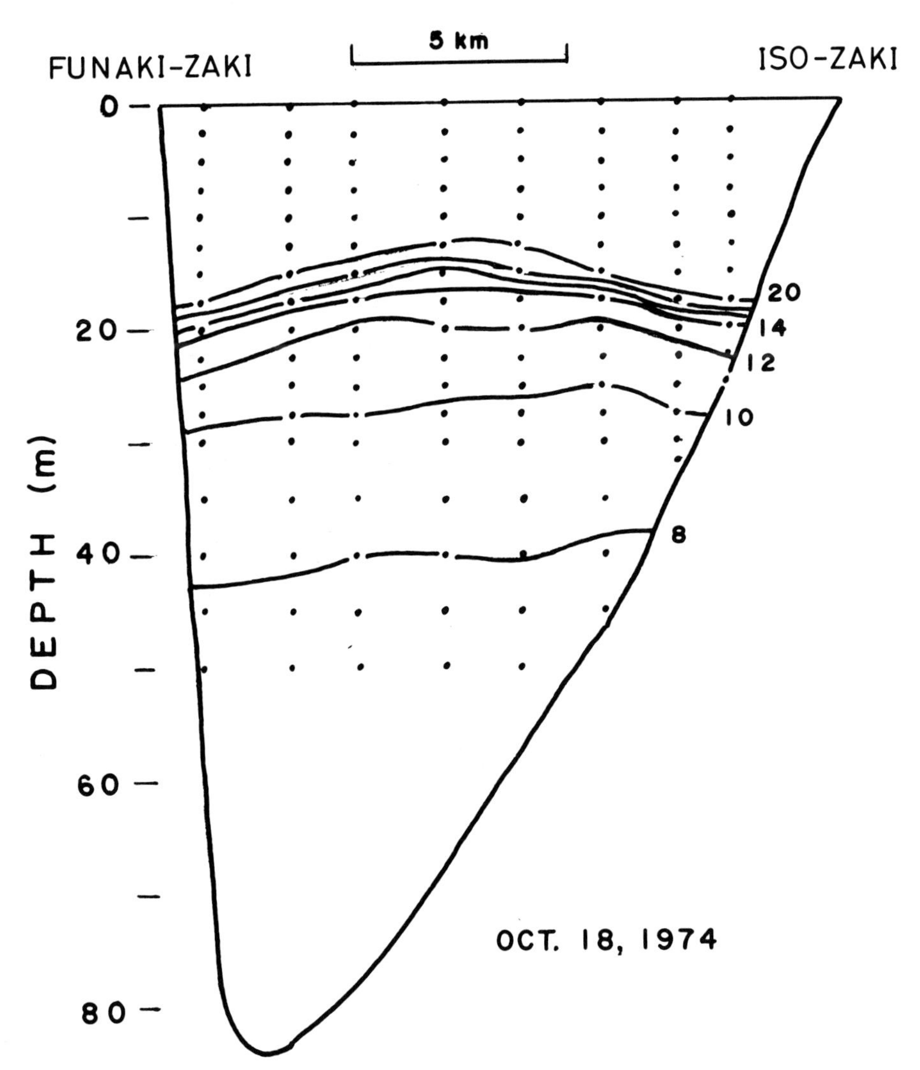

Fig. 75. Temperature distributions in a vertical section corresponding to the counter-clockwise gyre shown in Fig. 74.

Internal seiches in deep bay

In a deep bay, water currents will be seriously influenced by internal seiches developed in its main basin, although small gyratory movements can take place occasionally. Such an example observed in Shiozu Bay, situated at the north end of the main basin, is shown in Fig. 76. The water temperature showed periodic fluctuations, particularly at the bay head, with periods of 44 and 16 hours. The former coincides with the period of the fundamental oscillation to be developed

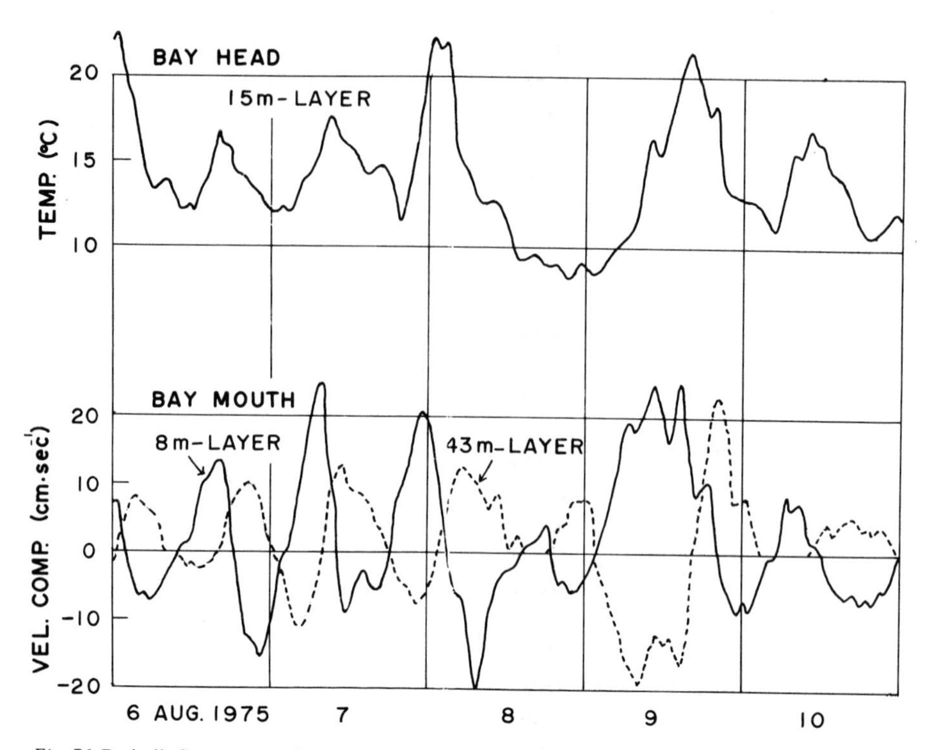

Fig. 76. Periodic fluctuations observed in water temperature (upper) and in longitudinal component of current velocity (lower) at Shiozu Bay. These oscillations are caused by intrusion of internal seiches developed in the main basin of Lake Biwa.
Positive: northward; negative: southward (Okamoto & Iwamoto 1977).

in the main basin, and the latter must be due to an oscillation modified in the bay.

Observed current velocities also show periodic variations with the same periods as those observed for the temperature. Typical configurations of flow velocities are given in the lower part of Fig. 76 as an illustration of the effort of internal seiches in Shiozu Bay. As is clearly seen in this figure, the upper layer water (at the depth of 8 m) flows northwards while the lower one (at 43 m) flows southwards, and vice versa.

The phases of temperature and of current velocity are not concurrent with each other but a quarter period of delay is seen for the temperature oscillations. This delay may be interpreted as follows: the temperature in the bay reaches to its maximum value only when the inflow of the warm upper-layer water with the simultaneous outflow of the cool lower-layer water has been completed.

Since the amount of river water flowing into the bay is negligible, the net flux must be nil in any vertical section. The boundary between the upper and the lower layer lies around the depth of the thermocline. The rate of water transport both for the upper and the lower flow is, respectively, estimated at 7×10^7 m^3 for a half period, which corresponds to about 10 percent of the whole bay water. It can be expected therefore that the water exchange may easily be conducted between the bay and the main basin. It has been pointed out that the thermocline is considerably raised with an upwelling phenomenon when the local wind blows from the bay head to the mouth (Okamoto et al. 1971).

Density currents

From autumn to winter, the water temperature gradually goes down both in the north and the south basins. The depression rates, however, are different from each other: since the south basin is much shallower than the north, the temperature drops more quickly in the south than in the north. As a result, the water temperature in the south is lowered by about 2–3°C below that of the north throughout these two seasons, from which a thermal front is formed at the boundary between both basins as shown in Fig. 77.

At the thermal front, the cooler water from the south sinks beneath the warmer water from the north, and then flows to the north along the sloped lake bottom. Fig. 78 gives an example for such observations. It is shown in this figure that the southward flows are observed in the upper layers at the depths of 1, 3, and 5 m, respectively, while, in the lower layer of 7 m depth, the northward flows are clearly recognized. This is the density current observed in Lake Biwa. The velocity of this current is usually in the range of a few cm s^{-1}, and the total amount of water to be transported from the south basin into the north has been estimated at about 5×10^7 m^3. Periodic velocity fluctuations visible in the drogue trajectories of Fig. 78 are caused by the seiche with a period of four hours.

Water-mass exchange

The gross amount of water volume in Lake Biwa is 2.75×10^{10} m^3 and the rate of the outflow through the River Seta is 0.52×10^{10} m^3 yr^{-1}, although the latter shows considerable variations from year to year. If we assume that the pre-existing lake water is pushed out into the River Seta by the inflowing river water; that is, if we ignore the mixing of both water-masses, then the whole lake water might be replaced within a period of 5.3 years. Since the mixing of the inflowing river water with the pre-existing lake water is not negligible, however, the actual exchange rate must be much smaller. According to the writer's estimation, 4.4 years are necessary until 50% of the lake water has been replaced, and 14.5 years for 90%. If the lake is once polluted, therefore, more than ten years would be necessary for the water to be refreshed again. This has actually become a great problem in several recent years.

One of the mechanisms of the water exchange between the north and the south basin has already been described in the section for density currents. There

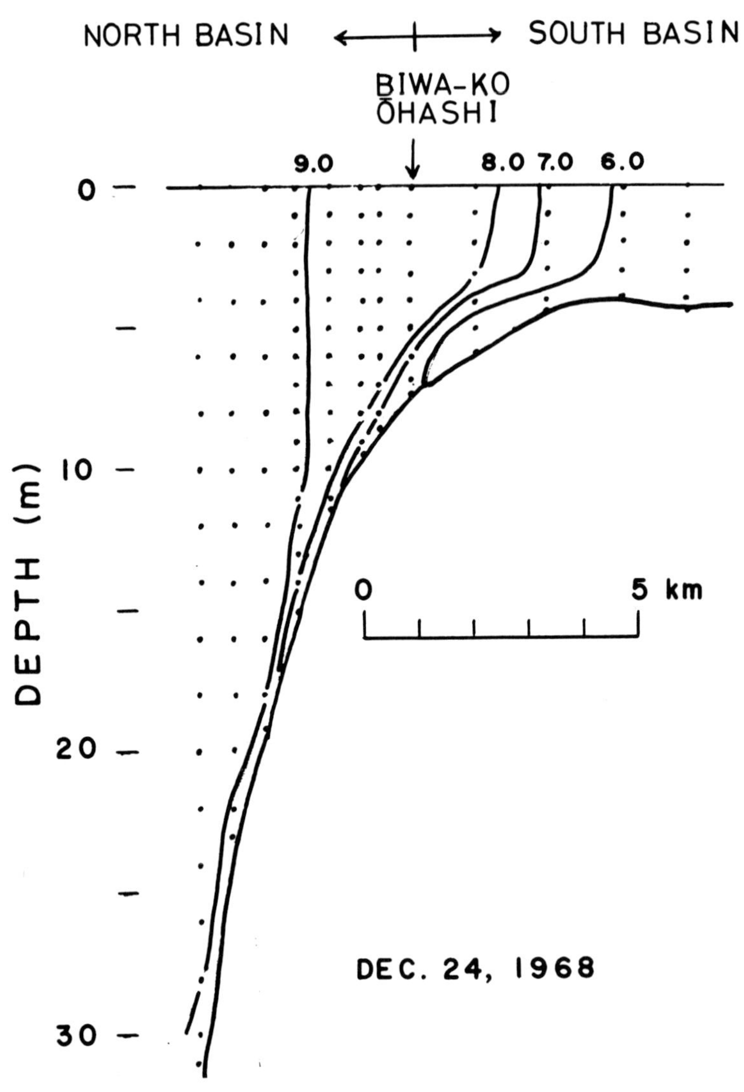

Fig. 77. Example of temperature distributions in a longitudinal section during winter time. A thermal front is formed at the boundary between the north and the south basin. The cooler water from the south sinks beneath the warmer one from the north and flows to the north along the sloped lake bottom (BIWA-KO ŌHASHI means Lake Biwa Big Bridge). (Okamoto 1971).

are two other mechanisms to be discussed here, that is, the internal seiches and the ordinary seiches. When the internal seiches are developed, 4×10^7 m^3 of hypolimnion water from the north basin intrudes into the bottom layer of the south during half of an oscillation period, accompanying the reverse flow of the surface water in the amount of 3×10^7 m^3. The balance, 1×10^7 m^3 is the amount to be flown out into the River Seta. This phenomenon often accom-

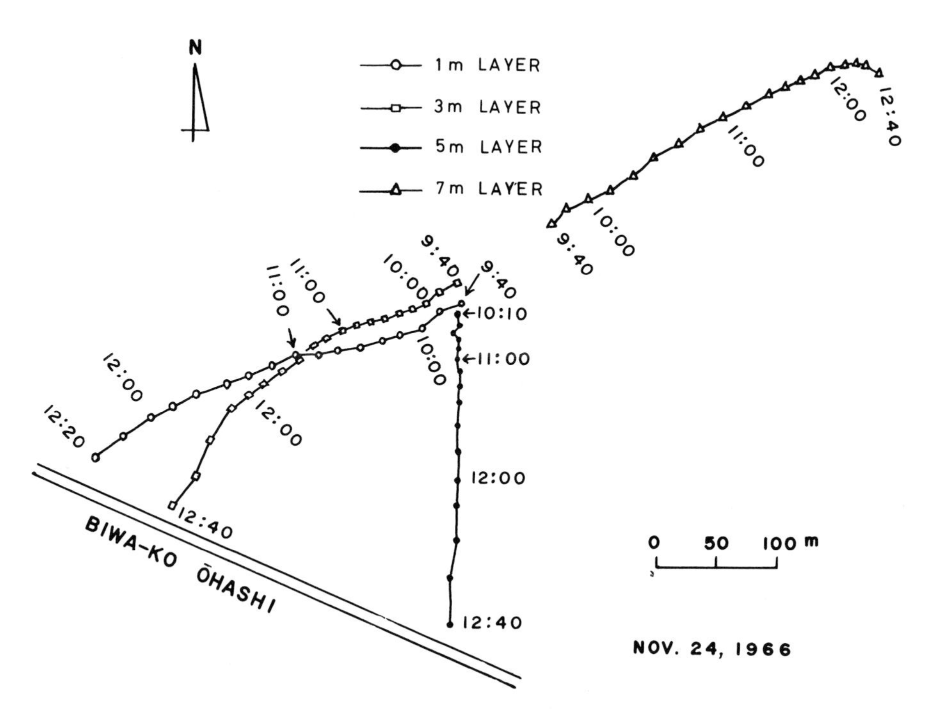

Fig. 78. Drogue trajectories around the boundary between the north and the south basin. A flow in the north-east direction, visible in the bottom layer of 7 m depth, is the density current given in Fig. 77 (BIWA-KO ŌHASHI means Lake Biwa Big Bridge) (Okamoto 1971).

panies a drastic change in temperature distributions of the south basin (Okamoto 1972).

Another type of oscillatory motion called 'Ordinary Seiches' has also been observed. In this case, the oscillation period is 4 hours or 70 minutes and the nodes appear around the boundary between both basins. When this type of oscillation is developed, 3×10^6 m^3 of water can be exchanged in each period, and a conspicuous oscillatory change is observed for the velocity fields in the south basin.

Internal waves and seiches

Seiichi Kanari

Introduction

It is well known that heating of the surface layer by solar radiation leads to remarkable growth of a thermocline in Lake Biwa during early summer to late autumn.

By such formation of a thermocline, the density of lake water becomes less in the upper layer (epilimnion) than that in the deeper layer (hypolimnion). In the main basin of the lake, the thickness of the epilimnion reaches to 15 m to 20 m during the stratification season, and the vertical density distribution exhibits almost two-layer stratification.

The stratification produces apparently reduced gravity in the lower layer, which makes it possible to introduce a large-scale vertical motion of the interface of the density stratification, due to comparatively weak external disturbances at the lake surface, such as wind and atmospheric pressure.

After the external disturbances have been removed, the interface is released from the displacement, tends to be restored toward its original position, and begins vertical free oscillations around its equilibrium level because of inertia of lake water.

Such oscillations in lakes are called internal seiches and have been studied in many European lakes and the Great Lakes in Canada and the United States.

The existence of internal seiches in small Japanese lakes has also been well known among Japanese limnologists. However, nobody could confirm their existence in the greatest lake in Japan until the 1960's.

Even in a report of the Kōbe Marine Observatory which carried out a systematic survey of Lake Biwa in 1926, not a line of description of internal seiches can be found.

In the 1960's, Okamoto and Morikawa for the first time noted long-term variations in the temperature profile of Lake Biwa and suggested the existence of long-period internal seiches, based on the observed temperature profile in the summer season.

In 1968, Kanari casted the first buoy station of a thermistor chain at Sugaura

Horie, S. Lake Biwa

ISBN 90 6193 095 2. Printed in The Netherlands

Bay, and in 1969 two other similar stations were anchored at Kido and Funaki. They have yielded a mass of invaluable information on internal waves and seiches in the lake through their continuous temperature records.

Since the foundation of the three buoy stations, the study of internal waves and seiches in Lake Biwa has remarkably advanced and the main features of internal seiches in Lake Biwa were nearly clarified by the year 1973 as is to be described below.

Bathymetric condition of the lake

Lake Biwa, the largest lake in Japan, has an area of 674.4 km^2, a length of about 68 km, and a maximum depth of 104.0 m as shown in Fig. 79. The basin consists of two basins, of which the main basin is called the North Basin (Hokko in Japanese) with the mean depth of 48 m* and the other sub-basin is called the South Basin (Nanko in Japanese) with the mean depth of less than 5 m*. The two basins are connected by a narrow channel at Katata. The maximum depth in the cross section at the Katata channel is about 8 m. The greatest depth at the 104.0 m-point of the North Basin is very close to the 80 m-contour line along the western shore, which is about 3 km off in the direction of NNW from the Funaki point and is just like a pin-hole. Except for the pin-hole, the maximum main depression of the North Basin is located at 7 km north of the Funaki point as shown by the 90 m-contour line in Fig. 79.

As is to be described later, the main thermocline is formed at the depth of 15 m in summer and 20 m in autumn. Hence, the South Basin may have virtually no connection with internal seiches, that is to say, the region in which internal seiches are active is confined within the North Basin.

Seasonal thermal condition of Lake Biwa

Owing to the difference of heat contents between the two basins, the seasonal thermal condition of the North Basin exhibits a great difference in comparison with that of the South Basin.

As shown in Fig. 80 (a)–(f), which was drawn from the past six years, regular observations conducted by the Ōtsu Hydrobiological Station of Kyoto University, the surface water temperature in the main basin is the lowest in March and the highest in August, in contrast to those in the sub-basin where the monthly variation of water temperature at the surface almost coincides with the variation at the bottom.

The amplitude of the seasonal temperature variation in the main basin clearly decreases with increasing depth. However, at the depth below 50 m, such variation becomes inconspicuous. This means that there exists a permanent hypolimnion with constant temperature of 7°C in Lake Biwa.

* In the present paper, calculation is based on the above-mentioned morphometric figure.

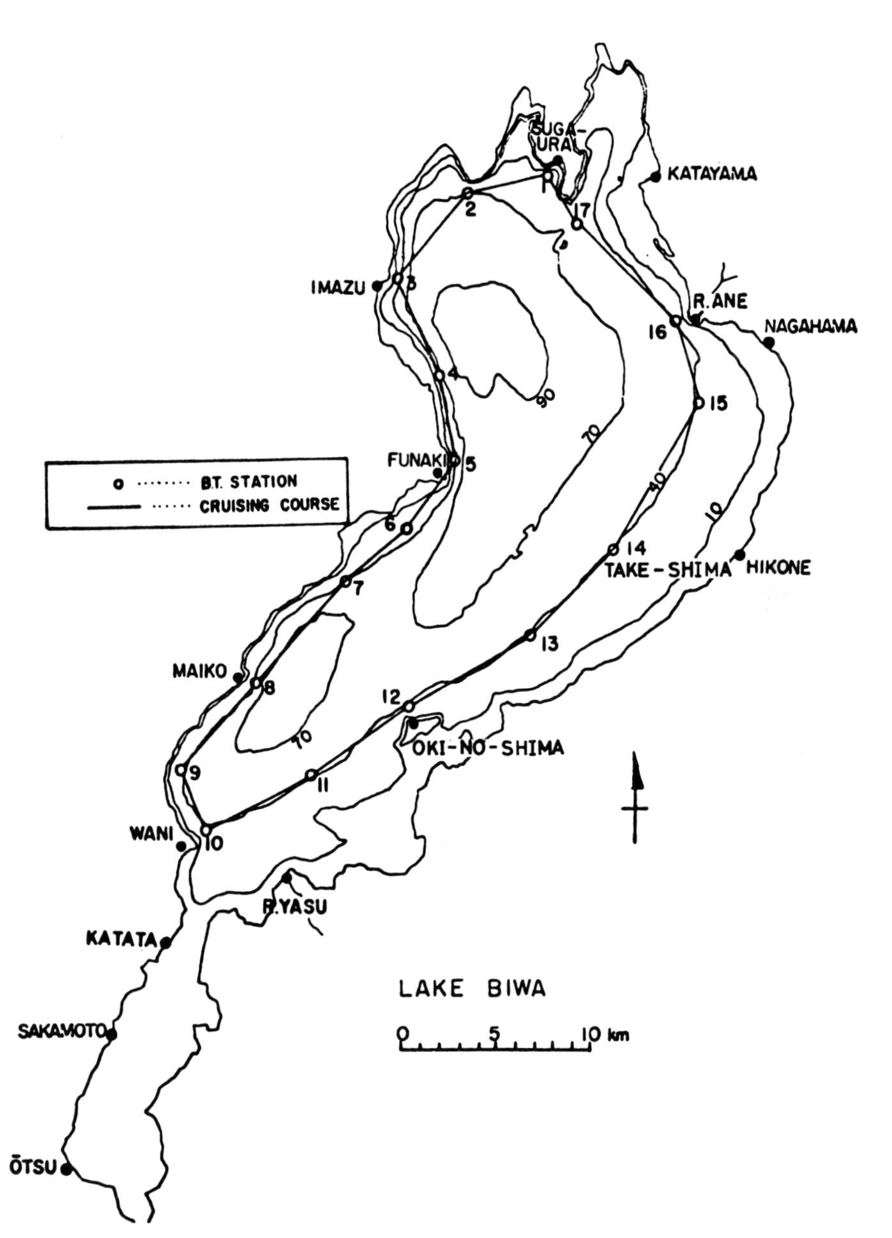

Fig. 79. Bathymetric map of Lake Biwa and location of survey stations. Black circles (●): thermistor-chain stations; open circles (○): round-cruise survey (after Kanari 1974b).

It should be noted that the annual surface temperature variation during the stratified season in the main basin may be due to the annual variation of solar radiation. However, variations at the 15 m layer are greater than those at the surface. For example, the temperature difference between those in 1966 and 1970

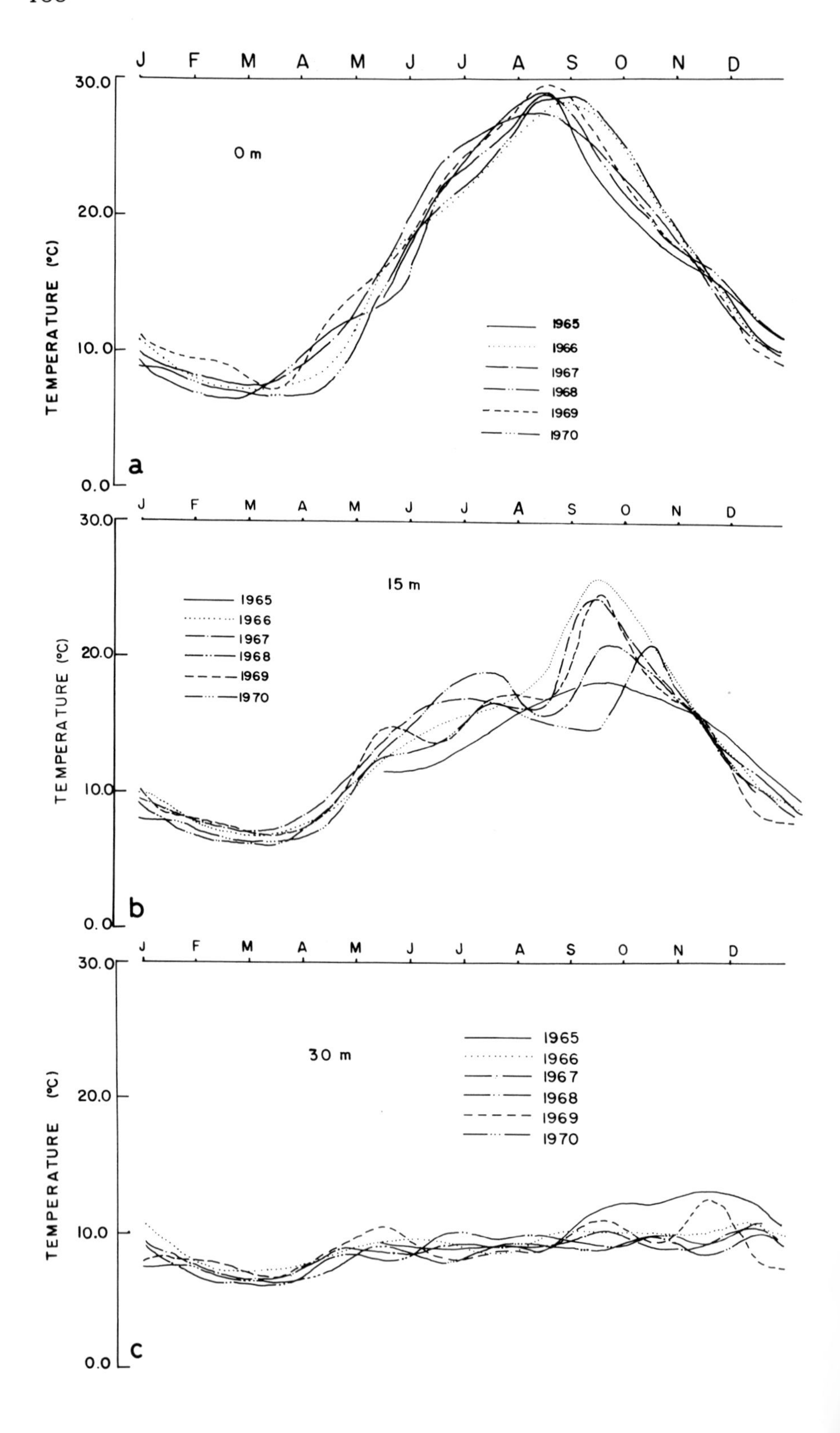
J F M A M J J A S O N D
30.0
20.0
10.0
0.0
TEMPERATURE (°C)
0 m
1965
1966
1967
1968
1969
1970
a
15 m
b
30 m
c

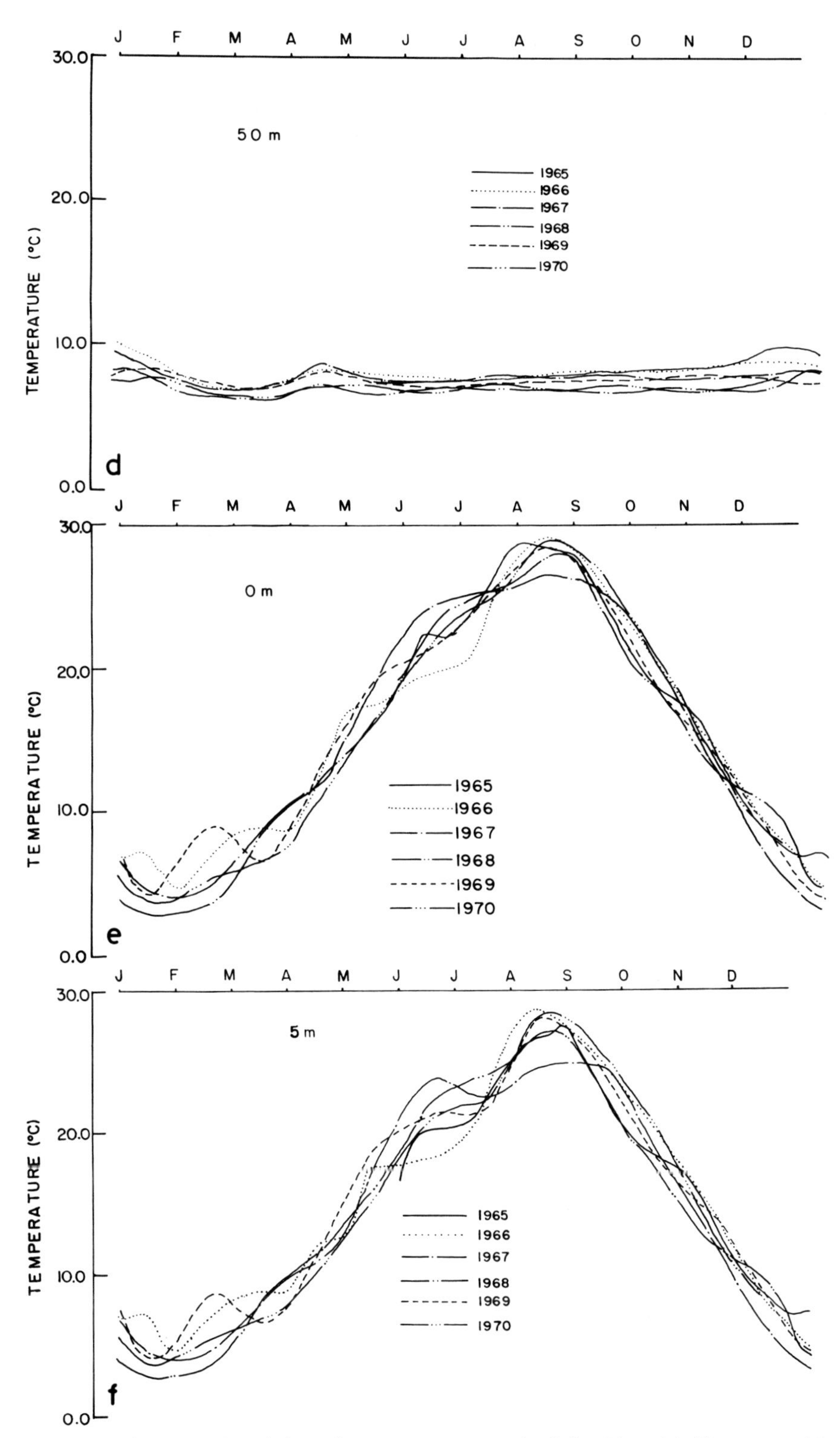

Fig. 80. (a)–(f). Seasonal variation of water temperature in Lake Biwa; (a)–(d) are monthly temperature changes at the depth of (a) 0 m, (b) 15 m, (c) 30 m and (d) 50 m, at the station Ie, which is at the center of the line connected with the points 7 and 12 of the North Basin in Fig. 79; and (e) 0 m, and (f) bottom of the South Basin (Kanari 1974c).

at the 15 m layer reaches to about 11°C in September, though the difference at the surface is only 2°C. Hence, such a large annual temperature variation at the 15 m layer may not be due to the annual change of solar radiation, but may be due to the large amplitude of internal seiches.

Density stratification

Fig. 81 shows the monthly variation of the vertical density anomaly defined by

$$\delta\rho = (1-\rho) \times 10^3 \ (\text{g/cm}^3),$$

which was calculated from the data of regular observations in 1970 shown in Fig. 80.

It can be seen from the figure that the stratification of Lake Biwa is formed in the season from spring to late autumn.

Assuming the depth of the steepest gradient in the vertical density distribution to be the position of the interface between the two layers, the seasonal change of thickness of the upper layer and the density difference between the two layers can be estimated as shown in Figs. 82 (A) and 82 (B). It is clear that the maximum density difference appears in August and its magnitude is 2.5×10^{-3} g/cm^3, and the minimum density difference is established in February to March.

When some external forces, such as a strong wind or a local atmospheric disturbance, act on the lake surface, a standing oscillation will be generated especially at the interface of the density stratified layers, being called internal seiches in a lake.

As the simplest case, the theoretical period of uninodal internal seiches in a two-layered rectangular lake of uniform depth can be given by the well-known formula [1]:

$$T = 2L\sqrt{\rho(h_1 + h_2)/\Delta\rho g h_1 h_2} \qquad [1]$$

where $\Delta\rho$ is the fractional density difference between the layers, ρ is the mean density, h_1 and h_2 are the thickness of the upper and the lower layers respectively, L is the longitudinal length of the lake, and g is the acceleration of gravity.

Since the longitudinal length and the mean depth of the North Basin are estimated roughly to be 50 km and 50 m respectively, one can calculate the uninodal period of internal seiches as the curve shown in Fig. 82 (C).

Buoy stations of thermistor-chain

The first temperature record of Lake Biwa was taken at the Funaki Station in October, 1968 by the writer, casting a thermistor-buoy station with double layer sensors.

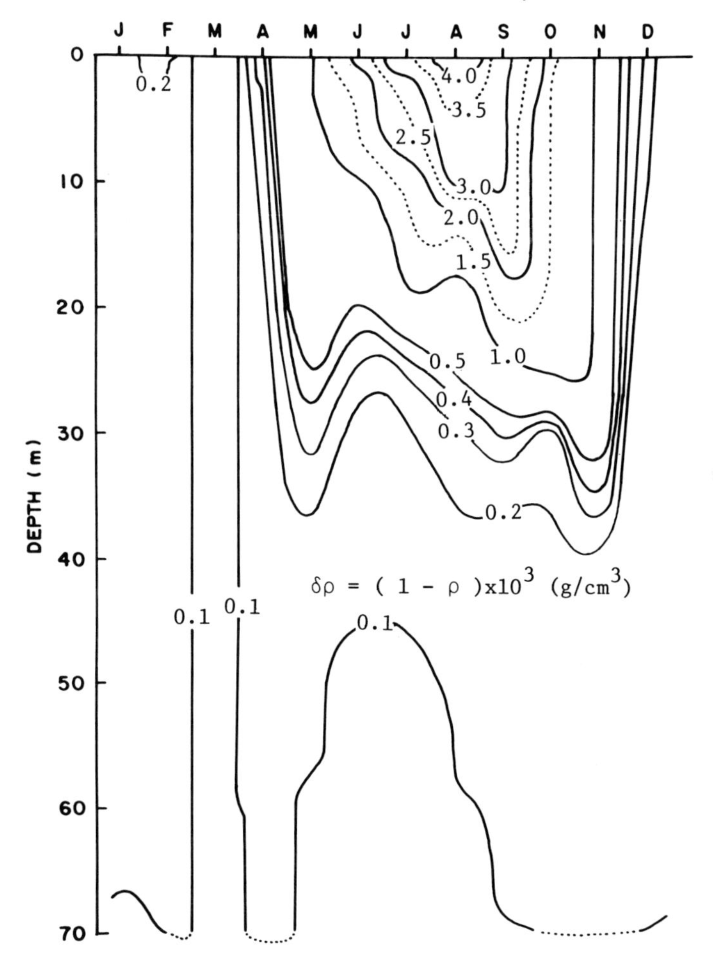

Fig. 81. Seasonal change of the vertical density-anomaly distribution in Lake Biwa. The numerals on equidensity-anomaly lines show the value of $\delta\varrho$ (after Kanari).

In 1969 to 1970, the writer established the other two stations both with the six layer sensors at Sugaura at the northern end and at Kido near the southern end of the main basin.

The structure of the buoy station is shown in Fig. 83.

The six thermistor sensors were arranged vertically so that the mean position of the interface of density discontinuity bears on the center of the thermistor array. The sub-surface buoy with 33 cm dia. has buoyancy of 10 kg, which keeps the thermistor sensors being held at their original levels even when the surface condition is wavy.

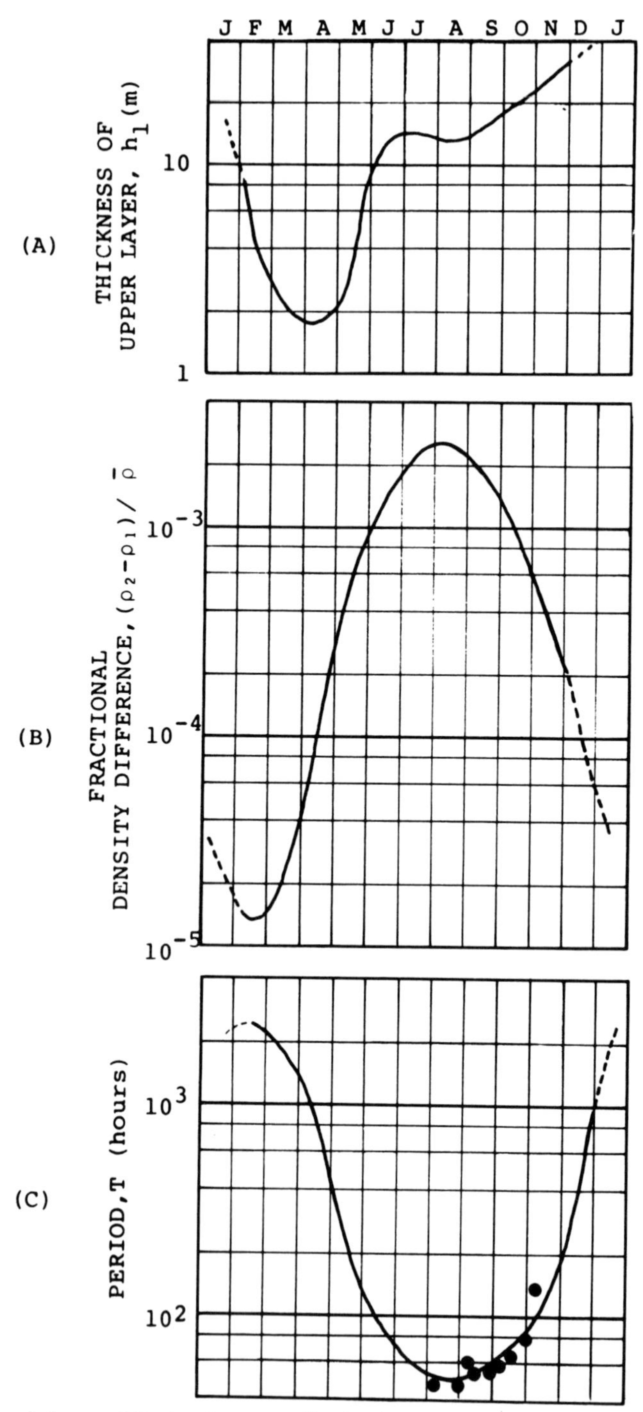

Fig. 82. Seasonal change of the two-layer stratification in Lake Biwa: (A) thickness of the upper layer, (B) fraction of the density difference between the upper and the lower layers, (C) periods of internal seiches calculated by formula (1) (thick line) and obtained from Fourier analysis as shown in Fig. 92 (black circles) (Kanari 1975b).

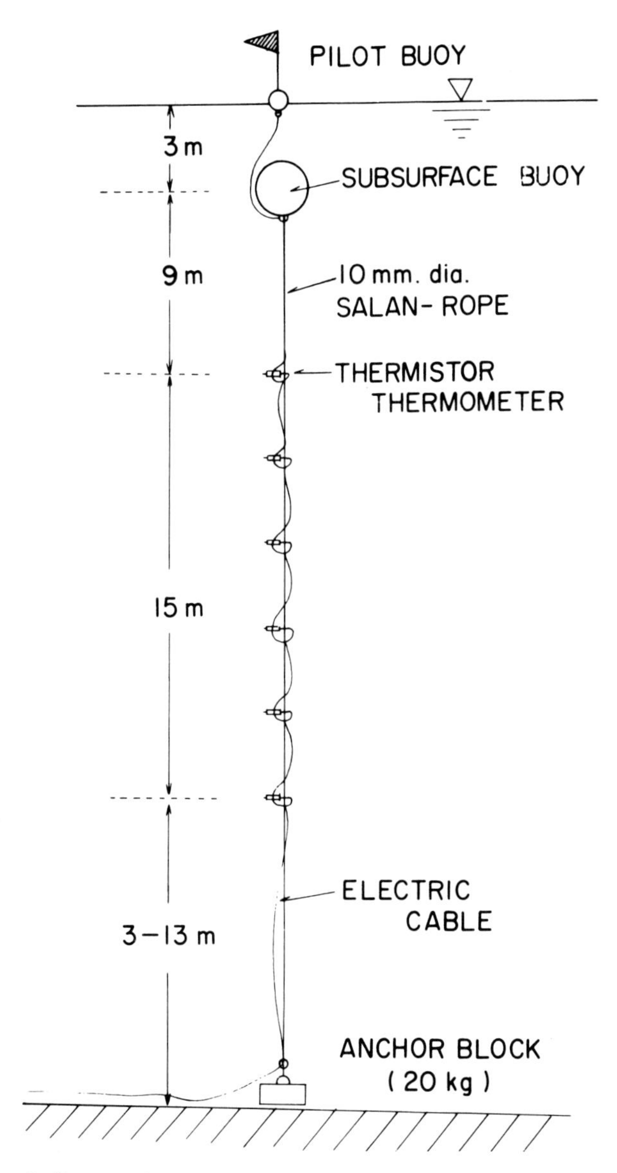

Fig. 83. Schematic diagram of the structure of a thermistor-buoy station (Kanari 1970a, 1974c).

Respective lines of thermistor sensors are connected with a 12-p electric cable which was settled in the lake bottom to the nearest shore enabling the shore station to take continuous records.

The measurable temperature range was designed to be 25°C (from 5°C to 30°C) with the accuracy of ±0.1°C.

The time constant of the thermistor sensors is about 5 seconds. However, the sampling interval of the recorder for the same channel was 30 seconds. Accordingly, the buoy system was able to record the temperature variation of the period longer than one minute without effective distortion.

Initial setup of internal seiches by wind

A large temperature oscillation as likely as a uninodal longitudinal internal seiche was recorded at the Sugaura Station in September 21 to 23, 1968.

The continuous temperature record of the 18 m layer was sampled at the interval of 30 minutes and then smoothed by a moving average of 8.5 h interval. The unduration of temperature starts at 0300 h in September 21 with the amplitude of about 5°C and the period of almost two and a half days. As shown in Fig. 84, the oscillation might be generated by a suddenly imposed wind. The upper panel in Fig. 84 shows the hourly N–S wind component averaged between Hikone and Funaki. From the figure, it is clear that the wind components from north to south blasted over 18 hours in September 21, removed the surface water in the main basin to the south. The southward transport of water above the thermocline would cause an ascent of the interface, which brings the thermistor sensor at the original interface into relatively colder water below the interface. This is the reason why the descending of water temperature corresponds to the strong north wind. The situation is schematically illustrated in Fig. 85.

At 1900 h in September 21, the wind calmed down and after that a virtually calm condition continued at least to 1200 h in September 23. However, the

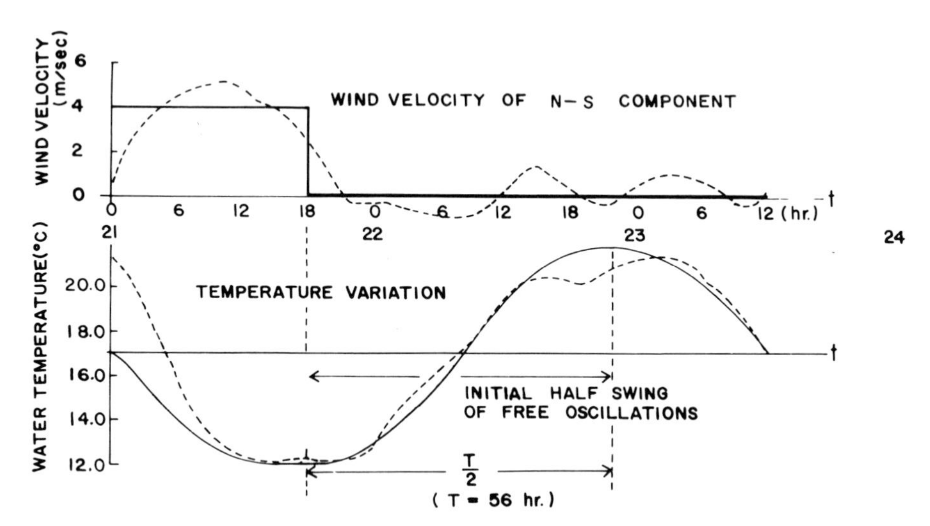

Fig. 84. The long-period temperature variation of the thermocline layer as compared with the N—S component of the wind velocity averaged between Hikone and Funaki. The broken lines show the plan of an idealized condition (Kanari 1970b).

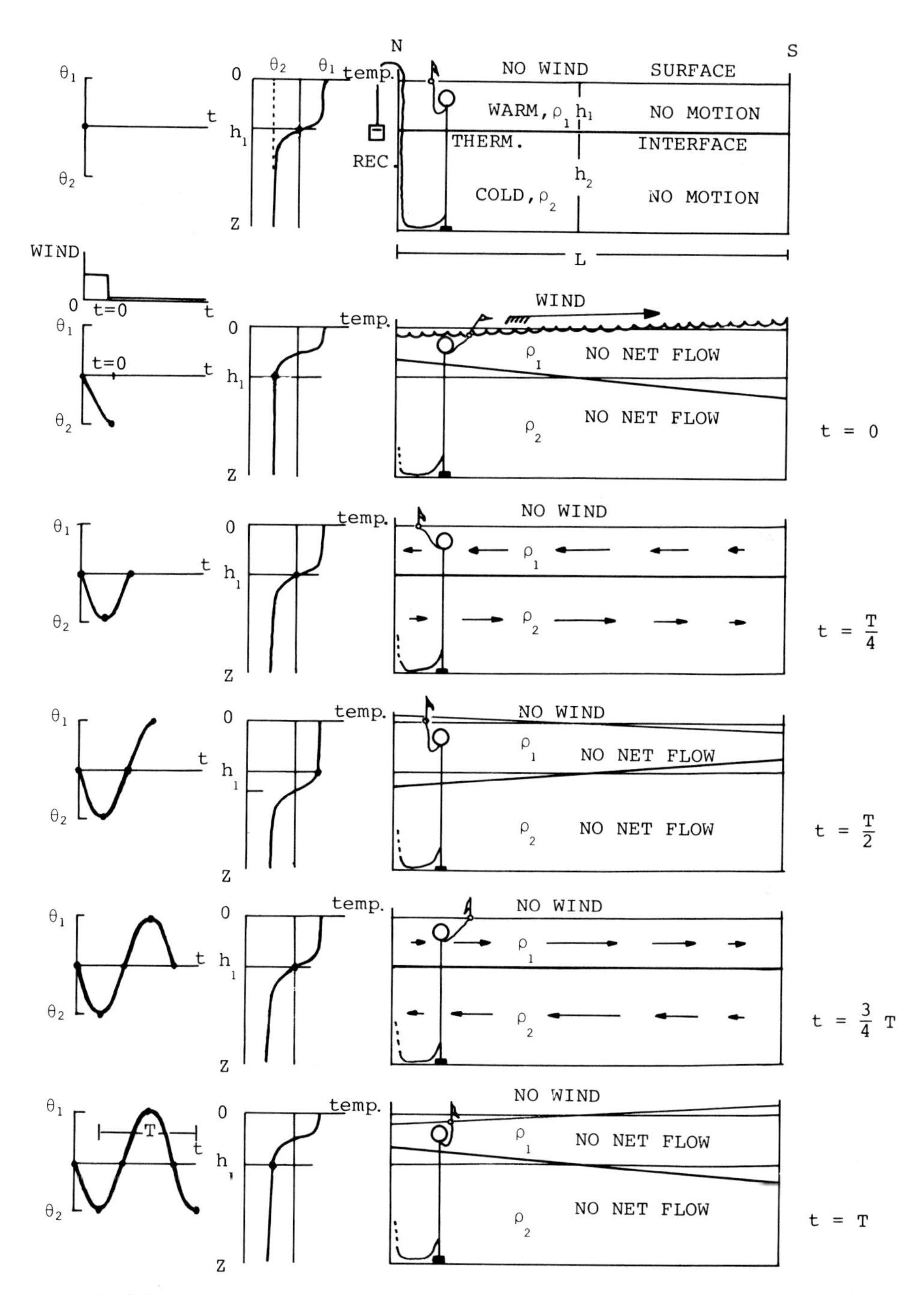

Fig. 85. Schematic representation of free oscillation of the interface generated by winds, and temperature change measured by a thermistor fixed at the level of the equilibrium interface.

temperature oscillation started at 1800 h, almost the same time as the cessation of the wind.

Assuming that the temperature variation is due to the free oscillation of the interface after the wind stress acted on the lake surface was removed, one can estimate the period of oscillation, because, as seen in Fig. 84, the time interval between the minimum temperature at 1800 hr in September 21 and the maximum temperature at 2130 h in September 22 (the broken line) corresponds to a half of the period of oscillation. Then, we can easily find the period of free oscillation to be 56 h.

The situation described above can be tested by Eq. [1]. The vertical density distribution observed off the Sugaura Station in September 19, 1968 is shown in Fig. 86. From the density distribution in Fig. 86, the thickness of the upper layer and the density difference between the upper and the lower layer can be estimated as $h_1 = 17.5$ m and $\Delta\bar{\rho} = 2.2 \times 10^{-3}$ (g/cm^3) respectively. Then, by taking $L = 50$ km and $h_2 = 32.5$ m (which was determined by subtraction of h_1 from the mean depth $\bar{H} \simeq 50$ m), the theoretical period of free oscillation given by Eq.[1] can be estimated to be 55.83 h, which is very close to twice the half swing period determined from the temperature record as shown in Fig. 84.

Experimental test for an estimate of vertical displacement of waters

At both ends of an elongate lake, the horizontal motion of waters is reduced and superseded by the vertical motion of waters. Consequently, if the vertical thermal eddy diffusion can be neglected in comparison with the vertical thermal advection, the temperature variation near the thermocline may approximately be expressed by the following equation.

$$\frac{\partial \theta}{\partial t} = -w \frac{\partial \theta}{\partial z} \qquad [2]$$

where w is the vertical velocity of waters and θ is the water temperature at a fixed depth z (downward positive). Let ζ_w be the vertical displacement of waters (upward positive). Then, the vertical velocity of waters, w, can be replaced by

$$w = -\frac{\partial \zeta_w}{\partial t}. \qquad [3]$$

From Eqs. [2] and [3], we can get the following equation:

$$\frac{\partial \zeta_w}{\partial t} = \frac{\left(\dfrac{\partial \theta}{\partial t}\right)}{\left(\dfrac{\partial \theta}{\partial z}\right)}. \qquad [4]$$

If the vertical displacement of waters is due to internal seiches and the vertical temperature gradient $\partial\theta/\partial z$ can be replaced by the time mean of $\partial\theta/\partial z$, then the

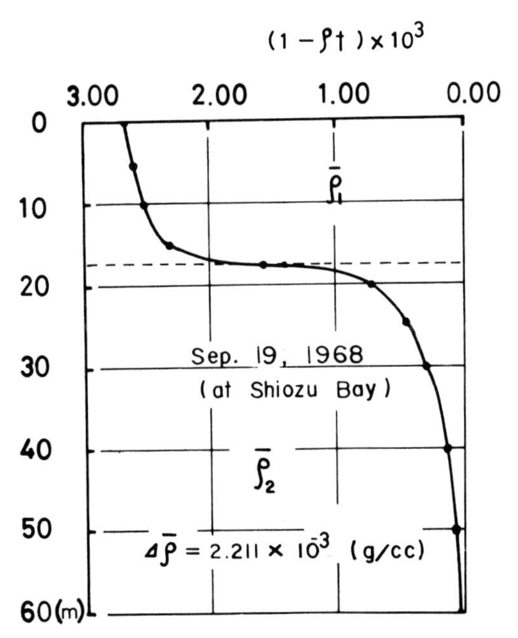

Fig. 86. Vertical density distribution in the main basin, observed in September 19, 1968 at the mouth of Shiozu Bay, 3 km east of the Sugaura Station (after Kanari 1970b).

Eq. [4] can be integrated as follows:

$$\zeta_w = \frac{1}{\left\langle \dfrac{\partial \theta}{\partial z} \right\rangle} [\theta(t) - \theta(t_o)] \qquad [5]$$

where $\langle \partial\theta/\partial z \rangle$ is the time mean of the vertical temperature gradient at the thermocline and $\theta(t_o)$ is the initial water temperature at an arbitrary time, $t = t_o$.

The validity of Eq. [5] was examined in October 2 to 3, 1969 at the bay of Sugaura, located in the north end of Lake Biwa, by making use of the instrumented neutrally-buoyant float (abbreviated as INBF below) which was designed by the writer. This float has a set of time-shared FM system, as shown in the lower half of Fig. 87, on the depth of the float and water temperature signals, which is able to eject supersonic FM signals toward the receiving system on the ship (the upper half of Fig. 87). Consequently, if we adjust the buoyance of the INBF so as to keep balance at the interface of density discontinuity, then the INBF may follow the vertical displacement of the interface and ejects momentarily information on the vertical position of the interface and water temperature just below the interface, as shown in Fig. 88.

In order to maintain the best receiving condition of the acoustic link, the horizontal motion of the INBF was restrained by a stainless steel wire of 1 mm dia., which was hung down from the survey ship, so that the INBF might be permissive only to vertical motions.

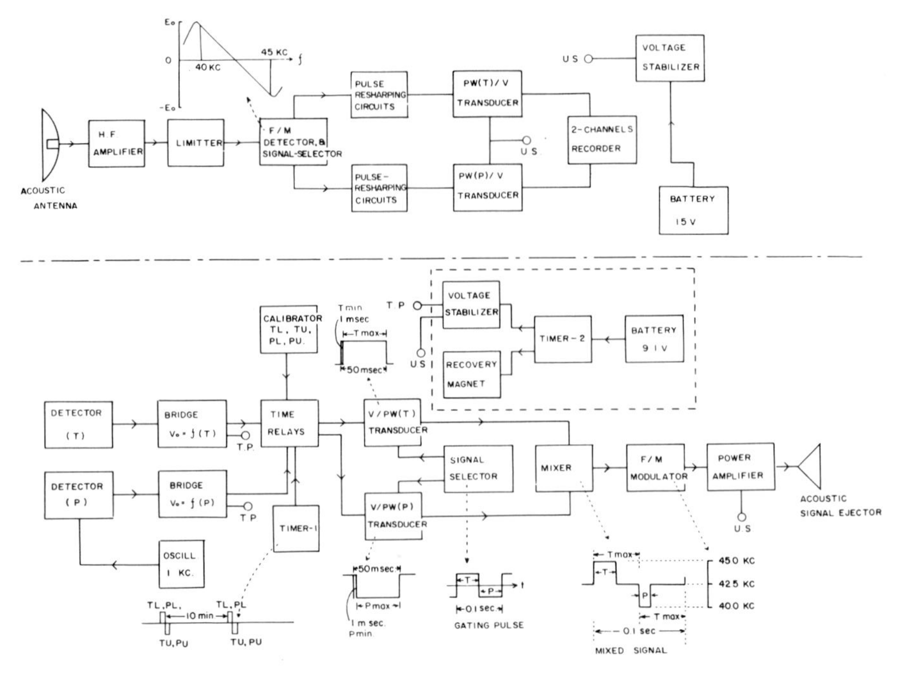

Fig. 87. The block diagram of the INBF system (lower) and the demodulation system of the acoustic FM signals (upper) (after Kanari).

The experimental test of the INBF was carried out simultaneously with the temperature recording by the six-layered thermistor chain at the Sugaura Buoy Station as shown in Fig. 83.

The vertical displacement of waters ζ_w was estimated from the temperature records of the thermistor chain by using the Eq. [5], which is shown in the middle of Fig. 89. On the other hand, the vertical displacement of the interface of density discontinuity was measured directly by demodulating underwater acoustic signals ejected from the INBF.

As shown in Fig. 89, the variation of the calculated vertical displacement of waters seems to have a good agreement with the trace of the INBF (ζ_F) which may represent the vertical displacement of the interface of density discontinuity, though there is a small discrepancy in detail. This result seems to assure us the validity of estimating the vertical displacement of waters due to internal seiches from temperature records by using Eq. [5].

Spectrum of water temperature records

A typical example of temperature records obtained by the six-layered thermistor-chain station is shown in Fig. 90, in which the long-period temperature variation with a large amplitude is conspicuous. Such multi-layered

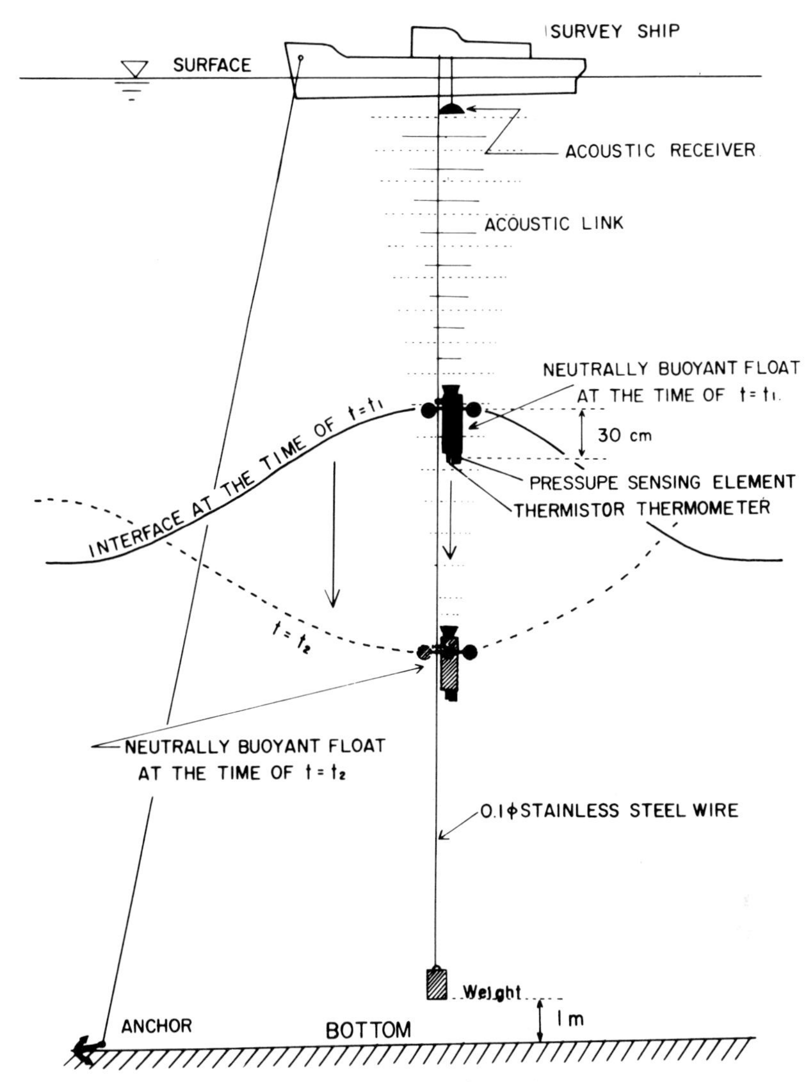

Fig. 88. Schematic representation of the plan of operation of the INBF (Kanari 1970a).

temperature records enable us to estimate the vertical displacement of waters by the method described in the preceding section.

An example of a pair of smoothed variation of the vertical water displacement obtained from simultaneous recording by the buoy stations at Kido and Sugaura is shown in Fig. 91.

As shown in Fig. 91, the large amplitude internal oscillation of a period of about 50 h shows almost anti-phase between the two stations.

The temperature records taken at the three stations, Sugaura, Kido and Funaki in Lake Biwa, during the past six years from 1968 to 1973 are listed in Table 25.

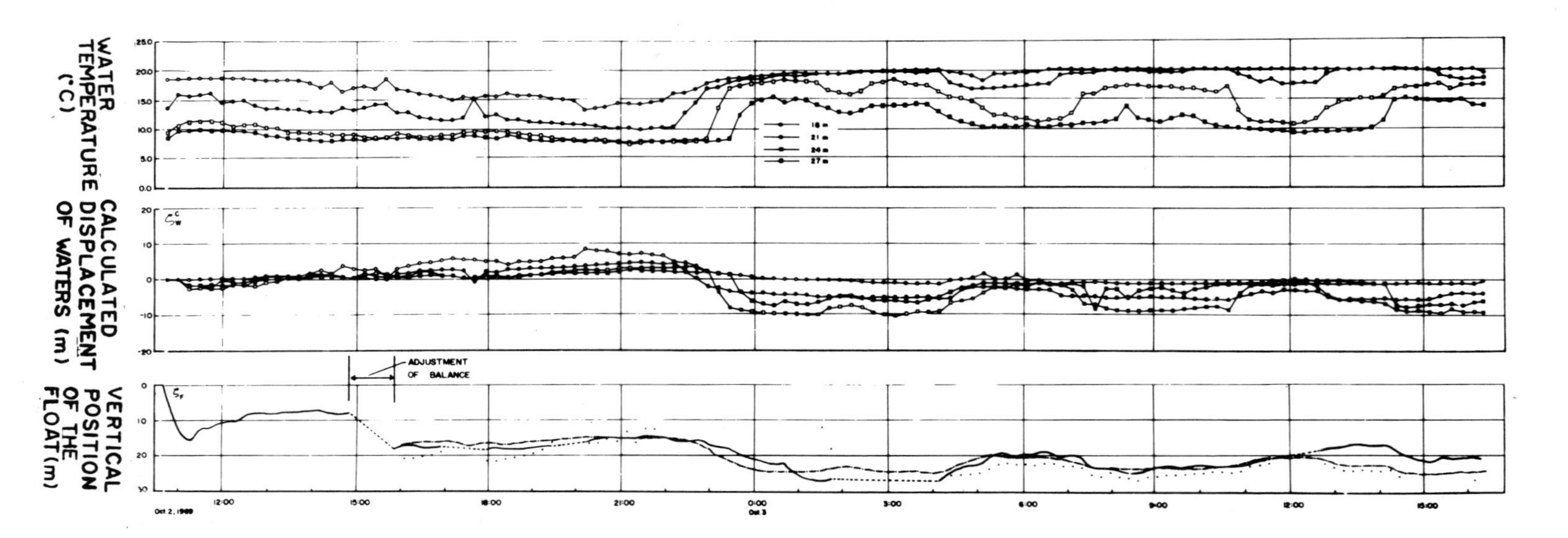

Fig. 89. Comparison between the calculated vertical displacement of waters (ζ_W) and the observed vertical displacement of the INBF (ζ_F) adjusted so as to be in neutral buoyancy at the interface. Upper: variation of water temperature at the depths of 18 m, 21 m and 27 m measured by the thermistor-chain at the bay of Sugaura. Middle: the vertical displacement of waters calculated from the variation of water temperature by using the Eq. [5]. Lower: trace of the vertical displacement of the INBF (full line), the vertical displacement of waters averaged over the four layers of 18 m, 21 m, 24 m and 27 m shown in the middle (dotted line), and estimated trace of the INBF, which was estimated from both of temperature records of the thermistor-chain and temperature changes measured by the INBF (dots) (Kanari 1970a).

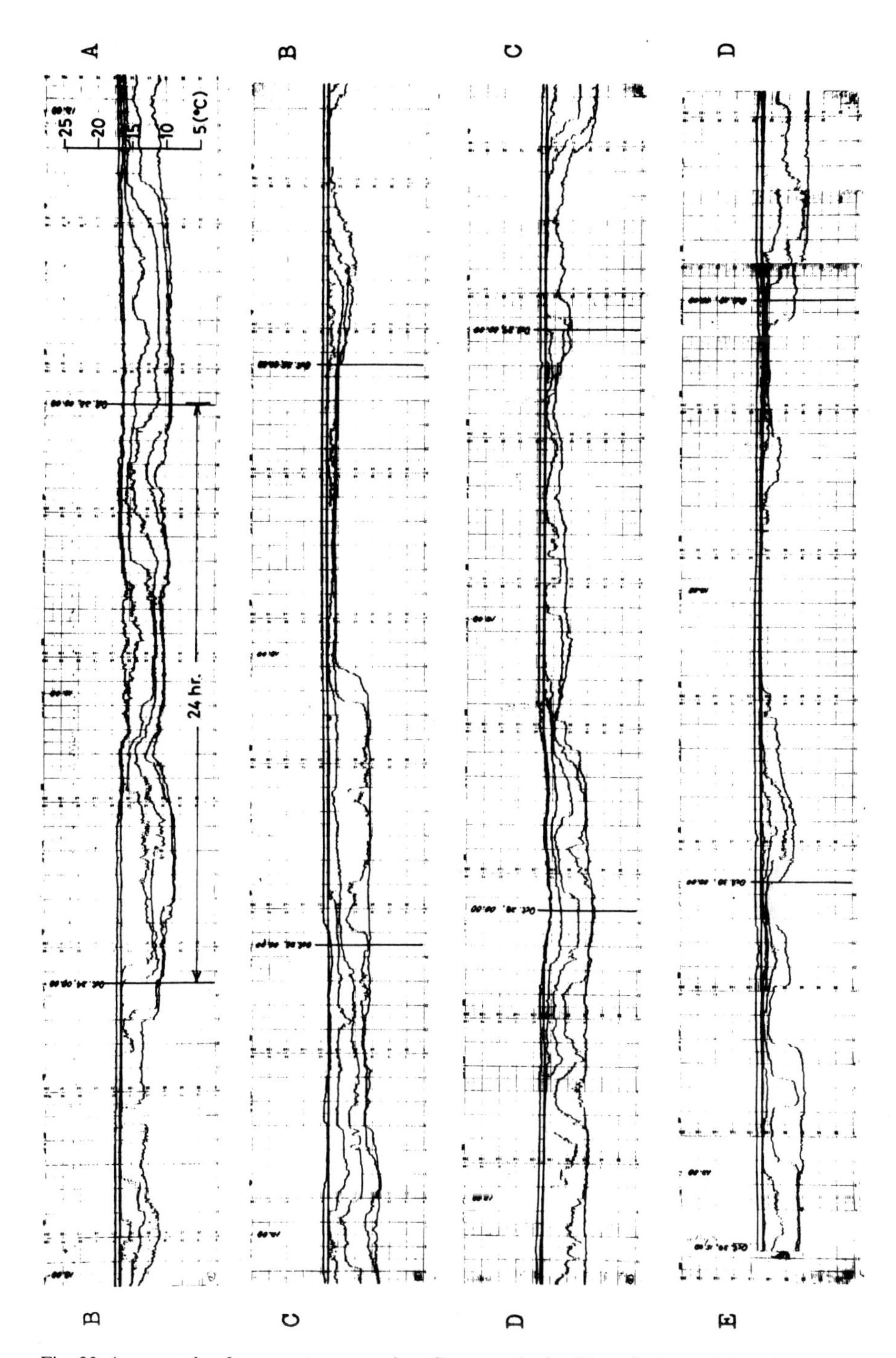

Fig. 90. An example of temperature records at Sugaura obtained by a six-layered thermistor-chain station. The left side of the top record B continues to the right of the second record B, and so on. The time of 00.00 h in each day is denoted by a vertical thick line (Kanari 1974b).

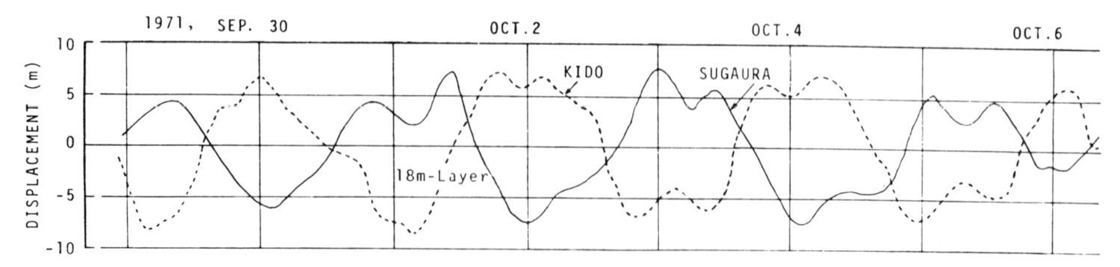

Fig. 91. Smoothed variations of vertical water displacement calculated from simultaneous temperature records at Kido and Sugaura (Kanari 1974b).

Table 25. The list of the observed records(Kanari 1974b).

Nos.	Year	Sugaura Station	Kido Station	Funaki Station
1	1968			Oct. 5–Oct. 31
2	1969	Sep. 9–Oct. 7		
3		Oct. 17–Oct. 30		
4	1970	Aug. 27–Sep. 12		
5		Oct. 23–Nov. 11		
6	1971		Sep. 11–Sep. 25	
7		Sep. 25–Oct. 9	Sep. 25–Oct. 9	
8			Oct. 9–Oct. 28	
9	1972	Aug. 5–Aug. 17	Aug. 5–Aug. 17	
10		Aug. 17–Aug. 30	Aug. 17–Aug. 28	
11		Oct. 13–Nov. 2		
12	1973		Aug. 20–Sep. 10	

The records of No. 2 to No. 12 in the table were analyzed by Fourier analysis with a sampling time interval of 30 minutes, and the results are shown in Figs. 92 (a) to (h).

The dominant peaks of the spectrum in Figs. 92 (a) to (h), which are indicated by arrows, seem to be the primary internal seiche in Lake Biwa, and their periods change seasonally depending on the change of density stratification of the lake.

The periods of the dominant peak, estimated from the above Fourier spectrum, are shown as black circles in Fig. 82 (c). The estimated periods seem to agree well with the theoretical periods calculated by Eq. [1] which is shown by the full line in Fig. 82 (c).

However, the results of the spectrum analysis suggest two inconsistencies. One is a dominant component of the spectrum at the Funaki Station as shown in Fig. 92 (h). If the internal seiche is a simple longitudinal one, there would be a nodal position and such a large amplitude oscillation could not be observed at the station. The other inconsistency is seen in phase differences $\delta\phi$ between the two spectra obtained simultaneously at the Sugaura and Kido Stations as shown in Figs. 92(f) and 92 (g). The phase differences at the period of dominant oscillation between those two stations show 159 degrees in the case of August 5 to 17, 1973 and 145 degrees in the case of September 25 to October 9, 1971 respectively. If the internal seiche in Lake Biwa is the simple longitudinal internal standing oscillation without the effect of the earth's rotation, then the phase difference must be 180 degrees. These problems included in the observed results will be explained well with the hypothesis that internal seiches in Lake Biwa are affected by the earth's rotation, as shall be described later.

Criterion of the existence of internal Kelvin waves

In 1967, Csanady discussed the water motion in the two-layered circular Great Lakes and provided the criterion for the existence of the Kelvin mode. According to Csanady, there is one Kelvin mode, provided that

$$\left(\frac{\gamma_e}{R}\right)^2 > K(K+1) \qquad [6]$$

holds for each azimuthal wave number K where γ_e denotes the equivalent radius of the circular lake, and R is Rossby's 'radius of deformation', which is given by the ratio of the wave velocity c to the Coriolis parameter $f=2\omega \sin \psi$, i.e.,

$$R=\frac{c}{f}. \qquad [7]$$

If we consider a circular lake with the size equivalent to Lake Biwa, then the equivalent radius may be estimated as $\gamma_e=14.4$ km, the radius of deformation for the internal standing oscillation with the period of 50 hr can also be estimated to be $R=6.5$ km. Consequently, the criterion for the Kelvin mode becomes

$$\left(\frac{\gamma_e}{R}\right)^2 = 4.9 > K(K+1). \qquad [8]$$

The condition [8] can be satisfied only for $K=1$, which suggests that there exists one Kelvin mode wave in Lake Biwa.

In any way, since the wave velocity of internal seiches may vary with the condition of stratification of the lake, the radius of deformation may also be affected by stratification.

The internal wave velocity in Lake Biwa may vary from 70 cm/sec in early summer to 40 cm/sec in late autumn. The condition [8] can be written

$$\left(\frac{\gamma_e}{R}\right)^2 = 2.8 \sim 8.5 > K(K+1), \qquad [9]$$

which means that there is a possibility of the existence of higher modes internal Kelvin waves according to seasons. However, their pattern of oscillation must be quite different from those in a circular basin because Lake Biwa has a rather elongated shape than a circular one.

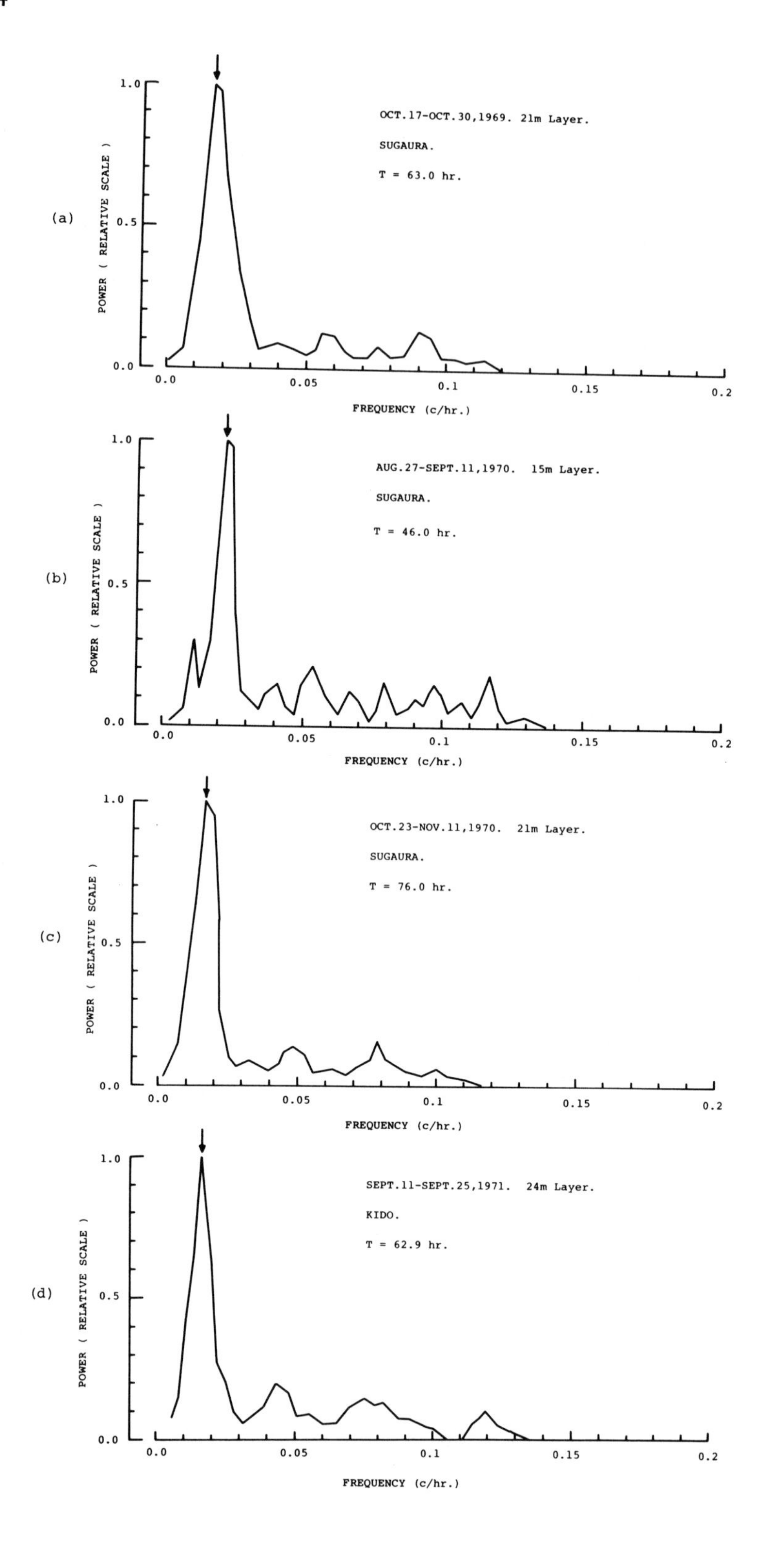
(a)
1.0
0.5
0.0
POWER (RELATIVE SCALE)
OCT.17-OCT.30,1969. 21m Layer.
SUGAURA.
T = 63.0 hr.
0.0
0.05
0.1
0.15
0.2
FREQUENCY (c/hr.)
(b)
1.0
0.5
0.0
POWER (RELATIVE SCALE)
AUG.27-SEPT.11,1970. 15m Layer.
SUGAURA.
T = 46.0 hr.
0.0
0.05
0.1
0.15
0.2
FREQUENCY (c/hr.)
(c)
1.0
0.5
0.0
POWER (RELATIVE SCALE)
OCT.23-NOV.11,1970. 21m Layer.
SUGAURA.
T = 76.0 hr.
0.0
0.05
0.1
0.15
0.2
FREQUENCY (c/hr.)
(d)
1.0
0.5
0.0
POWER (RELATIVE SCALE)
SEPT.11-SEPT.25,1971. 24m Layer.
KIDO.
T = 62.9 hr.
0.0
0.05
0.1
0.15
0.2
FREQUENCY (c/hr.)

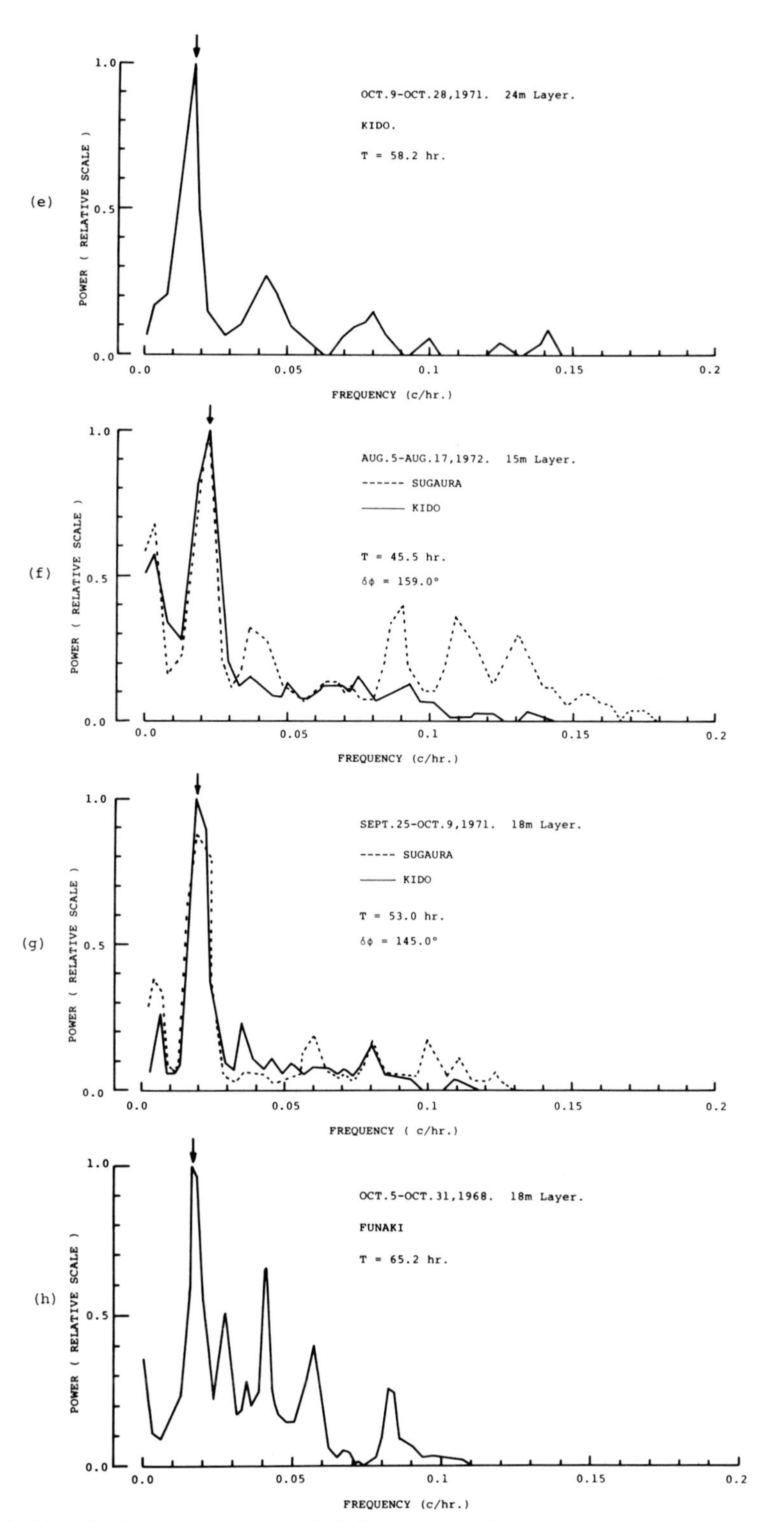

Fig. 92. (a) to (h). Power spectra of vertical displacement of waters (Kanari 1974b).

Approximate solutions in a two-layered rectangular lake

To examine the pattern of oscillation of internal seiches under the influence of the earth's rotation, let us consider a simple two-layered rectangular lake with a size equivalent to Lake Biwa, as shown in Fig. 93.

Let h_1 and h_2 be the equilibrium thickness of the upper layer and the lower layer with the constant density of ρ_1 and ρ_2 respectively, U_1, V_1, U_2, and V_2 be the volume transport vector components referring to the upper and the lower layers, and ζ_1 and ζ_2 be the elevations of the surface and the interface from the equilibrium positions respectively.

By assuming that motions in the respective layers can be expressed by the superposition of barotropic and baroclinic motions, the equations for the volume transport rate vectors in the barotropic and the baroclinic modes may be given by

$$\frac{\partial \mathbf{V}^*_m}{\partial t} + 2\Omega \times \mathbf{V}^*_m = -C^2_m \nabla_h \cdot \zeta^*_m, \qquad [10]$$

$$\frac{\partial \zeta^*_m}{\partial t} + \nabla_h \cdot \mathbf{V}^*_m = O \qquad [11]$$

where $\boldsymbol{\Omega}$ is the angular velocity vector of rotation of the earth, and $\mathbf{V}^*_m$, ζ^*_m, and C_m are the volume transport rate vector, the equivalent elevation and the equivalent wave velocity in the barotropic (m=1) and the baroclinic (m=2)

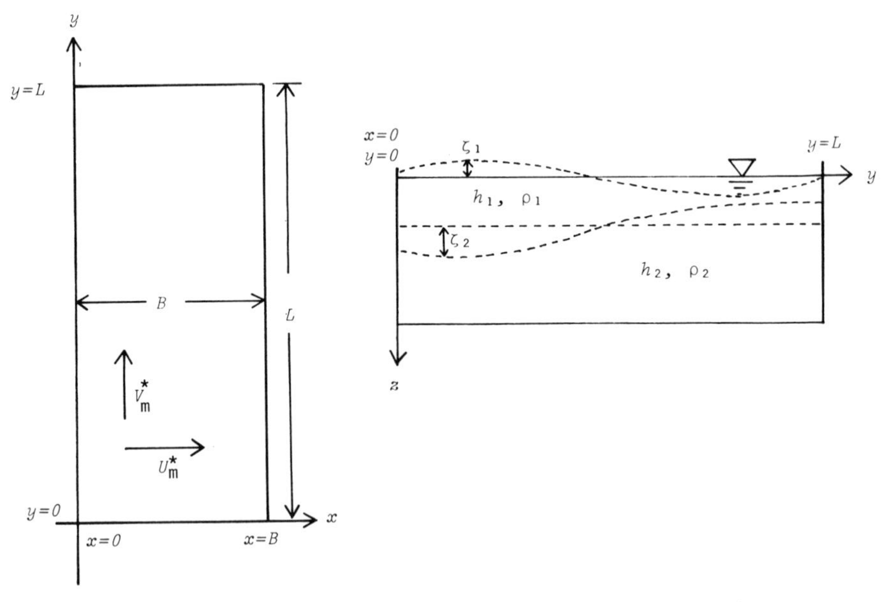

Fig. 93. Schematics of the two-layered rectangular Lake Biwa (Kanari et al. 1974).

modes respectively, $\nabla_h \equiv i\frac{\partial}{\partial x} + j\frac{\partial}{\partial y}$ and i, j are unit vectors in the x and y directions respectively.

Since $\mathbf{V}_m^* \equiv U_m^* \cdot i + V_m^* \cdot j$ and $2\mathbf{\Omega} \equiv 2\omega \sin\psi\mathbf{k}$, the Eqs. [10] and [11] can be written as follows

$$\frac{\partial U_m^*}{\partial t} - fV_m^* = -C_m^2 \frac{\partial \zeta_m^*}{\partial x}, \qquad [12]$$

$$\frac{\partial V_m^*}{\partial t} + fU_m^* = -C_m^2 \frac{\partial \zeta_m^*}{\partial y}, \qquad [13]$$

$$\frac{\partial \zeta_m^*}{\partial t} + \frac{\partial U_m^*}{\partial x} + \frac{\partial V_m^*}{\partial y} = 0 \qquad [14]$$

where $f \equiv 2\omega \sin\psi$ is the Coriolis parameter.

The quantities referring to the upper and the lower layers can be related with those in the barotropic and the baroclinic modes, that is,

$$U_1 \cong \frac{h_1}{H}(U_1^* - U_2^*),\ V_1 \cong \frac{h_1}{H}(V_1^* - V_2^*), \qquad [15]$$

$$U_2 \cong \frac{1}{H}(h_1 U_2^* + h_2 U_1^*),\ V_2 \cong \frac{1}{H}(h_1 V_2^* + h_2 V_1^*), \qquad [16]$$

$$\zeta_1 \cong \zeta_1^* - \left(\frac{C_2}{C_1}\right)\zeta_2^*,\ \zeta_2 \cong \frac{1}{H}(h_1 \zeta_2^* + h_2 \zeta_1^*), \qquad [17]$$

$$C_1^2 \cong gH,\ C_2^2 \cong \frac{(\rho_2 - \rho_1)}{\rho_2} g \cdot \frac{h_1 h_2}{H} \qquad [18]$$

where $H \equiv h_1 + h_2$.

For simplicity, let us assume that the equivalent elevation ζ_m^* can be given by the sum of geostrophic elevation ζ_{gm}^* and non-geostrophic elevation η_m^* and that the motion is in a geostrophic balance in the x direction, and that the non-geostrophic balance excels in the y direction, so that the Eqs. [12], [13] and [14] become

$$-fV_m^* \cong -C_m^2 \frac{\partial \zeta_{gm}^*}{\partial x}, \qquad [19]$$

$$\frac{\partial V_m^*}{\partial t} \cong -C_m^2 \frac{\partial \eta_m^*}{\partial y}, \qquad [20]$$

$$\frac{\partial \zeta_m^*}{\partial t} = \frac{\partial \eta_m^*}{\partial t} + \frac{\partial \zeta_{gm}^*}{\partial t} = -\left(\frac{\partial U_m^*}{\partial x} + \frac{\partial V_m^*}{\partial y}\right). \qquad [21]$$

Under the boundary conditions

$$U_m^* = 0 \text{ at } x = 0 \text{ and } x = B,$$

$$V_m^* = 0 \text{ at } y = 0 \text{ and } y = L,$$

$$\zeta_m^* \Big|_{y=0} = \eta_{0m} \text{ at } t = 0,$$

we can find a set of simple solutions:

$$U_m^* = -\frac{n\pi f}{2L}\eta_{0m} x(x-B) \sin\frac{n\pi y}{L} \cos\frac{2n\pi t}{T}, \qquad [22]$$

$$V_m^* = \frac{C_m}{2}\eta_{0m} \sin\frac{n\pi y}{L} \sin\frac{2n\pi t}{T}, \qquad [23]$$

$$\zeta_m^* = \frac{f}{C_m}\eta_{0m}\left(x - \frac{B}{2}\right) \sin\frac{n\pi y}{L} \sin\frac{2n\pi t}{T}$$

$$+\eta_{0m} \cos\frac{n\pi y}{L} \cos\frac{2n\pi t}{T} \qquad [24]$$

where n is the nodal number for the longitudinal oscillation and takes positive integers.

To transform the above equivalent quantities into the usual quantities related to the upper and the lower layers, the set of linear relationships from [15] to [17] should be used. However, the elevation in the barotropic mode ζ_1^* is usually very small as compared with the elevation in the baroclinic mode ζ_2^*. Accordingly, the following approximation is available:

$$\zeta_1 \cong \zeta_1^* \text{ and } \zeta_2 \cong \frac{h_1}{H}\zeta_2^*.$$

Taking the value of the parameter values in Lake Biwa,

$$L = 50 \text{ km},$$

$$B = 14 \text{ km},$$

$$f = 8 \times 10^{-5} \text{ 1/sec},$$

$$h_1 = 17.5 \text{ m},$$

$$H = 50.0 \text{ m},$$

$$\Delta\rho \cong \rho_2 - \rho_1 = 2.5 \times 10^{-3} \text{ g/cm}^3,$$

we obtain values of the wave velocities in the respective modes as $C_1 = 2{,}214$ cm/sec and $C_2 = 53$ cm/sec.

In the case of the barotropic elevation ζ_1^* in [24], the first term is negligibly small as compared with the second term, which means that the surface elevation may be almost unaffected by the earth's rotation. Accordingly, the surface elevation becomes usual longitudinal surface seiches given by

$$\zeta_1 \cong \zeta_1^* \cong \eta_{01} \cos \frac{n\pi y}{L} \cos \frac{2n\pi t}{T}. \qquad [25]$$

Such a surface oscillation has been treated in detail by Imasato et al. (1971).

Now, our major interest is in the behavior of the interfacial oscillation. The approximate solution of the internal elevation can be given by

$$\zeta_2 \cong \frac{h_1}{H}\zeta_2^* = \frac{h_1}{H}\eta_{02}\left\{\frac{f}{C_2}\left(x-\frac{B}{2}\right)\sin\frac{n\pi y}{L}\sin\frac{2n\pi}{T}t + \cos\frac{n\pi y}{L}\cos\frac{2n\pi}{T}t\right\}. \qquad [26]$$

Since $T=\dfrac{2L}{nC_2}$, the above solution can be rewritten in the following form.

$$\zeta_2(x,y,t) = \frac{h_1}{H}\eta_{02}\sqrt{\left[\frac{f}{C_2}\left(x-\frac{B}{2}\right)\sin\frac{n\pi y}{L}\right]^2+\cos^2\frac{n\pi y}{L}} \times \sin\left[X(x,y)-\frac{n^2\pi C_2}{L}t+\frac{\pi}{2}\right] \qquad [27]$$

where

$$X(x,y) \equiv \tan^{-1}\left[\frac{\frac{f}{C_2}\left(x-\frac{B}{2}\right)\sin\frac{n\pi y}{L}}{\cos\frac{n\pi y}{L}}\right]. \qquad [28]$$

The spatial dependency of the phase angle along the lake shore is shown in Fig. 94, in which the phase angles $X(0, y)$ and $X(B, y)$ are both shifted by 180° from the angle given by [28]. The phase distribution is quite different from those in the case of no rotation which is indicated by a dotted broken line in the same figure. Fig. 95 shows variation of the interfacial topography $\dfrac{\zeta_2(x,y,t)}{\eta_{02}}$ during a cycle of oscillation with the period $T=\dfrac{2L}{C_2}$ (for $n=1$). From the figure we can recognize a rotatory behavior of the interfacial elevation as expected in previous section.

The change of the topographic patterns can be transformed into the co-tidal and the co-range lines in a way similar to those used in tidal theory, as shown in Fig. 96.

As shown in Fig. 96, the internal oscillation for $n=1$ has one amphydromic point at the center of the lake.

For $n=2$, the two amphydromic points rise at $\left(\dfrac{B}{2},\dfrac{L}{4}\right)$ and $\left(\dfrac{B}{2},\dfrac{3}{4}L\right)$ respectively.

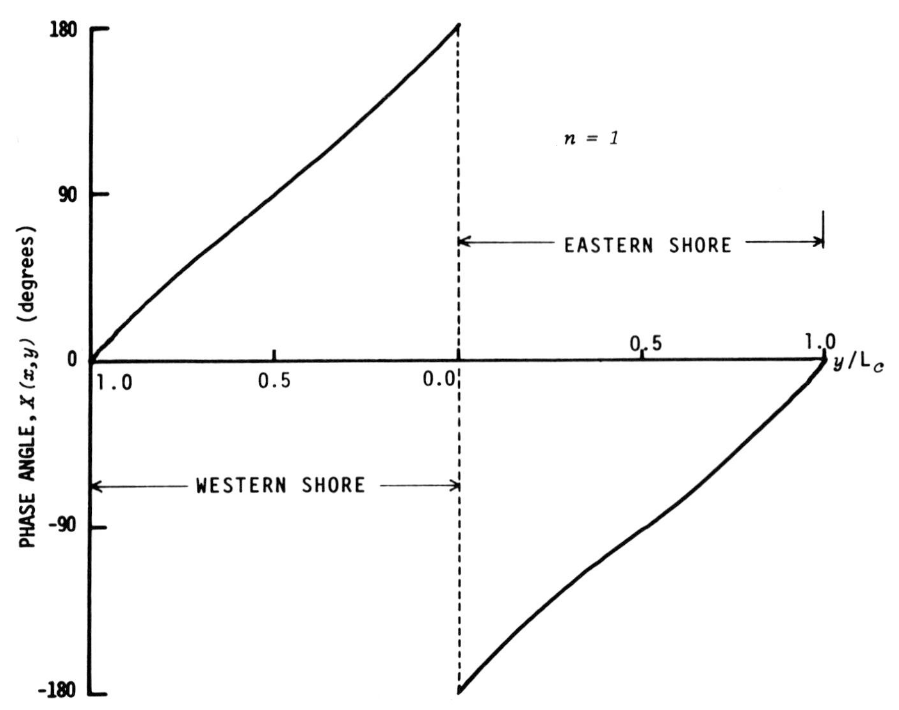

Fig. 94. Distributions of the phase angle along the western shore ($x = 0$) and the eastern shore ($x = B$) for $n = 1$. The phase angle is shifted by 180 degrees from those given by (28). $y/L_c = 0$ corresponds to the south end of the lake (after Kanari et al. 1974).

Confirmation of rotatory internal Kelvin waves by the round-cruising survey

For the purpose of confirming the existence of rotatory internal Kelvin waves suggested by the analytical solution for a rectangular lake, a series of daily Round-Cruising surveys (hereafter referred to as the RC survey) was carried out in August 9 to 14, 1973 by using an Electric Bathy-thermograph system (referred to as EBT).

Instrumentation of the EBT

The EBT system consists of a deck unit composed of a two-channel recorder and a pair of bridge circuits with a power supplier, and a submerged unit connected with the deck unit through a sealed cable of 100-meter long, including five lead wires. A thermistor and a pair of semi-conductor pressure sensors (one of which works as a thermal compensator) are mounted on the head of a submerging probe 27 cm long and 15 cm in diameter. The sensors constitute one arm of the respective electric bridge circuits in the deck unit as shown in Fig. 97, so that the

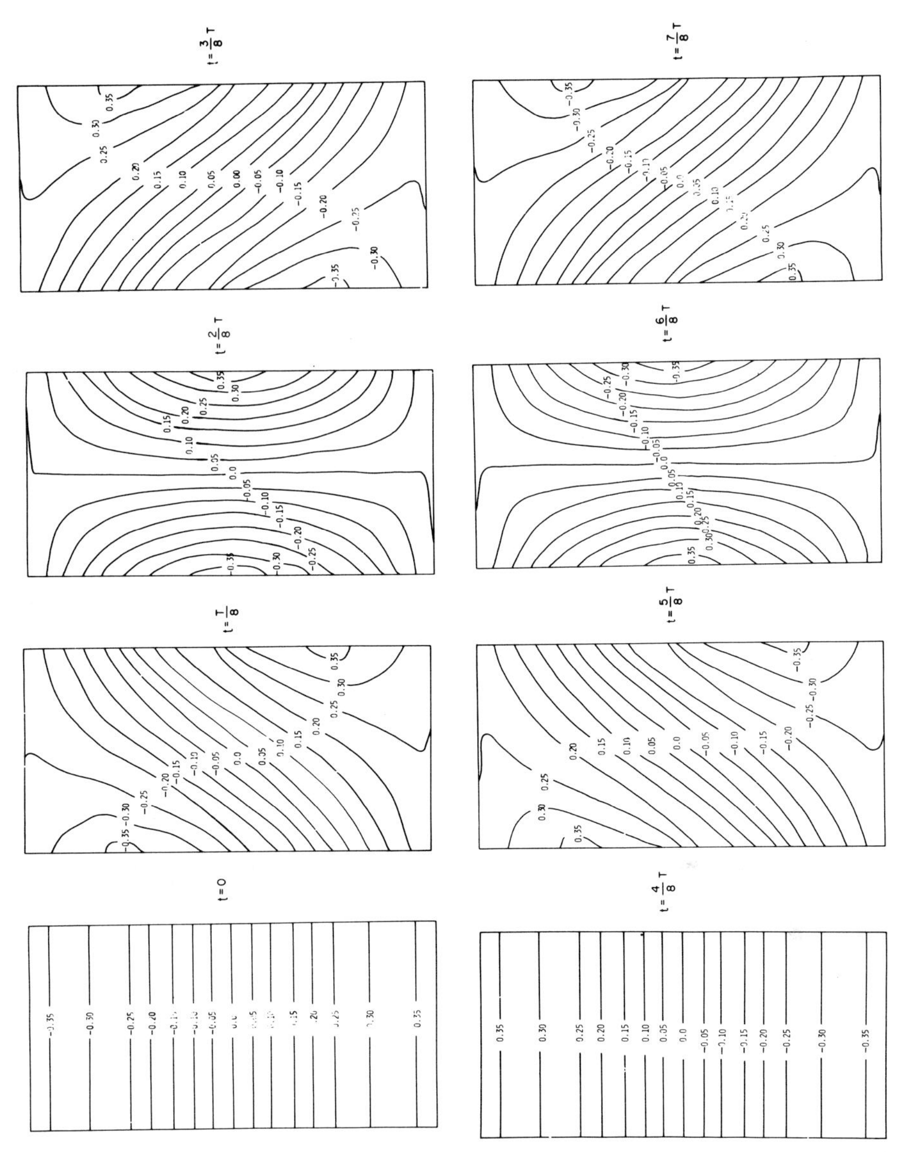

Fig. 95. Topographic changes of the interface during a cycle of internal oscillation with the period of T = 62.5 hr. The numerals show the relative elevation $\zeta_2\ (x,y,t)/\eta_{02}$.

bridge circuits would be balanced electrically at the given state of 15°C for the thermistor and of 1 atm (absolute pressure) for the pressure sensor.

The unbalanced voltages in the bridge circuits are approximately proportional to changes of temperature and of water pressure.

They are fed by the two-channel recorder, which takes the record as a pair of curves of depth and water temperature simultaneously with the lowering down

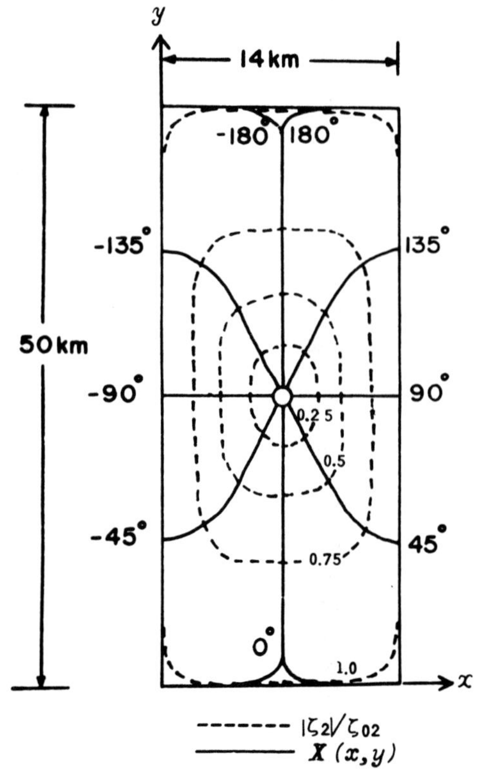

Fig. 96. Co-tidal lines and co-range lines of the interface due to the internal oscillation with the period of 62.5 hr (n = 1) in the rectangular Lake Biwa. The point of the maximum elevation transmits in a counterclockwise direction along the lake shore with the phase speed of 53 cm/sec. (after Kanari)

of the EBT-probe as shown in Fig. 98. Accordingly, the EBT-system enables us to record the vertical temperature profiles over the depth of 100 meters on the recording chart within several minutes.

RC survey with the EBT system

If rotatory internal Kelvin waves as described in the preceding section really exist in Lake Biwa, we may be able to catch various rotatory phases of oscillation through a series of successive surveys of the interface position along a closed survey line arranged along the contour line of 40 m depth which is shown in Fig. 79 by numbered open circles.

The daily RC survey was carried out by a high-speed survey boat equipped with the EBT-system. It took about four hours to complete one round-cruising, which was always made in the counterclockwise direction starting at Station No. 1 (Sugaura). Accordingly, the time intervals of successive cruising were about twenty hours.

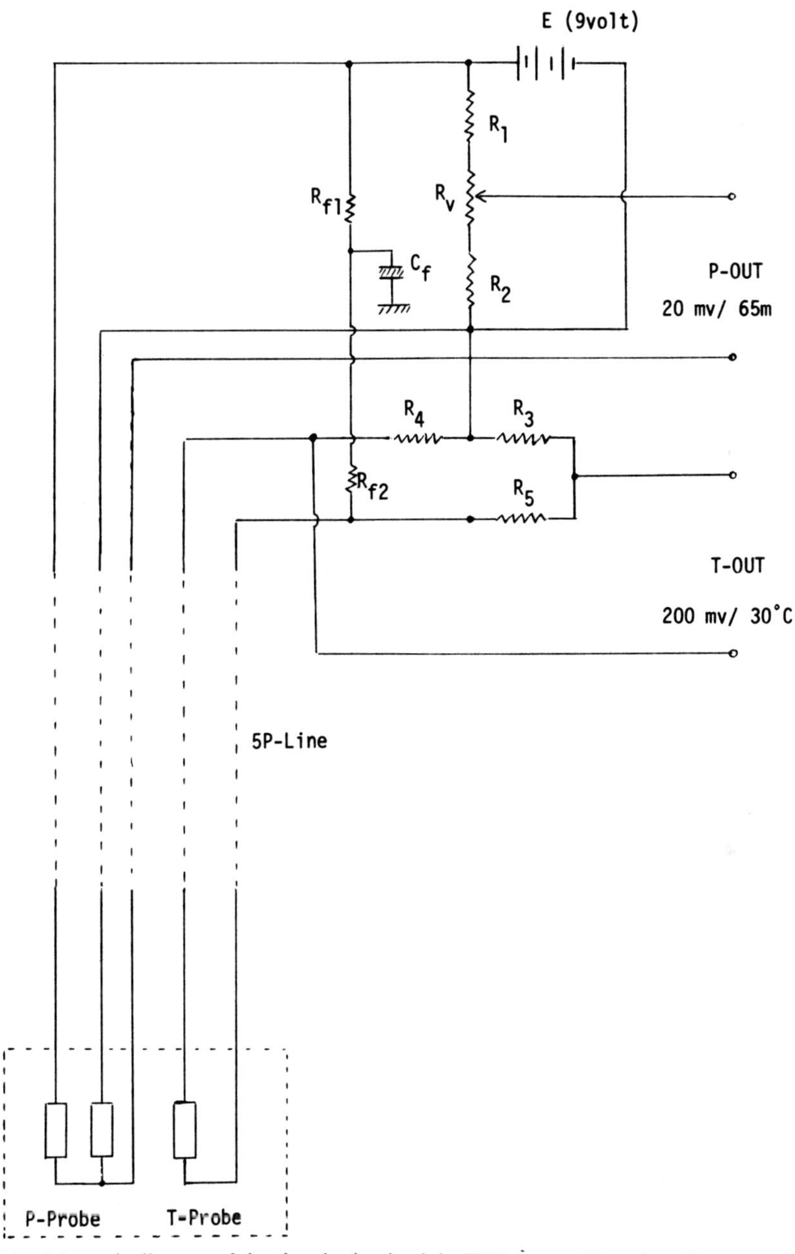

Fig. 97. Schematic diagram of the electric circuit of the EBT-system (Kanari 1974b).

The temperature profiles at seventeen stations over five days were obtained from reading EBT records, and the water temperature at the mean thermocline depth at respective stations was determined.

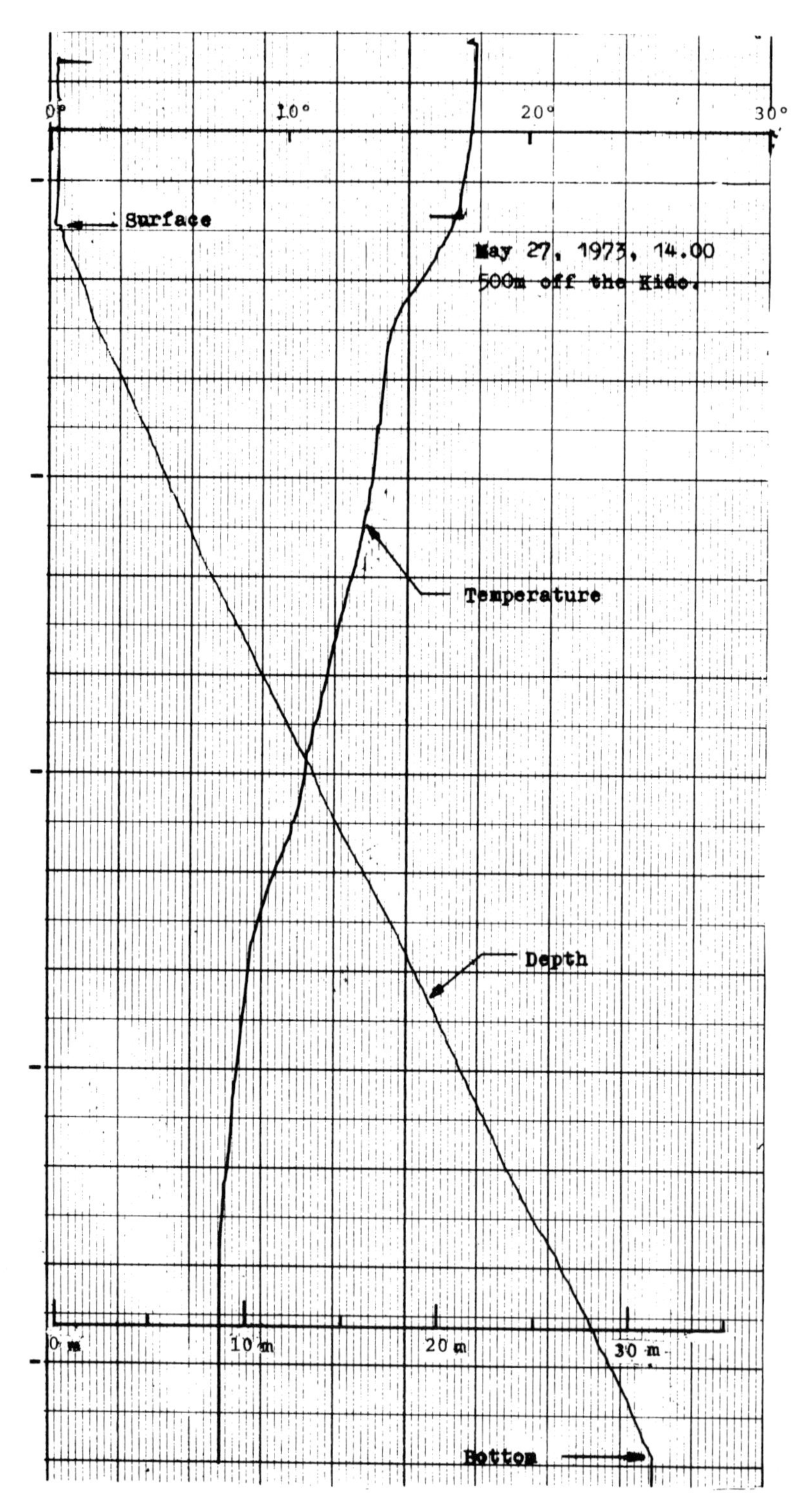

Fig. 98. An example of records of the EBT-system (Kanari 1974b).

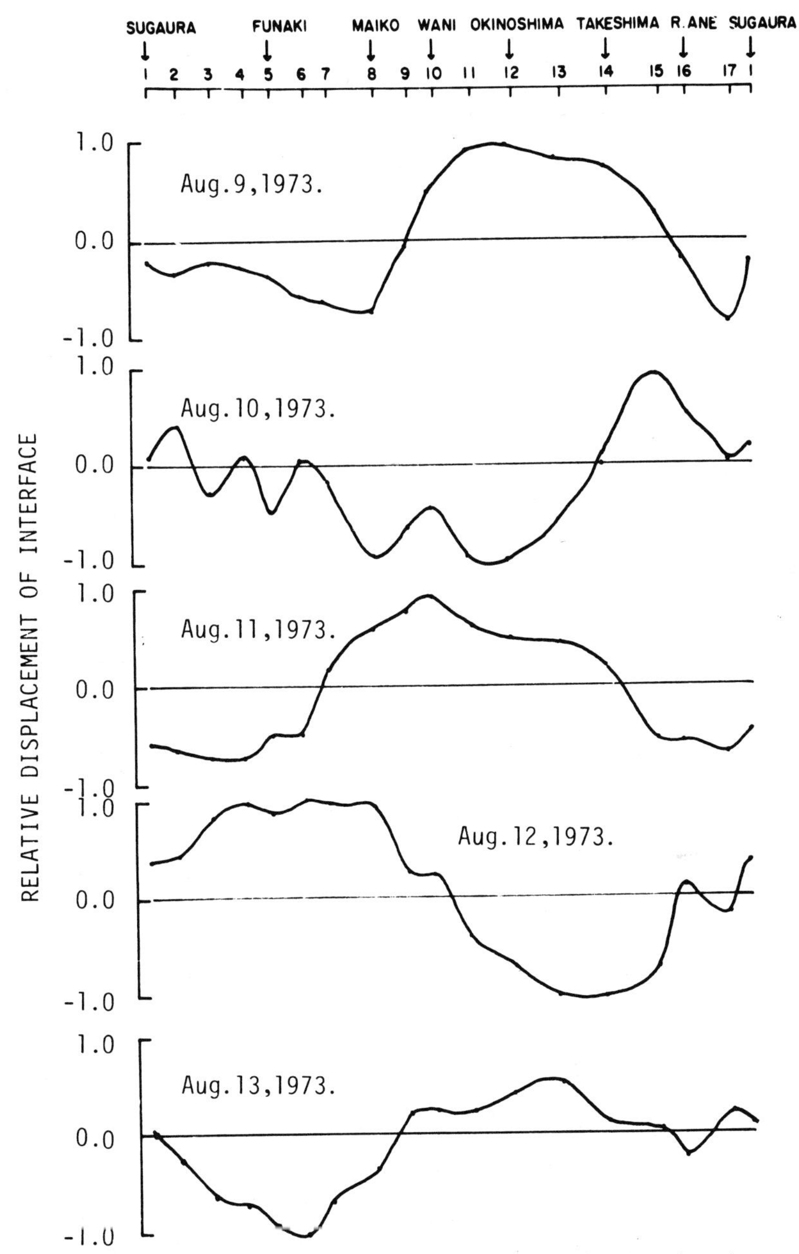

Fig. 99. Daily spatial distributions of the relative displacement of interface along the survey line (after Kanari 1974b).

The smoothed profiles of the depth along the survey line are shown in Fig. 99, though the amplitudes of temperature changes are small because of the succession of calm weather during this survey.

Spatial Fourier expansion of temperature profiles

The daily distributions of interface displacement as shown in Fig. 99 were expanded in Fourier series by means of the 12-point harmonic analysis method, in which the twelve values at equi-interval points were estimated by a linear interpolation. The obtained coefficients of sine and cosine terms are shown in Table 26.

By taking the first terms in both sine and cosine series and combining them into one, a sine term, including a phase angle against Station No. 1 for the respective distributions, it is as follows:

$1.21 \sin(2\pi s/L_c - 0.828\pi)$; August 9,

$1.26 \sin(2\pi s/L_c + 0.611\pi)$; August 10,

$1.46 \sin(2\pi s/L_c - 0.433\pi)$; August 11,

$2.32 \sin(2\pi s/L_c + 0.083\pi)$; August 12,

$0.92 \sin(2\pi s/L_c - 0.833\pi)$; August 13,

where L_c is the total length of the survey line and s is the distance from Station No. 1 measured in the counterclockwise direction along the survey line. The above sine functions are illustrated in Fig. 100.

The maximum (or minimum) amplitude point of the above sine curves should always occur at the same position, if the internal seiche be a longitudinal standing oscillation without the effect of the earth's rotation. However, as shown in Fig. 100, the wave form expanded along the survey line shows a certain shift of the maximum (also the minimum) amplitude point, and this fact clearly indicates that the wave is progressive along the lake shore counterclockwise.

Suppose that the obtained daily temperature profile, based on the largest amplitude term, is the reflection of the longest period internal progressive wave. Then, we can write the wave form of a unit amplitude along the shore line as

Table 26. Coefficients of harmonics (Kanari 1974b).

Date		Aug. 1973 9	10	11	12	13
	A_0	0.090	−0.040	0.010	0.160	−0.030
	A_1	−0.630	1.180	−1.430	0.610	−0.340
	A_2	0.140	−0.260	0.210	0.034	0.214
Cos	A_3	−0.070	0.010	0.026	−0.066	0.000
	A_4	−0.006	−0.060	0.026	0.020	−0.026
	A_5	0.041	0.048	−0.042	−0.018	0.048
	A_6	0.000	0.000	−0.010	−0.010	0.006
	B_1	−1.030	−0.440	0.300	2.240	−0.840
	B_2	0.265	−0.173	0.092	−0.196	0.162
Sin	B_3	−0.010	0.066	0.026	−0.094	−0.010
	B_4	0.034	0.011	−0.046	0.034	0.023
	B_5	−0.019	0.024	−0.028	−0.017	−0.011

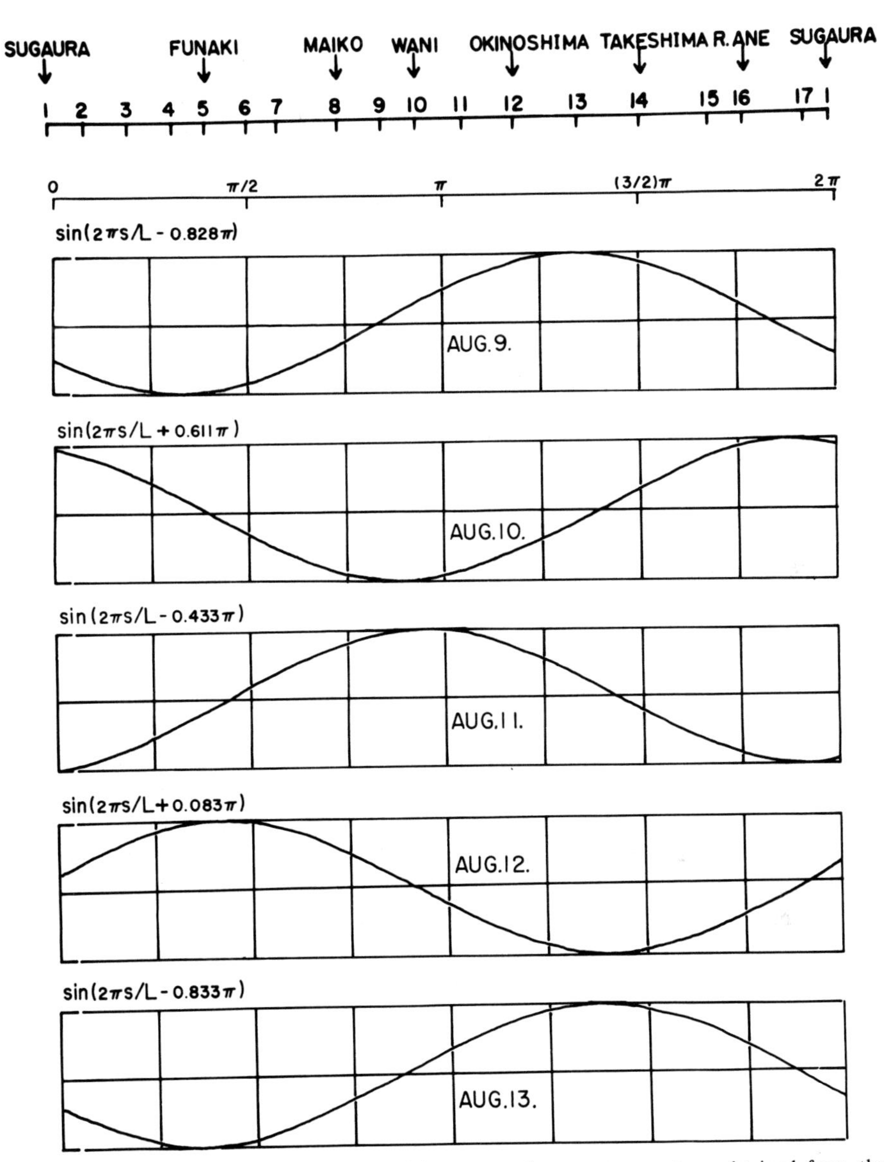

Fig. 100. Daily phase of composed sinusoidal waves of water temperature obtained from the principal component terms of the sine and cosine series in Table 26. The trace of the points of maximum temperature (black circles) shows a counterclockwise progression of wave forms along the lake shore (Kanari 1974b).

follows:

$$\zeta = \sin\psi = \sin(2\pi s/L_c - 2\pi t/T + \epsilon_0) \qquad [29]$$

where T is the period of oscillation which may be given by the formula [1], s and

L_c follow the same definition in the expanded since curves described above, and ϵ_0 denotes a constant arbitrary phase angle.

The wave given by [29] travels along the shoreline with the apparent phase velocity of L_c/T in the counterclockwise direction. Since $s=0$ at Station No. 1 (Sugaura) and T = 45.5 h which can be calculated by the formula [1] by use of the parameters of summer stratification, then the apparent phase angle at every time increment may be estimated by

$$\psi \Big|_{s=0} = \pi - 2\pi t/T \qquad [30]$$

as shown in Fig. 101, where $\epsilon_0 = \pi$.

In Fig. 101, the thick line shows the phase shift calculated by [30] and black circles with the dotted line show the phase shift obtained from the Fourier expansion. The tendency of the analyzed phase shift corresponds well with that of the calculated one. The result implies that the observed temperature profile resulted clearly from the internal seiche which travels along the shoreline with a period (45.5 h in August) approximately the same as that in the case of no rotation effect.

Here again, we can estimate the phase difference between Sugaura (Station No. 1) and Kido (Station No. 9) by Eq. [29].

Putting $t=0$, $s=0$ (Sugaura) and $s=0.43\ L_c$ (Kido), then we obtain

$$\delta\psi = \psi \Big|_{s=0.43L_c} - \psi \Big|_{s=0} = 0.86\pi = 154.8^\circ.$$

This result is very close to that shown in Fig. 92 (f), though the two observations were made independently. Besides, it should be noted that the phase difference

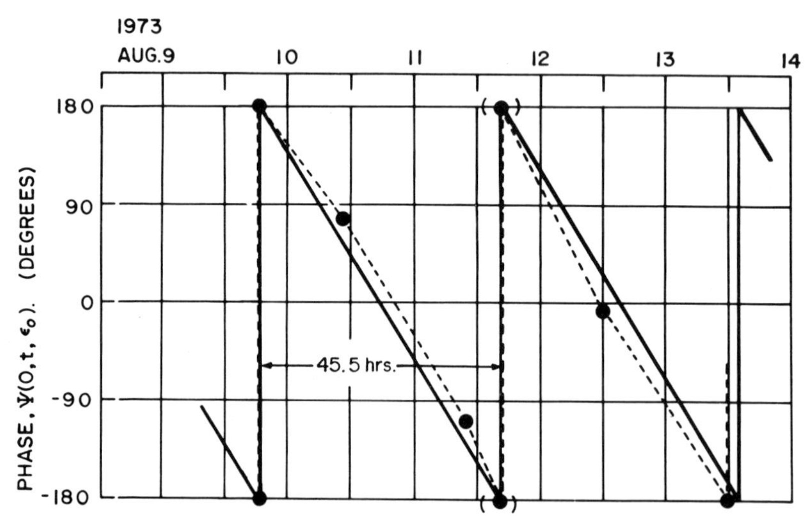

Fig. 101. Phase change of the internal oscillation estimated from the RC survey (black circles with the broken line) and the phase $\psi\,(0,t,\varepsilon_0)$ given by (29) (thick line). Bracketed black circles show the anticipated angles (Kanari 1975a).

$\delta\psi$ between any two points along the lake shore is virtually independent of the period of oscillation.

At an instant of arbitrary time, the spatial dependency of the phase angle $\psi(s, t_0)$ can be given by

$$\psi(s, t_0) = \frac{2\pi s}{L_c} + \text{const.}; \quad -\pi \leq \psi \leq \pi. \qquad [31]$$

The phase distribution $\psi(s, t_0)$ is given in Fig. 102, which exhibits almost the same pattern as Fig. 94.

Numerical experiments of the internal seiches in Lake Biwa

This section treats some behavior of internal seiches in Lake Biwa through a more realistic model than the rectangular model described in the preceding section. In this treatment a two-layer stratification was assumed. As shown in Fig. 103, the complicated real feature of the shoreline is taken into account, but the bottom topography is simplified as the uniform depth of 50 m for the North Basin and 5 m for the South Basin, in which the stratification is restricted within the North Basin. As an analogue of autumnal stratification, an equilibrium two-layer interface is set at the depth of 17.5 m. The model area is divided into 25 × 65 square meshes with a side of 1 km.

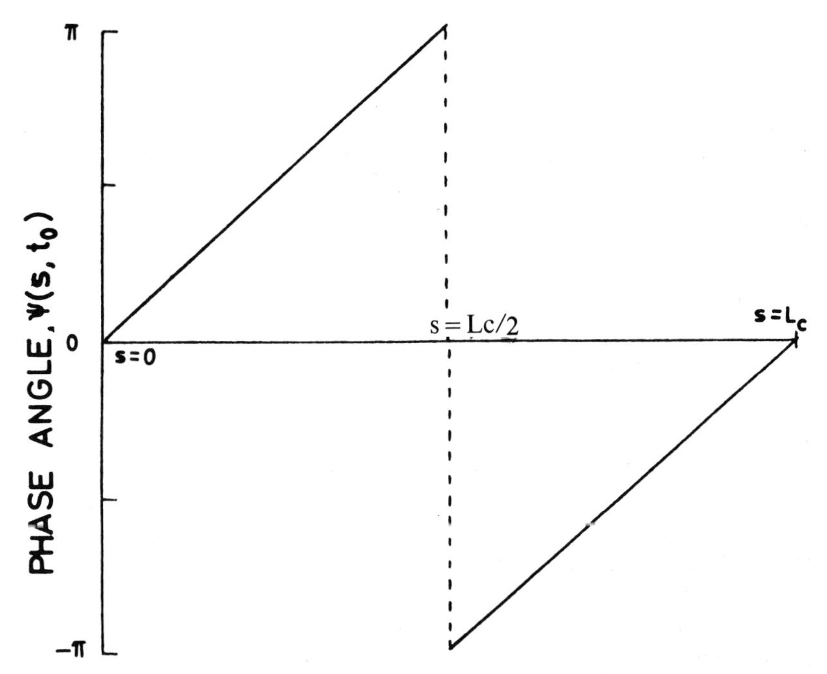

Fig. 102. Spatial dependency of the phase angle $\psi(s,t_0)$ given by (31), putting const. = 0.

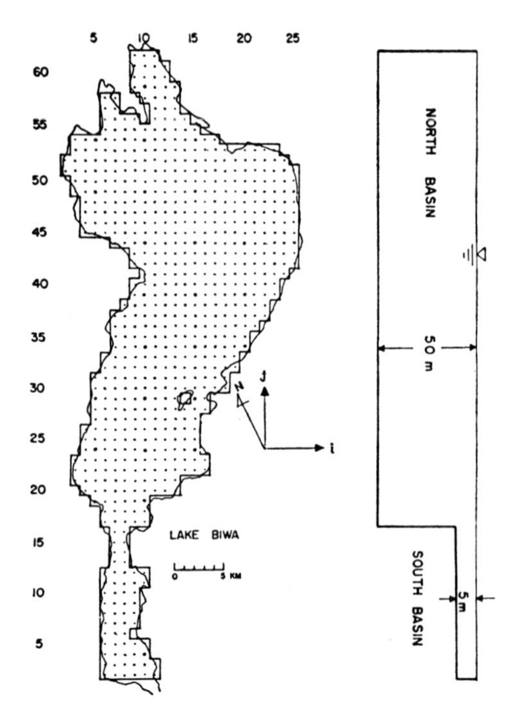

Fig. 103. Computational grid and the longitudinal cross-section of model Lake Biwa (Kanari 1974a).

Let x and y be the horizontal axes toward 26°30′ SE and 26°30′ NE respectively, and z be the vertical axis positive downward. Ignoring the inertia term, the equation of equivalent volume transport and continuity in a two-layered lake with a flat bottom can be given by the following equations (Csanady 1967):

$$\frac{\partial \mathrm{V}_m^*}{\partial t} + 2\Omega \times \mathrm{V}_m^* = -\mathrm{C}_m^2 \nabla_h \cdot \zeta_m^* + \mathrm{F}_m^*, \qquad [32]$$

$$\frac{\partial \zeta_m^*}{\partial t} + \nabla_h \cdot \mathrm{V}_m^* = 0 \qquad [33]$$

where $\nabla_h \equiv \partial/\partial x \cdot i + \partial/\partial y \cdot j$, i and j are the unit vectors referring to the x and y axes, and V_m^* is an equivalent volume transport rate vector which consists of a set of linear combinations of the volume transport rate vector of the earth's rotation, and ζ_m^* and F_m^* are the equivalent elevation and stress vectors corresponding to the definition of the flow vector V_m^*. Subscript m in each variable denotes the modal number, and the system represents a barotropic motion when $m = 1$ and a baroclinic motion when $m = 2$.

The equivalent wave velocity C_m in Eq. [32] can be determined as the roots of the following Stokes' equation:

$$C_m^4 - g(h_1+h_2)C_m^2 + \frac{\Delta\rho}{\rho} g^2 h_1 h_2 = 0. \qquad [34]$$

Neglecting the terms of $0(\Delta\rho^2)$, it follows that

$$C_m^2 = \begin{Bmatrix} g(h_1+h_2) & ; & m=1, \\ \Delta\rho g h_1 h_2/\rho(h_1+h_2); & & m=2. \end{Bmatrix} \qquad [35]$$

In the numerical calculation, the following stratification parameters are used: $h_1 = 17.5$ *m*, $h_2 = 32.5$ *m*, and $\Delta\rho/\rho = 2.5 \times 10^{-3}$.

Equation [32] takes the same form for the two modes; therefore, they can be treated as independent sets of two-dimensional equations. The numerical calculation for the two modes was carried out with different time increments of 20 s for the barotropic mode and 1,200 s for the baroclinic mode in the model lake.

Surface and internal seiches in Lake Biwa are mainly induced by wind stress. Unfortunately there are few meteorological stations around the shore, so that the exact distribution of wind stress over the lake surface cannot be determined. Therefore, it was assumed that a constant wind stress is suddenly imposed on the surface of the model lake that has no motion initially and that after a given duration the wind stress vanishes as suddenly.

As usually stated, the wind stress τ_s is proportional to the square of the wind velocity, that is,

$$\tau_s = \rho_a \gamma_a^2 \mathbf{W}|\mathbf{W}|,$$

where ρ_a is the density of air, γ_a^2 is the drag coefficient, and W is the wind velocity vector. In this study, the following four cases of wind stress are investigated: north wind, blowing for 7.5 h; north wind, blowing for 20 h; southwest wind, blowing for 7.5 h; east-southeast wind, blowing for 7.5 h, all with a uniform wind of 5 m s^{-1} and $\gamma_a^2 = 2.6 \times 10^{-3}$.

For the bottom stress, the following linear relationship (Yamada 1950) is assumed:

$$\tau_b = \rho K V - \beta \tau_s,$$

where K is a constant coefficient of the same dimension as that of the flow velocity, V is the velocity vector, and β is a nondimensional constant and takes the value of 1.0. In these experiments, the velocity coefficient K was chosen so that the bottom stress approaches that of the square-law relationship at the flow velocity of 1.0 cm s^{-1}.

The boundary condition at the shore is that the flow vector normal to the shoreline must vanish.

For the numerical calculations, the velocity vectors in both layers and elevations of the surface and the interface at all mesh points were printed out with a time interval of 15 min. At the same time, the flow velocity and the flow direction in both layers, the elevation of the surface, and the interface at 29 sampling points were punched out as the input data for Fourier analysis. In this analysis the time series in the stage of wind blow was excluded, which makes it possible to construct the spatial patterns of free oscillations in the form of the spatial distribution of amplitude and phase angle at the period of dominant oscillations.

As the result for surface seiches did not differ much from that for the homogeneous lake model treated by Imasato (1971, 1972), it will not be discussed further.

The spatial distribution of the phase angle and relative amplitude constructed from the result of Fourier analysis are completely different from the pattern that would be expected in the case without the effect of the earth's rotation. As shown in Fig. 104, which is an example of the interface oscillation in the case of a 20 h. north wind, the phase angle of high water is proportional to a distance along the lake shore in a counterclockwise direction with a period of 66.1 h., and an amphidromic point rises at about the center of the North Basin.

This means that internal waves with this period apparently progress along the shore in a counterclockwise direction with the apparent phase velocity of 1.5 km/h. There is always a phase difference of 180°C between two opposite shore points situated at the ends of a line through the amphidromic point. The

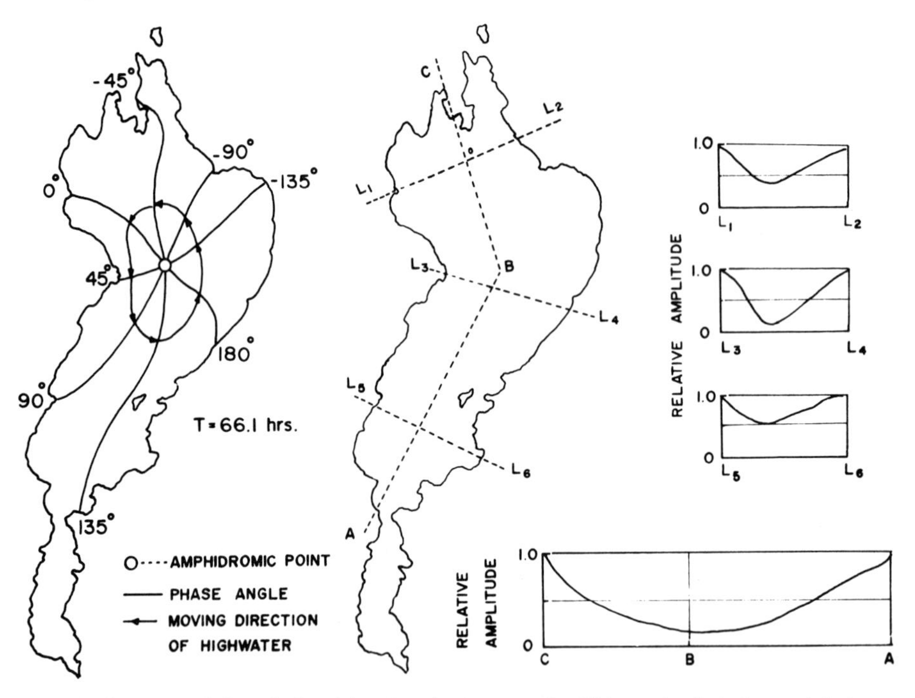

Fig. 104. Structure of the calculated interface in the case of a 20-h north wind; the spatial phase distribution (left), the longitudinal and the cross-sectional lines (center), and the relative amplitude distributions along them (right). In the left figure, maximum elevations at the initial stage of the free oscillation lie along the co-tidal line of −45° to 135° (Kanari 1975a).

large amplitude of the wave is concentrated toward the shore and decreases with the distance from the shore to the amphidromic point.

The internal oscillation starts with the high water to the left of the wind and the pattern of oscillation is essentially similar to the case of a 20-h north wind. The distributions of relative phase and amplitude along the shoreline for the respective cases are shown in Fig. 105, together with those in the case without the effect of the earth's rotation. There is a clear difference in phase distribution

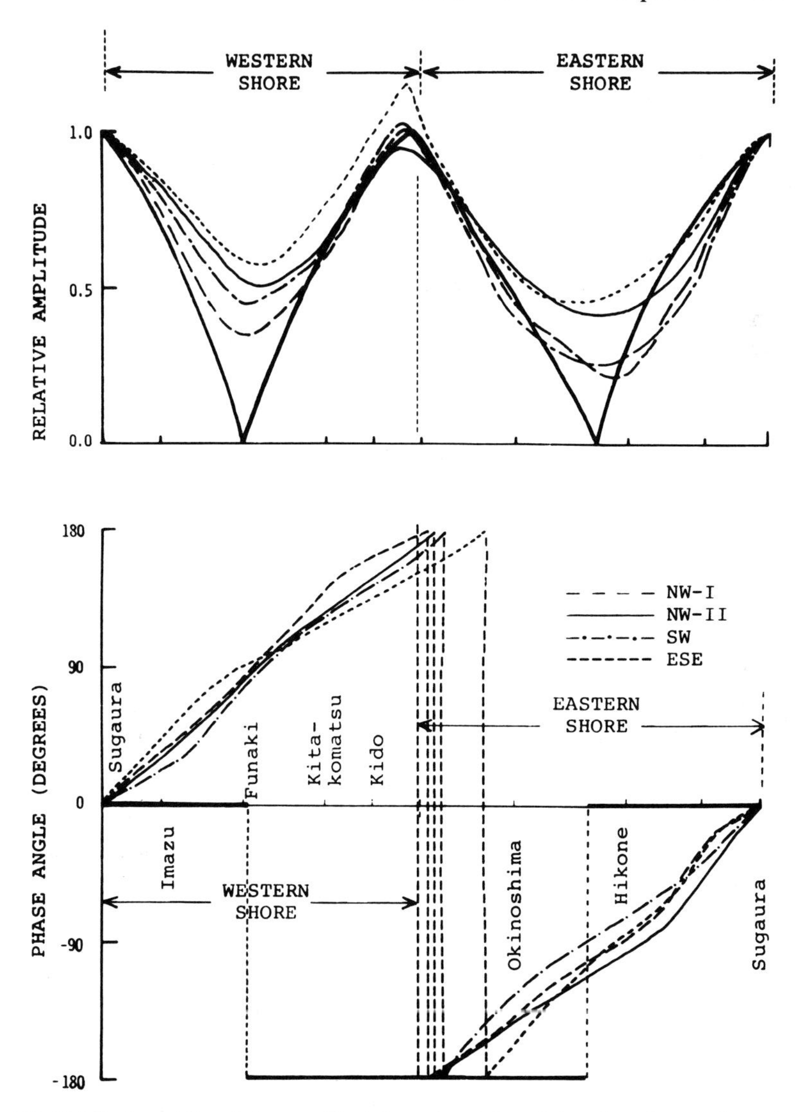

Fig. 105. Distribution of the relative amplitude (upper) and the relative phase (lower) along the lake shore in the case of four wind stresses. The heavy solid line shows those in the case without the effect of the earth's rotation (after Kanari 1975a).

Table 27. Phase differences between Kido and Sugaura (Kanari 1975a).

Numerical model			Observation			
Model	$\delta\Psi(0;s_o,t)$ (degs)	Period (h)	Date	$\delta\Psi(;s_o,t)$ (degs)	Period (h)	Remarks
N–7.5 h	167.0	66.1	25 Sep.–9 Oct.	145.0	53.0	Thermistor chain
N–20 h	150.0	66.1	5–17 Aug.	159.0	45.5	Thermistor chain
SW	148.0	66.1				
ESE	134.0	66.1	9–13 Aug.	154.8	45.5	Round cruise
Mean	149.7			152.9		

between the case with the effect of the earth's rotation and the case without it.

In the range of periods of higher harmonics, the number of amphidromic points increases; for example two amphidromic points appear at the period of 33.05 h. However, in the range below the inertia period, there seems to be no more rotatory oscillation, and the usual polynodal internal seiches appear (Kanari 1974a).

As described in section on Figs. 92 (f) and 92 (g), the phase differences of the internal oscillations between Kido and Sugaura, estimated from temperature records, were 145° in September 25–October 9, 1971 and 159° in August 5–August 17, 1972, and also 154.8° as estimated from the temperature data obtained by the Round-Cruising Survey as described in the previous section.

In the present numerical model, the phase differences between Kido and Sugaura are 167.0° ~ 150.0° for the case of NW-wind, 148.0° for SW-wind, and 134.0° for ESE-wind respectively.

Furthermore, the relative phase distribution along the lake shore shown in Fig. 105 shows a close agreement with the observed phase distribution (Fig. 102), and also with the theoretical phase distribution (Fig. 94) though it was a result of approximate solution. These results are collected in Table 27.

As described up to the present section, the results of a series of studies developed by Kanari have substantiated the existence of internal Kelvin waves in Lake Biwa. Such a wave is quite different in its oscillatory behavior from a simple longitudinal internal seiche in the case of no rotation effect of the earth.

Short-period internal waves in Lake Biwa

As a preliminary observation, a triangular thermistor array was set at Sugaura Bay in September 17, 1968. Its location is shown in Fig. 106. In the three stations, each thermistor sensor was set at the depth of 17.5 m where a sharp thermocline could be found in that season. The installation of the array system is almost the same as those described in previous section (Fig. 83) except for a single-layer thermistor.

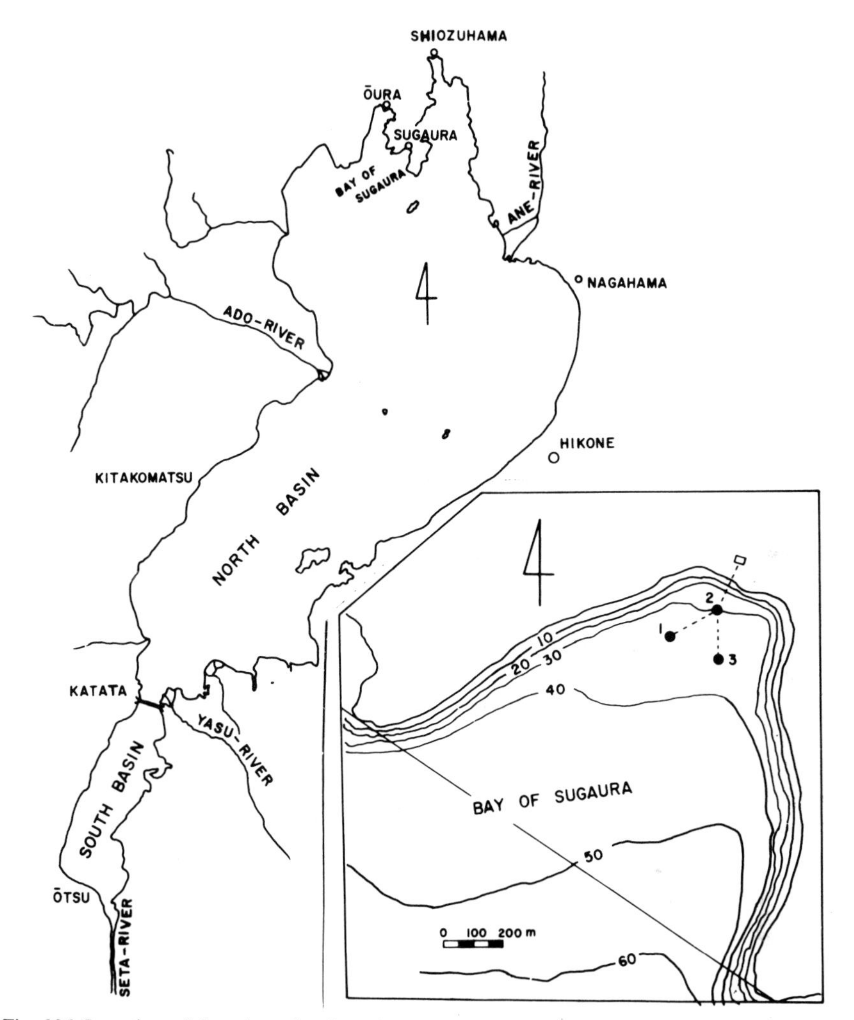

Fig. 106. Location of the triangular thermistor array at Sugaura Bay (Kanari 1970b).

Stability oscillation

According to a theory of internal waves, the high frequency cutoff of internal waves is given by the Brunt-Väisälä frequency, N.

$$N^2 = \frac{g}{\rho} \cdot \frac{d\rho}{dz}$$

where ρ is the local density of waters, g is the acceleration of gravity, and z is the vertical coordinate taken positive downward.

In the present observation, the vertical distribution of the stability period, $\frac{2\pi}{N(z)}$ is given in Fig. 107. As seen in this figure, the minimum stability period (corresponding to the maximum stability frequency) is found at the level of the thermocline and its period is about $1.7 \sim 2.0$ min. This means that stable internal waves of the period shorter than 1.7 min. may not be able to exist in the observed region in this season.

Fig. 108 shows some examples of short-period temperature fluctuations at the level of the thermocline drawn at random from the records of the thermistor array.

There can be seen many wave packets with the amplitude of about 1°C. Indeed, in the two lower records of Fig. 108, most periods of fluctuations seem to be longer than the stability period predicted by the theory of internal waves. However, in the two upper records of Fig. 108, the fluctuations, rather faster than those expected from the stability periods, sometimes appear and they must be a reflection of turbulent motions at the interface.

Spectrum of internal waves for $\omega < N$

The temperature records of the thermistor array, taken during the time from 0100 to 2100 h on October 8, 1968, were sampled at the interval of 1.0 minute. Long-period fluctuations, longer than the 4 h period, were filtered out from

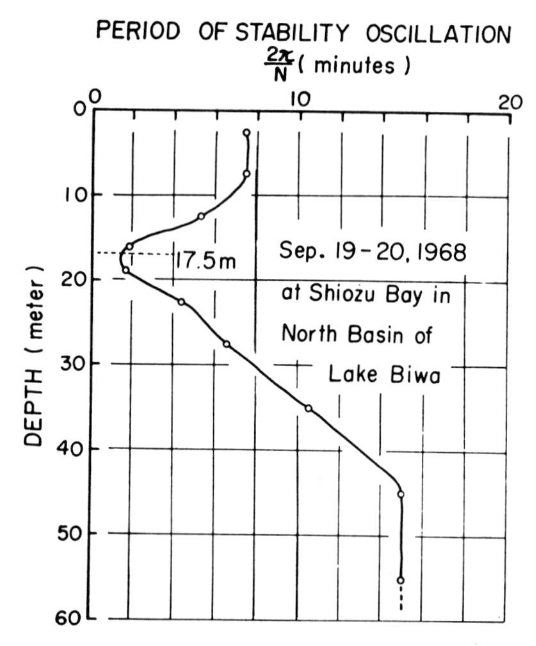

Fig. 107. Vertical distribution of the stability period $2\pi/N(z)$ of Lake Biwa in September (Kanari 1970c).

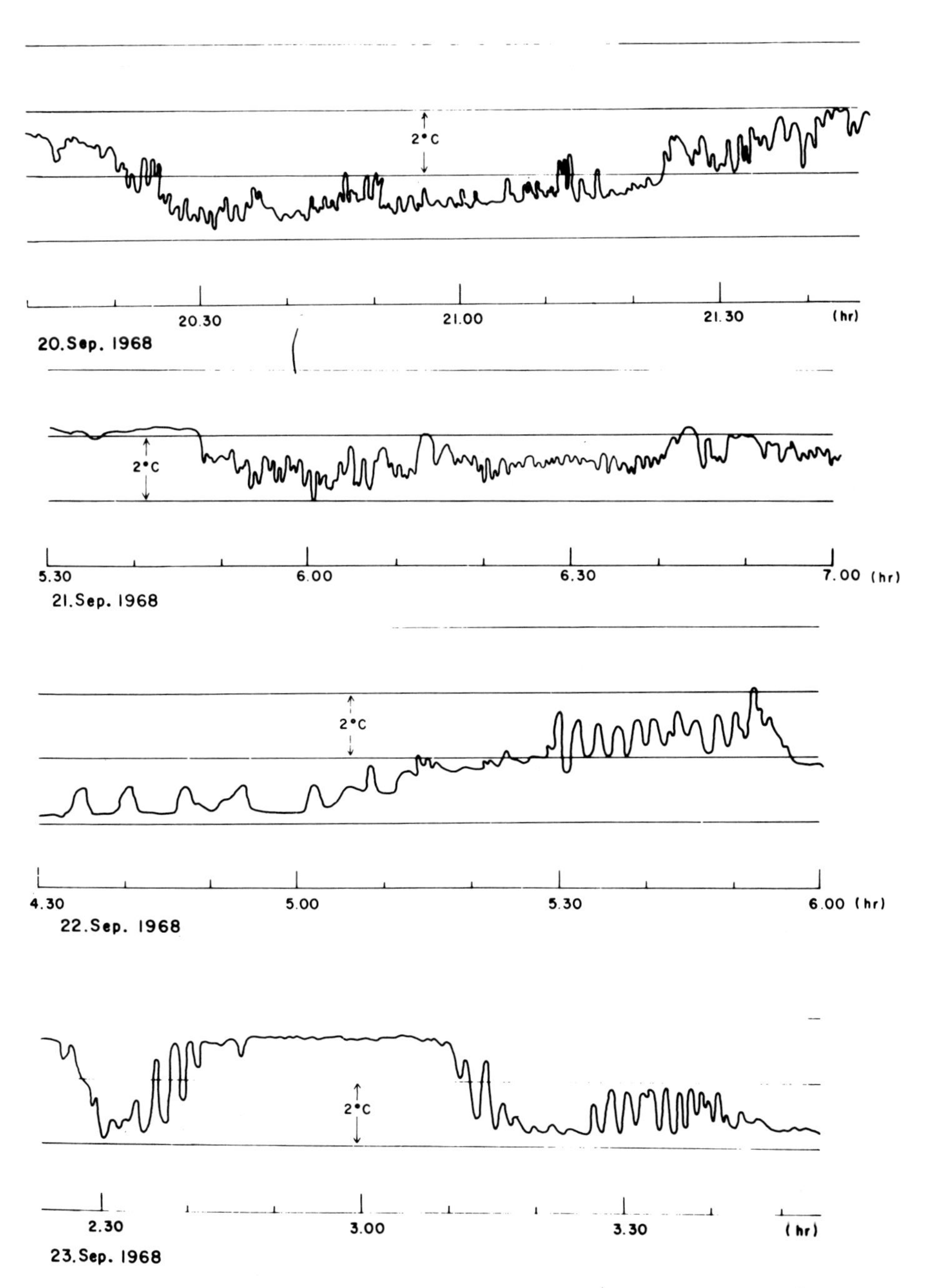

Fig. 108. Short-period fluctuations of temperature at the thermocline in Sugaura Bay. The above records were arbitrarily selcted from the records of the triangular thermistor array (Kanari 1970c).

the sampled data, and then the spectra of the temperature fluctuations with periods between 2 min. and 1.1 h were calculated by the method of Blackman and Tuckey. In this analysis, the degree of freedom was taken as 29.5, and the number of data was 1200. The calculated temperature spectra at the three stations, referred to as wave-1, wave-2 and wave-3, are shown in Fig. 109. The slopes of the spectra seem to be proportional to $\omega^{-2.5}$.

Phillips (1971) has derived the following spectral form for the oceanic internal waves:

$$\phi(z,\omega) = C(z) \cdot \frac{\sqrt{N(z)^2 - \omega^2}}{\omega^3} \qquad [36]$$

where C(z) is a parameter that depends on the depth considered and N(z) is the stability frequency at the same depth.

[36] clearly holds only in the frequency range of $N(z) > \omega$. In the frequency range of the present spectra, [36] is reduced approximately to

$$\frac{\phi(z, \omega)}{C(z)} \sim \omega^{-3.43}.$$

Recently, Eriksen (1978) discussed the form of an observed temperature spectrum and showed that not only the observed temperature spectra, but the observed temperature difference spectra used in the observation of the fine structure of the ocean are closely followed by the model spectrum formed by $P_m = Ak^{-\frac{5}{2}}$ throughout the frequency range of 0.1 cph to 20 cph, where k is the wave number and P_m is the power density spectrum of temperature in $°C^2$/cph. Eriksen's results fits rather well to the form of spectra in Lake Biwa.

Three-layers model

In the spring and early summer seasons, the density stratification of this lake is usually approximated by a three-fluid system rather than a two-fluid system. Even in summer to late autumn, the stratification should be approximated by the three-fluid system as far as short-period internal waves are concerned.

The writer (1968) estimated the Väisälä-frequencies in such a three-fluid system as $N_1 \simeq 10^{-2}$ (rad./sec) for the upper layer from the surface to the depth of 15 m, $N_2 \simeq 2 \times 10^{-2}$ (rad./sec) for the thermocline layer between the depths of 15 m and 25 m, and $N_3 \simeq 8 \times 10^{-3}$ (rad./sec) for the deep layer below the depth of 25 m to the bottom, where N_j j = 1, 2, 3) is defined by

$$N_j^2 = \frac{g}{\bar{\rho}} \left(\frac{d\bar{\rho}}{dz} \right)_j. \qquad [37]$$

By assuming that there is no flow as a basic state and neglecting the Coriolis effect, the basic equations (Lamb 1932) for the non-homogeneous fluid system

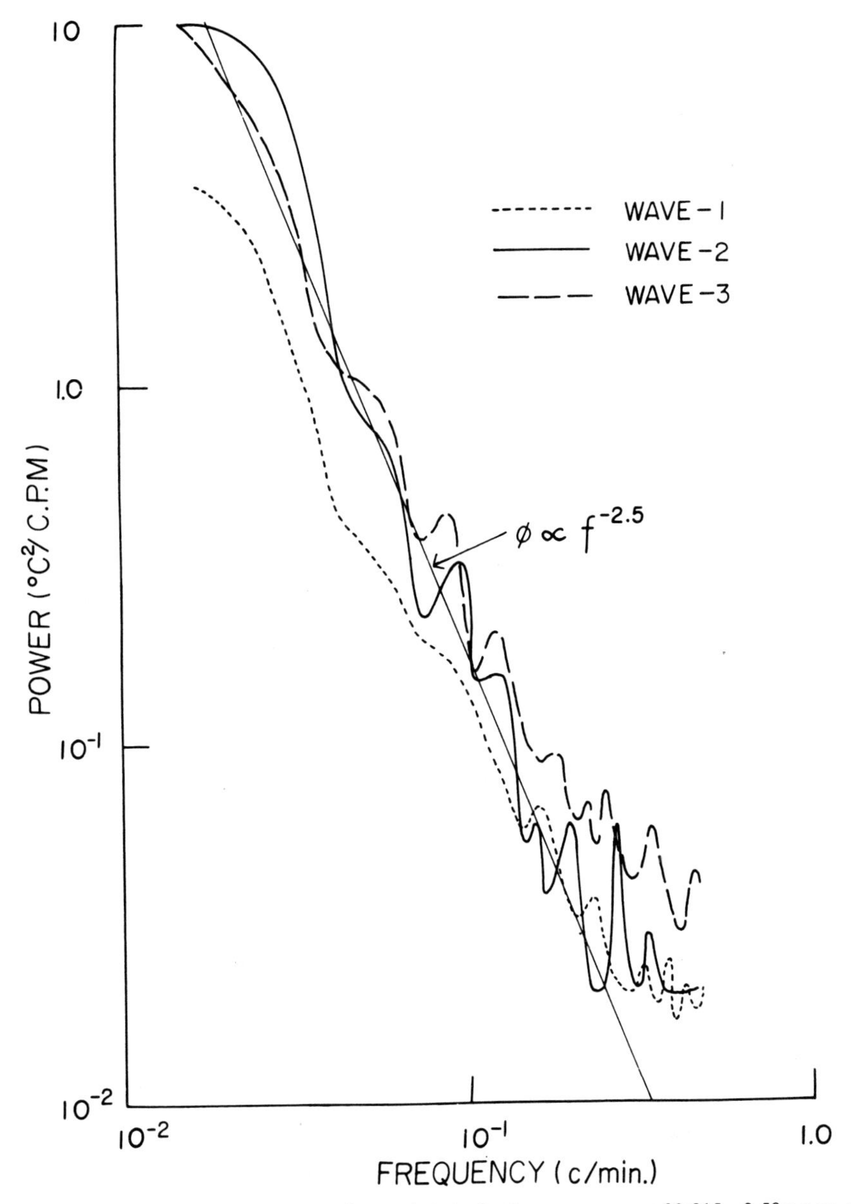

Fig. 109. Power spectra of temperature fluctuations in the frequency range of 0.015 ~ 0.50 c.p.m. at Sugaura Bay (Kanari 1970c).

become

$$\frac{\partial u}{\partial t} = -\frac{1}{\bar{\rho}}\frac{\partial p}{\partial x}, \qquad [38]$$

$$\frac{\partial v}{\partial t} = -\frac{1}{\bar{\rho}}\frac{\partial p}{\partial y}, \tag{39}$$

$$\frac{\partial w}{\partial t} = \frac{\rho}{\bar{\rho}}g - \frac{1}{\bar{\rho}}\frac{\partial p}{\partial z}, \tag{40}$$

$$\frac{\partial u}{\partial x} + \frac{\partial v}{\partial y} + \frac{\partial w}{\partial z} = 0, \tag{41}$$

$$\frac{\partial \rho}{\partial t} + w\frac{\partial \bar{\rho}}{\partial z} = 0, \tag{42}$$

where u and v are the horizontal velocity components referring to the horizontal coordinates x and y axes, w is the vertical velocity component referring to the axis z positive downward, p is the pressure and $\bar{\rho}$ is the density of water in the basic state assumed to be $\bar{\rho} = \bar{\rho}(z)$, and g is the acceleration of gravity.

The above five equations can be reduced to the following single equation in the vertical velocity w,

$$\frac{\partial^2}{\partial t^2}(\nabla^2 w) + N^2 \nabla_h^2 w = -\frac{N^2}{g}\frac{\partial^2}{\partial t^2}\left(\frac{\partial w}{\partial z}\right) \tag{43}$$

where

$$\nabla_h^2 \equiv \frac{\partial^2}{\partial x^2} + \frac{\partial^2}{\partial y^2},$$

$$\nabla^2 \equiv \nabla_h^2 + \frac{\partial^2}{\partial z^2},$$

$$N^2 = \frac{g}{\bar{\rho}}\frac{d\bar{\rho}}{dz}.$$

We assume a harmonic oscillation of the vertical velocity w as follows:

$$w(x, y, z, t) = W(z)\, e^{i(K_x x + K_y y - \omega t)} \tag{44}$$

where K_x and K_y are the horizontal wave numbers, ω is the angular frequency of waves, and $W(z)$ is the amplitude of the vertical velocity w. Substituting [44] into [43], we have

$$\frac{d^2 W}{dz^2} + \frac{K^2(N^2 - \omega^2)}{\omega^2} W + \frac{N^2}{g}\frac{dW}{dz} = 0 \tag{45}$$

where $K^2 \equiv K_x^2 + K_y^2$.

Here, we consider waves with the range of the angular frequency of the order of 10^{-4} (1/sec) and the horizontal wave number of the order of 10^{-5} (1/cm).

Since $N \simeq 10^{-2}$ (rad./sec), the magnitude of the respective terms in Eq. [45] become

$$\frac{N^2}{g}\frac{dW}{dz} \sim \frac{KN^2}{g\omega} = 10^{-8} \text{ (c.g.s.)},$$

$$\frac{K^2N^2}{\omega^2} W \sim \frac{K^2N^2}{\omega^2} = 10^{-6} \text{ (c.g.s.)},$$

$$\frac{d^2W}{dz^2} \sim \frac{K^2N^2}{\omega^2} = 10^{-6} \text{ (c.g.s.)},$$

$$K^2W \sim K^2 = 10^{-10} \text{ (c.g.s.)}.$$

Consequently, Eq. [45] can be approximated by the following equation:

$$\frac{d^2W}{dz^2} + \frac{K^2N^2}{\omega^2} W = 0. \qquad [46]$$

Kanari (1968) obtained the eigen solutions of the Eq. [46] with the following boundary conditions:

$$W = 0 \qquad \text{at } z = 0 \text{ and } z = H, \qquad [47]$$

$$\left.\begin{matrix} N(z) = N_1 \\ W = W_{\rm I}(z) \end{matrix}\right\} \qquad 0 \le z \le h_1, \qquad [48]$$

$$\left.\begin{matrix} N(z) = N_2 \\ W = W_{\rm II}(z) \end{matrix}\right\} \qquad h_1 \le z \le h_2, \qquad [49]$$

$$\left.\begin{matrix} N(z) = N_3 \\ W = W_{\rm III}(z) \end{matrix}\right\} \qquad h_2 \le z \le H, \qquad [50]$$

$$\frac{dW_{\rm I}}{dz} = \frac{dW_{\rm II}}{dz} \qquad \text{at } z = h_1, \qquad [51]$$

$$\frac{dW_{\rm II}}{dz} = \frac{dW_{\rm III}}{dz} \qquad \text{at } z = h_2 \qquad [52]$$

where N_1, N_2, and N_3 are assumed to be constant as shown in Fig. 110.

The solutions for respective layers are given by

(1) $0 \le z \le h_1$,

$$W_{\rm I}^{(n)}(z) = W_{\rm II}^{(n)}(h_1) \cdot \frac{\sin\left(\left[\frac{K}{\omega}\right]_n N_1 z\right)}{\sin\left(\left[\frac{K}{\omega}\right]_n N_1 h_1\right)}, \qquad [53]$$

(2) $h_1 \le z \le h_2$,

$$W_{\rm II}^{(n)}(z) = \frac{W_{\rm II}^{(n)}(h_1) \sin\left\{\left[\frac{K}{\omega}\right]_n N_2(h_2 - z)\right\} + W_{\rm II}^{(n)}(h_2) \sin\left\{\left[\frac{K}{\omega}\right]_n N_2(z - h_1)\right\}}{\sin\left\{\left[\frac{K}{\omega}\right]_n N_2(h_2 - h_1)\right\}}, \qquad [54]$$

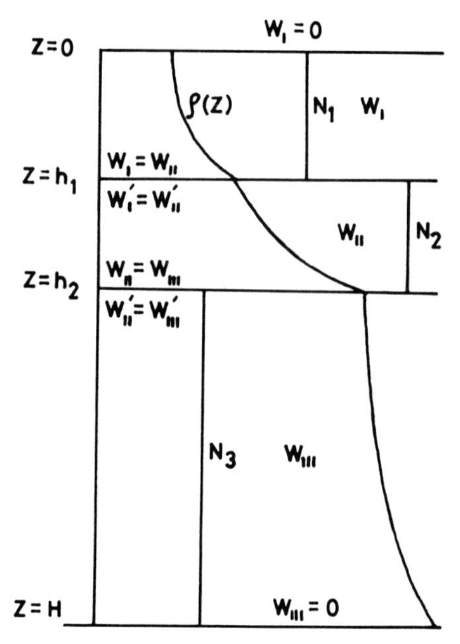

Fig. 110. Schematic representation of the three layer model. The prime of W shows differentiation with respect to z (Kanari 1968).

(3) $h_2 \leq z \leq H$,

$$W_{\mathrm{III}}^{(n)}(z) = W_{\mathrm{II}}^{(n)}(h_2) \cdot \frac{\sin\left\{\left[\frac{K}{\omega}\right]_n N_3 (H-z)\right\}}{\sin\left\{\left[\frac{K}{\omega}\right]_n N_3 (H-h_2)\right\}} \quad [55]$$

where n is the mode number and takes positive integers. The eigen value $\left(\frac{K}{\omega}\right)_n$ can be given by a series of roots of the following determinant equation which is derived from the conditions [51]~[52]:

$$\begin{vmatrix} \frac{N_1}{N_2}\cot\left(\frac{K}{\omega}N_1 h_1\right) + \cot\left(\frac{K}{\omega}N_2 D_1\right) & -\operatorname{cosec}\left(\frac{K}{\omega}N_2 D_1\right) \\ \operatorname{cosec}\left(\frac{K}{\omega}N_2 D_1\right) & -\left[\frac{N_3}{N_2}\cot\left(\frac{K}{\omega}N_3 D_2\right) + \cot\left(\frac{K}{\omega}N_2 D_1\right)\right] \end{vmatrix} = 0, \quad [56]$$

where $D_1 \equiv (h_2 - h_1)$ and $D_2 \equiv (H - h_2)$. Putting $N_1 h_1 \equiv \alpha$, $N_2 D_1 \equiv \beta$, $N_3 D_2 \equiv \gamma$, and $\frac{K}{\omega} \equiv \xi$, the Eq. [56] can be rewritten as

$$N_2^2 \cdot \frac{\tan \alpha\xi}{N_1} \cdot \frac{\tan \beta\xi}{N_2} \cdot \frac{\tan \gamma\xi}{N_3}$$

$$-\left(\frac{\tan \gamma\xi}{N_1}+\frac{\tan \beta\xi}{N_2}+\frac{\tan \alpha\xi}{N_3}\right)=0. \qquad [57]$$

Condition [56] also implies a relationship between $W_{II}(h_1)$ and $W_{II}(h_2)$. From the condition [51] and the solutions [53] and [54], we can get

$$\left(\frac{dW_I}{dz}\right)_{h_1}=W_{II}(h_1)\cdot\frac{\alpha\xi \cot \alpha\xi}{h_1}, \qquad [58]$$

$$\left(\frac{dW_{II}}{dz}\right)_{h_1}=W_{II}(h_2)\cdot\frac{\beta\xi}{D_1}\cdot\frac{1}{\sin \beta\xi}-W_{II}(h_1)\frac{\beta\xi}{D_1}\cot \beta\xi. \qquad [59]$$

Equating [58] and [59], we have

$$\frac{W_{II}(h_2)}{W_{II}(h_1)}=\frac{D_1\alpha}{h_1\beta}\sin \beta\xi \cot \alpha\xi+\sin \beta\xi \cot\beta\xi. \qquad [60]$$

This enables us to reduce the two arbitrary constants, $W_{II}(h_1)$ and $W_{II}(h_2)$ into one in the solutions [54] and [55].

Thus we can calculate the eigen solutions of the vertical velocity, give the parameters h_1, h_2, H, N_1, N_2, and N_3.

Fig. 111 shows the graphical solution of Eq. [57], in which the solid curve represents the function

$$y_1=\frac{\tan \gamma\xi}{N_3}+\frac{\tan \beta\xi}{N_2}+\frac{\tan \alpha\xi}{N_1}$$

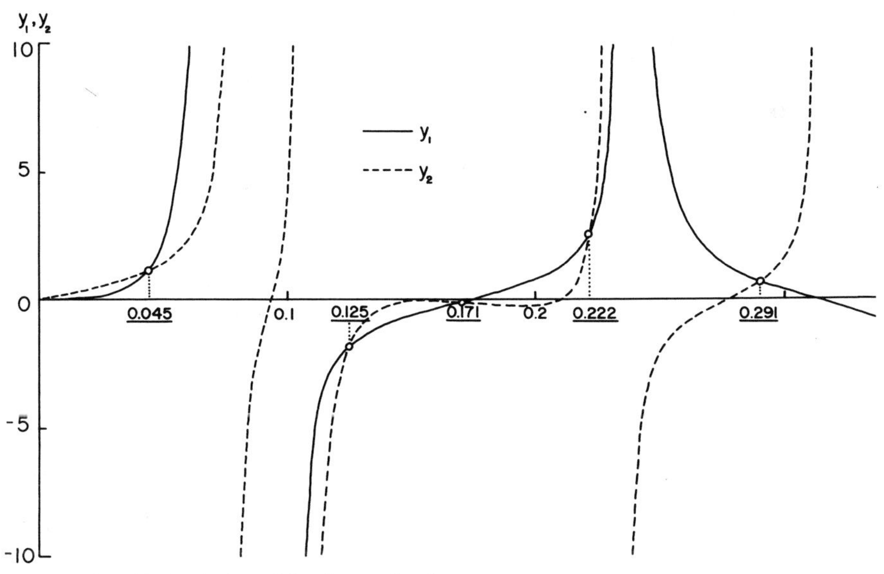

Fig. 111. Graphical solution of the eigen value ξ_n.

and the dotted curve represents the function

$$y_2 = N_2^2 \frac{\tan \alpha\xi}{N_1} \cdot \frac{\tan \beta\xi}{N_2} \cdot \frac{\tan \gamma\xi}{N_3}$$

for the case of $h_1 = 15$ m, $h_2 = 25$ m, $H = 50$ m, $N_1 = 10^{-2}$ (rad./sec), $N_2 = 2 \times 10^{-2}$ (rad./sec), and $N_3 = 8 \times 10^{-3}$ (rad./sec).

Consequently, the roots of ξ_n can be given by a series of ξ-values where the two functions intersect each other. The points of intersection are illustrated by open circles in the figure, except for the point $\xi = 0$.

The first five roots are as follows:

$$\xi_1 = 0.045,\ \xi_2 = 0.125,\ \xi_3 = 0.171,$$

$$\xi_4 = 0.223,\ \xi_5 = 0.291.$$

The roots determine the ratio of $W_{II}^n(h_2)$ and $W_{II}^n(h_1)$ and shown in Table 28. Thus the eigen solution in the term with the eigen value ξ_n is given by

(1) $0 \leq z \leq h_1$,

$$\frac{W_I^n(z)}{W_{II}^n(h_1)} = \frac{\sin N_1 \xi_n z}{\sin \alpha\xi_n}, \qquad [61]$$

(2) $h_1 \leq z \leq h_2$,

$$\frac{W_{II}^n(z)}{W_{II}^n(h_1)} = \frac{\sin N_2 \xi_n (h_2 - z) + A_n \sin N_2 \xi_n (z - h_1)}{\sin \beta\xi_n}, \qquad [62]$$

(3) $h_2 \leq z \leq H$,

$$\frac{W_{III}^n(z)}{W_{II}^n(h_1)} = \frac{A_n \sin N_3 \xi_n (H - z)}{\sin \gamma\xi_n} \qquad [63]$$

where $A_n = W_{II}^n(h_2)/W_{II}^n(h_1)$.

The calculated vertical distribution of $W^n(z)$ for the first five modes (n = 1, 2, 3, 4, and 5) are shown in Fig. 112, in which $W_{II}^n(h_1)$ is assumed to be unity.

As shown in Fig. 112, the maximum amplitude of the vertical velocity of the first mode occurs at the lower boundary of the thermocline ($z = 25$ m), while the second mode has two maxima at $z = 13$ m and $z = 33$ m but with anti-phase. For the higher mode, the solution has more maxima at other depth levels.

Table 28. Values of the first five roots and the ratio of $W_{II}^n(h_2)/W_{II}^n(h_1)$.

n	1	2	3	4	5
ξ_n	0.358	1.273	1.776	2.165	2.878
$\frac{W_{II}^n(h_2)}{W_{II}^n(h_1)}$	1.306	−0.927	−0.532	−4.737	0.759

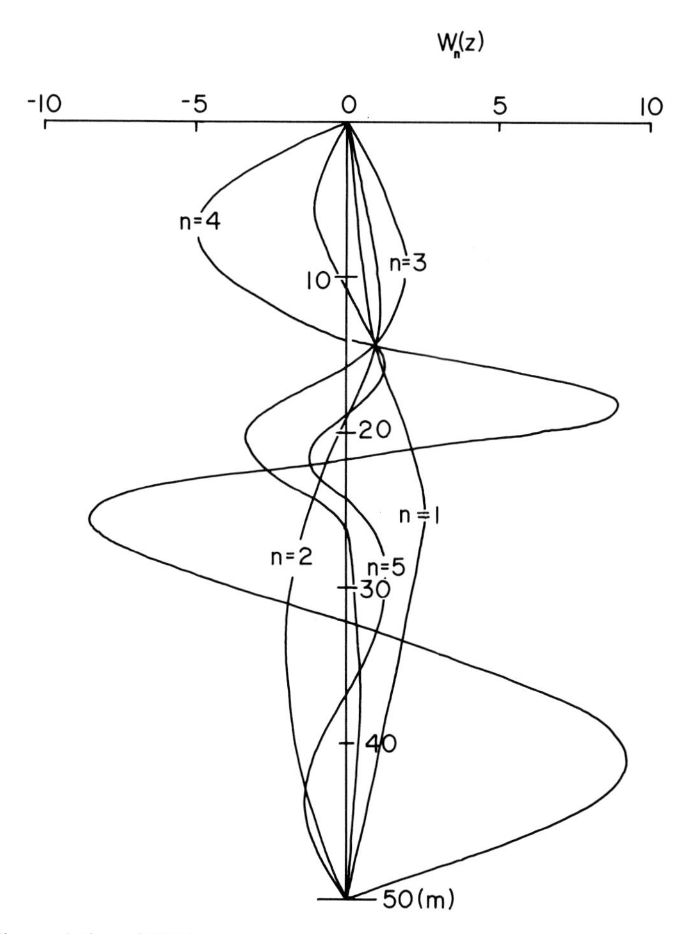

Fig. 112. Eigen solution of $W_n(z)$.

In real lakes, the vertical distribution of the vertical water velocity due to internal waves must be the result of superposition of many modes of oscillations.

The above solutions have been compared with the result of temperature observations, carried out by Kanari (1968), and it was found that the distribution of the vertical velocity estimated from the temperature variation corresponds to the second mode solution. However, the comparison is not complete because of an incomplete data processing, so that a more accurate data analysis is required.

Seiche

Norihisa Imasato

Seiche, its formulation

'Seiche' is the name originally given to periodic fluctuations of the water-level of a fairly long period and with the properties of a long wave in a closed lake basin. These periodic motions have been noticed in earlier times. According to Defant (1961), an old chronicle by Schulthaiss reported that seiche motions have been noticed in the Lake of Constance as early as in 1549 and also reported to have been observed in the Lake of Geneva already in the eighteenth century. Systematic and methodical observations of seiche motions were made first by Forel (1892) in the Lake of Geneva in 1869. Thereafter, a large number of contributions to the theory of seiche have been made by many scientists. These contributions have been summarized well by Defant (1961).

Seiche motions are the standing waves in a closed basin, in a bay or on a lake shelf. They are formed by the reflection of a progressive wave at a lake wall. They are governed by equations of motion and continuity (Eqs. 1) with appropriate boundary and initial conditions and the hydrostatic assumption in the vertical direction,

$$\frac{\partial u}{\partial t}+u\frac{\partial u}{\partial x}+v\frac{\partial u}{\partial y}=-g\frac{\partial \eta}{\partial x}+fv+\nu\frac{\partial^2 u}{\partial z^2}, \qquad [1\text{–}1]$$

$$\frac{\partial v}{\partial t}+u\frac{\partial v}{\partial x}+v\frac{\partial v}{\partial y}=-g\frac{\partial \eta}{\partial y}-fu+\nu\frac{\partial^2 v}{\partial z^2}, \qquad [1\text{–}2]$$

$$\frac{\partial \eta}{\partial t}=-\frac{\partial}{\partial x}\int_{-h}^{\eta}udz-\frac{\partial}{\partial y}\int_{-h}^{\eta}vdz \qquad [1\text{–}3]$$

where x is the coordinate axis positive to the east, y is that positive to the north t is time, u and v denote the velocity components in the directions of x and y, respectively, h the depth of the bottom below mean water level and η denotes the free surface, f denotes the Coriolis parameter, g the gravitational acceleration and ν the vertical kinematic viscosity, respectively. The boundary condition at a

Horie, S. Lake Biwa

ISBN 90 6193 095 2. Printed in The Netherlands

solid wall of the lake is

$$u_n = 0 \qquad [2]$$

where u_n is the velocity component normal to the solid wall.

It is impossible to get a general solution to the seiche motions because the basic equations are nonlinear and the actual basin generally has a very complicated geometry. Therefore many investigators have tried to get analytical solutions by making some approximations and assumptions, such as linearizing the equations of motion and assuming the seiche to oscillate along one dimensional 'Talweg' approximated by a well-known analytical function. The most representative example was given by Merian (1828) and it is briefly described below.

Seiche in a rectangular basin

Merian (1828) dealt with one-dimensional seiche motions in a rectangular closed basin with a uniform constant depth h and a length L, neglecting the viscous terms. The linearized equations governing the seiche motions are reduced from Eqs. 1 to Eqs. 3,

$$\frac{\partial u}{\partial t} = -g\frac{\partial \eta}{\partial x}, \qquad [3\text{–}1]$$

$$\frac{\partial \eta}{\partial t} = -\frac{\partial}{\partial x}\int_{-h}^{\eta} u dz = -\frac{\partial}{\partial x}(uh), \qquad [3\text{–}2]$$

and the boundary conditions are

$$u = 0 \text{ at } x = 0 \text{ and } x = L. \qquad [4]$$

According to Merian, the period T_n and the water surface displacement $\eta(x)$ of seiche with the n-nodal lines are given, respectively, as follows:

$$T_n = 2L/n\sqrt{gh}, \qquad [5\text{–}1]$$

$$\eta(x) = A \cdot \cos(2\pi n x/L) \cdot \cos(2\pi n t/T) \qquad [5\text{–}2]$$

where A is an arbitrally constant amplitude of the seiche.

Seiche in a basin with variable cross-section

Chrystal (1904) developed Merian's theory and gave some analytical solutions of seiche in a basin with a variable bottom profile represented by an analytical function. Defant (1918) developed a more refined and realistic method for determining the surface displacement profile and the period of seiche in a lake with any variable cross-section. But the seiche motion is considered to be one-dimensional yet, because the oscillation is assumed to be predominantly along the one-dimensional 'Talweg' normal to each variable cross-section. The volume continuity in the segment at x of the cross-section $S(x)$ and the width $b(x)$ at the surface is given by

$$\frac{\partial \eta}{\partial t} = -\frac{1}{b(x)} \cdot \frac{\partial}{\partial x}[uS(x)], \qquad [6]$$

which corresponds to Eq. [3–2]. Introducing an auxiliary variable

$$\xi = \int_0^t u dt, \qquad [7]$$

we have the linearized equation of motion as follows:

$$\frac{\partial^2 \xi}{\partial t^2} = -g\frac{\partial \eta}{\partial x}. \qquad [8]$$

Therefore, we have the basic equation for determining the period of seiche motion in the lake of variable cross-section as follows:

$$\xi = -\frac{1}{S(x)}\int_0^x \eta b(x) dx. \qquad [9]$$

The whole lake is broken up into segments of length Δx along the 'Talweg', cross-section $S_i(x)$ and width at the surface $b_i(x)$, i = 1 ... n, such that $L = n\Delta x$. These variables are deployed as shown in Fig. 113. We assume that the horizontal and vertical displacements ξ and η are simple harmonic in t, but perhaps quite complex in x in the form

$$\xi(x,t) = \xi_0(x) \cdot \cos S(2\pi t/T), \qquad [10–1]$$

$$\eta(x,t) = \eta_0(x) \cdot \cos S(2\pi t/T). \qquad [10–2]$$

When these two equations are substituted into Eq. [8] and the differential form is changed into the finite difference form, we have

$$\Delta\eta_0 = \frac{4\pi^2}{gT^2}\xi_0 \Delta x = \alpha\xi_0. \qquad [11]$$

If we assume that changes and displacements are linear from one segment to the next and introduce a second auxiliary variable q, we have a set of equations convenient for the practical computation as follows:

$$\xi_{i+1} = -\left(1 + \frac{\alpha b_{i+1}}{4S_{i+1}}\right)\left[q_i + \left(\eta_i + \frac{\alpha\xi_i}{4}\right)b_{i+1}\right]/S_{i+1}, \qquad [12–1]$$

$$\eta_{i+1} = \eta_i + \frac{\alpha(\xi_i + \xi_{i+1})}{2}, \qquad [12–2]$$

$$q_{i+1} = q_i + \frac{\eta_i + \eta_{i+1}}{2} b_{i+1} \qquad [12–3]$$

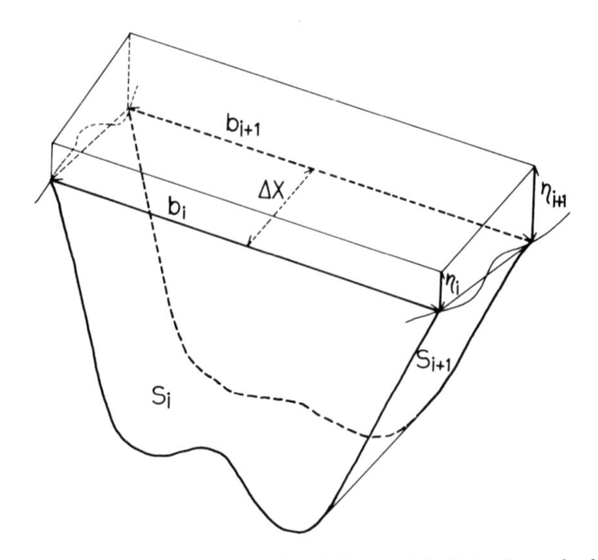

Fig. 113. A sliced segment and arrangement of variables used in Defant's methods.

The boundary condition at the one end of the 'Talweg' u (0) = 0 yields $\xi_0 = q_0 = 0$. On the other hand, η_0 must be set equal to an arbitrary value. First, an approximate value of T is formed by means of Merian's formula using the average depth of the basin. The result is an approximate value of α. Then, Eq. [12–1] yields ξ_1, which in turn gives η_1 from the second Eq. [12–2], and therefore the bottom Eq. [12–3] yields q_1. These values of ξ_1, η_1 and q_1 are substituted into Eqs. [12], and ξ_2, η_2 and q_2 are found. The same process is repeated until ξ_n, the value at the other end of the 'Talweg', is calculated. If the period of the seiche has been correctly guessed, ξ_n must be zero because the boundary condition $u(L) = 0$ must be satisfied. If we have $\xi_n = 0$, we can obtain not only the period but also the relative profiles of the vertical and the horizontal displacements $\eta_0(x)$ and $\xi_0(x)$ along the 'Talweg'. However, if we have not obtained $\xi_n = 0$, we must repeat the process for another estimated value of T.

The calculation has been very tedious in earlier times, but now the well-developed computer system enables us to perform it very easily. Therefore we can conveniently carry out the calculation for a fairly wide range of the period T by decreasing it by a small increment, plot ξ_n against T and draw a curve connecting each point of ξ_n. The points of intersection of this curve and the coordinate T give the exact periods of seiches including some higher modes. This method of Defant is a very convenient and fairly realistic method for estimating a first approximated characteristic of the longitudinal seiche oscillating along 'Talweg'.

Seiche in a basin of complicated geometry

Owing to the development of a large computer system, we have now means to understand more detailed characteristics of the horizontally two-dimensional seiche motion in an actual basin with a very complicated geometry. Integrating the basic equations (Eqs. 1 – 1 and 1 – 2) from the bottom $h(x, y)$ to the water surface $-\eta(x, y)$, we have the following equations as first approximation:

$$\frac{\partial M_x}{\partial t}+\frac{M_x}{h+\eta}\cdot\frac{\partial M_x}{\partial x}+\frac{M_y}{h+\eta}\cdot\frac{\partial M_x}{\partial y}=-g(h+\eta)\frac{\partial \eta}{\partial x}+fM_y+\frac{\tau_{ax}-\tau_{bx}}{\rho}. \quad [13\text{-}1]$$

$$\frac{\partial M_y}{\partial t}+\frac{M_x}{h+\eta}\cdot\frac{\partial M_y}{\partial x}+\frac{M_y}{h+\eta}\cdot\frac{\partial M_y}{\partial y}=-g(h+\eta)\frac{\partial \eta}{\partial y}-fM_x+\frac{\tau_{ay}-\tau_{by}}{\rho}, \quad [13\text{-}2]$$

$$\frac{\partial \eta}{\partial t}=-\frac{\partial M_x}{\partial x}-\frac{\partial M_y}{\partial y} \quad [13\text{-}3]$$

where $M_x=\int_{-h}^{\eta} u\,dz$ and $M_y=\int_{-h}^{\eta} v\,dz$. τ_a and τ_b denote the wind stress at the water surface and the bottom stress, respectively. They are usually expressed respectively as follows:

$$\tau_a=\rho_a\gamma_a^2|\bar{W}|\bar{W}, \quad [14\text{-}1]$$

$$\tau_b=\rho\gamma_b^2|\bar{U}|\bar{U} \quad [14\text{-}2]$$

where ρ_a is the density of air, $\bar{W}$ is the velocity vector of the representative wind, γ_a^2 the drag coefficient which is a function of $\bar{W}$ (Kunishi & Imasato 1966), ρ the density of water, $\bar{U}$ the vector of vertically averaged velocity and γ_b^2 the drag coefficient which is considered to be 2.3×10^{-3} after Hansen (1956).

The boundary condition at the solid wall is denoted by

$$M_n=0$$

where M_n is the velocity component normal to the solid wall. The partial differential equations are changed into a finite difference form. The whole basin is broken up into water columns with the variable depth $h_{i,j}$ and the square water surface of area ΔS^2, where $h_{i,j}$ means $h(i\Delta S, j\Delta S)$. Velocity components $u_{i,j}^m$ and $v_{i,j}^m$ and the elevation $\eta_{i,j}^m$ are calculated in turn step by step till a stable oscillation of seiche motions is attained (m denotes the m-th time step, i.e., $t=m\Delta t$).

Seiche in Lake Biwa

Beginning in 1911 with Nakamura and Honda, the history of studies on the seiche in Lake Biwa may be divided into three periods, i.e. from 1902 to 1938, from 1939 to 1969, and from 1970 to today. During the first period, the seiche in Lake Biwa was studied chiefly by the method of observations, although we must take note of a theoretical study of the seiche on the lake shelf developed by Nomitsu (1935). During the second period, no study was made on the seiche in Lake Biwa although the reason is not clear. In 1970, a study on the seiche in Lake Biwa was begun again by the writer by the method of numerical experiments on seiche in the complicated actual geometry of Lake Biwa. The characteristics of the longitudinal seiches of periods longer than 30 minutes may be said to have been fairly well understood through the writer's comparisons between the calculated and observed results. The writer must here emphasize that this situation is due to the rapid progress of a large computer system. These historical studies on the seiche in Lake Biwa are summarized in Table 29. The writer will describe the details in what follows. (Fig. 114).

Studies during the first period (1902–1938)

In 1911, Nakamura and Honda studied the seiche motion in Lake Biwa for the first time through observations and a hydraulic model experiment. They took observations at 13 sites in turn for about 2 days and found oscillations of periods of 231.2, 72.6, 30.5, 25.2 and 22.7 minutes. Unfortunately, however, they made no speculation about these observed oscillations. They not only took observations but also performed a hydraulic model experiment although it was a very simple one. They found two oscillations with periods of 71.9 and 30.5 minutes in their model basin. According to them, the oscillation of 71.9 minutes has two nodal lines, of which one is situated in the north basin and another in the south basin. They explained that the oscillation in the north basin was the seiche of the fundamental mode and the oscillation in the south basin was induced by the oscillation in the north basin. They also concluded that the oscillation with the period of 30.5 minutes is the higher mode of the oscillations mentioned above.

Suda et al. (Kobe Mar. Observ. 1926) took observations and detected oscillations of periods of 242.0, 71.0, 36 – 37, 30 and 25 – 22 minutes. Their explanations about these constituents are as follows: the characteristics of the 242.0 minutes oscillation could not be identified but this constituent was frequently accompanied with thunderstorms. The 71.0 and 36 – 37 minutes oscillations are the fundamental and its higher mode seiches in the north basin respectively and are the same as those identified by Nakamura and Honda (1911). The 30 minutes oscillation is the seiche of fundamental mode in the northern part of the north basin oscillating between Kaizu and Hikone. The oscillations with periods shorter than 25 minutes are local oscillations.

Table 29. List of periods reported in the previous studies on the seiches in Lake Biwa.

Investigator	Year		Period (minutes)					
			Mode-I	Mode-II	Mode-III	Mode-IV	Mode-V	Others
Nakamura & Honda	1911	Obs.	231.2	72.6			30.5	25.2, 22.7
		Ex.		71.9			30.5	
Suda et al (Kobe Mar.Observ.)	1926	Obs.	242.0	71.0		36–37	30.0	25–22, 15
Takaya	1931	Obs.	236.2					
Nomitsu	1935	Obs.	250.0	68.4			30.0	20, 12, 5
Takahashi	1935	Cal.	208.0	68.0				
Nomitsu et al.	1937	Obs.						18 – 15,5
Toyohara & Habu	1938	Obs.		66.0			32.0	
Takahashi & Namekawa	1938	Obs.	220.0					
Imasato	1970	Cal.	212.0	71.3	61.0	37.0	32.3	
Imasato	1971	Cal.	255.5	79.8	69.1			
		Obs.	243.9	74.1	65.1			
Imasato	1972	Cal.				38.7	31.9	51.1
		Obs.	232.3	74.1	66.4	40.2	32.6	
Imasato et al.	1973	Obs.	229.8	72.7	65.1	40.1	30.5	
Mean		Obs.	229.8 ± 12.2	72.7 ± 2.2	65.1 ± 1.5	40.1 ± 0.6	30.5 ± 0.8	
		Cal.	255.5	79.8	69.1	38.7	31.9	

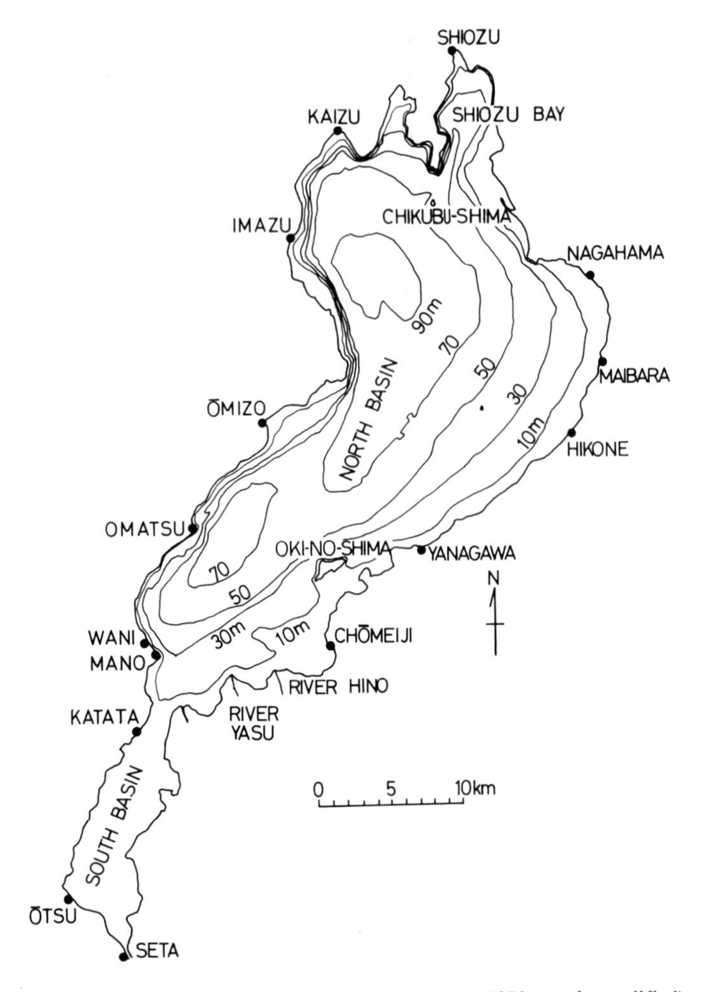

Fig. 114. Bathymetric map of Lake Biwa (Imasato 1970, partly modified).

Takaya (1931) discussed characteristics of the 236.2 minutes oscillation which he found to be predominant in the south basin. He made a histogram of the frequency of appearance of this constituent, studied monthly variations of the frequency of appearance, and concluded that this constituent was an oscillation forced by the seasonal wind in winter. But a careful examination of his histogram reveals that this oscillation is predominant in March and the wind is also the strongest in this month. Takahashi and Namekawa (1938) asserted that the seiche of this mode was a free oscillation in the south basin.

Nomitsu (1935) analyzed the records of displacement of the lake surface, so-called limnogram, during the period when the strong typhoon MUROTO passed over Lake Biwa from southwest to northeast and detected oscillations with the periods of 250.0, 68.4, 30.0, 20.0 and 5.0 minutes. He explained these oscillations as follows: the 250.0 minutes oscillation is the fundamental bay oscillation in the south basin. The 68.4 minutes oscillation is the uni-nodal seiche in the north basin and has loops at Nagahama and Katata. The oscillation of this period appears also in the south basin and it is a forced oscillation induced by the former in the north basin. The 30 and 20 minutes periods belong to oscillations in the northern part of the north basin. The 5 minutes oscillation is a shelf seiche on the lake shelf off Imazu. Nomitsu and his co-workers carried out some observations in order to prove his theory on the lake shelf seiche. Nomitsu et al. (1937) observed an 18 – 15 minutes oscillation off Ōmizo and a 5 minutes oscillation off Imazu and showed that they were in good agreement with Nomitsu's theoretical periods of the shelf seiche. Toyohara and Habu (1938) carried out observations simultaneously at Chōmeiji, Omatsu and Wani and detected 66 and 32 minutes oscillations. They identified the former oscillation as that of 68.4 minutes seiche reported by Nomitsu (1935). They also explained the latter oscillation as the shelf oscillation on the well-developed lake shelf off the river mouths of River Yasu and River Hino in the southeastern part of the north basin.

In 1935, a noteworthy study was made by Takahashi. He numerically found the solutions of the linearized inviscid equations and boundary conditions governing the one-dimensional seiche motions in the lake of variable cross-section. After tedious numerical calculations, he obtained periods of 208 and 68 minutes and clearly showed that they were the uni- and bi-nodal seiches in the whole basin of Lake Biwa. The period of 208 minutes is shorter than 220–250 minutes reported by the previous observations, but according to Takahashi, it agrees well with the observed period if friction is taken into account.

After 1938, the seiche in Lake Biwa had been out of the scientific interest of the investigators till 1970 when the writer began again numerical studies on characteristics of the seiche by making use of a large computer. Before the writer describes recent results of the seiche in Lake Biwa, it will be worth-while summarizing the results which have been mentioned above. The periods reported by many investigators during the first period may be classified into five modes, i.e. Mode I, II, III, IV and V as shown in Table 29. The seiche of Mode I is the 208 – 250 minutes oscillation and has been considered as the bay oscillation in the south basin or the uni-nodal seiche in the whole basin. Seiches of other modes except those of the shelf oscillations have been given no definite explanation.

Recent studies of seiche in Lake Biwa

Recently, we have been able to obtain numerical solutions of the nonlinear equations governing the seiche motions because of the rapid progress of a large computer system. In 1970, the writer calculated the periods of longitudinal seiches using Defant's method described in the previous section. They are 212.0, 71.3, 61.0, 37.0 and 32.3 minutes. The 212.0 minutes oscillation is the uni-nodal seiche in the whole basin, and it corresponds to the 208 minutes seiche obtained by Takahashi (1935). The 71.3 and 61.0 minutes oscillations are the bi- and tri-nodal seiches, respectively. The former corresponds to the 68 minutes seiche reported by Takahashi (1935) but the latter has been reported by no investigators. This failure seems to be due to the lack of a convenient method to extract the 71.3 and 61.0 minutes constituents independently from an observed record of lake surface displacement. It is very interesting that both of these two constituents have only one nodal line in the north basin. The 37.0 and 32.3 minutes oscillations are quadri- and quinque-nodal seiches in the whole basin.

Seiche motion is, of course, two-dimensional, and therefore its details cannot be revealed by any one-dimensional method such as Defant's one. The writer (1971, 1972) carried out numerical experiments by integrating the nonlinear equations of motion in order to understand the characteristics of the two-dimensional seiche motions in the actual and complicated basin of Lake Biwa. The whole basin is broken up into water columns of 1 km × 1 km square surface and variable depth. The external force inducing the seiche motion is the stress τ_a of wind which blows uniformly at the speed of $W = 5$ m/s (Eq. 14). The bottom stress τ_b is given after Miyazaki et al. (1961) as follows:

$$\tau_b = 0.0026|\bar{U}|U - 0.5\tau_a\gamma^2_b / \gamma^2_a$$

where $|\bar{U}|$ is the vertically averaged velocity vector. The initial conditions are $\eta = U = V = 0$ everywhere in the basin.

Calculation of $\eta^m_{i,j}$, $u^m_{i,j}$ and $v^m_{i,j}$ has been continued step by step from $t = 0$ to $t = 22$ hours, and the time series of calculated surface displacements are prepared at 64 mesh-points. The Fourier analysis is operated by the method of FFT (Fast Fourier Transform), yielding the calculated periods of the longitudinal seiches in Lake Biwa such as 255.5, 79.8, 69.1, 38.7 and 31.9 minutes, respectively. In the following, the characteristics of these five constituents of seiche will be briefly described.

The oscillation of the period of 255.5 minutes is the uni-nodal seiche in the whole basin as described by Takahashi (1935) and Imasato (1970) and it is obvious that this constituent corresponds to the seiche of Mode I shown in Table 29. As is clearly shown by the solid curve in Fig. 115, it has only one nodal line in the southern end of the north basin at about 1.5 km north of the narrowest channel which the previous investigators considered as the mouth of the south basin. The amplitude in the north basin is about 0.1 times as large as that in the south basin, and perhaps this should be the reason why the previous investigators could not detect this constituent in limnograms obtained in the north basin.

The oscillations of the periods of 79.8, 69.1 and 38.7 minutes are the bi-, tri- and quadri-nodal seiches in the whole basin, respectively. Their spatial distributions of nodal curves and amplitudes are shown in Figs. 116, 117 and 118. It is very interesting that the bi- and tri-nodal seiches differ in period only by 10.7 minutes. Unfortunately, however, this interesting fact has made it impossible for the previous investigators to detect these two constituents separately on limnograms.

The writer (1971, 1972) (Imasato et al. 1973) took observations at Ōtsu and Hikone (Fig. 114). Calculating the amplitude spectrum of the observed surface displacements by means of the FFT method, the writer clearly showed that these constituents actually and ordinarily exist in Lake Biwa. Fig. 119 is an example of such an observed amplitude spectrum. In this figure, arrows show the dominant peaks corresponding to each constituent of seiche. The mean observed periods corresponding to these constituents are 229.8 ± 12.2, 72.7 ± 2.2, 65.1 ± 1.5 and 40.1 ± 0.6 minutes, respectively, and they agree well with the calculated periods.

According to the writer (1972), the oscillation of the period of 31.9 minutes (Mode V) consists of the following four constituents: the first is the longitudinal quinque-nodal seiche in the whole basin, the next is the uni-nodal seiche oscillating between Kaizu and Maibara in the northern part of the north basin. the third is the bay oscillation in Shiozu Bay and the last is the shelf seiche on the lake shelf off river mouths of Rivers Yasu and Hino in the southern part of the north basin. The last constituent of this mode corresponds to the 32 minutes seiche reported by Toyohara and Habu (1938). These characteristics of oscillations are shown in Fig. 120. The left panel shows those of the first longitudinal constituent and the right panel shows those of the others. The calculated period of this constituent, 31.9 minutes, agrees well with the observed period 30.5 ± 0.8 minutes.

In Fig. 121, the distributions of velocity amplitude of each constituent of seiches are plotted (solid curves) along a line, although no detailed discussion is given here about them. In each panel, distributions of the lake surface elevation and the bottom profile are also plotted by broken and dotted curves, respectively.

As described above, the characteristics of seiches of periods longer than about 30 minutes have been clarified fairly well through observations and two-dimensional numerical experiments. We can easily find many dominant constituents of shorter periods in the amplitude spectrum but they seem to be local oscillations in a fairly small area in the basin and unfortunately their characteristics have not been well known. If we want to know them precisely through a numerical experiment and observations, we must adopt a mesh interval much smaller than 1 km.

Formation and decay of seiche

Some investigators have studied the formation and decay of seiche motions in Lake Biwa. Okada et al. (1914) enumerated the origins of seiche motions as follows:

1. Impacts of falling raindrops on the surface of the lake,
2. Accumulation of rainwater on one portion of the lake,
3. Impulsive action of winds on the surface,

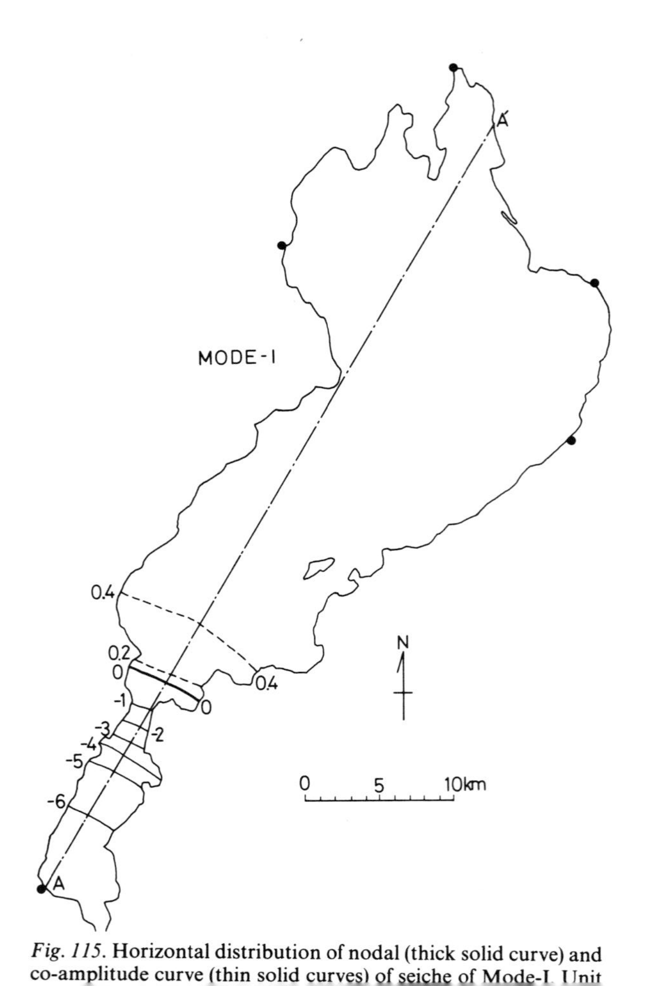

Fig. 115. Horizontal distribution of nodal (thick solid curve) and co-amplitude curve (thin solid curves) of seiche of Mode-I. Unit

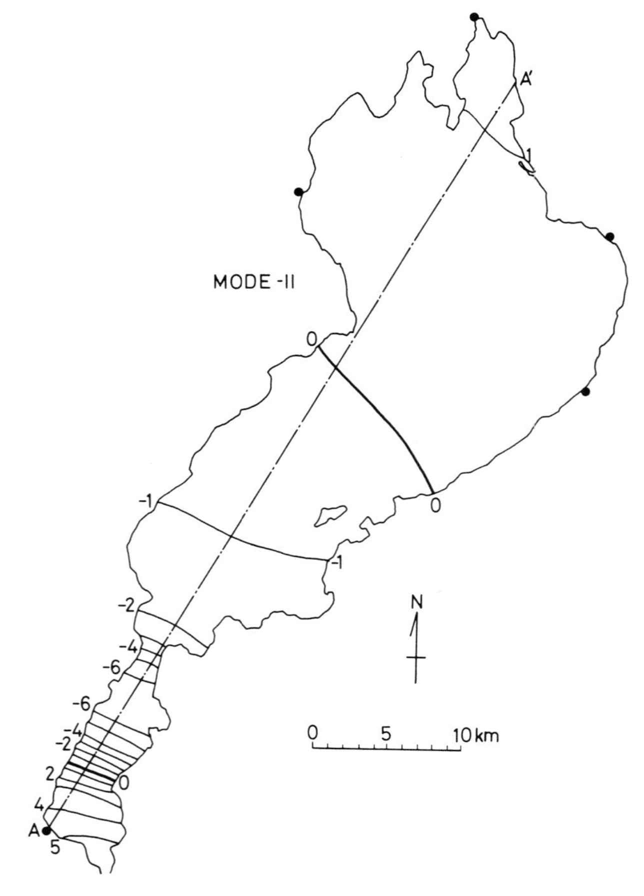

Fig. 116. Horizontal distribution of nodal (thick solid curves) and co-amplitude curves (thin solid curves) of seiche of Mode-II

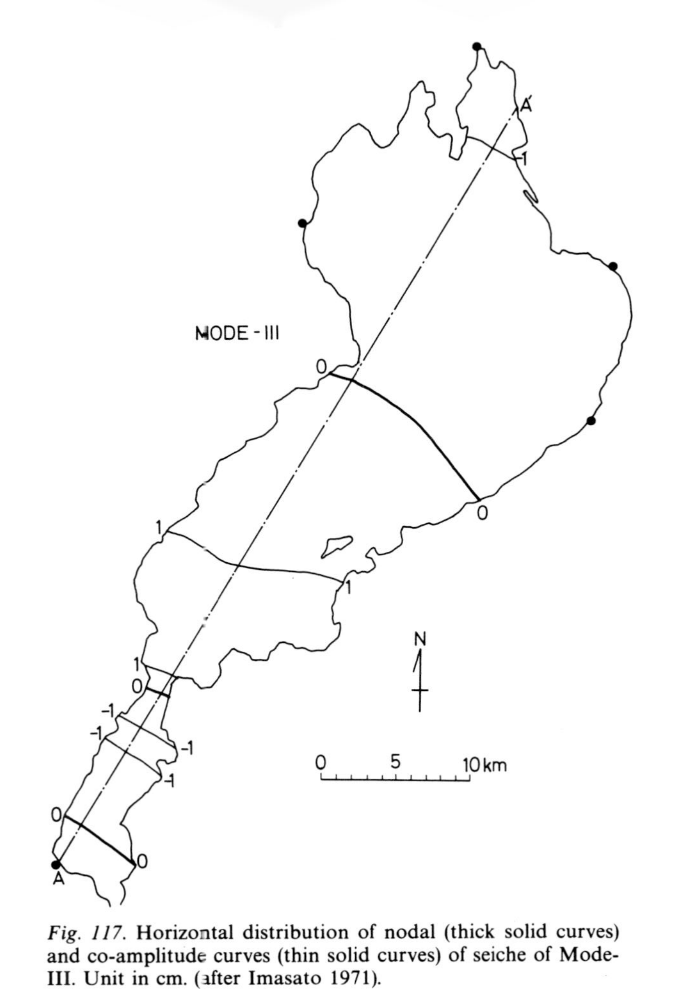

Fig. 117. Horizontal distribution of nodal (thick solid curves) and co-amplitude curves (thin solid curves) of seiche of Mode-III. Unit in cm. (after Imasato 1971).

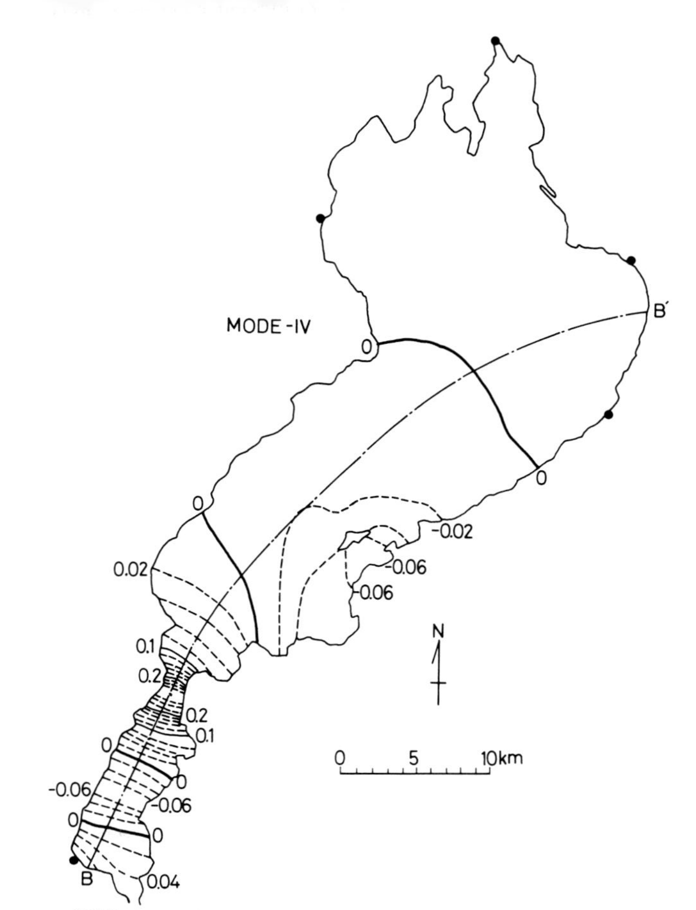

Fig. 118. Horizontal distribution of nodal (thick solid curves) and co-amplitude curves (thin solid and broken curves) of seiche of Mode-IV. Unit in cm. (after Imasato 1972).

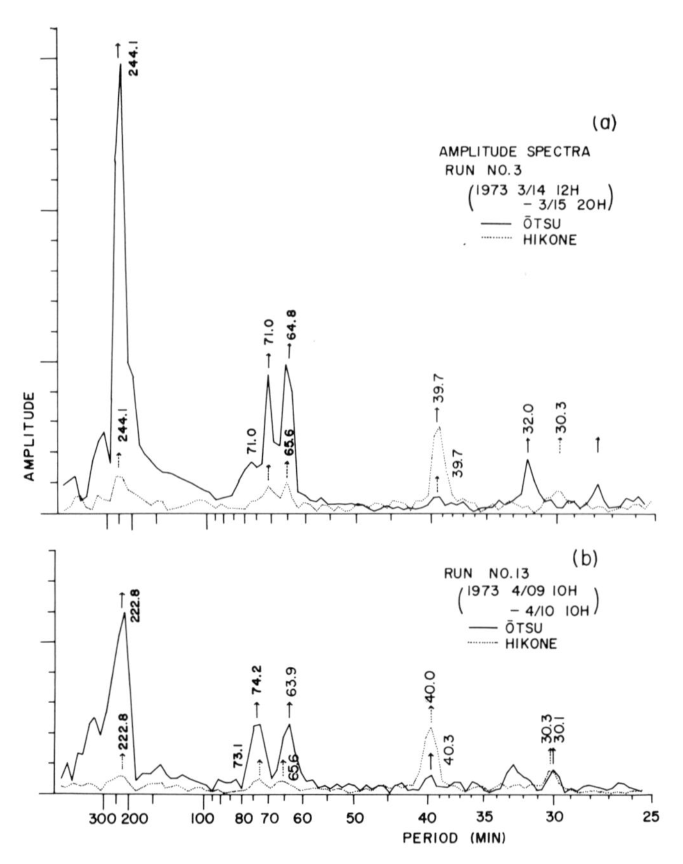

Fig. 119. Amplitude spectra of water surface displacements at Ōtsu (solid curve) and Hikone (dotted curve). Arrows indicate the spectral peaks of each constituent of seiche (Imasato et al. 1937).

4. Release of electric attraction of thunder-clouds,
5. Sudden changes of the barometric pressure.

They applied their theory to the 30 minutes seiche in the north basin and concluded that the parts contributed to the amplitude of the seiche by the barometric change, the rainfall and the impulsive action of wind are 38.4, 35.5 and 26.1%, respectively. It must be mentioned here that all of the above phenomena are accompanied by the wind blowing over the lake.

Suda et al. (Kobe Mar. Observ. 1926) stated that the 242.0 minutes constituent is frequently accompanied by thunderstorms. They also gave some qualitative discussions about external forces which induce the seiche motion. As has already been mentioned, Takaya (1931) made a histogram of wind speed and of the frequency of appearance of Mode I seiche, and concluded that the origin of seiche motion is the seasonal wind in winter.

Imagine that a fairly large amount of energy is given to a portion of lake water during a short time by any reason. Then, it will be dispersed and make

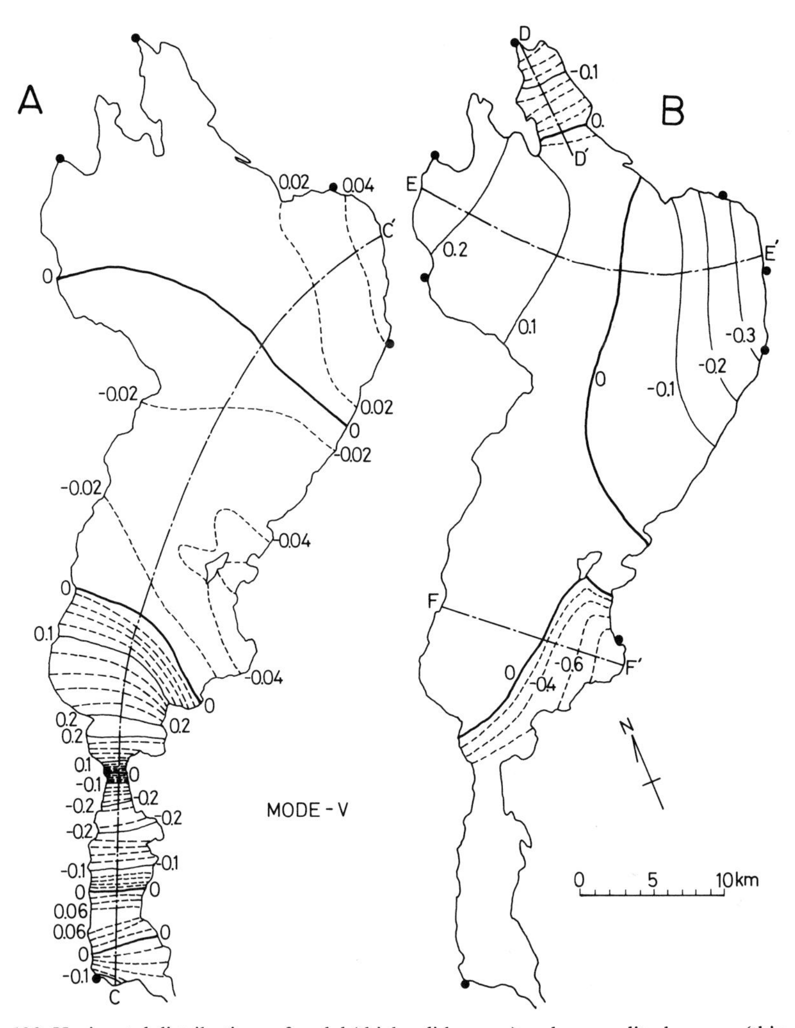

Fig. 120. Horizontal distributions of nodal (thick solid curves) and co-amplitude curves (thin solid and broken curves) of seiches of Mode-V.
A: the longitudinal quinqe-nodal seiche in the whole basin, B: the bay oscillation in Shiozu Bay, the uni-nodal seiche oscillating between Kaizu and Maibara and the shelf seiche in the southern part of the north basin. Unit in cm. (after Imasato 1972).

progressive waves of various scales. Oscillations except seiche will disappear soon through their breaking on the lake wall. However, seiche will continue to be alive because seiche is the standing wave and is induced just by the reflection of the progressive wave. The writer thinks that Takaya's histogram proves that the wind blowing over the lake is coherent with the appearance of seiches not only in winter but also in every month of the year. According to the writer (1972), the amplitude of each constituent of seiches in Lake Biwa varies with the wind speed and the wind direction. Seiche motions always appear in Lake Biwa

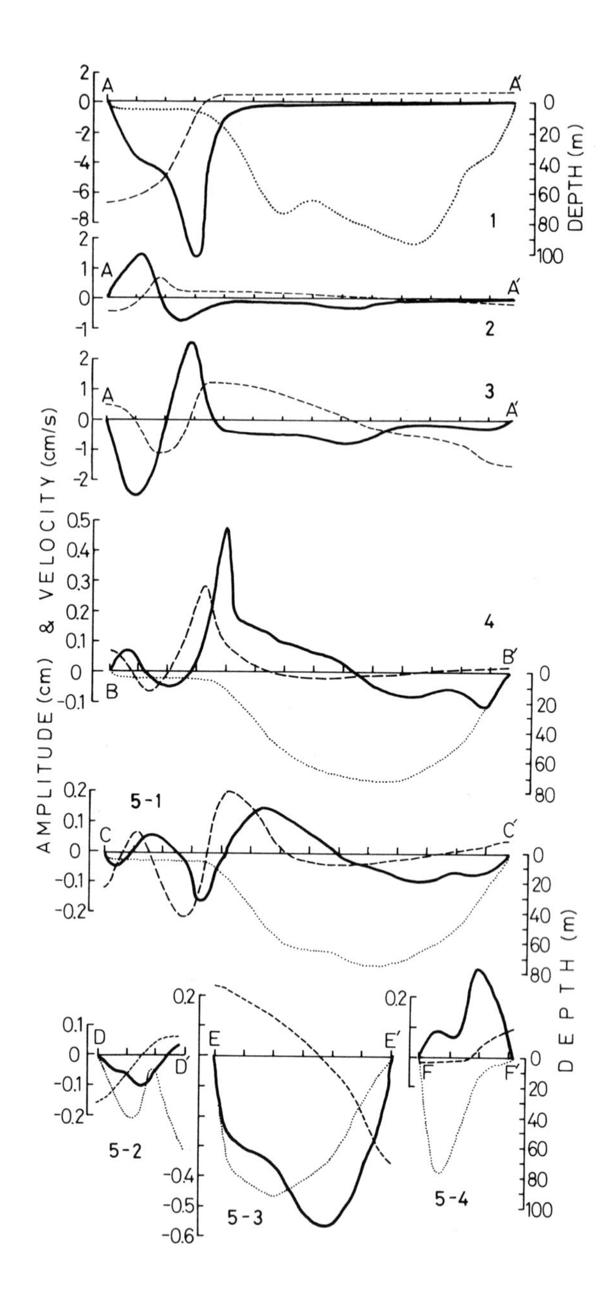

Fig. 121. Profiles of lake bottom (dotted curve), amplitude (broken curve) and velocity (thick solid curve) of seiche. Profiles in panels 1, 2 and 3 are drawn along the line $\overline{AA'}$ in Figs. 115, 116 and 117, respectively. Profiles in panel 4 are drawn along the line $\overline{BB'}$ in Fig. 118, those in panel 5–1 along $\overline{CC'}$ in Fig. 120-A and those in panels 5–2, 5–3 and 5–4 along $\overline{DD'}$, $\overline{EE'}$ and $\overline{FF'}$ in Fig. 120-B, respectively (after Imasato 1971 and 1972).

and the wind always blows over the lake, too. Therefore it may be concluded that the most prevailing external force which is able to make a movement of lake water and to induce seiche motions is the wind stress although other phenomena such as the discharge of a large amount of river-water are able to induce a seiche.

On the other hand, we have some unsolved problems in the decay process of seiche. Seiche continues to lose its energy due to the bottom and lateral frictions, but we know little of the eddy viscosity governing the energy loss, although Nomitsu (1935) stated that the eddy viscosity is a function of velocity. Recently the writer (1979) studied the decay of seiche motion due to the nonlinear interaction. If one or two constituents of seiche transfer energy nonlinearly to a progressive wave, this progressive wave should lose its energy rapidly through breaking on the solid wall of a lake, and as a result seiches will decay more rapidly than in the case they lose energy only through friction.

According to Hasselmann et al. (1963) and Tukey (1963), the bi-spectrum $B(\omega_1,\omega_2)$ is the measure of nonlinear interactions among three component waves of which angular frequencies ω_1, ω_2 and ω_3 satisfy the relation $\omega_1+\omega_2+\omega_3=0$. If these three waves do not interact nonlinearly with each other, we have $B(\omega_1,\omega_2)=0$. The writer calculated the bi-spectrum of observed and calculated water surface displacements in Lake Biwa and examined which three component waves nonlinearly interact with each other. An example of the bi-spectrum of surface displacements of Lake Biwa is shown in Fig. 122, where panel A shows the power- and bi-spectra observed at Ōtsu and panel B shows the spectra obtained in the numerical experiment. In each panel, the top is the power spectrum and the bottom is the bi-spectrum. In the figure, an arrow denotes a frequency of the constituent of seiche, and a narrow belt around the arrow and bounded by two broken lines denotes the frequency band where the constituent of seiche makes a dominant peak and has significant power-spectral densities. From this figure, it is easily seen that a dominant nonlinear-interaction occurs among the constituents of seiche. Especially, in the bi-spectrum at Ōtsu, it dominates at the frequency of seiche of Mode I. According to the writer (1979), if we are concerned only with the constituents of periods longer than 30 minutes, it does not happen that all of the three waves nonlinearly interacting are constituents of the seiche and it therefore follows that at least one of the three waves is a progressive wave. An example illustrating this situation is panel B of Fig. 122 obtained in the numerical experiment. In the power-spectrum (upper panel), we find a dominant peak at the period of 50.2 minutes, but this is not a constituent of seiche but a progressive wave. The bi-spectrum has the dominant peak at (ω_1,ω_2) corresponding to the constituents of Mode I and Mode III. It indicates that the wave of the period of 50.2 minutes interacts nonlinearly with the two constituents of seiche of Mode I and Mode III to make the dominant peak at 50.2 minutes in the power spectrum. This fact indicates that the seiches transfer their energy to a progressive wave which decays relatively rapidly through breaking on the lake wall. Consequently, the seiches lose energy more rapidly than in the linear system.

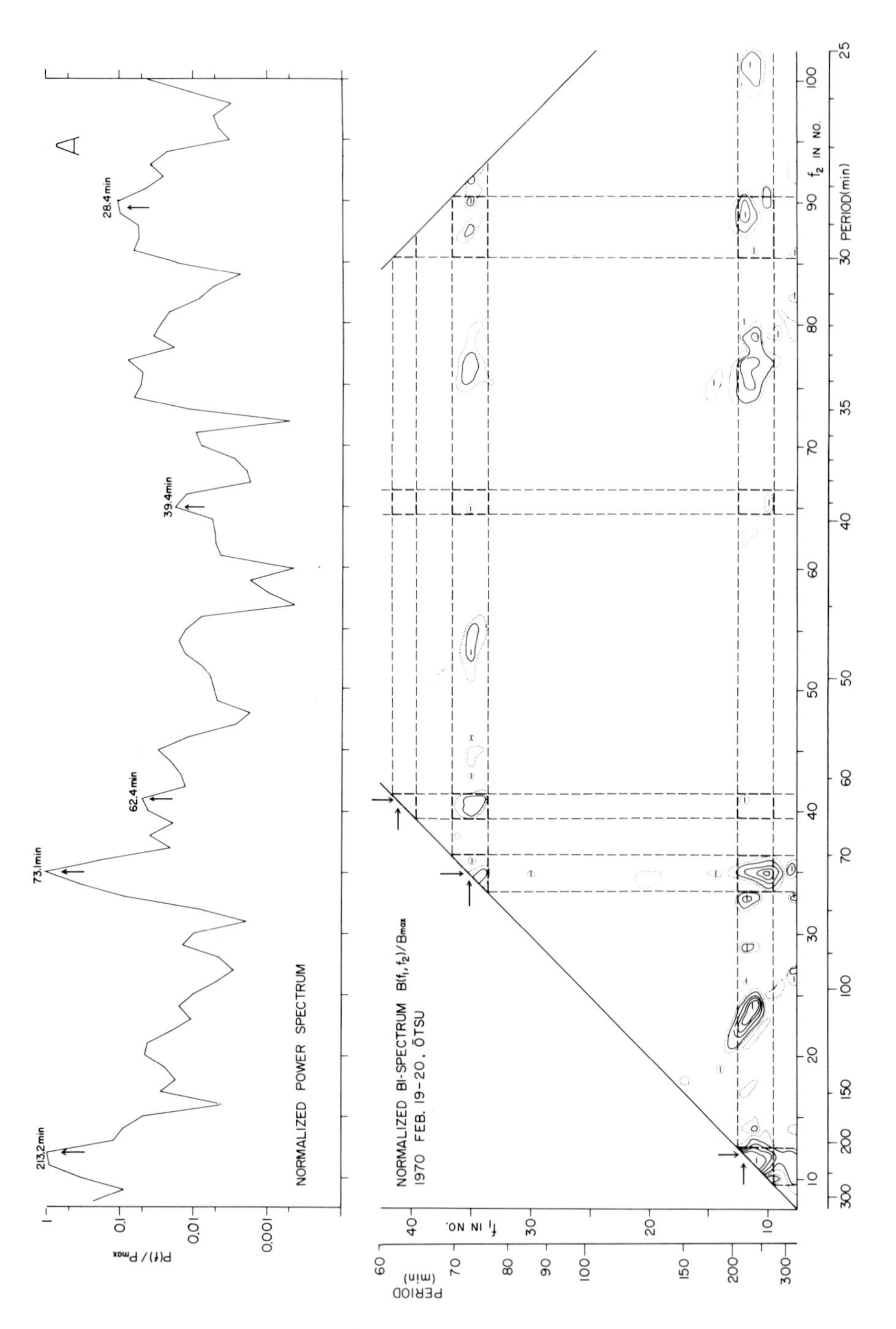
A
NORMALIZED POWER SPECTRUM
213.2min
73.1min
62.4min
39.4min
28.4min
P(f)/Pmax
NORMALIZED BI-SPECTRUM B(f1, f2)/Bmax
1970 FEB. 19-20, ŌTSU
f1 IN NO.
PERIOD (min)
f2 IN NO.
PERIOD(min)

Fig. 122A.

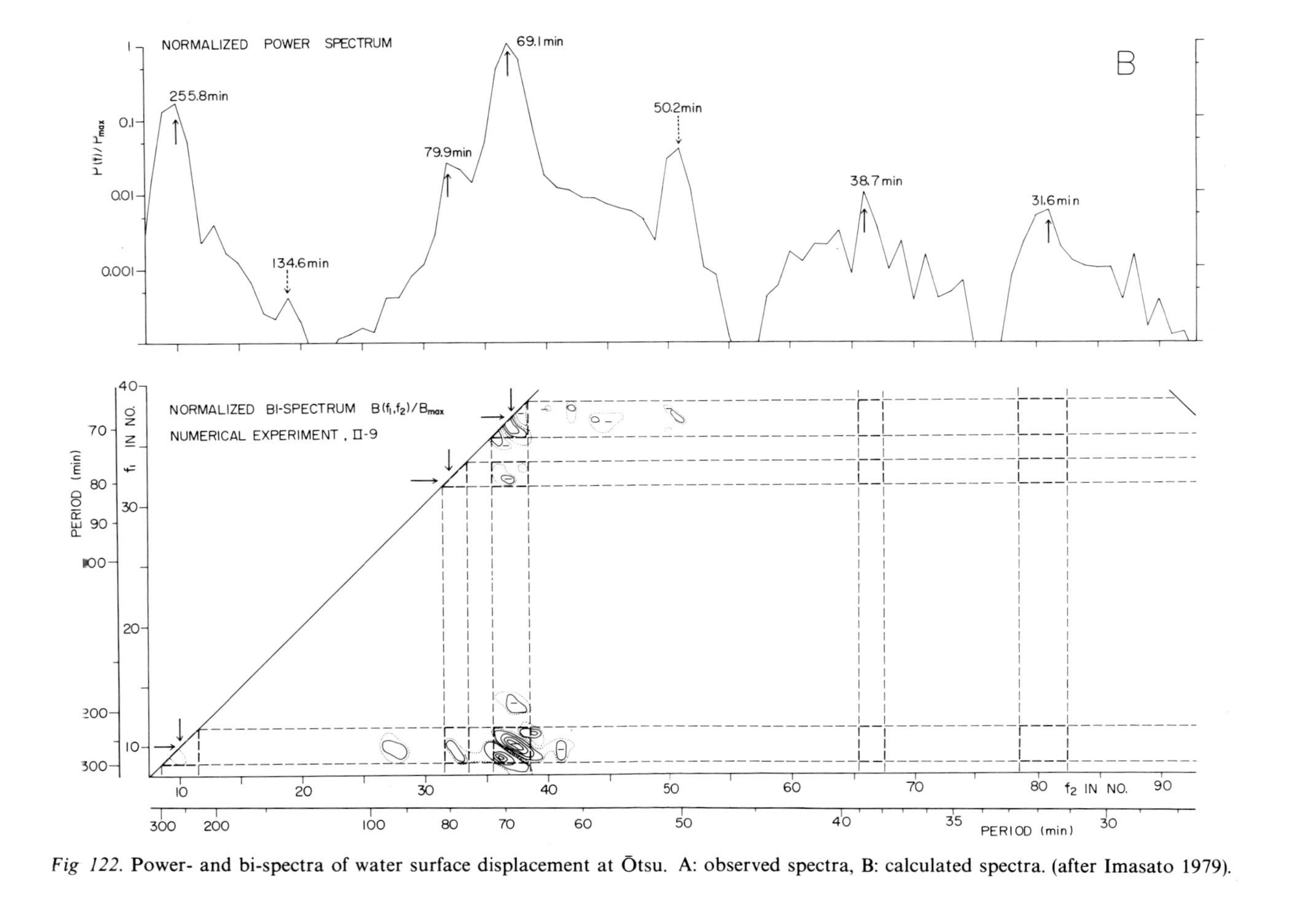

Fig 122. Power- and bi-spectra of water surface displacement at Ōtsu. A: observed spectra, B: calculated spectra. (after Imasato 1979).

Fig. 122B.

Table 30. Magnitude of nonlinearity of seiche motion in Lake Biwa *(after Imasoto 1979).*

		D_{max}
Seiche	Mean observed value at Ōtsu	0.837
	Calculation	0.704
Wind-wave	Lake Biwa	9.369
	Wind-tunnel	16.84

Table 30 shows the magnitude of nonlinearity of seiche motions in Lake Biwa in terms of nondimensional bi-spectral density $D_{max} = B_{max}(\Delta f)^2/(P_{max}\Delta f)^{3/2}$, where B_{max} is the bi-spectral density of the most dominant bi-spectral peak, P_{max} is the power-spectral density largest in the power densities of the three waves interacting with each other and Δf is $1/\Delta t$. According to this table, the value of D_{max} of the observed seiche in Lake Biwa is 0.837, which is a little larger than that in the numerical experiment, and is about 5–9% of that of the fully developed wind-waves.

Inorganic chemical aspects

Mutsuo Koyama, Munetsugu Kawashima and
Takejiro Takamatsu

Chemistry of hydrosphere around the lake

Environmental background

As is well known, chemical characteristics of the lake water and sediment reflect geological, climatological, and limnological situations, such as natures of soil and rocks in and around the lake, the rate and amount of precipitation, the inflow and outflow through rivers, the extent of variation of the temperature, and the depth of the basin. In addition to these, impacts of human activities on the lake water have become much more evident in recent decades than before, as is common in a civilized country. Therefore, it is worthwhile to describe briefly the present environmental background of the Lake Biwa in order to understand general features of the lake water.

It is customary to divide the lake into two parts as is shown in Fig. 123, because of their different limnological aspects. The northern part of the lake with the mean depth of 44.7 m accumulates almost ninety percent of the lake water. The southern part of the lake with the mean depth of 2.8 meters carries the lake water from the north to the River Seta which is the only natural outflow.

It is said that in the northern part there are three gyres of the lake water which play a significant role in transporting dissolved materials within the lake. Another mass movement of the lake water is seiche which sometimes transports as much as 40 million tons of water from the southern to the northern part according to I. Okamoto (Okamoto & Yagi 1969). The fact that the material transport from the southern to the northern lake occurs was proven by one of the water research groups (Fujinaga et al. 1966). They found more than one third of the materials such as nitrogen compounds, silicate ions, and phosphorus compounds in the South Lake is continuously transported to the North Lake by back-streaming water.

Horie, S. Lake Biwa

ISBN 90 6193 095 2. Printed in The Netherlands

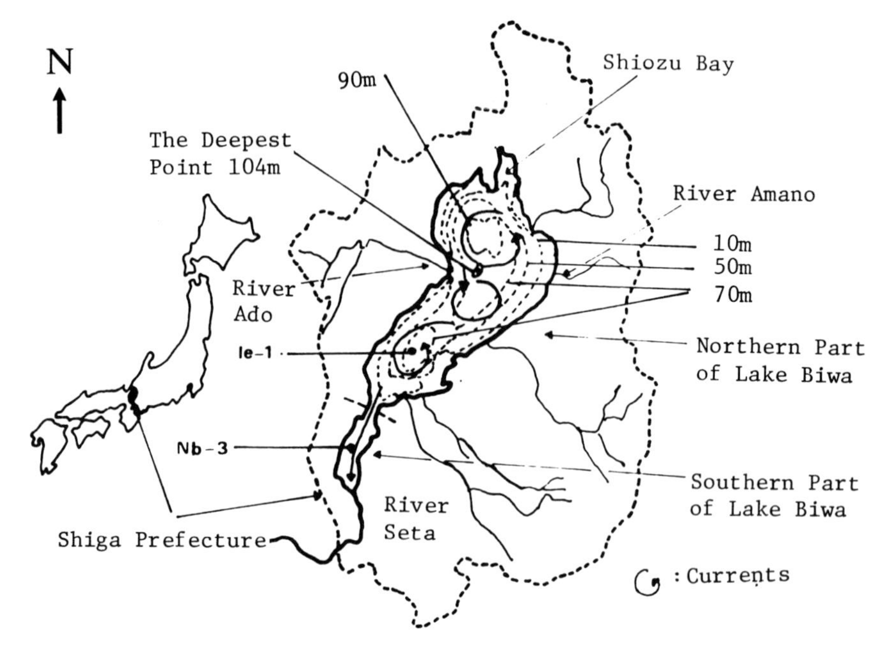

Fig. 123. Map of Lake Biwa (Koyama et al. 1975).

Inflow to the lake

More than one hundred rivers flow into the lake, transporting approximately one-third of rain or snow precipitated in the watershed of the lake, which is about five times as wide as the area of the lake itself. Although those rivers differ in water quality from one to another, average concentrations of dissolved chemical species are similar to those of rivers throughout Japan. For example, the concentration of calcium ions seldom exceeds 10 ppm and that of silicate ions is around 12 ppm. In Table 31 are shown concentrations of several nutrients in representative rivers flowing into the lake.

As is obvious, small rivers flowing into the southern part of the lake contain much more nutrition than those around the northern part where the urbanization is much less significant.

Y. Saijo estimated the annual inputs of several kinds of nutrients into the lake from various sources, as is presented in Table 32. It is surprizing that considerable portions of nutrients are supplied from sediments.

Chemistry of lake water

Since Lake Biwa is a warm monomictic lake, it has one vertical circulation period in winter and one stratification period which begins in the early spring

Table 31. Concentration* of nutrients in rivers flowing into Lake Biwa (Koyama 1974).

Name of river	Total-P	Nitrate-N	Total-N	Ca^{2+}	Mg^{2+}	Colorimetric SiO_2	Cl^-
River Ado	0.01	0.074	0.22	5.3	1.8	8.0	4.6
River Amano	0.04	0.19	0.31	27.0	3.5	8.3	6.3
River Echi	0.01	0.14	0.31	7.7	1.9	9.4	3.7
River Yasu	0.04	0.09	0.60	10.1	2.3	11.0	9.8
Rivers flowing into South Lake	0.02–1.9	0.04–0.83	0.60–10.5	—	—	—	9.5–363
Average of Rivers in Japan	0.02	0.22	—	10.4	3.6	10.1	7.1

*Concentration: ppm.

Table 32. Material balance of nutrients in Lake Biwa (Saijo et al. 1966).

	Nitrogen		Phosphorus	
	tons/year	(%)	tons/year	(%)
Inflow				
river water	1,188	55	65	33
precipitation	582	27	51	26
dissolution from sediment	390	18	82	41
total	2,160	100	198	100
Outflow				
river water	1,680	78	79	40
sedimentation	480	22	119	60
total	2,160	100	198	100

and lasts to the late fall in the northern part. In the southern part, however, stratification is very feeble and the vertical mixing of water takes place throughout the year. These physical characteristics have an important role on chemical species in the lake water. Therefore, the concentrations of chemically and biologically active substances vary to a considerable extent from season to season and from place to place. The averages of those are summarized in Table 33.

Vertical distribution of chemical species

In Figs. 124 and 125, typical vertical distributions of chemical species in both the circulation and the stratification periods are shown.

As is obvious, distributions of chemical species such as oxygen, nitrate, filtrable-phosphate, total-phosphate and dissolved silicate ions have close correlations with those of temperature. This sort of correlation is also demonstrated in Fig. 126, which represents the cross-sectional profiles of distributions of those chemical components.

On the other hand, concentrations of such ions as chloride, magnesium and calcium are almost constant from the surface to the bottom of the lake.

Horizontal distribution of chemical species

In the pelagic part of the northern part, the horizontal distribution of chemical components in surface water is considerably homogeneous, probably because of the mixing of water by gyres.

However, definite concentration gradients are observed for most of the chemical species in the vicinity of the boundary between the northern and the southern parts and also in the littoral areas.

Table 33. Probable concentrations of elements in the lake water, planktons and sediment (Koyama 1974).

Element	Lake water (mg/l)	Plankton (mg/g)	Sediment (mg/g)
Na	5.0	0.85	8.5
K	3.8	5.5	30
Ca	9.0	4.8	3
Mg	2.4	—	8
Cl	6.0	0.8	—
Br	0.013	0.002	—
Si	0.5	150	250
P	0.01	4	1.5
N	0.2	38	5
V	0.0003	—	0.02
Mn	0.001	0.05–0.5	1.0—10.0
Fe	0.03	4.0	50
Co	0.00003	0.002	0.018
Cu	0.0025	0.068	0.046
Zn	0.031	0.26	0.13
As	0.00001	0.014	0.03—0.5
Al	0.026	1.20	110
Sc	4×10^{-6}	0.001	0.014

Seasonal variation of chemical states of nutrients

Although concentrations of nutrients in the lake water vary from time to time, some trends of variation can be observed. Especially noticeable is the seasonal variation of colorimetrically measurable silicate ions which vary with a certain periodicity as is shown in Fig. 127. Namely, the concentration of the ions begins to decrease in the late fall, marks a minimal value during spring when the bloom of diatom becomes usually maximal and begins to increase in the early summer which is the rainy season in Japan. It can be calculated that as many as sixty thousand tons of silicate ions are precipitated from the lake water to the bottom annually. It should be noticed that there exists a kind of silicate ions that is filtrable with the use of a Millipore Filter but not colorimetrically measurable. The kind of silicate ions which are considered to be soluble oligomers are dominant during the period of fixation of silicate ions by diatoms.

On the other hand, the way in which phosphate ions change is not clear. An example is shown in Fig. 128. As a general trend, however, the fraction of the filtrable phosphate ions begins to increase in autumn. This trend is clearly seen in the bottom layer water of the northern part.

Seasonal variations in concentrations of nutrients are also correlated with biological activities in the following fashion. In Figs. 129 and 130, the concentration of organic nitrogen, total nitrogen, organic phosphate, or total phosphate is plotted against that of chlorophyll-a which is taken as a measure of biological activity.

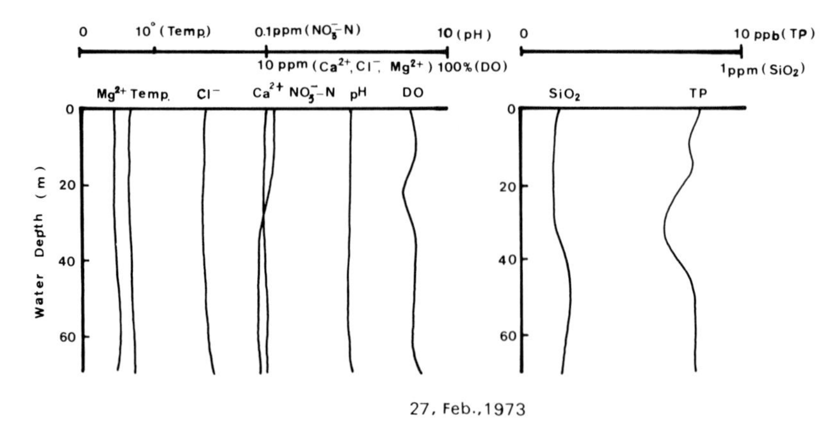

Fig. 124. Vertical distribution of chemical species in the northern part of Lake Biwa St. Ie-1(water depth 70 meters) (after Koyama 1974).

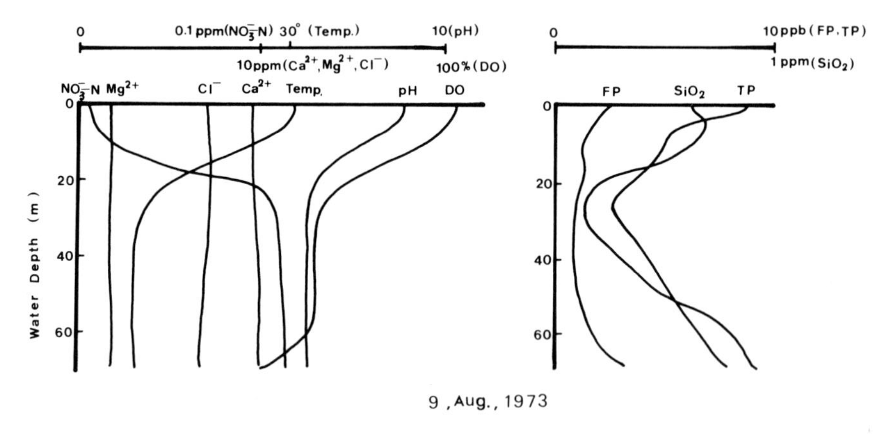

Fig. 125. Vertical distribution of chemical species in the northern part of Lake Biwa St. Ie-1 (water depth 70 meters) (after Koyama 1974).

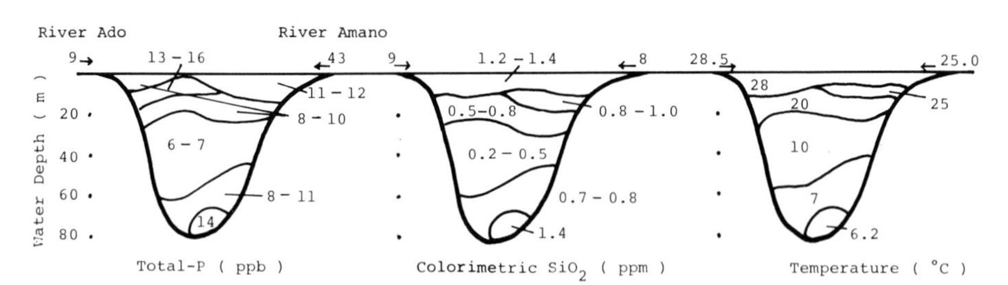

Fig. 126. Cross-sectional distribution profiles of chemical components and temperatures in the northern part of the lake (along the line between Rivers Amano and Ado) (after Koyama 1974).

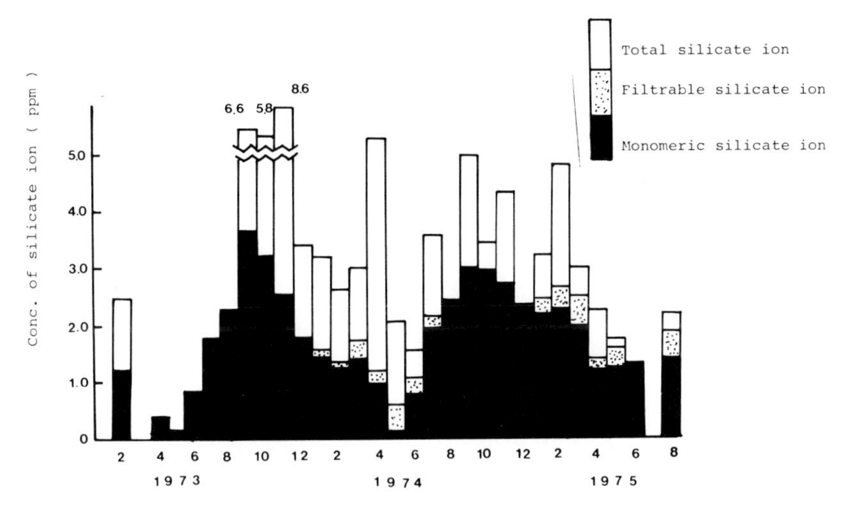

Fig. 127. Monthly variation of silicate ion concentration in the southern part of Lake Biwa (Nb-3, 0 meter) (Koyama et al. 1975).

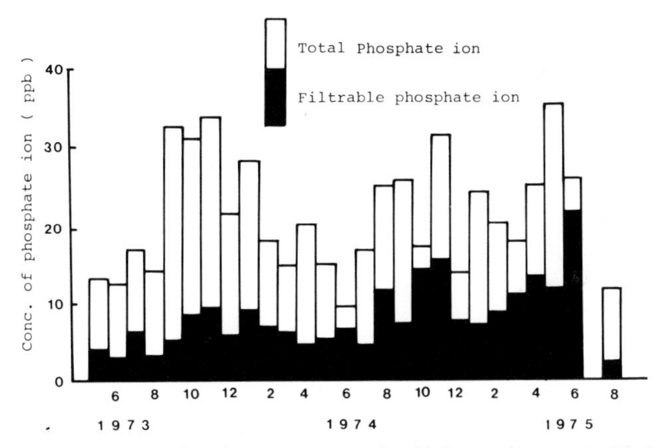

Fig. 128. Monthly variation of phosphorus concentration in the southern part of Lake Biwa (Nb-3, 0 meter) (Koyama et al. 1975).

It can be clearly observed that the data points which seem to be randomly scattered can be classified into several groups within each of which a definite linear correlation is established. Apparently, the data of the southern part of the lake constitute groups different from those of the northern part of the lake. It may also be noted that the data from different regions constitute different groups. These facts seem to demonstrate that there exist differences and/or successions of phytoplanktonic species which respond to the change of trophic and seasonal circumstances. Another point which should not be overlooked is that all regression lines in these figures converge to 0.008 ppm with respect to the total phosphate ion concentration and at 0.2 ppm with respect to the total

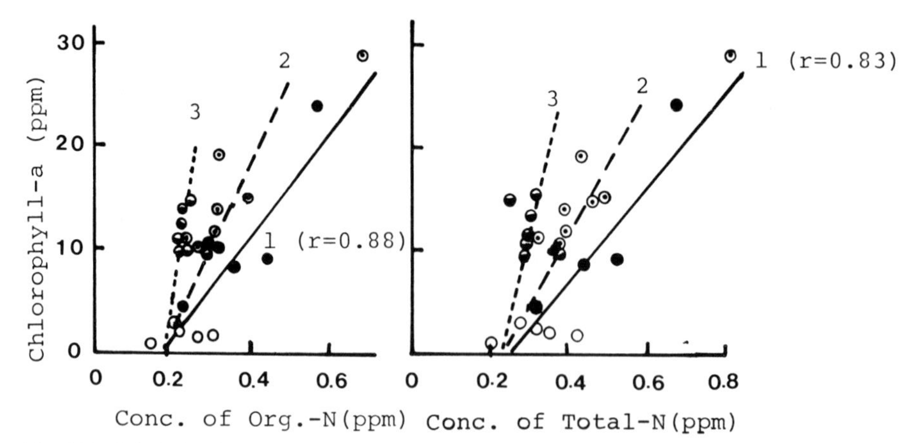

Fig. 129. Correlation between chlorophyll-a and nitrogen compounds (Koyama et al. 1975).
1: 16, July, 1975 (o ; northern, ● ; southern)
2: 13, September, 1975 (⊙ ; southern)
3: 13, September, 1975 (◒ ; northern)

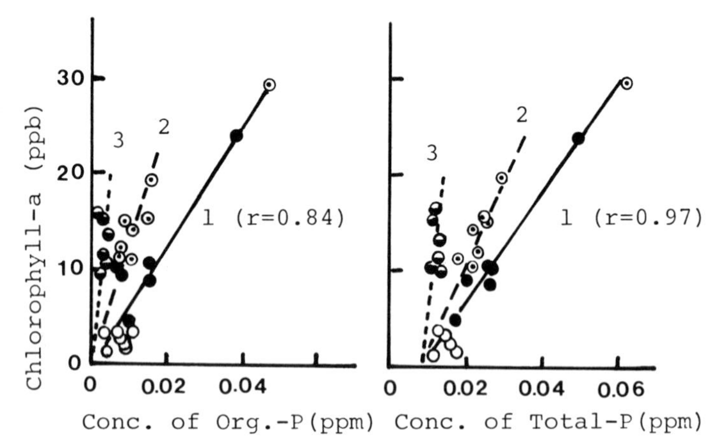

Fig. 130. Correlation between chlorophyll-a and phosphorus compounds (Koyama et al. 1975).
1: 16, July, 1975 (o ; northern, ● ; southern)
2: 13, September, 1975 (⊙ ; southern)
3: 13, September, 1975 (◒ ; northern)

nitrogen concentration when extrapolated to the zero value of chlorophyll-a concentration. This fact suggests that biological species quite different from those which are present now would appear in Lake Biwa, when the concentrations of total phosphate and total nitrogen become extremely different from the present levels. This kind of deduction can be applied to the past history of the lake water in which such an extremely oligotrophic condition would have very likely been present. The evidence of a paleotrophic situation could be visualized in connection with microfossils with the concentrations of nutrients contained in the deep boring core.

Seasonal distribution of trace metal ions

Vertical circulation and stratification of the lake water have important effects on supplying oxygen to the bottom and on transporting dissolved elements to the upper layer. A typical example is the vertical distribution of manganese that is shown in Fig. 131. Most of manganese ions in the circulation period are present in the form of particulate matters in the lake water and the concentration gradient of the ion is feeble. On the other hand, the manganese concentration of the bottom layer increases up to several hundred $\mu g/l$ during the stratification period and a large concentration gradient can be observed from the bottom to the middle layer of the lake water. These facts suggest that water-soluble manganous ions are continuously supplied from the sediment of reducing condition and a part of the ions is removed by precipitation in the form of oxidized manganese compounds. If once manganese dioxide which has strong affinity with manganous ions is formed, the compound will scavenge a large part of manganous ions dissolved in the lake water.

Other transition elements which are sensitive to reduction and oxidation in natural environments behave in a manner more or less similar to manganese depending on their redox potentials.

In Table 34 are shown the concentrations of several elements in particulate matters taken from various depths of the lake water in different seasons.

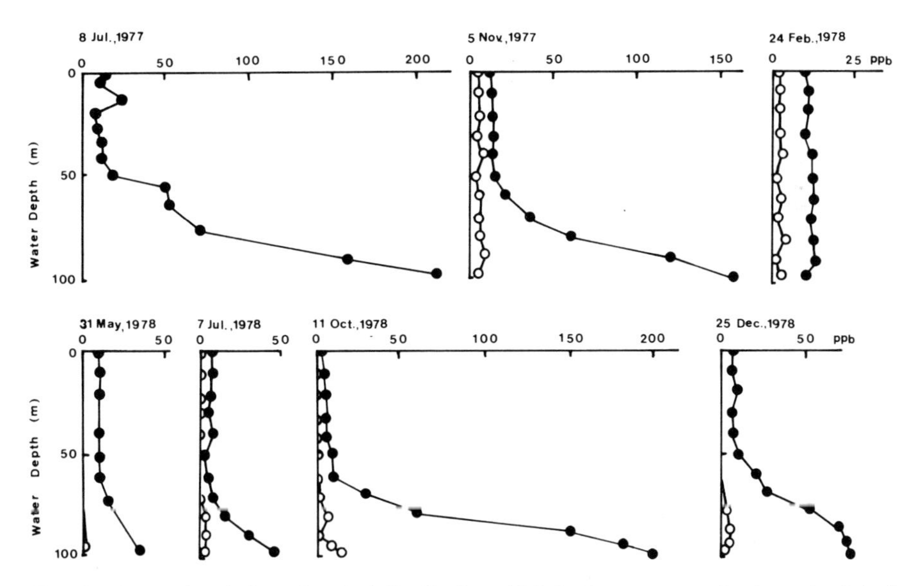

Fig. 131. Seasonal variation of vertical distribution of Mn in the water near the deepest point of Lake Biwa (Kawashima et al. 1978).
●——●: total manganese
○——○: filtrable manganese (Mn^{2+})

Table 34. Concentrations of several elements in the form of particulates in the lake water.

Water depth (in m)	Element							
	Na	Mn	Fe	Co	Cr	Zn	Sc	Sb
5	21	4.8	29	0.030	1.1	1.6	3.9×10^{-3}	0.15
40	9.5	5.7	17	0.030	0.31	2.4	2.7×10^{-3}	0.08
70	23	75	31	0.054	0.35	1.9	4.4×10^{-3}	0.04

Sampling station: Ie–1, water depth 70 m.
Sampling date: 19 July, 1976.
Concentration: μg/1 lake water.

Chemistry of sediments

Sediments near the bottom surface

The analysis of sediment cores near the bottom surface is important in finding out the relation between sedimentary environments and chemical composition of sediments and the recent environmental history preserved in the core — especially of pollution problems and the chemical and biochemical processes taking place after sediments have deposited. By accumulating this kind of information, data obtained from long sedimental core samples which have deposited from several hundred thousand years before up to the present could be understood more precisely by taking into account what had vanished or what had been preserved before the sediments were fixed in composition.

In reality, the composition of the bottom surface sediment is profoundly influenced by the sedimentary environments. The most influential is the water depth which is directly related to the movement of water, hence, the supply of material.

Several examples of distribution profiles of chemical compositions in relation to the water depth are shown together with oxidation reduction potentials (Eh) in Figs. 132, 133, 134 and 135. These elements were determined by means of neutron activation, X-ray fluorescence and atomic absorption spectrometry. Eh values were measured with a couple of electrodes (Pt wire vs. SCE) and converted to potentials of NHE. For the sake of simplicity, classification of elements according to their distribution profiles is shown in Fig. 136.

Distribution of Eh

The Eh value measured on the core taken at the place of large water depth drops sharply to approximately 80 m volt just below the bottom surface and gradually shifts to the less negative direction and approaches to 100 m volts on going from the bottom surface to 40 cm below the surface.

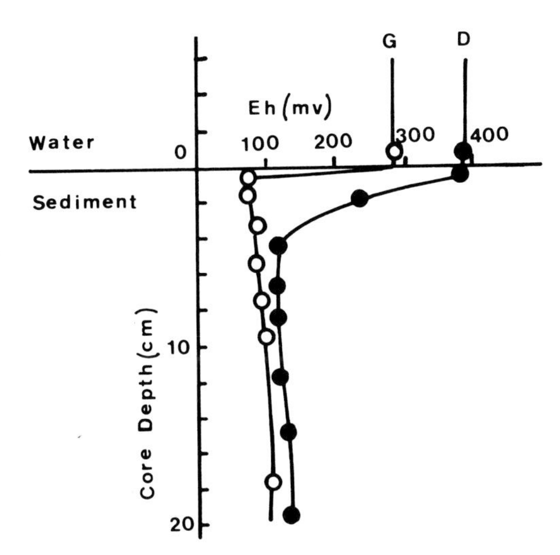

Fig. 132. Vertical distribution of Eh value in the bottom water and in the sediment cores at station G and D (Kawashima et al. 1978).
Eh values shown here are based on NHE. Water depth of stations: G 97 m, D 4 m.

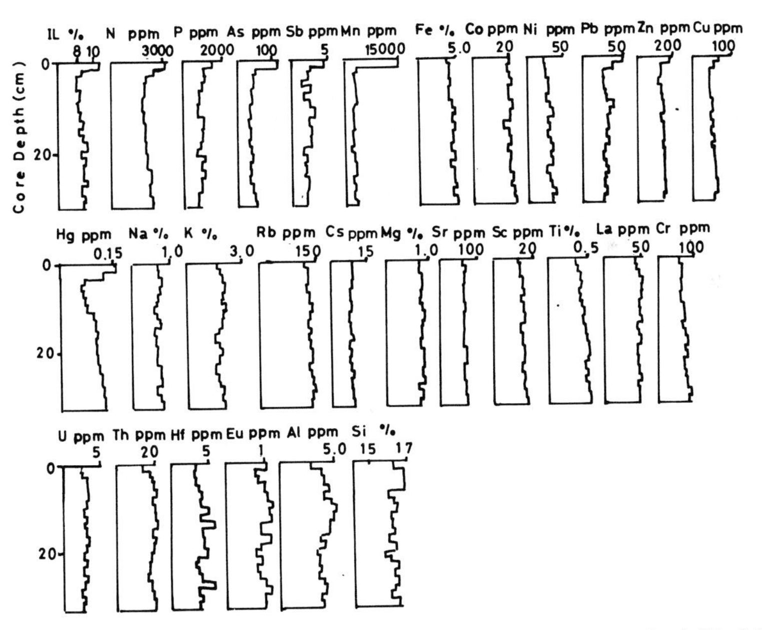

Fig. 133. Vertical distribution of elements in the sediment core at St. G (water depth 97 m) (Kawashima et al. 1978).

At the place with a small water depth, i.e., 4 m, the region of a positive potential extends to a deeper part and the minimum potential becomes less negative. However, the potential in the steady region approaches to 100 m volt which is almost the same as in the sediments taken at the place with a large water depth.

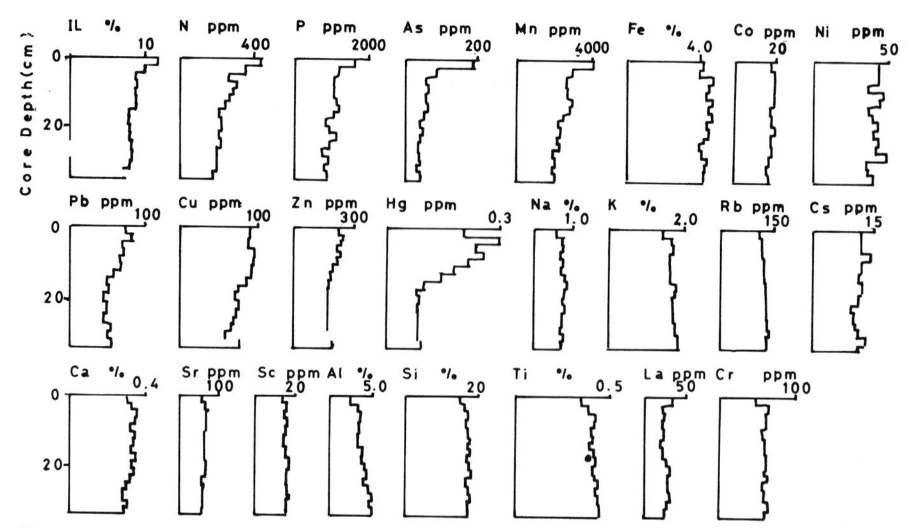

Fig. 134. Vertical distribution of elements in the sediment core at St. E (water depth 75 m) (Kawashima et al. 1978).

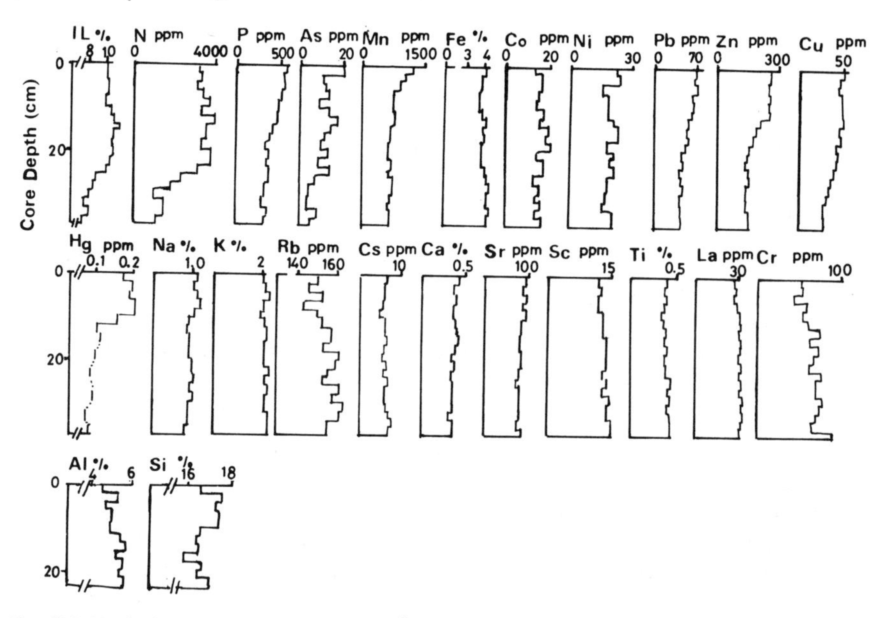

Fig. 135. Vertical distribution of elements in the sediment core at St. D (water depth 4 m) (Kawashima et al. 1978).

Distribution of inorganic ions

Manganese, phosphorus and arsenic. Among twenty-five inorganic ions measured with 40 cm core samples, manganese, phoshorus, and arsenic ions give distribution patterns closely correlated with those of Eh along the core. These ions are found to be particularly concentrated in manganese nodules which are sometimes found in the size of a finger nail. Since a number of minute manga-

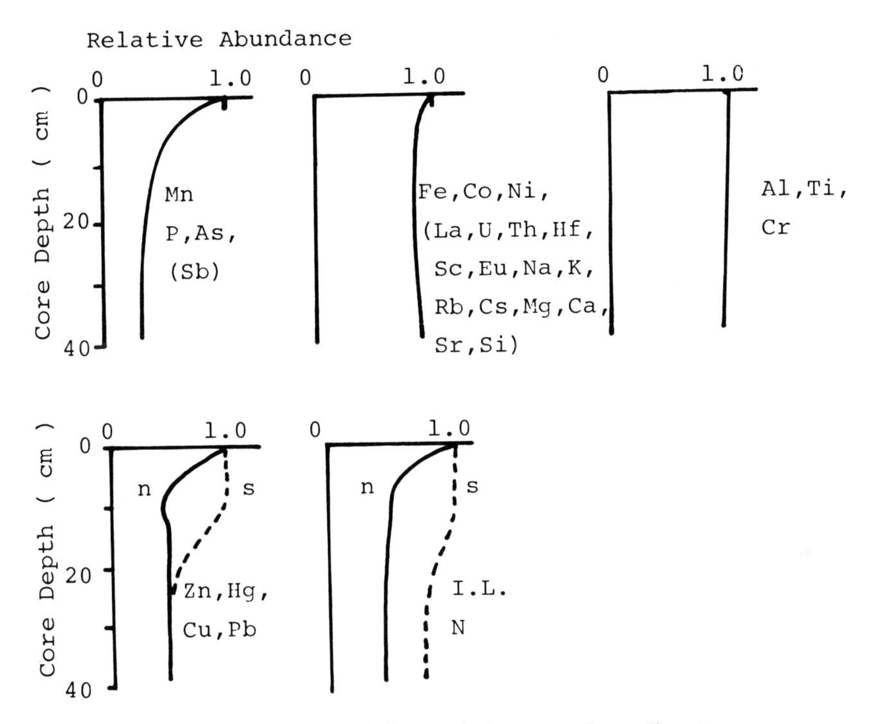

Fig. 136. Typical distribution patterns of elements in bottom surface sediment cores.
n: North Lake; s: South Lake; I.L.: ignition loss; N: total nitrogen. Ignition losses were corrected for all the distribution patterns except for nitrogen.

nese nodules are found in a microscopic view of the bottom surface sediments, a large part of manganese in the oxidized layer of sediments is considered to be present in the form of nodules.

There are frequent discussions in the literature that the accumulation of phosphorus in surface sediments is the evidence of the recent accelerating eutrophication. It is certainly true that most of the phosphorus carried into the lake will be deposited on the lake bottom. However, the contribution of the recent eutrophication to the phosphorus content in the sediment should be evaluated by taking into consideration the natural process of concentrating the element in the bottom surface of the lake. Let us assume the average excess concentration of phosphorus in the surface layer sediments with thickness of 1 cm to be 800 ppm. The total amount of phosphorus in the surface layer of the lake is calculated to be 5,000 tons. The annual rate of accumulation of phosphorus on the bottom surface has been estimated in the 1960's by Saijo. According to his calculation, the net deposition of phosphorus ions which are brought by incoming water, namely, rainfall and inflowing rivers, is estimated to be 40 tons/year. If the deposition of phosphorus which has dissolved out from the bottom sediments is taken into account, the mean deposition rate amounts to 120 tons/year.

The period required for sedimentation of 1 cm thickness is calculated to be ten years by using the annual sedimentation rate of 1 mm/year in Lake Biwa which is commonly accepted. If so, the amount of phosphorus precipitated annually becomes 500 tons/year on average which is four to twelve times as much as what Saijo estimated. Which figure one might take, a simple process such as the successive sedimentation of phosphorus in which no movement of phosphorus ions in the sediment is taken into account, fails in explaining the excess phosphorus concentration in the bottom surface sediments.

The fact that the significant decrease of the phosphorus concentration in sediment cores is usually observed along with that of the Eh value suggests that the sediment beneath the oxidized layer becomes the supply source of phosphorus ions and the oxidized layer is playing a role of trap during the diagenesis process.

Although this kind of calculation is only crude, the contribution of pollution to the phosphorus accumulation in the surface sediment is estimated to be less than at most one half of the total amount in view of the fact that the pollution around the lake became evident since after the early 1960's. In the case of manganese, it is difficult to find the artificial source of 10,000 tons of manganese around the lake. Therefore, the direct effect of human activity could be excluded as the main cause of the accumulation of manganese. Most of the manganese in the surface sediments is considered to be concentrated by the natural biochemical and physicochemical process. However, as eutrophication advances, an indirect effect such as the following may occur: the reduction of manganese dioxide proceeds in surface sediments as the precipitation of organic substances is increased and, consequently, manganous ions exude to the lake water in a greater quantity from sediments. And eluted manganous ions are reoxidized in the epilimnion in which oxygen is abundant, precipitate as manganous dioxide, and accumulate on the bottom surface. In other words, eutrophication accelerates to concentration of manganese in the bottom surface up to a certain limit. When this limit is exceeded, manganese exudes mostly to lake water because of the exhaustion of oxygen in the bottom water.

Such a process is what is really taking place in the lake as a natural phenomenon, in view of the fact that the deeper the water depth is, the more concentrated manganese is in the surface and the thinner the oxidized layers are.

It should be mentioned that another important reason of accumulation of these elements in pelagic sediments is that precipitation of coarse clay minerals seldom takes place in pelagic parts of the lake, consequently decreasing chances to be diluted with materials which are poor in phosphorus, arsenic, manganese and other heavy metal ions.

Iron, cobalt and nickel. Distributions of iron, cobalt, and nickel are little changed from the top of the core to the depth of 40 cm along the cores. However, concentrations of these ions in the upper layer increase by factors of 1.1–1.2 if the ignition loss is corrected for. This fact implies that these ions are more concentrated in the surface layers than in the deeper layers of sedimental core samples, even though not as evident as those of manganese, arsenic or phosphorus. The relation between the water depth and the concentration in sediments for these ions are again similar to those held for manganese and phosphorus, that is, the deeper the water depth is, the more concentrated these ions are in the core.

Distributions of mercury, copper, zinc, lead and sulfide ions. In analyzing distribution patterns of these metal ions in recent sediments, two important factors should be taken into account, that is, pollution and chemical nature of ions. It is very probable that these ions, especially mercury and lead, are contaminated by the activity of mankind. Sources of pollution are pesticides containing mercury which have been sprayed over rice fields around the lake for nearly ten years from the early 1960's and lead compounds used for anti-

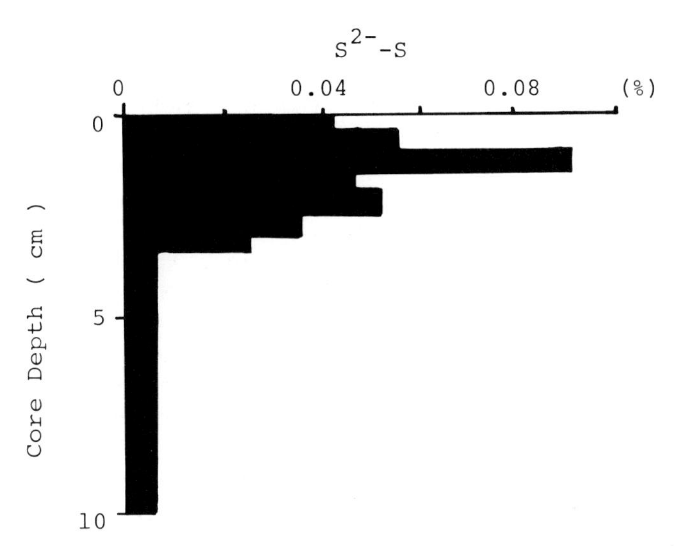

Fig. 137. Vertical distribution profile of S^{2-} S along a core. Samling Station E (water depth 75 m) (Kawashima et al. 1978).

knocking additives in gasoline. Although both have been ceased to be used, the effect may be preserved in recent sediments. As for the chemical natures of these ions, they can be classified into the type of ions which form difficultly soluble compounds with sulfide ions in the naturally occurring pH range, i.e., 6.0–7.0. Distribution patterns pictured in Fig. 136 are similar to that of labile sulfide ions as shown in Fig. 137. However, if these ions once form sulfide compounds, they would have the least chance to be dissolved out of the sedimental layers in natural environments. If metallic mercury or dialkylmercury are produced by chemical reduction or by biochemical processes, mercury will have chances to be diffused out of the deeper sedimental layers to the upper layers.

The writers are at present of the opinion that the surface enrichments of mercury and lead are possibly caused by pollution. If lake sediments which are proven to be unpolluted are available, behaviors of these elements in natural environments will be much more clearly understood.

Organic geochemistry

Kazuko Ogura

Photosynthetic products and fecal pollution

Organic compounds in lakes are derived mostly from aquatic biological products and partly from terrestrigenous materials. Primary productions in Lake Biwa have been studied by a few researchers. However, organic substances in water other than those by biological production are scarcely known. Therefore, dissolved and particulate organic substances, which are expected to be in greater abundance than the standing crop of phytoplankton, will not be discussed below.

In this section, with basic data on biological production, the distribution of organic substances in surface sediments, both major substances such as carbohydrate and trace substance such as sterol, is mainly described.

Gross production by phytoplankton, investigated by Saijo et al. (1966) in 1963—1964, is about 110 g-C/m^2 in one year, both in the North Basin and the South Basin. That of the whole lake is estimated to be 67,200 t-C in one year. Gross productions of macrophyte and epibenthic algae throughout the year are 1,050 t-C and 63 t-C respectively and, thus, correspond to the 1/67 and 1/1000 of that by phytoplankton. The product by phytoplankton is almost all degraded promptly; it is estimated that only 5% is incorporated into the sediments as nitrogen. The further decomposition for nitrogenous organic compounds in the sediments makes the nitrogen elute to the water as inorganic compounds. Consequently, only 1% of the former product remains as organic nitrogen in the sediments during 150 years (Saijo et al. 1966.)

Standing crops of phytoplankton, macrophyte and epibenthic algae are 2,009 t-C, 1,050 t-C and 1.2 t-C in summer, respectively and 2,275 t-C, nil and 5.6 t-C in winter, respectively. Fig. 138 shows the horizontal distribution of chlorophyll *a* in the surface water in summer and winter as investigated by Saijo et al. The stagnation is maintained in summer when the depth of thermocline is 20–15 m, corresponding to that of the euphotic zone. In winter, the lake water is mixed up and chlorophyll *a* is distributed uniformly in the water column.

The distribution of organic carbon in the surface sediments is shown in Fig. 139. Fig. 140 and Fig. 141 also show those of phaeopigment and carbohydrate, respectively, which are both derived from photosynthetic products.

Horie, S. Lake Biwa
© 1984, Dr W. Junk Publishers, Dordrecht/Boston/Lancaster
ISBN 90 6193 095 2. Printed in The Netherlands

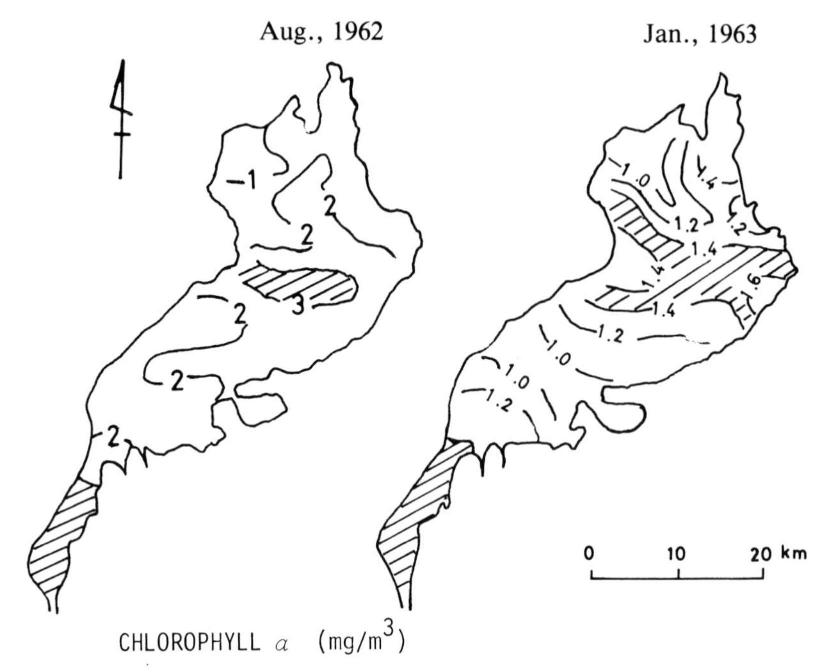

Fig. 138. Distribution of chlorophyll *a* in the water of production layer at the time of both stagnation of water (August, 1962) and mixing of water (January, 1963) (after Saijo & Sakamoto 1970).

Only one study is available as regards the composition of organic substances in the sediments of Lake Biwa studied by Handa (unpublished). In the core sample taken from the North Basin, carbon of carbohydrate, protein including amino acid, humic acid and lipid, contained in total organic carbon, vary 27 to 25%, 37 to 15%, 37 to 57% and 1.2 to 0.2%, respectively, from the top to the layer of 14 cm in depth.

The distribution of the organic carbon and carbohydrate, except for that of phaeopigment have features similar to that of chlorophyll *a* in the water as shown in Fig. 138. The quantities of chlorophyll *a* in the water reflect that of organic substances in surface sediments. This should be taken as evidence for the fact that the organic carbon and carbohydrate are mostly derived from phytoplankton, which is produced in the upper layer. However, the similarity of features in the distribution of chlorophyll *a* in the water to those of organic carbon and carbohydrate can be accounted for by the transportation mechanisms of lake currents. As noted by Saijo et al., in the North Basin, photosynthetic productivity is increased by the nutrients which are transported by lake currents, especially, at the junction of two major gyres. On the other hand, the high concentration of chlorophyll *a* at the South Basin can be explained by the shallowness (mean depth: 2.8 m). When the data of surface sediments are examined in detail, the distributions of organic carbon and carbohydrate seem to be related to that of the water depth, as expected from differences in the size

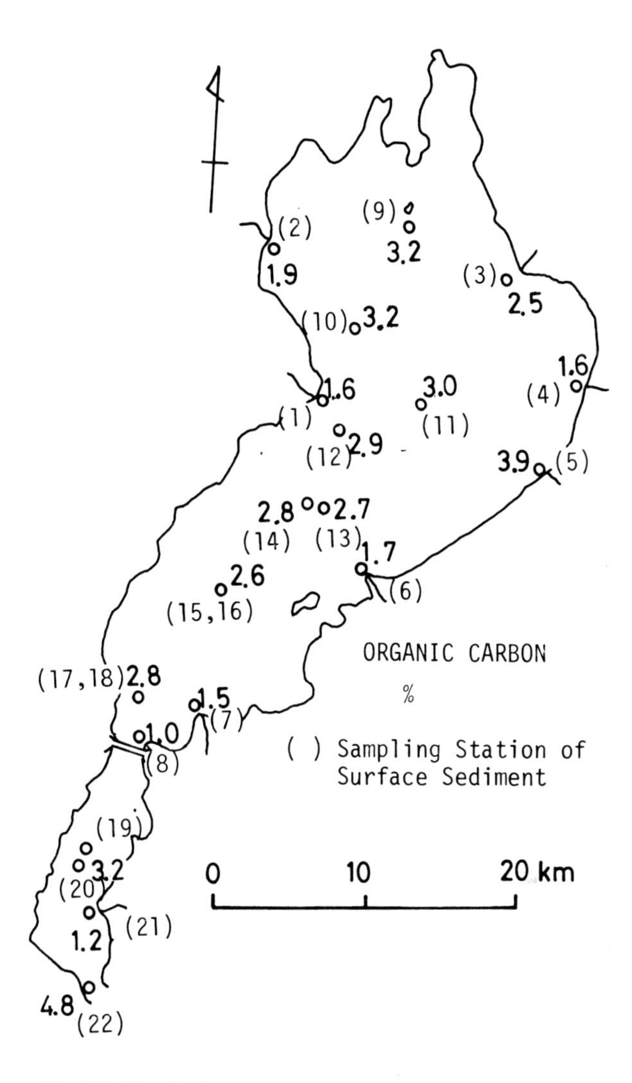

Fig. 139. Distribution of organic carbon in surface sediments.

of particles in the sediments according to the water depth. It seems that the distribution of organic substances in surface sediments depends mostly on the transportation mechanisms of the lake water rather than the influence of the productivity of the upper column. These phenomena can very well be explained by the distribution of trace polluted materials, such as coprostanol which is a sterol of mammals' feces.

There are three major gyres in the North Basin. The area around the South Basin is densely populated so that the pollutant has been introduced into the southern part of the lake.

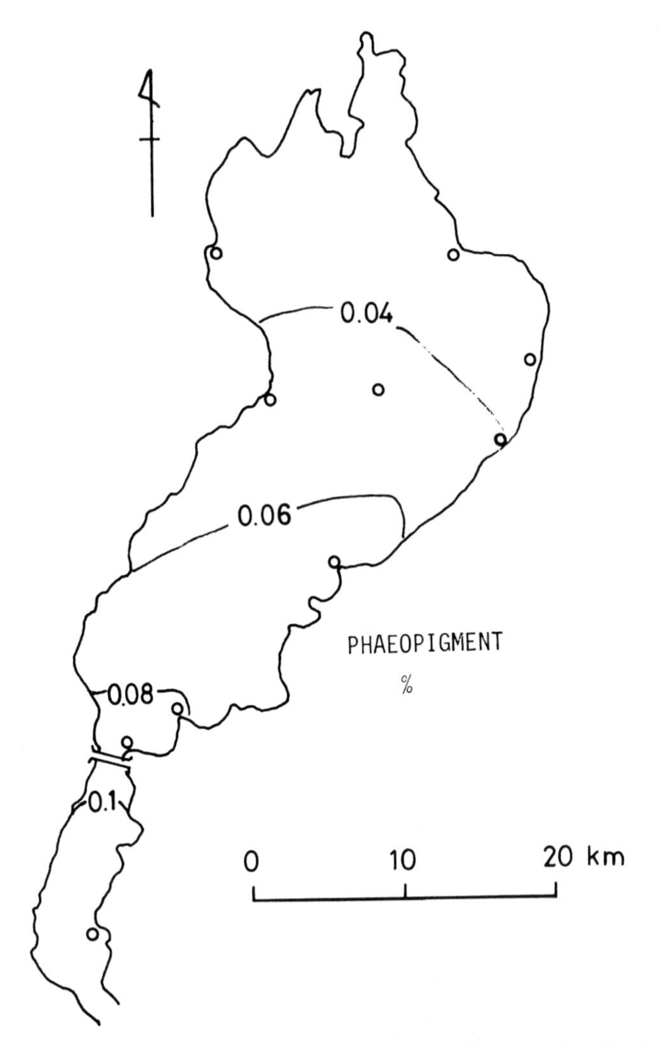

Fig. 140. Distribution of phaeopigment in surface sediments (After Handa).

Coprostanol in surface sediments is distributed in the whole lake, in the range from 0.1 μg/g to 13.6 μg/g, as shown in Table 35 and Fig. 142 (Ogura 1975). In the table, coprostanol-C/TOC (normalized coprostanol) shows the degrees of pollution in the sediments. The movements of the pollutant from the South Basin to the North Basin can be explained by the distribution of normalized coprostanol. High concentrations of coprostanol at the entrance to the South Basin from the north one, St. 17 and St. 18, show that the pollutant is moving from south to north through the bottom surface, either depending on the difference of water depth between the two basins, or possibly with water, when the south wind blows or by other actions such as seiche. The pollutant, once

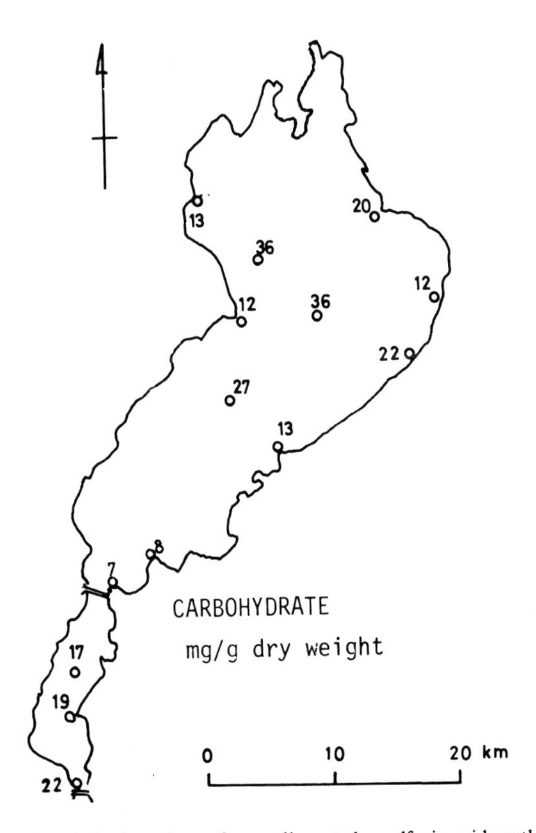

Fig. 141. Distribution of carbohydrate in surface sediments by sulfuric acid method (After Handa).

introduced into the northern part, slips down to the deepest part, St. 14 and St. 15, and is partly distributed on the whole lake, carried by lake currents. The most polluted area in the North Basin except for St. 17 and St. 18, is St. 9 where coprostanol is thought to be derived from the Chikubu-shima Island. But the second most polluted area is at the junction of the two currents, St. 12. At the North Basin, the normalized coprostanol in the mouths of rivers, is so small (0.2×10^{-4}), so that the high value at St. 12 (2.2×10^{-4}) can be explained by the lake currents and it is estimated that the 2/7 of the total organic carbon in the sediments is derived from polluted sediments in the South Basin, on the basis of normalized coprostanol at St. 21 (7.1×10^{-4}), which is the most polluted area in this study.

Compared with other areas in Japan, the normalized coprostanol of the lake is not so small; for example, in the mud of River Tama, which is the most polluted urban river by domestic wastes, the value is about 1×10^{-3} and in the sediments of a very small lake in a hot spring – the Nikko-Yumoto town (1/16000 of Lake Biwa in water volume), the value is 8×10^{-4} at the out-fall of the effluent of Treatment Plant of Sewages and 2×10^{-4} at other areas.

Table 35. Concentrations of coprostanol in dry weight of surface sediments and ratios of both coprostanol and total sterols to total organic carbon.

Station No.	Coprostanol (μg/g)	Coprostanol-C/ TOC ($\times 10^{-4}$)	Total sterols-C/ TOC ($\times 10^{-4}$)
1	0.1	0.05	4.2
2	0.4	0.18	6.6
3	0.4	0.13	9.7
4	0.2	0.11	7.4
5	1.1	0.24	8.3
6	0.3	0.14	9.9
7	0.4	0.23	9.9
8	0.3	0.24	10.2
9	9.2	2.40	75.0
10	2.8	0.63	38.7
11	2.4	0.67	22.9
12	7.5	2.17	49.5
13	3.5	1.11	36.0
14	2.6	0.79	47.9
15	4.6	1.51	52.5
16	5.6	1.85	51.5
17	8.2	2.65	73.6
18	13.6	3.91	74.1
19	2.9	0.75	13.9
20	2.7	0.70	16.8
21	10.2	7.14	30.7
22	6.2	1.08	15.8

Table 36. Percentages of each sterol in total sterols (Ogura, 1975).
A: Coprostanol; B: Cholesterol*; C: Brassicasterol; D: Campesterol*; E: Stigmasterol; F: β-sitosterol*.

Station No.	A	B	C	D	E	F
1	1.3	17.7	5.0	16.5	15.2	44.0
2	2.7	26.8	6.7	18.8	12.3	32.9
3	1.4	10.8	4.9	14.0	11.9	57.0
4	1.4	18.6	5.7	13.6	10.0	50.7
5	2.9	28.7	5.2	14.1	12.3	36.8
6	1.5	17.6	5.9	14.2	8.3	52.5
7	2.3	21.1	8.0	16.6	11.4	40.6
8	2.4	22.2	3.2	13.5	7.1	51.6
9	3.2	25.3	8.2	21.9	8.2	33.2
10	1.6	28.1	13.0	21.4	10.0	25.8
11	2.9	39.5	17.1	12.0	3.5	24.8
12	4.3	26.0	8.5	22.6	8.1	30.5
13	3.0	25.1	10.3	21.8	10.8	28.9
14	1.6	43.3	8.2	24.0	5.9	17.0
15	2.9	23.5	9.4	21.8	10.7	31.6
16	3.6	24.7	8.2	22.6	9.4	31.4
17	3.6	31.4	9.9	16.0	10.3	28.9
18	5.2	27.7	10.9	20.6	9.3	26.3
19	5.4	23.1	8.5	22.6	12.4	28.0
20	4.2	19.6	7.1	17.9	13.7	37.5
21	23.3	20.8	5.0	13.0	8.2	30.1
22	6.8	25.3	5.7	17.4	11.9	32.8

* Including corresponding stanol.

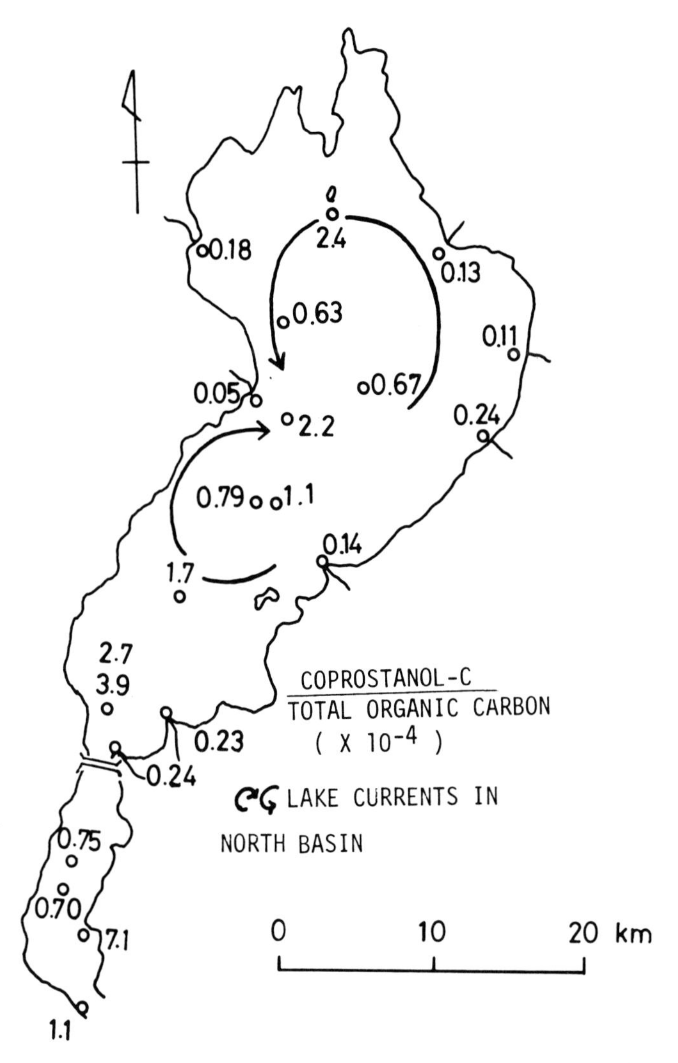

Fig. 142. Distribution of ratios of coprostanol to total organic carbon (normalised coprostanol) in surface sediments.

Polluted materials such as coprostanol which has low solubility are easily adsorbed into fine particles such as clay minerals or refractory substances, so that they are widely distributed with clay minerals in the water systems.

Other sterols have also been determined from the surface sediments, as listed in Table 36. Almost all the sterols found in the lake sediments are not so specific in this district when compared with those in other recent sediments (Lake Haruna and Lake Yu [Yu-no-ko]).

The characterization of the sediments can be estimated from the relative abundance of each of the sterols in total sterols as shown in Table 36. In the sediments at the mouths of rivers, β-sitosterol is relatively abundant, showing the contribution of soils from neighboring areas. Brassicasterol which is derived from diatoms is relatively abundant in the sediments of St. 10 and St. 11 and very little so in the sediments at the mouths of rivers both in the North Basin and the South Basin. Brassicasterol in sediments shows the productivity in the water column of the lake.

The concentration of total sterols is comparable to other lakes such as Lake Yu (Yu-no-ko) and Lake Haruna which have two times mixing and are both located more northern than Lake Biwa is in Japan.

Phytoplankton

Masami Nakanishi

Composition of phytoplankton

The phytoplankton community in Lake Biwa is composed of more than ninety species of algae including twelve of Cyanophyceae, forty-one of Chlorophyceae, six of Chrysophyceae, twenty-seven of Bacillariophyceae and two of Dinophyceae (Table 37). From the viewpoint of the algal composition, Lake Biwa is characterized by the existence of a few endemic species such as *Pediastrum Biwae Negoro (Negoro 1954a). Further, it is worth noting that Melosira solida* and *Stephanodiscus carconensis* in diatoms which are rare species in the world occur abundantly in Lake Biwa and that desmids including a rare species, *Staurastrum dorsidentiferum*, are rich in their number of species and in their quantities. *Ceratium hirudinella*, found commonly in the whole water body, is also an endemic form, the so-called 'Lake Biwa Type' (Negoro 1959). It is likely that the very long history of Lake Biwa is reflected in its characteristic composition of the algae that differs considerably from that in other Japanese lakes. *Pediastrum*, *Staurastrum* spp., *M. solida* and *S. carconensis* are found commonly in 200 m core samples of Lake Biwa (Kadota 1974; Mori, Sh. 1974).

Though most of the planktonic algae distribute widely in the whole water of Lake Biwa, it is said that there are some differences in the algal composition between the main basin (North Basin) which has a mean depth of 44.7 m and a surface area of 617.6 km^2 and the sub-basin (south basin) which has 2.8 m and 56.8 km^2 respectively. According to Negoro (1956 and 1971), *M. solida* and *S. carconensis* develop only in the main basin and *Melosira italica* and *Melosira granulata* are the most common and abundant ones in the sub-basin.

At present, eutrophication is in progress in Lake Biwa. This is clear in the sub-basin, but even the main basin that has been typically oligotrophic is shifting to a eutrophic water body. We are afraid of the effect of eutrophication on the growth of some valuable species that have been developing for an extraordinarily long period under oligotrophic conditions. In fact, at least in the latest 13 years, the *S. carconensis* population seems to have been decreasing. And in the sub-basin, some of the main or dominant algal species in the period 1935–

Horie, S. Lake Biwa

ISBN 90 6193 095 2. Printed in The Netherlands

Table 37. The list of planktonic algae in Lake Biwa. After Negoro, 1971.
● = Endemic species; ○ = characteristic species.

Cyanophyceae
Microcystis aeruginosa Kützing
Aphanothece clathrata W. et G.S. West
Aphanothece nidulans P. Richter
Aphanocapsa elachista W. et G.S. West
var. *conferta* W. et G.S. West
Chroococcus dispersus (Keissler) Lemmermann
Merismopedia elegans A. Braun
Merismopedia tenuissima Lemmermann
Oscillatoria tenuis Agardh
Phormidium tenue (Menegh.) Gomont
Lyngbya limnetica Lemmermann
Anabaena macrospora Klebahn
var. *crassa* Klebahn
Anabaena spiroides Klebahn
var. *crassa* Lemmermann
Chlorophyceae
●*Pediastrum biwae* Negoro
●*Pediastrum biwae* Negoro
var. *triangulatum* Negoro
●*Pediastrum biwae* Negoro
var. *ovatum* Negoro
Pediastrum duplex Meyen
var. *cohaerens* Bohlin
Pediastrum tetras (Ehrenberg) Ralfs
Sphaerocystis schroeteri Chodat
Eudorina elegans Ehrenberg
Volvox aureus Ehrenberg
Oedogonium sp.
Dictyosphaerium pulchellum Wood
Oocystis sp.
Coelastrum microporum Nageli
Coelastrum cambricum Archer
Kirchneriella lunaris (Kirchner) Möbius
Kirchneriella contorta (Schmidle) Bohlin
Quadrigula lacustris (Chodat) G.M. Smith
Selenastrum gracile Reinsch
Lagerheimia citriformis (Snow) G.M. Smith
Ankistrodesmus falcatus (Corda) Ralfs
Ankistrodesmus falcatus (Corda) Ralfs
var. *tumidus* (W. & G.S. West) G.S. West
Micractinium pusillum Fressenius
Tetraedrom minimum (A. Braun) Hansgirg
Crucigenia rectangularis (A. Braun) Gay
Crucigenia fensestrata Schmidle
Scenedesmus denticulatus Lagerheim
Scenedesmus arcuatus Lemmermann
○*Staurastrum dorsidentiferum* W. et G.S. West
var. *ornatum* Grönbl
○*Staurastrum limneticum* Schmidle
var. *burmense* W. et G.S. West
Staurastrum pingue Teiling
Staurastrum tohopekaligense Wolle
Staurastrum sebaldi Reinsch
var. *productum* W. et G.S. West
Straurastum asterias Nygaard
●*Staurastrum biwaensis* Hirano
Cosmocladium constrictum (Archer) Joshua
Arthrodesmus convergens Ehrenberg
Xanthidium hastiferum Turner
var. *javanicum* (Nordst.) Turner
Spondylosium moniliforme Lundell
Hyalotheca dissiliens (Sm.) Brebisson
var. *tatrica* Raciborski
Closterium aciculare Tuffen West
var. *subpronum* W. et G.S. West
Closterium acerosum (Schrank) Ehrenberg
Spirogyra sp.
Chrysophyceae
Dinobryon cylindricum Imhof
Dinobryon bavaricum Imhof
Dinobryon divergens Imhof
Mallomonas fastigata Zacharias
Synura uvella Ehrenberg
Botryococcus braunii Kützing
Bacillariophyceae
○*Melosira sólida* Eulenstein
Melosira italica (Ehr.) Kützing
Melosira granulata (Ehr.) Ralfs
Melosira granulata (Ehr.) Ralfs
var. *augustissima* Müller
Melosira varians Agardh
Melosira undulata (Ehr.) Kützing
var. *Normanni* Arnott
○*Stephanodiscus carconensis* Grunow
○*Stephanodiscus carconensis* Grunow
var. *pusilla* Grunow
Coscinodiscus lacustris Grunow
Rhizosolenia longiseta Zacharias
Rhizosolenia eriensis H.L. Smith
Atteya Zachariasi J. Brun
Tabellaria fenestrata (Lyngb.) Kützing
Fragilaria crotonensis Kitton
Fragilaria capcina Desmazieres
Asterionella formosa Hassall
Synedra ulna (Nitzsch) Ehrenberg
Synedra acus Kützing
Synedra rumpens Kützing
Synedra beroliensis Lemmermann
Bacillaria paradoxa Gmelin
Gyrosigma acuminatum (Kütz.) Rabenhorst
Cymatopleura solea (Bréb.) W. Smith
Cymatopleura elliptica (Bréb.) W. Smith
var. *constricta* Grunow
Surirella biseriata Brébisson
Surirella robusta Ehrenberg
var. *splendida* (Ehr.) van Heurck
Surirella Capronii Brébisson
Dinophyceae
○*Ceratium hirundinella* (O.F. Müller) Schrank
Peridinium africanum Lemmermann

1939 (Yamaguti 1960), such as *Synura uvella*, *Volvox aureus*, and *Aphanocapsa elachista*, have become rare species. Thus, in Lake Biwa, some of the planktonic algae that have lived for long time seem to be decreasing or diminishing. On the other hand, at the end of May of 1977, 'Akashio' (red tides) was brought about by a sudden propagation of *Ulogrena americana* that had never been observed in Lake Biwa.

Seasonal change in planktonic algae

The seasonal succession of the phytoplankton in the main and sub-basins of Lake Biwa is to be outlined below.

Negoro (1956) who examined samples taken from the main basin in 1950 and 1952 concluded that the basic pattern of seasonal change in the dominant species of phytoplankton was as follows: *Stephanodiscus carconensis* during January-March, *Melosira solida* in April and December, *Asterionella formosa* in May, *Atteya Zachariasi* in June and October, *Ceratium hirundinella* in July and *Pediastrum Biwae* in August and September. However, this basic pattern was broken at least in 1965 because of an extreme decrease in the population densities of *S. carconensis* and *A. Zachariasi*. Now, the dominant species of phytoplankton community in the main basin seems to change seasonally with the pattern that is shown in Fig. 143A. This pattern is summarized as follows: *M. solida* during January-April and November-December, *A. formosa* in May, *Staurastrum dorsidentiferum*, *Closterium aciculare* and/or *P. Biwae* during June-October. In addition to this basic pattern, an exceptional one may be found in some year when *S. dorsidentiferum* and/or *C. aciculare* propagate remarkably from October to January. In such a year, the population density of *M. solida* tends to decrease and the dominant species during October-December and January-April become *S. dorsidentiferum* and/or *C. aciculare* in place of the diatom (Nakanishi 1976). Such an exceptional pattern by propagation of these desmids occurred in the whole water body from the end of 1965 to the beginning of 1966 and from 1971 to 1972 in the last twelve years.

The pattern of seasonal change of the phytoplankton in the sub-basin varies in a complicated manner from year to year and from place to place so that it is very difficult to find any basic pattern in the sub-basin. According to the data obtained for the period 1965–1974 (cf. the regular limnological survey of the Ōtsu Hydrobiological Station, Kyoto University, e.g. Mori, Sy. (1978)), the respective main species of *P. Biwae, S. dorsidentiferum, C. aciculare* and *M. granulata* differ a great deal not only in the degree of propagation but also in the time of propagation from year to year. This fact may suggest that physical, chemical and biological factors in the water of the sub-basin are subject to relatively irregular seasonal changes every year which are strong enough to influence the growth of these planktonic algae. As an example, seasonal changes in the main algae at a station in the northern part of the sub-basin during 1973—

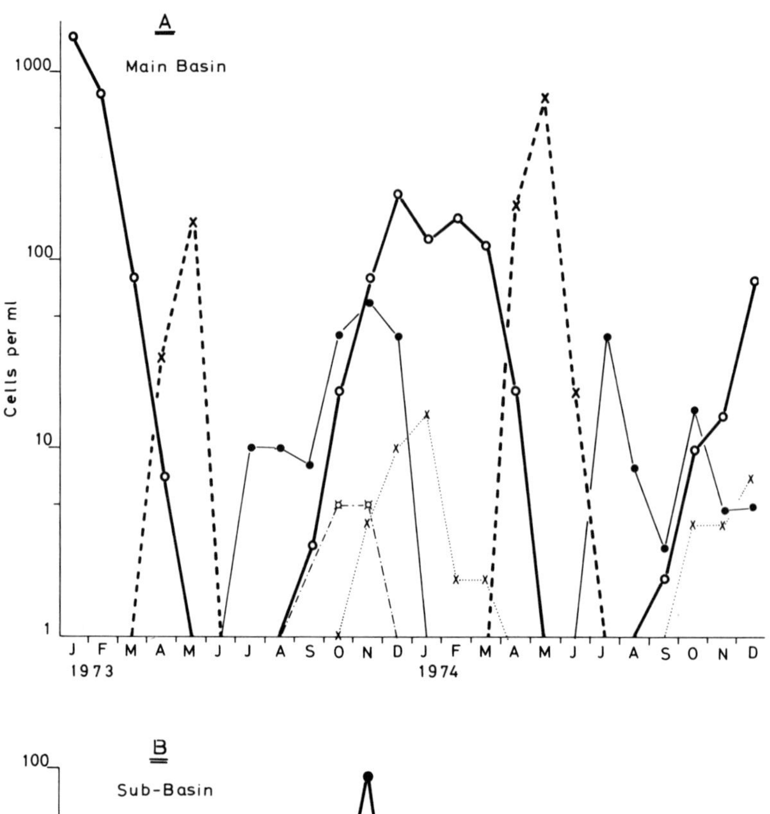

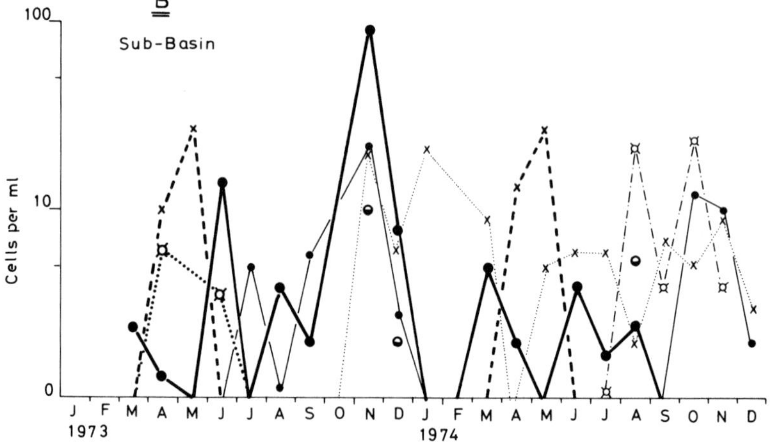

Fig. 143. A and B. Examples of seasonal changes in the dominant species of planktonic algae in pelagic areas of the main basin (A) and the sub-basin (B) of Lake Biwa. Cell number (only *P. biwae* was counted as colony) are averages from the surface to 10 m deep in the main basin and from the surface to the bottom (ca. 4 m deep) in the sub-basin. The data are taken from the regular limnological survey conducted by the Ōtsu Hydrobiological Station, Kyoto University.

M. solida: —○—; *M. italica*: —●—; *M. granulata*: ◒; *A. formosa*: --×--; *F. crotonensis*: ···¤···; *S. dorsidentiferum*: —●—; *C. aciculare*: ···×···; ***P. biwae***:—·¤·—.

1974 are shown in Fig. 143B. The regular pattern of rise and fall through the two years was observed only for *A. formosa.*

Primary production of phytoplankton

The available information is too poor to enable us to understand the primary production of phytoplankton in a large-scale water body like Lake Biwa. Therefore, this section is confined to a rough description of some parameters such as daily compensation depth, chlorophyll *a* amounts and primary production rates in pelagic areas of both basins. Physical and chemical parameters or water temperature and nutrient concentration related to the primary production are not described below as they are treated elsewhere in this book.

According to the measurement by a submarine photometer of light attenuation through lake water, the daily compensation depth varies seasonally from 7 to 18 meters below the surface in the pelagic area of the main basin and from 1.5 to more than 4 meters in the pelagic area of the sub-basin (under clear conditions, 15 per cent of the surface light intensity attenuate on the bottom at the depth of ca. 4 meters). The compensation depth of the main basin tends to be high in winter (December-February) and low in early summer (May-June) and early autumn (September-October). This seasonal variation is related mainly to chlorophyll *a* amounts. On the other hand, the compensation depth of the sub-basin which is shallow is strongly influenced by bottom sediments whirled up by winds and turbid currents that are brought about by precipitation. Thus, in the sub-basin, light attenuation into water depends on the weather rather than on chlorophyll *a* amounts.

Concentration of chlorophyll *a* in the pelagic area of the main basin is relatively low and does not differ so much in its distribution among places. In contrast, it is quite different among places in the sub-basin and becomes noticeably high as one proceeds to the south (Fig. 144). In addition, the writer presents Fig. 145 that shows the horizontal distribution of chlorophyll *a* concentration in the sub-basin. The distribution pattern of chlorophyll *a* amounts in the sub-basin seems to be influenced mainly by the water movement and the trophic status. Low concentrations of chlorophyll *a* in the northern part of the sub-basin must be due to inflow of the main basin water with relatively low chlorophyll *a* concentration. As we move to the south in the sub-basin, we find the effect of the main basin water to become less, while the effect of propagation of phytoplankton by enrichment of lake water through urban sewages and rivers becomes greater. Consequently, chlorophyll *a* concentration forms a rising gradient from the north to the south in the sub-basin. Mean concentration of chlorophyll *a* within the euphotic zone in the main basin varies seasonally from 0.5 to 5 $mg.m^{-3}$ except in a year when *S. dorsidentiferum* and/or *C. aciculare* are extraordinarily propagated. In December, 1971, when *S. dorsidentiferum* increased greatly, it came up to 20 $mg.m^{-3}$ (Mitamura, unpublished). In the sub-basin it varies widely from 1 to 30 $mg.m^{-3}$.

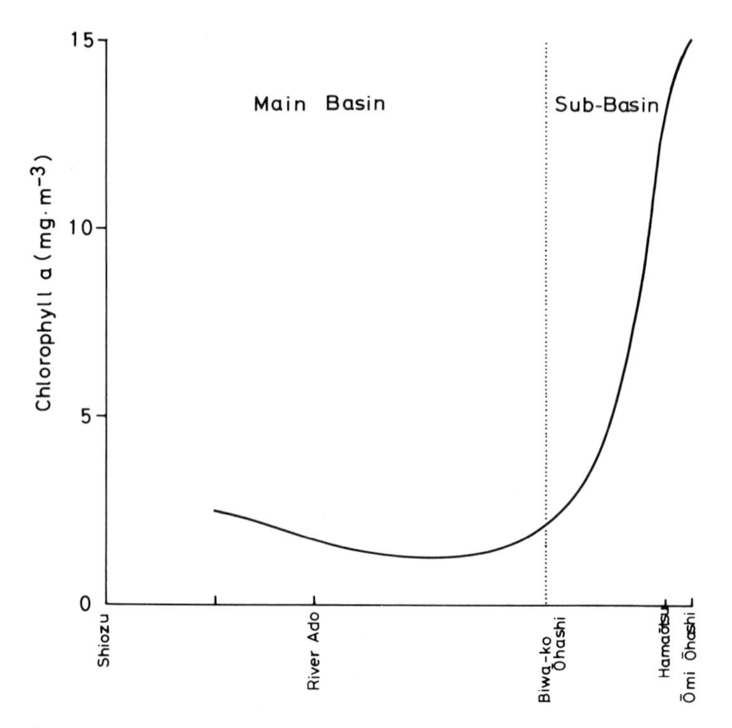

Fig. 144. The pattern of successive changes in chlorophyll *a* amounts of pelagic surface water from the main basin to the sub-basin. Data collected on 17 October 1976, by the Kinki Chihō Kensetsu-Kyoku, Kensetsu-Shō (Ministry of Construction).

The seasonal change in chlorophyll *a* amounts depends mainly on the population densities of *S. dorsidentiferum*, *C. aciculare* and *P. Biwae*, all of which fluctuate in a great degree from year to year. This makes it very difficult to find any regular pattern of seasonal change in chlorophyll *a* amounts as well as that in algal composition.

We do not have a good example to explain the pattern of seasonal change in the rate of primary production of phytoplankton in the pelagic areas of both basins. Here, as an example, we describe the pattern of seasonal change in primary production rates in Shiozu Bay which is located in the northernmost part of the main basin and whose water quality is almost the same as that in the pelagic area of the main basin.

Fig. 146 shows chlorophyll *a* amounts and the monthly gross and net production rates during the period July 1971 – June 1973 within the euphotic zone and the whole water column in Shiozu Bay at the depth of ca. 40 meters. Two peaks are found in the primary production rates over the year: one in summer with small amounts of chlorophyll *a* and other in late autumn with fairly large amounts of chlorophyll *a*. The first peak in summer is caused by high photosynthetic activity of phytoplankton which with light being saturated is 3.7 C mg.chl.*a* $mg^{-1}.h^{-1}$ and the second peak depends on the population densities

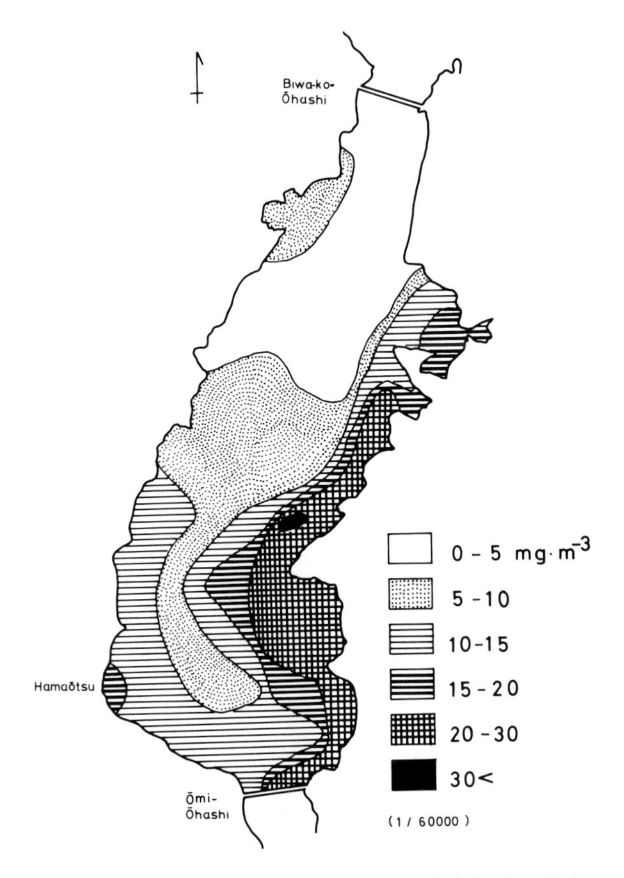

Fig. 145. The horizontal distribution of chlorophyll *a* amounts in 0.5 m layer below surface water in the sub-basin. Data collected on September 3 and 4, 1976 (Biwa-ko Ōhashi means Lake Biwa Big Bridge and Ōmi-Ōhashi means Ōmi Big Bridge) (Yamano & Matsue unpublished).

of planktonic algae of *S. dorsidentiferum* and/or *C. aciculare* rather than their photosynthesis which is 0.5–1.0 Cmg.chl.*a* $mg^{-1}.h^{-1}$. It seems that the second peak is not always found every year because the degree of propagation of these algae differ a great deal from year to year. The primary production rate at the second peak exerts a great influence on variations in annual primary production. In Shiozu Bay, for instance, the annual gross and net production within the euphotic zone were 340 C $g.m^{-2}.yr^{-1}$ and 210 C $g.m^{-2}.yr^{-1}$ respectively from July, 1971 to June, 1972 when *S. dorsidentiferum* and *C. aciculare* were extraordinarily propagated, while they were 140 C $g.m^{-2}.yr^{-1}$ and 120 C $g.m^{-2}.yr^{-1}$ in a later year which had less propagation.

In order to examine whether some changes can be found in primary production rates and chlorophyll *a* amounts in the latest 13 years, we plot in Fig. 147 sporadic data obtained in 1963–1964 (Saijo et al. 1966) and in 1976 at one pelagic station in the main basin at the depth of ca. 60 meters and at another

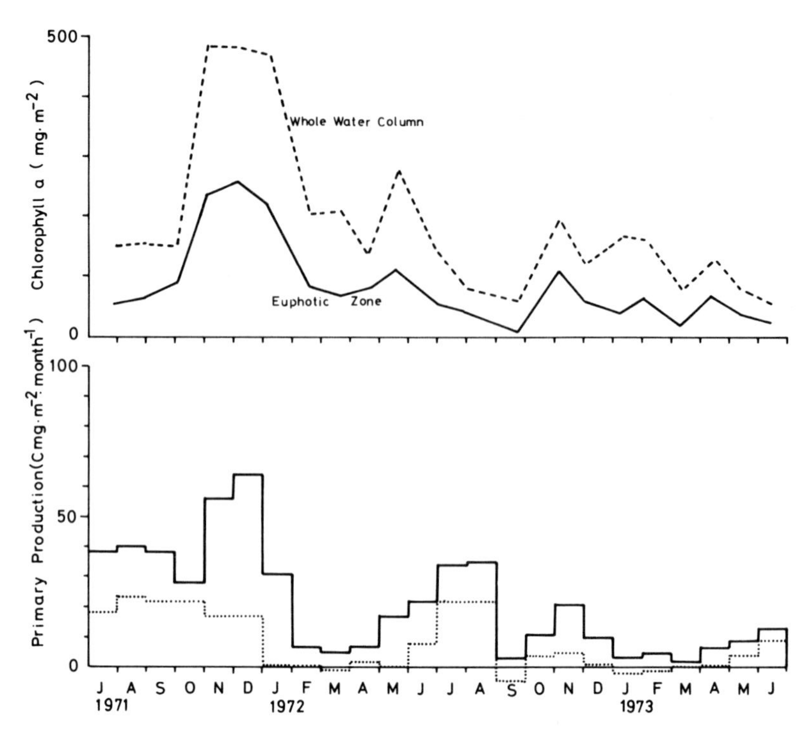

Fig. 146. Seasonal changes in chlorophyll *a* amounts and monthly gross and net primary production in Shiozu Bay, Lake Biwa (Nakanishi 1976). Solid and dotted lines in the lower figure present monthly gross production and monthly net production in the whole water column, respectively.

one in the northern part of the sub-basin at the depth of ca. 4 meters. It seems that there are no noticeable differences in daily primary production and chlorophyll *a* amounts within the euphotic zone between 1963–1964 and 1976.

This fact may suggest that, even though eutrophication had gone on during this period, its effect on changes in these parameters were not very serious. It arouses our interest that the pH values in the main basin are quite different in seasonal change between the latest 13 years (1964–1976) and around 1951 (Fig. 148). The range of seasonal change in the pH value was between 7.0 and 7.3–7.4 in 1937 and 1951, but between 7.1 and 8.2–8.8 after 1964. It is worth nothing that the pH values became extremely high in summer after at least 1964. High values of pH in summer seem to correspond with high rates of primary production (Figs. 146 and 147). If photosynthetic activity or phytoplankton activity influences changes in the pH value of lake water, it may be possible to some extent to discuss, on the basis of seasonal changes in the pH value, how primary productivity of phytoplankton stood in around 1951 as against in 1964 and 1976. It is concluded on this assumption that the primary productivity was very low before 1951 as compared with that after 1964. Further, judging from the above-mentioned remarkable changes in the dominant species that occurred

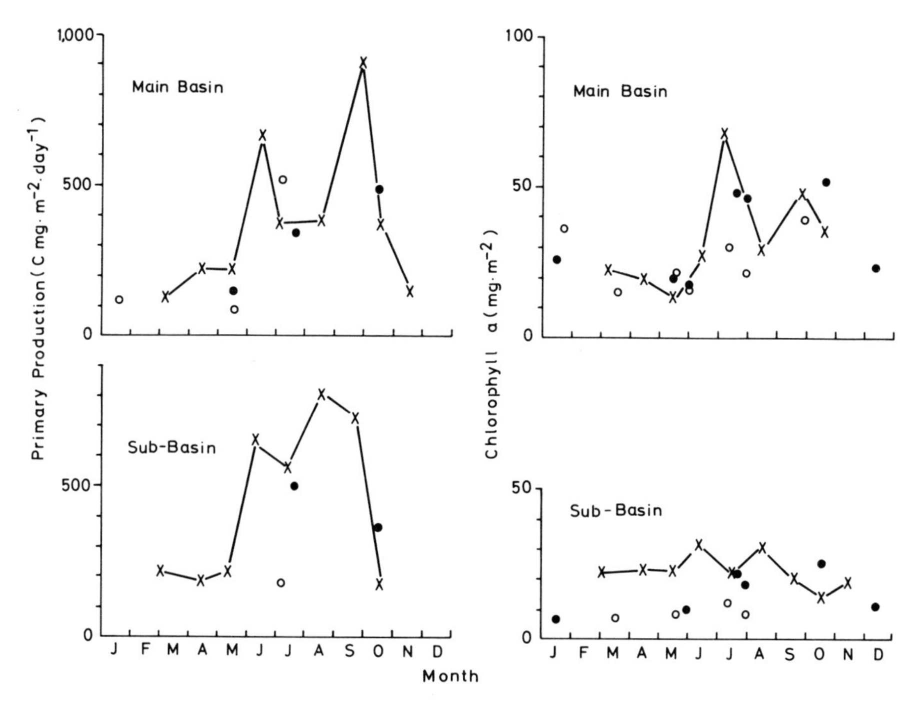

Fig. 147. Daily primary production and chlorophyll *a* amounts, within the euphotic zone, obtained in 1963–1964 (Saijo et al. 1966) and 1976.
1963: ○; 1964: ●; 1976: —×—.

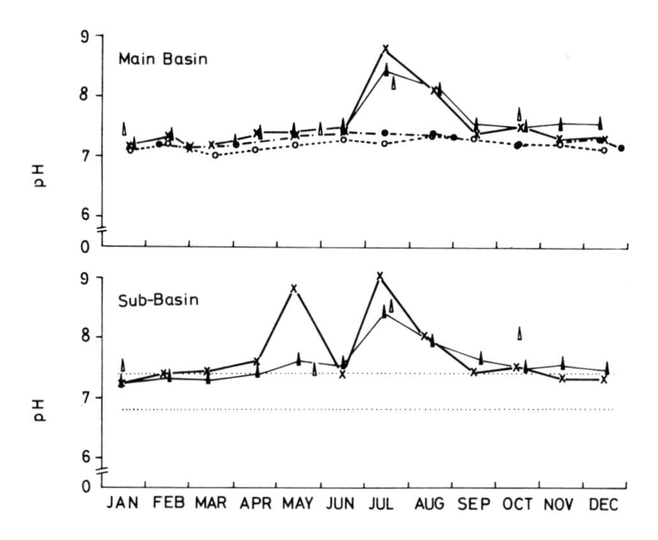

Fig. 148. Comparison of seasonal changes in the pH value in pelagic areas of both basins over the latest 30 years. The data are taken from Sugawara (1938), Negoro (1956), Yamaguti (1960) and the regular limnological survey conducted by the Ōtsu Hydrobiological Station, Kyoto University. The dotted line in the figure of the sub-basin presents the range of seasonal change in the pH value during 1933–1939.
1937: -- ○--; 1951: —·—●—·—; 1964: △; 1967: —▲—; 1976: —×—.

between around 1950 and 1964, it seems that noticeable changes occurred during this period not only in the productivity but also in the structure of the phytoplankton community. Though it is uncertain what brought about such changes, the one of the causes might have been progress in eutrophication by sudden increases in human activities around Lake Biwa since a little before 1964 when economic advances started there. It is very difficult to discuss this matter in the sub-basin in the absence of useful data on the seasonal change in the pH value which were measured before 1964 (except for the period 1933–1939 (Yamaguti 1960)). However, it seems that we can assert that the primary production processes in the sub-basin have changed in the same way as in the main basin.

In passing, let us remark on the levels of the daily primary production rates and chlorophyll *a* amounts within the euphotic zone though the data are poor. The daily primary production and chlorophyll *a* amounts vary seasonally from 100 to 1,000 C $mg.m^{-2}.day^{-1}$ and from 15 to 100 chl.*a* $mg.m^{-2}$ respectively in the main basin and from 200 to 800 C $mg.m^{-2}.day^{-1}$ and from 10 to 30 chl.*a* $mg.m^{-2}$ respectively in the northern part of the sub-basin, except in a year when desmids are subject to sudden propagation. It is expected that these parameters have higher values in the southern part of the sub-basin because it is more eutrophic.

Algae

Shinobu Mori and Shoji Horie

Introduction

Negoro (1956) listed 35 species of algae which constitute the phytoplankton of the main basin of Lake Biwa as representing the composition of phytoplankton of that lake, consisting of two parts, the main basin (north lake) and the sub-basin (south lake). They are two Cyanophyceae, fifteen Bacillariophyceae, eight Desmidiaceae, three Chrysophyceae, one Dinoflagellata, five Chlorophyceae and one Heterokontae.

Composition

As far as our investigations are concerned, main plankton of Lake Biwa in nature are as follows.*

Cyanophyceae (Myxophyceae)

Aphanocapsa elachista W. et G.S. West var. *conferta* W. et G.S. West (Negoro 1954; Fig. 149 – 1)

Aphanothece nidulans P. Richter (Negoro 1954; Fig. 149 – 2)

Bacillariophyceae (Diatomeae)

Melosira varians C.A. Agardh (Negoro 1954; Fig. 149 – 3)

Melosira granulata (Ehr.) Ralfs var. *angustissima* Müller (Hustedt 1930; Negoro 1954; Fig. 149 – 4)

Melosira solida Eulenstein (Negoro 1954; Fig. 149 – 5)

Stephanodiscus carconensis Grun. (Negoro 1954; Fig. 149 – 6)

Stephanodiscus carconensis Grun. var. *pusilla* Grun (Negoro 1954; Fig. 149 – 7)

Melosira undulata (Ehr.) Kützing var. *normanni* Arnott (Hustedt 1930; Negoro 1954; Fig. 149 – 8)

Gyrosigma acuminatum (Kütz.) Rabh. (Hustedt 1930; Negoro 1954; Fig. 149 – 9)

Cymatopleura solea (Brébisson) W. Smith (Hustedt 1930; Negoro 1954; Fig. 149 – 10)

Cymatopleura elliptica (Brébisson) W. Smith var. *constricta* Grun.

* Negoro (1954) is included in Shiga Prefectural Fisheries Experimental Station (1954).

Horie, S. Lake Biwa

ISBN 90 6193 095 2. Printed in The Netherlands

(Hustedt 1930; Negoro 1954; Fig. 149 – 11)
Surirella robusta Ehr. var. *splendida* (Ehr.) van Heurck (Hustedt 1930; Negoro 1954; Fig. 149 – 12)
Synedra ulna (Nitzsch) Ehrenberg (Hustedt 1930; Negoro 1954; Fig. 149 – 13)
Melosira solida Eulenstein (Negoro 1954; Fig. 149 – 14)
Surirella biseriata Brébisson (Hustedt 1930; Negoro 1954: Fig. 149 – 15)
Fragilaria capucina Desmazières (Hustedt 1930; Negoro 1954; Fig. 149 – 16)
Asterionella formosa Hassall (Negoro 1954; Fig. 150 – 17, 18)
Surirella Capronii Brébisson (Hustedt 1930; Negoro 1954; Fig. 150 – 19)
Attheya Zachariasi J. Brun. (Hustedt 1930; Negoro 1954; Fig. 150 – 20 – 26)

Desmidiaceae

Closterium aciculare T. West var. *subpronum* W. et G.S. West (Negoro 1954; Fig. 151 – 27 – 29)
Staurastrum dorsidentiferum W. et G.S. West var. *ornatum* Grönblad (Negoro 1954; Fig. 151 – 30)
Staurastrum dorsidentiferum W. et G.S. West var. *ornatum* Grönblad (a summer type) (Negoro 1954; Fig. 151 – 31)
Staurastrum paradoxum Meyen (Negoro 1954; Fig. 151 – 32)
Staurastrum limneticum Schmidle var. *burmense* W. et G.S. West (Negoro 1954; Fig. 151 – 33)
Staurastrum biwaensis Hirano (Negoro 1954; Fig. 151 – 34)
Cosmocladium constrictum Archer (West and Carter 1923; Negoro 1954; Fig. 151 – 35)
Hyalotheca dissiliens (Smith) Brébisson (West and Carter 1923; Negoro 1954; Fig. 151 – 36)
Xanthidium antilopaeum (Brébisson) Kützing (Negoro 1954; Fig. 151 – 37)

Chrysophyceae

Dinobryon cylindricum Imhof (Negoro 1954; Fig. 152 – 38)
Dinobryon bavaricum Imhof (Negoro 1954; Fig. 152 – 39, 40)
Dinobryon cylindricum Imhof (Negoro 1954; Fig. 152 – 41, left) a cell
Dinobryon bavaricum Imhof (Negoro 1954; Fig. 152 – 41, right) a cell
Mallomonas fastigata Zacharias (Negoro 1954; Fig. 152 – 42)

Dinoflagellata

Ceratium hirundinella (O.F. Müller) Schrank (Negoro 1954; Fig. 152 – 43 – 46)

Chlorophyceae

Pediastrum biwae Negoro (Negoro 1954; Fig. 153 – 47) (Pl. 9)
Pediastrum biwae Negoro var. *triangulatum* Negoro (Negoro 1954; Fig. 153 – 48) (Pl. 9)
Pediastrum biwae Negoro var. *ovatum* Negoro (Negoro 1954; Fig. 153 – 49) (Pl. 9)

Eudorina elegans Ehrenberg (Negoro 1954; Fig. 153 – 50 – 52)
Volvox aureus Ehrenberg (Smith 1920; Negoro 1954; Fig. 153 – 53)
Sphaerocystis schroeteri Chodat (Negoro 1954; Fig. 154 – 54)
Oedogonium sp. (Negoro 1954; Fig. 154 – 58, 59)

Heterokontae

an aggregated colony of *Botryococcus braunii* Kützing (Negoro 1954; Fig. 154 – 55)
a solitary colony of *Botryococcus braunii* Kützing (Negoro 1954; Fig. 154 – 56, 57)

Among the species listed above, both *Melosira solida* and *Stephanodiscus carconensis* of Diatomeae are only found in Lake Biwa in Japan, but they occur in North America. Further, *Pediastrum biwae* of Chlorophyceae is also found only in Lake Biwa. In addition to such a fact, their abundance makes them the most important constituents of the plankton of Lake Biwa. Next to these, *Ceratium hirundinella* of the main basin is also noticeable that it should be called as the 'Lake Biwa type' in various respects. As for *Oedogonium* sp., it is probably regarded as a species endemic to Lake Biwa. Three species of *Staurastrum,* namely, *S. dorsidentiferum* var. *ornatum, S. limneticum* var. *burmense* and *S. paradoxum*, are abundantly found in the main basin of Lake Biwa, though the former two are not very common.

How many species of plankton constituents exist in Lake Biwa? According to Negoro (1968), the number differs between the main basin and the sub-basin. In the former there are found about seventy species; the phytoplankton are nearly as many as the zooplankton. The sub-basin receives almost all the plankton that go down by the flow of lake water from the main basin. Moreover, it is believed that about fifty species of plankton occur only in the sub-basin. (There are about thirty species of phytoplankton and about twenty species of zooplankton.)

The previous descriptions are so far concerned only with net-plankton. If nannoplankton through the main and the sub-basin are taken into consideration, about twenty more plankton species are added (in this case, they belong to almost all phyto-nannoplankton).

Endemic phytoplankton algae

Many endemic species inhabit Lake Biwa. Among them, so far as the plankton are concerned, Negoro (1959) listed the following species:

The endemic or semi-endemic phytoplankters of Lake Biwa as mostly stated already:

Pediastrum biwae Negoro
Staurastrum biwaensis Hirano
Staurastrum dorsidentiferum W. et G.S. West var. *ornatum* Grönbl.
Staurastrum limneticum Schmidle var. *burmense* W. et G.S. West
Oedogonium sp.
Always sterile. Probably endemic!
Melosira solida Eulenstein
Stephanodiscus carconensis Grunow
Ceratium hirundinella (O.F. Müller) Schrank
The endemic form ('Lake Biwa type')!

Desmids in Lake Biwa — biogeography —

Hirano (1968) gave a detailed description of Desmids. They are *Closterium aciculare* T. West, *Arthrodesmus convergens* Ehrenb. forma *curta* Turner, *Xanthidium hastiferum* Turner var. *curvispinosum* Okada, *Staurastrum arctiscon* (Ehrenb.) Lund., *St. biwaensis* Hirano, *St. dorsidentiferum* W. & G. S. West var. *ornatum* Grönblad, *St. gracile* Ralfs, *St. hantzii* Reinsch var. *japonicum* Roy et Biss., *St. leptocladum* Nordst., *St. leptodermum* Lund. var. *capitatum* Hirano, *St. limneticum* Schmidle var. *burmense* W. et G.S. West, *St. longiradiatum* W. et G.S. West, *St. paradoxum* Meyen, *St. pingue* Teiling, *St. sebaldi* Reinsch var. *ornatum* Nordst., var. *productum* W. et G.S. West, *St. sonthalianum* Turner, *St. subborgesenii* Grönblad, *Cosmocladium constrictum* (Archer) Joshua, *C. saxonicum* De Bary, *Spondylosium ellipticum* W. et G.S. West, *Sp. lütkemülleri* Grönblad, *Sp. moniliforme* Lund., *Sphaerozosma vertebratum* (Bréb.) Ralfs, *Onychonema laeve* Nordst. and *Gymnozyga moniliformis* Ehrenb.

Both northern and southern elements are involved in the desmid flora of this lake. The following species have hitherto been known to be distributed over the northern countries: *Closterium aciculare* T. West, *Staurastrum arctiscon* (Ehrenb.) Lund., *St. dorsidentiferum* W. et G.S. West var. *ornatum* Grönblad, *St. pingue* Teiling, *St. borgesenii* Grönblad, *Cosmocladium constrictum* (Arch.) Joshua, *C. saxonicum* De Bary, *Spondylosium ellipticum* W. et G.S. West, *Sp. lütkemülleri* Grönblad and *Sp. moniliforme* Lund.

It is of great interest that such species which have hitherto been thought to inhabit only the tropical Asian districts are distributed also in Lake Biwa: they are: *Arthrodesmus convergens* Ehrenb. forma *curta* Turner, *Staurastrum longiradiatum* W. et G.S. West, *St. leptocladum* Nordst., *St. sonthalianum* Turner and *Staurastrum leptodermum* Lund. var. *capitatum* Hirano, for which a similar one has been reported from the tropical African lakes. It is also very interesting to find species which have been known to be mainly from the tropical Asian district: *Ichtyocercus longispinus* (Borge) Krieger, *Pleurotaenium kayei* (Archer) Rabenh., *Pl. ovatum* Nordst., *Pl. subcoronulatum* (Turner) W. et G.S. West, *Euastrum ansatum* Ralfs var. *javanicum* (Gutw.) Krieger, *E. exile* Joshua, *E. flammeum* Joshua, *E. indicum* Krieger var. *capitatum* Krieger, *E. turgidum* Wall., *Micrasterias alata* Wall., *M. foliacea* Bail., *M. lux* Joshua, *Cosmarium australe* (Racib.) Lütkem., *C. javanicum* Nordst., *C. otus* Krieger, *C. pandriforme* Turner, *Xanthidium hastiferum* Turner var. *javanicum* (Nordst.) Turner, *Arthrodesmus convergens* Ehrenb. forma *curta* Turner, *A. curvatus* Turner, *Staurastrum acanthastrum* W. et G.S. West, *St. asterias* Nygaard, *St. columbetoides* W. et G.S. West, *St. leptocladum* Nordst, *St. saltans* W. et G.S. West, *St. subsaltans* W. et G.S. West, *St. tauphorum* W. et G.S. West var. *sumatranum* Krieger, *St. unguiferum* Turner.

The Kinki district, where Lake Biwa is involved, is situated in the central part of the Japanese Islands and has a large water system centering around the lake. Therefore there are included species of both north and austral Asian origin (after Hirano 1968).

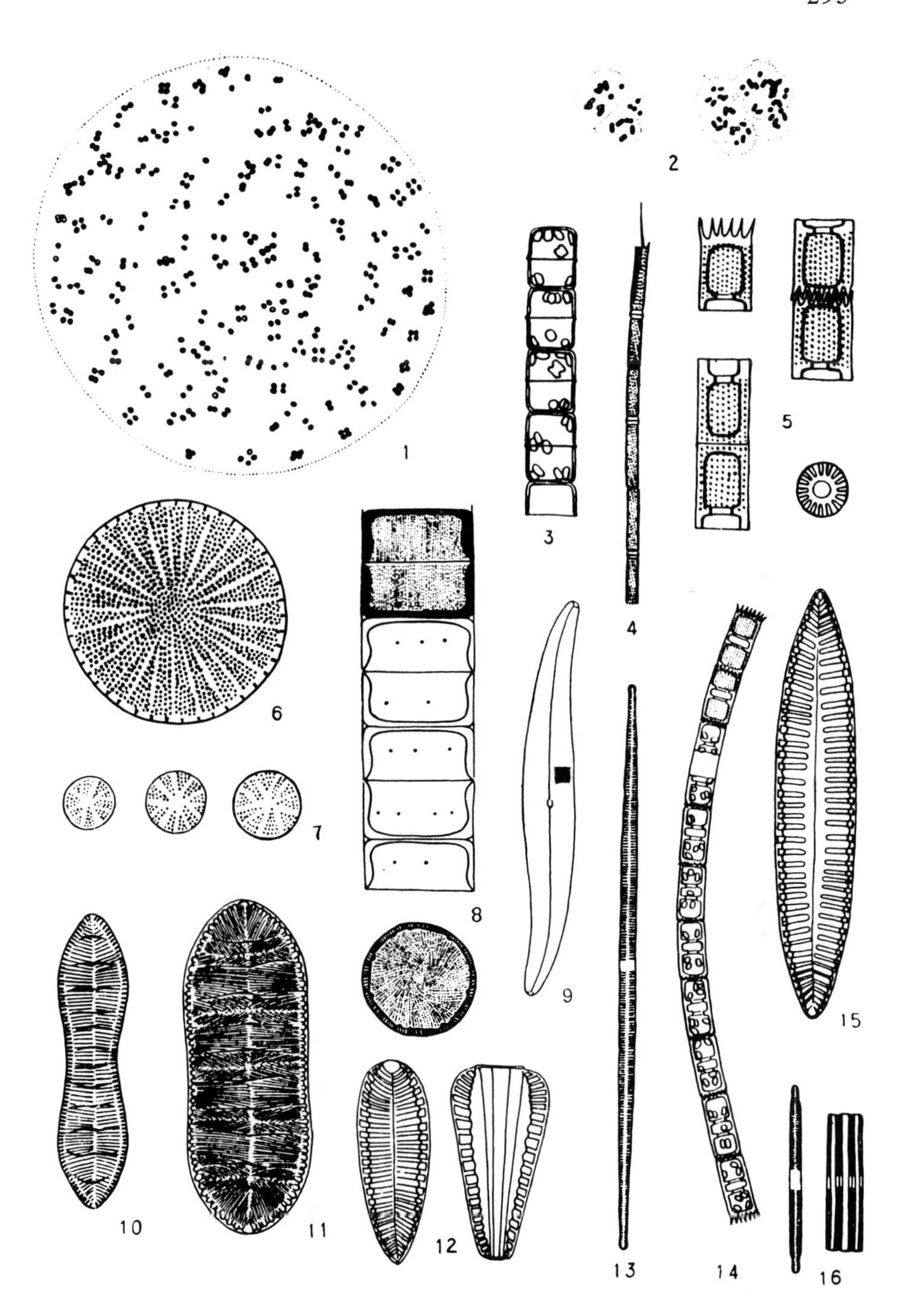

Fig. 149. Algae in Lake Biwa.

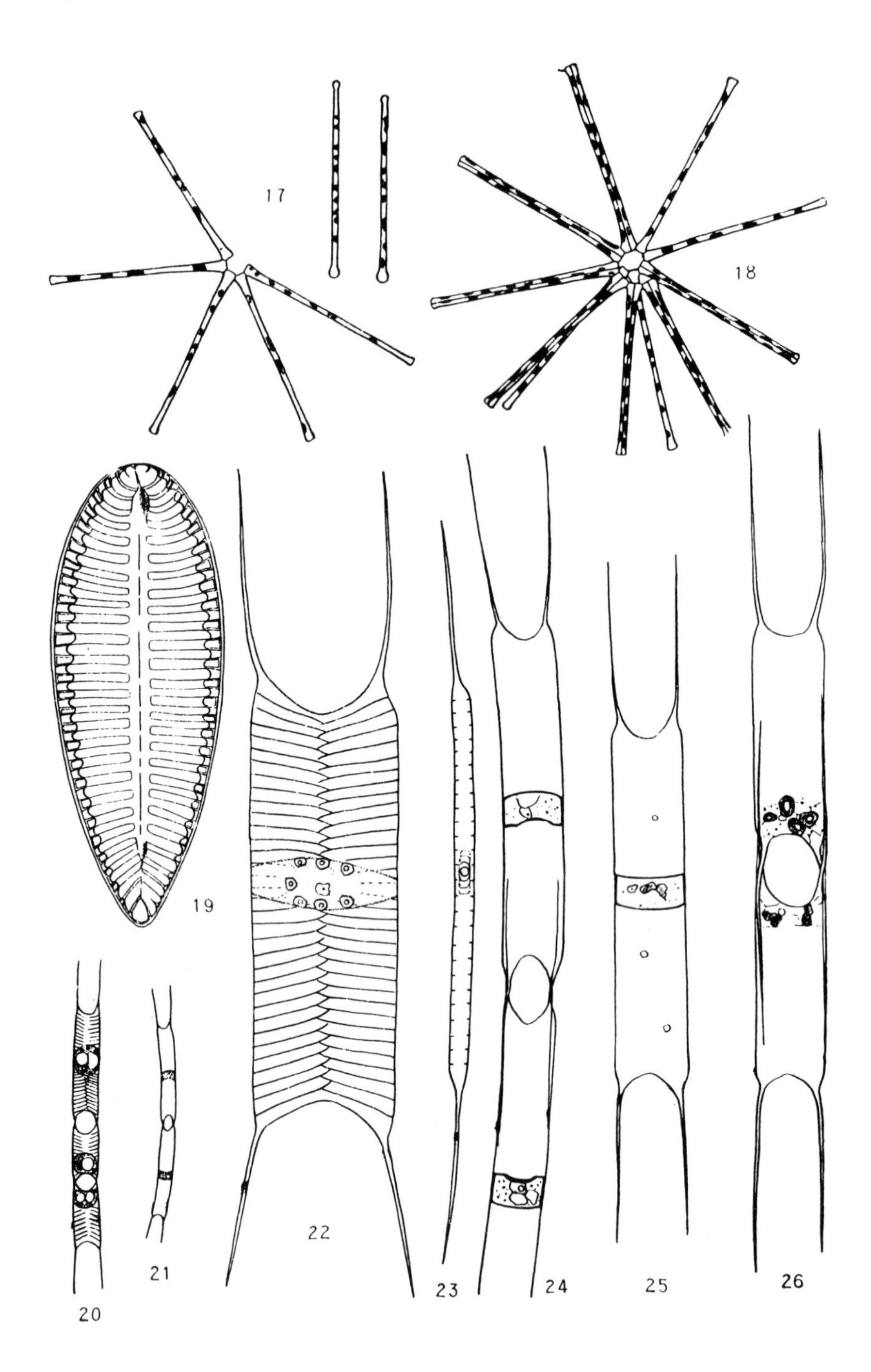

Fig. 150. Algae in Lake Biwa.

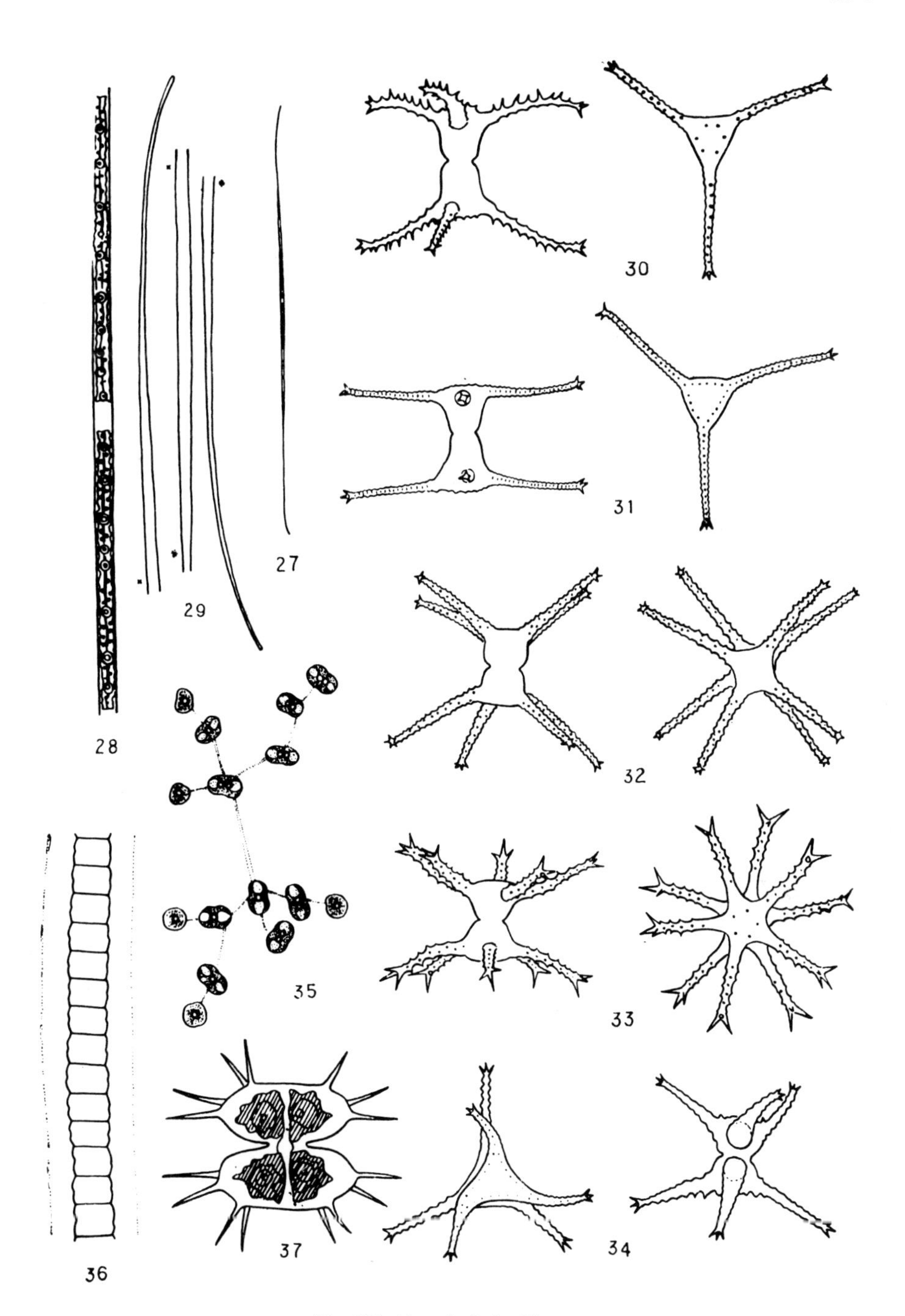

Fig. 151. Algae in Lake Biwa.

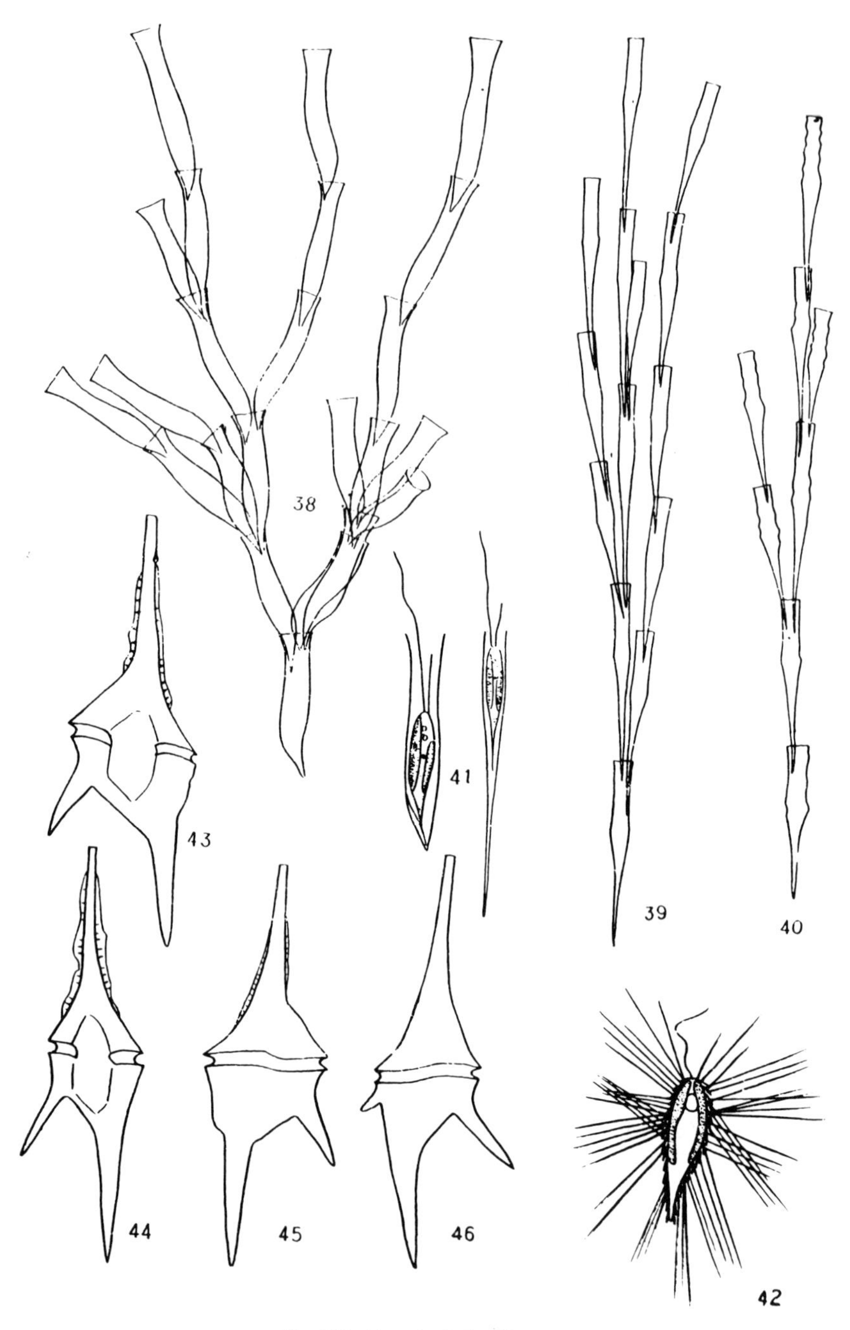

Fig. 152. Algae in Lake Biwa.

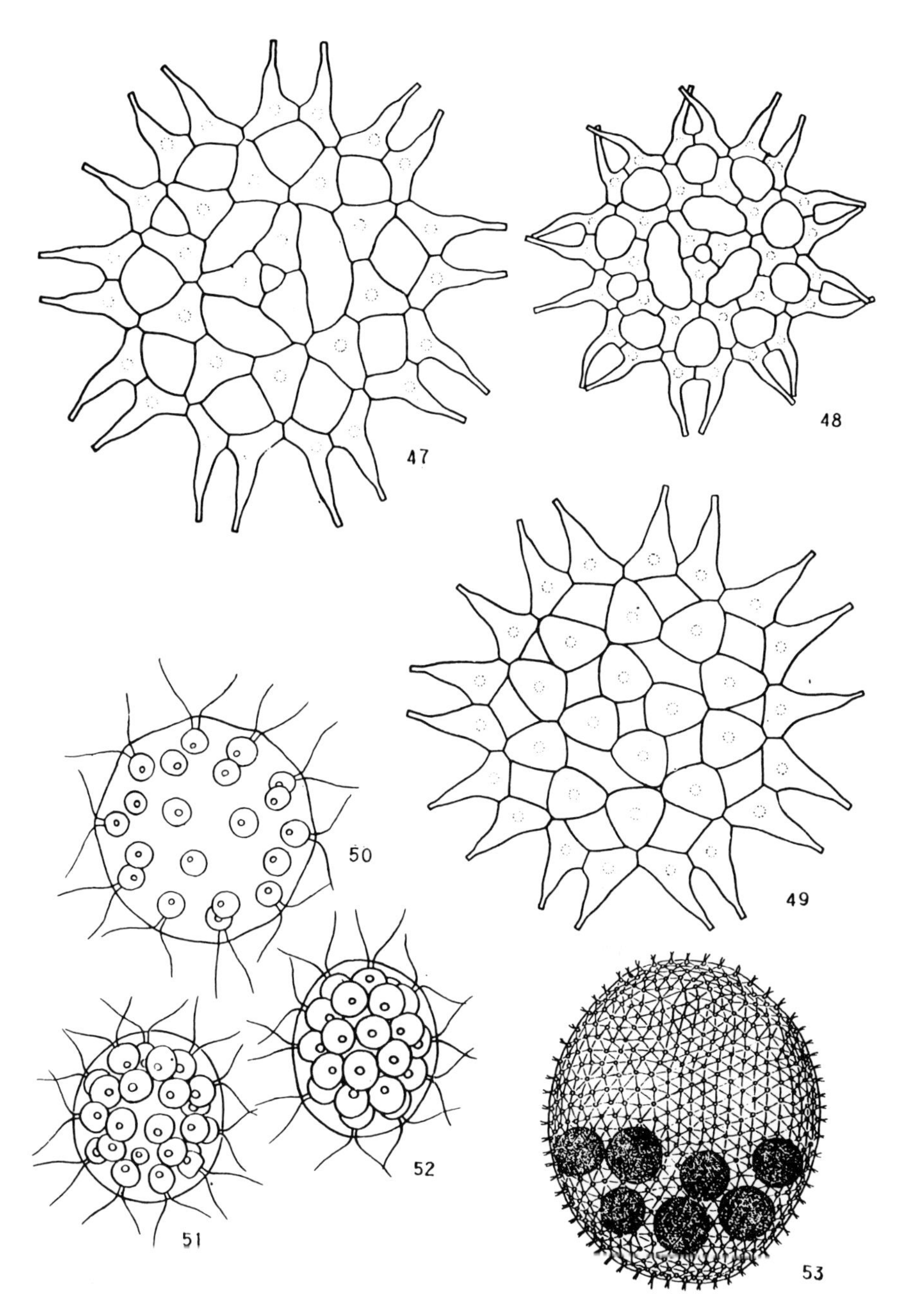

Fig. 153. Algae in Lake Biwa.

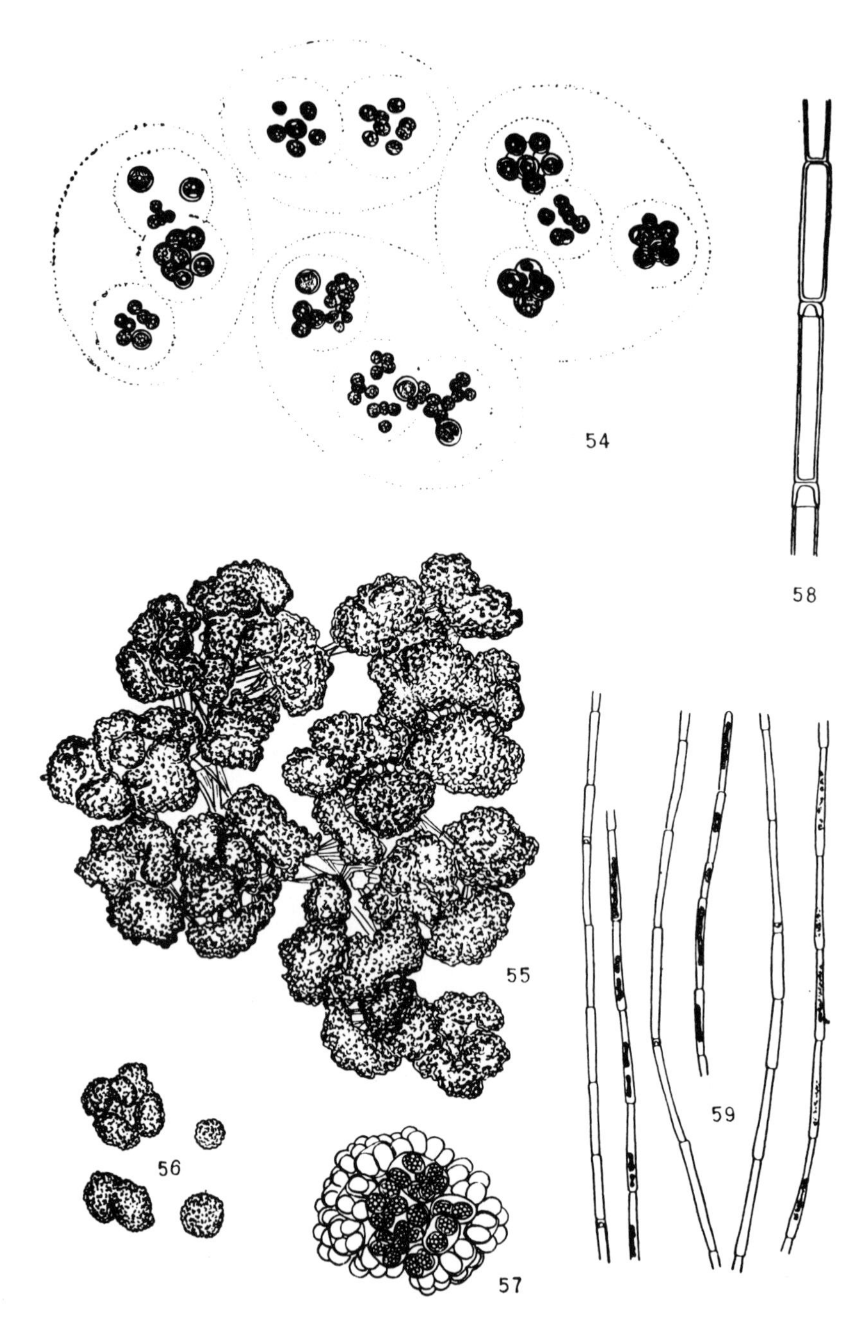

Fig. 154. Algae in Lake Biwa.

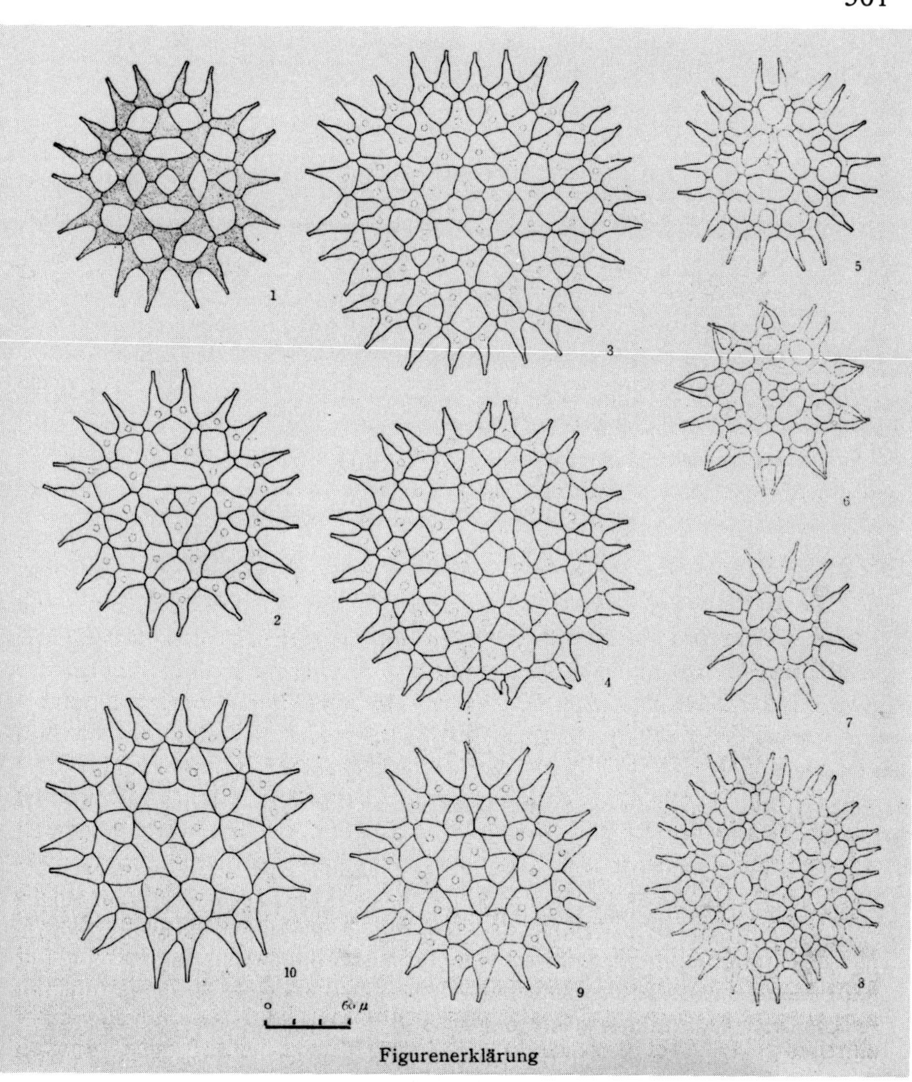

Plate 9. *Pediastrum biwae* found in Lake Biwa.
1–4: *Pediastrum biwae* Negoro.
5–8: *Pediastrum biwae* var. *triangulatum* Negoro.
9, 10: *Pediastrum biwae* var. *ovatum* Negoro (after Negoro 1954a).

Aquatic macrophytes

Isao Ikusima

Introduction

This section is a review of the studies conducted by many researchers on aquatic macrophytes in Lake Biwa and its surroudings. The emphasis of this section will be placed on the floristic compositional studies of the lake, species changes caused by a progress of eutrophication and by a dense pure stand of exotic submerged plants, and a production ecology of aquatic macrophytes.

Floristic composition of aquatic macrophytes

The studies on the morphology and identification of aquatic macrophytes were made by Miki in Pond Ogura, Pond Mizoro, and also in Lake Biwa and swamps and ponds near Lake Biwa. General knowledge on the autecology of some species was obtained through his studies. In Lake Biwa and its surroundings, at least 57 species or about 80 percent of freshwater macrophytes in Japan were found (Miki 1927, 1929, 1937). The rich and diversified condition may be attributed to the presence of other lakes, many small ponds and marshes which are connected to the lake by streams or channels. Lake Biwa is of a very old origin and its environment varies from one site to another. However, the lake has only two endemic species, *Potamogeton biwaensis* Miki and *Vallisneria asiatica* var. *biwaensis* Miki* (Miki 1934).

In 1935, 395 stations located at the southern part of the sub-basin (south basin) of Lake Biwa were surveyed by Yamaguti. It was found that the standing crop for 0–2 m, 2–4 m and 4–6 m depths was 38.6%, 58.8% and 2.6% of the total standing crop, respectively. About 10 dominant species including *Vallisneria biwaensis, Hydrilla verticillata, Potamogeton maackianus, Vallisneria denseserrulata* and *Ceratophyllum demersum* were living in the lake (Yamaguti 1938). He also surveyed the sub- and main basins of the lake from 1938 to 1941. Observations were made on the relation between the dominant species and

* *Vallisneria biwaensis* (Miki) Ohwi
Scientific names used in 'Flora of Japan' by J. Ohwi (1972) are applied to all aquatic macrophytes in this section.

Horie, S. Lake Biwa

ISBN 90 6193 095 2. Printed in The Netherlands

water depth or nature of the bottom, and from this study he concluded that in the sub-basin the shallower part within 5 m depth was occupied by submerged plants, while in the main basin the lower limit of *Chara, Nitella* and members of green algae was about 10 m and 8–12 m, respectively. The list of aquatic macrophytes is given in Table 40. It is seen from this table that more than 71 species were living in Lake Biwa and its surrounding areas, where 61 of those species were living in connecting ponds and channels, and 37 species of those were living in small ponds in the surrounding areas. Only 34 species were living inside the lake (Yamaguti 1943). Unfortunately, the rich flora found in small ponds and marshes surrounding the lake has disappeared when those areas were modernized and industrialized.

Many species of aquatic macrophytes were found in Imazu and Takashima

Table 40. Floristic composition of aquatic macrophytes in Lake Biwa (after Yamaguti, 1943).

Brasenia schreberi J.F. Gmelin
Nuphar japonicum DC.
N. oguraense Miki
N. subintegerrimum (Casp.) Makino
Nymphaea tetragona Georgi
Ceratophyllum demersum L.
C. demersum L. var. *pentacanthum* Kitagawa
Ranunculus nipponicus Nakai var. *major* Hara
Callitriche verna L.
Elatine triandra Schkuhr
Rotala mexicana Cham. et Schltdl.
Trapa japonica Flerov
T. incisa Sieb. et Zucc.
T. macropoda Miki
Myriophyllum oguraense Miki
M. spicatum L.
M. ussuriense Max.
M. verticillatum L.
M. brasiliense Camb.
Ludwigia ovalis Miq.
Nymphoides indica (Linn.) O. Kuntze
N. peltata (Gmel.) O. Kuntze
Limnophila sessiliflora Blume
Trapella sinensis Oliver var. *antennifera* (Lév.) Hara
Utricularia dimorphantha Makino
U. japonica Makino
U. multispinosa (Miki) Miki
U. pilosa (Makino) Makino
U. tenuicaulis Miki
Potamogeton anguillanus Koidz.
P. biwaensis Miki
P. crispus L.
P. dentatus Hagst.
P. distinctus A. Benn.
P. fryeri A. Benn.
P. maackianus A. Benn.
Potamogeton octandrus Poiret
P. octandrus P. var. *miduhikimo* (Makino) Hara
P. malaianus Miq.
P. malainoides Miki
P. nipponicus Makino
P. oxyphyllus Miq.
P. pectinatus L.
P. perfoliatus L.
Najas graminea Del.
N. indica (Willd.) Chamisso
N. marina L.
N. minor All.
N. oguraensis Miki
Sagittaria pygmaea Miq.
S. trifolia L.
Caldesia reniformis (D. Don) Makino
Alisma canaliculatum A. Br. et Bouché
A. plantago-aquatica L. var. *orientale* Samuels
Ottelia alismoides (L.) Persoon
Hydrocharis dubia (Blume) Backer
Vallisneria asiatica Miki
V. biwaensis (Miki) Ohwi
V. denseserrulata (Makino) Makino
Hydrilla verticillata (L.f.) Casp.
Zizania latifolia Turcz.
Phragmites communis Trin.
Scirpus tabernaemontani Gmel.
Eleocharis kuroguwai Ohwi
Lemna paucicostata Hegelm.
Spirodela polyrhiza (L.) Schleid.
Marsilea quadrifolia L.
Azolla japonica Franch. et Sav.
Salvinia natans Allioni
Chara spp.
Nitella spp.

located at the western part of the lake. Most of these species occurred abundantly in the shallow littoral region at the depth of 2–5 m. The highest standing crop was found at 6–7 m depth and the limit of distribution was 10 m depth and at the deepest places only *Hydrilla verticillata* was observed (Uéno 1945).

In Lake Sone (Sone-numa), one of the lagoons on the east coast of Lake Biwa, there were 12 species of aquatic macrophytes which were the same species as those living in Lake Biwa. *Trapa japonica* was the dominant and *Nelumbo nucifera* was found in a poor growing condition (Yamaguti 1956).

The aquatic macrophytes of Lake Yogo located at the north of Lake Biwa were surveyed in 1951. The well developed submerged vegetation was found in 2–2.5 m of water at the northern part of the lake and the standing crop ranged from 107–467 g m^{-2} in dry weight. The dominant species which grew only on the muddy bottom was *Hydrilla verticallata* and *Potamogeton maackianus* (Yamaguti 1955a).

Non-planktonic water blooms of the pollen grain of *Pinus densiflora* were scattered on the water surface in the sub-basin of Lake Biwa from early to mid-May. In early September, water blooms due to the white male flowers of *Vallisneria biwaensis* occurred. These water blooms were periodically observed for a long time. Large floating masses of the pollen grain and the male flower were carried by water currents from the main basin to the sub-basin and extended over the vast area off Ōtsu city (Yamaguti 1956).

For the measurement of standing crop of submerged macrophytes, 108 stations were selected along the shoreline of the lake in 1953. From data of the Shiga Prefectural Fisheries Experimental Station (1954) the seasonal maximum standing crop was calculated at 340–430 g fresh weight m^{-2} at 0–3 m depth, 130 g m^{-2} at 3–4 m depth and 60 g m^{-2} at 4–5 m depth. The rough estimates of the total standing crop in the main and sub-basins were 1.17×10^4 and 7.9×10^3 metric tons in fresh weight, respectively. About 40% of the total standing crop was harvested in the sub-basin, the area of which covers only 8.5% of the total lake (Ikusima et al. 1962).

An intensive survey in Lake Biwa which includes 88 stations at intervals of 1–2 km was conducted during the period of 1961–1966 and 21 species were listed. The dominant species of aquatic macrophytes were *Potamogeton crispus, P. maackianus, P. malaianus, Najas marina, Hydrilla verticillata, Vallisneria biwaensis, V. denseserrulata, Ceratophyllum demersum, Myriophyllum spicatum* and *Elodea nuttallii.* The geographical distribution of the vegetation was mapped on 1:10,000 scale. The area covered by submerged vegetations was 718.5 ha in the main basin and 60.2 ha in the sub-basin, respectively. The vertical distribution of standing crop at main habitats in the lake is shown in Fig. 155. The total standing crop in the lake was estimated at 2.62×10^3 ton in dry matter.

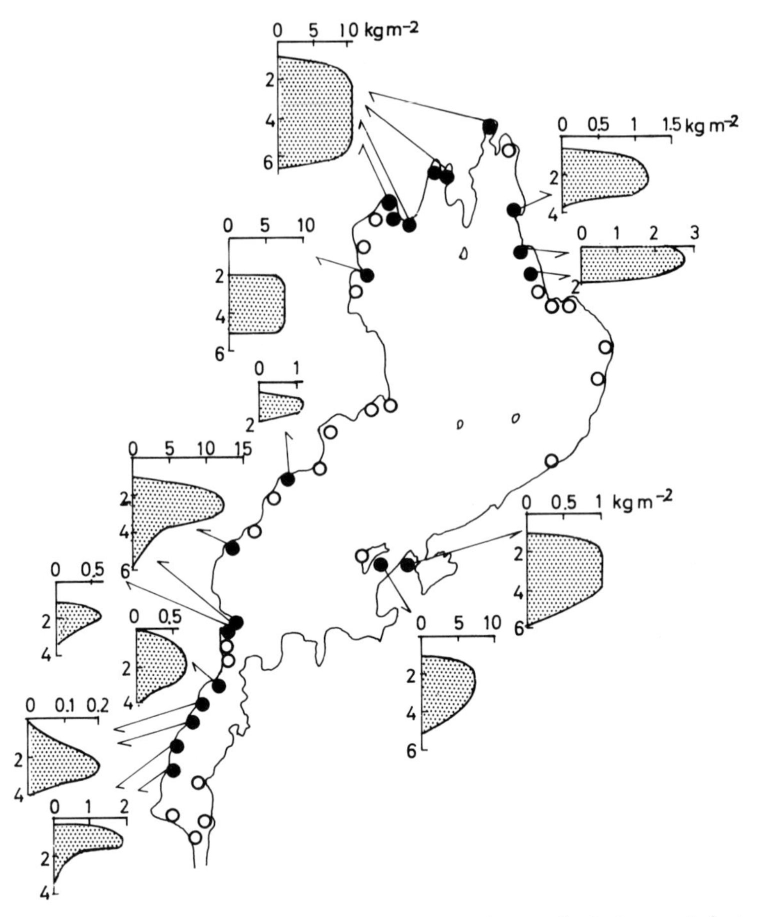

Fig. 155. Vertical distribution of standing crop (kg fresh weight per m²) of submerged plants at well-developed stands (●). (○) indicates areas where submerged plants were found (after Ikusima).

Some naturalized exotic aquatic macrophytes in Lake Biwa

An explosive expansion of *Elodea nuttallii** broke out in the lake. The plant is native in North America and was first noticed in November 1961 at Shiozu Bay located at the northern end of the lake. This was its first record in Japan. Soon after that year the adventive distribution of *Elodea nuttallii* was found in 3 bays situated at the northern part of the lake and in the main fishing ports of Oki-no-shima Island and Katata. Only male flowers were first found at the northeastern

* *Elodea occidentalis* (Pursh) St. John was used in the papers by Ikusima and Kabaya (1965) and Ikusima (1965, 1967), but *Elodea nuttallii* (Planch) St. John has been used since Mizushima (1967) pointed out that *E. occidentalis* was mis-applied in 1920.

locations in the summer of 1965, so that this extreme rapid spread of the submerged plant may have been caused by its rapid vegetative multiplication. The plant could increase throughout the year by the fragmentation of shoots under suitable habitat conditions. In 1965 or thereabout almost all littoral regions were occupied by the plant, which developed into dense pure stands at 2 – 5 m depth. The standing crop reached up to 700 – 1,000 g m^{-2} in dry weight which was 3 – 5 times larger than that of the native plants. The role of this plant in the primary production became more and more important as the dense infestation developed (Ikusima & Kabaya 1965).

In 1969, 8 years after *Elodea nuttallii* had been initially found, almost all shallow waters up to 5 m deep in Lake Biwa were occupied by the plant and were covered by vast dense mats which blocked small bays. Changes of specific composition, coverages and standing crop of macrophytes in the presence of extensive pure colonies of *Elodea nuttallii* were surveyed. Listed in the order of standing crop are species, *Elodea nuttallii, Hydrilla verticillata, Vallisneria biwaensis, Ceratophyllum demersum* and *Potamogeton malaianus,* the standing crops of these five species accounting for 98% of the total standing crop, with the standing crop of *Elodea nuttallii* alone covering up to 81% of the total. The area of about 30% of the total basin of 0 – 7 m depth was covered with the submerged plant and the total standing crop was estimated at 3.97×10^3 ton in dry weight (Shiga Prefectural Fisheries Experimental Station 1972).

When the eutrophication of the lake progressed, the vigorous adventive spread and growth of *Elodea nuttallii* gradually decreased and the equilibrium state was maintained, while the other exotic submerged plant, *Egeria densa** gradually increased and established its pure stands in the *Elodea nuttallii* stands. *Egeria densa* had not yet appeared in 1966, but in 1971 a small stand of the plant was found by Ikusima and Tachibana at the depth of 2 m near the bay of Chikubushima Island situated at the northern part of the lake. In 1973, dense stands of *Egeria densa* were found in a native submerged community developed at shallow waters from Onoe to Hayasaki at the northeastern shore in the main basin.

A survey conducted in 1974 indicated that *Egeria densa* was found at almost all shorelines of the lake with *Elodea nuttallii* and particularly the critical infestation of *Egeria densa* extended along the western shore in the sub-basin (Nagai 1975).

About 5.8% of the total area of the sub-basin was covered by *Egeria densa*. The maximum standing crop reached 411 g m^{-2} in dry weight and the annual net production of *Egeria densa* was approximately 1 kg dry matter m^{-2} $year^{-1}$ in a dense stand. The large amount of nitrogen which is equivalent to 22% of the total inflowing nitrogen from the main basin to the sub-basin was considered to be captured by *Egeria densa,* so that the nitrogen circulation through the plant could not be neglected (Tanimizu & Miura 1976). In the dense *Egeria densa* stand, there was no certain evidence to prove that the plant had never been limited in its vigorous growth by the deficiency of nitrogenous nutrients even in

* *Egeria densa* Planch. [*Elodea densa* (Planch.) Casp.].

the most productive period. However, the deficiency of nitrogenous nutrients was probably not so great and seemed to be covered well, mainly by the excretion of shrimps (Miura et al. 1976).

Eichhornia crassipes, the exotic free floating plant, was frequently seen in the accessory small lakes such as Dai Naka-no-ko,* Nishi-no-ko,** Lake Sone (Sonenuma)* and some eastern parts of the sub-basin for the last 5 – 7 years. The outbreak and the extent of its colonies varied from year to year. *Myriophyllum brasiliense*, the exotic aquatic plant having both submerged and emerged types, established pure stands in small areas of the western and eastern parts of the lake. The colony was well developed at Matsunoki Naiko***, the water surface of which was occupied by the plant and the native plant *Trapa japonica* was excluded by the plant.

The exotic plants, except *Eichhornia crassipes,* can be survived during the cold winter period making thick mats of the shoot. Their rapid growth by means of the vigorous vegetative spread of shoots in early spring may change the status of many native species and the exotic plants tend to form extensive pure stands more easily than the native plant.

Light-dependent photosynthesis and production

The relation between net photosynthesis and light intensity of the dominant species of submerged plants was obtained. The maximum net photosynthesis under saturated light conditions amount to a level of 35 mgO_2 $g^{-1}hr^{-1}$ (equivalent to 5 mg CO_2 dm^{-2} leaf area hr^{-1}), and this is only 1/5 of that of terrestrial plants (Ikusima 1965). The photosynthetic rate generally differed less between species than within species. Within species, the rate was affected by the season, the age of shoots and the position in a stand or a community (Table 41).

An attempt was made to calculate the daily photosynthesis in submerged plants at various depths and under different water conditions on the basis of the photosynthesis-light intensity curve, the attenuation of light in water and in the dense shoot of a stand, and the record of light intensity above the water level with reference to transparency and light extinction coefficients in water and within plant covers. The rate of daily gross production of *Elodea nuttallii* stand in summer was nearly proportional to the biomass where the shoot biomass was less than 300 g m^{-2}, but tended to saturate rather rapidly at greater biomass values, reaching a maximum of about 25 g dry matter m^{-2} land area day^{-1} (Fig. 156, Ikusima 1970).

The daily gross production of a H-type stand (apparently corresponding to the herb-type recognized in terrestrial stands) was slightly higher than that of G-type (corresponding to the grass-type). Moreover, considering losses due to

* Its open waters became extremely small after the 2nd World War.
** Nish-no-ko is situated south of Dai Naka-no-ko at present.
*** Other name of Matsunoki Naiko is Yotsugawa Naiko (Table 80).

Table 41. Rates of net photosynthesis and dark respiration of submerged plants in Lake Biwa (after Ikusima 1970).

Species	Water temperature (°C)	Light-saturated photosynthesis (mg O_2 $g^{-1}hr^{-1}$)	Dark Respiration (mg O_2 $g^{-1}hr^{-1}$)	Date
Elodea nuttallii				
(apical portion)	10±1	6.6	0.8	1964.4.
Hydrilla verticillata				
(apical portion)		7.2	1.9	
Vallisneria biwaensis	27±2			1964.7.
(leaf)		7.8	1.3	
Elodea nuttallii				
(apical portion)		7.7	1.4	
(basal portion)		0.9	0.8	
Elodea nuttallii				
(apical portion)	7±1	6.9	0.5	1965.1.
Elodea nuttallii				
(apical portion)		11.5	1.8	
(basal portion)	24.0±0.5	4.1	1.5	1966.7.
Hydrilla verticillata				
(apical portion)	27.8±0.6	14.6	1.4	1969.7.

respiration, the efficiency of net production of a small height G-type stand was slightly higher than that of a tall height H-type stand. In nature, the plant which winters with the G-type may have the ability to reveal a high growth rate and to make a thick pure stand for a short period from spring to early summer as observed in *Elodea nuttallii* (Ikusima 1972).

The mean daily net production during the growing season of *Elodea nuttallii* in Lake Biwa was estimated to be 8 g dry matter $m^{-2}day^{-1}$, where the ratio (net production of a stand/gross production of a stand) was assumed to be 40%.

In addition to the productivity, the ratio of daily compensation depth for aquatic macrophytes (Z_c)/transparency of water (Tr) was elucidated. Fig. 157 illustrates the dependence of the Z_c/Tr ratio on the relative light intensity on a monthly mean basis, where Z_c is the compensation depth of the upper part of shoot, in *Elodea nuttallii* and the relative light intensity is evaluated on an assumption that if fine, cloudy and rainy days continued for a whole month the value would be 100, 50 and 25%, respectively. When the bad weather continues the photosynthesis is restricted by insufficient light intensity and the compensation depth would become shallower. On the other hand, if the photosynthetic activity of the plant was high and the respiration rate was low, the compensation depth would become deeper. Physiological properties of the plant as well as environmental factors such as the amount of radiation, the day-length, weather etc., usually vary with the season, so that the compensation depth also changes with the season. Using the weather record based on a fairly long

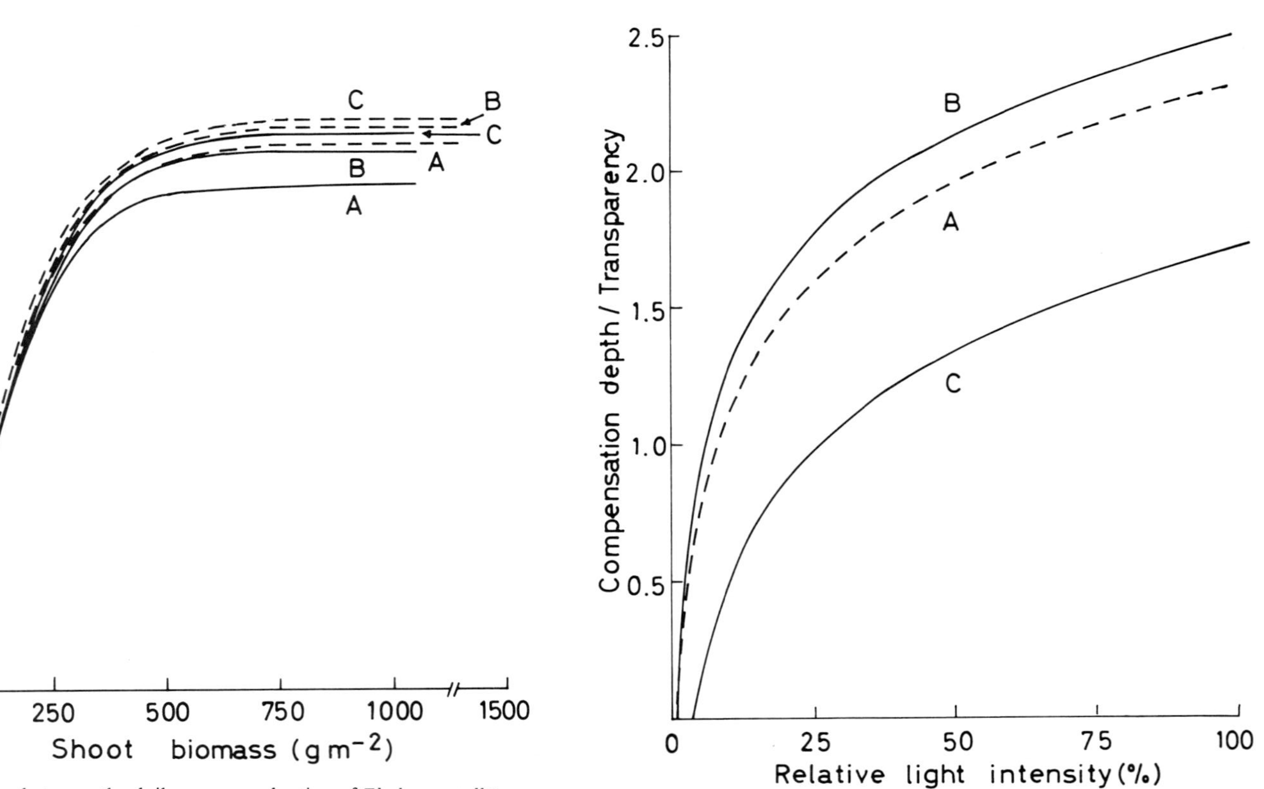

Fig. 156. Relation between the daily gross production of *Elodea nuttallii* stand and the shoot biomass of the stand. Solid lines refer to the G-type stand, while broken lines to the H-type stand. Curves A, B and C correspond to different grades of water turbidity equivalent to the light extinction coefficients of water of 1.0, 0.5 and 0.3 (1/m) (Ikusima 1970).

Fig. 157. Daily compensation depth/transparency ratio in *Elodea nuttallii* in relation to relative light intensity in January (A), April (B) and July-August (C) (Ikusima 1967).

period, the monthly mean of the daily compensation depth/transparency ratio of *Elodea nuttallii* in the lake was computed at 2.0, 2.4 and 1.4 in January, April and July-August, respectively. In July-August, the ratio of *Hydrilla verticillata,* which had been found down to about 10 m, ranged from 0.7 on a rainy day to 1.3 on a fine day, while that of *Elodea nuttallii* ranged from 1.0 to 1.6 (Ikusima 1967). There was a little difference in the ratio between these two species, so that it may be expected that in vigorous growing season *Elodea nuttallii* can survive at the deeper place than *Hydrilla verticillata.* At the first period of *Elodea nuttallii* invasion of Lake Biwa, the plant settled in new habitats at the deeper place where the native plants such as *Hydrilla verticillata* had not been found and the cover of *Elodea nuttallii* increased without any competition with native aquatic macrophytes.

Zooplankton

Tetsuya Narita and Kunihiro Okamoto

Introduction

In a lake food chain the zooplankton play the most important role of the secondary producers as a key industry organism. The herbivorous zooplankton graze on the phytoplankton and the bacteria, meanwhile are exposed to the predation by fishes and carnivorous zooplankton. As the study on the roles of zooplankton in the lake community is still on the way, in this section the faunal composition will be treated.

Zooplankton composition

The zooplankton fauna in Lake Biwa was first studied by T. Kawamura (1918). Since then, there have been taxonomical studies of Cladocera by M. Ueno and of Rotifera by K. Yamamoto and ecological studies of the crustacean zooplankton by K. Kikuchi. More than a hundred species of zooplankton have been so far recorded from this lake — 7 species of Copepoda, 18 of Cladocera, 80 of Rotifera and 14 of Protozoa (Table 42). Recently, probably due to eutrophication, Rotifera and Protozoa have increased. As a whole, the zooplankton of this lake is composed mostly of the common species of the temperate zone.

In Diaptomidae, *Eodiaptomus japonicus* (Fig. 158) is the dominant species and it is endemic to Japan (Kiefer 1932), but occurs commonly in lakes and ponds of southern Japan. Another species, *Sinodiaptomus valkanovi** is very rare. In Cyclopidae, *Mesocyclops leuckarti* is present dominant in pelagic water. However, according to Kikuchi et al. (1942), this species was rare in pelagic and occurred in the littoral area while *Cyclops strenuus* was common in the pelagic water at the end of the 1920's. Though *C. strenuus* had been reported from the lake till the end of 1940's, Ito, in his intensive study on copepods, did not find the

* Though Mori, Sy. (1945) recorded *Sinodiaptomus chaffanjoni* from the sub-basin, only *S. valkanovi* occurs in Japan. It is know that *S. chaffanjoni* occurs in the northern China (Ito & Miura 1973 and personal communication).

Horie, S. Lake Biwa
© 1984, Dr W. Junk Publishers, Dordrecht/Boston/Lancaster
ISBN 90 6193 095 2. Printed in The Netherlands

Table 42. The list of zooplankton in Lake Biwa.

CRUSTACEA

Copepoda

Sinodiaptomus valkanovi Kiefer
Eodiaptomus japonicus (Burckhardt)
Macrocyclops fuscus (Jurine)
Mesocyclops leuckarti (Claus)
Cyclops vicinus Uljanin
Eucyclopus serrulatus (Fischer)
Paracyclops fimbriatus (Fischer)

Cladocera

Sida crystallina (O.F. Müller)
Diaphanosoma brachyurum (Lieĕvin)
Holopedium gibberum Zaddach
Daphnia biwaensis Uéno
D. longispina O.F. Müller
Scapholeberis mucronata (O.F. Müller)
Simocephalus vetulus (O.F. Müller)
Moina macrocopa Straus
M. dubia De Guerne et Richard
Bosmina longirostris (O.F. Müller)
Bosminopsis deitersi Richard
Camptocercus rectirostris (Schoedler)
Graptoleberis testudinaria (Fischer)
Alona guttata G.O. Sars
A. rectangula G.O. Sars
Chydorus sphaericus (O.F. Müller)
Monospilus dissar G.O. Sars
Leptodora kindti (Focke)

ROTIFERA

Rotaria rotatoria (Pallas)
R. neptunia (Ehrenberg)
Floscularia ringens Hudson et Gosse
Sinantherina socialis (Linne)
Conochilus unicornis Rousselet
Conochiloides natans (Seligo)
Collotheca mutabilis (Bolt)
C. cornuta (Dobie)
Filinia longiseta (Ehrenberg)
Hexarthra mira (Hudson)
Testudinella patina (Hermann)
Pompholyx complanata Gosse
P. sulcata Hudson
Synchaeta oblonga Ehrenberg
S. stylata Wierzejski
S. tremula (O.F. Müller)
Polyarthra trigla (Ehrenberg)
P. vulgaris Carlin
P. euryptera Wiezejski
Ploesoma hudsoni (Imhof)
P. truncatum (Levender)
Chromogaster ovalis (Bergendal)
C. testudo Lauterborn
Trichocerca ierinis (Gosse)
T. cylindrica (Imhof)
T. longiseta (Schrank)
T. cristata Harring
T. bicristata (Gosse)
T. capucina (Wierzejski)
T. chattoni (Wierzejski)
T. elongata (Gosse)
Trichocerca (Diurella) tigris (O.F. Müller)
T. stylata (Eyeferth)
T. birostris (Minkiewcz)
T porcellus (Gosse)
T. brachyura (Gosse)
Asplanchna priodonta Gosse
A. sieboldi (Leydig)
A. multiceps (Schrank)
Brachionus falcatus Zacharias
B. quadridentatus Hermann
B. urceolaris O.F. Müller
B. angularis Gosse
B. budapestinensis Daday
B. calyciflorus Pallas
B. forficula Wiezejski
B. dimidiatus (Bryce)
B. diversicornis (Daday)
Platyias patulus (O.F. Müller)
P. quadricornis (Ehrenberg)
Keratella cochlearis (Gosse)
K. quadrata (O.F. Müller)
K. valga (Ehrenberg)
Kellicottia longispina (Kellicott)
Notholca striata (O.F. Müller)
N. foliacea (Ehrenberg)
N. labis Gosse
Anuraeopsis fissa (Gosse)
Epiphanes senta (O.F. Müller)
Euchlanis dilatata Ehrenberg
Dipleuchlanis propatula (Gosse)
Lepadella ovalis (O.F. Müller)
L. acuminata (Ehrenberg)
Lophochalis salpina (Ehrenberg)
Mytilina ventralis (Ehrenberg)
Trichotria pocillum (O.F. Müller)
T. tetractis (Ehrenberg)
Colurella adriatica (Ehrenberg)
C. obtusa Gosse
C. bicuspidata Ehrenberg
Lecane luna (O.F. Müller)
L. flexilis (Gosse)
L. ludwigii (Eckstein)
L. lunaris (Ehrenberg)
L. pygmaea Daday
L. stenroosi Meissner
L. bulla (Gosse)
L. crenata Harring
L. quadridentata (Ehrenberg)
L. hamata (Stokes)

Table 42. Continued

PROTOZOA	*D. lanceolata* Penard
Arcella vulgaris Ehrenberg	*D. brevicolla* Cash
A. discoides Ehrenberg	*D. globulus* (Ehrenberg)
Centropyxis aculeata (Ehrenberg)	*Cyphoderia ampulla* Ehrenberg
Difflugia acuminata Ehrenberg	*Tintinnidium fluviatile* Stein
D. oblonga Ehrenberg	*Vorticella* sp.
D. biwae Kawamura	*Strombidium* sp.
D. pyriformis Perty	

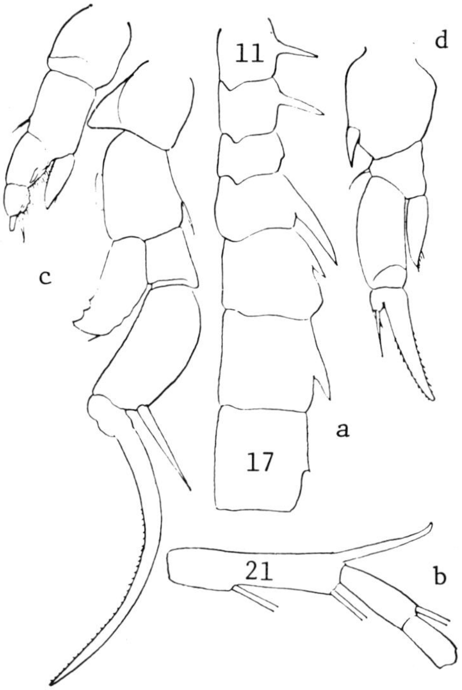

Fig. 158. Eodiaptomus japonicus (Bruckhardt).
a) right antenna ♂ showing spine and processes of segments 11–17.
b) apex of right antenna ♂.
c) Leg 5 ♂.
d) Leg 5 ♀.
Ito (In Uéno, 1973)

species and only *Cyclops vicinus* was found in the lake (Yamamoto et al. 1966). In Cladocera, *Daphnia biwaensis*, which was firstly described as a geographical race (*D. pulex biwaensis*, Ueno 1937), was recorded as an endemic species of this lake by Ueno (1973). It differs from *D. pulex* in the number of pecten on the claw (Fig. 159).

The zooplankton composition of the main basin, which is now mesotrophic, is fairly different from the sub-basin which is eutrophic. In the monthly collections

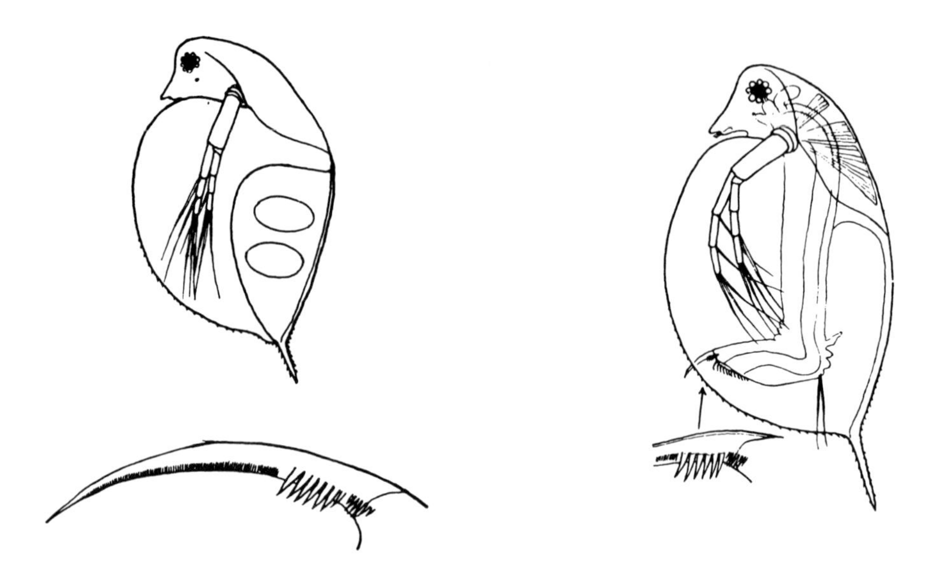

Fig. 159. Morphological difference between *Daphnia pulex* (right) and *Daphnia biwaensis* (left) (Uéno 1973).

of the zooplankton from April 1974 to April 1975, 8 to 19 species (mostly 10 to 13) occurred in the main basin and 7 to 23 species (mostly 13 to 17) in the sub-basin (Fig. 160).

In the main basin, *E. japonicus* is the most dominant species. This species occurs in the epilimnion and is rare in the hypolimnion in the warm season and moves to the bottom layer (hypolimnion) in winter to overwinter in the copepodite stages (Fig. 162). This species shows apparent diurnal vertical migration when the water temperature is nearly uniform, while it aggregates mostly in the epilimnion in summer when the thermal stratification is established (Narita & Mori, Sy. 1973) (Fig. 161). The younger stages of the species occur in rather shallower layers, and as they grow they occupy deeper layers (Table 43). *Mesocyclops leuckarti* occurs commonly in the epilimnion, and shows notable seasonal change in number. *Cyclops vicinus* occurs in the hypolimnion throughout the year, but not as much as the former species, and also shows seasonal change in number. These two cyclopoids seem to segregate their inhabiting layers; that is, *Mesocyclops* occupies epilimnion while *Cyclops* occupies hypolimnion (Table 44). In Cladocera, *Daphnia longispina* is dominant species in the pelagic epilimnion in the warm season (Fig. 162). *Daphnia biwaensis* was known to occur in the hypolimnion and hardly collected in the epilimnion (Kikuchi et al. 1942). In the writers' collections of 1974–75, the writers did not find this species at all even in deeper layers and only found *D. longispina, Bosmina longirostris* occurs in the meta- and hypolimnion in summer just as in other lakes, while it occurs commonly in a shallow water in ponds.

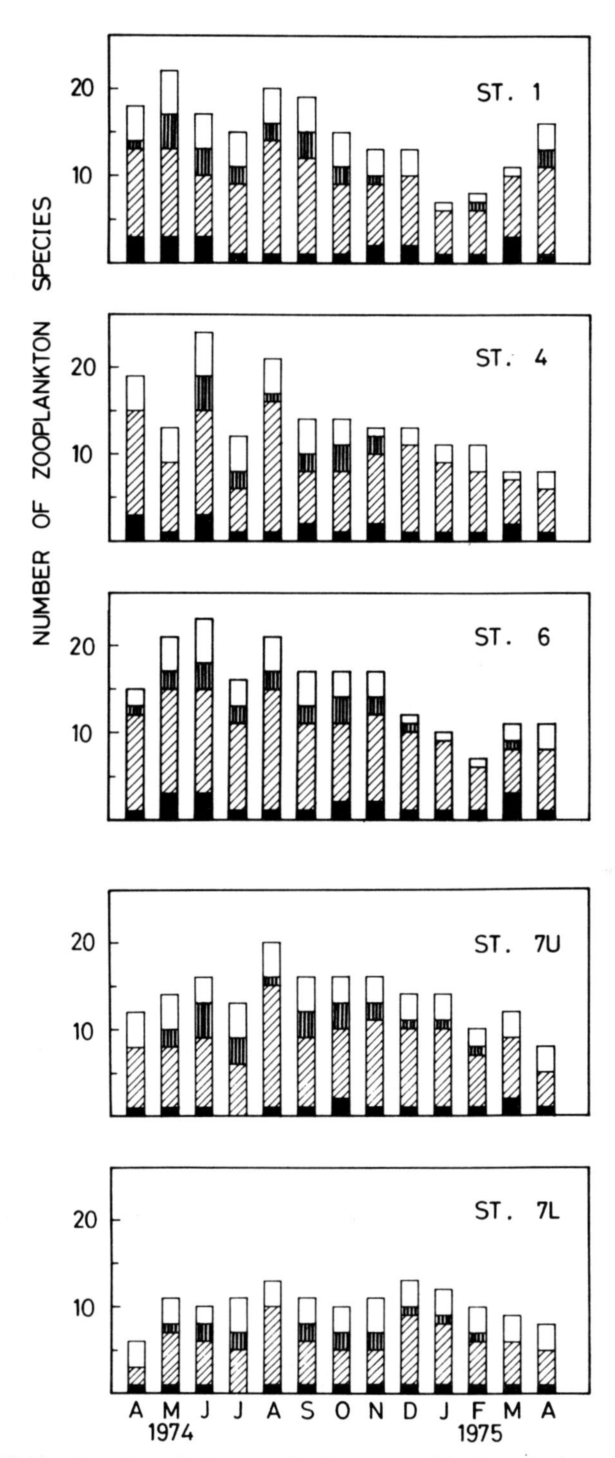

Fig. 160. Number of zooplankton species that occurred in the collection of 1974 – 1975. (U: upper layer, 0–20 m, L: lower layer, 20–40 m).

Protozoa, Rotifera, Cladocera, Copepoda

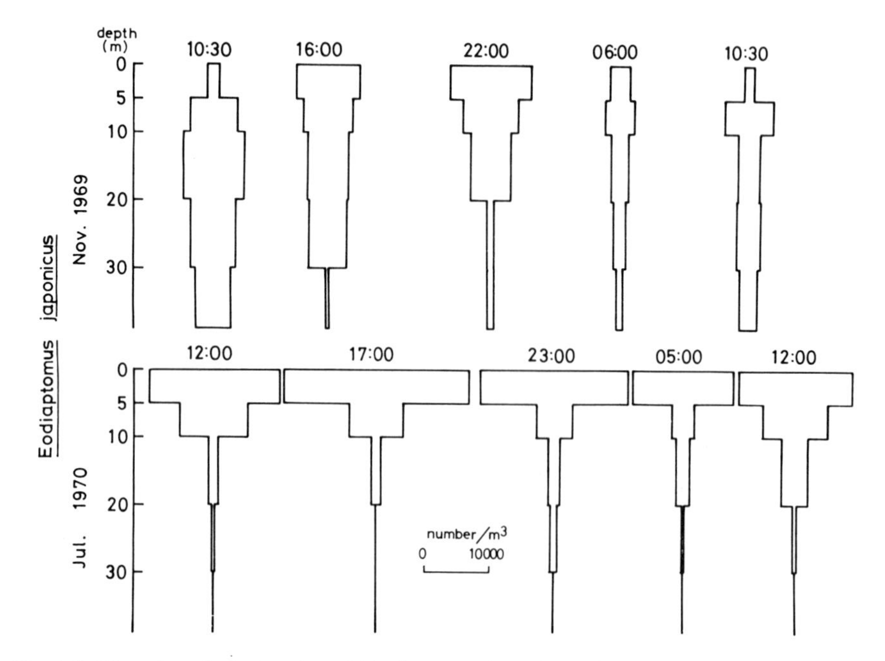

Fig. 161. Diurnal vertical migration of *Eodiaptomus japonicus* (after Narita & Mori, Sy. 1973).

Table 43. Vertical distribution of different stages of *Eodiaptomus japonicus* (N/liter) (December, 1926). (After Kikuchi et al., 1942).

Stage	Depth (in m)								Average depth
	0	2	5	10	15	20	25	30	
Nauplii	2.0	2.8	8.8	6.4	0.4	0	0	0	7.1 m
0.4–0.5 mm	0.1	0.1	0.5	1.0	0.6	0	0	0	10.4
0.5–0.6	0	0	0.1	0.2	1.8	0.1	0	0	14.5
0.7–0.85	0	0	0	0.2	1.9	0.2	0	0	15.0
♂(0.9–1.0)	0	0	0	0.1	0.4	0.7	0.2	0	18.9
♀(1.2)	0	0.1	0	0.2	1.7	0.4	0.3	0.3	16.4

In Rotifera, *Kellicottia longispina* is dominant and occurs throughout the year. *Keratella cochlearis, K. quadrata* and *Polyarthra vulgaris* are common members but are rarely found in a large number in the main basin.

In the sub-basin, *Bosminopsis deitersi* and *Chydorus sphaericus* are the common crustaceans in the littoral and the embayment (Fig. 162). *Scapholeberis mucronata*, which commonly occurs in the paddy fields around the lake, is sometimes found among aquatic plants in the embayment. Among rotifers, *Synchaeta stylata, Polyarthra vulgaris* and *Keratella cochlearis* are the dominant species. *Brachionus angularis* and *Filinia longiseta* are the common rotifer in eutrophicated embayment.

Table 44. Vertical distribution of *Mesocyclops leuckarti* and *Cyclops vicinus* (number/m^3) (After Yamamoto et al., 1966).

	1963							1964				
	Jun. 16	Jul. 23	Aug. 18	Sep. 22	Oct. 22	Nov. 24	Dec. 8	Jan. 12	Feb. 7	Mar. 17	Apr. 18	May 16
Mesocyclops leuckarti												
0–2	920	1132	283	1062	1274	142	0	70	0	0	0	0
2–5	3303	1415	1510	708	472	94	0	0	0	94	0	0
5–10	5095	4302	2038	849	283	142	0	0	0	28	226	0
10–15	1076	2038	1181	2406	283	142	57	0	0	28	453	0
15–20	771	906	736	170	57	142	57	0	57	28	113	0
20–30	—	28	57	113	14	142	14	0	0	42	170	28
30–50	—	0	0	0	0	14	0	0	7	0	0	0
50–70	—	0	0	0	0	0	0	0	7	0	0	0
Total (n/m^2)	46459	43019	25481	22503	7219	4396	710	140	565	1122	5660	280
Cyclops vicinus												
0–2	0	0	0	0	0	0	0	0	0	0	0	0
2–5	0	0	0	0	0	0	0	0	0	0	0	0
5–10	0	0	0	0	0	0	0	0	0	0	0	0
10–15	0	0	0	0	0	0	0	0	0	0	283	0
15–20	0	0	0	0	57	0	0	0	0	0	113	0
20–30	—	0	42	28	99	0	28	40	0	0	85	28
30–50	—	120	99	99	78	28	28	10	0	0	57	78
50–70	—	45	71	42	7	7	4	10	0	0	14	78
Total (n/m^2)	0	3300	3820	3100	2975	700	920	800	0	0	4250	3400

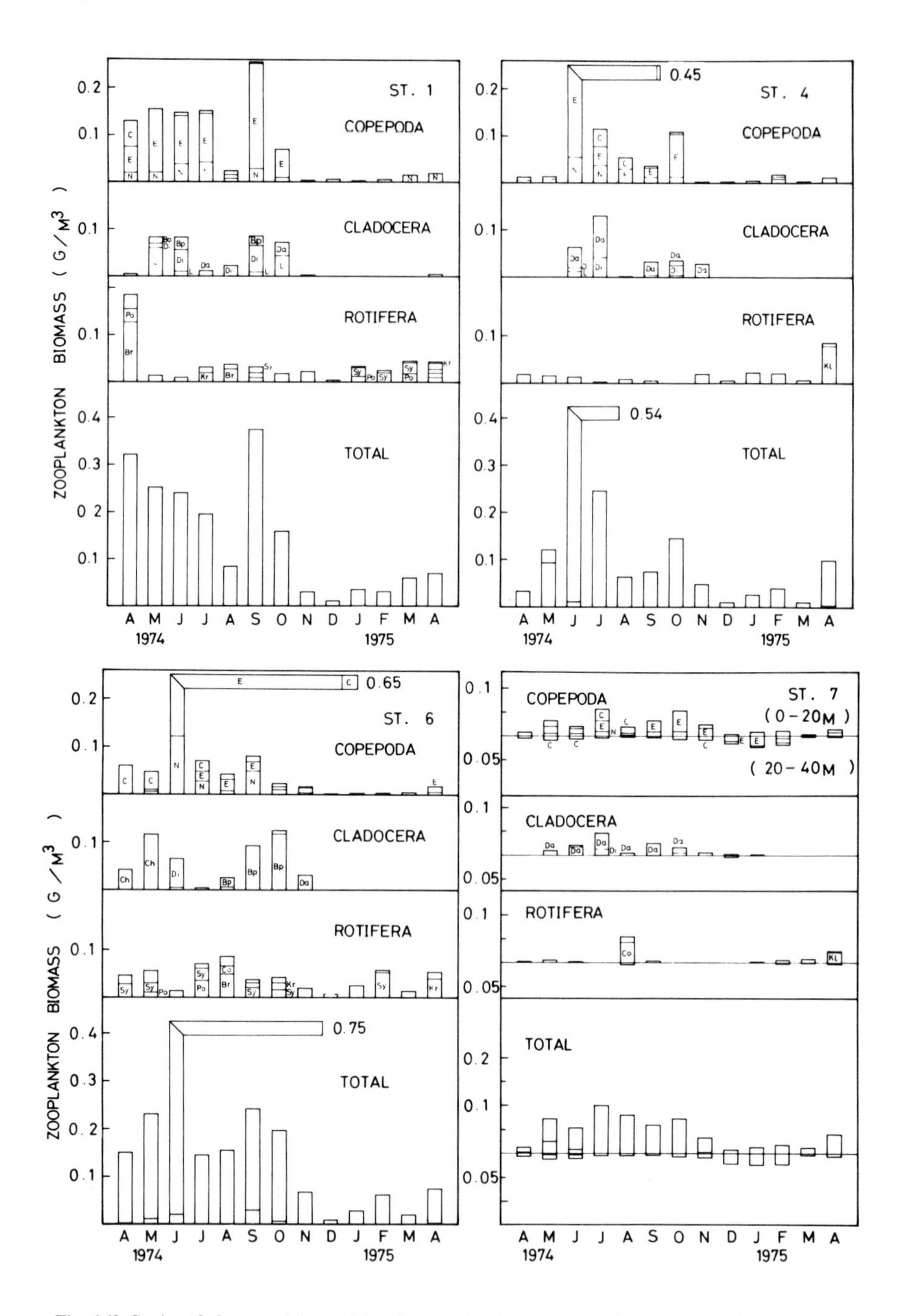

Fig. 162. Seasonal change of dry weight biomass density at three points of sub-basin (St. 1, 4 and 6) and one point of main basin (St. 7).

E:	*Eodiaptomus*	C:	*Mesocyclops* and *Cyclops*	N:	nauplii of copepods
Da:	*Daphnia*	Di:	*Diaphanosoma*	L:	*Leptodora*
Bo:	*Bosmina*	Bp:	*Bosminopsis*	Ch:	*Chydorus*
Sy:	*Synchaeta* spp.	Po:	*Polyarthra* spp.	Br:	*Brachionus* spp.
Kr:	*Keratella* spp.	Co:	*Conochirus unicornis*	Kl:	*Kellicottia*

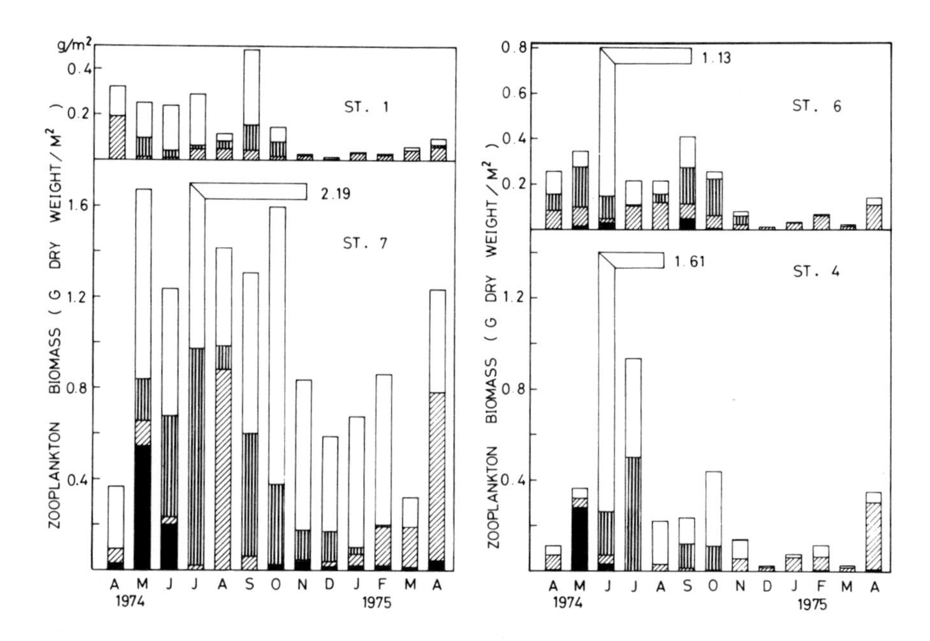

Fig. 163. Dry weight biomass of zooplankton at three points of sub-basin (St. 1, 4 and 6) and one point of main basin (St. 7) (dry weight g/m²).

The total dry weight biomass changes depending on the depth and the season. The annual mean of the biomass in the collections was 1.2 g dry weight/m² in the main basin and 0.16 to 0.32 g dry weight/m² in the sub-basin (Fig. 163). The average of the biomass density was 0.03 g dry weight/m³ in the main basin and 0.11 – 0.16/m³ in the sub-basin (Fig. 162). As usual with lakes in the temperate zone, the zooplankton biomass is large in summer and is small in winter in this lake. As a whole, crustaceans composed a large percentage of the biomass even in the sub-basin and the rotifers have rather small biomass in spite of their large number. In average, about 48 – 57% of the biomass was composed of the copepods and their nauplii, 20 – 24% composed of cladocerans 15 – 25% of rotifers and 0.2 – 7% of protozoans.

Benthos

Eiji Harada

Introduction

As is noted for other ecological assemblies of organisms living in Lake Biwa, the benthos also comprises a rich variety of species and includes many endemic species, if compared to those of other freshwater areas in Japan. The mode of life varies greatly among these species that occupy various habitats afforded in the lake.

The study of benthic organisms in Lake Biwa started, in its reality as biological limnology, with the works of Annandale (1916a,b, 1918, 1921, 1922), Annandale and Kawamura (1916), Stephenson (1917) and Tattersall (1921). Since then, the knowledge on the benthos of Lake Biwa has been accumulated. The research works that have been undertaken so far, however, are mostly concerned with larger animals, and the micro- and meio-fauna have been left rather untouched. It is only recently that these minute organisms of the benthos have received much attention in connexion with the pollution of the lake and in regard to their contributions to the nutrient cycle and the energy flow in the lake. In this section, therefore, the macro-zoobenthos in Lake Biwa is principally and briefly reviewed.

Taxonomic composition

The known species of the macro-zoobenthos in Lake Biwa, other than molluscs and fish that are treated separately in other sections, number more than 178, of which 152 have been identified to the species. The larval insects dominate in number of species, followed by molluscs. The remaining groups are generally small in number of species. Two groups, Tubificidae and Chironomidae, should be mentioned as having been acknowledged to have much greater number of species than shown in Table 45, since the taxonomic work has not yet been sufficiently advanced, though quite numerous forms have already been found for these groups. Only the species identified to the species are given in the list of Table 46.

Horie, S. Lake Biwa

ISBN 90 6193 095 2. Printed in The Netherlands

Table 45. Taxonomic composition of macro-zoobenthos in Lake Biwa (based on Kato, 1962; Takao, 1964; Mizuno & Tetsukawa, 1965; Kumode, 1965; Mori, Sy. 1970b; and the unpublished records compiled by Ōtsu Hydrobiological Station).

Taxonomic groups	Number of species confirmed	Number of species identified	Number of endemic species
Porifera	6	6	0
Platyhelminthes	3	2	1
Ectoprocta	4	4	0
Mollusca	44	44	18
Oligochaeta	6	4	1
Hirudinea	6	3	1
Amphipoda	3	3	1
Isopoda	2	2	0
Decapoda	4	4	0
Hexapoda	101	84	6

Table 46. List of the species identified of zoobenthos in Lake Biwa, excluding molluscs and fish (based on Kato, 1962; Takao, 1964; Mizuno & Tetsukawa, 1965, 1966; Kumode, 1965; Tsuda et al., 1966; Mori, Sy. 1970b, and the unpublished records compiled by the Ōtsu Hydrobiological Station).

Pyrophyta
Dinophyceae
Peridiniales

Ceratium hirundinella (O.F. Müller)

Euglenophyta
Euglenophyceae
Euglenales

Euglena viridis
Phacus triqueter (Ehr. & Duj)
Phacus pleuronectes (O.F. Müller) Duj.
Anisonema acinus Duj.

Chlorophyta
Chlorophyceae
Volvocales

Eudorina elegans Ehr.

Protozoa
Sarcodina
Testacida

Arcella vulgaris Ehr.
Difflugia biwae Kawamura
Euglypha tuberculata
Centropyxis aculeata (Ehr.) Stein
Cyphoderia ampulla Ehr.

Ciliata
Peritrichida

Vorticella campanula Ehr.

Oligotrichida

Strombidium gyrans Stokes
Halteria grandinella (O.F. Müller)

Hypotrichida

Balladyna parvula Kowalewski

Table 46. Continued

Porifera	
Demospongia	
Haplosclerina	
	Spongilla lacustris (Linne)
	Spongilla clementis Annandale
	Spongilla semispongilla (Annandale)
	Heteromeyenia baileyi var. *petri* (Lauterborn)
	Ephydatia müllleri (Lieberkühn)
	Ephydatia mülleri var. japonica (Hilgendorf)
Platyhelminthes	
Turbellaria	
Tricladida	
	Dugesia japonica Ichikawa et Kawakatsu
	Bdellocephala annandalei Ijima et Kaburaki
Aschelminthes	
Rotatoria	
Bdelloidea	
	Rotaria rotatoria (Pallas)
	Philodina roseola Ehr.
Flosculariacea	
	Conochilus unicornis Rousselet
	Testudinella patina (Hermann)
Collothecacea	
	Collotheca cornuta (Dobie)
Ploima	
	Synchaeta tremula (O.F. Müller)
	Trichocerca cylindrica (Imhof)
	Brachionus angularis Gosse
	Keratella cochlearis (Gosse)
	Keratella valga (Ehr.)
	Euchlanis dilatata Ehr.
	Dipleuchlanis propatula de Beauchamp
	Lepadella oblonga (Ehr.)
	Mytilina ventralis (Ehr.)
	Colurella colurus (Ehr.)
	Lecane flexilis (Gosse)
	Lecane hamata (Stokes)
	Lecane hastata (Murray)
	Monostyla bulla Gosse
	Monostyla closterocerca Schmarda
	Monostyla pygmaea Daday
Ectoprocta	
Phylactolaemata	
	Plumatella emarginata Allman
	Pectinatella gelatinosa Oka
	Lophopodella carteri (Hyatt)
	Fredericella sultana (Blumenbach)
Annelida	
Oligochaeta	
Archioligochaeta	
	Criodrilus bathybates Stephenson
	Kawamuria japonica Stephenson
	Branchiura sowerbyi Beddard
	Limnodrilus grandisetosus Nomura

Table 46. Continued

Hirudinea	
Rhynchobdellida	
	Ancyrobdella biwae Oka
	Whitmania pigra (Whitman)
Gnathobdellida	
	Mimobdella japonica Blanchard
Arthropoda	
Crustacea	
Amphipoda	
	Anisogammarus annandalei (Tattersall)
	Kamaka biwae Uéno
	Orchestia japonica (Tattersall)
Isopoda	
	Tachea chinensis Thielemann
	Asellus hilgendorfii Bovallius
Decapoda	
	Palaemon paucidens de Haan
	Macrobrachium nipponense (de Haan)
	Paratya compressa (de Haan)
	Procambarus clarkii (Girard)
Hexapoda	
Trichoptera	
	Ecnomus omiensis Tsuda
	Platyphylax yokouchii Iwata
	Dipseudopsis stellata MacLachlan
	Nemataulius admorsus (MacLachlan)
	Limnephilus fusovittatus Matsumura
	Goera japonica Banks
	Gumaga okinawaensis Tsuda
	Molanna moesta Banks
	Georigium japonicum (Ulmer)
	Triplectides magnus (Walker)
	Triaenodes yamamotoi Tsuda
	Athripsodes biwaensis (Tsuda et Kawayama)
	Setodes biwae Tsuda
	Mystacides azurea (Linne)
	Mystacides longicornis Linne
	Oecetis morii Tsuda
	Oecetis yukii Tsuda
Ephemeroptera	
	Ecdyonurus kibunensis Imanishi
	Ecdyonurus yoshidae Takahashi
	Cinygma hirasana Imanishi
	Siphlonurus sanukensis Takahashi
	Polymitarcis shigae Takahashi
	Potamanthus kamonis Imanishi
	Choroterpes trifurcata Uéno
	Ephemera lineata Eaton
	Ephemerella longicaudata Uéno
Plecoptera	
	Neoperla nipponensis (MacLachlan)
Coleoptera	
	Eubrianax granicollis Lewis
	Metaeopsephenus japonicus Matsumura
	Hydrochus japonicus Sharp
	Coelostoma stultum (Walker)
	Enochrus simulans simulans (Sharp)
	Enochrus vilis (Sharp)

Table 46. Continued

Coleoptera *continued*	*Enochrus haroldi* (Sharp)
	Berosus signaticollis punctipennis Harold
	Regimbartia profunda (Sharp)
	Sternolophus rufipes (Fabricius)
	Noterus japonicus Sharp
	Laccophilus difficilis Sharp
	Ilybius apicalis Sharp
	Cybister tripunctatus orientalis Gschwendtner
	Gyrinus curtus Sharp
	Dineutus orientalis (Modeer)
	Heterocurus asiaticus Nomura
Megaloptera	*Parachadiodes japonicus* MacLachlan
Odonata	*Cercion calamorum* Ris
	Cercion hieroglyphicum Brauer
	Calopteryx atrata Selys
	Stylurus oculatus (Asahina)
	Stylurus annulatus (Djakonov)
	Gomphus melaenops Selys
	Gomphus oculatus Asahina
	Gomphus postocularis Selys
	Trigomphus interruptus Selys
	Trigomphus ogumai Asahina
	Nihonogomphus viridis Oguma
	Onychogomphus viridicostus Oguma
	Ictinogomphus clavatus Fabricius
	Sieboldius albardae Selys
	Aeschnophlebia longistigma Selys
	Anax parthenope julius Brauer
	Macromia amphigena Selys
	Epophthalmia elegans Brauer
	Orthetrum albistylum speciosum Uhler
	Orthetrum triangulare melania Selys
	Crocothemis servilia Drudy
	Deielia phaon Selys
	Sympetrum frequens Selys
	Sympetrum darwinianum Selys
	Sympetrum risi risi Bartenef
	Sympetrum infuscatum Selys
	Sympetrum eroticum eroticum Selys
	Pseudothemis zonata Burmeister
Neuroptera	*Sisyra nikkoana* Navas
Diptera	*Chironomus salinarius* Kieffer
	Chironomus plumosus Linne
	Tokunagayusurika akamusi (Tokunaga)
Lepidoptera	*Nymphula vittalis* Bremer
	Nymphula interruptalis Pryer
	Nymphula turbata Butler
Hemiptera	*Ilyocoris exclamationis* Scott
	Ranatra chinensis Mayr
	Ranatra unicolor Scott
	Gerris paludum insularis Motschulsky

Life types

Major life types recognizable in the macrobenthos are the free-living and the sessile life. Most of the species of the macro-zoobenthos found in Lake Biwa are free-living.

Crustaceans and insect larvae are mostly bottom inhabitants and make up a prominent part of the epifauna. Shrimps and prawns also frequent submerged plant growths, but are only scarcely represented in the nekton (Harada 1966). A gammarid often gathers and clings to the remains of woods and weeds on the bottom, and also spends pelagic life to be found as a member of the plankton (Narita 1976). Insect larvae of the epifauna are further attributable to either of the types of life of crawling, nest-carrying or nest-spinning (Takao 1964). Hirudineans crawl on the surface of stones and pebbles, and tricalds also occur attached to wood pieces, tins, polyethylene bags and anything hard lying on the bottom. Oligochaetes are dominating members of the infauna and the burrowing forms of insect larvae, particularly chironomids, also are well represented in that (Kato 1962; Takao 1964).

The representative sessile animals found in Lake Biwa are sponges and ectoprocts. They are living attached to various kinds of hard substrates, like rocks, boulders, reeds, fishing traps and even shells of living molluscs (Mizuno & Tetsukawa 1965). Many of the protozoans and rotifers are also found on these substrates (Wakamatsu 1963; Kondo 1963; Iima 1963).

Distribution and abundance

The shallow littoral area of Lake Biwa is composed of different substrates, with or without aquatic vegetation, and affords various habitats to benthic animals, whereas the profundal region is rather uniform in substrate and configuration of the bottom and is devoid of the growth of macrophytes. Generally speaking, insect larvae and hirudineans dominate in the littoral zone, while oligochaetes are the dominant in the profundal region (Tsuda 1962).

In the fringing areas of the shallow south basin, where reeds and aquatic plants grow densely, larvae of Orthoptera and Hemiptera occur abundantly and the large population of *Asellus hilgendorfii* is characteristic (Ito 1964). Shrimps, prawns and crayfish inhabit these areas in quantities, and sponges, particularly *Spongilla lacustris* which is the most dominant among the sponges in the lake, and ectoprocts, particularly *Pectinatella gelatinosa*, are also abundantly found there (Mizuno & Tetsukawa 1965). Various heteropterans and coleopterans are inhabiting there. Hirudineans and insect larvae dominate on the shallow bottoms of pebbles and stones in the south basin. Frequent occurrence of *Anisogammarus annandalei* characterizes these areas (Ogino 1963; Ito 1964). The muddy level bottom over the central part of the south basin supports benthic faunas with predominating populations of oligochaetes and chironomids (Suzuki & Mori 1967).

Table 47. Densities and biomasses of benthic animals at various habitats in Lake Biwa (based on Ogino, 1963; Harada, 1966; Suzuki & Mori, 1967). Densities are given in number per m², and biomasses in wet weight in gr per m².

Animal group	Habitat			
	North Basin Profundal, Mud			
	Winter		Summer	
	Density	Biomass	Density	Biomass
Oligochaetes	311–622	2.7–20.9	44–650	0.4–11.3
Hirudineans	0–44	0–0.89	0–89	0–1.5
Chironomids	0–89	0–0.04	0–44	0–1.5
Other Insects				
Gammarids	0–800	0–20.9		
Prawns	0–0.99	0–0.31		

Animal group	North Basin Littoral, Mud		Littoral, Pebbles	
	Summer		Summer	
	Density	Biomass	Density	Biomass
Oligochaetes	0–1890	0–10.6	0–8	0–0.89
Hirudineans	0–44	0–0.20	0–20	0–1.68
Chironomids				
Other Insects	0–89	0–0.73	2–126	0.002–1.36
Gammarids			0–36	0–1.13
Prawns			0–126	

Animal group	South Basin Shallow, Mud			
	Winter		Summer	
	Density	Biomass	Density	Biomass
Oligochaetes	0–444	0–16.9	0–311	0–3.6
Hirudineans				
Chironomids	0–178	0–13.3	0–89	0–1.2
Other Insects				
Gammarids				
Prawns				

Animal group	South Basin Littoral, Pebbles	
	Summer	
	Density	Biomass
Oligochaetes	0–6	0–0.01
Hirudineans	2–48	0–9.45
Chironomids		
Other Insects	0–94	0–0.11
Gammarids	0–154	0–0.45
Prawns		

The reed belts fringing the north basin of Lake Biwa sustain the faunas similar to that in the south basin, but a sponge, *Spongilla clementis*, occurs and the next dominant ectoproct, *Plumatella emarginata*, is abundant there (Mizuno & Tetsukawa 1965). *Palaemon paucidens* is a representative animal inhabiting submerged plant growths, such as those of *Potamogeton*, *Najas*, *Vallisneria* or *Elodea*, during the warmer months from spring to early winter (Harada 1966). *Paratya compressa* is comparatively rare in these areas in the north basin. *Palaemon paucidens* also inhabits the shallow pebble bottoms, as well as artificial banks of stones, where *Macrobrachium nipponense* is relatively abundant. On the shallow bottoms of pebbles, stones and rocks, *Ecdyonurus yoshidae*, *Potamanthus kamonis*, *Neoperla nipponensis* and *Metanopsephenus japonicus* are, among the insects, characteristically found. Hirudineans are abundant and *Anisogammarus annandalei* also occurs abundantly. Chironomids are prominent members of the faunas on the muddy and sandy bottoms from the littoral to the profundal region. Oligochaetes predominate in the faunas of the profundal muddy bottoms. *Bdellocephala annandalei* is almost exclusively found on the profundal bottoms and hirudineans are represented characteristically by *Ancyrobdella biwae* there (Annandale 1922, Gose 1966). *Anisogammarus annandalei* is a notable constituent of the profundal faunas, too (Narita 1976). *Palaemon paucidens* also populates the profundal bottoms in winter (Harada 1966).

The representative estimates of the densities and biomasses of benthic animals at various habitats in Lake Biwa are shown in Table 47. Benthic animals vary markedly between places and with season both in density and biomass, but no general trends of difference in macro-zoobenthos have ever been discussed and confirmed.

Molluscs

Syuiti Mori

Introduction

About 60 species (including subspecies) of molluscs are living in Japanese freshwaters, among which 44 species are found in Lake Biwa, including 19 endemic species. This means that about 2/3 of the total are living in Lake Biwa and 1/3 are found only in Lake Biwa.

This is a peculiar feature of this lake incomparable to other Japanese lakes. This seems to be caused by its extraordinary long geological history (five million years) and its large surface area (674.4 km^2), which means there has been enough time for differentiation of organisms and enough kinds of habitats allowing to differentiate various types in mode of life. Furthermore, the temperature of waters deeper than 40 m is maintained at about 7°C throughout the year, whereas the surface temperature fluctuates between 7°C and 28°C (or even 30°C) within the year, and this makes it possible for many kinds of molluscs originated in cold waters as well as in warm waters to coexist.

Kinds of molluscs

Specific names of molluscs found in Lake Biwa are listed in Table 48, in which the species marked with* shows an endemic species (modified from Mori, Sy. 1970).

On some note on biology of several important species

The following descriptions are based on the materials reported in Hayashi et al. (1966) and Mori, Sy. (1971).

Hyriopsis (*Nipponohyria*) *schlegeli* (v. Martens) (Plate 10)

This is an endemic bivalve in Lake Biwa and is very important economically, i.e., this is used as a mother mussel for artificial production of freshwater pearls and

Horie, S. Lake Biwa

ISBN 90 6193 095 2. Printed in The Netherlands

Table 48. List of molluscan species living in Lake Biwa (Mori, Sy. & Miura 1980).

Class Gastropoda		
Ord.		Mesogastropoda
	Fam.	Viviparidae
		Cipangopaludina japonica (v. Martens)
		Cip. chinensis malleata (Reeve)
	*	*Heterogen longispira* (Smith)
		Sinotaia quadrata histrica (Gould)
	Fam.	Valvatidae
		Valvata (*Cincinna*) *japonica biwaensis* Preston
	Fam.	Bithyniidae
		Parafossarulus manchouricus japonicus (Pilsbry)
	Fam.	Thiaridae
		Semisulcospira libertina (Gould)
	*	*Sem. decipiens decipiens* (Westerlund)
	*	*Sem. decipiens multigranosa* Boettger
	*	*Sem. decipiens reticulata* Kajiyama et Habe
		Sem. kurodai Kajiyama et Habe
	*	*Sem. nakasekoae* Kuroda
	*	*Sem. niponica* (Smith)
		Sem. reiniana (Brot)
Ord.		Basommatophora
	Fam.	Lymnaeidae
		Austropeplea ollula (Gould)
		Fossaria truncatula (Müller)
		Radix auricularia japonica (Jay)
	*	*Rad.* (*Biwakoia*) *onychia* (Westerlund)
	Fam.	Physidae
		Physa acuta Draparnaud
	Fam.	Planorbidae
		Camptoceras hirasei Walker
		Culmenella prashadi (Clench)
		Polypylis hemisphaerula nitidella v. Martens
		Gyraulus chinensis hiemantium (Westerlund)
	*	*Gyr. biwaensis* (Preston)
	*	*Gyr. amplificatus* (Mori)
	Fam.	Ancylidae
		Pettancylus nipponica (Kuroda)
		Pet. japonica Habe et Burch
Ord.		Stylommatophora
	Fam.	Succineidae
		Oxyloma hirasei (Pilsbry)
Class Bivalvia		
Ord.		Palaeoheterodonta
	Fam.	Unionidae
	*	*Lanceolaria oxyrhyncha* (v. Martens)
		Unio (*Nodularia*) *douglasiae nipponensis* v. Martens
	*	*Unio (Nod.) biwae* Kobelt
		Inversidens brandti (Kobelt)
	*	*Inv. reiniana* (Kobelt)
		Inv. japanensis (Lea)
		Pseudodon (*Obovalis*) *omiensis* (Heimburg)
	*	*Hyriopsis* (*Nipponohyria*) *schlegeli* (v. Martens)
	*	*Cristaria plicata clessini* (Kobelt)
		Anodonta lauta v. Martens
	*	*Ano. calipygos* Kobelt

Table 48. Continued

Ord.	Heterodonta
Fam.	Corbiculidae
*	*Corbicula sandai* Reinhardt
	Cor. (*Corbiculina*) *leana* Prime
Fam.	Sphaeriidae
*	*Pisidium kawamurai* Mori
	Pis. cinereum lacustre Woodward
*	*Sphaerium japonicum biwaense* Mori

the value of production of pearls accounts for about 1/3 of the total value of fishery production in this lake. Larval mussels are mostly found at the sandy or sandy mud bottom 0.5–3 m deep, but adults are living at the muddy bottom down to 30 m deep, most abundantly at about 10 m deep. Their stomach content is composed 60–70% of detritus, 30–37% of organisms and a few percent of inorganic materials. Their spawning season is spring, from April to June. The number of glochidia seen in an individual breeding pouch is about 300,000–400,000. The duration from egg to glochidium is 13 days at about 16°C and the period of glochidium parasitic on fish is 11–25 days.

Corbicula sandai Reinhardt (Plate 11)

This is also an endemic, but very common species in this lake and very important economically as foodstuffs of local people. The young likes the sandy bottom 2 to 8 m deep, and the adults are living on a sandy or sandy mud substratum ranging over a wider area than the young down to 40 m deep, but mostly distributed at shallower places than 10 m. 40% of their stomach content is debris and 60% is inorganic materials. The breeding season is from June to October and the number of eggs in a female (length 2.5 cm) is about 35,000 to 45,000. This mussel is dioecious, and fertilization takes place in water. Fertilized eggs lay first on the bottom, develop into trochophora after 20 hours, young shells are formed after 65 hours, and then get into the bottom after 86 hours (at 20–28°C).

There is a fact which should be emphasized concerning the biology of this bivalve. There are two main strains in Japanese corbiculan mussel; one is the *Corbicula japonica* strain, which is living in brackish waters; being dioecious, fertilization takes place in water and has a planktonic larval stage. The other is the *Corbicula leana* strain (Plate 12), which is living in freshwaters, being hermaphrodite, fertilization takes place in a breeding pouch and the young is produced into water after formation of young shell and it gets instantly into the bottom, so that it has no planktonic stage Hurukawa 1953; Miyazaki 1957). Considering these facts, the mode of life of *Corbicula sandai* is rather similar to that of *Corbicula japonica* and it is possible to presume from this fact that Lake

Biwa has once been connected with brackish waters and *Corbicula sandai* had been derived from *Corbicula japonica* Prime.

This species had been widely distributed all over the lake up to a few years ago, but it almost disappeared from the South lake since about 1971, and instead, *Corbicula leana* appeared and took the habitat which had species is undoubtedly caused by a recent rapid progress of eutrophication in the South undoubtedly caused by a recent rapid progress of eutrophication in the south lake. *Corbicula leana* is a species living widely in Japanese freshwaters which are usually more polluted than Lake Biwa. At about year 1971 Japan was in the midst of her rapid economic and industrial growth and around Lake Biwa the area along the southern coast had been subjected to remarkably rapid growth economically and industrially.

Unio (Nodularia) biwae Kobelt (Plate 13)

This is an endemic species too, and is very common all over the lake and is important as a foodstuff for local people. This is living at nearly the same place as *Corbicula sandai*, though it can live at more muddy places and cannot live at places more than 20 m deep. The breeding season is from April to September and the number of eggs in a mussel is about 150,000 to 240,000. This is the most active bivalve in the lake and moves about most at night.

Heterogen longispira (Smith) (Plate 14)

This is an endemic gastropod and very important for fishing. It lives on all kinds of substrata, but especially prefers sandy mud substrata, and is found most frequently at the depth of 2 to 10 m and rarely at places deeper than 30 m. Only 17% of the stomach content is organic materials, which is much less than *Corbicula sandai* (about 40%) or *Hyriopsis schlegeli* (more than 90%). This may be due to the difference in the mode of eating behavior, i.e., this gastropod takes food by creeping on the bottom mud, so that it naturally takes in a lot of inorganic materials, whereas bivalves take suspending materials, which have a larger percentage of organic materials than in mud. This is hermaphrodite and viviparous, and its breeding season covers spring till autumn. The time when it is active is highly characteristic, that is, it moves about actively from 01:00 to 15:00, but mostly remains inactive from the late afternoon to the early night.

Recently in the South lake, it was replaced by *Sinotaia quadrata histrica* (Gould) (Plate 15). The latter species is a common species distributed in small waters all over Japan, and has been also a common species at a river mouth or in a small bay of Lake Biwa where water is somewhat polluted, and it had never appeared in the central part of the lake. However, it has extended its distribution area toward the central part of the South lake since about 1971, the time corresponding to the period of rapid growth of Japanese economy and the time when a rapid progress of eutrophication took place especially in the southern part of the lake.

Semisulcospira decipiens group (Plate 16)

There are three subspecies (Watanabe 1970a and b; regards as species) in this group as listed in Table 48. These are all endemic in this lake and are the most numerous species among all molluscs living in this lake. In the North lake, subspecies *decipiens* is found abundantly on the littoral bottom shallower than 3 m and *reticulata* is living mostly at the places deeper than 8 m, whereas *multigranosa* occupies the habitat of intermediate depth (5 – 8 m). However, there is a strange phenomenon in the South lake (the deepest depth is 7.5 m), that is, *multigranosa* nearly disappears there and *reticulata* comes to be distributed at the depth of 3 – 7.5 m, and shares the bottom habitat nicely with *decipiens* (Watanabe 1970a)

Plate 10. Hyriopsis (*Nipponohyria*) *schlegeli* (v. Martens) (see p. 331).

Plate 11. Corbicula sandai Reinhardt (see p. 333).

Plate 12. Corbicula (*Corbiculina*) *leana* Prime (see p. 333).

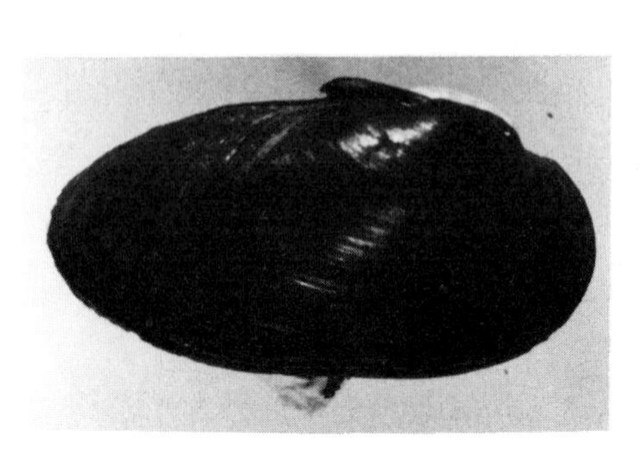

Plate 13. Unio (*Nodularia*) *biwae* Kobelt (see p. 334).

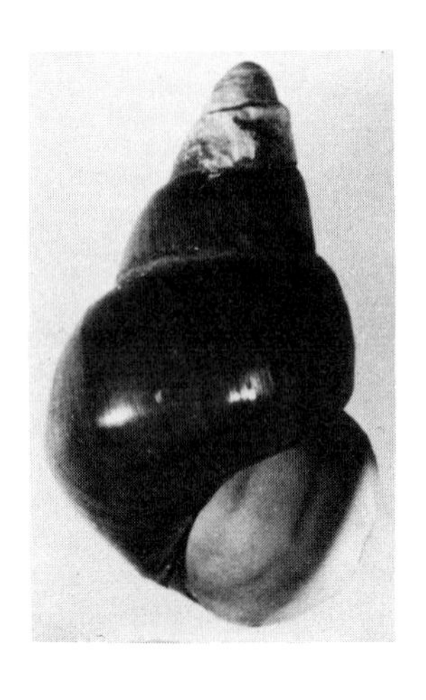

Plate 14. Heterogen longispira (Smith) (see p. 334)

Plate 15. Sinotaia quadrate histrica (Gould) *(see p. 334).*

Plate 16. Semisulcospira decipiens reticulata Kajiyama et Habe *(see p. 335).*

Fishes

Toshinobu Tokui and Hiroya Kawanabe

Introduction

There are two distinctive characteristics of the fish fauna in Lake Biwa. First, the species richness of fishes in Lake Biwa is greatest in Japan. Over sixty species, including subspecies live there, which represent one half of the freshwater fish fauna of Japan. Nearly one-half of these belong to Cyprinidae. Secondly, there are many endemic fishes whose affinities lie with species of the Asian Continent. Furthermore, there still remain in the lake the survivors of sea origin and glacial ages. Why do so many kinds of fish and endemic forms exist in Lake Biwa? One reason is that Lake Biwa is a large, deep lake that is blessed with many different habitats permitting co-existence of closely related species. There is another reason. The appearance of Lake Biwa may date back to Pre-Quaternary Period, at which time the Japanese Islands were probably connected to the Asian Continent. Lake Biwa is the third oldest lake in the world after Lake Baikal and the Caspian Sea. Throughout the long period since its formation, new species or subspecies would have been formed and specialized into living in the new environments created in Lake Biwa.

In this paper, the sections of Introduction, Species of fish, Endemic species and important species in fishery were written by T. Tokui and the sections of Life of fishes in major habitats and Endemism and speciation in fishes by H. Kawanabe.

Species of fish

Table 49 shows the species of fish which live in Lake Biwa and its affluents (Shiga-ken 1978, partly modified). The scientific name with an asterisk indicates an endemic species or subspecies. The following forms are of exotic origin: *Salmo gairdneri*, *Ctenopharyngodon idellus*, *Hypophtalmichthys molitrix*, *Rhodeus ocellatus ocellatus*, *Channa argus*, *Micropterus salmoides*, and *Lepomis macrochirus*.

Horie, S. Lake Biwa

ISBN 90 6193 095 2. Printed in The Netherlands

Table 49. Fishes of Lake Biwa and of its affluents (After Shiga-ken, 1978).

Petromyzonidae
Lampetra mitsukurii (Hatta) = *Lampetra* (*Lethenteron*) *reissneri* (Dybowski)
Salmonidae
Salvelinus pluvius (Hilgendorf) = *Salvelinus leucomaenis* (Pallas) f. *pluvius* (Hilgendorf)
+ *Salmo gairdneri* Richardson = *Salmo* (*Salmo*) *mykiss* Walbaum
Oncorhynchus rhodurus Jordan et McGregor = *Salmo* (*Oncorhynchus*) *masou macrostomus* Günther f. *ishikawai* Jordan et McGregor
Plecoglossidae
Plecoglossus altivelis Temminck et Schlegel
Osmeridae
Hypomesus transpacificus nipponensis McAllister = *Hypomesus transpacificus* McAllister f. *nipponensis* McAllister
Cyprinidae
Pungtungia herzi Herzenstein
Gnathopogon elongatus elongatus (Temminck et Schlegel)
* *Gnathopogon elongatus caerulescens* (Sauvage) = *Gnathopogon caerulescens* (Sauvage)
Squalidus gracilis (Temminck et Schlegel)
Squalidus japonicus Sauvage = *Squalidus japonicus japonicus* (Sauvage)
Squalidus biwae (Jordan et Snyder) = *Squalidus chankaensis biwae* (Jordan et Snyder)
Hemibarbus barbus (Temminck et Schlegel) = *Hemibarbus labeo* (Pallas)
Hemibarbus longirostris (Regan)
Sarcocheilichthys variegatus (Temminck et Schlegel)
⊙ *Abbottina rivularis* (Basilewsky) = *Pseudogobio* (*Abbottina*) *rivularis* (Basilewsky)
Pseudogobio esocinus (Temminck et Schlegel) = *Pseudogobio* (*Pseudogobio*) *esocinus* (Temminck et Schlegel)
Biwia zezera (Ishikawa) = *Pseudogobio* (*Biwia*) *zezera* Ishikawa
Pseudorasbora parva (Temminck et Schlegel)
Hemigrammocypris rasborella Fowler = *Aphyocypris* (*Hemigrammocypris*) *rasborella* (Fowler)
Tribolodon hakonensis (Günther) = *Leuciscus* (*Tribolodon*) *hakonensis* Günther
Moroco steindachneri (Sauvage) = *Phoxinus lagowski* Dybowski f. *steindachneri* Sauvage
Moroco jouyi (Jordan et Snyder) = *Phoxinus lagowski* Dybowski f. *oxycephalus* (Sauvage & Dabry)
+ *Ctenopharyngodon idellus* (Cuvier et Valenciennes) = *Ctenopharyngodon idella* (Valenciennes)
Opsariichthys uncirostris (Temminck et Schlegel)
Zacco platypus (Temminck et Schlegel)
Zacco temmincki (Temminck et Schlegel)
Ischikauia steenackeri (Sauvage)
+ *Hypophthalmichthys molitrix* (Cuvier et Valenciennes) = *Hypophthalmichthys molitrix* (Valenciennes)
Cyprinus carpio Linnaeus
Carassius auratus langsdorfi Temminck et Schlegel = *Carassius gibelio langsdorfi* (Valenciennes)
* *Carassius auratus grandoculis* Temminck et Schlegel = *Carassius carassius grandoculis* (Temminck et Schlegel)
* *Carassius auratus cuvieri* Temminck et Schlegel = *Carassius cuvieri* Temminck et Schlegel
Acheilognathus lanceolata (Temminck et Schlegel) = *Rhodeus* (*Acheilognathus*) *lanceolatus* (Temminck et Schlegel)
Acheilognathus limbata (Temminck et Schlegel) = *Rhodeus* (*Acheilognathus*) *limbatus* (Temminck et Schlegel)

Acheilognathus rhombea (Temminck et Schlegel) = *Rhodeus* (*Paracheilognathus*) *rhombeus* (Temminck et Schlegel)
    Acheilognathus cyanostigma Jordan et Fowler = *Rhodeus* (?) *cyanostigma* (Jordan et Fowler)
    Acheilognathus tabira tabira Jordan et Thompson = *Rhodeus* (?) *tabira* Jordan et Thompson
  + *Rhodeus ocellatus ocellatus* (Kner) = *Rhodeus* (*Rhodeus*) *ocellatus* (Kner)

Cobitidae
    Misgurnus anguillicaudatus (Cantor) = *Cobitis* (*Misgurnus*) *anguillicaudatus* Cantor
    Lefua echigonia Jordan et Richardson = *Lefua costata costata* (Kessler)f. *echigonia* Jordan et Richardson
    Leptobotia curta (Temminck et Schlegel)
    Cobitis delicata Niwa = *Cobitis* (*Niwaella*) *delicata* Niwa
    Cobitis taenia striata Okada et Ikeda = *Cobitis* (*Cobitis*) *taenia* Linnaeus f. *striata* Ikeda
    Cobitis biwae Jordan et Snyder = *Cobitis* (*Cobitis*) *biwae* Jordan et Snyder

Siluridae
  * *Parasilurus lithophilus* Tomoda = *Silurus* (*Parasilurus*) *lithophilus* Tomoda
    Parasilurus asotus (Linnaeus) = *Silurus* (*Parasilurus*) *asotus* Linnaeus
  * *Parasilurus biwaensis* Tomoda = *Silurus* (*Parasilurus*) *biwaensis* Tomoda

Bagridae
    Liobagrus reini Hilgendorf
    Pelteobagrus nudiceps (Sauvage) = *Pseudobagrus* (*Pelteobagrus*) *fulvidraco* (Richardson)

Anguillidae
    Anguilla japonica Temminck et Schlegel

Oryziatidae
    Oryzias latipes (Temminck et Schlegel)

Gasterosteidae
    Gasterosteus aculeatus microcephalus (Girard) = *Gasterosteus aculeatus* Linnaeus f. *leiurus* Cuvier

Channidae
  + *Channa argus* (Cantor) = *Channa* (*Ophicephalus*) *argus* Cantor

Centrarchidae
  + *Micropterus salmoides* (Lacépède)
  + *Lepomis macrochirus* Rafinesque

Eleotridae
    Odontobutis obscura (Temminck et Schlegel)

Gobiidae
    Chaenogobius annularis Gill = *Chaenogobius* (*Chaenogobius*) *annularis* Gill
  * *Chaenogobius isaza* Tanaka = *Chaenogobius* (*Chaenogobius*) *isaza* Tanaka
    Rhinogobius brunneus (Temminck et Schlegel)
    Rhinogobius flumineus (Mizuno)

Cottidae
    Cottus pollux Günther = *Cottus* (*Cottus*) *hilgendorfi* Steindachner et Döderlein
    Cottus reinii (Hilgendorf) = *Cottus* (*Cottus*) *hilgendorfi* Steindachner et Döderlein

---

* = Endemic; ⊙ = Introduced; + = Exotic species.

Endemic species and important species in fishery

First, the writers describe two important species for the fishery *Oncorhynchus rhodurus* and *Plecoglossus altivelis*. The writer then describes some endemic species. Information is based on the reports by Miura (1966), Miyadi et al. (1976), Nakamura (1969, 1975), Shigaken (1978) and Tomoda (1961).

Oncorhynchus rhodurus Jordan et McGregor (Plate 17)

This species is the sole salmon in Lake Biwa and native to this lake. In the spawning season, from the middle of October to the end of November, the fish run into Rivers Ado, Ane, Inukami, Chinai, Shiozu-okawa and Ishida. A majority of the anadromous shoals is caught by means of weirs and traps built on the river. The fish captured are applied to artificial hatching. The fry hatched in winter are released into the lake in spring. The relation between the released number and the recovery in recent years is shown in Table 50. The young, before the release, are parr-marked and have red spots on their sides. After coming to the lake, they lose their parr marks and red spots and assume a brilliant silver color. In the spawning season the adult develops a nuptial coloration, becoming dark brown with vertical bars or blotching. It is reported that *O. rhodurus* likes a layer of water with a temperature of about 13°C. Therefore, the fish tend to inhabit the water mass below 20–25 m in summer and come up to the surface in winter.

Small individuals live on large planktons, while large ones feed on smaller fish such as *Plecoglossus altivelis* and *Chaenogobius isaza*. After three or four years of life in the lake, they run into rivers to spawn. The normal growth is 18 cm in one year, 29 cm in two years, 37 cm in three years, and 43 cm in four years, respectively. All adults perish after the spawning.

Table 50. Number of Released Fry and Amount of Catch of *Oncorhynchus rhodurus* (Shiga-ken, 1978).

Year	Number of Fry (thousands)	Catch (kilograms)
1967	242	20,794
1968	150	23,429
1969	156	21,816
1970	230	23,109
1971	264	21,031
1972	541	39,492
1973	340	52,897
1974	739	39,106
1975	1290	22,000
1976	967	35,966
1977	644	20,464

Source: Shigaken, 1978.

Plecoglossus altivelis Temminck et Schlegel (Plate 18)

This species is originally amphidromous. Since the passage to the sea from Lake Biwa was blocked by the sluice at Nangō-Arai Dam, the lake population of *P. altivelis* has been isolated completely from the sea population. *P. altivelis* of Lake Biwa may be classified into two principal groups: the stream population and the lake population . The former members run into inlet streams to grow whereas the latter members, which are smaller in size than the former, live in the lake for the most part of their lives and run into the lower parts of the streams only in the spawning season.

The spawning occurs in autumn at the lakeside or in inlet streams. The larvae hatched in the stream get washed down immediately into the lake, the number of floaters being greatest at dark and showing a slight increase at dawn. They become dispersed widely over the lake and, in the early stages, they tend to stay in the surface layer of the lake in daytime. After the yolk sacs are absorbed, they move to the bottom layer in the daytime and rise to the surface only at night. Although they live on rotifers, wheel-animalcules, in the earliest stages, they soon begin to feed on floating crustaceans.

In the middle fry stage, it is possible to distinguish the stream population from the lake population on morphological grounds. The stream population begins to run into inlet streams from February or March.

At the time when the fish come close to inshore waters and travel upstream, they are captured en masse as seed-fish. The annual catch for the release is nearly 300 tons (75–100 million individuals), which corresponds to 70 per cent of all domestically released fry of this species. The stocked seed fish are released in most parts of the country, from Hokkaido in the north to Okinawa Prefecture in the south. The latest production of the seed-fish is shown in Table 51. The population which remains in the lake feeds on algae which grow on gravel or pebbles along the lake margins, or on zooplanktons in the center of the lake. The lake population usually occupies the mid-water layer. In the mid-summer, however, they come up to the surface and often leap above the water.

The number of *P. altivelis* in Lake Biwa fluctuates greatly from year to year. If

Table 51. Production of Seed Fry of *Plecoglossus altivelis* (Shiga-ken 1978).

Year	Fry harvest (tons)	Year	Fry harvest (tons)
1967	198*	1973	348
1968	210	1974	325
1969	353	1975	311
1970	310	1976	255
1971	312	1977	278
1972	293		

* Approximately 4 tons per one million fry.

there is no typhoon-induced inundation in the spawning season (autumn), the number of *P. altivelis* present at the beginning of winter on the whole reflects the number of the spawning fish. Reduction of food organisms in winter results in a decline in the number of *P. altivelis* and the decline appears to be further influenced by competition with *Chaenogobius isaza*, which utilizes same food sources. It is reported that the population of *P. altivelis* diminishes when conditions favor increase in the number of *Chaenogobius isaza*.

Gnathopogon elongatus caerulescens (Sauvage) (Plate 19)

Gnathopogon elongatus caerulescens, a small-sized cyprinid, is endemic to Lake Biwa and the River Yodo. It is an allied subspecies of *Gnathopogon elongatus elongatus*, which is widely distributed over the southern part of Japan. It usually schools in the middle layer of the offshore areas and feeds on zooplankters. From late autumn to early winter they gather at depths of 60–100 meters in the northern and the western parts of the lake to overwinter. In early spring they gradually move toward the shore to spawn. By the end of March, they appear in the zone of emergent vegetation (reeds and water oats) in the eastern and the southern parts of the lake. Some shoals enter the lagoons or creeks of Lake Biwa. The spawning season is from April to May. Spawning is accomplished by one female and a number of males. Eggs are deposited on the stalks of sprouting water oats and submerged willow roots. After spawning most adults leave the shallow waters, but some remain in the lagoon. The young grow to 7–11 cm in length by November, 12 cm in the second year, and about 13 cm in the third year. The annual catch of this fish reaches about 300 tons.

Opsariichthys uncirostris (Temminck et Schlegel) (Plate 20)

This species is restricted to Lake Biwa, the River Yodo System, Lake Mikata (Fukui Prefecture), and the River Hasu which flows into Lake Mikata. In recent years, it has been transferred accidentally to many places, mixed with the fry of *P. altivelis* of Lake Biwa, and propagated everywhere. This is the only predacious cyprinid. The mouth is well adapted to catching the prey. Usually, the fish swim actively in the mid-layer and the surface zones and they have a habit of jumping high out of water. In May the fish gradually come shorewards, especially around the mouth of inflowing streams where they actively chase and feed on the fry of *P. altivelis*.

The spawning season is from June to August. The spawning is carried out by a sexual pair, on the sandy or pebbled bottom of the stream and of the shallow lake margin. The spawning fish are usually three or four years old in the male and two or three years old in the female. The young grow to 6–7 cm in length in the first year, reaching 13–14 cm in two years, 18–20 cm (male) or 16–18 cm (female) in three years, and 22 cm (male) in four years. The yearly catch of *Opsariichthys uncirostris* is about 100–150 tons in Lake Biwa. An allied species is distributed to the Asian Continent.

Carassius auratus grandoculis Temminck et Schlegel (Plate 21)

Carassius auratus grandoculis inhabits the bottom layers of the central part of Lake Biwa, feeding on zooplankters and benthic animals. This is in contrast to *C. auratus cuvieri*, which lives in surface and middle layers and feed on phytoplankters. Both subspecies are considered to be differentiated within Lake Biwa. Their subspeciation would have been impossible without isolation through habitat segregation within the lake.

C. a. grandoculis has fewer gill-rakers than *C. a. cuvieri* and its intestinal length is two to five times the body length. The spawning season is from April to July, the peak occurring in May. In the spawning season, a large number of this subspecies visit lagoons or shallow regions in the southern part of the lake and spawn upon water weeds and other floating objects. The fry hatched swim in the shallow layer among the reeds till August. They feed mainly on protozoans such as *Difflugia* until they grow to more than 18 mm in body length. Then they feed mainly on semi-sessile crustaceans and, when grown to about 20 cm, eat benthic or adhesive algae. The adult is about 25–30 cm long, though some are known to grow over 40 cm in length. *C. a. grandoculis* requires two years to mature in Lake Biwa.

The 'funa-zushi', a kind of fish-cheese, is made mainly of the females of this fish caught in the coastal region of Lake Biwa and large gravid females fetch a high price from the makers of 'funa-zushi'.

Carassius auratus cuvieri Temminck et Schlegel (Plate 22)

This subspecies was previously known only from Lake Biwa and the tributaries of the River Yodo but in recent years it has been reported to breed in most parts of the country. These fish move around in shoals in the offshore surface water. As they live mainly on phytoplankters, their gill-rakers are numerous and their intestines are five to seven times as long as their body length. This is a remarkable morphological differentiation from other forms of the crucian carp.

The spawning season is from late March to middle July; May and June are the months of the most active spawning. Their spawning grounds, i.e. lagoons, have become limited considerably because of land reclamation in recent years. Spawning is most actively pursued before dawn just after a rise of water level due to heavy rainfall. The fish spawn upon water plants floating near the surface. The spawning is carried out by three- or four-year old males and four- or five-year old females.

The fry inhabit the environs of floating algae and reed beds in lagoons, and feed initially on zooplankters, and later on sessile algae. When they grow up to 2–7 cm in length, they enter the lake. The standard growth is 9–11 cm in the first year, 15 cm in two years, 23–25 cm in three years, and up to 30 cm in 4–6 years. The variety of this subspecies bred in ponds by fish culturists of Osaka Prefecture, is called 'Herabuna'. The yearly catch of *Carassius auratus*, including *C. a. cuvieri*, *C. a. grandoculis* and *C. a. langsdorfi*, amounts to 600–700 tons.

Parasilurus lithophilus Tomoda (Plate 23)

This species inhabits only Lake Biwa and Lake Yogo in the north of Lake Biwa. The eyes of this species protrude from the side of the head, so that they can be seen clearly from the ventral side of the fish. This unique character distinguishes the species from the two other species of catfish found in Lake Biwa. This species was described as a new species in 1961 (Tomoda 1961). The specific name, *lithophilus*, means stone loving. In fact, these fish are reef dwellers, and they prey on other fish and small animals at night. The spawning season is from late May to late June. The spawning takes place on sand and gravel bottoms at depths of 2–4 m. The fish are estimated to grow to 15–20 cm in length in a year and reach 30–40 cm when fully matured at two years. Females sometimes grow to 60 cm in length. This species is highly valued as it is the most palatable among the three species of the genus *Parasilurus* in Lake Biwa.

Parasilurus biwaensis Tomoda (Plate 24)

Parasilurus biwaensis was described as a new species by Tomoda (1961). The adult of this species is large: the male grows up to 50–70 cm and the female 70–100 cm in total length. This fish resembles the ordinary catfish, *Parasilurus asotus*, in many respects. However it has the following characteristics which distinguish it from other species of *Parasilurus*: the upper lobe of the caudal fin being longer than the lower lobe, the large head, and a large number of vertebrae. The fish inhabit the bottom and rise to the surface or the middle layer while chasing small fish. The spawning season is from late June to middle August. They spawn on the flat bottom 0.5 m deep near the lake shore when the water becomes muddy after a heavy rainfall. This species is unpalatable and is sold cheaply on the market.

Chaenogobius isaza Tanaka (Plate 25)

Chaenogobius isaza is an endemic species of Lake Biwa, living near the bottom deeper than 30 m in the daytime and rising close to the surface at night. The spawning season is from early April to late June. The spawners deposit eggs under small stones in shallow gravelled areas. The male protects the eggs until they hatch. The young grow to 3.5–4.5 cm long by October at the end of the growing season. They start growing again in April and reach 5–6 cm in length by the next October. Though most females spawn in the second year and die, some spawn in the first year. Effects of population density on their growth are apparent: when large numbers are caught, the fish are invariably of small average weight. It is said that *C. isaza* of under one year of age and the fry of *Plecoglossus altivelis* compete with each other for food and living space. The yearly catch of *C. isaza* from Lake Biwa is 200–500 tons.

Life of fishes in major habitats

In discussing the habitats of fishes the following ecological division of a lake community will be respectively applied: i.e., littoral and profundal zones (not the benthal system in the strict sense but the bentho-pelagial one) and epi- and hypolimnion (the pelagial system in the strict sense). In examining the life of fishes in Lake Biwa, we need to subdivide the littoral zone roughly into 1) rocky shores, 2) sandy shores or inlet streams, as well as 3) bays, lagoons or hydrophyte zone; and to add a transitional zone (bentho-pelagial system) between the littoral and the profundal zones, which is especially developed in the lake.

Table 52 shows the seven major habitats of fishes in each of them.

For spawning grounds, most fishes select the hydrophyte zone or sandy shores. There are no spawners in rocky shores or the pelagial zone; in other words, floating egg spawners are lacking and almost all eggs are adhesive (Miyadi et al. 1976).

In the hydrophyte zone, most fishes attach their eggs to the surface of floating or submerged plants (Table 53) during spring and summer (Hirai 1970). In the case of *Carassius gibelio langsdorfi*, the males of which are virtually absent, the species develops its embryos by being stimulated by sperms of any other species of fish. Bitterlings and *Sarcocheilichthys variegatus* place their eggs inside mussels.

In sandy shores or in the sandy bottom of inlet streams, many fishes lay their eggs with adhesive membranes between sand or sandgrains, as mesopsammons. But *Salmo* (*Oncorhynchus*) *masou macrostomus* f. *ishikawai* lays its non-adhesive eggs down deeply under gravels, while gobies do so under stones guarded by males.

Among rare autumnal spawners are *S.* (*O.*) *masou macrostomus* f. *ishikawai*, *Plecoglossus altivelis*, *Rhodeus* (*Paracheilognathus*) *rhombeus*, etc.

The hydrophyte zone is a very secure nursery ground for most fishes including hydrophyte and sandy-shore spawners. The food relation of juveniles is shown in Fig. 164 (Hirai 1971). Earlier in the season, when the total population of

Table 52. Habitats of main fishes in Lake Biwa (Miura et al. 1966 & Miyadi et al. 1976).

Spawning grounds	Nursery grounds for juvenile	Habitats of adult fish
LITTORAL ZONE		
Rocky shore		
		P. altivelis
		S. variegatus
		R. tabira
		R. lanceolatus
		S. asotus
Shady shore and/or inlet		
S. masou macrostomus f. *ishikawai*	*Z. platypus*	*Z. platypus*
P. altivelis	*O. uncirostris*	*S. variegatus*
L. hakonensis	*S. variegatus*	*P. esocinus*
Z. platypus	*P. esocinus*	*P. zezera*
Z. temmincki	*P. zezera*	*R. lanceolatus*
O. uncirostris	*R. tabira*	*R. tabira*
P. esocinus	*C. taenia* f. *striata*	*R. cyanostigma*
P. parva	*S. biwaensis*	*R. brunneus*
C. taenia f. *striata*		*C. hilgendorfi*
S. biwaensis		
P. nudiceps		
C. isaza		
R. brunneus		
C. hilgendorfi		
Hydrophyte zone (bays and lagoons)		
Z. platypus	*Z. platypus*	*Z. platypus*
I. steenackeri	*Z. temmincki*	*Z. temmincki*
A. rasborella	*I. steenackeri*	*I. steenackeri*
G. caerulescens	*G. caerulescens*	*P. parva*
S. chankaensis biwae	*S. chankaensis biwae*	*P. zezera*
S. japonicus japonicus	*S. japonicus japonicus*	*A. rasborella*
P. parva	*S. variegatus*	*C. gibelio langsdorfi*
C. gibelio langsdorfi	*P. parva*	*C. carassius grandoculis*
C. carassius grandoculis	*C. gibelio langsdorfi*	*R. ocellatus* f. *smithi*
C. cuvieri	*C. carassius grandoculis*	*R. lanceolatus*
R. ocellatus f. *smithi*	*C. cuvieri*	*R. rhombeus*
R. lanceolatus	*R. ocellatus* f. *smithi*	*R. cyanostigma*
R. rhombeus	*R. lanceolatus*	*C. taenia* f. *striata*
R. cyanostigma	*R. rhombeus*	*S. asotus*
R. tabira	*R. cyanostigma*	*P. nudiceps*
S. asotus	***R. tabira***	*R. brunneus*
R. brunneus	*C. taenia* f. *striata*	
	S. asotus	
	P. nudiceps	
	C. annularis	
	R. brunneus	
TRANSITIONAL ZONE		
S. lithophilus	*P. altivelis*	*S. chankaensis biwae*
	G. caerulescens	*S. japonicus japonicus*
	S. variegatus	*C. carassius grandoculis*
	C. carassius grandoculis	*S. lithophilus*
	S. lithophilus	
	R. brunneus	
	C. hilgendorfi	

Table 52. Continued

PROFUNDAL ZONE		
		for hibernation
		G. caerulescens
		S. chankaensis biwae
		S. japonicus japonicus
		S. variegatus
		P. esocinus
		P. zezera
EPILIMNION		
	S. masou macrostomus f. *ishikawai*	*S. masou macrostomus* f. *ishikawai* (winter only)
	C. isaza	*P. altivelis*
	R. brunneus	*L. hakonensis*
		O. uncirostris
		G. caerulescens (summer only)
		C. cuvieri
		S. biwaensis
		C. isaza (winter only)
HYPOLIMNION		
	S. masou macrostomus f. *ishikawai*	*S. masou macrostomus* f. *ishikawai*
	O. uncirostris	*P. altivelis*
	C. isaza	*L. hakonensis*
		C. isaza

juveniles is small and the biomass of total food organisms is rather large, planktonic crustacean feeders, planktonic rotifer feeders and phytal crustacean (and rotifer) pickers are roughly separated from one another in the food relation. Later in the season, the same relationship tend to hold, but the juvenile population is rather large relative to food biomass. Many fishes feed together on phytal crustaceans, especially *Camptocercus*. Larger juveniles of *Carassius carassius grandoculis* and some species of bitterlings turn partly to *phytal algae,* and *Rhinogobius brunneus* catches chironimid larvae and fry of *C. carassius grandoculis* which is relatively numerous in the season. Juvenile of *C. carassius grandoculis* are numerous in the submerged plant belt; the more abundant the floating plant material, the higher the density of juveniles (Hirai 1972). These phenomena also apply to some bitterlings, especially *Rhodeus* (*Rhodeus*) *ocellatus* f. *smithi*, but *Gnathopogon caerulescens*, *Zacco platypus* and *R. brunneus* are widely distributed in lagoons regardless of the amount of hydrophyte present.

Along the sandy shore of the open lake side, juveniles of *Z. platypus* and *Opsariichthys uncirostris* inhabit shallow areas, where they catch planktonic crustacea carried by water currents (Sunaga 1970), whereas juvenile of *G. caerulescens, C. carassius grandoculis, Pseudogobio* (*Biwia*) *zezera,* etc. are distributed in somewhat deeper areas.

The littoral zone is another important habitat for many of adult and subadult fishes. Fig. 165 shows the seasonal change on the food relation of fishes in a lagoon (Maki 1964). Zooplankters and phytal animals are eaten mainly by small

Table 53. Seasonal changes of number of fry and juvenile of fishes in a hydrophyte zone of Lake Biwa caught by trap (after Hirai 1970).

	April		May			June			July			August
Z. platypus	0	0	0	4	0	3	23	37	188	646	86	8
I. steenackeri	0	0	0	0	0	0	0	0	0	18	17	1
G. caerulescens	11	19	77	36	4	1	2	7	0	4	0	0
P. parva	0	0	0	0	15	21	19	15	27	8	0	0
C. carassius grandoculis	29	172	505	348	74	110	290	107	2448	497	160	8
R. ocellatus f. *smithi*	0	0	0	10	8	867	180	759	829	320	203	93
R. lanceolatus	0	0	0	0	9	33	27	56	39	10	2	3
R. rhombeus	0	6	53	42	27	22	70	20	0	0	0	0
R. cyanostigma	0	0	0	0	8	200	59	139	51	30	9	3
R. tabira	0	0	0	0	0	12	31	222	74	5	5	0
P. nudiceps	0	0	0	0	0	0	0	4	10	104	655	20
R. brunneus	139	432	205	199	152	283	574	1689	767	359	1847	110

fishes, while aquatic plants are grazed by *Ischikauia steenackeri* nearly through the year. Phytal algae and aquatic insects, especially chironomids, are other main food items in the lagoon; the former is utilized from summer to early winter and the latter from spring to autumn.

Around rocky shores, a few of *P. altivelis* graze algae grown on stones; *S. variegatus*, *Rhodeus* (?) *tabira*, *Rhodeus* (*Acheilognathus*) *lanceolatus*, etc. feed on various organisms, and *Silurus* (*Parasilurus*) *lithophilus* catches some fishes.

In pelagial zones, a thermocline develops from May to October at depth between 5 and 25 m, and disappears completely in winter and early spring. Thus the water temperature is higher than 25°C in the epilimnion and about 8°C in the hypolimnion in summer, while it is about 6°C in the whole water column during winter.

In summer, *Carassius cuvieri* (phytoplankton filterer), *G. caerulescens* and *P. altivelis* (zooplankton feeders), as well as three piscivorous species i.e. *Leuciscus* (*Tribolodon*) *hakonensis*, *O. uncirostris* and *Silurus* (*Parasilurus*) *biwaensis* are found in the epilimnion; they sail around actively in search of food. In contrast, two species of zooplankton feeders i.e. *P. altivelis* and *Chaenogobius* (*Chaenogobius*) *isaza*, and two species of piscivores i.e. *L.* (*T.*) *hakonensis* and *S.* (*O.*) *masou macrostomus* f. *ishikawai*, are active in the hypolimnion. In winter, however, all fishes, including those restricted to the hypolimnion during summer, utilize to the entire pelagial layers, but a warm-water plankton feeder, *G. caerulescens*, disappears.

In the transitional zone, there are *C. carassius grandoculis, Squalidus chankaensis biwae* and *Squalidus japonicus japonicus,* all of which are zooplankton feeders, and *S.* (*P.*) *lithophilus,* a piscivore, living there during summer.

Many of the warm-water fishes keep themselves to the profundal zone in winter for 'hibernations'. Among them are *G. caerulescens, S. chankaensis biwae, S. japonicus japonicus, I. steenackeri, Pseudogobio* (*Pseudogobio*) *esocinus, P.* (*B.*) *zezera* and *S. variegatus.*

The food relations of fishes in the pelagial region in a broad sense are shown in Fig. 166 (Shiga Pref. Fish Exp. St. 1942; Sunaga 1964).

It is worth noting that most of these pelagial fishes are either endemic or nearly so. Conversely speaking, all of the endemic or nearly endemic species are restricted to the pelagial region.

Endemism and speciation in fishes

In Lake Biwa, as shown in the section of 'Endemic species and important species in fishery', there are six endemic species or subspecies in the strict sense (Table 49). Each of them has been differentiated *in situ* from its closely related species distributed widely in Japan including the area around the lake. *C. carassius grandoculis* as well as *C. cuvieri* has been derived from *Carassius carassius buergeri,* and *G. caerulescens* from *Gnathopogon elongatus elongatus S.*

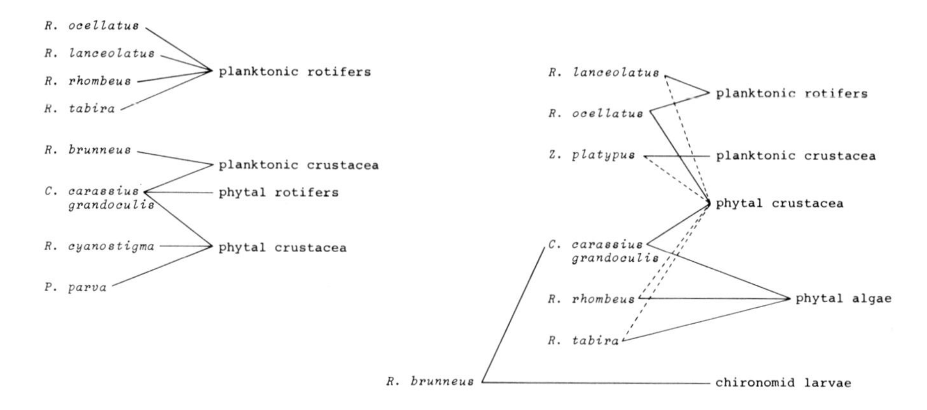

Fig. 164. Food relation of juvenile fishes in a nursery ground, hydrophyte zone in Lake Biwa. Left: until mid-June; right: since mid-June (redrawn from Hirai 1971).

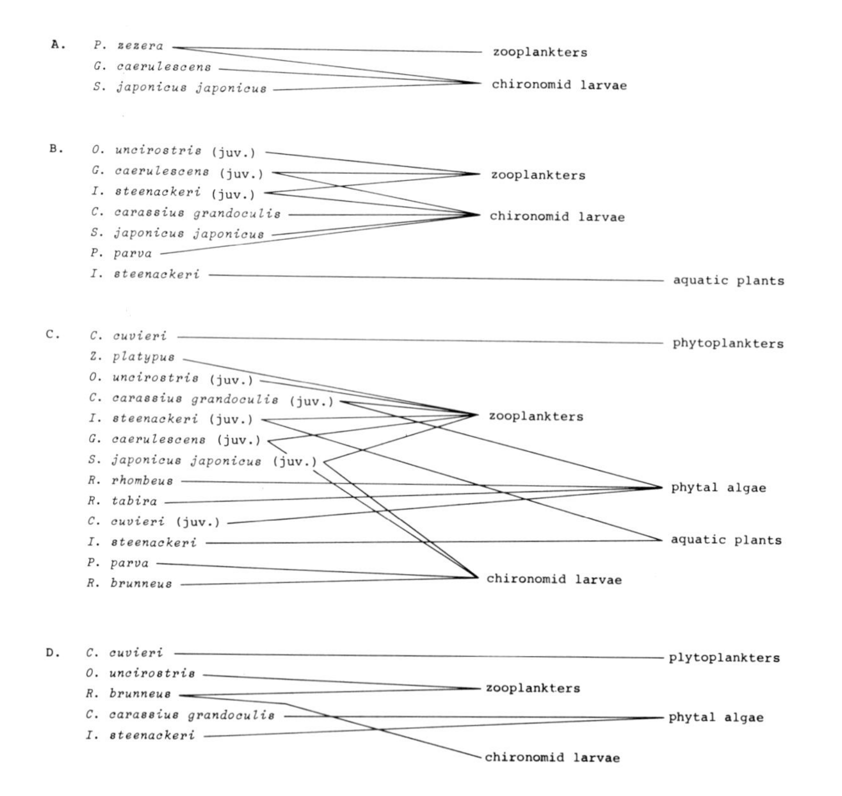

Fig. 165. Food relation of fishes in a lagoon of Lake Biwa.
A: spring; B: summer; C: autumn; D: winter (redrawn from Maki 1964).

Fig. 166. Food relation of adult main fishes in pelagic region of Lake Biwa.
A: spring; B: summer; C: autumn; D: winter (redrawn from Shiga Pref. Fish Exp. St. 1942; Sunaga 1964).

(*P.*) *lithophilus* as well as *S.* (*P.*) *biwaensis* seems to be specialized from *Silurus* (*Parasilurus*) *asotus*, and *C.* (*C.*) *isaza* is differentiated from *Chaenogobius* (*Chaenogobius*) *annularis*.

There are also five endemic forms in Lake Biwa: i.e. *Salmo* (*Oncorhynchus*) *masou macrostomus* f. *ishikawai,* two forms of *Plecoglossus altivelis,* and lacustrine forms of *Cottus* (*Cottus*) *hilgendorfi* and *Rhinogobius brunneus.*

Along the Pacific coast of the western Japan, a landlocked trout *Salmo* (*Oncorhynchus*) *masou macrostomus* f. *stricto* lives in the upper reaches of rivers, and a few individuals descend to the sea coast and then ascend back to the streams. In recent years, comparative studies have been made intensively between the trout and the Lake Biwa trout, and several differences have been discovered in morphological, ecological and physiological characters (Kato 1973, 1975; Honjoh 1977; Yoshiyasu 1973). The former spends 1 or 2 years in streams before descending to the sea, while the latter stays only 1 to 3 months in inlet streams and then descends to the lake. *P. altivelis* is originally an amphidromous fish, and its landlocked form lives only in Lake Biwa save the transplanted populations. The landlocked type is clearly different from the amphidromous type both ecologically and physiologically (Azuma 1969, 1973; Shimadzu 1954; Kawanabe 1975, 1976), but the close similarity of morphological features of the two types indicates that the invasion period seems to have occurred recently, possibly in some of interglacial periods. This landlocked stock, moreover, consists of at least two populations which are probably genetically isolated (Fig. 167). One has an 'amphidromous habit' between the lake (instead of the sea) and its inlets. The other is a completely lacustrine form and ascends inlets up to 2 or 3 km upstreams from the river mouth at the spawning season, though some individuals never ascend inlets but spawn in the coastal pebble bottom of the lake. *P. altivelis* is the sole member of subfamily Plecoglossinae, which is a derivative of Osmeridae, suggesting the lacustrine type of the fish in Lake Biwa may have returned to its ancestral type in migration.

C. (*C.*) *hilgendorfi*, distributed to most parts of Japan, can be divided into 3 forms by morphological characters especially in their larval stages; i.e., amphid-romous, fluvial and lacustrine ones (Kurawaka 1976). The last form is endemic to Lake Biwa and is living not only on the shallow bottom but also in a deeper part. *R. brunneus* is one of the most variable species distributed widely in the Far East and consists at least of 6 forms in Japan (Mizuoka 1962, 1967, 1974, 1976; Nishijima 1968; Mizuno 1960, 1971). An orange-coloured form living in the lake produces very small eggs and its juveniles to live even in the offshore waters, all individuals maturing in 1 year.

It is also worth-while noting that all derived species, subspecies or forms mentioned above clearly show the pelagic mode of life in the lake as compared to their closely related or ancestral ones.

There are some nearly endemic species, subspecies or forms in the lake, all of which have closely related forms in the Chinese Continent and/or the Korean Peninsula. *O. uncirostris* is restricted in its distribution to Lake Biwa and the

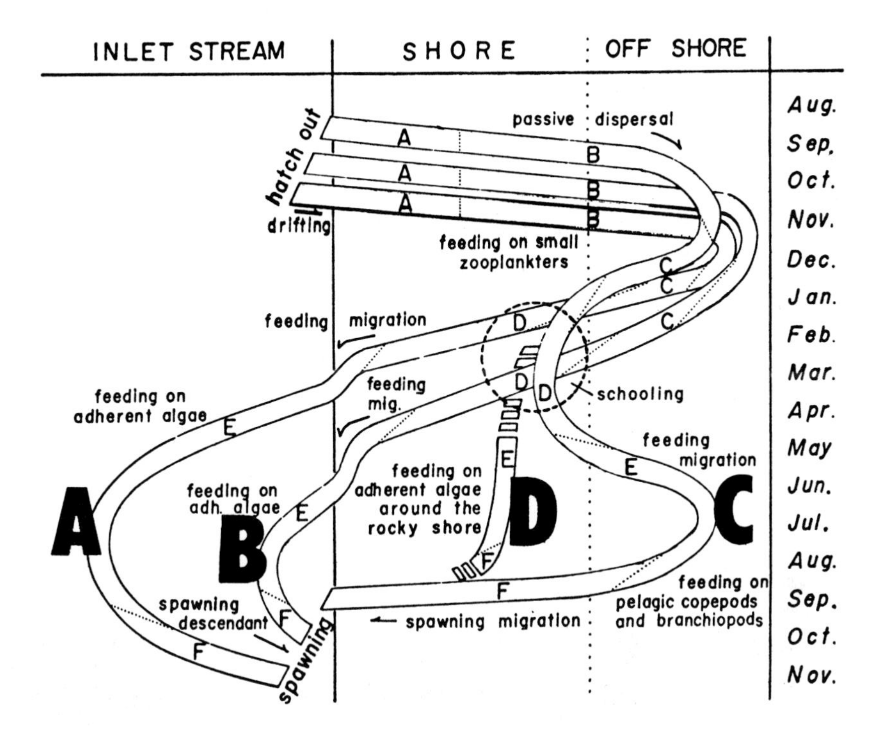

Fig. 167. Schematic diagram representing the main events in the life history of different populations of *Plecoglossus altivelis* in Lake Biwa.
Large letter A–D: early ascending population,.late ascending population, lake population with small size, lake population with large size, respectively.
Small letter A–F: the phase of development (Azuma 1973).

outlet, River Yodo, and Lake Mikata along the Japan Sea Coast in Japan but is widely distributed on the Continent. Continental populations of the fish are believed to be separate species e.g. *Opsariichthys bidens* and *O. amurensis*, but they may be different forms of the same species. The population living in Mikata exhibits intermediate morphological characters between the Continental and Lake Biwa populations. *Ischikauia steenackeri* is known only from Lake Biwa, River Yodo and several ponds in Nara Prefecture, which were connected to the River Yodo system. Other species of this genus are found in the southern part of the Chinese Continent, Formosa and Hainan Island. Two species of the genus *Squalidus*, *S. chankaensis biwae* and *S. japonicus japonicus*, is distributed to in Lake Biwa and the Yodo, as well as the Kiso, the Nagara and the Ibi river system. Related subspecies of both species are found in wide areas of the Chinese Continent and the Korean Peninsula.

Fossil evidence indicates that many freshwater fishes have reached Japan from the continent in some geological periods. For instance, fossils of two cyprinid fishes, *Xenocypris* and *Distoechodon*, have been discovered at Gifu and Shizuoka Prefectures lying about 100 and 250 km east of the lake at late

Miocene, but such xenocyprinean fishes are completely lacking in the Japanese freshwater fauna at present. (Uyeno 1965; Suginohara & Uyeno 1967).

The ancestors of relict species in the lake had probably arrived from the continent through a land bridge in the Pliocene or earlier and lived in a wide range of Japan, at least in its western part, for a time, and then were eliminated except in a large, deep and stable lake — Lake Biwa. *O. uncirostris* and *I. steenackeri* have piscivorous and weed-grazing habits respectively and the main habitat of *Squalidus chankaensis biwae* and *S. japonicus japonicus* is the transitional area between the littoral and profundal zones. These fishes were probably eliminated from other freshwater systems in Japan but could continue to survive only in Lake Biwa.

Plate 17 (see p. 342).

Plate 18 (see p. 343).

Plate 19 (see p. 344).

Plate 20 (see p. 344).

Plate 21 (see p. 345).

Plate 22 (see p. 345).

Plate 23 (see p. 346).

Plate 24 (see p. 346).

Plate 25 (see p. 346).

Bacteria

Nobuhiko Tanaka and Hajime Kadota

Introduction

Organic substances produced by primary producers in the lake and those supplied to the lake from terrestrial sources are metabolized through activities of various organisms, and finally mineralized to inorganic substances such as CO_2, NH_4^+ and PO_4^{3-} which can be utilized again for the primary production. In the processes of decomposition and mineralization of organic substances fungi, especially heterotrophic bacteria, play a very important role. In the lake ecosystem the bacteria also contribute to the secondary and higher production by serving their own cells as foods for protozoa, zooplankton and some benthic animals. In this section some ecological and physiological aspects of bacteria in Lake Biwa are described.

Distribution of bacteria in Lake Biwa

It has been reported that population densities of heterotrophic bacteria in pelagic waters of Lake Biwa counted by cultural methods were $10^2 - 10^3$ and $10^3 - 10^4$ per ml in the north and south basins, respectively. However, in some littoral or neritic waters there existed $10^4 - 10^5$ heterotrophic bacteria per ml. Very high population densities of bacteria were often found in waters of the southern half of the south basin. The horizontal distributions of bacteria in surface water were in general correlated with those of the concentration of organic matter, the primary productivity, etc. Fig. 168 shows the distribution patterns in surface and 5 m deep waters of heterotrophic bacteria in the whole area of Lake Biwa in summer (Kadota et al. 1966). As illustrated in this figure bacteria-rich water mass was found in a pelagic part of the north basin in summer. The presence of such bacteria-rich water mass may be attributable to the horizontal circulation of lake water reported by Morikawa et al. (1966). The direct microscopic counting of bacteria in water of Lake Biwa usually gave counts 10 to 1,000 times as many as those obtained by conventional cultural

Horie, S. Lake Biwa

ISBN 90 6193 095 2. Printed in The Netherlands

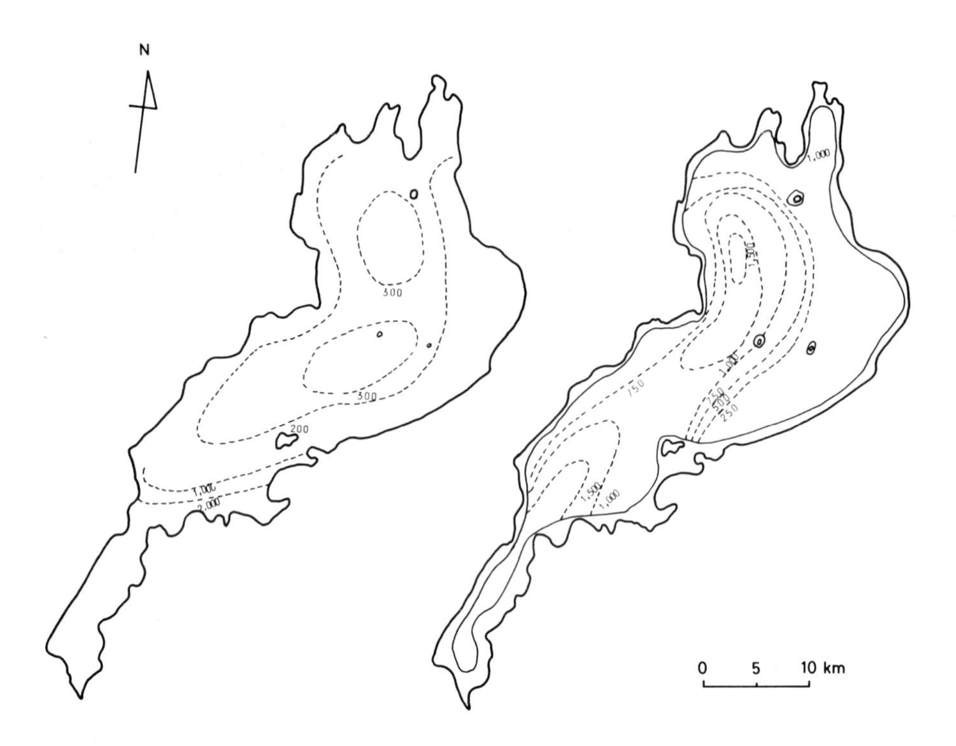

Fig. 168. Horizontal distribution of heterotrophic bacterial in surface (left) and 5 m deep (right) water in Lake Biwa in July 1963 (bacterial number per ml) (Kadota et al. 1966).

methods. The difference in bacterial count between the two methods might be dependent upon the environmental conditions, e.g., the concentration of organic matter; the difference was generally larger in oligotrophic water than in eutrophic one.

Fig. 169 shows the vertical distribution pattern of bacteria obtained by using the direct microscopic method in a pelagic area of Shiozu Bay (Tanaka et al. 1974a). The maximum bacterial count was obtained at the depth just above the thermocline in the water column in summer. The population densities of bacteria in the epilimnion were in general higher than those in the metalimnion (Kadota et al. 1966). As illustrated in Fig. 169, the population maxima of bacteria were found at the depths of 1.5–3 meters and 10–15 meters in summer and at the 1.5–3 meter depth in winter, respectively. The maximum peak of bacterial count found at the 10–15 meter deep layer in summer probably reflected the accumulation of particulate organic matter near the depth of thermocline. The particulate organic matter might be utilized by bacteria as substrates for their growth as well as the solid surface to be attached to by them. The peak found at the depth of 1.5–3 meters throughout the year might be related to other factors, probably the supply of soluble organic substrates excreted by phytoplankton (Tanaka et al. 1974a).

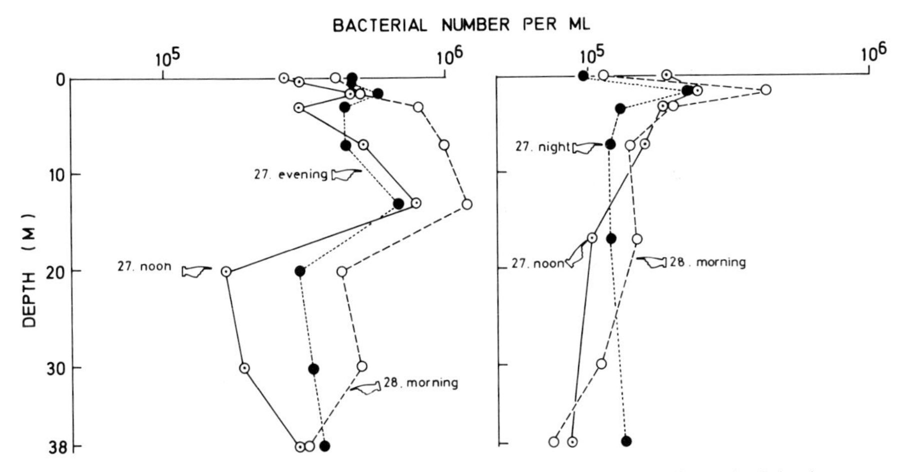

Fig. 169. Vertical distribution of bacteria, counted by the direct microscopic method, in the water column in Shiozu Bay of Lake Biwa in November 1969 (right) and July 1971 (left). Diurnal changes in the distribution were shown (Tanaka et al. 1974a).

The population densities of bacteria in surface water of Lake Biwa varied seasonally at different stations (Fig. 170, Tanaka et al. unpublished). The population densities of bacteria were generally high in littoral and eutrophic areas (Stations 1, 2, 3, 5 and 6 in Fig. 170), and relatively low in pelagic and oligotrophic areas (Stations 4 and 7). Although regular seasonal changes were not found, the patterns of the seasonal variations in bacterial populations at the stations examined were fairly similar, e.g., the population densities of bacteria tended to increase from April to June, to decrease from July to August, and to increase again from September to December. No remarkable change was found from January to April. From these data it is suggested that the variations found in the bacterial number during the period examined were related to some climatic conditions, such as the frequency of rainfall and the water temperature in the Lake Biwa region.

According to the data obtained by Tanaka et al. (1977), the major groups of bacteria found in water of Lake Biwa were *Flavobacterium*, *Vibrio*, Enterobacteriaceae, *Pseudomonas*, *Achromobacter* and *Aeromonas*. These groups were found in the ratios of 10–60, 5–20, 0–10, 0–10, 0–5 and 0–5%, respectively. It was also found that 20–70% of the bacterial strains isolated did not grow when transplanted to new media.

Structure of bacterial community in water of oligotrophic and eutrophic areas in Lake Biwa

The concentration of organic matter in water of Lake Biwa varied very much depending on location of the station employed. The COD value of water in

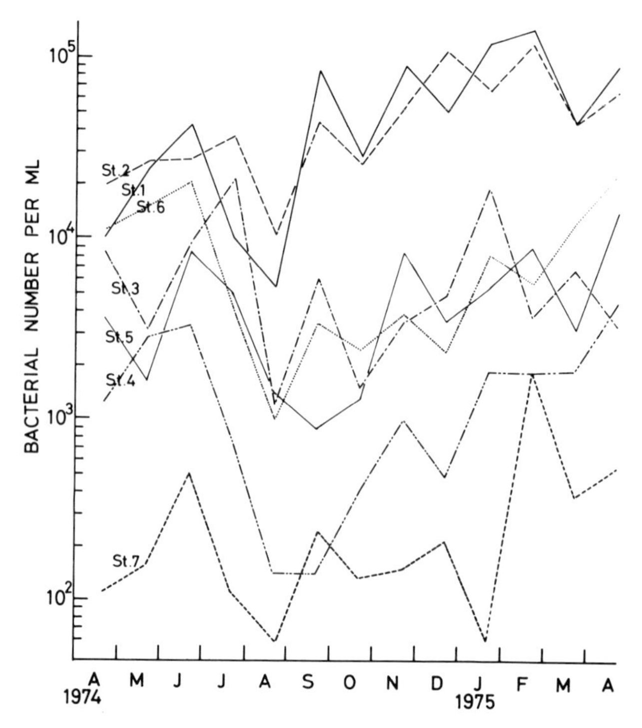

Fig. 170. Seasonal variations in population densities of bacteria in surface water at various stations in Lake Biwa (Tanaka et al. unpublished).

pelagic area of the north basin was approximately 2 mg $O_2/1$ and that in the south basin was higher than 5 mg $0_2/1$. The amount of particulate organic matter in water was also very different between oligotrophic and eutrophic waters. In response to the difference in environmental conditions, especially concentration of organic matter, the population densities of heterotrophic bacteria and the structure of bacterial community varied. Fig. 171 illustrates the relationships between the concentration of peptone in the agar medium and population densities of bacteria counted with water samples collected at several stations in oligotrophic and eutrophic areas in Lake Biwa (Tanaka et al. 1977). As shown in this figure, the bacterial group which can form a colony on the agar medium containing peptone in low concentration was predominant in the oligotrophic area, and the bacterial group which can be counted using conventional medium containing peptone in high concentration was predominant in the eutrophic area.

The amount of small sized seston ($<10\ \mu$) was five to ten times larger in the south basin than in the north basin. Bacterial populations which attached to different sized particles of seston were found to be different in the response to concentration of peptone in the counting medium. These facts suggest that the

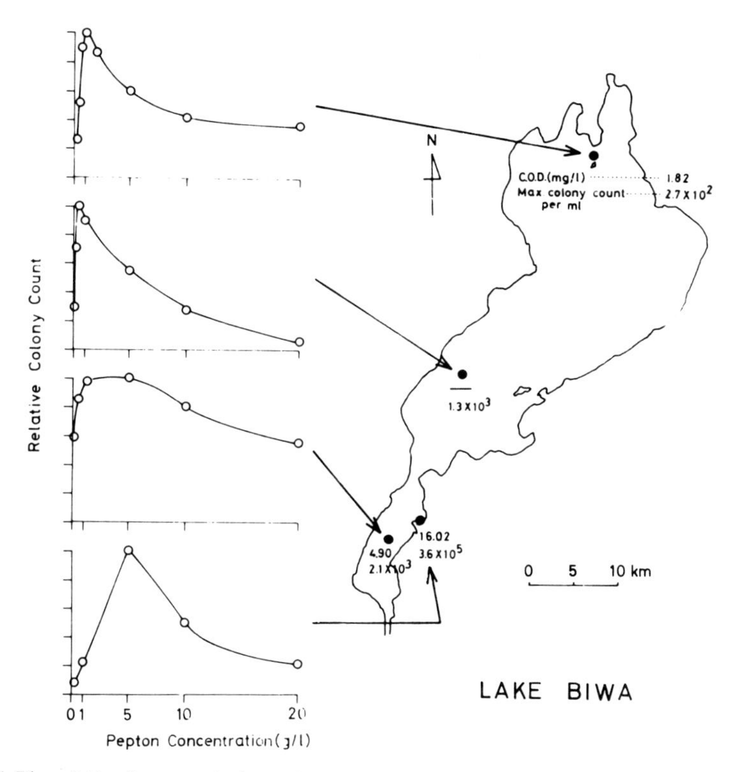

Fig. 171. The relation between the bacterial count and the concentration of peptone in agar media (Tanaka et al. 1977).

difference in structure of bacterial community in water between oligotrophic and eutrophic areas is attributable to the differences in concentration of dissolved organic matter in water and in size composition of seston between oligotrophic and eutrophic areas.

Uptake and mineralization by bacteria of dissolved organic matter in water of Lake Biwa

According to Kadota and Tanaka (1975), the production rate of bacteria in water of the north basin, which was determined at *in situ* temperature by using the method reported by Romanenko (1963, 1964), was about 440 mg C/m^2 water column (40 m deep)/day. This rate was about 2/3 of the primary production rate by the natural phytoplankton population. This suggests that the bacterial production is important in the ecosystem of Lake Biwa. It was also found with water in the north basin of Lake Biwa that the mineralization rates of glucose by bacteria in summer were 3 – 8 mg C/m^3/day and 1 mg C/m^3/day in the epilimnion and the hypolimnion, respectively.

Table 54. Bacterial production in water column at Station 1 in Shiozu Bay, Lake Biwa (November 30 (Kadota & Tanaka 1975)).

Depth (m)	Bacterial number (count/ml)	Bacterial biomass (mg C/m^3)	Bacterial production							
			Production rate (mg C/m^3/hour)				Generation time (hour)			
			11°C	18°C	25°C	*in situ*	11°C	18°C	25°C	*in situ*
3	5.68×10^5	21.3	0.368	0.729	1.036	0.467	58	29	21	50
20	4.42×10^5	16.6	0.403	0.741	1.133	0.492	41	22	15	36

Organic matter present in the lake water is constituted from autochthonous and allochthonous ones. The excretion of organic substances from phytoplankton during photosynthesis contributes very much to the autochthonous production of organic matter in the water. Tanaka et al. (1974a, b, 1975a, b) found that 1 to 11% of the photosynthetic product was released from the algal cells mainly as the form of glycollic acid in water of Lake Biwa. It was also found that the glycollate-utilizing bacteria were dominant in the heterotrophic bacterial population in the photosynthetic layer in summer. The kinetic parameters for glycollic acid uptake by microorganisms in the lake waters, which were measured by the method of Wright and Hobbie (1965), are shown in Table 55 (Tanaka et al. 1975a). As shown in Table 55, the maximum velocity for the glycollic acid uptake by microbial population varied within the range of 0.55–4.52 mg C/m^3/day in September and 0.19 – 0.70 mg C/m^3/day in November. These rates corresponded at most to about 15% of the total values of organic matter taken up by bacteria in the photosynthetic layer.

Table 55. Kinetic parameters for the uptake of glycollate by microorganisms at various depths in the water column at Station 1 in Shiozu Bay, Lake Biwa (Tanaka et al. 1975a).

Date	Depth (m)	V ($\times 10^{-4}$mg/1/hr)	T (hr)	Sn + Kt (μg/1)	Water temperature (°C)
Sep. 21, 1972					
	0	1.61	1798	290	22.7
	1	1.92	166	32	22.4
	3	5.15	233	120	22.0
	7	7.69	261	201	21.8
	13	7.20	322	232	21.6
	25	0.94	1679	159	11.7
Nov. 30, 1972					
	0	0.82	5026	414	12.9
	1	0.98	5029	493	12.9
	3	1.19	5443	648	12.9
	7	0.79	5863	462	12.9
	13	0.32	1220	39	12.9
	17	below the limit of measurement			12.9

Sedimentation and its significance in lake metabolism

Yatsuka Saijo, Mitsuru Sakamoto and Yoshimasa Toyoda

Introduction

It has long been recognized that the process of sedimentation plays an important part in lake metabolism. The study of sedimentation related to organic matter production in lakes was first attempted by Hogetsu et al. (1952) in Lake Suwa and later by Saijo (1956), Ohle (1962) and T. Koyama et al. (1975) and so on. Most of these studies were conducted in rather small and shallow lakes. But, because of strong influences from the shore and the bottom, such lakes were often unsuitable for investigating the process of sedimentation and decomposition of autochthonous organic substances. From 1963 to 1964, an opportunity was offered for a study of primary production as well as sedimentation in Lake Biwa (Saijo et al. 1966 Toyoda et al. 1968).

Seasonal changes in the rate of deposition near the bottom

For the sampling of deposits, one pair of large mouthed glass bottles was suspended every ten meters below an anchored buoy at a station located near the center of the lake with the depth of 60 m. Deposits accumulated in these bottles were collected after the bottles had been left for a week to a month.

Fig. 172 shows seasonal changes of the rate of deposition at 50 m depth, i.e., ten meters above the bottom surface. A remarkable change of deposition was observed through the period of investigation. The maximum values, 3–6 $g/m^2/day$ as dry weight, were observed in April and May 1963 and June 1964, whereas the minimum values, about 0.1 $g/m^2/day$, were found from November to the beginning of March. Some lower values, 0.4 to 0.5 $g/m^2/day$, were also obtained from July to September. To elucidate the reason for such seasonal changes, the monthly average values of daily outflow of this lake are plotted in Fig. 172. Here, due to the lack of reliable data on inflow, the values of outflow were employed in place of inflow, assuming that both values vary similarly. As seen from this figure, the rate of deposition was generally high in the season

Horie, S. Lake Biwa

ISBN 90 6193 095 2. Printed in The Netherlands

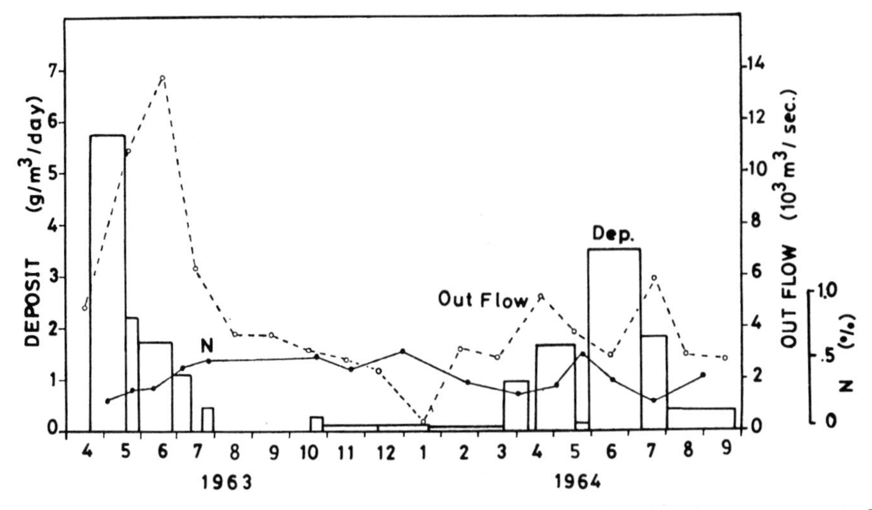

Fig. 172. Seasonal changes in depositing rate, nitrogen content in deposit and average amount of the monthly outflow (Toyoda et al. 1968).

when the rate of outflow was high and vice versa. This suggests a considerable influence of inflow to the deposition, that is to say, the greater part of deposits was derived from allochthonous substances which had been supplied by inflow, even in such a big lake as Lake Biwa. Furthermore, seasonal changes of organic nitrogen contents in deposit seem to support the above explanation. As illustrated in Fig. 172 an inverse relationship was found between the rate of deposition and the content of organic nitrogen in deposit, i.e. when the rate of deposition was low, the nitrogen content of deposit was high, and vice versa. These phenomena seem also to indicate that the larger part of deposit is composed of allochthonous inorganic substances when the rate of deposition is high.

From these rates of deposition at 50 m, the yearly deposition rate of Lake Biwa can be calculated as 208 $g/m^2/year$. If this value is plotted in a figure (Fig. 173) of Saijo (1956) and Saijo et al. (1966) showing the relation between the Secchi disc depth and the yearly rate of deposition as dry weight in Japanese lakes, Lake Biwa occupies almost the same position as Lake Aoki, a typical oligotrophic lake having the maximum depth of 58 m.

Vertical distribution of the rate of deposition

Seasonal changes in the vertical distribution of the rate of deposition are illustrated in Fig. 174. Vertical changes of water temperature are added as reference. As can be seen, remarkable seasonal changes of the pattern were observed for the vertical distribution of deposition. In autumn, the maximum

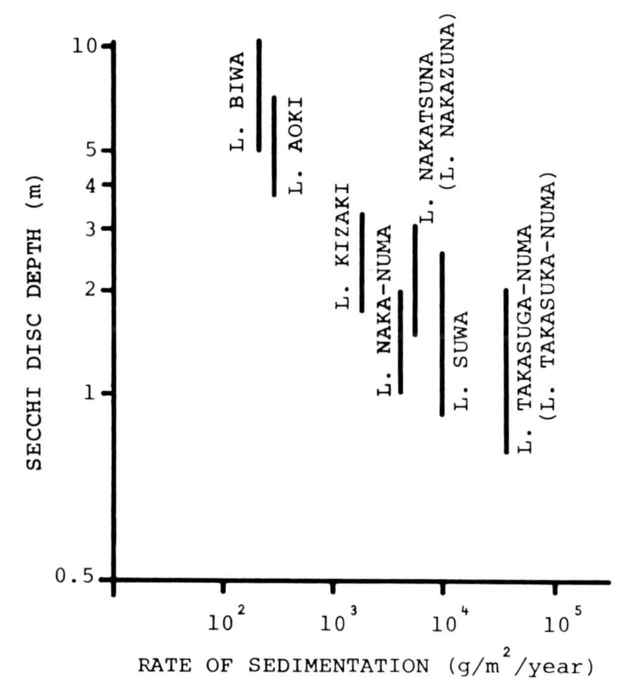

Fig. 173. Relation between the Secchi disc reading and the rate of deposition in some Japanese lakes (after Saijo et al. 1966).

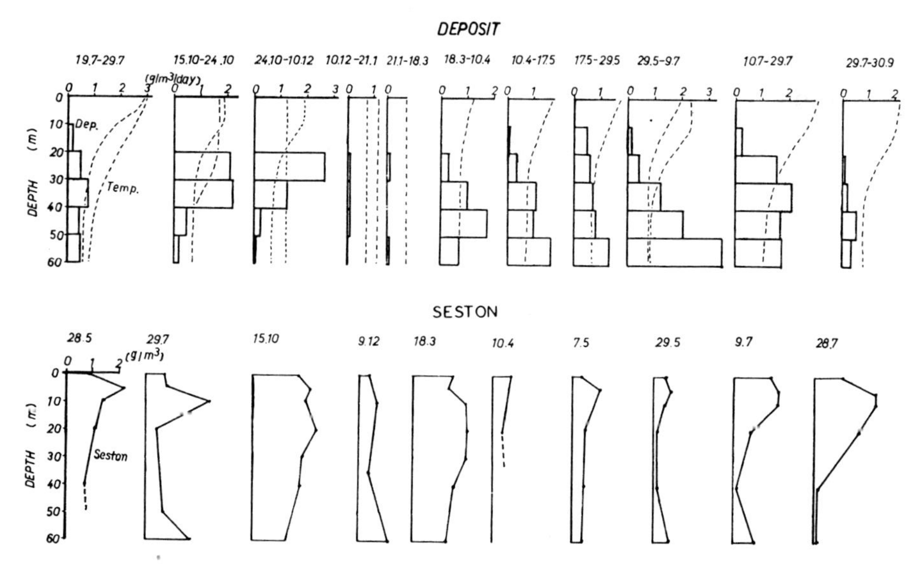

Fig. 174. Seasonal changes in the vertical distribution of depositing rate and seston (Toyoda et al. 1968).

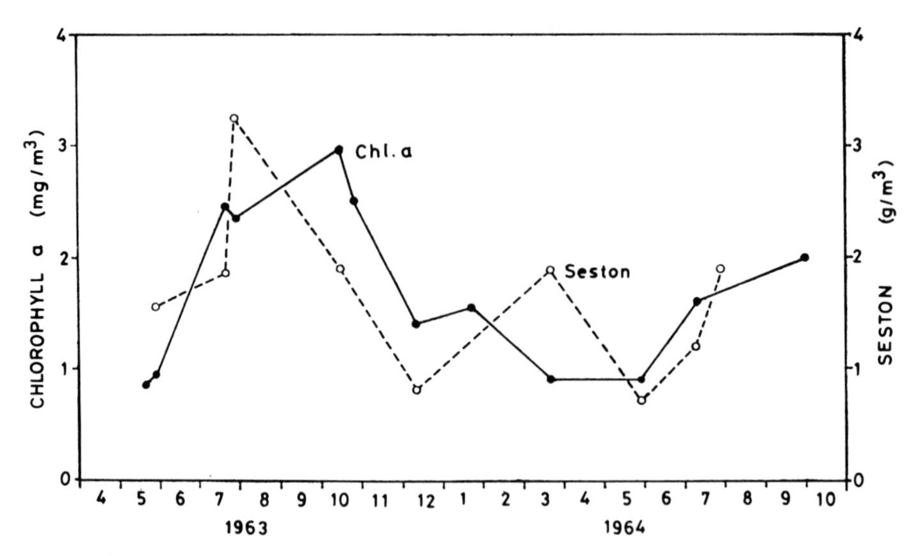

Fig. 175. Seasonal changes in the total amount of chlorophyll *a* and seston in the euphotic layer (0– 20 m depth) (Toyoda et al. 1968).

values of deposits were obtained in the upper layer and decreased with depth, while from spring to early summer, the minimum values were found in the upper layers. As to the explanation of the origin of deposits, the vertical distribution of seston is illustrated in Fig. 174. The amount of seston was generally high in near surface layers and decreased with depth. It is an interesting fact that there was almost no similarity in the vertical distributions of deposit and of seston. This seems to suggest the difference in the characters of deposit and seston. When suspending particles have been transported in the lake through inflowing rivers, their larger part sinks rather rapidly to the bottom but a few smaller particles remain in water as seston with autochthonous substances for considerably long periods. Such a phenomenon might be conspicuous after a heavy rainfall. Therefore the sample of deposit accumulated in the suspended bottles includes some autochthonous substances as well as a large amount of allochthonous substances, whereas seston is composed of a larger part of autochthonous organic substances and a minor amount of allochthonous matters. The relation between seston and chlorophyll *a* in sample waters seems to support the above explanation. As illustrated in Fig. 175 a rather good correlation was found between chlorophyll and seston for their total amounts in the euphotic layer (0 – 20 m).

Nitrogen and phosphorus in deposits

Seasonal variations in the vertical distributions of nitrogen and phosphorus in deposits are presented in Fig. 176. In the case of nitrogen, the maximum value of

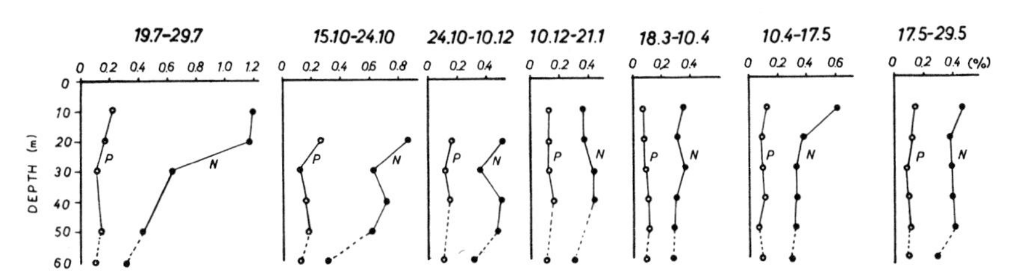

Fig. 176. Seasonal changes in the vertical distribution of nitrogen and phosphorus contents in deposit (Toyoda et al. 1968).

1.2% was found at 10 m and 20 m depth in July 1963, then decreased with depth, and the values of 0.4 and 0.3% were obtained at 50 m and 60 m, respectively. From October to December, similar patterns were observed in the vertical distribution of nitrogen, though the difference of nitrogen contents between the upper and lower layers were not so large as that in July. From December to April, the nitrogen contents were consistently low and no appreciable differences were found between the deposits and the surface of bottom sediments.

Such seasonal variations in the vertical distribution of nitrogen contents can be explained from seasonal changes of thermal stratification of water and the production of phytoplankton. Namely, in July and October 1963, the chlorophyll *a* contents were the highest through the period of investigation and a marked thermal stratification was observed. Therefore, it can be surmised that a large amount of organic substances was produced near the surface and sank slowly subject to decomposition. On the other hand, from December to January and from March to April, the standing crop of phytoplankton was very low and the production of organic substances was quite limited. Furthermore, the lake water was mixed through all layers because of the lack of thermal stratification. The nearly constant low nitrogen values in winter and spring seem to derive from such circumstances.

Regarding phosphorus, its vertical changes were not so conspicuous as those of nitrogen. There was no appreciable difference in phosphorus contents between the sampled deposit and the bottom sediment. Such a difference in the vertical distribution of content between the phosphorus and nitrogen might be caused by the difference in the decomposition rates of both elements. In the case of nitrogen, the process of decomposition goes rather slowly and continues from the euphotic layer until after it has been buried in bottom sediments. As regards the phosphorus, on the contrary, the process of decomposition goes so rapidly that most of the organic phosphorus in particles is destroyed into soluble form already in the upper part of the euphotic layer. The remaining phosphorus, which is mainly composed of inorganic phosphorus like silicate minerals, does not receive any more appreciable decomposition either during the sinking in water or after being buried in bottom sediments.

Vertical distributions of nitrogen and phosphorus in a core sample

A core sample of bottom sediment (18 cm length) was taken at the same station in order to investigate changes of organic nitrogen and phosphorus after depositing to the bottom. The results of the chemical analysis are shown in Fig. 177. The highest values of nitrogen (0.26%) were observed at the surface of sediment and decreased with depth, but below 5 – 6 cm, the nitrogen contents were almost constant. These changes seem to indicate the rate of decomposition of organic nitrogen in sediments. Accordingly, if the age of each layer in the sediment can be determined, it may be possible to estimate the yearly rate of decomposition in sediments, assuming that there was no marked change in the supply of autochthonous and allochthonous substances during that period.

Negoro (1954b) observed no appreciable change of diatom composition in a core (45 cm length) of bottom sediment. Furthermore, summarizing the prevailing data, the writers found that the transparency of this lake had been substantially constant for the past 60 years (until 1954). From these facts, it can be assumed that there was no appreciable change in organic matter production in this lake.

For the phosphorus contents of bottom sediments, a similar pattern was observed in the depth distribution as seen in Fig. 177 but the change was not so

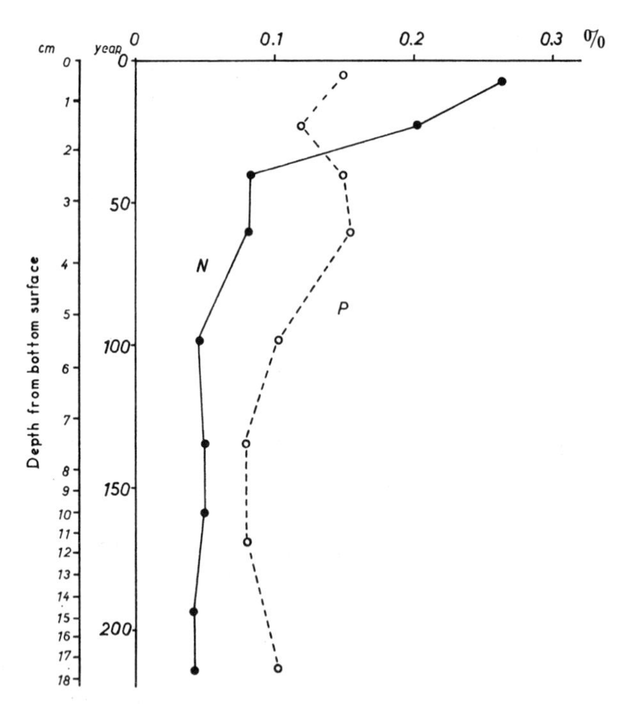

Fig. 177. Vertical distribution of the nitrogen and the phosphorus (percentage for dry weight) in bottom sediments and its relation to the age of sediments (Toyoda et al. 1968).

clear as that of nitrogen. Therefore, it was difficult to estimate the liberation rate from the sediments.

The age of the sediments layer was estimated from the average rate of deposition near the bottom, 208 g/m^2/year as dry weight. To examine whether or not this annual rate of deposition can be used as the basis for calculating the age of sediments, the amount of outflow from 1963 to 1964 was compared with the change of yearly outflow from 1875 to 1964, and the value was found to be not so far from its average value for the past 88 years. Then, using the above yearly rate of deposition, the period required for accumulating the sediment of 18 cm thickness from the surface was calculated at about 220 years. From this rate of sedimentation and the vertical distribution of organic nitrogen in the core sample, the yearly rate of nitrogen liberation from the sediment was calculated at 380 mg N/m^2/year.

Nitrogen metabolism in the lake

To discuss the significance of the depositing process in lake metabolism, the results of observation for nitrogen in the north basin of Lake Biwa are summarized in Fig. 178 (Saijo & Sakamoto 1970). The amount of assimilated nitrogen was estimated from the results of ^{14}C productivity measurement. The supply of allochthonous matters were calculated from the data of Kobayashi (1952) and the excretion from zooplankton was determined by incubation experiments. Other parameters were estimated from available data.

It is readily apparent from Fig. 178 that approximately 90% of the organic matter produced by phytoplankton in the trophogenic layer is decomposed within that layer, about 1/3 as a result of grazing by zooplankton and the remainder by other decomposition processes. Two-thirds of the 10 percent which reach the tropholytic layer are further subjected to degradation as they

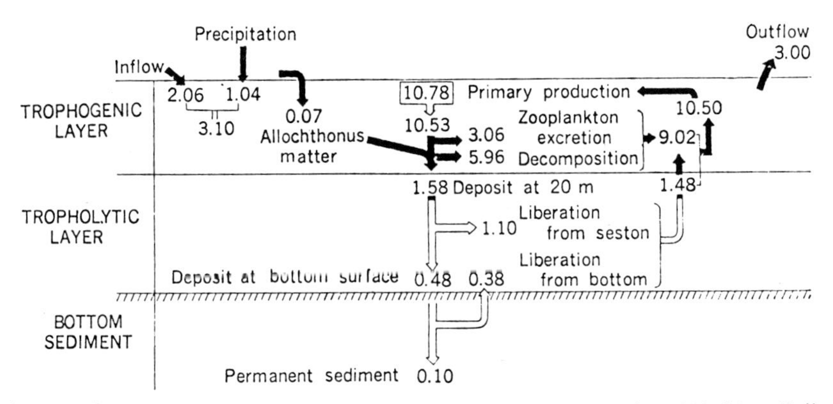

Fig. 178. Nitrogen cycle in the north basin of Lake Biwa. Amounts in g N/m^2/yr. (Saijo & Sakamoto 1970).

settle through it, the remaining 1/3 being deposited on the lake bottom. In addition, the latter material gradually decomposes over a long period, only about 1/5 of which ever being buried as permanent bottom sediment. Thus, such accumulation barely represents 1 percent of the total organic production of the lake.

The above results suggest that for lakes of the same general type as Biwa's north basin the greater part of the decomposition of organic material (into the inorganic substances which support primary production) take place in the producing, trophogenic layer. The so-called tropholytic layer, in spite of its original meaning, accounts for only a small part of the total decomposition. Furthermore, the supply of nutrients into water by decomposition of bottom sediment is far less than previously believed and cannot be said to be of really major importance.

References

Annandale, N., 1916a. Zoological results of a tour in the Far East. Introduction. Mem. Asiat. Soc. Bengal 6: 1-11.

Annandale, N., 1916b. Zoological results of a tour in the Far East. The Mollusca of Lake Biwa, Japan. Mem. Asiat. Soc. Bengal 6: 41-74.

Annandale, N. and Kawamura, T., 1916. The sponges of Lake Biwa. Jour. Coll. Sci., Imp. Univ. Tokyo 39: 1-27.

Annandale, N., 1918. Zoological results of a tour in the Far East. Sponges. Mem. Asiat. Soc. Bengal 6: 195-216.

Annandale, N., 1921. Zoological results of a tour in the Far East. The viviparous water-snail of Lake Biwa, Japan. Mem. Asiat. Soc. Bengal 6: 399-401.

Annandale, N., 1922. The macroscopic fauna of Lake Biwa. Annot. Zool. Japon 10: 127-153.

Azuma, M., 1969. Some considerations on the variation of the Ayu-fish in Lake Biwa, with special references to their stadia of development. Jap. Jour. Michurin Biol. 5: 165-172.

Azuma, M., 1973. Studies on the variability of the landlocked Ayu-fish, *Plecoglossus altivelis* T. et S., in Lake Biwa. IV. Consideration on the grouping and features of variability. Jap. Jour. Ecol. 23: 255-265.

Birge, E. A., 1915. The heat budgets of American and European lakes. Trans. Wis. Acad. Sci. Arts Lett. 18: 166-213.

Chrystal, 1904. Some results in the mathematical theory of seiches. Proc. Roy. Soc., Edinb. 25: 328-337.

Csanady, G. T., 1967. Large-scale motion in the Great Lakes. Jour. Geophys. Res. 72: 4151-4162.

Defant, A., 1918. Neue Methode zur Ermittlung der Eigenschwingungen (Seiches) von abgeschlossenen Wassermassen (Seen, Buchten, usw.). Ann. Hydrogr. Mar. Meteorol. 46: 78-85.

Defant, A., 1961. Physical Oceanography, 2: Oxford, London, New York, Los Angeles, Paris, Frankfurt, Pergamon Press, 598 pp.

Endoh, S., Onishi, Y. and Imasato, N., 1976. Studies on the currents in Lake Biwa-ko, III. Baroclinic responses to the wind stress with a curl. Pre-print, Oceanogr. Soc. Japan, 1976, 163-164.

Eriksen, C. C., 1978. Measurements and models of fine structure, internal gravity waves, and wave breaking in the deep ocean. Jour. Geophys. Res. 83: 2989-3009.

Forel, F. A., 1892. Le Léman: monographie limnologique. 1: Lausanne, F. Rouge, 543 pp.

Fujinaga, T., Morii, F., Saijo, Y., Horie, S., Hori, T. and Fuse, S., 1966. Interim report of water chemistry. In: Biwa-ko Seibutsu-shigen Chōsa-dan (BST) Chūkan Hōkoku. Osaka, Kensetsu-shō, Kinki Chihō Kensetsu-kyoku. 106-217.

Fujita, T., 1977. Determination of turbulent flux by the bulk method using measurements at two levels. Papers Meteorol. Geophys. 29: 1-15.

Gose, K., 1966. Food habit of *Bdellocephala Annandalei* of Lake Biwa. Jap. Jour. Ecol., 16, 78-79.

Hansen, W., 1956. Theorie zur Errechnung des Wasserstandes und der Strömungen in Randmeeren nebst Anwendungen. Tellus 8: 287-300.

Harada, E., 1966. Shrimps and prawns. In: Biwa-ko no Seibutsushigen o Megutte. Biwa-ko Seibutsu-shigen Chōsa-dan (BST). 14-18.
Hasselmann, K., Munk, W. and Macdonald, G., 1963: Bispectra of ocean waves. In: Proc. Symp. Time Series Analysis, ed. Murray Rosen Blatt, New York, London, John Wiley and Sons, 125-139.
Hayashi, K., Mori, Sy., Higashi, S. and Kawanabe, H., 1966. Interim report of mollusca. In: Biwa-ko Seibutsushigen Chōsa-dan (BST) Chūkan Hōkoku. Osaka, Kensetsu-shō, Kinki Chihō Kensetsu-kyoku. 607-707.
Hikone Local Meteorological Observatory, 1969. Shiga-ken Bōsai Kishō-Yōran, 459 pp.
Hirai, K., 1970. Ecological studies on fry and juvenile of fishes at aquatic plant area in a bay of the Lake Biwa, I. On the distribution of fish larvae. Bull. Fac. Educ., Kanazawa Univ. No. 19 (Nat. Sci.) S: 93-105.
Hirai, K., 1971. Ecological studies on fry and juvenile of fishes at aquatic plant area in a bay of the Lake Biwa, II. On the food habits of fish larvae. Bull. Fac. Educ., Kanazawa Univ. No.20 (Nat. Sci.) S: 59-71.
Hirai, K., 1972. Ecological studies on fry and juvenile of fishes at aquatic plant areas in a bay of Lake Biwa, III. Relationship of the food habits to the habitat of Nigorobuna (*Carassius carassius grandoculis*) larvae. Jap. Jour. Ecol. 22: 69-93.
Hirano, M., 1968. Desmids. In: Shiga-ken Shokubutsushi, ed. S. Kitamura, Osaka and Tokyo, Hoiku-sha, 248-274.
Hirose, H., Yamagishi, T., Akiuama, M., Ioriya, T., Imahori, K., Kasaki, H., Kumano, S., Kobayashi, H., Takahashi, E., Tsumura, K. & Hirano, M., 1977. Illustrations of the Japanese freshwater algae. Tokyo, Uchidarokakuho Pub. Co., Ltd., 933 pp.
Hogetsu, K., Kitazawa, Y., Kurasawa, H., Shiraishi, Y. and Ichimura, S., 1952. Fundamental studies on the biological production and metabolism of inland waters, mainly of Lake Suwa. Suisan Kenkyu Kaihō No. 4 41-127.
Honjoh, T., 1977. Studies on the culture and transplantation of Amago Salmon, *Oncorhynchus rhodurus*. Rept. Gifu Pref. Fish. Exp. St. No. 22, 1-103.
Hurukawa, M., 1935. An ecological studies on the Bivalve "Seta-Shijima", Corbicula sandai Reinhardt of the Lake Biwa – I. Bull. Jap. Soc. Sci. Fish., 19, 88-90.
Iima, N., 1963. Study on the protozoan fauna of the Lake Biwa. Biol. Jour., Nara Women's Univ., No. 13, 5-6.
Ikusima, I., Furukawa, M. and Ikeda, J., 1962. The summer standing crop of rooted aquatic plants in Lake Biwa. Jour. Coll. Arts Sci., Chiba Univ. 3: 483-494.
Ikusima, I., 1965. Ecological studies on the productivity of aquatic plant communities, I. Measurement of photosynthetic activity. Bot. Mag. 78: 202-211.
Ikusima, I. and Kabaya, H., 1965. A new introduced aquatic plant, *Elodea occidentalis* (Pursh) St. John in Lake Biwa, Japan. Jour. Jap. Bot. 40: 57-64.
Ikusima, I., 1967. Ecological studies on the productivity of aquatic plant communities, III. Effect of depth on daily photosynthesis in submerged macrophytes. Bot. Mag. 80: 57-67.
Ikusima, I., 1970. Ecological studies on the productivity of aquatic plant communities, IV. Light condition and community photosynthetic production. Bot. Mag. 83: 330-341.
Ikusima, I., 1972. Organic matter production of plants in water, I. Freshwater macrophytes. Seitaigaku-kōza, No. 7, Tokyo, Kyōritsu Shuppan K.K., 98 pp.
Imasato, N., 1970. Study of seiche in Lake Biwa-ko, (I) - On the numerical calculation by Defant's method.-Spec. Contr. Geophys. Inst., Kyoto Univ., No. 10, 93-103.
Imasato, N., 1971. Study of seiche in Lake Biwa-ko, (II) - On a numerical experiment by non-linear two-dimensional model.- Contr. Geophys. Inst., Kyoto Univ., No. 11, 77-90.
Imasato, N., Kanari, S. and Kunishi, H., 1971. A numerical experiment of water motion in Lake Biwa-ko. - On the two-dimensional one - layer model. - Disaster Prev. Res. Inst. Ann., Kyoto Univ. No. 14B 451-464.
Imasato, N., 1972. Study of seiche in Lake Biwa-ko (III) - Some results of numerical experiments by nonlinear two-dimensional model. - Contr. Geophys. Inst., Kyoto Univ., No. 12, 63-75.

Imasato, N., Tanaka, K. and Kunishi, H., 1973. Study of seiche in Lake Biwa-ko (IV). - Observation with a new portable long period water level gauge. - Contr. Geophys. Inst., Kyoto Univ., No. 13, 65-72.

Imasato, N., 1979. A study on nonlinear interaction and decay of seiches in Lake Biwa. Jap. Jour. Limnol. 40: 115-122.

Ito, K. & Okamoto, I., 1974. Time variation of water temperature in Lake Biwa-ko, (VIII) - Heat budget in both basins. - Jap. Jour. Limnol. 35: 127-135.

Ito, S., 1964. Biological investigation in Yama-no-shita Bay, Ogoto, Lake Biwa. Tansui Seibutsu, No. 9, 43-44.

Ito, T. and Miura, Y., 1973. Copepoda. In: Freshwater Biology of Japan, ed. M. Ueno, Tokyo, Hokuryukan, 434-454.

Jacobs, W. C., 1942. On the energy exchange between sea and atmosphere. Jour. Mar. Res. 5: 37-66.

Japan Meteorological Agency. Geophys. Rev. No. 762, 1963 Feb. 62 pp (1964), No. 788, 1965 Apr. 47 pp (1966), No. 793, 1965 Sep. 113pp (1967), No. 888, 1973 Aug. 56 pp (1974), No. 910, 1975 Jun. 56 pp (1976).

Japan Meteorological Agency, 1971. Climatic Table of Japan Part 1. Monthly normals of atmospheric pressure, temperature, relative humidity, precipitation and sunshine duration, 1941-1970. 56 pp.

Japan Meteorological Agency, 1972. The monthly normals of temperature and precipitation at climatological stations in Japan, 1941–1970. Technical Data Series, No. 36, 209 pp.

Kadota, H., Hata, Y. and Miyoshi, H., 1966. Interim report of micro-organisms. In: Biwa-ko Seibutsushigen Chōsa-dan (BST) Chūkan Hōkoku. Osaka, Kensetsu-shō, Kinki Chihō Kensetsu-kyoku, 370-405.

Kadota, H. and Tanaka, N., 1975. Bacterial production and decomposition of organic matter. In: Productivity of Communities in Japanese Inland Water, JIBP Synthesis, 10, ed. Sy. Mori and G. Yamamoto, Tokyo, Univ. Tokyo Press, 12-14.

Kadota, S., 1974. A quantitative study of microfossils in a 200-meter core sample from Lake Biwa. Paleolimnology of Lake Biwa and the Japanese Pleistocene, 2, ed. S. Horie, 236-245.

Kanari, S., 1968. On the studies of internal waves in Lake Biwa, (I). Disaster Prev. Res. Inst. Ann., Kyoto Univ., No. 11, B, 179-189.

Kanari, S., 1969. On the studies of internal waves in Lake Biwa, (II). – On the instrumented neutrally-buoyant float. - Disaster Prev. Res. Inst. Ann., Kyoto Univ., No. 12, B, 669-680.

Kanari, S., 1970a. On the studies of internal waves in Lake Biwa, (III). - On the measurement of the vertical displacement of waters by using an instrumented neutrally-buoyant float. - Disaster Prev. Res. Inst. Ann., Kyoto Univ., No. 13, A, 601-608.

Kanari, S., 1970b. Internal waves in Lake Biwa, (I). - The responses of the thermocline to the wind action. Bull. Disaster Prev. Res. Inst., Kyoto Univ. 19: Pt. 3, 19-26.

Kanari, S., 197oc. On the study of internal waves in Lake Biwa. Suion no Kenkyu 14: No. 3, 2-16.

Kanari, S., 1974a. On the study of numerical experiments of two layer Lake Biwa. Jap. Jour. Limnol. 35: 1-17.

Kanari, S., 1974b. Some results of observation of the long-period internal seiche in Lake Biwa. Jap. Jour. Limnol. 35: 136-147.

Kanari, S., 1974c. Long-period internal waves in Lake Biwa. Suion no Kenkyu 17: No. 5 2-14.

Kanari, S., Imasato, N. and Kunishi, H., 1974. On the studies of internal waves in Lake Biwa (IV) Disaster Prev. Res. Inst. Ann., Kyoto Univ., No. 17, B, 249-258.

Kanari, S., 1975a. The long-period internal waves in Lake Biwa. Limnol. Oceanogr., 20, 544-553.

Kanari, S., 1975b. The long-period internal waves in Lake Biwa. Verh. Internat. Verein. Limnol. 19: 97-103.

Kato, F., 1973. On the sea-run form of *Oncorhynchus rhodurus* obtained in Ise Bay. Jap. Jour. Ichthyol. 20: 107-112.

Kato, F., 1975. Large specimens of *Oncorhynchus rhodurus* growth up in a dam and rivers in Fukui Prefecture. Jap. Jour. Ichthyol. 22: 183-185.

Kato, Y., 1962. Thirty species of Chironomidae larvae from Lake Biwa. Tansui-Seibutsu, No. 8, 37-47.

Kawamura, T., 1918. Nihon Tansui Seibutsugaku, Jō. Tokyo, Shōkabo, 362 pp.

Kawanabe, H., 1975. On the orgin of Ayu-Fish (Pisces, Osmeridae) in Lake Biwa (preliminary notes). Paleolimnology of Lake Biwa and the Japanese Pleistocene, 3, ed. S. Horie, 317-320.

Kawanabe, H., 1976. A note on the territoriality of Ayu, *Plecoglossus altivelis* Temminck et Schlegel (Pisces: Osmeridae) in the Lake Biwa stock, based on the 'relic social structure' hypothesis. Physiol. Ecol. Japan 17: 395-399.

Kawashima, M., Nakagawa, T., Nakajima, M., Shiota, A., Taniguchi, T., Itasaka, O., Takamatsu, T., Matsushita, R., Koyama, M. and Hori, T., 1978. Vertical distribution and chemical properties of elements in Lake Biwa sediments. - Distribution of Manganese, Phosphorus and Arsenic. - Mem. Fac. Educ., Shiga Univ., Nat. Sci., No. 28, 13-29.

Kiefer, F., 1932. Versuch eines Systems der Diaptomiden (Copepoda Calanoida). Zool. Jahrb. Abteil Syst., Ökol. Geogr. Tiere, 63: 451-520.

Kikuchi, K., Enokida, Y. and Tateno, H., 1942. Seasonal change in the vertical distribution of the plankton in Lake Biwa. Jap. Jour. Limnol. 12: 63-72.

Kobayashi, J., 1952. On the chemical character of irrigation waters. Shiga-ken Nōsei Shiryō, No. 1, 81 pp.

Kobe Marine Observatory, 1926. Biwa-ko Chōsa Hōkoku, I. Bull. Kobe Mar. Observ., No. 8, 104 pp.

Kobe Marine Observatory, 1928. Biwa-ko Chōsa Hōkoku, II. Bull. Kobe Mar. Observ., No. 15, 59 pp.

Kodama, R., 1966. On the wind system in Shiga Prefecture. Jour. Meteorol. Res. (Kishōchō Kenkyu-Jihō) 18: 301-304.

Kodama, R., 1973. On the weak winds in Shiga Prefecture. Jour. Meteorol. Res. (Kishōchō Kenkyu-Jihō) 25: 333-337.

Kondo, J. and Watabe, I., 1969. Seasonal changes of vertical profile of water temperature and evaporation from deep lakes. Rept. Nat. Res. Cent. Disaster Prev., No. 2, 75-88.

Kondo, K., 1963. The pollution biological class determined by the weed Aufwuchs of the Lake Biwa. Biol. Jour., Nara Women's Univ., No. 13, 3-4.

Kotoda, K., 1975. Lake and People. Chiri, 20: No. 6, 35-44.

Kotoda, K. and Mizuyama, T., 1975. Water balance of the Lake Biwa and its catchment area. Pub. l'Assoc. Internat. Sci. Hydrol. Symp. Tokyo 1975, No. 117, 21-28.

Kotoda, K., 1977. A method for estimating lake evaporation based on climatological data. Bull. Environmental Res. Center, Univ. Tsukuba No. 1 53-65.

Koyama, M., 1974. Variation in chemistry. In: Biwa-ko no Dōtai, ed. T. Fujinaga, Tokyo, Jiji Tsushin-sha, 109-135.

Koyama, M., Hori, T., Itasaka, O., Kawashima, M. and Hori, T., 1975. Analysis of the present situation of Lake Biwa. Proc. Internat. Congr. Human Environ. (HESC). Science for better environment. Kyoto, ed. Science Council of Japan. 391-397.

Koyama, T., Matsunaga, K. and Tomino, T., 1975. Geochemical studies on the cycle of carbon and nitrogen in a Mesotrophic lake. Nitrogen fixation and nitrogen cycle. In: JIBP Synthesis, ed. H. Takahashi, Jap. Comm. Internat. Biol. Program, Tokyo, Univ. Tokyo Press 12: 115-123.

Kumode, M., 1965. Benthic animals in the rocky areas on the northeast coast of Lake Biwa. Tansui-Seibutsu, No. 10, 10-12.

Kunishi, H. and Imasato, N., 1966. On the growth of wind waves by high-speed wind flume. Disaster Prev. Res. Inst. Ann., Kyoto Univ., No. 9, 667-676.

Kunishi, H., Okamoto, I. and Sato, H., 1967. Observation of water circulation in Lake Biwa-ko. Disaster Prev. Res. Inst. Ann., Kyoto Univ., No. 10, B: 321-329.

Kurawaka, K., 1976. Study of speciation in fish. Dr. Thesis, Fac. Agricul., Kyoto Univ.

Lamb, H., 1932. Hydrodynamics. Cambridge, Univ. Press, (Sixth edition), 738 pp.

Maki, I., 1964. The relationship between fishes and their food in a bay of the Lake Biwa, Japan. Physiol. Ecol. 12: 259-271.

Matsuoka, T., 1970. Biwa-ko no Sōgō Kaihatsu. Otsu, Kensetsu-shō, Biwa-ko Kōji Jimusho, 287 pp..

Merian, J. R., 1828. Über die Bewegung tropfbarer Flüssigkeiten in Gefässen. Nach J. R. Merian bearbeitet von Karl Vonder Mühl. Math. Ann. 27 (1886): 575-600.

Miki, S., 1927. Oekologische Studien über die Sumpf- und Wassergewächsel sowie ihre Formationen im Ogura-Teiche. Kyoto-fu Shiseki Meishōchi Chōsa-kai Hōkoku. No. 8, 81-145.

Miki, S., 1929. Oekologische Studien vom Mizoro-Teiche. Die Entstehung der Schwimminseln und die Schwankung des Sauerstoffs der Kohlensäure und der Wasserstoffonenkonzentration in Laufe des Tages und der Jahreszeiten (Vorläufige Mitteilung). Kyoto-fu Shiseki Meishō Tennen Kinenbutsu Chōsa Hōkoku. No. 10, 61-145.

Miki, S., 1934. On fresh water plants new to Japan. Bot. Mag., 48: 326-337.

Miki, S., 1937. The water phanerogams in Japan, with special reference to those of Prov. Yamashiro. Kyoto-fu Shiseki Meishō Tennen Kinenbutsu Chōsa Hōkoku. No. 17, 127 pp.

Miura, T., 1966. Ecological notes of the fishes and the interspecific relations among them in Lake Biwa. Jap. Jour. Limnol. 27: 49-72.

Miura, T., Sunaga, T., Kawanabe, H., Maki, I., Azuma, M., Tanaka, S., Hirai, K., Narita, T., Tomoda, Y., Mizuno, N., Nagoshi, M., Takamatsu, S., Shiraishi, Y., Onodera, K., Suzuki, N. and Yanagishima, S., 1966. Interim report of fish. In: Biwa-ko Seibutsushigen Chōsa-dan (BST) Chūkan Hōkoku. Osaka, Kensetsu-shō, Kinki Chihō Kensetsu-kyoku. 711-906.

Miura, T., Kawakita, A., Iwasa, Y. and Tanimizu, K., 1976. Studies on the submerged plant community in Lake Biwa, II. Macro-invertebrates as an important supplier of nitrogenous nutrients in dense macrophyte zone. Physiol. Ecol. Japan 17: 587-591.

Miyadi, D., Kawanabe, H. and Mizuno, N., 1976. Coloured illustrations of the freshwater fishes of Japan. Osaka, Hoiku-sha, 462 pp.

Miyazaki, I., 1957. Nimai-gai to sono Yōshoku. Tokyo, Isana Shobō, 158 pp.

Miyazaki, M., Ueno, T. and Unoki, S., 1961. Theoretical investigations of typhoon surges along the Japanese coast. Oceanogr. Mag. 13: 51-75, 103-117.

Mizuno, N., 1960. Study on a freshwater Goby, *Rhinogobius similis* Gill, with a proposition on the relationships between land-locking and speciation of some freshwater Gobies in Japan. Mem. Coll. Sci., Univ. Kyoto, Ser. B, 27: 97-115.

Mizuno, N., 1971. Fishes in Hii River. In: Hii-gawa, Kando-gawa Suikei no Seibutsu ni kansuru Sōgō Chōsa, Shimane-ken, 65-87.

Mizuno, T. and Tetsukawa, T., 1965. Distribution of fresh-water sponges and Bryozoa in Lake Biwa. Jap. Jour. Limnol. 26: 134-145.

Mizuno, T. and Tetsukawa, T., 1966. Interim report of attached animals. In Biwa-ko Seibutsushigen Chōsa-dan (BTS) Chūkan Hōkoku. Osaka, Kensetsu-shō, Kinki Chihō Kensetsu-kyoku. 496-517.

Mizuoka, S., 1962. Studies on fluvial variations in the Gobioid fishes' 'Yoshinobori'. I. Two types in the Ota River. Educ. Stud. Fac. Educ., Hiroshima Univ., Part 2, 10: 71-95.

Mizuoka, S., 1967. Studies on fluvial variations in the Gobioid fish 'Yoshinobori'. IV. Distributions and variations in color pattern and the pectoral fin ray numbers. Bull. Fac. Educ., Hiroshima Univ., Part 3, No. 16, 43-52.

Mizuoka, S., 1974. Studies on the variation of *Rhinogobius brunneus* (Temminck et Schlegel). III. On the six types of body color and color pattern in the San in district, the Hokuriku district, the San yo district, and the Gotō Islands. Bull., Fac. Educ., Hiroshima Univ., Part 3, No. 23, 31-40.

Mizuoka, S., 1976. Electroporetic studies on some Gobioid fishes. On *Rhinogobius flumineus, Rhinogobius giurinus* and the six types of *Rhinogobius brunneus*. Bull. Fac. Educ., Hiroshima Univ., Part 3, No. 25, 23-32.

Mizushima, M., 1967. On the scientific name of 'ko-kanada-mo'. Jour. Jap. Bot. 42: 48.

Mori, Sh., 1974. Diatom succession in a core from Lake Biwa. Paleolimnology of Lake Biwa and the Japanese Pleistocene, 2, ed. S. Horie, 247-254.

Mori, Sy., 1945. Yearly succession of plankton community in the sub-basin of Lake Biwa from 1934 to 1937. Seiri-Seitaigaku Kenkyugyoseki, Kyoto Univ., No. 53, 1-28.

Mori, Sy., 1970. List of animal and plant species living in or on Lake Biwa. Mem. Fac. Sci., Kyoto Univ., Ser. Biol. 3: 22-46.

Mori, Sy., 1971. Molluscs in Lake Biwa. In: Biwa-ko Kokutei Kōen Gakujutsu Chōsa Hōkokusho, Otsu, Shiga-ken. 301-312.

Mori, Sy., 1978. Seventh report of the regular limnological survey of Lake Biwa (1973), II. Phytoplankton. Mem. Fac. Sci., Kyoto Univ., Ser. Biol., 7, No. 2, 1-9.

Mori, Sy. and Miura, T., 1980. List of plant and animal species living in Lake Biwa. Mem. Fac. Sci., Kyoto Univ., Ser. Biol. 8: 1-33.

Morikawa, M. and Okamoto, I., 1960. The surface currents of Lake Biwa-ko deduced from drift-bottle experiments. Jap. Jour. Limnol. 21: 173-186.

Morikawa, M. and Okamoto, I., 1962. On the currents of Lake Biwa-ko. Mem. Fac. Lib. Arts Educ., Shiga Univ., No. 12, 21-30.

Morikawa, M., Okamoto, I. and Horie, S., 1966. Interim report of hydrology. In: Biwa-ko Seibutsushigen Chōsa-dan (BST) Chūkan Hōkoku. Osaka, Kensetsu-shō, Kinki Chihō Kensetsu-kyoku. 53-105.

Nagai, K., 1975. The distribution and ecology of aquatic macrophytes. In: Biwa-ko Suisei Shokubutsu Jittai Chōsa Hōkokusho, Osaka, Toshi Kagaku Kenkyu-sho, 1-32.

Nakamura, M., 1969. Cyprinid fishes of Japan. - Studies on the life history of Cyprinid fishes of Japan. - Spec. Publ. Res. Inst. Natur. Resources, No. 4, 455 pp.

Nakamura, M., 1975. Keys to the freshwater fishes of Japan fully illustrated in colors, 5 ed., Tokyo, Hokuryukan, 260 pp.

Nakamura, S. and Honda, K., 1911. Seiches in some lakes of Japan. Jour. Coll. Sci., Imperial Univ. Tokyo 28: (Art 5): 1-95.

Nakanishi, M., 1976. Seasonal variations of chlorophyll *A* amounts, photosynthesis and production rates of macro- and microphytoplankton in Shiozu Bay, Lake Biwa. Physiol. Ecol. Japan 17: 535-549.

Narita, T. and Mori, Sy., 1973. Biomass and ingestion rate of zooplankton in Shiozu Bay of Lake Biwa. In: Nihon Rikusui Gunshū no Seisan-ryoku ni kansuru Kenkyu. Cont. JIBP-P F: 24-28.

Narita, T., 1976. Occurence of two ecological forms of *Anisogammarus annandalei* (Tattersall) (Crustacea: Amphipoda) in Lake Biwa. Physiol. Ecol. Japan, 17, 551-556.

Negoro, K., 1954a. *Pediastrum Biwae* spec. nov., eine neue planktische Grünalge aus dem Biwasee. Acta Phytotax. Geobot., 15: 135-138.

Negoro, K., 1954b. Diatom shell composition of profundal bottom sediments in Lake Biwa. Kagaku 24: 527-528.

Negoro, K., 1956. The phytoplankton of the main basin of Lake Biwa-ko. Jap. Jour. Limnol. 18: 37-46.

Negoro, K., 1959. Some noticeable plankton-algae of Lake Biwa and its effluent. Algal indicators of a water system. Mem. Coll. Sci., Kyoto Univ., Ser. Biol. 26: 311-314.

Negoro, K., 1968. Phytoplankton of Lake Biwa. In Flora Ohmiensis (Shiga-ken Shokubutsushi). Ed. S. Kitamura, Osaka & Tokyo, Hoiku-sha, 275-330.

Negoro, K., 1971. Plankton in Lake Biwa. In: Biwa-ko Kokutei Kōen Gakujutsu Chōsa Hōkokusho, Otsu, Shiga-ken. 245-274.

Nishijima, S., 1968. Two forms of the gobioid fish *Rhinogobius brunneus* from Okinawa-jima, Ryukyu Islands, Zool. Mag. 77: 397-398.

Nomitsu, T., 1935. Surface fluctuations of Lake Biwa caused by the Muroto typhoon. Mem. Coll. Sci., Kyoto Imp. Univ., Ser. A,18: 221-238.

Nomitsu, T., Habu, K. and Nakamiya, M., 1937. Proper oscillations of lake-shelves. Mem. Coll. Sci., Kyoto Imp. Univ., Ser. A, 20: 3-17.

Ogino, Y., 1963. Study on the benthic fauna of Lake Biwa. Biol. Jour., Nara Women's Univ., No. 13, 7-9.
Ogura, K., 1975. Coprostanol in the surface sediments of Lake Biwa as an index of fecal pollution. Paleolimnology of Lake Biwa and the Japanese Pleistocene, 3, ed. S. Horie, 263-276.
Ohle, W., 1962. Der Stoffhaushalt der Seen als Grundlage einer allgemeinen Stoffwechseldynamik der Gewässer. Kieler Meeresforsch. 18: 107-120.
Ohwi, J., 1972. Flora of Japan, (revised edition), Tokyo, Shibundo, 1560 pp.
Okada, T., Fujiwhara, S. and Maeda, S., 1914. On thunderstorms as a cause of seiches. Proc. Tokyo Mathematico-physical Soc. 2nd Ser. 7, 1913-14, 210-221.
Okamoto, I. and Morikawa, M., 1961. Water circulation in Lake Biwa-ko as deduced from the distribution of water density. Jap. Jour. Limnol. 22: 193-200.
Okamoto, I. and Yagi, K., 1969. Time variation of water temperature in Lake Biwa-ko, (III) - Drastic change of temperature accompanied with the intrusion of deep water - Jap. Jour. Limnol. 30: 108-118.
Okamoto, I., 1971. Water currents in Lake Biwa. In: Biwa-ko Kokutei Kōen Gakujutsu Chōsa Hōkokusho, Otsu, Shiga-ken. 177-213.
Okamoto, I., Kitamura, K. and Tomoda, N., 1971. Current observations in Shiozu Bay, Lake Biwa-ko, (II). - Distribution of the current velocity at the bay mouth. - Mem. Fac. Educ. Shiga Univ., Nat. Sci., No. 21, 75-87.
Okamoto, I., 1972. Time variations of water temperature in Lake Biwa-ko (VI). - Extraordinary change of temperature caused by the typhoon - Jap. Jour. Limnol. 33: 29-35.
Okamoto, I. and Yamashita, M., 1973. Time variation of water temperature in Lake Biwa-ko, (VII). - Oscillation characteristics of the internal seiche in the main basin - Mem. Fac. Educ., Shiga Univ., Nat. Sci., No. 23, 44-54.
Okamoto, I. and Iwamoto, N., 1977. Current observation in Shiozu Bay, Lake Biwa-ko (V). - Automatic continuous records for current velocity and water temperature - Mem. Fac. Educ., Shiga Univ., Nat. Sci., No. 27, 44-54.
Okuda, S., Nakajima, C. and Yokoyama, K., 1970. Aerial survey method for water temperature and flow in Lake Biwa-ko. In: Koshō Butsurigaku no Kenkyu (Ministry of Education), 59-63.
Okuda, S. (ed.), 1978. Synoptic observations and monitorings of current velocity in Lake Biwa. Lake Biwa Institute Rept. No. 9. 163 pp.
Okumura, Y. and Yamamoto, A., 1976. Observations of lake currents by the radio- tracking system of drifting buoys - Bull. Osaka Electro-Communication Univ. No. 12, 191-204.
Penman, H. L., 1963. Vegetation and hydrology. Tech. Comm., No. 53, Commonwealth Agr. Bureaux, Farnham Royal Bucks, England, 124 pp.
Phillips, O. M., 1971. On spectra measured in an undulating layered medium. Jour. Phys. Oceanogr. 1: 1-6.
Romanenko, V. I., 1963. Potential capacity of aquatic microflora for heterotrophic assimilation of carbon dioxide and for chemosynthesis. Microbiol. 32: 569-574.
Romanenko, V. I., 1964. Heterotrophic assimilation of CO_2 by bacterial flora of water. Microbiol. 33: 610-614.
Saijo, Y., 1956. Chemical studies in lake metabolism. Jour. Chem. Soc. Japan 77: 917-936, 1184-1196.
Saijo, Y., Sakamoto, M., Toyoda, Y., Kadota, H., Miyoshi, H., Horie, S., Kawanabe, H. and Tsuda, M., 1966. Interim report of lake metabolism. In: Biwa-ko Seibutsushigen Chōsa-dan (BST) Chūkan Hōkoku. Osaka, Kensetsu-shō, Kinki Chihō Kensetsu-kyoku. 406-466.
Saijo, Y. and Sakamoto, M., 1970. Primary production and metabolism of lakes. Some normal and specialized examples from Japan. In: Profiles of Japanese Science and Scientists, 1970, ed. H. Yukawa, Tokyo, Kodansha, Ltd., 208-225.
Shiga Prefectural Fisheries Experimental Station, 1942. Biwa-ko Jūyō Gyozoku Tennen Shiryō Chōsa Hōkoku. ed. K. Araki, Hikone, Shiga Pref. Fish. Exp. St., 80 pp.

Shiga Prefectural Fisheries Experimental Station, 1954. A report of Investigation of counterplans for low water level in Lake Biwa. Hikone, Shiga Pref. Fish. Exp. St., 1-11 & 1-40.

Shiga Prefectural Fisheries Experimental Station, 1972. Report on the littoral community of Lake Biwa. Hikone, Shiga Pref. Fish. Exp. St., 121 pp.

Shiga-ken, 1978. Shiga no Suisan. 86 pp.

Shimadzu, T., 1954. Stocking results of Ayu-fish (*Plecoglossus altivelis*) in Nuru-River, Gumma Prefecture, 1950 to 1952. Bull. Freshwater Fish. Res. Lab. 3: No. 2, 25 pp.

Stephenson, J., 1917. Zoological results of a tour in the Far East. Aquatic oligochaeta from Japan and China. Mem. Asiat., Soc. Bengal 6: 85-99.

Sugawara, K., 1938. The seasonal variation of the occurrence of phytoplankton and the circulation of silicon in Lake Biwa. Jap. Jour. Limnol. 8: 434-445.

Suginohara, M. and Uyeno, T., 1967. Miocene fossil cyprinid fishes from Mitake, Gifu Prefecture. Jour. Geol. Soc. Japan 73: 101.

Sunaga, T., 1964. On the seasonal change of food habits of several fishes in the Lake Biwa. Physiol. Ecol. 12: 252-258.

Sunaga, T., 1970. Relation between food consumption and food selectivity in fishes. I. Food selectivity of juvenile Hasu (*Opsariichthys uncirostris* Teminck et Schlegel) in Lake Biwa. Jap. Jour. Ecol. 20: 129-137.

Suzuki, N. and Mori, Sy., 1967. First report of the regular limnological survey of Lake Biwa (Oct.1965–Dec.1966), IV. Benthos. Mem. Coll. Sci., Kyoto Univ., Ser. Biol. 1: 78-94.

Takahashi, A., 1959. Analysis of water temperature in lakes. Suion no Kenkyu 3: 63-74.

Takahashi, R., 1935. Level changes of Lake Biwa caused by the typhoon of Sept. 21, 1934. Bull. Earthq. Res. Inst., Tokyo Imp. Univ., Suppl. 2, 40-58.

Takahashi, T. and Namekawa, T., 1938. Study of seiche in the south basin of Lake Biwa-ko (I). Umi to Sora 18: 256-261.

Takao, S., 1964. Study on the community of aquatic insects of Lake Biwa in 1963. Biol. Jour., Nara Women's Univ., No. 14, 3-4.

Takaya, S., 1931. Study of oscillation with a long period in Lake Biwa-ko. Umi to Sora 11: 186-191.

Tanaka, N., Nakanishi, M. and Kadota, H., 1974a. Nutritional interrelation between bacteria and phytoplankton in a pelagic ecosystem. In: Effect of the Ocean Environment on Microbial Activities, ed. R. R. Colwell and R. Y. Morita, Baltimore, London, Tokyo, Univ. Park Press, 495-509.

Tanaka, N., Nakanishi, M. and Kadota, H., 1974b. The excretion of photosynthetic product by natural phytoplankton population in Lake Biwa. Jap. Jour. Limnol. 35: 91-98.

Tanaka, N., Nakanishi, M. and Kadota, H., 1975a. Distribution and activity of glycollate-utilizing bacteria in the water column in Lake Biwa. Bull. Jap. Soc. Sci. Fish. 41: 251-256.

Tanaka, N., Nakanishi, M. and Kadota, H., 1975b. Seasonal variation of glycollate-utilizing bacteria in the water column of Lake Biwa. Bull. Jap. Soc. Sci. Fish. 41: 1129-1134.

Tanaka, N., Ueda, Y., Onizawa, M. and Kadota, H., 1977. Bacterial populations in water masses of different organic matter concentrations in Lake Biwa. Jap. Jour. Limnol. 38: 41-47.

Tanimizu, K. and Miura, T., 1976. Studies on the submerged plant community in Lake Biwa. I. Distribution and productivity of *Egeria densa*, a submerged plant invader, in the south basin. Physiol. Ecol. Japan 17: 283-290.

Tattersall, W. M., 1921. Zoological results of a tour in the Far East. Mysidacea, Tanaidacea and Isopoda. Mem. Asiat. Soc. Bengal 6: 405-433.

Tomoda, Y., 1961. Two new catfishes of the genus *Parasilurus* found in Lake Biwa-ko. Mem. Coll. Sci., Univ. Kyoto, Ser. B, 38: 347-354.

Toyoda, Y., Horie, S. and Saijo, Y., 1968. Studies on the sedimentation in Lake Biwa from the viewpoint of lake metabolism. Mitt. Internat. Verein. Limnol. No. 14, 243-255.

Toyohara, Y. and Habu, K., 1938. On the seiches of two-step lake-basin. Jap. Jour. Limnol. 8: 242-249.

Tsudo, M., 1962. Comments on benthos in biological resources research in Lake Biwa. Tansui Seibutsu, No. 8, 1-2.
Tsuda, M., Iwao, Y. and Watanabe, T., 1966. Interim report of attached protozoans. In Biwa-ko Seibutsushigen Chōsa-dan (BST) Chūkan Hōkoku. Osaka, Kensetsu-shō, Kinki Chihō Kensetsu-kyoku, 535-543.
Tukey, J. W., 1963. What can data analysis and statistics offer today? In: Ocean Wave Spectra, Proc. Conf., London, Sydney, Toronto, Paris, Tokyo, Mexico City, Prentice-Hall Inc., 347-351.
Ueno, M., 1937. Branchiopoda. Fauna Nipponica 9: Tokyo Sanseidō. 135 pp.
Ueno, M., 1945. The littoral communities at the northwestern part of the main basin. The littoral communities of Lake Biwa 7. Seiri-Seitaigaku Kenkyugyoseki, Kyoto Univ., No. 30, 1-19.
Ueno, M., ed. 1973. Freshwater Biology of Japan, Tokyo, Hokuryukan, 760 pp.
Uyeno, T., 1965. On a cyprinid fish from a Pleistocene bed in Shizuoka Prefecture, Japan, and 'fossil species' problem. Rept. Jap. Soc. Syst. Zool. No. 1: 27-29.
Wakamatsu, K., 1963. The pollution biological class determined by the stone Aufwuchs of the Lake Biwa. Biol. Jour., Nara Women's Univ., No. 13, 1-2.
Watanabe, N. C., 1970a. Studies on three species of *Semisulcospira* in Lake Biwa. I. Comparative studies of shell form and habitat. Venus 29: 13-30.
Watanabe, N. C., 1970b. Studies on three species of *Semisulcospira* in Lake Biwa. II. Comparative studies of radulae. Venus 29: 93-98.
Wright, R. T. and Hobbie, J. E., 1965. The uptake of organic solutes in lake water. Limnol. Oceanogr. 10: 22-28.
Yamada, H., 1950. Theoretical estimation of meteorological high water. Rept. Res. Inst. Fluid Eng., Kyushu Univ. 6: No. 2, 22-33.
Yamaguti, H., 1938. A survey of the larger aquatic plants in the southern basin of Lake Biwa. Ecol. Rev. 4: 17-26.
Yamaguti, H., 1943. The littoral communities of Biwa-ko (Lake Biwa), 1. Aquatic vegetation. Jap. Jour. Limnol. 13: 92-104.
Yamaguti, H., 1955a. Bottom deposits and higher aquatic plants of Lake Yogo, north of Lake Biwa. Jap. Jour. Limnol. 17: 81-90.
Yamaguti, H., 1955b. Non-planktic water-bloom occurring in Lake Biwa. Jap. Jour. Limnol. 17: 141-148.
Yamaguti, H., 1956. Aquatic vegetation of Soné-numa, a small marshy lake close to Lake Biwa-ko. Jap. Jour. Limnol. 18: 93-109.
Yamaguti, H., 1960. The phytoplankton of the southern basin of Lake Biwa-ko. Jap. Jour. Limnol. 21: 315-326.
Yamamoto, G. and Kondo, J., 1964. Evaporation from Lake Towada. Jour. Meteorol. Soc. Japan, Ser. II 42: 85-96.
Yamamoto, G., Cheinm, B. T., Yasuda, N. and Kondo, J., 1972. Evaporation from deep lakes in Japan. Jour. Meteorol. Soc. Japan, Ser. II 50: 423-430.
Yamamoto, K., Ito, T., Mizuno, T. and Yamaji, I., 1966. Interim report of zooplankton. In: Biwa-ko Seibutsushigen Chōsa-dan (BST) Chūkan Hōkoku. Osaka, Kensetsu-shō, Kinki Chihō Kensetsu-kyoku. 469-495.
Yoshimura, S., 1935-1936. A contribution to the knowledge of deep water temperatures of Japanese lakes. Part I. Summer temperatures. Jap. Jour. Astr. Geophys. 13: 61-120.
Yoshiyasu, K., 1973. Starch-gel electrophoresis of hemoglobins of freshwater salmonid fishes in southwest Japan -. II. Genus *Oncorhynchus* (salmon). Bull. Japan Soc. Sci. Fish. 39: 97-114.

Radiometric age on lacustrine deposits

Susumu Nishimura

Introduction

Since the study of sediments of Lake Biwa is concerned with reconstructing the change of surroundings of the lake from its beginning approximately a few million years ago up to the present day, the provision of a chronological framework is a basic necessity. The development of physical dating techniques has therefore had an immense impact on the study of sediments of Lake Biwa since these techniques provide essential dates which can be used to establish the temporal relationship between the surrounding conditions, and thus to build up a world-wide chronology.

Physical methods of age determination are based on the decay of radioactive isotopes and variations in the direction and intensity of the earth's magnetic field.

Radioactive elements are unstable, the atomic nuclei decaying to a chemically different species with the emission of high energy radiation, typically alpha particles, beta particles and gamma rays — for example, the radioactive potassium isotope (^{40}K), referred to as the parent isotope, with the emission of beta particles.

Radioactive decay is a spontaneous process which is unaffected by the external environment and thus provides an ideal basis for age determination. The rate of decay at any specified time is proportional to the number of nuclei (N) present at that time:

$$dN/dt = -\lambda N.$$

where λ is the decay constant. The number of parent nuclei decreases progressively with time and the number (N) present at time t is given by

$$N = N_0 \exp(-\lambda t)$$

where N_0 is the initial number of parent nuclei at time $t=0$. Similarly, the number of stable daughter nuclei increases progressively with time and, assuming that no daughter nuclei are present at time $t=0$, the number (D) present at time t is given by

Horie, S. Lake Biwa

ISBN 90 6193 095 2. Printed in The Netherlands

$$D = N(\exp(\lambda t) - 1).$$

Alternatively, the radioactive decay process can be specified in terms of half-life (τ) which is thc time required for one half-life of a given number of parent nuclei to decay to stable daughter nuclei. The half-life is related to the decay constant (λ) by the equation

$$\tau = \frac{\ln 2}{\lambda} = \frac{0.693}{\lambda}.$$

Radiocarbon (^{14}C) which decays to nitrogen with a half-life of 5,730 years is continually formed in the upper atmosphere due to the cosmic ray bombardment of nitrogen atoms. It then becomes uniformly distributed throughout the atomosphere, the biosphere and the oceans.

Hence, the time that has elapsed since death can be determined from the measurement of the ^{14}C content of the dead sample (N) and the ^{14}C content of living material (N_0). Because of small variations in the ^{14}C content of living material in the past, there is a significant deviation between the radiocarbon age, obtained from the above measurements, and the true calender age. Therefore, in order to obtain a truly absolute chronology corrections, provided by measurement on samples of known age, must be applied to the radiocarbon ages. The most suitable types of samples for radiocarbon dating are charcoal and well-preserved wood, although leather, cloth, peat, shell and bone can also be used. Because of the comparatively short half-life of ^{14}C, radiocarbon dating is not applicable to samples with ages greater than approximately 50,000 years, the remaining ^{14}C concentration being too small for accurate measurement.

By careful selection of samples and the use of sophisticated measurement techniques, the potassium-argon and the spontaneous fission-track method can both be applied to volcanic deposits of lake sediments. In addition, various isotopes in the uranium decay series can be used for the age determination of lake sediments and bones.

When a volcanic material is formed, its initial concentration of argon ($\lambda = 0.58 \times 10^{-10}\ y^{-1}$) is usually zero since any ^{40}A, which had accumulated prior to the volcanic eruption, would have been expelled when the rock was in a molten state. Subsequently, the concentration of ^{40}A increases as a result of the decay of the radioactive potassium isotope, ^{40}K, using the modified equation

$$D = N(\exp(\lambda t) - 1),$$

the time that has elapsed since the formation of the volcanic deposits can be calculated from measurements of the concentration of ^{40}K (N) and ^{40}A (D).

Because of the high abundance of potassium in the earth's crust and the high sensitivity of the measurement techniques for argon, volcanic rocks formed as recently as 100,000 years ago can be dated by this method.

The spontaneous fission of the uranium isotope ^{238}U results in the formation of two heavy nuclei with very large kinetic energies. These energetic nuclei disrupt the crystal lattice of the minerals in which the ^{238}U occurs and produce damage tracks which are visible under the microscope after etching. Again, any damage tracks previously created are removed when the volcanic rocks are in a molten state so that the number of tracks present at the time of formation of a volcanic deposit is zero. Subsequently, tracks are created by spontaneous fission of ^{238}U and counting the number of tracks present in a rock sample provides a value for D, the concentration of the accumulated daughter product. The concentration of ^{238}U (N) is determined by counting the number of additional tracks produced by thermal neutron induced fission during a controlled thermal neutron irradiation in a nuclear reactor.

The method can be used to date natural glass. While, by separating out specific minerals (e.g. zircon), the age determination of volcanic ash is also possible.

In addition to radiocarbon dating, a number of other radioactive methods can be used to determine the age of sediments. These methods are based on the decay of intermediate radioactive isotopes in ^{238}U and ^{235}U radioactive series which have half-lives in the range of 10^4—10^5 years. Lal and Somayajulu (1975) also proposed that it would be very useful to determine precisely the chronology of lake sediments using cosmogenic ^{32}Si, and ^{210}Pb and ^{226}Ra radio-nucleide belonging to the ^{238}U decay series.

When clay or rocks are fixed to approximately 700°C and allowed to cool in the earth's magnetic field, they acquire thermoremanent magnetism. The evidence from paleomagnetism indicates that for certain periods of the geological past the earth's magnetic field was reversed with respect to the present direction (Cox 1969). The present normal polarity epoch is termed the Brunhes and this was preceded by the Matuyama reversed epoch, the transition occurring around 0.7 million years ago. During an epoch there are occasional occurrences of polarity events associated with reversals with respect to the primary direction of the epoch and lasting for 5,000 to 20,000 years. The polarity data for the past 2.5 million years, presented by Cox (1969), has been required principally from the measurement magnetism of volcanic lavas, the age determination of the lavas being achieved using the K–A method. These changes are synchronous throughout the world and therefore if any events were detected in the thermoremanent magnetism of volcanic ashes, they should provide approximate dates for these structures.

The polarity data has been used for the age determination of lake sediments. In this case, it is the direction of the detrial or depositional remanent magnetism that is measured; the sediments include magnetic particles which inherit a remanence from their parent rocks and which tend to be aligned by the earth's magnetic field as they settle (Harrison 1966). These studies on sediments of Lake Biwa are discussed by N. Kawai in another section of this book.

Radioactive ages on Lake Biwa sediments

^{14}C *Dating*

Because of its convenient half-life and the common and world-wide occurrence of materials containing carbon, the radiocarbon method has been extensively applied to the field of archaeology and anthropology. More recently the method has been extended to study the rate of sedimentation of ocean or lake sediments, late Pleistocene geology and climate, and also the terrestrial ages of meteorites. The excellent agreement was obtained between the age of known historical samples and radiocarbon age. However, there are many radiocarbon dates that are not in agreement with historical ages and caution is needed in interpreting results. Some of these errors are possibly related to cumulative errors in historical chronology; ages greater than 4,000 years are subject to large errors as this marks the first astronomical fix. The following data on the sediments of Lake Biwa are obtained by Isotopes A Teledyne Co., USA (Horie et al. 1971). These results are

0.8 ± 0.05 m depth; $1{,}430 \pm 95$ years,
4.5 ± 0.15 m depth; $3{,}650 \pm 105$ years,
11.5 ± 0.2 m depth; $14{,}980 \pm 460$ years

and also shown in Fig. 179.

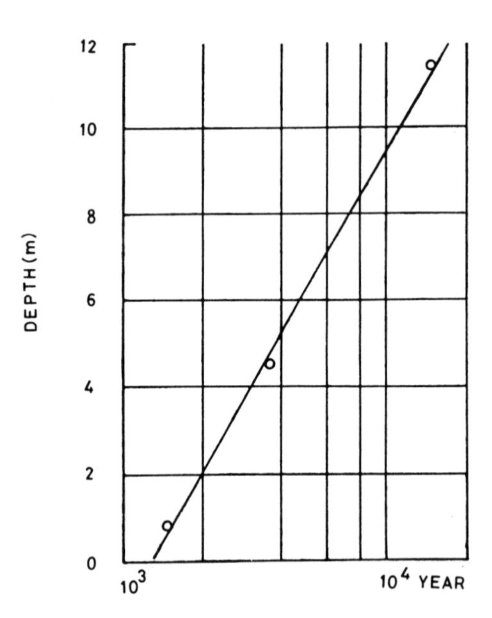

Fig. 179. Sedimentation rate in the center of Lake Biwa inferred from the C^{14} data on the core sample. (Horie et al. 1971).

Fission-track dating

In Lake Biwa, there accumulated abundant sediments named the Kobiwako Group. A 200 m core sample has been collected from this lake. This sample is mainly composed of clay sediments containing some volcanic materials. Such volcanic ashlayers, if identified correctly, will serve the purpose of the key-layers for the regional geologic chronology. They are at least thirty volcanic seams, as shown in Fig. 180-a. The features and radiogenic ages of these volcanic ashes

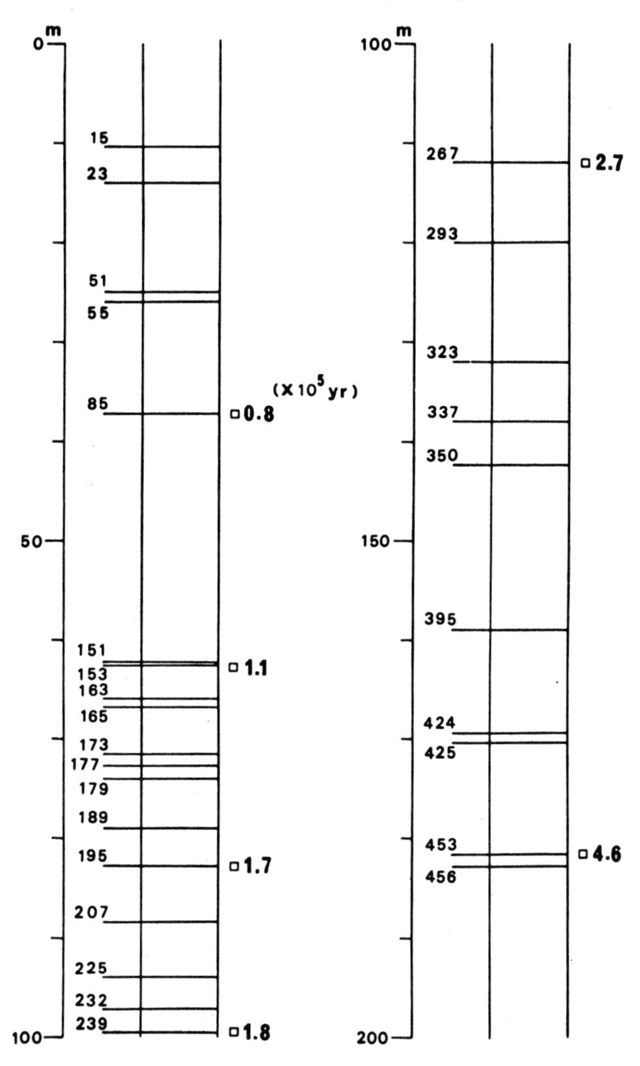

Fig. 180–a. Column of the 200 m core sample in Lake Biwa and sampling horizons of volcanic ashes for the fission-track dating (horizontal lines indicate volcanic seams). (Nishimura & Yokoyama 1975.)

are very important as a foundation of all other works concerning the Quaternary geochronology. The volcanic ashes were chosen as the objective materials and the fission-track technique was applied to them. Since the volcanic materials in this core had deposited under water, there is no doubt that the crystals in the volcanic layers were unaltered after the volcanic ejection and deposition.

The recent development of the fission-track technique for dating minerals and glasses offers a new tool for the purpose of dating tuffs and ash-layers. There are usually at least one or often more grains of zircons in tuffs and ash-layers; they can be used for the fission-track method of dating.

Zircon, because of its high uranium impurity content, its fragment occurrence in common rocks and its high temperature stability, is suitable for the application of the fission-track dating method. The developments by Krishnaswami et al. (1974) and Nishimura (1975) of a suitable etchant for revealing tracks in zircon has made it possible for us to determine zircon ages.

The method depends on the spontaneous fission of ^{238}U atoms in mineral or glass, taking place at a constant rate and leaving fission-tracks, as mentioned above. Once formed, the fission-tracks disappear if the material is heated above a critical temperature. The fission-track age, T years, can be represented by the following equation (Fleischer & Price 1964):

$$T = \frac{1}{\lambda} \ln \left[1 + \frac{\lambda}{\lambda_f} \frac{\rho_s}{\rho_i} \frac{\phi_\sigma}{\eta} \right]$$

where ρ_s is the fossil fission-track density (cm^{-2}), ρ_i is the induced track density by the bombardment with thermal neutrons (cm^{-2}), λ is the decay constant for uranium (y^{-1}), λ_f is the fission decay constant for ^{238}U (we used 6.85×10^{-17} y^{-1}; Nishimura 1975), σ is the thermal neutron cross section for fission ^{235}U (cm^2), ϕ is the thermal neutron dose (cm^{-2}), and η is the isotope ratio $^{235}U/^{238}U$. If T is smaller than 10^9 y, the equation can be written.

$$T = 6.12 \times 10^{-8} \, \phi \frac{\rho_s}{\rho_i}.$$

The volcanic ash layers in the 200 m core sample are chiefly composed of volcanic glass flakes and pumice grains. They are roughly divided into two types from the petrological viewpoint. Some of the volcanic ashes are andesitic and others are more acidic, perhaps rhyolitic. The former contains two pyroxenes, hornblende and iron ore, as the main heavy minerals in the volcanic glass. The latter generally contains orthorombic pyroxene, hornblende, biotite and quartz as crystalline components in the volcanic glass and pumice grains. In addition to the above-mentioned minerals, they contain other crystalline minerals such as apatite, zircon, feldspar, etc. The samples from which Nishimura and Yokoyama (1973, 1975) could successfully obtain the fission-track ages, are BB85, BB153, BB195, BB239, BB267 and BB453.

Zircon was separated from these samples using the standard separating method with heavy solutions and isodynamic separator. Zircon required 1 to 2 min. in concentrated H_3PO_4 at 450° – 500°C or 4 to 10 hr. in 1 : 1 mixture volume of 52% HF and 98% H_2SO_4 at 190°C in a teflon capsule within a screw type stainless container for etching. After the sample had been etched and washed, it was ready for counting.

For the reactor run, a small amount of zircon was packed in a plastic container and irradiated in a reactor of Kyoto University. Neutron flux was also obtained by the fission-track method.

Ages of zircon from six ash-layers in the 200 m core sample of Lake Biwa are given in Table 56 and in Fig. 180–b. Zircon has so high a uranium impurity content that the young age of this mineral can be measured with fairly good accuracy. It is found that there are wide variations in uranium concentrations among zircon in each of the samples of Lake Biwa, which leads us to believe that the crystals with suitable uranium concentrations and uniform distributions can be selected from samples.

Current studies of the effect of high temperature and high pressure on track storage time show that zircon is the most retentive of all the minerals we have studied. The tracks will survive for a million years in zircon at 300°C. The accuracy of these data is such that their sampling errors are within 20%.

^{210}Pb Dating

^{210}Pb, a member of the ^{238}U radioactive series with a halflife of 22.3 years is a useful method for geological processes over the past one hundred years. ^{210}Pb is involved in sedimentary processes through the following series of events. The noble gas nucleide ^{222}Rn diffuses out of the earth's crust. ^{222}Rn decays in the atomosphere and ^{210}Pb is formed. Lead together with any ^{210}Pb is removed from the atomosphere with precipitation. The lead in water reservoirs is quickly scavenged to the sediment floor. As the atomosphere-delivered ^{210}Pb activities in the sediments decay along a half-life of 22.3 years, the sedimentation rate is then determined. The chronological work of Lake Biwa with ^{210}Pb was reported by Matsumoto (1975) as follows.

Table 56. Fission-track ages of zircon from the core sample of Lake Biwa (Nishimura & Yokoyama 1974, 1975).

Sample No.	Depth (m)	Spontaneous fission-track density (cm^{-2})	Induced fission-track density (cm^{-2})	Neutron dose (cm^{-2})	Fission-track age (years BP.)
BB 85	37	1.7×10^4	7.2×10^6	0.55×10^{15}	8.0×10^4
BB 153	62	2.1×10^4	1.3×10^7	1.11×10^{15}	1.1×10^5
BB 195	82	1.5×10^4	6.0×10^6	1.1×10^{15}	1.7×10^5
BB 239	99	6.7×10^4	2.5×10^7	1.10×10^{15}	1.8×10^5
BB 267	110	2.9×10^4	7.2×10^6	1.1×10^{15}	2.7×10^5
BB 453	181	5.4×10^4	7.9×10^6	1.1×10^{15}	4.6×10^5

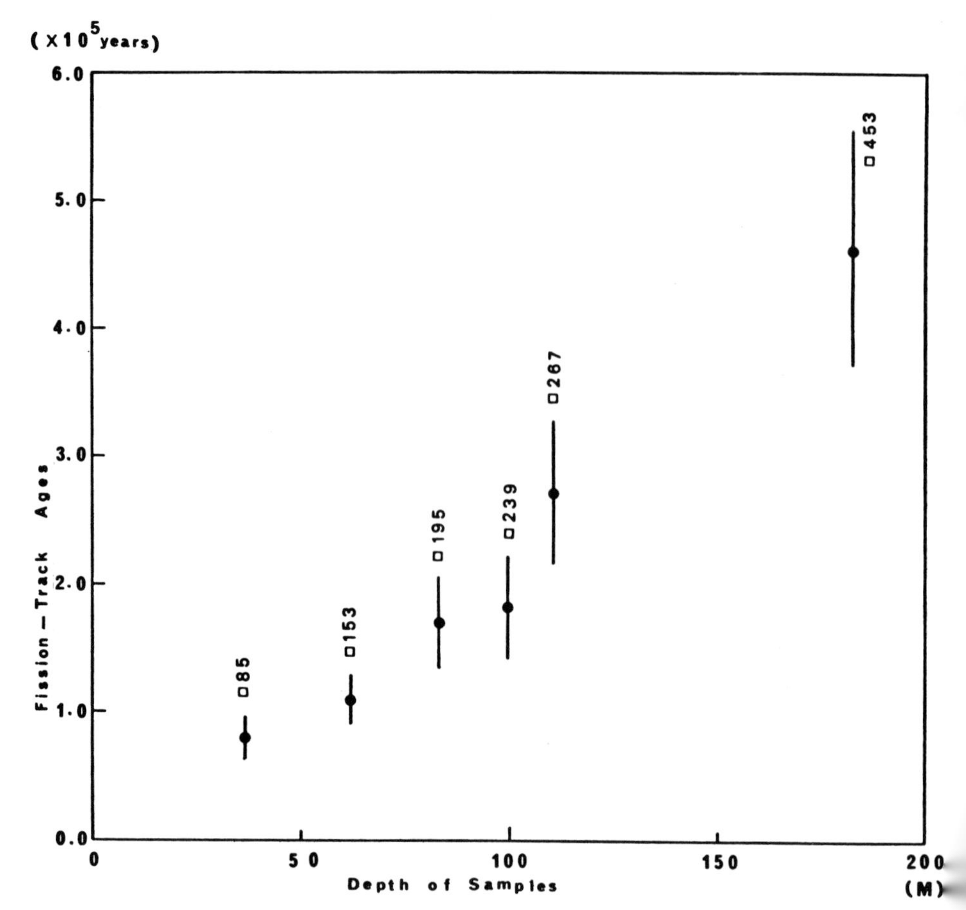

Fig. 180–b. The relation between fission-track ages and horizons of volcanic ashes in 200 m core sample in Lake Biwa. (Nishimura & Yokoyama 1975.)

Three core samples, 2P, IG and 3G, were collected from Lake Biwa by a Horie sampler and a manual type grab sampler. Core 2P was collected in November 1973 near the center of the northern part of north basin of Lake Biwa from its 90 m water depth. Core 1G was collected in the center of the southern part of the north basin from its 70 m water depth; and core 3G was in the center of the south basin in the shallow area of 5 m water depth in August 1973 (Fig. 181).

Immediately after their collection, the cores were cut into portions, 2–3 cm long, and dried at 110°C. The weight loss was recorded so as to measure the water content in the sediments.

The dried sample, typically 3–4 g, was leached with 30 ml of 6N HCl and 3 ml of H_2O_2 on a hot plate for 1 hr. Pb^{2+} equivalent to 4 mg PbO_2 was added for the extraction of ^{210}Pb. The supernatant was analyzed for ^{210}Pb activity. The

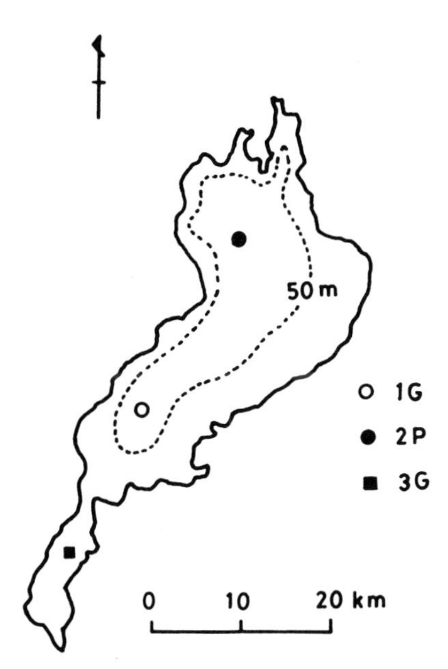

Fig. 181. Bathymetric map of Lake Biwa and sites of core used in this investigation. (Matsumoto 1975.)

leachate was taken to dryness and solids were dissolved in 5% HCl with excess KI. The lead was extracted by methyle-iso-propyl ketone and back-extracted by 5% HCl. The solution was taken to dryness and solids were dissolved in 1.5 ml HCl. The solution is then passed through an anion exchange column (Dowex 1 × 8, 50–100 mesh). The column is washed with three volumes of distilled water. Finally, the lead is electrodeposited on a platinum disc as PbO_2. The electro-deposition of lead was carried on for 60 minutes in 4N HNO_3 under the conditions of 70 mA and 3V. The deposited platinum disc was dried at 110°C to constant weight, and prepared for radioactive assay. The overall recovery of lead by this procedure varied 50 to 80%. The ^{210}Pb radioactive assay was made utilizing the growth of the ^{210}Bi daughter from ^{210}Pb decay. The samples were counted in a low background gas flow beta-counter after sufficient growth of ^{210}Bi. The counting efficiency (23%) for ^{210}Bi was ascertained by counting a standard ^{210}Pb source. A blank of ^{210}Pb through the analytical procedure could be detected.

From the results, Matsumoto concluded as follows: The unsupported ^{210}Pb values, defined as $^{210}Pb_{exc.}$ were determined by subtracting the contribution due to radium-supported ^{210}Pb in the detritus minerals from the total observed ^{210}Pb analyses in the deep strata of the sediments, where the $^{210}Pb_{exc.}$ have completely decayed. If the collected core is long enough, we can obtain a reasonable background level of supported ^{210}Pb. The $^{210}Pb_{exc.}$ value in the surface sediments of core 2P is 13 dpm/g of the dried sediments, compared to the

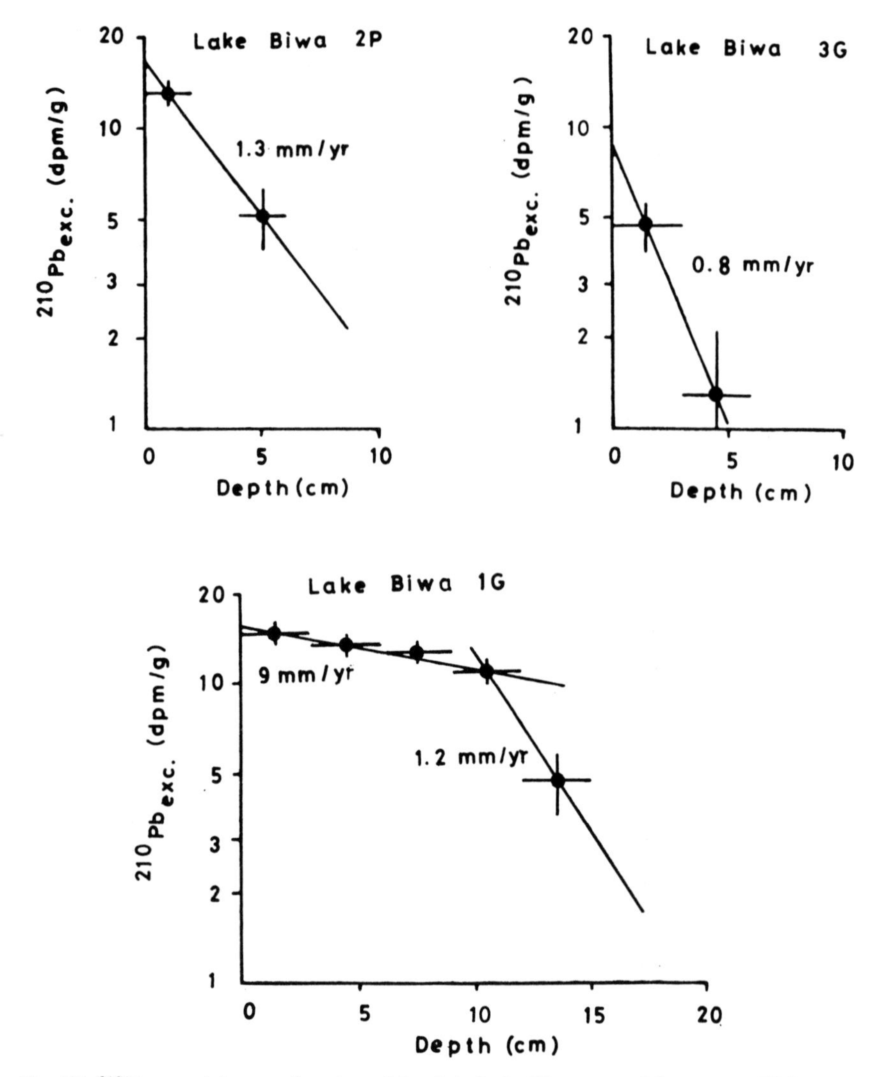

Fig. 182. $^{210}Pb_{exc.}$ activity as a function of depth in Lake Biwa cores. (Matsumoto 1975.)

supported ^{210}Pb of 3.8 dpm/g. In core 3G, $^{210}Pb_{exc.}$ is about 5 dpm/g, compared to the supported ^{210}Pb of 2.4 dpm/g. In core 1G, the core length of 15 cm did not permit us to obtain a reasonable background level of supported ^{210}Pb, and therefore we assumed that the supported ^{210}Pb value in core 1G is 3.1 dpm/g, which is the average of those obtained in cores 2P and 3G.

The values of $^{210}Pb_{exc.}$ are plotted against depth in Fig. 182. The sedimentation rates are calculated from the slope of the straight line. The sedimentation rate in core 2P from the $^{210}Pb_{exc.}$ data is 1.3 mm/yr. The sedimentation rate in core 3G is 0.8 mm/yr., but the calculated rate could be significantly affected by a background correction, since the observed $^{210}Pb_{exc.}$ values are relatively small. The $^{210}Pb_{exc.}$ value in the surface sediments of the core 1 G is about 15 dpm/g, and $^{210}Pb_{exc.}$ gradually decreases with depth to the 10 cm level, but abruptly decreases at more than 10 cm. The changes in the slope represent the changes in the sedimentation rates, and the rate obtained is 9 mm/yr. at the level shallower than 10 cm and 1.2 mm/yr. at the level deeper than 10 cm.

Paleomagnetic study of the Lake Biwa sediments

Naoto Kawai

Introduction

For more than the three recent decades there has been a growing interest in the post-depositional remanent magnetism (PDRM) of sedimentary strata that have been formed in both the sea and lakes. Recently, Kent (1973) experimentally examined a hypothesis that the remanent magnetic vector of such a stratum represents faithfully the three elements of the past geomagnetic field under which each PDRM occurred. Consequently, if one were able to obtain a vertical core penetrating such strata and succeed in preparing a series of continuous sections, PDRM of each of which is measurable, he could certainly clarify a long geomagnetic history since the time when sedimentation started in each geologic basin.

It was Johnson et al. (1948) who published for the first time a study along the above-mentioned line of thought. They successively utilized varved clay as their samples in 1948. Since then, a number of papers have successively been published by investigators such as Nagata et al. (1949), Kawai (1951), Griffiths et al. (1960), Opdyke (1972), Denham and Cox (1971), and Creer et al. (1972).

When Horie succeeded in coring out a long boring column from Lake Biwa in 1971, he asked the present writer to make use of it for the palaeomagnetic study. Several small blocks were tested with a sensitive astatic magnetometer in Osaka University. It was then found that each had a weak but stable permanent magnetism of which the direction and intensity were easily determinable without sacrificing accuracy.

Worthy of mention in the case of Lake Biwa sediments is that the core thus obtained is quite uniform everywhere in the column, looking like dark grey lode composed mainly of fine clay particles and the gyttja, although there exist about thirty thin layers of volcanic ashes sandwiched in the entire length of the core (197.2 m). This shows that the sedimentary layers, except the above-mentioned ashes, have been stratified not only continuously but also gradually in the whole cycle of deposition in the lake bottom. The mean rate of deposition is estimated,

Horie, S. Lake Biwa

ISBN 90 6193 095 2. Printed in The Netherlands

as is to be described later, to be 0.0426 cm/yr. The value is several hundred times greater than that of the sediments formed in the deep-sea bottom. Consequently, it is possible to compile a continuous palaeomagnetic record with a very high resolution power, when the sections prepared from the lake core will all be measured from the top to the bottom and arranged in the descending order.

This paper reports the writer's magnetic investigation of the lake sediments. Its last section shows similar magnetic data that we obtained from two deep sea cores collected from the Meranesian Basin in the Central Pacific. As the age of the lake sediments and that of the ocean sediments overlap in the middle part of the time span, it is possible to compare the magnetic data at different parts of the earth's surfaces. This comparison will shed light upon the question as to whether or not the magnetic aspects that the writer has obtained are world-wide phenomena.

The geomagnetic record of the uppermost layers

On the uppermost part of the lake or ocean sediments there lies an oozy layer made of many small particles still movable in the free space that remains in the sediments. This layer bears, of course, no permanent magnetic moment or what we call natural remanent magnetism, NRM. As the depth increases below this layer, the gap in which particles make Brownian motions becomes reduced so effectively that even the contained magnetic particles soon become fixed and embedded within the sediments, regardless of the existence of the applied geomagnetic field. The layer having such magnetic particles and stable NRM is important for the present study.

It, however, is difficult to assume theoretically the boundary above which we have unstable NRM and below which we have stable NRM. At the present stage of our study, the best estimate of the depth of the boundary seems to be deducible from actual magnetic observations.

The palaeomagnetic data obtained from the uppermost part of the core (see the curve a in Figs. 183 and 184) should represent the recent secular geomagnetic variation, although we do not know how old the variation is.

On the other hand, the historical secular variation called archaeomagnetic variation is obtainable independently by using baked earths of known historical ages. The declination and inclination of the field in the vicinity of Lake Biwa have been studied by Hirooka (1971) and the present writer as shown by the curve b in Figs. 183 and 184.

The curves a and b resemble each other considerably. The inclination curve obtained from the sediments shows a peak value at the depth of about 5.3 m, whereas in the curve from the archaeomagnetic study there appears a similar peak in the past about 1,300 BP. Besides, in the shape of both curves we can easily confirm a good agreement traceable to the left-hand side of the abscissa. The youngest archaeomagnetic data of inclination is about 100 years old. It,

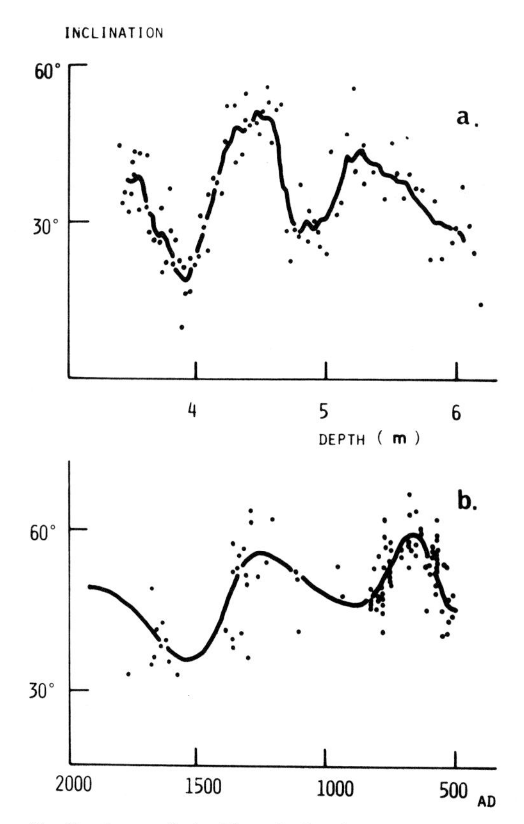

Figure 183. Change of inclination. a: Lake Biwa, b: Southwest Japan (after Hirooka, 1971) (after Nakajima & Kawai 1974).

therefore, is likely that the very recent geomagnetic field not older than 100 years has been preserved as PDRM in the uppermost layer of the obtained core.

The resemblance of the two curves is seen not only in the inclination change but also in the declination change as shown in Figs. 183 and 184. The upper part of the core is still soft and compressible, while the lower part has become reasonably harder. It may have received a significant compaction under the gravitational load. The curve showing the inclination change, therefore, is much deformed from its true state. Furthermore, we have no exact data on the possible change in the sedimentation rate of the lake basin. In this case, one must determine the age of sediments on an assumption that the maxima or minima of the inclination or declination has the same age of the respective values in the archaeomagnetic data.

As a result of this comparison, the relation between the age and the sedimentation rate has been obtained as shown in Table 57. Besides the directional change, the intensity variation has been determined as shown in

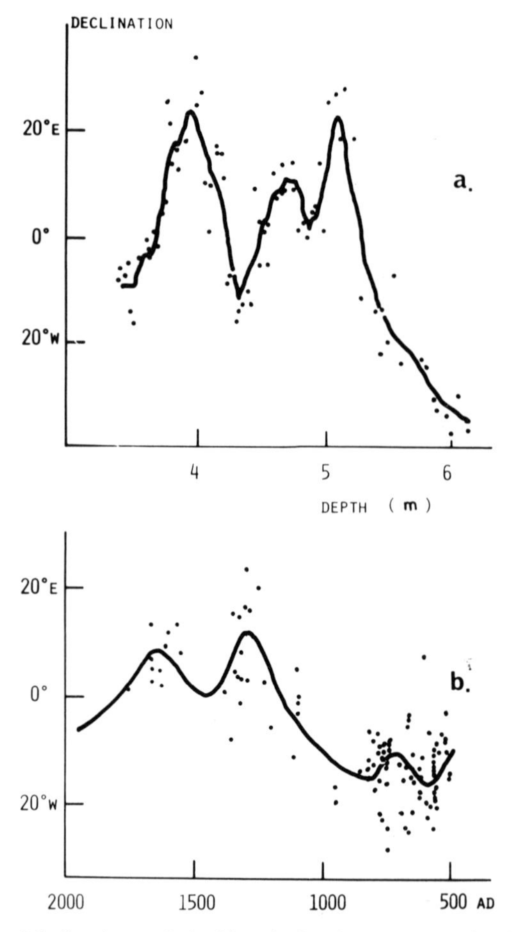

Figure 184. Change of declination. a: Lake Biwa, b: Southwest Japan (after Hirooka, 1971) (after Nakajima & Kawai 1974).

Table 57. Relation between the age and the sedimentation rate (Nakajima & Kawai 1974).

Depth (m)	Age (years BP.)	Rate of Deposition (mm/yr)
3.4	0	1.25
3.9	400 (I_{min}, D_{max})	2.00
4.2	550 (D_{min})	2.00
4.5	700 (I_{max})	2.00
4.6	750 (D_{max})	1.14
5.0	1100 (I_{min})	1.50
5.3	1300 (I_{max})	
	mean	1.65

Fig. 185 in which the three upper curves are the archaeomagnetic intensity data obtained from various parts of the world (Bucha 1971).

The above-mentioned fact suggests that the geomagnetic secular change can be traced faithfully back to the remote past of the Quaternary period with almost no gap in the record, when the NRM of an obtained column will be measured.

Stability test

NRM of each sample was measured by an astatic and a spiner magnetometer. Before measurements were taken, the stability of NRM was examined by means of the progressively increasing AC field of demagnetization. It was made clear

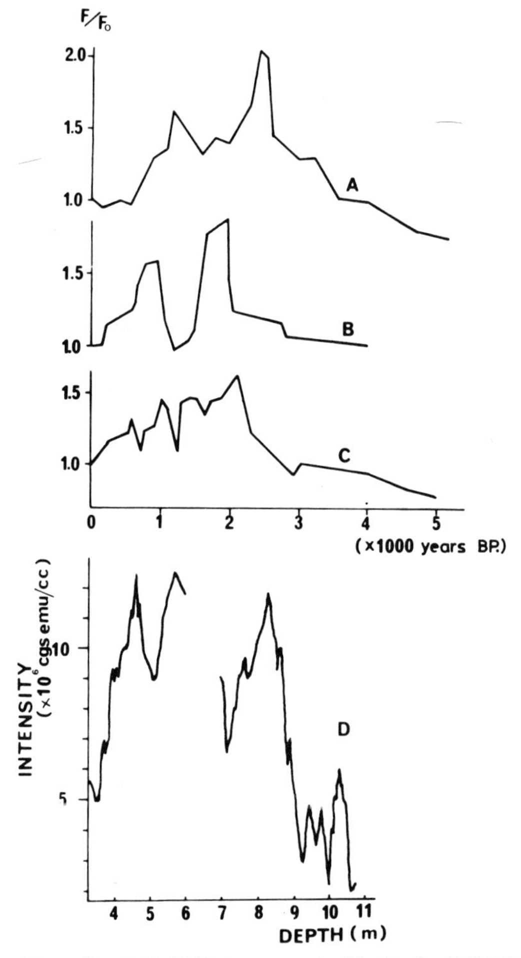

Figure 185. Change of intensity. A, B, C: Data summarized by Bucha (1971) from Czechoslovakia, America, and Japan, respectively. D: Lake Biwa (after Nakajima & Kawai 1974).

that the stability of NRM was considerably high. The mean value of the half destructive demagnetization field was as high as 400 Oe in many specimens (Fig. 186). In stable specimens, no remarkable change of direction was found on the increasing alternative field up to 600 Oe (Fig. 187–a). But in some of the specimens a small change of the direction was observed in the initial stage of the application of the increasing alternative field up to 100 Oe (Fig. 187–b). So, we have decided to use the demagnetization field with a peak field of 100 Oe for magnetic cleaning and every sample had been treated before measurements were made.

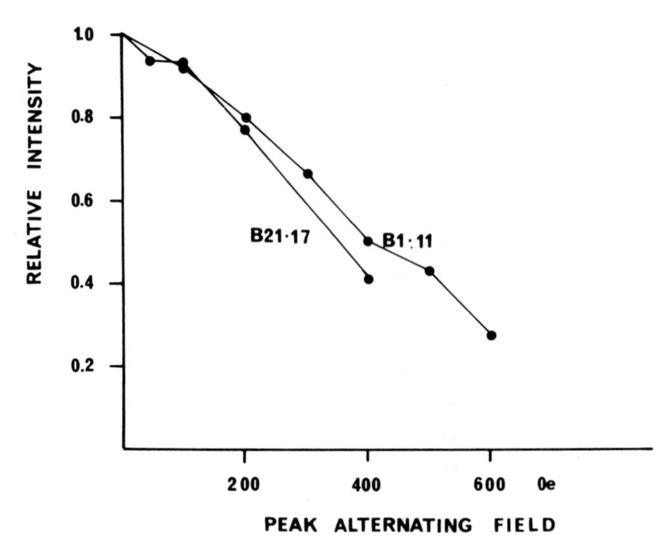

Figure 186. Two examples of AF demagnetization (intensity change) (after Yaskawa et al. 1973).

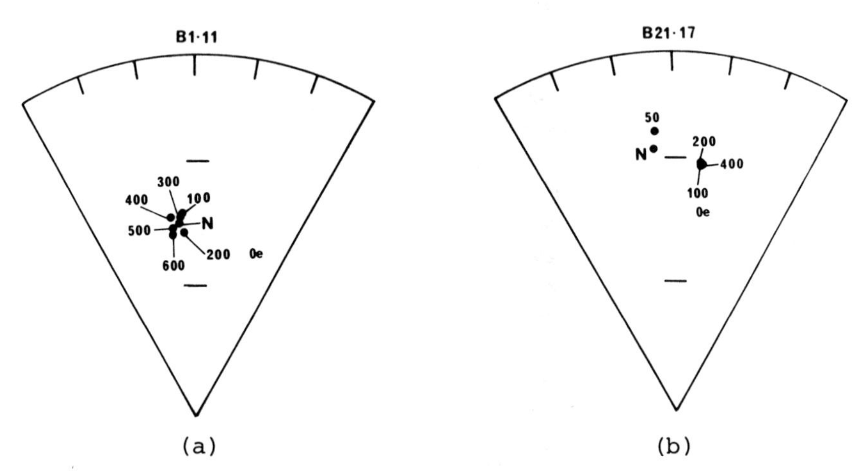

Figure 187. Two examples of AF demagnetization (directional change) (after Yaskawa et al. 1973).

It is hardly possible to extract ferromagnetic particles responsible for the remanent magnetism of the sediments. The difficulty is due both to the scarcity of particles and their very small mean diameters. They can hardly be moved in a non-uniform magnetic field, since the attractive force is extremely reduced because of the small particle size, however stronger the magnetic force may be. Furthermore, it becomes almost impossible to separate them from small-sized non-magnetic organic matters that coexisted. So, we still do not know of what materials or in what shape each particle is made up.

Four geomagnetic events and many excursions in the Brunhes normal epoch

In the 200 m core, about 40 sampling horizons were chosen in such a manner that they may spread over the entire core as uniformly as possible with a nearly equal separation by approximately 5 m each, then three specimens were cut off successively along the core axis from each horizon, and measured after the magnetic cleaning using the AC field mentioned above.

Although most of the inclinations we observed have positive values and fluctuate smoothly from point to point in the diagram in Fig. 188, there appear at least four regions A, B, C and D in which one positive value jumps suddenly to a negative one. When more specimens were taken and examined carefully at each region, many transient inclinations (represented by square dots) were discovered. The geomagnetic field itself seems to have changed the sign so frequently in the geologic time.

It has long been considered that one normal geomagnetic field has persisted in

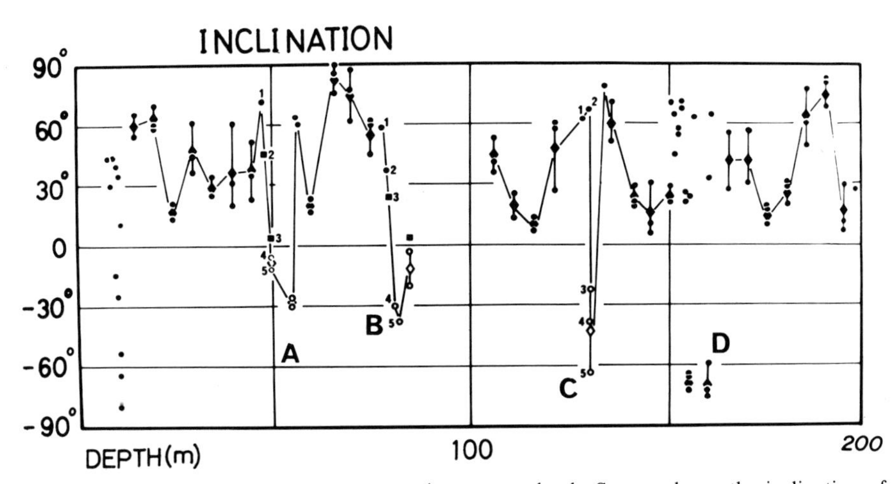

Figure 188. Inclination of permanent magnetism versus depth: Square shows the inclination of transient magnetization. Hollow circle shows each inclination of the reverse fields of which the polarity was confirmed also from the declination check. The mean value is shown with rhombus (after Kawai et al. 1972).

the entire Brunhes epoch since 0.69 million years ago. Smith and Foster (1969) discovered from the sediments collected from the Atlantic and Caribbean basin one abnormal event called the Blake event in which the normal field was broken up and replaced by a reverse field (since 114,000 years BP. till 108,000 years BP.). Unfortunately, no direct evidence was found as to whether the reversal of the field is a world-wide phenomenon or not.

The age of the youngest of the four events, which we observed and which are to be described later in this paper, is so close to the age of the Blake event that it is likely that a world-wide geomagnetic reversal occurred.

The effect of the reversal seems to have been discovered at different parts of the earth surface. The second geomagnetic short reversal is called the Biwa I event, while the third and the fourth event are called the Biwa II event and the Biwa III event respectively.

Besides the four geomagnetic events, other types of magnetic anomalies were discovered from the continuous sampling and measurements we carried out in the upper part of the core down to 60 m depth. A cube with its edge length 2 cm was used as one sample, and the gap that occurs between two adjoining samples was reduced as small as possible as shown in Fig. 189. As shown in Figs. 190 and 191, both inclination and declination of NRM during this type of anomaly made sudden flips from the respectively normal positions for a very short interval less than 1 or 2 millenius. When the virtual pole positions during the shift were plotted on an equal area projection, the pole position made a long distance excursion from the point near the north pole towards the south pole through the equator, as shown in Fig. 192. The deviation of the pole soon became reduced.

Next to be mentioned is the fact that the intensity of the NRM during some excursions drops abruptly down to approximately one tenth of that of the field in the normal epoch.

The youngest excursion that we found from Lake Biwa, is at the depth of 13 m, and 18,000 years old, while the second youngest one, is found at the depth of 26 m and 49,000 years old. More than 5 similar geomagnetic short aspects have been observed in the core down to the depth of 60 m. Results of the magnetic measurement from the upper part of the core downwards are shown in Fig. 193. Both the intensity and inclination changes with the depth are shown. The mean value of both intensity and inclination is obtained out of 5 vertically adjoining specimens and plotted in Fig. 194.

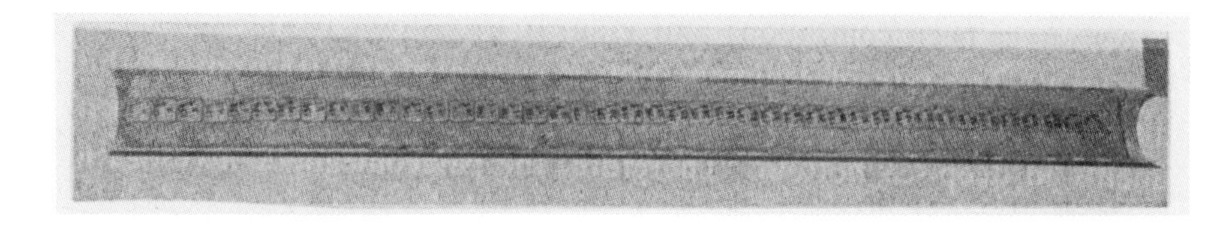

Figure 189. A vertical section of a core with plastic cubes inserted into the core.

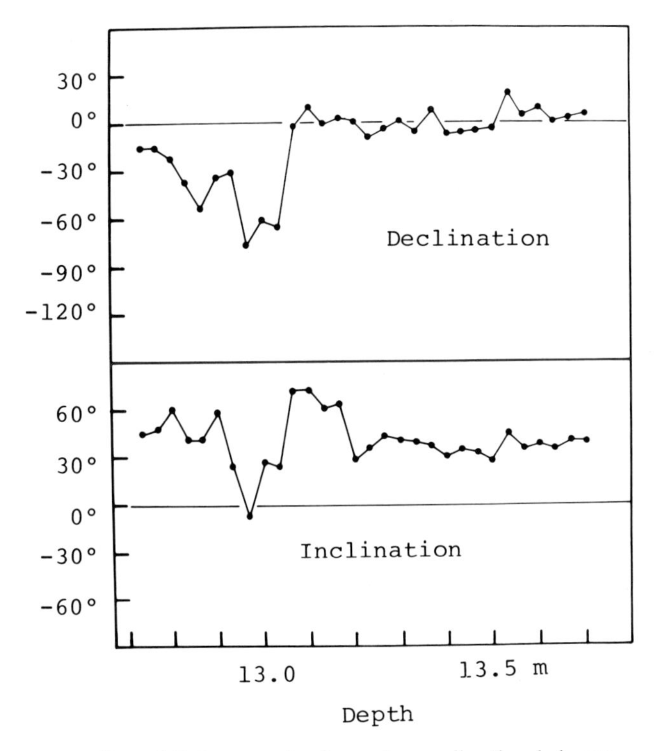

Figure 190. An example of conspicuous directional change.

Depth vs age relation in the core

Although we can date sedimentary layers in the uppermost part of the core on the basis of the archaeomagnetic data, the dating becomes difficult as soon as the depth increases beyond 6 m below the surface. The most reliable method is to utilize the ash layers containing heavy minerals called zircon. The crystals, when separated, can be dated by means of the fission-track technique.

Nishimura and Yokoyama (1973, 1975) could apply this method to the six ash layers in the sediments. The zircon crystal grains were separated by using heavy solution and an isodynamic separator. After etching the surface of each crystal they counted the number of spontaneous fission-track per unit surface area. This number, together with the total amount of the uranium impurities in the crystal (obtainable independently), gives the authentic time length since the crystals met their quenching from high temperature (from the time of the eruption of each volcanic ash in the case of sediments).

The depth *vs* the fission-track age relation thus obtained is shown in Fig. 195. On the other hand, the age is obtainable from the state of compression under the gravitational load in the prolonged duration. Kanari and Takenoya (1975) estimated the age-depth relation as shown by the smooth curve in Fig. 195. It

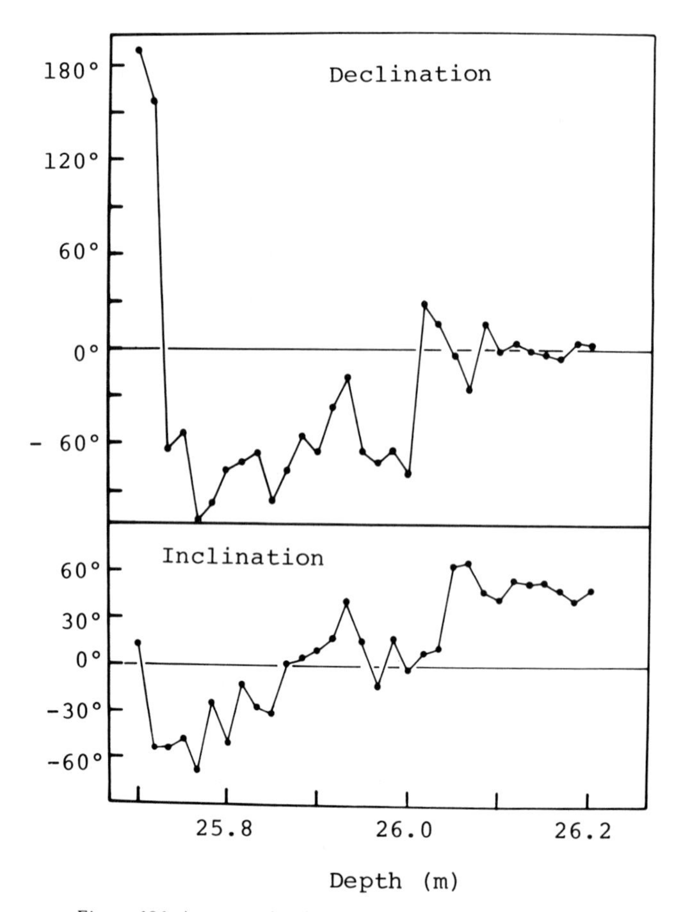

Figure 191. An example of conspicuous directional change.

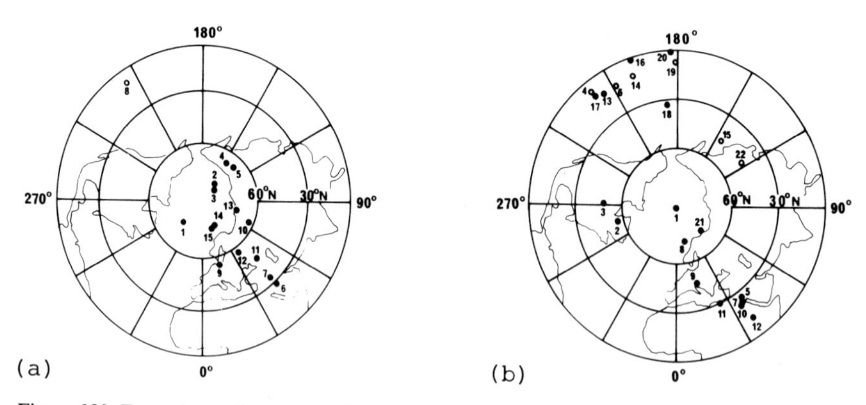

Figure 192. Excursions of geomagnetic north pole (after Yaskawa et al. 1973).

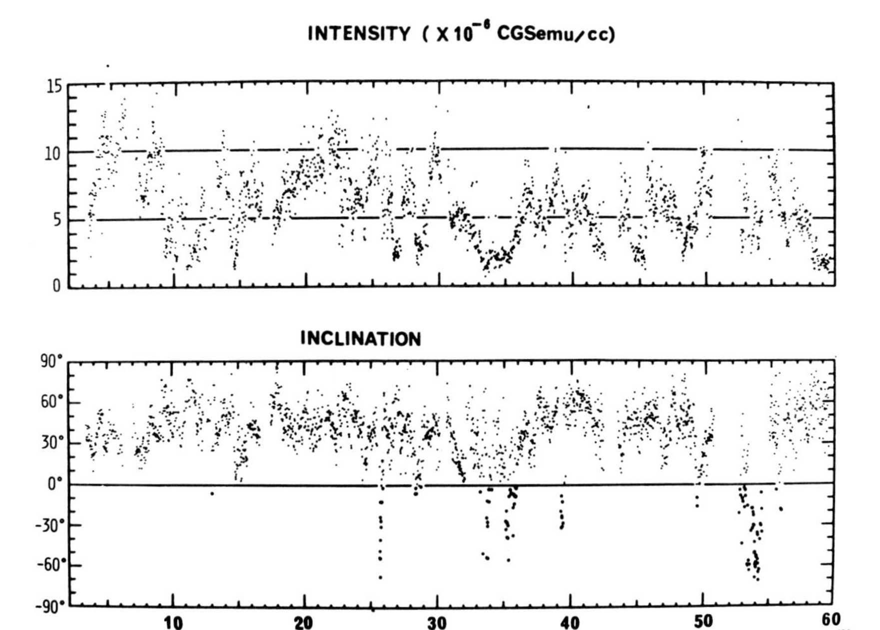

Figure 193. Results of magnetic measurements (after Kawai et al. 1975).

then becomes possible to make both the extrapolation and interpolation of this curve in order to obtain the ages of the four geomagnetic events and many excursions.

The age of the youngest geomagnetic event turns out to be 100,000 years BP. according to the above-mentioned estimation, while the ages of Biwa I, Biwa II, and Biwa III are found to be 160,000 years BP., 310,000 years BP. and 380,000 years BP. respectively.

As mentioned in the previous section, the age estimated for the youngest event and the age of the Blake event determined by Smith and Foster (1969) become quite close to each other.

It is interesting now to compare the magnetic data obtained by Wollin et al. (1971) from the sediment as precise as possible as shown in Fig. 196. The event appears at an age of 100,000 years BP. in the ocean sediment, while it appears in the lake sediments at the depth of 50 m.

When Wollin et al. made the magnetic observation, the resolution power of the ocean record was too weak for them to recognize each magnetic anomaly in their record as the geomagnetic event in the past. However, it is now possible to estimate these inclination anomalies to be those corresponding to the events clarified from the Lake Biwa sediments as compiled by Nakajima et al. (1973) and Yaskawa (1973).

On the other hand, one geomagnetic anomaly (a big inclination shift) was discovered from an observation of an upper layer of the ocean core collected

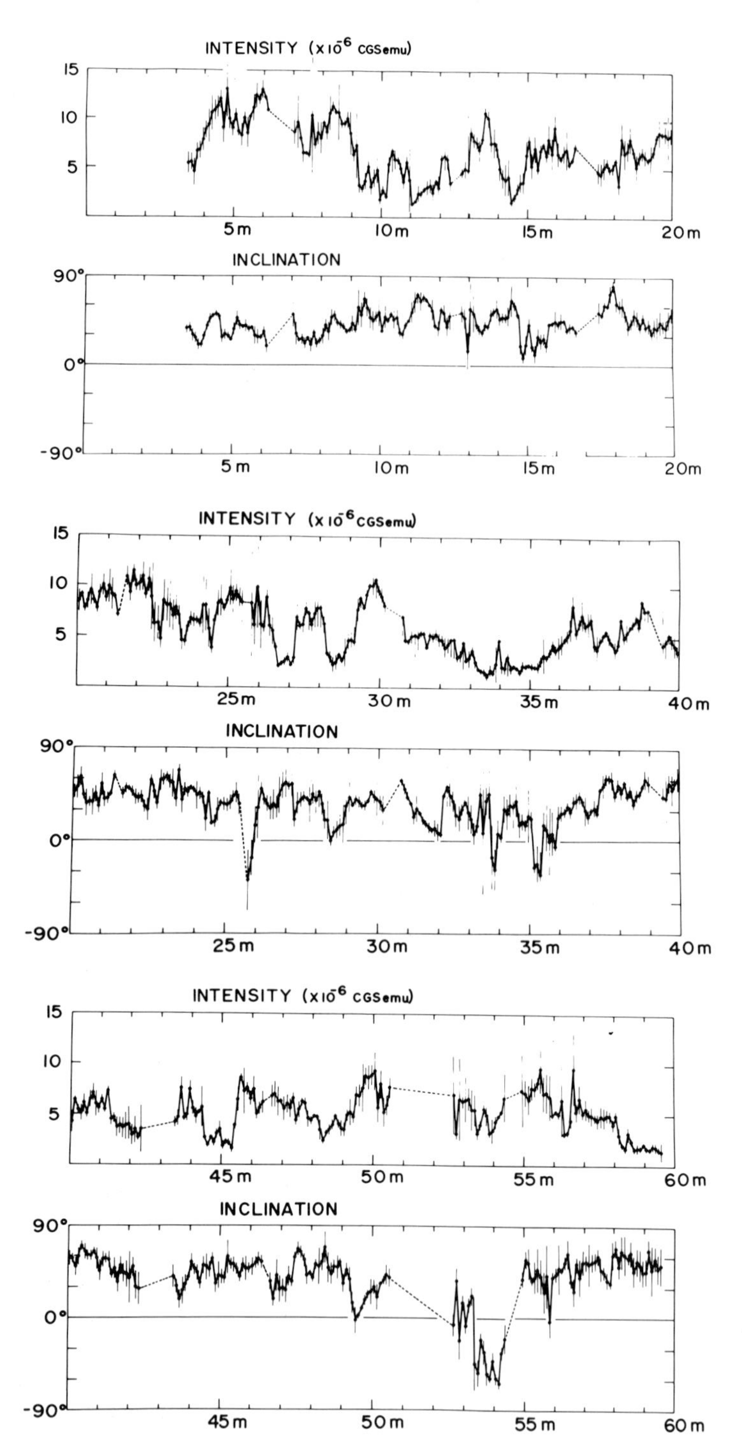

Figure 194. Intensity and Inclination *vs* depth. The curves were obtained based on the 5-pt. moving average (after Kawai et al. 1975).

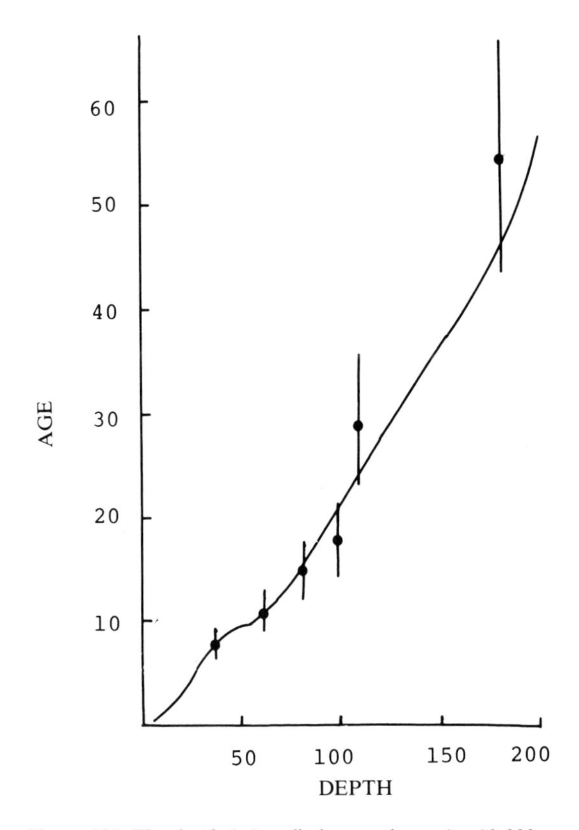

Figure 195. The depth (m) *vs* fission-track age (× 10,000 years).

from the Meranesian Basin in the Central Pacific (Kawai et al. 1976). The estimated age of anomaly, 0.33 million years BP., is very close to that of the Biwa II event. If the two ages should be the same, the Biwa II event is indeed a world-wide geomagnetic phenomenon, too.

As these geomagnetic events and excursions are world-wide phenomena, their respective ages that have been determined and such anomalies should appear simultaneously in any geological basin, and each stratum should be used as a standard bed or a key bed with which the stratigraphy of the geological basin can be constructed. The magnetic information we gathered from the lake basin is very useful in the Quaternary geology.

Comparison of palaeomagnetic study with geochemical, isotope, and pollen analysis

Various elements consisting of such organic materials as carbohydrate, protein, and lipid were extracted by Koyama, Handa, Ishiwatari and Ogura from the same boring core with which we studied palaeomagnetism (Koyama T. 1972, 1976;

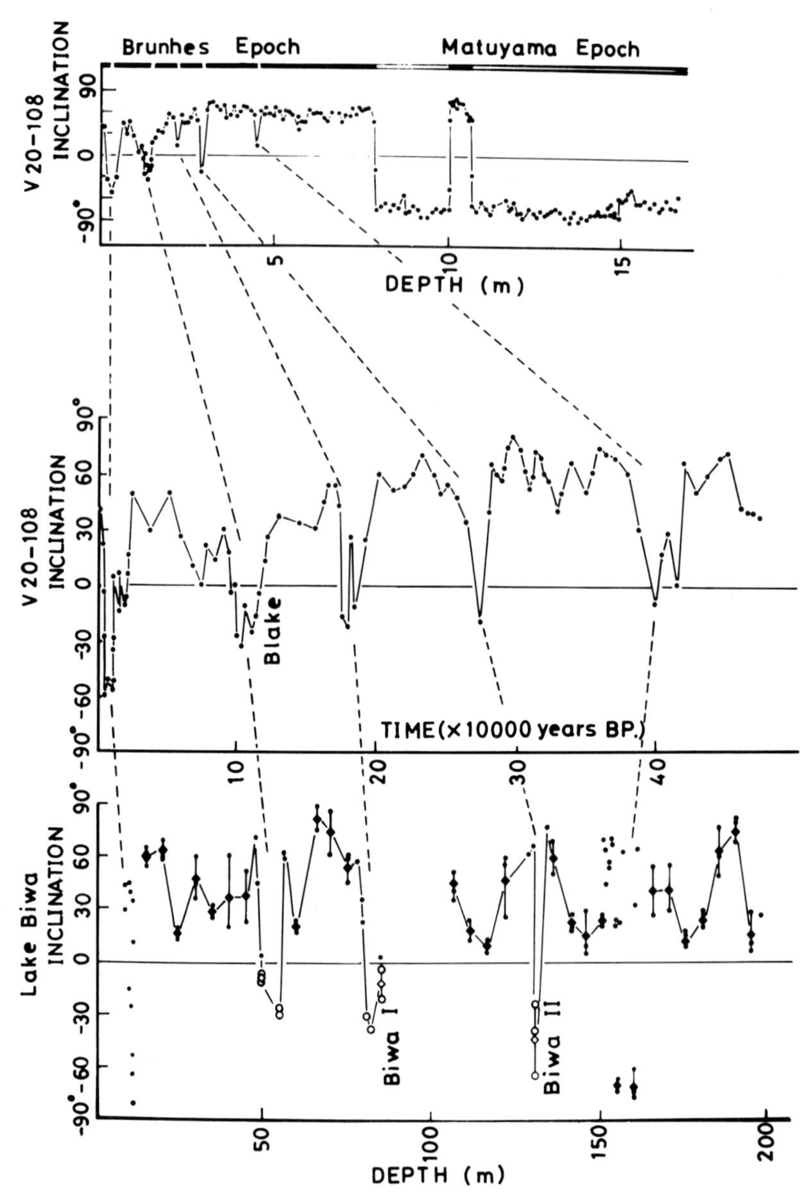

Figure 196. Correlation of magnetic inclination between lake sediments and deep-sea sediments (after Nakajima et al. 1973).

Handa 1973, 1975; Ogura & Ishiwatari 1976). Then, they undertook respective chemical analysis whose results are now available to be compared with each other as well as with the palaeomagnetic data.

On the other hand, the carbon isotope ^{13}C, especially the change of its concentration as the depth of the core increases was investigated intensively by Nakai (1972). The isotope is very important, since it indicates the paleoclimatic environment. Independently of the above-mentioned chemical analysis, Fuji carried out his own pollen analysis using the same sediments in order to know the palaeoclimatic change in the geological time in which the upper 200 m layer of the lake sediments had been stratified (Fuji & Horie 1972, 1977; Fuji 1973a).

Of the various elements in the carbon, organic carbon at various depths in the core is shown in the curve b in Fig. 197, while carbohydrate carbon and protein carbon are shown by the curves c and d respectively in the same diagram. On the other hand, the geomagnetic inclination change with depth is represented by the

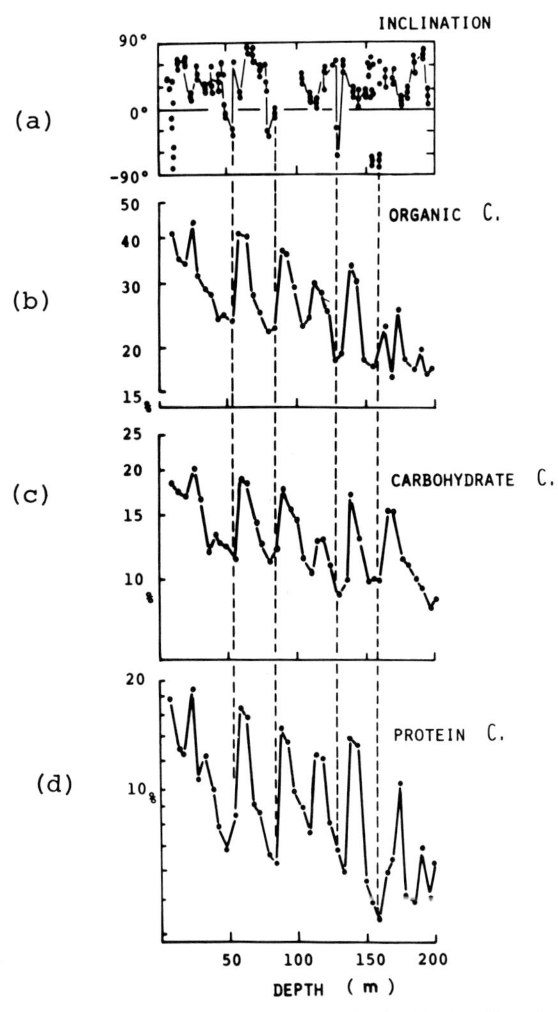

Figure 197. Carbon contents of various carbonates *vs* depths. Broken line shows the correlation between changes of magnetic inclination and of these contents (Kawai 1974).

curve a. It is interesting and important to point out the fact that each element rapidly decreases in the organic materials every time the geomagnetic event recurs. The four vertical lines drawn in the diagram demonstrate that the coincidence of each geomagnetic event with the abrupt drop of the carbon concentration is so evident. This implies that the photosynthetic reaction must have been suppressed in the lake and its vicinity when the event had occurred. the event had occurred. The synthetic reaction becomes in general active when the climate turns out to be hotter under higher insolation, while it becomes sluggish when the climate turns out to be colder under less insolation.

An alternative explanation is possible, too. The organic materials produced in the process of photosynthesis is so unstable that the synthesized materials are easily decomposed until they reach the lake bottom from the places where they were formed. The putrefaction is generally much enhanced when the mean atomospheric temperature is higher than the normal value.

It, therefore, is difficult now to determine in which way (cooler or hotter) the climate changed during and after the geomagnetic event.

Recently Ogura (1978) obtained an interesting result from her chemical analysis of lipid organic compounds. The concentration of a particular element becomes exceedingly high in a limited part of the core in which each geomagnetic event is clearly traceable, though other coexisting organic elements such as carbohydrate, protein and total carbon decrease abruptly as stated before.

Lipid carbon, deriving from phytoplanktons, zooplanktons, plants, benthos, fishes, etc. is said to become extremely condensed in the soil when living things are going to face either a glacial or polar environment. On the contrary, it becomes rarefied when they are going to confront an interglacial or equatorial environment.

As Lake Biwa is located in the middle latitude zone of the northern hemisphere, it is likely that the palaeoenvironment during each geomagnetic event was more glacial than interglacial. A cooler climate should have predominated during and after the geomagnetic event.

The cause of the cooler climate has, however, not been elucidated.

The question remains unsolved as to whether or not the coincidence of the geomagnetic event and the cooler climatic environment is a general, world-wide, and time independent fact. In order to clarify the question, declination and intensity change of NRM that we obtained from the Meranesian basin are compared with the recent data of palaeoclimatic observation by Shackleton and Opdyke (1976) as shown in Fig. 198. They used oxygen isotope ^{18}O extracted from $CaCO_3$ in the sediments in estimating palaeo-temperature. A reverse-normal geomagnetic chart was also obtained from their palaeomagnetic observation. They used this chart to determine the age of the sediments. For example, the Matuyama-Brunhes boundary (M-B boundary) is estimated to be 0.69 million years BP. and the upper and lower Jaramillo Boundary to be 0.89

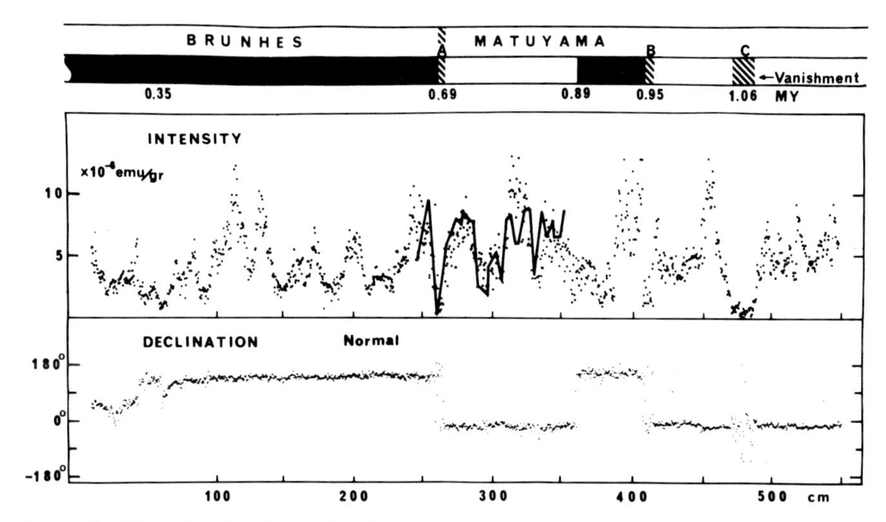

Figure 198. Figure showing the relationship between geomagnetism and climatic changes. Solid line indicates the climatic variation obtained by Shackleton and Opdyke (1973). By the sequence of dots the changes of geomagnetic field is shown (after Kawai et al 1977).

and 0.95 million years BP., although the intensity change in the entire core was not measured.

The main materials which the sediments consist of are fossils of foraminifera whose main chemical composition is $CaCO_3$. According to our palaeomagnetic observation such geomagnetic boundaries, which they used as the standards of geological age, indicate the geomagnetic events at which each intensity decreased not only abruptly but also so conspicuously as if the field had vanished (Kawai et al. 1976, 1977; Sueishi et al. 1977). The weakened field persisted as long as more than 10,000 years in the interval of the boundary. At the end of the interval the field intensity recovers and increases abruptly. The field appears as though reversed when the increased intensity has a direction opposite to that of the field prior to its abrupt decrease. On the other hand, even the same great geomagnetic anomaly is apt to be overlooked when the increased field is accidentally directed in the same direction as that of the decreased field.

During the prolonged interval when the field was weakened, an oscillation of either the declination or inclination continues as we have observed in the case of the Biwa geomagnetic events. So, the geomagnetic aspect we encountered in the reverse-normal magnetic boundary can be said to be a geomagnetic event, too.

It is interesting now to confirm in the geomagnetic boundary a new fact that extreme weakening of the field with or without the field reversal and climatic depression had taken place simultaneously.

The ice age recurred at least at the Upper Jaramillo and the M-B transition boundary (Fig. 198). The geomagnetic intensity drop and climatic depression occurred not only in the boundary but also in the epoch between the reversals.

Intensity drops can be seen at the time of Reunion, Lower Olduvai, Upper Olduvai, Lower Jaramillo and M-B boundary. Similar drop without reversal of the field direction occurred at the time of the Reunion event and at a special event whose age is estimated to be about 1.06 million years BP.

Chemical composition and clay minerals of sediments

Kazuo Shigesawa

Introduction

The Kobiwako (Paleo-Biwa) Group is widely distributed in Shiga and Mie Prefectures, and it is divisible into eight formations: Shimagahara, Iga-Aburahi, Sayama, Gamō, Yōkaichi, Zeze, Katata and Takashima in ascending order. They are mainly composed of lacustrine sediments in the old Lake Biwa (so-called Kobiwako). The writer already reported on the clay minerals of Shigaraki Clay and the tuffaceous sediments from the Kobiwako Group (Shigesawa 1957, Shigesawa & Nishikawa 1963). The former belongs to the lowest Kobiwako Group – perhaps to the Shimagahara Formation – and the latter is collected from the Gamō Formation. The clay mineral which is characteristic of the former is kaolinite and that of the latter is halloysite. Besides, the writer has investigated the chemical and mineral compositions of the recent bottom sediments of Lake Biwa and several samples from the 200 m core sample of Lake Biwa. In this section, the results of these investigations are outlined.

Samples and experimental methods

The localities of samples are shown in Fig. 199. The samples from L1 and L2 were collected from the lowest part of the Kobiwako Group, and those from L3 were collected from the Gamō Formation, the middle part of the Kobiwako Group. Samples Nos. and their localities are as follows (Table 58):

Table 58. Sample number and locality.

	Sample No.		Locality*
11	dark grey clay	L1	Shimagahara, Ayama-gun,
12	black clay		Mie Prefecture
13	elutriated black clay		Shimagahara Formation

* Both Shimagawara and Shimagahara mean same location but have a different pronunciation.

Horie, S. Lake Biwa

ISBN 90 6193 095 2. Printed in The Netherlands

Table 58 (continued)

	Sample No.		Locality
21	yellowish white clay	L2	Shigaraki-cho, Kōga-gun, Shiga Prefecture Iga-Aburahi Formation
22	grey white clay, containing quartz grains		
23	dark grey clay		
24	elutriated dark grey clay		
31	tuffaceous sediments	L3	Hino-cho, Gamō-gun, Shiga Prefecture Gamō Formation
32	tuffaceous sediments		
33	tuffaceous sediments, containing quartz and feldspar grains		
34	elutriated tuffaceous sediments		
41	recent bottom sediments		Collected from the recent bottom sediments of Lake Biwa
42	elutriated from No. 41, grain size less than 7.5		
43	elutriated from No. 41, grain size less than 3.5		
44	elutriated from No. 41, grain size less than 2.0		
51	the 200 m core sample, original sample number, No. 38 (ca. 20 m depth)		Collected from the 200 m core sample of Lake Biwa
52	the 200 m core sample, original sample number, No. 75 (ca. 34 m depth)		

Some of the samples were elutriated to remove sandy materials, and especially the recent bottom sediments were carefully elutriated to collect finer parts of the samples.

In order to determine the chemical composition and the characteristics of clay minerals, the chemical analysis and the examination by X-ray powder diffraction were performed.

Results

The results of the chemical analysis are listed in Table 59, and X-ray powder diffractograms are shown in Fig. 200. As the X-ray diffractograms of Nos. 11–13 and those of No. 21, 22, 24 are similar, the former diagrams are omitted in Fig. 200.

In the chemical analysis, no remarkable differences are observed in the silica content among the raw (not elutriated) samples, but rather remarkable differences are found in the contents of Al_2O_3 and Fe_2O_3. Nos. 41, 43, 51 and 52, which are collected from the recent bottom sediments or the 200 m core sample of Lake Biwa, have less Al_2O_3 content and more Fe_2O_3 content than those of the other samples. The results of the X-ray examination characterize the clay minerals of Nos. 11–13 and Nos. 21–24 as kaolinite and those of Nos. 31–33, collected from the tuffaceous sediments, as halloysite. Characteristics of the clay minerals of Nos. 41–43, 51 and 52 cannot be clearly determined; the clear peaks of non-clay minerals, especially quartz and feldspar, are found.

Table 59. Analytical data

Sample No.	11	12	13	21	22	23	24	31	32	33	41	43	51	52
SiO_2	62.94	64.64	48.14	51.90	53.95	54.84	50.19	52.12	58.33	58.10	57.76	54.27	61.38	61.30
TiO_2	–	–	1.03	1.01	0.29	0.54	0.60	tr.	tr.	0.87	1.03	0.83	1.05	1.01
Al_2O_3	23.85	21.61	35.01	32.09	23.79	27.51	30.76	23.73	23.14	22.54	14.97	16.62	15.99	15.88
Fe_2O_3	0.85	1.15	1.17	1.61	1.68	2.90	1.71	3.36	2.87	1.84	7.17	8.39	7.39	7.88
CaO	0.52	0.33	0.54	0.44	1.31	0.22	0.05	1.48	1.10	1.20	0.69	0.54	1.05	0.94
MgO	0.32	0.25	0.28	0.11	0.35	0.20	0.38	0.05	0.16	0.34	1.05	1.93	2.39	1.26
MnO	–	–	tr.	–	–	0.01	–	0.03	tr.	tr.	0.94	–	0.53	0.29
K_2O	1.23	1.35	0.44	–	0.99	2.59	2.03	1.72	0.84	2.30	2.08	–	2.17	1.28
Na_2O	0.53	0.58	0.40	–	3.21	1.08	1.25	1.19	1.01	3.18	2.28	–	1.39	0.64
$H_2O^{(+)}$	9.78*	10.20*	13.27*	6.10	8.30	7.97	12.74	8.02	4.70	4.07	9.56	10.22	6.94*	8.81*
$H_2O^{(-)}$				4.08	6.04	1.27	0.85	9.01	7.40	6.31	3.22	3.77		
Total	100.02	100.11	100.28	97.34	99.91	99.13	100.56	100.71	99.55	100.75	100.75	96.57	100.28	99.29
Analyst	The Government Research Institute for Ceramics			Shigesawa		Shigesawa and Okada	Shigesawa	Shigesawa and Nishikawa			Shigesawa			

Note: *Ignition loss

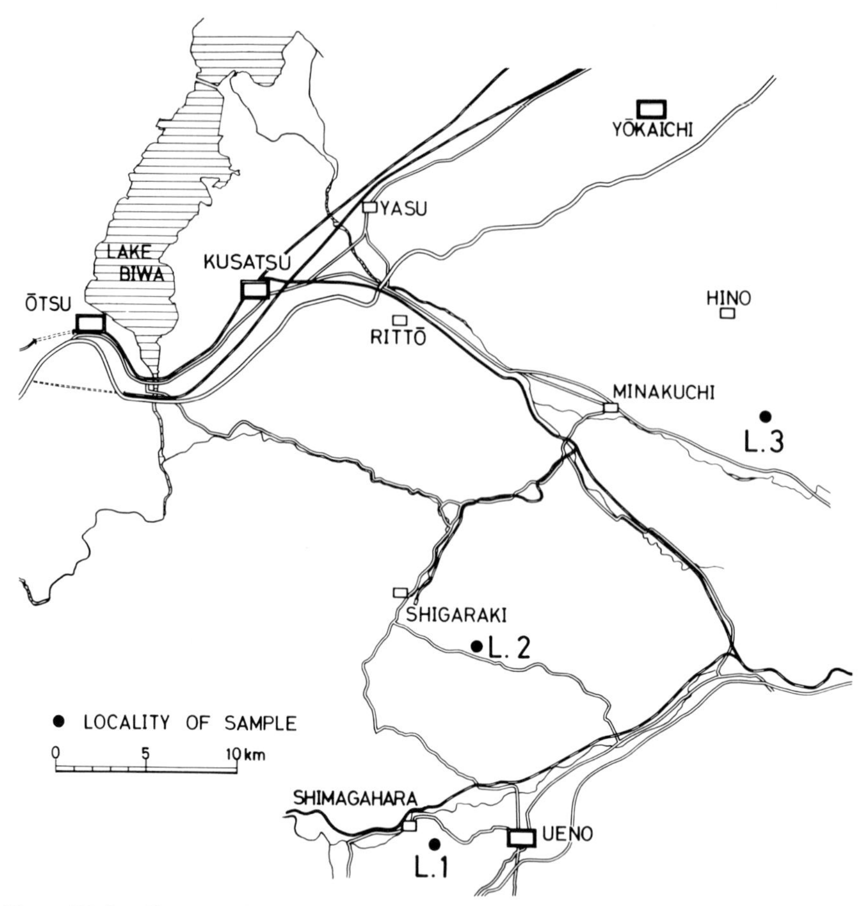

Figure 199. Locality map of sampling sites.

collected from the tuffaceous sediments, as halloysite. Characteristics of the clay minerals of Nos. 41-43, 51 and 52 cannot be clearly determined; the clear peaks of non-clay minerals, especially quartz and feldspar, are found.

Consideration

The sediments of L1 and L2 are composed of alternating strata of sand, sandy clay and clay, which are mainly derived from decomposed or weathered granitic rocks, and sometimes they contain thin tuffaceous beds. Meta-halloysite is found in these tuffaceous materials, but in the other sediments kaolinite is the characteristics of clay minerals, and it is a major clay mineral in almost all clay materials. In these districts, kaolinite is always found in clay or sandy clay materials which contain quartz and feldspar grains — granitic rock forming minerals — and therefore kaolinite may be derived from granitic rocks, and the weathering products of granitic rocks may contain kaolinite in these districts.

In L3, the sediments are composed of alternating strata of sand, sandy clay and tuffaceous clay. Tuffaceous clay beds are several centimeters to several meters thick, and these rather thick tuffaceous beds show that active volcanism

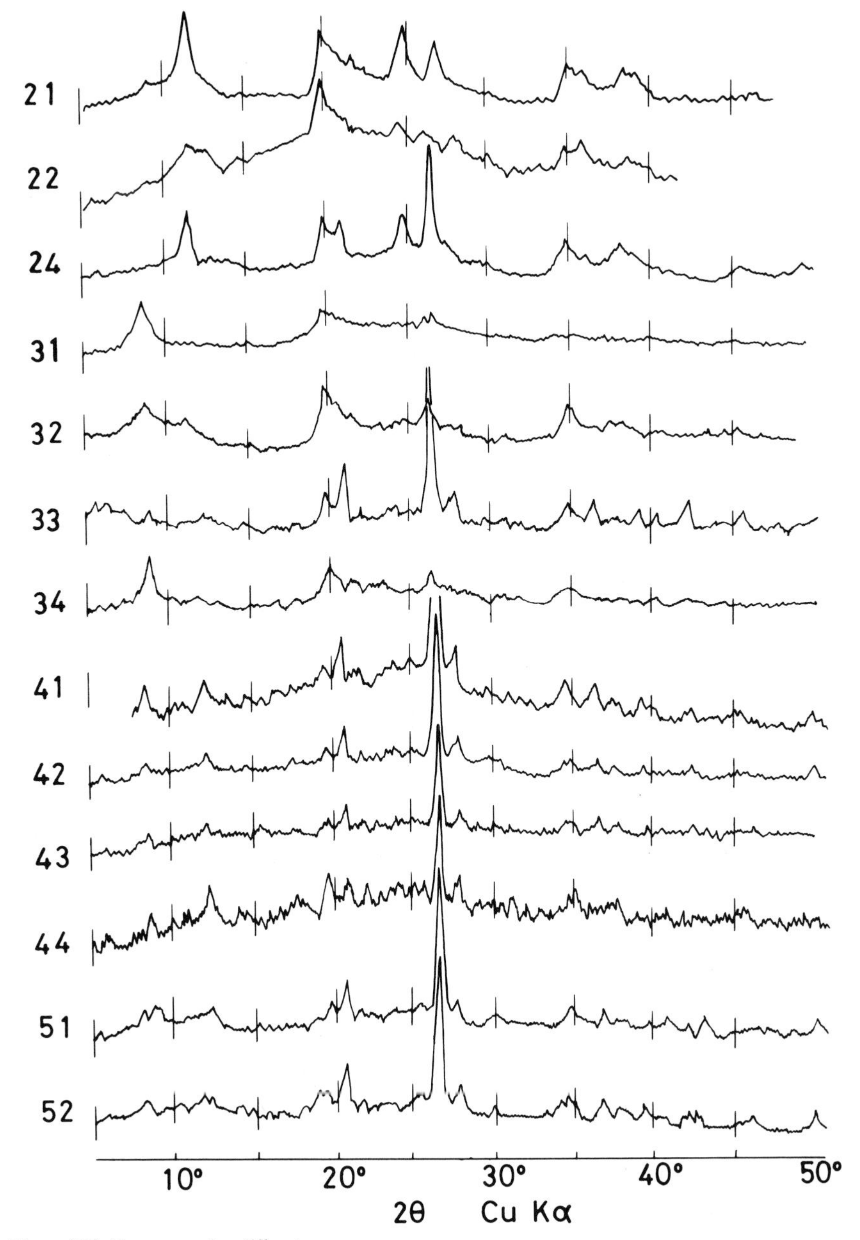

Figure 200. X-ray powder diffractograms.

took place while the sedimentation was performed in this district. Samples Nos. 31, 32 and 33 are collected from these tuffaceous beds. In these samples, halloysite is found, which is a main clay mineral, and kaolinite cannot be detected in tuffaceous materials. Some of the tuffaceous beds are mainly composed of halloysite and glassy materials as shown in X-ray powder diffractograms. Therefore, it can be thought that rather thick tuffaceous sediments and halloysite in these sediments are the characteristics in this district, and halloysite may be the weathering product of the tuff.

From the X-ray powder diffractograms of samples Nos. 41–44, 51 and 52, the existence of kaolinite, chlorite and illite in those samples can be detected, but the reflection peaks of non-clay minerals are notable, especially in Nos. 41–43. Of these reflection peaks, those of clay minerals are weak and not so clear as in the other samples described above. Therefore, the recent bottom sediments and the core sample contain less clay minerals than the sediments of the Kobiwako Group. Furthermore, the recent bottom sediments and the core sample have larger Fe_2O_3 content than those of the Kobiwako Group. Thus, it can be thought that the recent sediments and the core sample are mixtures of weathered or decomposed materials and fragments of granitic rocks and those of Palaeozoic rocks. This finding accords with the geological environment.

Conclusion

Generally speaking, the lower formations of the Kobiwako Group are distributed in the southern districts, that is, Shimagahara (Ayama-gun, Mie Prefecture), and Shigaraki (Kōga-gun, Shiga Prefecture), and these are mainly composed of clay, sandy clay and sand, which are derived from weathered or decomposed granitic rocks. And nearly every clay mineral of these sediments is kaolinite. Accordingly, it can be thought that kaolinite is an important clay mineral of the weathering products of granitic rocks in the lower Kobiwako Group. In the Gamō Formation — may be the middle Kobiwako Group — halloysite derived from tuffaceous materials is characteristic, and much of tuffaceous sediments shows active volcanism. The recent bottom sediments and the core sample are different from the other samples in chemical and mineral compositions, and these differences may be due to differences in the geological environment.

Isotopic studies

Nobuyuki Nakai

Introduction

The stable isotopic methods are often discussed in connection with the possibility that they may yield temperature indicators.

Recently, development of a technique for obtaining paleotemperature data by measuring oxygen isotopic compositions ($^{18}O/^{16}O$) of foraminifera in deep sea sediments has led to the confirmation of the pleistocene glacial cycles (Emiliani 1966a, 1966b, 1972, Emiliani & Shackleton 1974, Shackleton and Opdyke 1976). As is well known, this method is based on the temperature dependence of oxygen isotope fractionation between precipitates and H_2O when precipitation occurs under isotopic exchange equilibrium. Fortunately, in the case of marine sediments, the isotopic composition of H_2O in which $CaCO_3$ was precipitated can be assumed to be identical with that of the present sea water. On the contrary, the stable isotope paleotemperature scale which measures oxygen isotopic ratios cannot be applied to fresh water sediments such as Lake Biwa sediments, because the oxygen isotopic composition of lake water was changed widely by evaporation and supply of water in the past geologic time and because fossil carbonate itself can hardly be found in lacustrine environments. Therefore, the present writer proposed using a carbon isotope method for the cored sediments obtained from Lake Biwa.

To study the climatic and environmental change of Japanese Islands in the past, 200 m long cored sediments were obtained in 1971 from the bottom of Lake Biwa, 65 m below its present water level. The cored column is composed mainly of clayey sediments, including about thirty thin volcanic ash layers, fossil pollen grains, fossil diatoms and other organic residues. Except for volcanic ash layers, the clayey sediments have a rather homogeneous appearance throughout the core column and seems to have been deposited continuously. The cored sediments were analyzed chronologically by the ^{14}C, fission-track and constant sedimentation rate methods (Horie et al. 1971; Nishimura & Yokoyama 1973, 1975). According to Kanari's estimation based on the suitable curve-fitting method applied to several sets of data derived by the ^{14}C and fission-track

Horie, S. Lake Biwa

ISBN 90 6193 095 2. Printed in The Netherlands

methods, the deepest horizon of the cored column (197.2 m below the bottom surface) showed the age of 544.2×10^3 BP. (Kanari & Takenoya 1975).

The following sections will discuss the climatic fluctuation (Nakai 1972, 1975; Nakai & Shirai 1977, 1978) deduced from a carbon isotopic study of Lake Biwa sediment column in connection with vertical profiles of fossil diatom numbers (Mori, Sh. 1974), climates estimated by palynological studies (Fuji & Horie 1977) and geomagnetic inclinations from paleomagnetic studies (Kawai et al. 1972) for the same cored sediments. In addition to the studies of climatic changes by means of the 200 m cored sediments, the detailed climatic fluctuation has been estimated by measuring stable oxygen isotopic ratios of speleothem $CaCO_3$ in samples collected in Gifu Prefecture near Lake Biwa and the results will be reported for the past 37×10^3 years.

The experimental procedure

Sediment samples for carbon isotope analyses were taken at intervals of five meters from the cored sediment column of 200 meter length. A few hundred milligrams each of air-dried samples, which were pretreated with a dilute HCl solution to remove carbonate, were used for determining organic carbon contents of the sediment samples and the $^{13}C/^{12}C$ ratios. This analytical procedure involves the complete combustion of organic residues to CO_2, the purification of the produced CO_2, the manometric measurement of the purified CO_2 in a constant-volume system and the final mass spectrometry of CO_2 as described in the writer's previous papers (Nakai 1960, 1972).

Stalactite and flowstone samples were collected from the Otaki Cave, Gifu Prefecture, about 75 km east of Lake Biwa. Calicium carbonate samples were drilled out of the cross-section of stalactite and the vertical section of flowstones at intervals of every 1 to 5 mm. All of the carbonate was mineralogically identified to be calcite by X-ray analysis. The calcium carbonate samples were then analyzed for stable oxygen isotope ratios ($^{18}O/^{16}O$) as reported by Epstein et al. (1953), and for age by the ^{14}C method.

The isotopic data for both of carbon and oxygen are presented in this paper as per mil deviation from the PDB-Chicago standard (fossil belimnite, $^{13}C/^{12}C = 0.0112372$ by Craig 1957) and SMOW (Standard Mean Ocean Water, $^{18}O/^{16}O = 0.0019934$ by Craig 1961), respectively. This deviation is defined by

$$\delta^{13}C \text{ or } ^{18}O\ (\text{‰}) = \frac{R(\text{Sample}) - R(\text{Std.})}{R(\text{Std.})} \times 1000$$

where R denotes the $^{13}C/^{12}C$ or $^{18}O/^{16}O$ ratio for the sample and the standard. A positive value suggests that the sample material is relatively more abundant in heavy isotope than the standard is. On the other hand, the negative value shows that it is relatively poorer in heavy isotope than the standard is.

Stable carbon isotope ratio and its distribution in nature

There are two kinds of ^{12}C and ^{13}C as stable carbon isotopes. Both of these occur in nature and the abundance is 98.892% of ^{12}C and 1.108% of ^{13}C in the limestone of Nier (1950) which is often referred to. The isotopic composition varies in nature through various chemical reactions and physical phenomena. The main phenomena producing isotope fractionation are:

1. the isotope exchange equilibrium reaction,
2. the kinetic process that depends on differences in reaction rates of isotopic molecular species,
3. isotopic fractionation due to other physico-chemical effects.

Isotopic exchange equilibrium

Isotopic exchange equilibrium is equivalent to chemical equilibrium in chemical reaction. When a certain element, which is carbon in this case, exists in two or more chemical compounds which are different from each other, isotopic exchange reaction occurs in the element. On the occasion of isotope exchange, the fractionation of isotopes between the chemical compounds is decided upon according to the difference in their molecular structures and temperatures. The system of CO_2–HCO_3^-, for example, is expressed by the following equation.

$$^{13}CO_2 + H^{12}CO_3^- \rightleftharpoons {}^{12}CO_2 + H^{13}CO_3^-$$

As is shown in Table 60, the difference of ^{13}C value between them shows the fixed value in compliance with the temperature.

HCO_3^- is always relatively more abundant in ^{13}C than that of CO_2.

Table 60. Fractionation of carbon isotopes between gaseous CO_2 and aqueous HCO_3^- in equilibrium (after Deuser & Degens 1967).

Temperature (°C)	$\delta^{13}C_{HCO_3^- - CO_2}$ (‰)
0	9.2
10	8.1
20	7.3
30	6.8

Isotopic unidirectional reaction

The second main phenomena producing isotope fractionations are kinetic effects because light isotopic species differs from a heavy one in the reaction rate. The isotope fractionation introduced in the course of this reaction may be considered in terms of the ratio of rate constants for isotopic molecules. For two competing isotopic reactions of chemical compounds A to B,

$$A_1 \xrightarrow{k_1} B_1 \text{ (for light isotope)}$$

$$A_2 \xrightarrow{k_2} B_2 \text{ (for heavy isotope),}$$

the ratio of rate constants, k_1/k_2, for the reaction of light and heavy isotopic species are generally larger than unity. This suggests that ^{12}C-molecules react faster than ^{13}C-molecules do, resulting in a 'light' product during the uni-directional reaction. The photo-synthetic reaction affords an example.

Some other factors

In addition to the above-mentioned cases of chemical and biochemical reactions, there is another fractionation of isotopic compositions according to the physico-chemical processes, such as evaporation, condensation, precipitation, and diffusion.

The present data on the distribution of $\delta^{13}C$ values in nature, especially in connection with the following discussions, are given in Fig. 201. The atmospheric CO_2 which is a material for photosynthesis indicates relatively constant $\delta^{13}C$ values ranging from -7 to $-9‰$. The values of organic carbon for organisms synthesized as a consequence of photosynthesis show -8 to $-36‰$

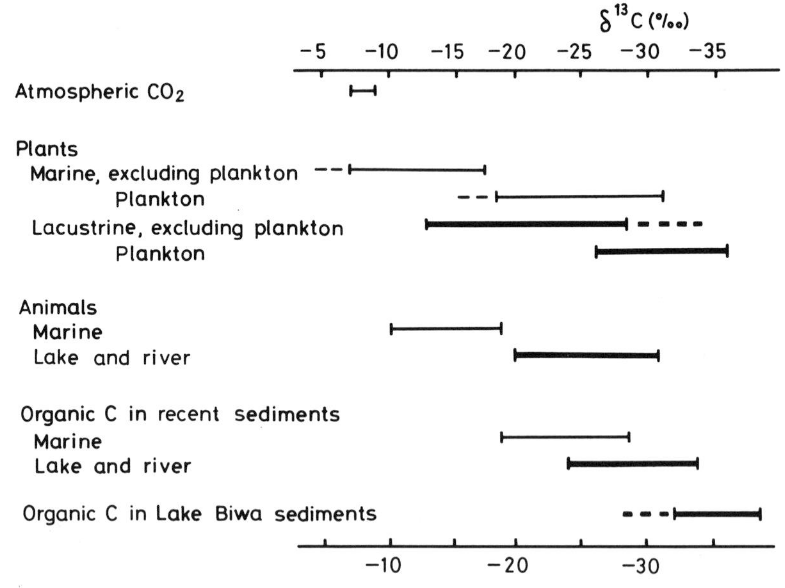

Figure 201. Distribution of $\delta^{13}C$ values in nature, relative to PDB.

more ^{13}C-depleted than those of the atmospheric CO_2. Marine and lacustrine plants excluding plankton range from -8 to $-18‰$ and from -13 to $-27.5‰$, respectively; marine phytoplanktons from -18 to $-29‰$ and lacustrine phytoplanktons from -26 to $-35‰$. It follows from these results of many workers that a) there is a difference between marine and lacustrine $\delta^{13}C$ values, b) marine and lacustrine planktons are respectively composed of the carbon isotope ratios depleted in ^{13}C more than any other plant. These organism residues are accumulated in the bottom sediments and their carbon isotopic compositions may contain useful information on the past depositional environment. Both in the marine and in the fresh water, isotopic compositions of organic carbon contained in it generally indicate the intermediate values between those in the plankton and in the plant excluding plankton, which are rather close to the plankton values.

According to the Degens' detailed results (1969) given in Fig. 202, $\delta^{13}C$ values in the organic carbon undergo changes through oceanic latitudes (probably changes of relative kinetic effect in the photosynthetic reaction by water temperature) besides the marine and lacustrine difference. As is to be described in a later section, Sackett et al. (1965) reported that living marine planktons consist of a wide range of $\delta^{13}C$ values depending on the water temperature at which they are growing.

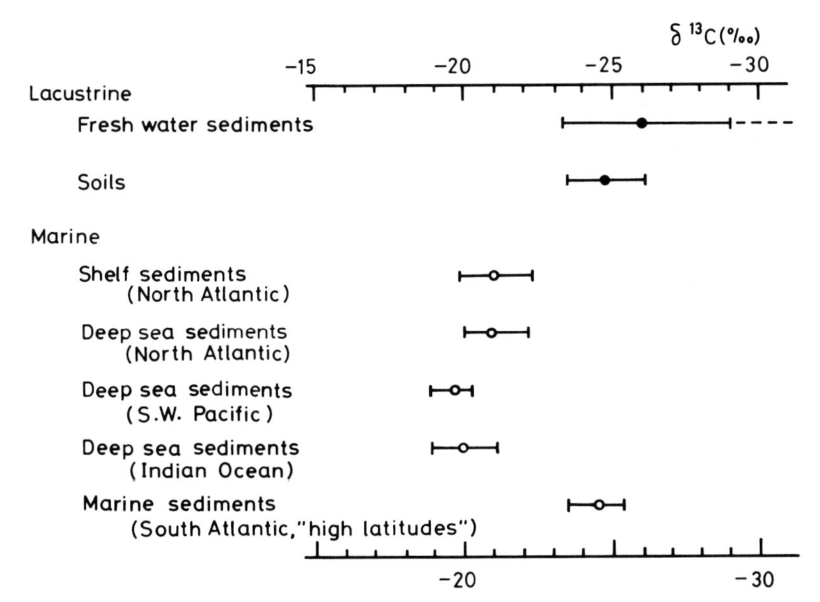

Figure 202. $\delta^{13}C$ values of organic carbon in recent sediments relative to PDB (after Degens 1969 and others).

Carbon isotope compositions of Lake Biwa sediments and their climatic indication

Vertical distribution

The carbon in organic materials in the sediment samples other than those of the bottom surface has $\delta^{13}C$ values ranging from -31.8 to -37.9‰, and total organic carbon contents range from 0.3 to 1.2% in dry samples, as shown in Fig. 203. The $\delta^{13}C$ values of the bottom surface sediments show -27.4 to -28.4‰. The ^{12}C-enriched values of the sediments indicated in Lake Biwa may possibly be explained by a large contribution of plankton residues to sedimentary organic materials as seen in Fig. 201. As has been reported by Craig (1953) and Wickman (1952), the $\delta^{13}C$ values of terrestrial plants and organisms show -12.2 to -29.1‰. However, recent investigations indicate that freshwater planktons are richer in ^{12}C and that their $\delta^{13}C$ values range from -26.4 to -34.5‰. Sackett et al. (1965) have studied marine plankton samples from the Atlantic and the Pacific Oceans; they showed that there is a large difference between the isotopic composition of organic carbon in marine planktons and that of other marine organisms. They pointed out that the $\delta^{13}C$ values of 25 marine planktons ranged from -18 to -31‰, considerably enriched in ^{12}C relative to the range from -8 to -18‰ for those of other marine organisms. From these facts, it is evident that the $\delta^{13}C$ values of organic carbon in the freshwater environment are smaller than in the marine environment, and that plankton is richer in ^{12}C than other organic materials in fresh and marine environments.

To understand the ^{12}C enrichment in plankton, the chemical form of inorganic carbon, which organisms utilize through photosynthesis, and the carbon isotope fractionation among CO_2, HCO_3^- and $CO_3^=$ must be considered. An experimental study by Deuser and Degens (1967) on the carbon isotope fractionation in the CO_2–HCO_3^-–$CO_3^=$ system has shown that the difference in carbon isotopic composition ($\delta^{13}C_{HCO_3^-}$–$\delta^{13}C_{CO_2}$) is from 9.2 to 6.8‰ at temperatures of 0°C to 30°C under the condition of an isotopic exchange equilibrium, as is shown in Table 60. These fractionations, of course, depend on the equilibrium temperature. Under the conditions of lake water, however, the $CO_3^=$ ion can be neglected. CO_2 dissolved in water has the lightest carbon. If planktons utilize exclusively molecular CO_2 during photosynthesis, as has been suggested in growth experiments of diatoms by Degens et al. (1968), the $\delta^{13}C$ values of plankton should be lighter by 7 to 9‰ than in other organisms which take up other chemical species of inorganic carbon, such as bicarbonate ions. Thus, the ^{12}C-enriched values of the sediments in Lake Biwa may possibly be explained by the large contribution of plankton residues to sedimentary organic materials.

Fig. 203 shows the change in the $\delta^{13}C$ values; the organic carbon can be found all through the core samples. The isotopic composition of organic carbon

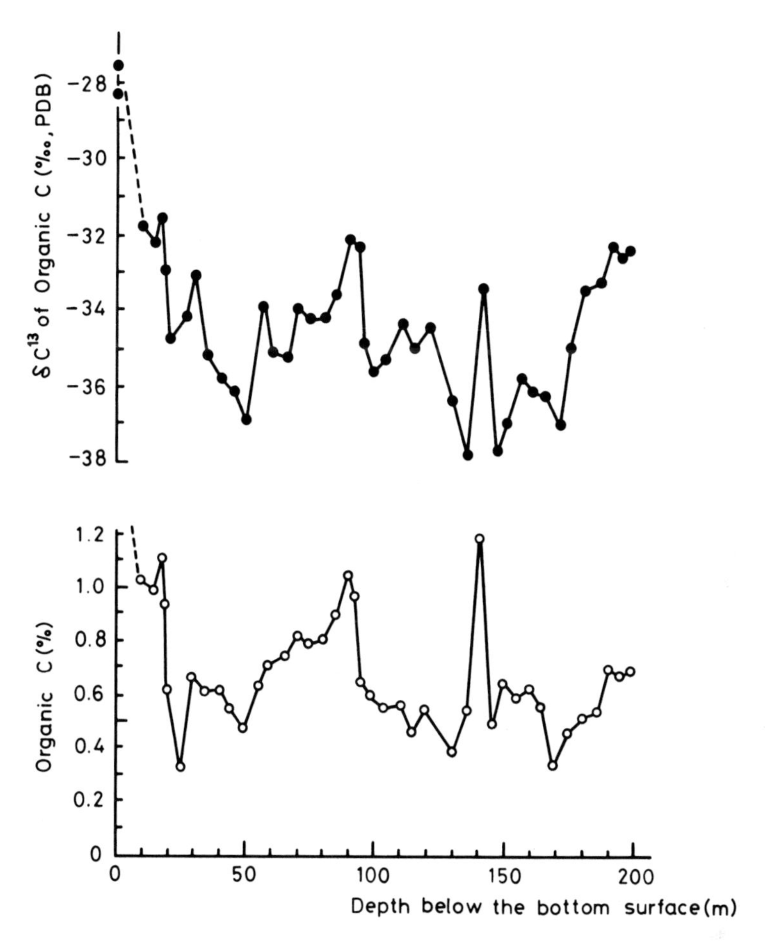

Figure 203. The vertical variation of organic carbon contents and the $\delta^{13}C$ values of sediments from Lake Biwa (Nakai 1972, 1975).

becomes enriched in ^{13}C with an increase in the organic carbon contents. The isotopic variation in organic carbon in the core column is undoubtedly controlled with kinetic isotope effect by temperature at which the plankton grew. In a warmer climate, the production rate of organic materials such as plankton in the lake and their accumulation rate in the sediments are relatively higher, resulting in relatively higher ^{13}C enrichment than those in a colder climate. The paleoclimatic variation estimated by a palynological investigation (Fuji & Horie 1977) also shows an excellent correlation with that by the $\delta^{13}C$ values as will be shown later. A general feature of these results obtained so far seems to be a tendency toward larger $\delta^{13}C$ values when organic carbon contents are larger, indicating that the climatic condition is rather temperate or warm.

Climate dependence of carbon isotopic compositions

Sackett et al. (1965) reported $\delta^{13}C$ values of living marine phyto- and zooplanktons taken from the Atlantic and the Pacific Oceans at high to low latitudes. They found the apparent water temperature dependence of $\delta^{13}C$ values as shown in Table 61. This temperature dependence shows that planktons from low temperature regions at 0°C are depleted in ^{13}C by about 6‰ relative to those from tropical regions at 25°C. These results give the temperature coefficient of 0.25‰/°C on the average for the photosynthetic process. Degens et al. (1968) calculated the temperature coefficient to be 0.35‰/°C on the average on the basis of their results in laboratory growth experiments of marine phytoplanktons.

Pearman et al. (1976) and Wilson and Grinstead (1977) have attempted to obtain paleoclimatic information from $\delta^{13}C$ values of tree rings and reported the same trend in the temperature dependence of isotopic compositions as that of planktons. Pearman et al. (1976) had taken samples out of the cross-section of two King Billy pine trees which grew in a natural forest in mountains of northern Tasmania and determined the $\delta^{13}C$ values to discuss their temperature dependence from 1895 to 1970. By adding the air temperature data recorded at weather stations in this area, the results are illustrated in Fig. 204 which gives ^{13}C histograms for each tree. They found the temperature coefficient of 0.24 ±0.11 and 0.48±0.12‰/°C for trees A and B, respectively. This temperature dependence also gives the same trend as that of planktons mentioned above. From these facts for the temperature dependence of stable carbon isotopic compositions, one can expect the general tendency toward larger $\delta^{13}C$ values of photosynthetic products when the climate is warm or temperate.

For the same 200 m core samples from Lake Biwa as those used for $\delta^{13}C$ measurements of organic materials, Mori, Sh. (1974) studied the species composition of fossil diatoms and reported the vertical profile of total fossil diatom cell numbers. The relationship between organic carbon contents of sediments and numbers of fossil diatom cells in sediments is shown graphically in Fig. 205. The linear relationship observed between them tells us that the population of diatom, which has been one of the main producers of organic materials, indicates the relative productivities in the past lake water. A few points, which deviate by exceptionally large distance from the straight line in the figure, belong

Table 61. Temperature dependence of $\delta^{13}C$ in living planktons from the Atlantic and the Pacific Ocean (Sackett et al. 1965).

Water temperature (°C)	Averaged $\delta^{13}C$ (‰)	Number of samples
25±2	−21.7	9
14.4	−24.0	1
0±2	−27.9	12

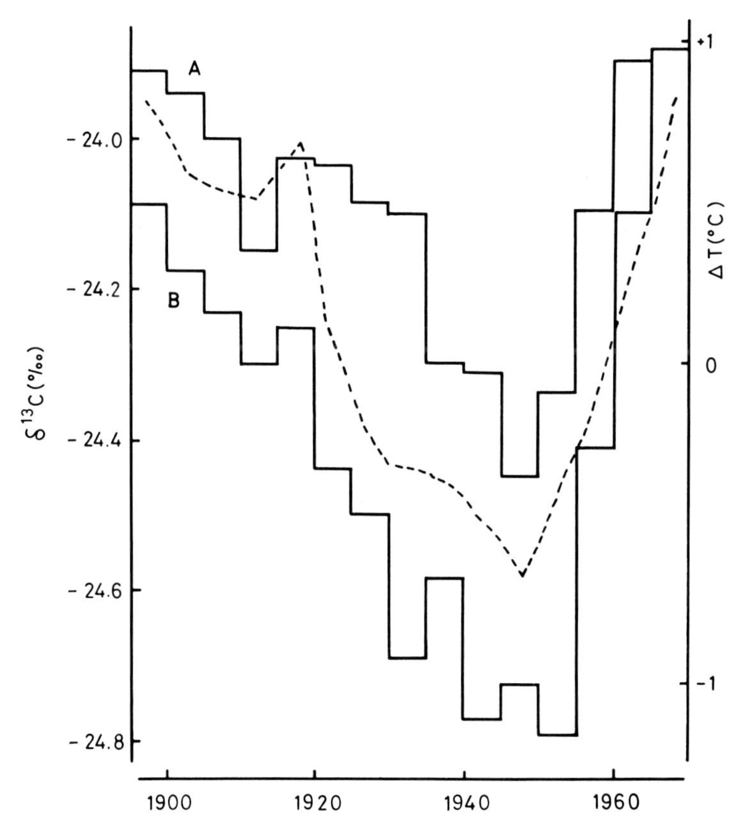

Figure 204. $\delta^{13}C$ of tree rings from Tasmanian trees A and B, and smoothed temperature deviation (ΔT) from the long term average of February maximum temperature for six Tasmanian stations (after Pearman et al. 1976).

to the sediment samples in the periods of paleogeomagnetic reversals which have been measured by Kawai et al. (1972). The vertical profile of total fossil diatom cell numbers is as shown in Fig. 206A by the absolute age scale; the fluctuation pattern observed throughout the entire core column gives an excellent agreement with that of $\delta^{13}C$ values (Fig. 206B). The periods of abundant diatom cells correlate well with the periods of larger $\delta^{13}C$ values.

Fuji and Horie (1977) established the paleotemperature curve of the 200 m core sample on the basis of the palynological treatment of fossil pollen grains. They found 10 temperate and 12 cold periods during the last 560×10^3 years. The temperate and cold climatic periods have been designated as t − 1, t − 2, ..., and c − 1, c − 2, ..., respectively, from the young to old periods. In Fig. 206B, the climatic periods deduced from the palynological study are indicated on the $\delta^{13}C$ fluctuation curve. As shown in this figure, an excellent correlation can be found between the $\delta^{13}C$− climatic fluctuations and the palynological paleoclimate. $\delta^{13}C$ of organic materials gives relatively high values in the palynologically

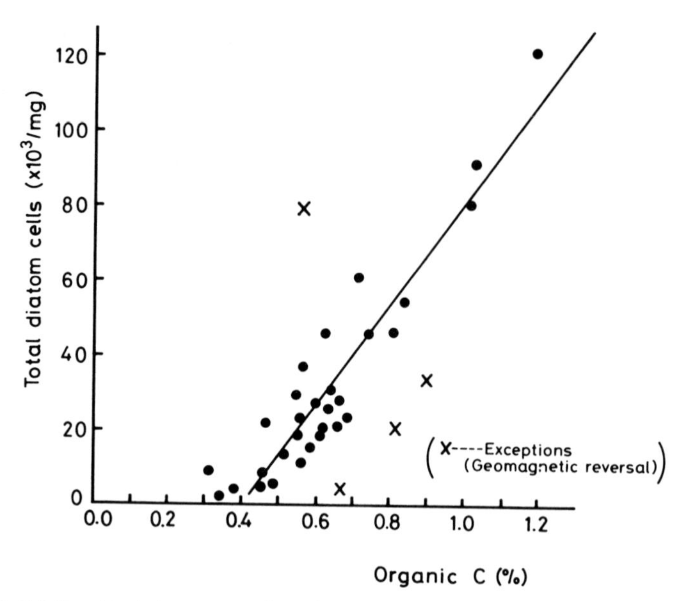

Figure 205. Relationship between organic carbon contents and total diatom cells of the sediment column from Lake Biwa (after Nakai & Shirai 1977).

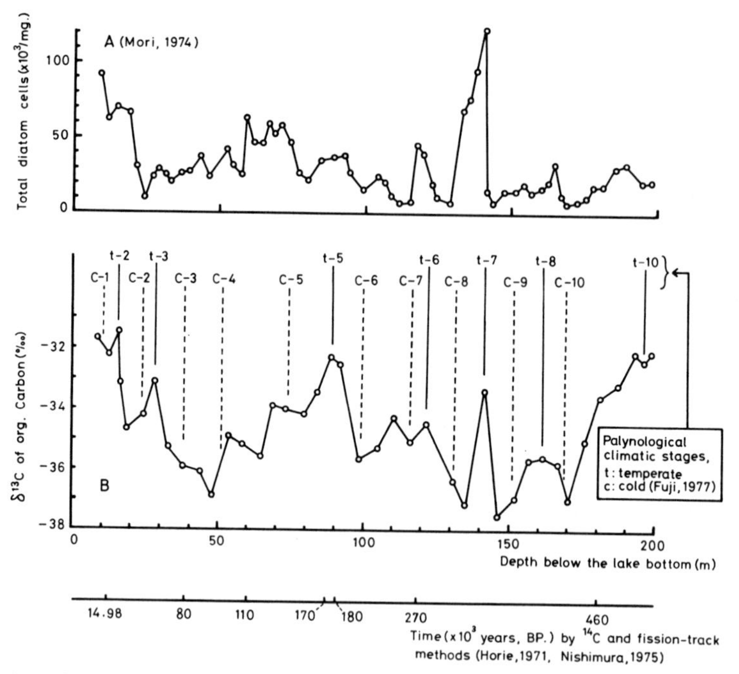

Figure 206. The vertical profile of total fossil diatom cells (A) and δ^{13}C values of organic carbon (B) in the Lake Biwa cored column by depth in meters and by absolute age (after Nakai & Shirai 1977).

temperate climate, t − 1 to t − 10, and relatively low values in the cold climate, although some exceptions are found in the periods t − 4, c − 11 and c − 12. In other words, in a warm or temperate climate, the production rate of organic materials in lake water and their accumulation rate in the bottom sediments are relatively high, resulting in a relatively high ^{13}C enrichment than in a cold climatic period.

From a comparison of the carbon isotopic results with the fossil diatom and the fossil pollen studies, it is clearly concluded that the carbon isotope fluctuation of fossil organic materials through the cored column undoubtedly reflects the change in the past temperature and paleoclimate.

Relation of $\delta^{13}C$ values for organic materials to paleomagnetic polarity change

In the previous section, we found from Figs. 206A and B that changes in the relative productivities of diatoms in the past lake were closely connected with those in the $\delta^{13}C$ values of organic products at the time. In Fig. 207, $\delta^{13}C$ values

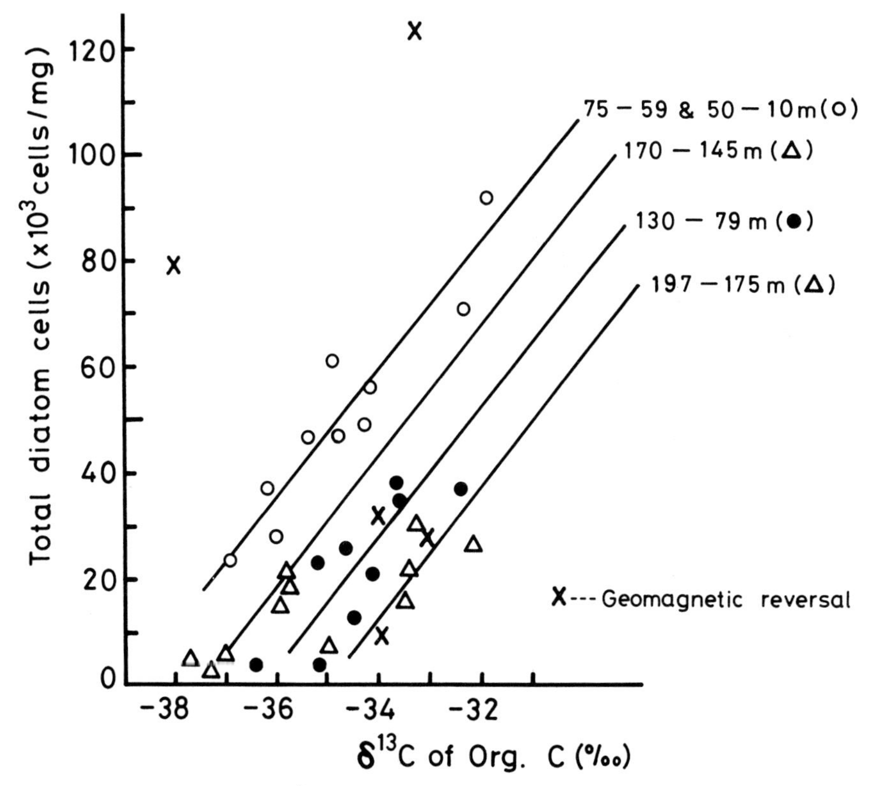

Figure 207. Relationship between $\delta^{13}C$ values of organic matters and total diatom cells in the sediment column from Lake Biwa. (Nakai & Shirai 1977).

of organic carbon in the sediment core are plotted against the numbers of total fossil diatom cells. This figure appears to show no correlation between them. However, if sediment layers with limited thickness are separated, an excellent linear relation can be found between them in each of the layers. This means that $\delta^{13}C$ values of sedimentary organic carbon and the past productivity of the lake have different linear relations in different periods. As shown in this figure, four straight lines can be drawn and these linear relationships make up different lines before and after the geomagnetic reversal period reported by Kawai et al. (1972). In other words, $\delta^{13}C$ values and the productivities at the time show a linear relationship during a normal geomagnetic period and the $\delta^{13}C$ values of fossil organic materials can be used as an indicator of the past productivity and temperature fluctuation within the limited time interval. This may indicate that short reversed polarity events in the Brunhes normal polarity epoch which has continued for the last 690×10^3 years have some effects upon the $\delta^{13}C$ values of photosynthetic products. It may suggest that the fluctuation of cosmic ray intensities caused supposedly by geomagnetic field changes had direct effects on the appearance and disappearance of organism species, resulting in different metabolic fractionations of carbon isotopes, or on $\delta^{13}C$ values of materials in photosynthesis, atmospheric CO_2.

Fig. 208 shows a comparison between variations of paleogeomagnetic inclination and of $\delta^{13}C$ values of organic carbon with depth in sediments from Lake Biwa. Through the paleogeomagnetic studies, the existence of three short

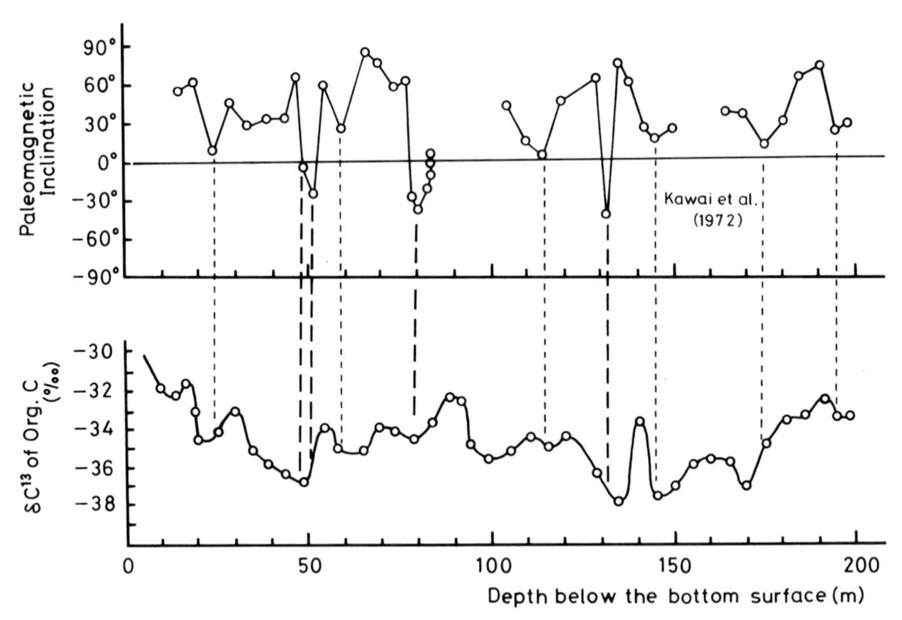

Figure 208. Vertical variations of geomagnetic inclination and $\delta^{13}C$ values of organic carbon for Lake Biwa sediments.

reversed polarity events including the Blake event and two geomagnetic excursions was recognized in the 200 meter core by Kawai et al. (1972). A detailed comparison of the geomagnetic polarity fluctuation with $\delta^{13}C$ fluctuation throughout the cored column reveals that the magnetically reversal and the lower inclination epochs correspond to the smaller or minimum $\delta^{13}C$ periods. It suggests that the climatic fluctuation may be related to some of the magnetic events. The relation among the $\delta^{13}C$ value, climate and geomagnetic polarity changes becomes a serious problem, but it will be solved only through detailed measurements and analyses in the future in co-operation with geophysicists and cosmophysicists.

Climatic fluctuation deduced from $\delta^{13}C$ measurements

Considering the $\delta^{13}C$ fluctuation and the absolute ages in Fig. 206B, one can examine glacial ages based on the $\delta^{13}C$ values for their correspondence with the European and North American glacial ages, the existence of which was certified from the viewpoint of glacial geology, paleontology and others. The last glacial age, Würm or Wisconsin, corresponds to the $\delta^{13}C$ drops at the depth of 10 to 30 m in the Lake Biwa sediment column. The earlier cold period with the lowering $\delta^{13}C$ values between the depth of 60 and 70 m may be considered to be the Riss glacial age, and that between the depth of 95 and 105 m to be the Mindel glacial age. The cold period of the minimum $\delta^{13}C$ value at the depth of about 135 m may possibly correspond to the Günz glacial age. The drops of $\delta^{13}C$ values at the depth of 45 to 50 m, 145, and 170 m correspond to none of the glacial ages recognized in Europe and North America. However, the climatic estimation based on measurements of $\delta^{18}O$ of ice sheets from Greenland and Arctic Canada indicated a cold stage around 90×10^3 years BP. during the Sangamon interglacial stage found between the Würm and the Riss glacial age (Johnsen et al. 1972; Paterson et al. 1977). This cold stage chronologically agrees well with the carbon isotope cold stage at the depth of 45 to 50 m in the Lake Biwa sediment column.

Paleoclimatic changes during the last 37×10^3 years deduced from oxygen isotope compositions of speleothems

In addition to the studies in climatic changes by means of the 200 m cored column, the climatic estimation has been attempted by measuring oxygen isotope ratios of cave $CaCO_3$ for the last glacial age to Holocene. Fortunately, speleothems can be collected from the Otaki Cave, a cave in the Gujohachiman region about 75 km east of Lake Biwa. The isotopic compositions were expressed in per mil deviation from the standard as explained in the previous section.

Hendy and Wilson (1968) found temperature dependent fluctuations in the $^{18}O/^{16}O$ ratios of calcite deposited on speleothems in New Zealand. They calculated the deposition temperature of calcite on an assumption that $^{18}O/^{16}O$ ratios of speleothems in a non-tropical region with the maritime climate would give the same fluctuations in magnitude as those in the tropical. The present writer, however, tried a direct calculation of the deposition temperature by a method different from Hendy's as follows.

In an isotopic equilibrium of oxygen isotopes between calcite and water in which calcite was deposited, the isotopic fractionation is a function of the temperature, increasing by 0.24‰ when the temperature of deposition decreases by 1°C. The isotopic equilibrium fractionation, α, can be expressed as a function of temperature (°K) according to Eq. [1] (O'Neil et al. 1969),

$$\delta^{18}O(CaCO_3) - \delta^{18}O(H_2O) = 100 \ln \alpha^{CaCO_3}_{H_2O} = 2.78(10^6 T^{-2}) - 3.39 \quad [1]$$

Though the $\delta^{18}O$ values of calcite can be measured, it is impossible to get those of the water in which calcite was deposited in the geologic time. Thus, we have to estimate the isotopic composition of the water. The $\delta^{18}O$ values of the water passing through the speleothem surface are in turn determined by those of the average atomospheric precipitation in the region. Dansgaard (1964) reported that the annual average value of the oxygen isotopic composition of rain for coastal regions at middle to high latitudes varies linearly with the mean annual air temperature. The $\delta^{18}O$ composition of rain changes by 0.695‰ when the temperature changes by 1°C as represented by the following equation:

$$\delta^{18}O(H_2O)SMOW = 0.695t - A \quad [2]$$

where A is a constant and t the mean annual temperature in the centigrade scale. Dansgaard (1964) assigned the value of 13.6 to A for the North Atlantic coast and Greenland but our estimation for Central Japan indicates it to be 15.2 instead of 13.6.

By eliminating the $\delta^{18}O$ of water from equations [1] and [2], the isotopic equilibrium temperature can be calculated. The ranges of the $\delta^{18}O$ values of speleothems and the calculated temperatures are presented in Table 62. The calculated temperature ranges from 10.6 to 15.0°C. By connecting results obtained from two flowstones and a stalactite which overlap each other

Table 62. Oxygen isotopic compositions of speleotherms from the Otaki Cave and the estimated isotopic temperatures (Nakai & Shirai 1977).

Speleothem	$\delta^{18}O$ of $CaCO_3$ (‰ relative to SMOW)	Isotopic temperature (°C)
Flowstone I	+23.7 ~ +25.2	11.7 ~ 15.0
Stalactite	+23.2 ~ +24.3	10.6 ~ 13.0
Flowstone II	+23.2 ~ +24.0	10.6 ~ 12.3

chronologically, a paleotemperature curve can be drawn for the last 37×10^3 years, as shown in Fig. 209A. In this figure, only one point around 10×10^3 years BP. is separated widely from other points at similar ages so that this point was eliminated in drawing a smoothed curve. The paleoclimatic fluctuation curve has a feature indicating that the late Würm glacial stage (the last glacial age) and the Holocene warm time can apparently be seen before and after about 10×10^3 years BP., respectively. During the late Würm glacial stage, stadials, extremely cold ages, can be found in the ages of 17 to 18×10^3 years BP. and 30×10^3 years BP., and interstadial, relatively warm age, in the age of 20 to 27×10^3 years BP. These stadial and interstadial ages chronologically agree well with the climatic estimation from the sea-level changes (Minato 1966; Fairbridge

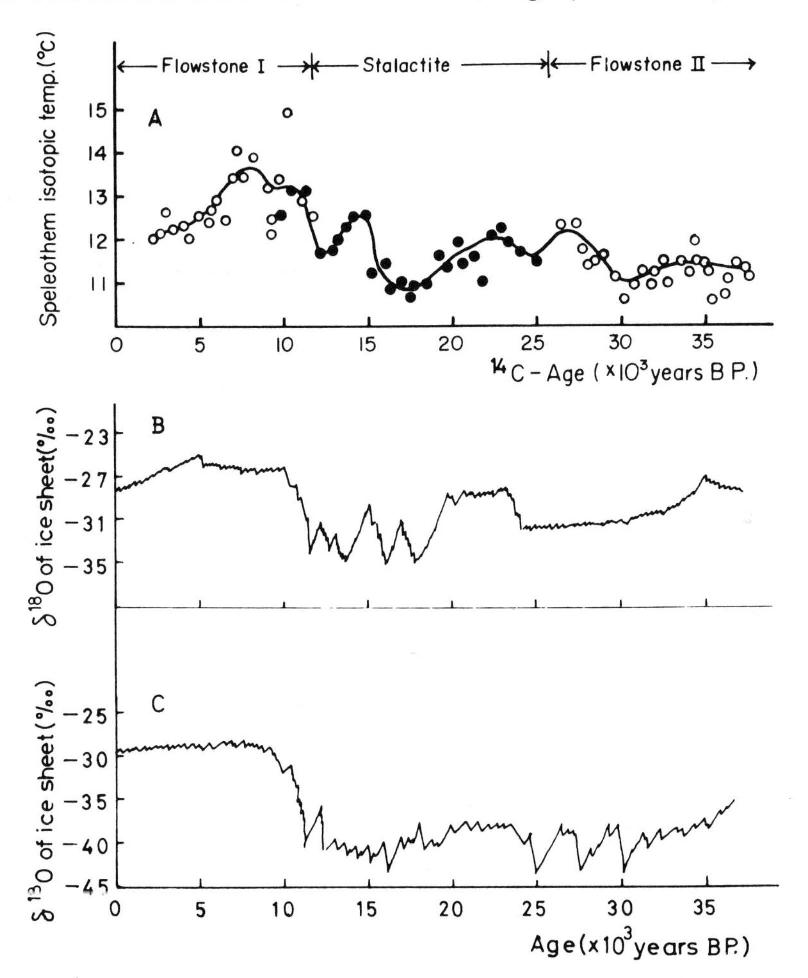

Figure 209. Temperature fluctuation during the last 37,000 years estimated by $\delta^{18}O$ measurements of speleothems from the Otaki Cave, Central Japan (A) and of ice sheets from Arctic Canada (B) (Paterson et al. 1977) and Greenland (C) (Johnsen et al. 1972).

1961) and from $\delta^{18}O$ of ice sheets (Paterson et al. 1977; Johnsen et al. 1972). The climatic variations of ice sheets from Arctic Canada and Greenland are shown in Figs. 209B and 209C. During Holocene we can find a temperature maximum in the age of 7 to 8×10^3 years BP., and the temperature decreases continuously toward the present time. The above-estimated temperatures show those of water in the cave in the strict sense. Even for the past temperature in the cave the fluctuation could be found as shown in the figure. The time resolution of this method is much better than that estimated by $\delta^{18}O$ of foraminifera from deep sea cores, having an average of one point/450 years.

As reported above, the carbon and oxygen isotope methods have been applied to Lake Biwa sediments and speleothems, respectively, and the paleo-temperature variations have been estimated for Central Japan. The comparison between the climatic profiles by isotopic and several other methods not only gave a close agreement to each other but also suggested that unknown short cold periods could be found by these isotopic methods applied to the lake sediments and speleothems.

Grain size variation

Atsuyuki Yamamoto

Introduction

Bottom sediments in the central part of Lake Biwa consist mainly of fine-grained gyttja brought in by rivers from the surrounding basin, partly of planktonic remains and sometimes of volcanic ash. Since sediments that have arrived at the river mouths as a load of river water are subjected to prompt hydraulic sorting, leaving coarser grains behind, only finer grains of the sediments are dispersed into the lake. Thus, sediments transported and dispersed by water as far as the central part of the lake are characterized as sufficiently fine particles of very small settling velocity. In this way, the granulometric constitution of sediments is considered to represent a resultant of the dynamic processes in water, such as suspension, transportation and settling in water, which sedimentary grains have passed through, and to reflect physical environments around the lake, which bring forth the sediment supply at the river mouths as well as the hydraulic conditions on river inflow and lake water movement. Particularly, the grain size distribution of sediments on the lake bottom is expected to be closely related to the precipitation as an element of climate, which induces the dispersion of sediments into the lake. Thus, an investigation of the vertical variations of grain size in bottom sediment layers should be an indispensable branch of the paleolimnological or paleo-climatological study of the lake.

It seems meaningful, too, to take note of the bulk density of sediments as a physical parameter as to the assembly of grains, because the increase in the bulk density of a particular sediment layer, which is under consolidation beneath the bottom surface, is time-dependent and, accordingly, the vertical profile of bulk density passing through loose sediment layers is readily available to provide a scale for the age of deposition if only absolute dates for some layers are obtained.

In Lake Biwa, several borings into bottom sediments have been conducted at a site of about 65 m in water depth between Ōmi-Maiko and Oki-no-shima Island in the central area of the lake. In 1967 a 12 m-long core sample was taken there (Horie et al. 1971), and in 1971 a 197.2 m-long continuous core sample

Horie, S. Lake Biwa

ISBN 90 6193 095 2. Printed in The Netherlands

(so-called 200 m core) was obtained as the very first one in lakes the world over (Horie 1972). The present paper describes the analysis of some features of grain size variations in these core samples and then applies it to estimating the paleoclimatic change, especially paleoprecipitational change, during the last 500,000 years. Before proceeding to this work, however, we have to introduce the method of scaling ages along the core.

The objective and the method of grain size analysis

Main objective

The main objective of grain size analysis of core sediments lies presently in estimating the variation of paleoprecipitation which, along with paleotemperature, is an element indispensable to paleoclimatological study and necessary for finding the existence of pluvial periods. The estimation is based on the following inferences.

Bottom sediments accumulated in the central basin of the lake generally ought to be constituted by evenly fine grains. However, in the period when a heavy precipitation occurred very frequently in the basin, discharges of river water into the lake must have increased, accompanied by an increase in flow velocity as well as the intensification of turbulence in water. This must have made river water transport not only a great deal of silty grains but also much coarser grains than usual in suspension into the lake. Since the increment of suspended sediments augments the density of river water, enlarging the density difference between river water and lake water, it could have caused turbidity currents at the river mouth, by which a sediment load in turbid river water is brought in along the bottom within the body of lake water. So in past pluvial periods, it is inferred, such coarse grains might have been brought in by turbidity currents onto the central bottom as never observed in arid periods. Now, suppose that the lake has not experienced any notable change in size and form these several hundred thousand years. Then, at the same site of the central part of the lake where only very fine sediments are able to arrive, the coarsened grain size of bottom sediments suggests that age must have been so heavily rainy a period that strong turbidity currents were frequently generated. Thus, this inference leads us to estimate the presence of past pluvial periods. This is the feature of paleoprecipitation based on the grain size analysis of the core (Yamamoto 1974a, b). Despite its importance, our knowledge of paleoprecipitation compares unfavorably with that of paleotemperature.

Method of grain size analysis

With the 200 m core sample, 5 cm-long slender pillars at each 25 cm interval are cut off over the whole length of the core. The grain size analysis has been applied to them by means of the settling analysis with the photo-extinction method. The procedure is as follows: after measuring by a pycnometer the mean grain density for each sample of several hundred mg, fractions coarser than 4.5 ϕ (44 μm) are separated through the wet sieving. The remainder which is finer fractions dispersed by sodium hexa-meta phosphate in water is analyzed by the photo-extinction method using a Hitachi PSA II to obtain their grain size distribution. Grain sizes are wholly expressed on the phi-scale in terms of the equivalent diameter of a sphere having a density of 2.6 g/cm^3, which is the density averaged throughout the core. To represent the grain size distribution, Inman's statistical measures are used here (Inman 1952).

Age-scaling based on the consolidation model

Before describing results of the grain size analysis, it seems useful to explain the composition of the age scale used hereafter.

The original principle

In a chronological study of a core sample, it is necessary to establish the rationale of interpolating between the discrete values of absolute dates.

A sediment layer, once deposited on the surface of the lake bottom, begins to contract its thickness through the consolidation process under the cumulative load of sediments continually depositing above. Suppose that, taking a vertical column of sediments having a unit section, a sediment layer of thickness $d\zeta$ with bulk density ρ_B, which is situated at the depth ζ beneath the bottom surface at present, is originally that which had deposited on the surface t years ago, forming a layer of thickness dS with bulk density ρ_{BO}. During these t years, the layer has experienced contractions of Δ per unit thickness through consolidation, so that $d\zeta = dS(1-\Delta)$. Since the contraction is due to the squeezing of water, the equation of mass conservation then is $\rho_B(1-\Delta) = \rho_{BO} - \rho\Delta$, where ρ is the density of water. Thus, the integration of dS over the last t years, i.e., the total thickness if not yet consolidated, is

$$S = \int_0^\zeta \frac{d\zeta}{1-\Delta} = \int_0^\zeta \frac{\rho_B - \rho}{\rho_{BO} - \rho} d\zeta, \qquad [1]$$

While, assuming the sedimentation rate h(t) to be constant ($= h_0$),

$$S = \int_0^t h(t)dt = h_0 t. \qquad [2]$$

Hence, equating [2] to [1] yields:

$$t = \frac{1}{h_0(\rho_{BO} - \rho)} \int_0^\zeta (\rho_B - \rho) d\zeta. \qquad [3]$$

Equation [3] ensures the age-scaling along the core sample, when the vertical profile of bulk density and reference ages are given.

Bulk density ρ_B obtained from weighing core samples cut by 1 m in length enables us to express the profile of ρ_B as a function of depth ζ (Fig. 210). Using ^{14}C-dates as reference age values, this method of age scaling has been successfully applied to core samples taken from Lake Kizaki (Horie et al. 1980) and Osaka Bay (Yamamoto 1977a) and to the 12 m core from Lake Biwa (Yamamoto 1973).

An empirically modified age-scale

According to Eq. [3], a preliminary age-scale for the 200 m core was composed in reference to the ^{14}C-dates of three layers shallower than 12 m in depth (Yamamoto et al. 1974). For the deeper half of the core, however, it has been recognized that ages on the scale are overestimated as compared with several fission-track dates obtained later (Nishimura & Yokoyama 1975). In re-composing the age-scale by mobilizing all nine available absolute dates of ^{14}C and fission-track, the power law is adopted to relate the age t to the integral I, where $I = \int_0^\zeta (\rho_B - \rho)d\zeta$, instead of that of proportionality between t and I, which has been adopted in the form of Eq. [3] under the condition of a constant sedimentation rate. The better fit of the power law to the 9 reference dates is the reason why the assumption of Eq. [3] was abandoned. Thus, the age-scale is empirically modified as shown in Fig. 211 and is readily available for the chronological treatment of grain size variation. Yet, to anticipate our further argument, the age-scaling must inevitably consider the variability of the sedimentation rate.

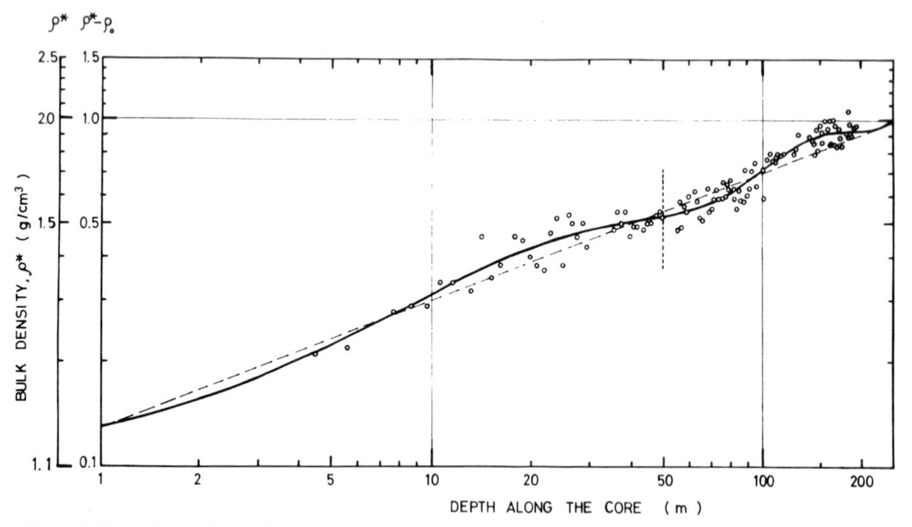

Figure 210. Bulk densities of the core sample for every meter. The increasing densities with depth show the consolidation effect of bottom sediments. Bulk densities as a function of the depth are approximated by a polynomial, of which the curve (solid line) undulates around the curve of a power formula (broken line) (Yamamoto, 1974b).

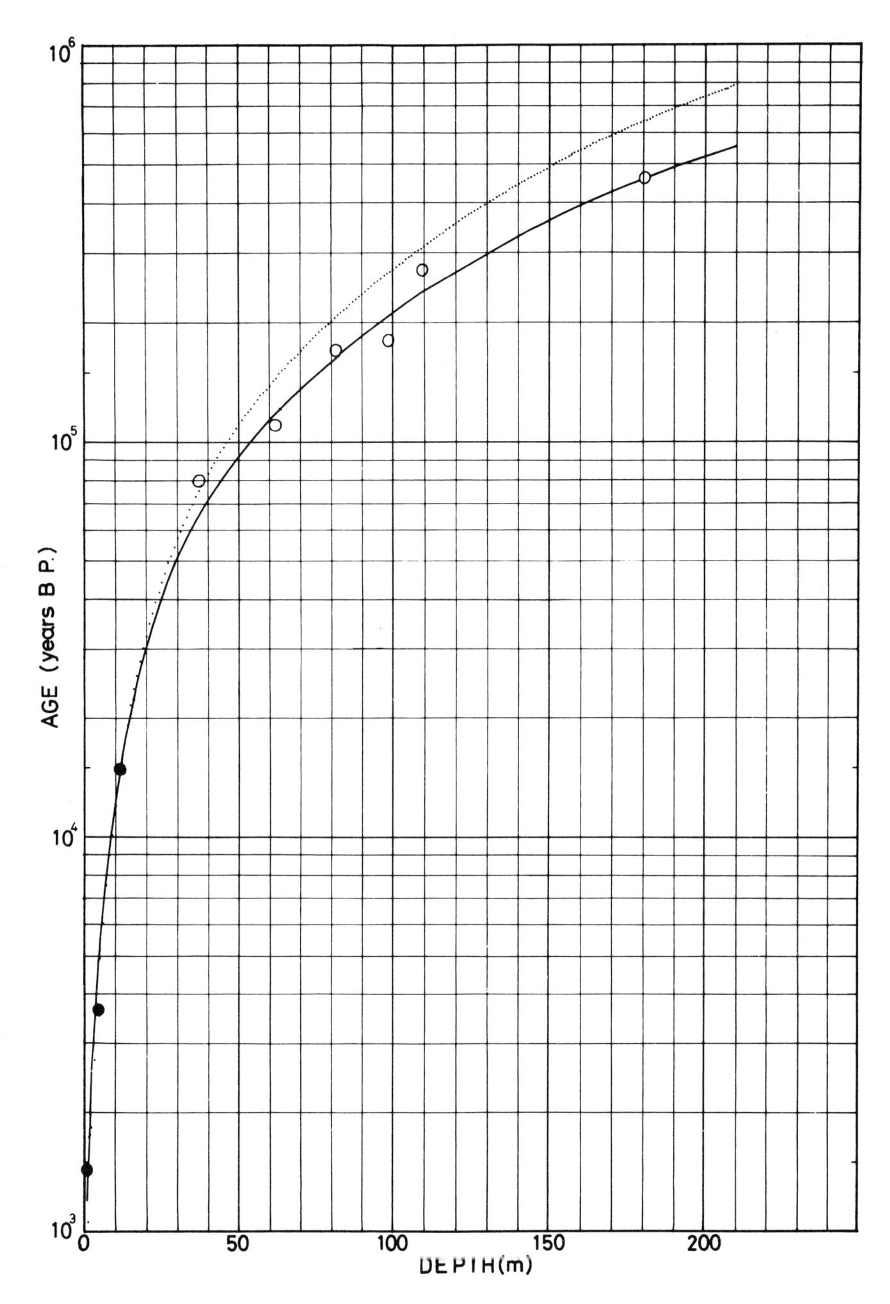

Figure 211. An empirically modified age scale (solid line), with nine reference ages including three ^{14}C-dates (solid circles) and six fission-track dates (open circles). Such an overall fit partially causes minor deteriorations. Dotted line is the former age scale referred to the ^{14}C-dates alone. (Yamamoto et al. 1974.)

Some features of grain size variations

Variations in density and grain size of sediments throughout the 200 m core, obtained from the analysis at intervals of 25 cm, exhibit some interesting features. The vertical variations of granulometric parameters for the core samples are shown in Fig. 212 for the shallower half and in Fig. 213 for the deeper half, where five parameters have been chosen: (a) grain density, (b) fractions finer than 8 ϕ (2 μm), (c) fractions coarser than 4.5 ϕ (44 μm), (d) Inman's phi deviation measure σ_ϕ, and (e) phi median diameter. As to the rest of Inman's statistical measures of grain size distribution, their curves are shown in Fig. 214 only for the uppermost 40 m part of the core, though complete data are compiled in the data series of I to IV (Yamamoto 1974c; Yamamoto & Higashihara 1975, 1976, 1978).

The grain density averaged over the whole core is about 2.64 g/cm^3, while the average over every quarter part of the core changes slightly from 2.66 g/cm^3 at the lowest part through 2.68 and 2.62 to 2.63 at the uppermost one. These values are common to assemblages of particles. However, as seen in Fig. 212 for the upper half of the 200 m core and in Fig. 213 for the lower half, the grain density varies greatly and sometimes reveals peaks higher than 2.8 g/cm^3 and troughs lower than 2.2 g/cm^3. Following the curve from the deeper layer to the

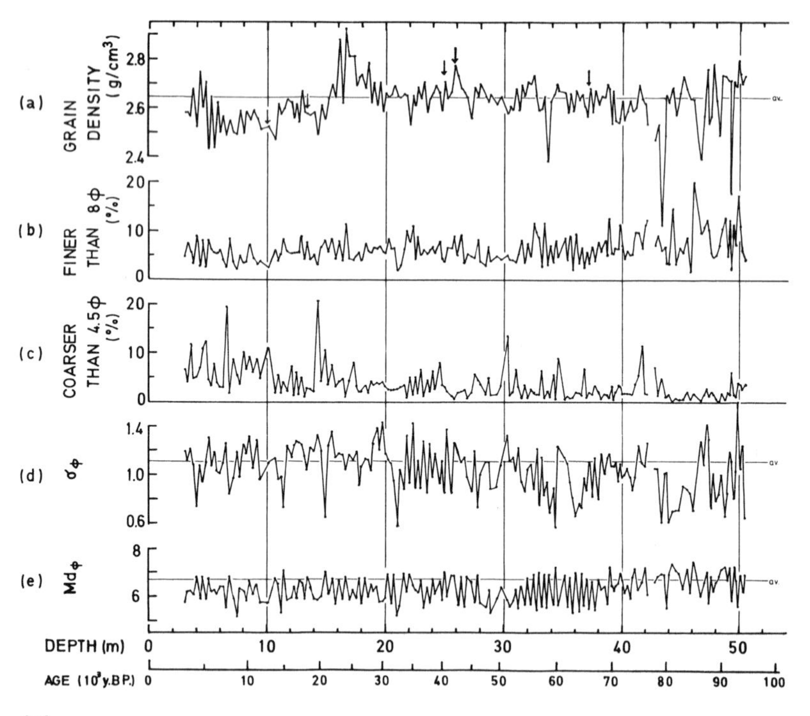

Figure 212.

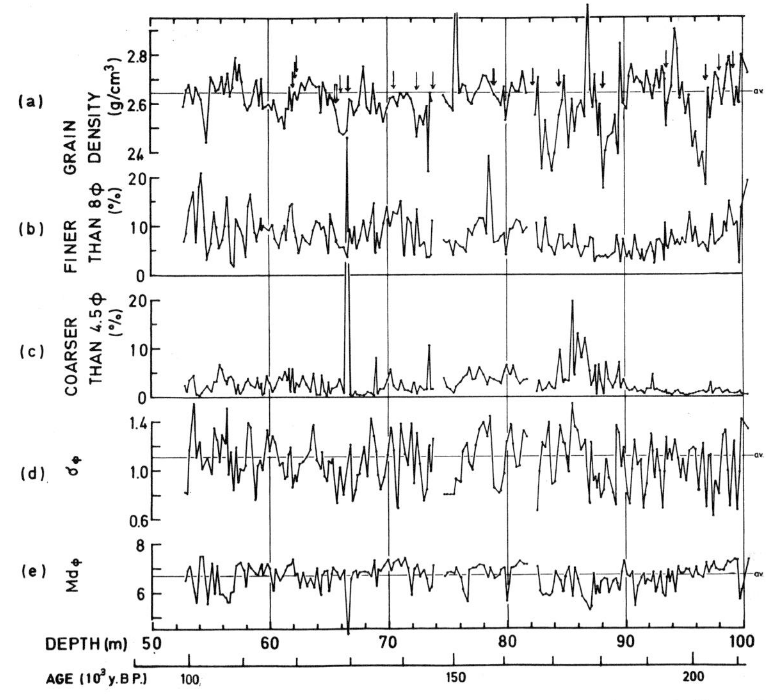

Figure 212. Vertical variations of the density and grain size parameters, in the shallower half of the 200 m core (after Yamamoto 1975 & 1976).
(a) mean density of grains,
(b) contents of fractions finer than 8 ϕ,
(c) contents of fractions coarser than 4.5 ϕ,
(d) phi deviation measure, σ_ϕ,
(e) phi median diameter, Md_ϕ.
Arrow marks upon the curve (a) indicate the locations of the layers of volcanic ash.

shallower, we find a characteristic pattern in density variations, namely, the density abruptly decreases from a high peak to a conspicuously low minimum after fluctuating within somewhat high values and then slowly recovers though accompanied by fluctuations. Moreover, it is to be noted that such an abrupt lowering of density occurs within a short period of 1 to 3 thousand years. For example, the average density is high at the depth of 19 to 16 m (28 to 23×10^3 years BP.), and the highest peak of 2.88 g/cm^3 appears at 16.5 m depth (23.8×10^3 years BP.) among them. From this peak with an extraordinarily high value the density rapidly descends to a low minimum value of 2.49 g/cm^3 within a lapse of about four thousand years. Such a change of density is often found at other horizons of the core, e.g., from 2.70 g/cm^3 at 34.65 m to 2.38 g/cm^3 at 33.9 m, and so on. Two illustrations of the rapid descent near 16 m and near 125 m in depth are shown in Fig. 215.

For the grain size distribution of sediments, both the phi median diameter Md_ϕ and the phi mean diameter M_ϕ average at about 6.5 (11 μm) throughout the whole core and vary in the range of 5 to 7. Of the sorting, the phi deviation measure σ_ϕ, indicating that the smaller the value, the better the sorting, varies from 0.6 to 1.5 with the mean value of 1.1. The fifth percentile ϕ_5 as a representative of the larger sizes of grains averages at about 4.8 ϕ (36 μm). Thus, it is confirmed that the bottom sediments in the central area of Lake Biwa mainly consist of very fine silt of about 10 μm in size since at least 0.5 million years past. Fractions finer than 8 ϕ (4 μm) are over several wt. %. On the other hand, though the fractions coarser than 4.5 ϕ (44 μm) are less than 5 wt. % at most layers, sometimes conspicuous peaks of coarser fractions more than 10% emerge. As inferred above, the content of coarser fractions might be regarded as an indirect measure of the feature of paleoprecipitation at that time, although the extraordinary peak of 53 wt. % in coarser fractions at the 66.4 m layer which originates only in sand-rich volcanic ash is naturally to be excluded. Around the high peak of coarser fractions, there seems to exist a bunch of relatively high peaks. In the curve of coarser fractions smoothed by moving averages (Fig. 216b), bunches of high peaks stand out in relief as protrusions and such periods are considered as pluvials.

Through our inspection of the curves of grain density and grain size parameters, the following characteristics of the variation become evident:

1. maxima and minima in the density curve correspond well to those in the finer fractions, respectively;

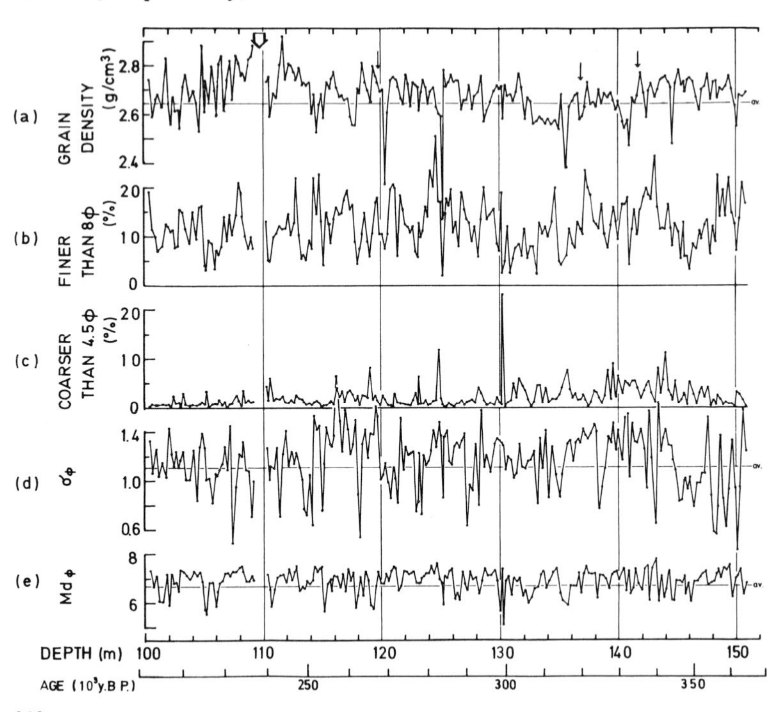

Figure 213.

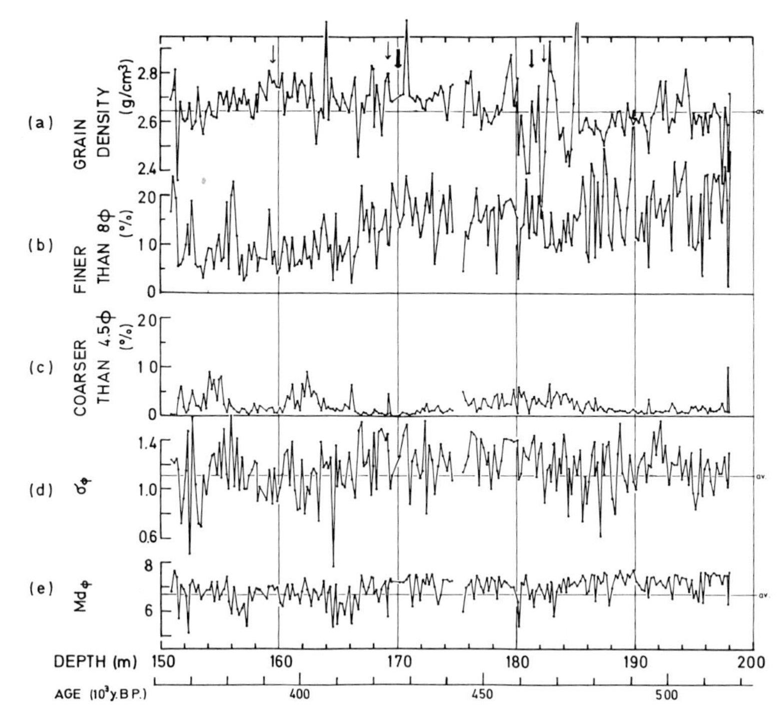

Figure 213. Vertical variations of the density and grain size parameters, in the deeper half of the 200 m core (after Yamamoto 1975 & 1976).
(a) mean density of grains,
(b) contents of fractions finer than 8 ϕ,
(c) contents of fractions coarser than 4.5 ϕ,
(d) phi deviation measure, σ_ϕ,
(e) phi median diameter, Md_ϕ.
Arrow marks upon the curve (a) indicate the locations of the layers of volcanic ash.

2. markedly low minima in density correspond well to conspicuous maxima in the coarser fractions;

3. thus, the density that mounted before rapidly decreases along with the rapid coarsening of sediments;

4. subsequently, with a rapid decrease of density, there emerges an abrupt change of σ_ϕ towards poorer sorting.

However, note that, e.g., at the layers of 97.35 m to 97.1 m, there are no conspicuous peaks of coarser fractions corresponding to the rapid decrease in density, but that their very low contents remain unchanged regardless of the evident occurrence of the characteristic change of density. In this duration, however, the finer fractions tend to decrease slowly and Md_ϕ shows the tendency to coarsen. Accordingly, σ_ϕ as the mean moves towards poorer sorting. And the tendency to coarsen certainly occurs with the passage of time within the fractions finer than 4.5 ϕ.

Turning our eyes not only to the central tendencies but also to the tail in the

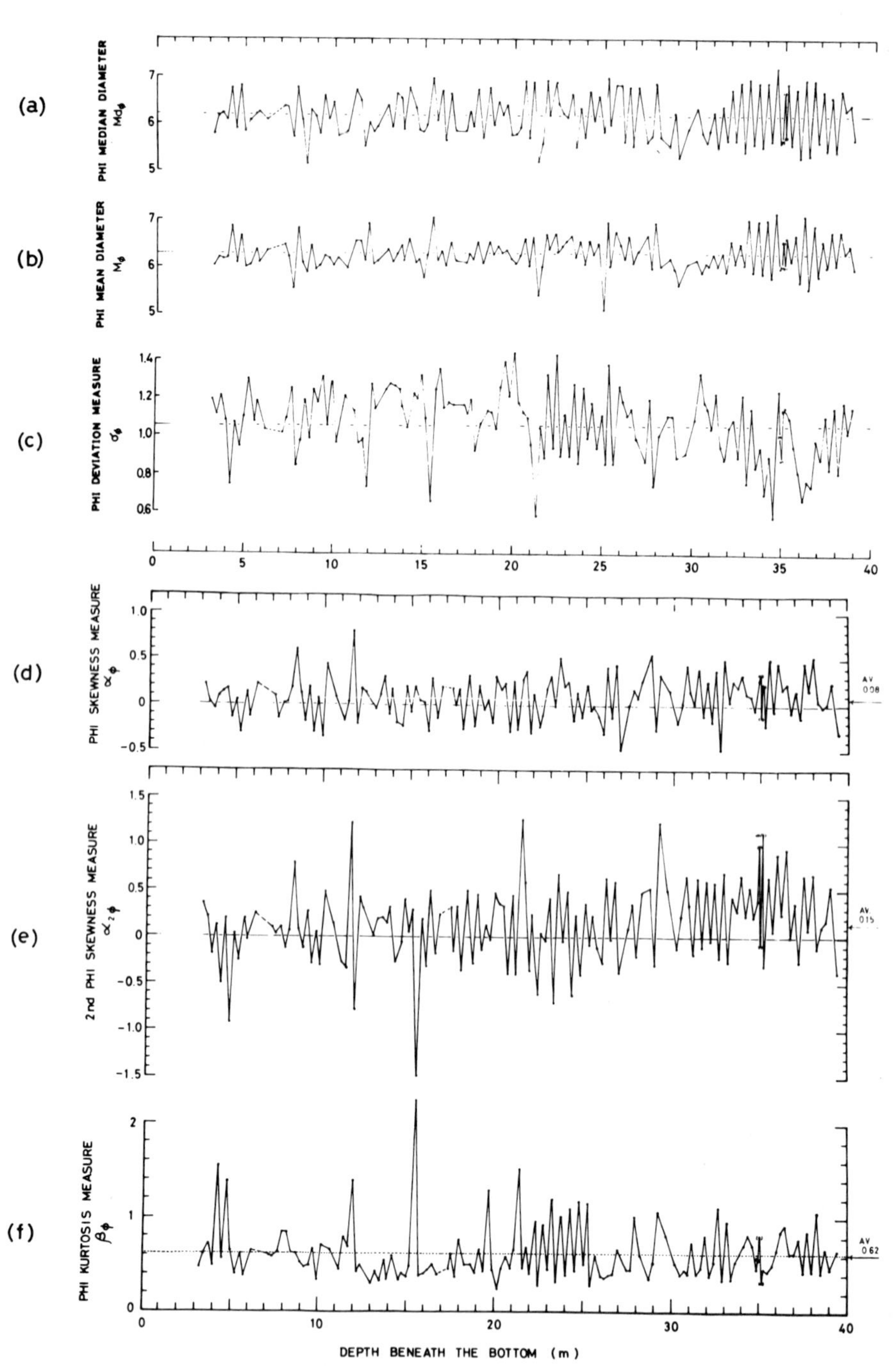

Figure 214. Vertical variations of Inman's descriptive measures of grain size distribution, in the uppermost 40 meters of the 200 m core (Yamamoto 1974b).

(a) phi median diameter, Md_ϕ
(b) phi mean diameter, M_ϕ
(c) phi deviation measure, σ_ϕ
(d) phi skewness measure, α_ϕ
(e) 2nd phi skewness measure, $\alpha_{2\phi}$
(f) phi kurtosis measure, β_ϕ

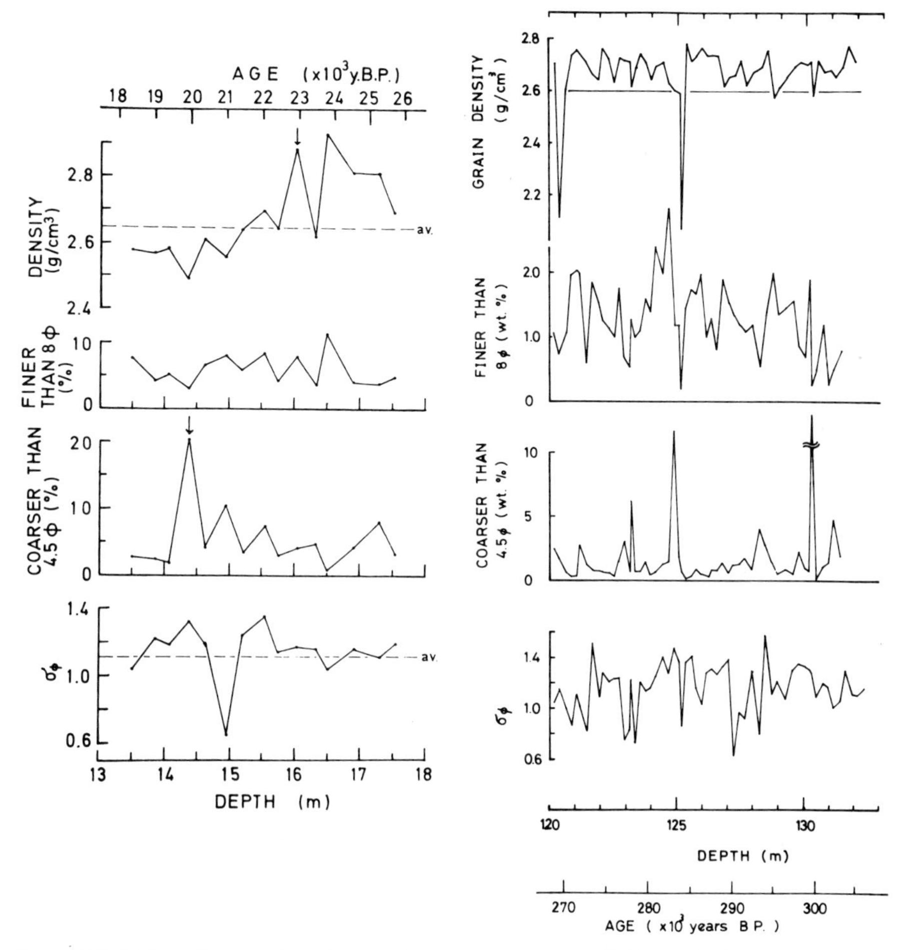

Figure 215. Enlarged parts of vertical variations, near 16 m and 125 m in depth, showing a characteristic pattern of granulometric variations following after the rapid lowering of density (after Yamamoto 1974a, Yamamoto 1976).

grain size distribution, it is noteworthy in the instance of layers at the depth of 16.5 m to 15 m where the density rapidly decreases that the solitary peaks of $\alpha_{2\phi}$ and β_ϕ arise at the 15.5 m layer due to the rush-in of grains near the coarsest size before the peak occurred in the fractions coarser than 4.5 ϕ. Besides, Md_ϕ, σ_ϕ and α_ϕ show a look of finegrained sediment at that layer of 15.5 m. This situation suggests that a substantial coarsening has started at the 15.5 m layer.

The characteristic pattern of variations is illustrated for two specific cases in Fig. 215. Furthermore, the σ_ϕ variations seem to be very useful in our arguments, because, in the boring site where bottom sediments consist mainly of very fine ones, shifts of σ_ϕ towards the poorer sorting are chiefly due to coarser grains being carried.

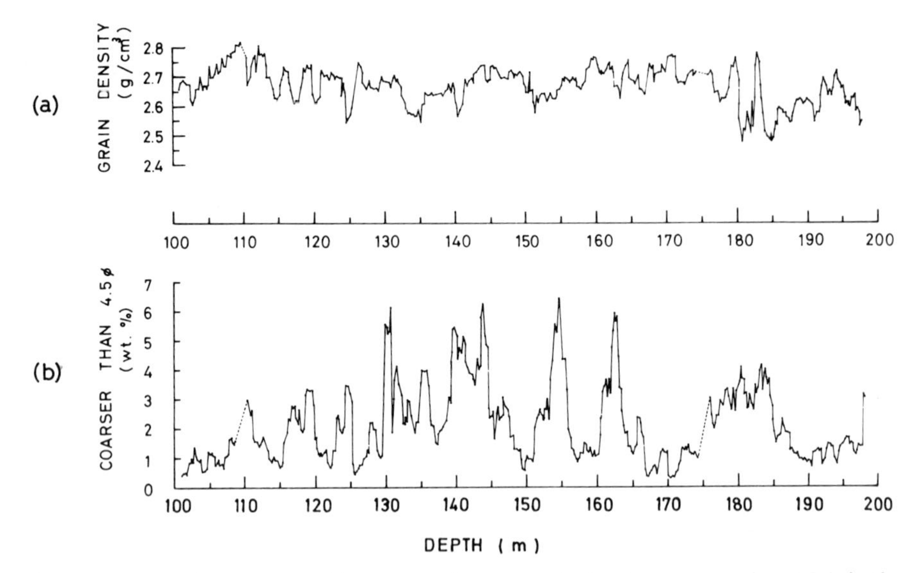

Figure 216. Smoothed curves of (a) grain density and (b) the fractions coarser than 4.5 ϕ, in the deeper half of the 200 m core. Gross amountings in (b) imply the existence of pluvial periods (after Yamamoto 1976).

Paleoprecipitation and climatic change

In line with the view that a temporary increase of very coarse grains in the bottom sediments can be attributed to an increase of precipitation at that time, one may use the contents of fractions coarser than 4.5 ϕ as an index of the frequency and abundance of paleoprecipitation. Thus, paleoprecipitation is redrawn on the empirically modified age scale, in Fig. 217 and Fig. 218, where several pluvial periods are exposed: at the times of about 470 to 450, 400, 380, 330, 300, 260, 180, 160, 120, 75, 60, 45, 30–20, 10 ($\times$ 10^3 years BP.).

In Fig. 217(c) and (d) covering the whole Würm Glacial Age, a comparison of variations of paleoprecipitation since 90×10^3 years BP. (in the uppermost quarter part of the core) with Dansgaard's climatic curve which shows variations of temperature in terms of δ^{18}O in an ice core from Greenland on the time axis based on the ice flow model (Dansgaard et al. 1969) leads us to a working hypothesis that there exists the following regularities that relate the variation characteristic of the granulometric parameters to climatic changes (Yamamoto 1974a, 1976):

1. conspicuous peaks of coarser fractions correspond well to the stage of rapidly lowering temperature;

Figure 217. Comparison of paleoprecipitational change with Dansgaard's climatic curve. Climatic curve (c) is based on δ^{18}O in ice core from Greenland (Dansgaard et al. 1969). Grain density curve (a), smoothed curve of phi deviation measure (b) and variations of coarser fractions as a measure of paleoprecipitation (d) are drawn up on the empirically modified age scale. (Yamamoto 1976.)

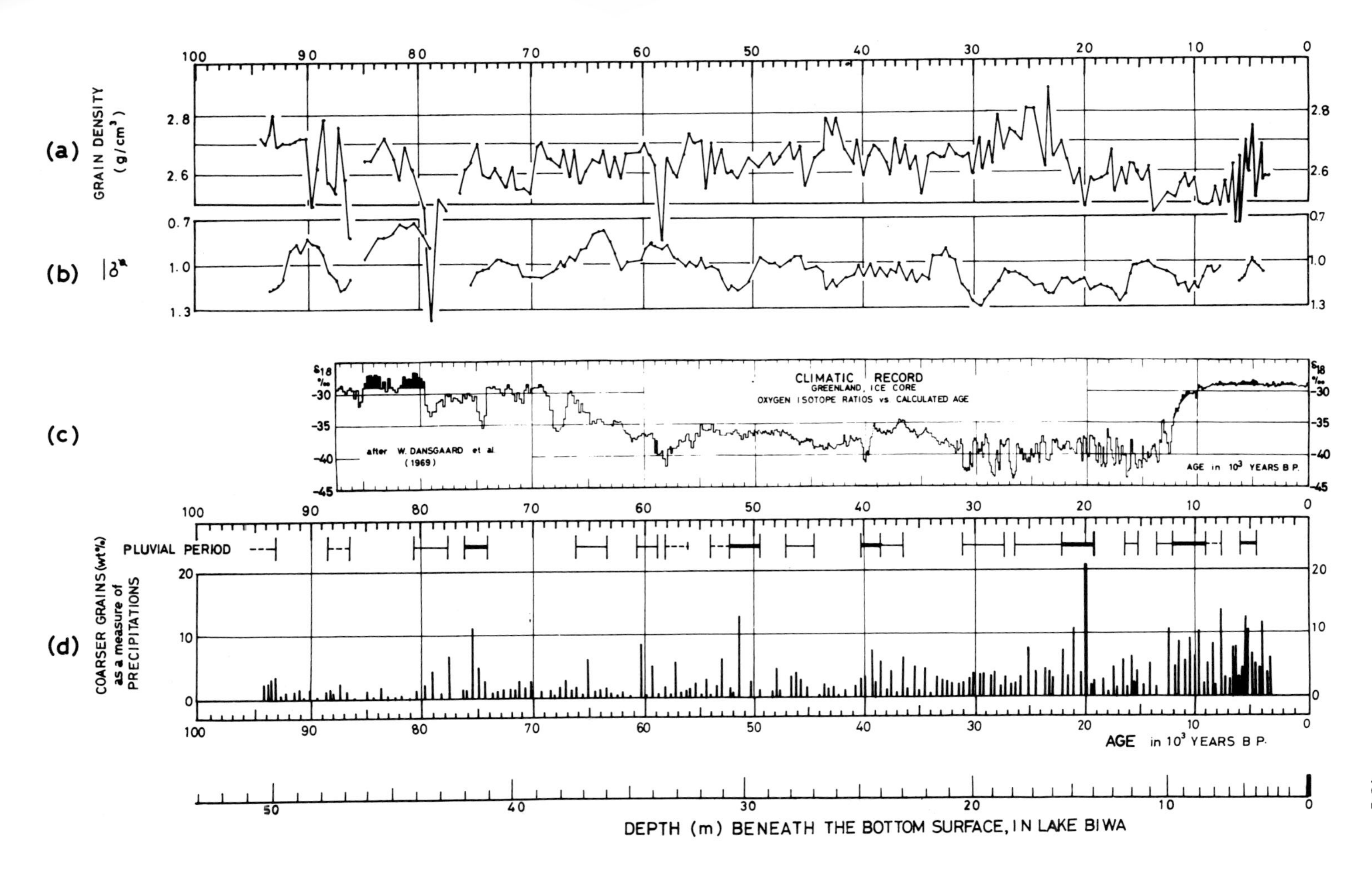
(a)
GRAIN DENSITY
(g/cm³)
(b)
(c)
CLIMATIC RECORD
GREENLAND, ICE CORE
OXYGEN ISOTOPE RATIOS vs CALCULATED AGE
after W. DANSGAARD et al.
(1969)
AGE in 10³ YEARS B.P.
(d)
COARSER GRAINS (wt%)
as a measure of
PRECIPITATIONS
PLUVIAL PERIOD
AGE in 10³ YEARS B.P.
DEPTH (m) BENEATH THE BOTTOM SURFACE, IN LAKE BIWA

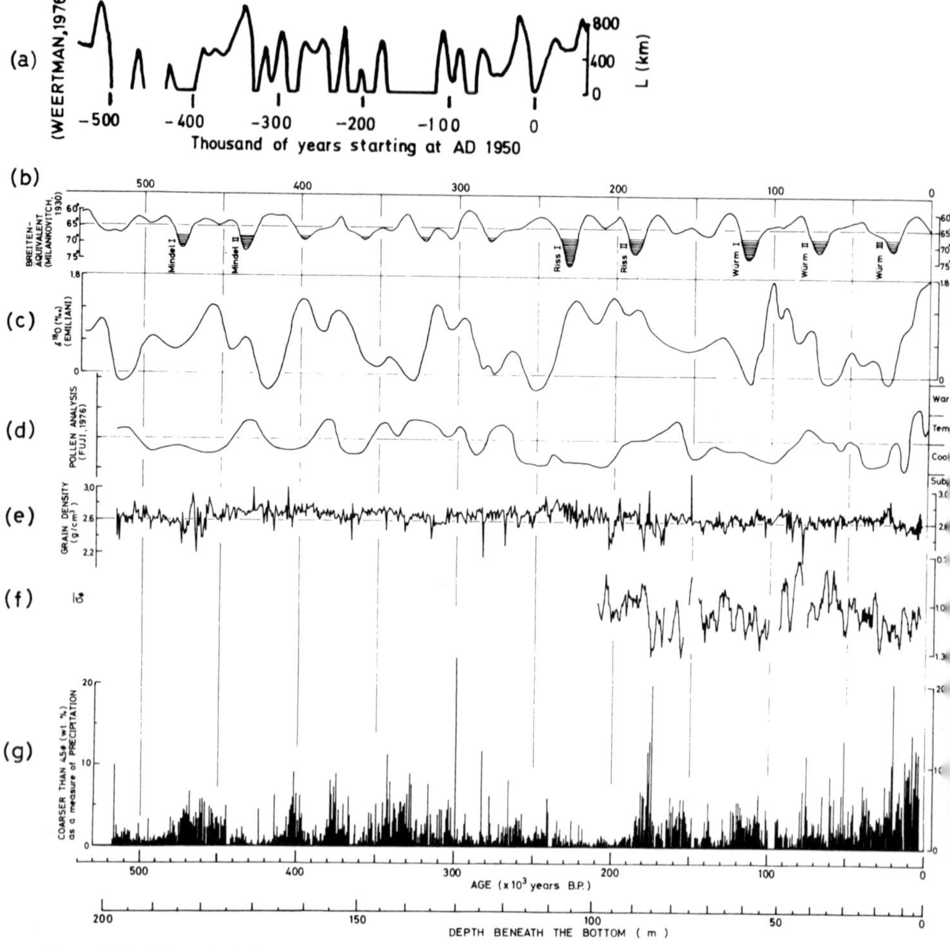

Figure 218. Paleoprecipitational change during the last 520,000 years and some comparisons (after Yamamoto 1976).
(a) Time variations of half width L of ice sheet, calculated using glacier mechanics and Milankovitch solar radiation variations. This curve is redrawn after J. Weertman's article (1976).
(b) Milankovitch's insolation curve.
(c) Emiliani's climatic curve.
(d) Fuji's climatic curve based on the pollen analysis of the same 200 m core sample, redrawn on the empirically modified age scale in common with curves (e), (f) and (g).
(e) Variations of grain density.
(f) Smoothed curve of the phi deviation measure.
(g) Paleoprecipitational change estimated from coarser fractions.

2. a pronounced pluvial period seems to occur mostly in a severely cold period;

3. individual fluctuations of grain density correspond fairly to those of temperature, although not always parallel;

4. hence, a period of rapidly decreasing density from a high maximum,

namely, a period of rapidly coarsening grain sizes or a period of remarkable increasing σ_ϕ (tending towards poor sorting) is regarded as equivalent to a period of steeply cooling climate.

If this working hypothesis is applicable to the whole core going back to 520,000 years BP.,* main pluvials preceding the Würm Glacial Age seen in Fig. 218, such as those at the depth of 55–65 m, 75–90 m, 125–140 m, 155–165 m and 175–185 m, might correspond to cool or rapidly cooling periods, too. First, an attempt is made to estimate climatic changes going back to 220×10^3 years BP. by applying the hypothesis that a period of remarkably coarsening sediments in accordance with a rapid decreasing grain density from a high value corresponds to a stage of severely cooling climate. Then, it is inferred, on the one hand, that the prominent pluvial period from 183 to 175×10^3 years BP. (ca. 89 to 86 m in depth) may be the very beginning stage of pronounced cooling, probably of the Riss Glacial Age. On the other hand, the period of 220 to 205 $\times 10^3$ years BP., when there were no bottom currents capable of bringing such coarse grains as sand, might pertinently be regarded as that of a rather dry climate. Thus, it follows from this view of density variations (1) that, after a fairly calm (or warm) period, a cycle of characteristic variations of density exists prior to the beginning of the Riss Glacials, i.e., fluctuating within the range of fairly high values → very rapid and deep lowering → slow recovering, and represents the preliminary starting (or preceding change) towards the real cooling of climate, (2) that within the period of the 2nd cycle of density variation, the real cooling and pluvials commences, and (3) that in the 3rd cycle, the pluvial climax, which seems to be representative of the catastrophic change, comes to an end. From this point of view, it is supposed that the repetition of rapid decrease in density, starting at ca. 105×10^3 years BP. (55 m), ca. 330×10^3 years BP. (140 m), and ca. 490×10^3 years BP. (190 m) and commonly accompanying some pluvials, may suggest the beginning of the coming glacial age.

In Fig. 218, instead of Dansgaard's curve, Fuji's climatic curve is drawn on the basis of the pollen analysis at rough intervals of 5 m along the same core. A comparison of the density and paleoprecipitation curves with Fuji's climatic curve seems, in general, to support the above-noted hypothesis, with some exceptions, and to suggest that the curve of the sorting σ_ϕ possibly resembles the climatic curve, if smoothed by moving averages in longer terms. In this figure, although Milankovitch's insolation curve is shown (Milankovitch 1930), it seems difficult to find any direct correspondence with paleoprecipitation. However, a comparison with Weertman's curve of ice-sheet extension, which has been computed from Milankovitch's solar radiation and the glacier mechanics of ice sheets acting as a plastic flow, seems to exhibit a good correspondence to the paleoprecipitational variation. It appears to favor the hypothesis.

Now, taking a look at the outline of interrelations among the results from various analyses as shown in Fig. 219, one finds that variations of grain density,

*According to the empirically modified age-scale as mentioned above (Fig. 211), the date of the deepest layer of the 197.2 m long core is almost 520,000 years BP.

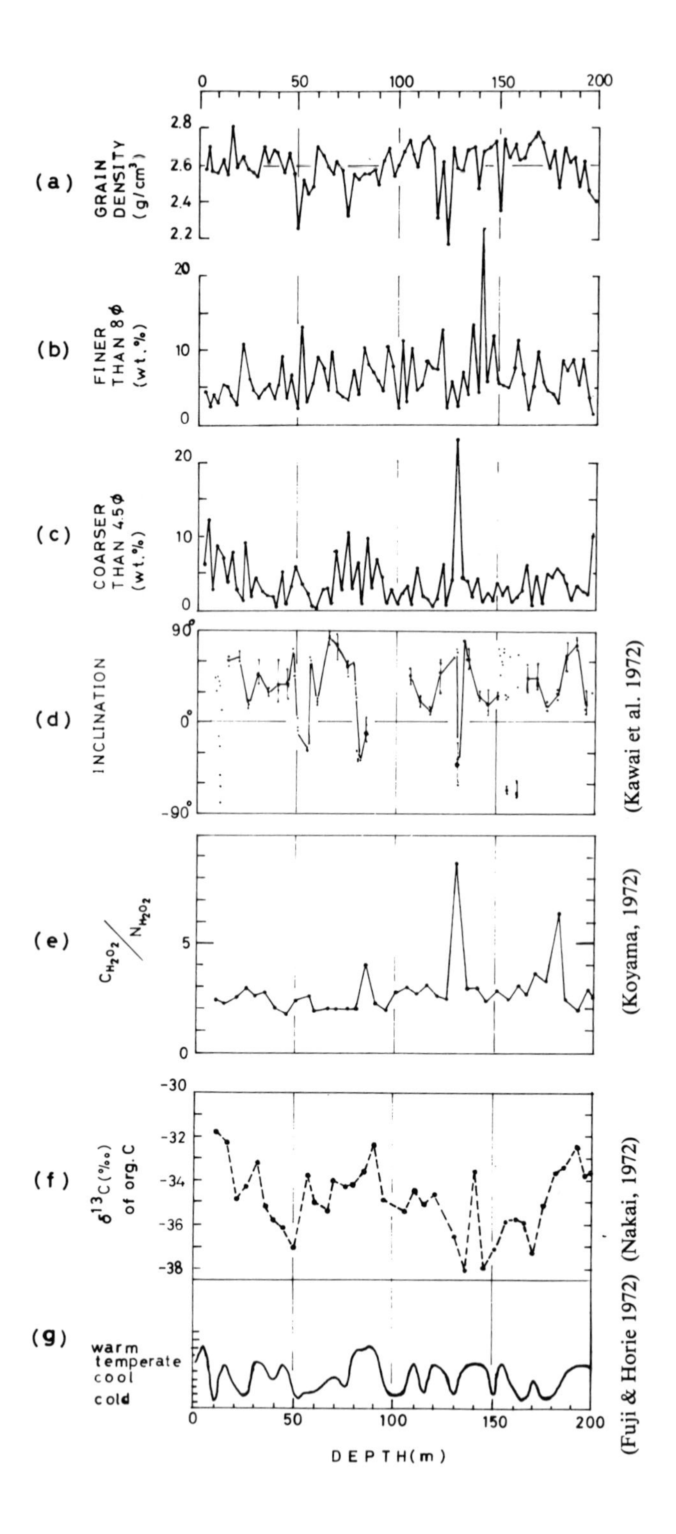

(a)
GRAIN DENSITY (g/cm³)
(b)
FINER THAN 8φ (wt.%)
(c)
COARSER THAN 4.5φ (wt.%)
(d)
INCLINATION
(Kawai et al. 1972)
(e)
(Koyama, 1972)
(f)
δ13C(‰) of org.C
(Nakai, 1972)
(g)
warm
temperate
cool
cold
(Fuji & Horie 1972)
DEPTH(m)

fractions finer than 8 ϕ and fractions coarser than 4.5 ϕ analyzed at the interval of 2.5 m, expose the presence of conspicuous pluvial periods, roughly at the depth of 180 m, 130 m, 80 m, 55 m, and so on, which might be closely related to the beginning of remarkable Glacial Ages, if characteristic variations of grain density are taken into account. In this figure, the period at the depth from 140 m with a finers' peak up to 130 m with a coarsers' peak very much impresses us with the clearness of change. Considering the simultaneous occurrence of the geomagnetic reversal (Kawai et al. 1972) and the extraordinary peak of the C/N ratio (Koyama 1972), this period should have been the ages of high significance to the environments around Lake Biwa, whether it was the beginning of Glacial Age or not. It is very interesting to observe that the geomagnetic events occurred three times at least in synchronization with the marked cooling periods estimated from grain size variations, respectively. Moreover, their concurrence with distinct peaks of the C/N ratio might be a clue to the conditions of river inflow loaded with sediments. Two kinds of climatic curves are shown in the figure: one is the $\delta^{13}C$ curve (Nakai 1972) which is regarded as the variations of lake water temperature and the other is the climatic curve based on the pollen analysis indicating climatic temperature in the areas neighboring the lake (Fuji & Horie 1972). Both climatic curves, it seems, allow us to conclude that the fire periods pointed out were the cooling or cooled periods.

Some considerations

As to the premise in estimating the Paleoprecipitation

In the inference that the coarseness of bottom sediments depends upon the precipitational change, the constancy of size and shape of the lake has been assumed at first. There might be a thought that ascribes the grain size variations at a given site in the central area of the lake to the change of the distance from the shore and, so, to the change of size and form of the lake. In order to support this thought, however, the requirement must be satisfied that the lake level change of more than 20 meters should have occurred very frequently. Such evidence seems to be unverified. In this connection, we may note that depending on the change of lake size alone seems to be less realistic than assuming the rough constancy of Lake Biwa during the last 500,000 years. Nevertheless, the second assumption of turbidity currents bringing in coarse sediments is not always complete because it is not certain if such strong turbidity currents capable of

Figure 219. A general survey of various changes throughout the whole 200 m core. Note that the periods of rapid decrease in grain density correspond well to those of conspicuous change in other curves prepared from independent analyses. (a) grain density, (b) fractions finer than 8 ϕ, (c) fractions coarser than 4.5 ϕ, (d) inclination of permanent magnetism (Kawai et al. 1972), (e) ratio of total carbon to nitrogen, treated with hydrogen peroxide (Koyama 1972), (f) $\delta^{13}C$ of organic carbon (Nakai 1972), (g) paleoclimatic curve based on the pollen analysis (Fuji & Horie 1972) (Yamamoto 1976).

bringing in coarse sediments at 130 m-layer might have actually occurred. To answer these questions needs more information based on hydrology, hydraulics and paleogeography (Yamamoto 1974a).

As to the variation of grain density

It is striking that the density of bottom sediments in lakes varies over a wide range of values reflecting the climate in the period of deposition. As the mean density of sediment grains depends on the assorting of different grains, sediments of extraordinarily high density should contain more grains of heavy minerals than in the ordinary clastic sediments of 2.6 to 2.7 g/cm^3. Likewise, sediments of extraordinarily low density should consist of plenty of light minerals or some organic matters. The local parallelism between the density curve and the curve of fractions finer than 8 ϕ suggests that the cause of mounting density should lie in very fine particles of sediments and, hence, the possibility has been pointed out that what controls the mounting of density could be minute particles of iron or iron oxides (Yamamoto 1974c). If minute spherules of iron truly play the role of giving rise to the density variation, then the concentration of such spherules in sediments must be related to the mechanism of climatic change.

In considering the causes of decreasing density of sediments, except one due to some organic matter, it seems unreasonable to overlook the influences of volcanic ash on the density of sediments as a mixture of different clastics. Arrow marks (↕) in Figs. 212 and 213 indicate the locations where layers of volcanic ash interlie among layers of ordinary clastic sediments, representing a bulk layer with a bold arrow and a thin layer with a fine arrow. By a close inspection of the density curves, the arrows (↕) are found to agree with the minima or stages of decreasing density, in spite of there being no plots for the layers consisting of volcanic ash alone, except at the 66.4 m layer. The density of volcanic ash itself found in this core is 2.2 to 2.4 g/cm^3, being fairly low when compared with ordinary sediments. Therefore, low density values in sediments are mostly realized when ordinary clastics are assorted with volcanic ash in different proportions. Now, suppose that volcanic ashes once accumulated on the surrounding basins on land around Lake Biwa are eroded and transported far out very slowly so as to settle down on the lake bottom with various time lags from the time of eruption, then there will be a very long interval between the ashfall by the eruption and the settlement on the lake bottom of ordinary clastic sediments lacking in volcanic ashes. In consideration of an inference that there may have intermittently existed such geologic times when volcanic eruptions frequently occurred on a world-wide scale (Lamb 1971; Budyko 1974), it seems not so absurd to regard the volcanic ashfall as one of the main causes of density lowering in sediments. The finding that a rapid decrease of density corresponds to a cooling of climate implies naturally a possibility that a persistent obstruction of solar radiation due to the volcanic ashes frequently ejected over the globe may have produced a strong effect on the climate, not to say a radical effect on climatic changes.

As to the closely analyzed results and the periodicity seen therein

Now, it seems meaningful to note the results of detailed grain size analysis which has been carried out using 5 cm-long samples taken without any skip along the core, both for the whole 12 m core and the part near 130 m in depth of the 200 m core.

First, with respect to the 12 m core, Fig. 220 shows variations smoothed by 5-

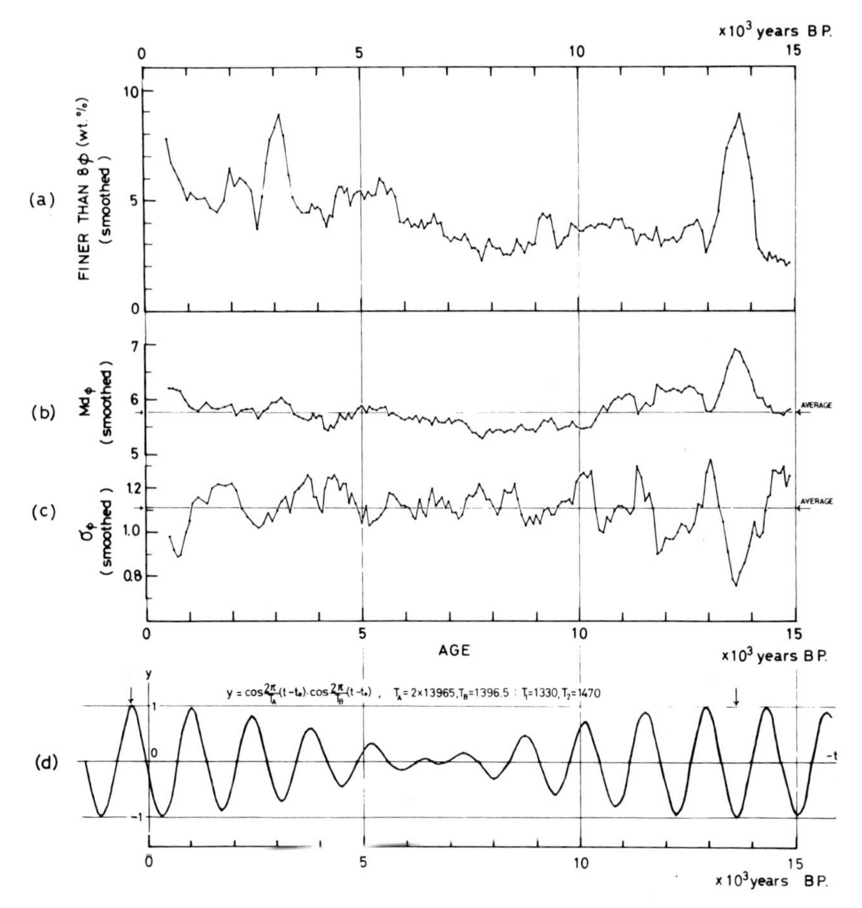

Figure 220. Granulometric change along the 12 m-long core sample taken from Lake Biwa, covering the whole Post-glacial Age. Note that the smoothed curve of phi deviation measure σ_ϕ reveals a periodicity of about 1,400 years and seems to be simplified by the curve of beat oscillation (the lowest curve) (Yamamoto 1977b).

point moving averages of three grain-size parameters, namely, (a) fractions finer than 8ϕ, (b) Md_ϕ and (c) σ_ϕ. Though this figure does not show important curves of grain density and fraction coarser than 4.5 ϕ, it could be deduced from the hypothesis presented in the previous section that the time of ca. 13×10^3 years BP. when a steep increase of σ_ϕ followed by a rapid decrease in the finers might be a severely cooling period accompanied by pluvials. This period seems to correspond well to the famous cold period (Older Dryas) in the Late Glacial Age. The time of ca. 10.5×10^3 years BP. with another steep increase followed by a coarsening of Md_ϕ, even though not so conspicuous, corresponds also to another cold period (Younger Dryas). Although comparisons in Postglacial Age are not described here, it is necessary to mention in passing that the grain-size parameters have seemingly varied in accordance with the cooling of climate since the pre-historic times, too (Horie & Yamamoto 1977).

In this connection, it is very interesting to observe that the smoothed σ_ϕ curve seems to be periodic of about 1,400 years and morphologically to have a beat of oscillations as if the composite of two simple harmonic motions with a little different periods (Fig. 220(c) and (d)).

Next, turning to the 200 m core, the writer shows the results of detailed analysis of samples by 5 cm in length in Fig. 221, using only two parameters, namely, those of (a) fractions coarser than 4.5 ϕ and (b) grain density. The coarsers' curve is regarded as representing a somewhat detailed structure of paleoprecipitational change and needs to be elaborately inspected. In the present context, however, our interest lies in density variations since the density curve looks like being of periodic character.

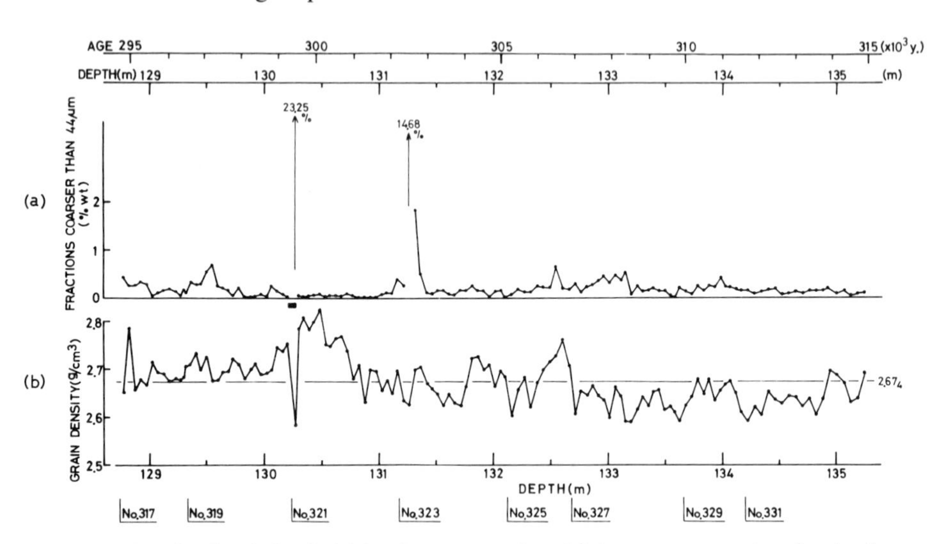

Figure 221. Details of variation in (a) fractions coarser than 4.5 ϕ as a representative of grain size parameters (upper) and (b) grain density (lower), in depths from 129 m to 135 m along the 200 m core. A block mark ■ at the depth of about 130.3 m indicates a geomagnetic reversal named Biwa II event (Yamamoto 1977b).

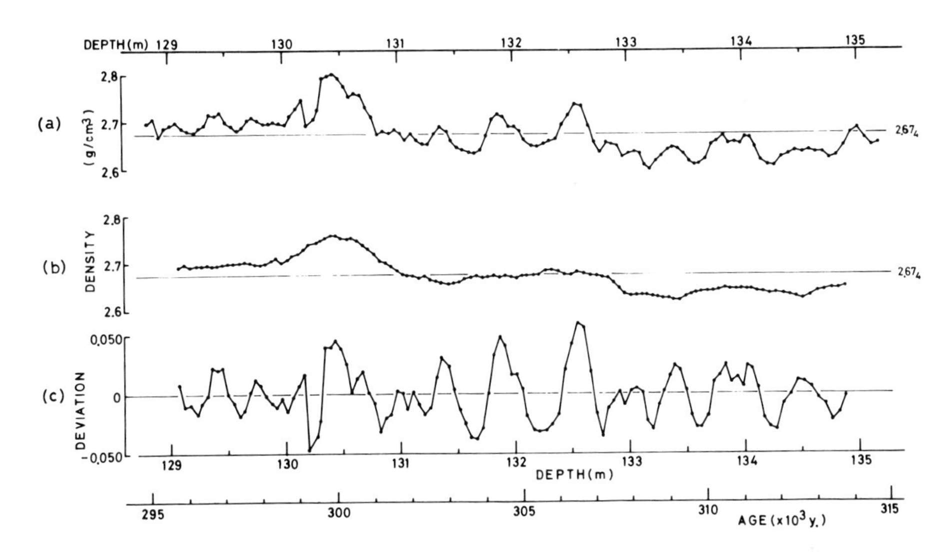

Figure 222. Curves of density variation, smoothed by (a) 3-point and (b) 13-point moving averages. Curve (c) represents the deviations of (a) from (b), revailing a periodicity of about 1,500 years (after Yamamoto 1977b).

To clarify the situation, two smoothed curves of density by 3- and 13-point moving averages are presented in Fig. 222. (a) The former curve is suggestive of the periodicity in density variations and (b) the latter exposes a step-by-step mounting of density as a part of the secular variation. Subtracting (b) from (a) yields the curve (c) in Fig. 222. Though not harmonic-analyzed, the curve (c) manifests the density oscillating with the period of about 1,500 years. A rough comparison of Fig. 221(a) with Fig. 222(c) shows that it is likely that pluvial periods emerge more or less every cycle of density variation. It is very noticeable that periodicities of Fig. 220(c) and Fig. 222(c) approximate each other (Yamamoto 1977b).

Sea level changes in the Quaternary in Japan

Yô Naruse and Yoko Ota

Historical review

Though R.A. Daly's glacial control theory had already been introduced into Japan as early as the 1920's, the geomorphological and geological significance of the glacial eustasy to the Quaternary development of coastal areas was not understood until 1950. For it has long been believed in Japan that the glacial eustasy, if present, might be hidden away in the tectonic eustasy in a tectonically active coast such as the Japanese coast. Therefore, a regional transgression or regression was attributed exclusively to a regional crustal movement (Yabe 1929; Otuka 1948). For the first time, the theory of the glacial eustasy was applied in the 1950's either to explain the post-glacial (the Yurakucho) transgression in South Kanto (Sugimura 1950) or to explain the development of the submarine topography around Kanto (Kaizuka 1955). During the 1960's the age-depth relation of the late Quaternary marine formation came to be elucidated as an increase of the data on radiocarbon measurement, and the attitude of the sea level change from the latest Pleistocene to the Holocene time was amply clarified. Since then, the upland surface developed in the rear of the present coastal plain and the constituting late Pleistocene formations (the Shimosueyoshi Formation and its correlatives) have been investigated in detail. Then, it has been thought that, judging from the progress of the transgression in relation to the climatic change, the Shimosueyoshi Transgression was a noticeable transgression prior to the post-glacial one and that, therefore, it was caused by the rising sea level in the Last Interglacial (Kanto Loam Research Group 1960). In the 1970's, the fission-track method has been applied to the late Pleistocene period prior to 40,000 years BP., and the above correlation has been supported radiometrically. Now, aspects of the sea level changes during the middle to early Pleistocene time have gradually been elucidated on the basis of

This section is devoted to introducing extremely deep coring operations in 1982–83 on Lake Biwa, with a discussion of the problem of possible invasion by ocean water in ancient times (Shoji Horie).

Horie, S. Lake Biwa

ISBN 90 6193 095 2. Printed in The Netherlands

the stratigraphy in relation to climatic changes and radiometric ages of formations of the days.

The sea level change in the Holocene

The sea level change in the Holocene, one of the problems that have been discussed recently in Japan, has been studied actively by many researchers. Previous studies and related problems on the Holocene sea level change have been summarized by Iseki (1977).

In what follows, special attention is to be paid to the sea level change in the last 10,000 years, covering many different opinions as to the sea level height of 'Climatic Optimum' and the sequence of the sea level after that.

Post-glacial Transgression of ca. 6,000 years BP.

It is obvious that there was a large-scale transgression in the Japanese Islands sometime in the Holocene, as indicated by the following facts:

1) The Holocene marine terraces are found along almost all the shoreline, surrounding the Pleistocene terraces and hilly lands. The Numa Terrace in the southern tip of the Bōsō Peninsula in the Kanto area is the most famous one.

2) Holocene coastal plains of various extents are distributed in the lower course of relatively large rivers, and consist of thick marine formation which buries the irregular relief of bedrock topography.

3) Many shell mounds consisting of marine shells, remains of the Jōmon age, are found on the above-mentioned marine terraces and coastal plains.

This transgression is called the Yurakucho Transgression, after the name of a marine Holocene bed of the Tokyo area or the Numa Transgression after the Numa coral bed which consists of the Numa Terrace in the Bōsō Peninsula. It is also called the Jōmon Transgression after the age of the remains. Hereafter, we use the name of the Numa Transgression to represent the post-glacial transgression. In the last 20 years, various kinds of samples obtained from the constituents of Holocene terraces and coastal plains or remains on them have been subjected to ^{14}C age measurements. Their radiometric age shows that the Numa Transgression culminated at about 5,500 to 6,500 years BP., almost simultaneously throughout the Japanese Islands, as indicated by the following examples:

Listed from the north to the south — 5,500–6,500 years BP. (peat and marine shells from the Tsugaru Plain, Umitsu 1976), 5,000–6,600 years BP. (marine shells from the Miura Peninsula, Matsushima 1976), 5,600–7,800 years BP. (corals or shells from South Kanto, Yonekura 1975), 5,500 years BP. (marine shells from the Nōbi Plain, Furukawa 1972; Iseki 1977), 6,000–7,000 years BP. (marine shells and woods from the Osaka Plain, Kajiyama & Itihara 1972),

6,000–6,300 years BP. (marine shells from the Osaka Plain, Maeda 1976) and 6,000–7,000 years BP. (corals from Kikai-jima Island, Ota et al. 1978; Nakata et al. 1978).

Moreover, it is noted that the temperature at the time of the transgression was higher than both before and after the event. According to the palynological data, the temperature about 6,000 years ago was 2–3°C higher than at the present, resulting in a 200–300 m rise of the timber line (Fuji 1966, 1975a; Fujii & Fuji 1967).

Littoral molluscan fauna found along the Pacific coast of Japan suggests that the water temperature was at most about 5°C higher than at the present. This means that the water condition of that time seems to correspond to that of 1–6°C lower in latitude (Matsushima & Ohshima 1974). The Numa coral bed also indicates a warmer water temperature in the time of its formation.

Consequently, it is obvious that the Numa Transgression does not represent the regional subsidence of the Japanese Islands in the Holocene, but has originated from the eustatic sea level rise of Climatic Optimum. Therefore, it is correlated with the Flandrian transgression.

The sea level height at the time of the Numa Transgression

The present height of the marine terrace formed by the Numa Transgression is usually less than several (5–6) meters, but in the seismo-tectonic areas along the Pacific coast, it reaches to 25 m at the southern tip of the Bōsō Peninsula and the Ōiso area, 13 m at the Muroto Peninsula and Kikai-jima Island. Even along the Japan Sea coast, it reaches to 10 m at some places. Moreover, the height of the Holocene terrace is different from place to place even in each individual area and the mode of vertical deformation deduced from the height distribution of the Numa Terrace is very similar to that of the late Pleistocene terrace. Therefore, it can be said that the tectonic movement has continued with a uniform mode in each area. It means that the present height of the Numa Terrace must be the sum of the vertical tectonic movement and the sea level lowering in the Holocene, as noted by previous works. Thus, to estimate the former sea level height of the Numa Transgression, it is necessary to separate the effect of the tectonic uplift from the present height of terraces. The first attempt to do this was made by Sugimura and Naruse (1954, 1955). They noted a positive correlation between the amount of the seismic uplift of the 1923 earthquake and the height of the Numa Terrace and devised the following formula: $y=6+11x$ where y is the height of the Numa Terrace and x the coseismic uplift in 1923. Thus, they concluded that the sea level lowering since the time of the Numa Transgression must be 6 m.

Many other estimations have been undertaken since then. Ota et al. (1976) assumed in their study of Sado-ga-shima Island, for example that the sea level in the Numa Transgression had to be 2 m higher than the present one because the height difference between the Numa Terrace and the 1802 earthquake terrace was 2 m whatever may be the amount of the coseismic uplift, which was not the same from place to place. Other studies of many different areas reported that the sea level at that time could be about 5 m higher (Fuji 1975a; Sakaguchi 1963; Fujii & Fuji 1967; Maeda 1976) or 3–4 m higher (Kaseno et al. 1972; Umitsu 1976; Furukawa 1972; Nakata et al. 1978). Some scholars insisted that the height was almost the same as today (Kajiyama & Itihara 1972) or 0–3.5 m higher (Hasegawa 1975).

As these estimations are usually based on the present height of a Holocene terrace or marine Holocene formation, the effect of the tectonic movement may still remain. However, it is known that the upper limit of a marine Holocene formation with a shell bed is 2 m high in the Nōbi Plain where the tectonic uplift was not recognized (Iseki 1977). In addition, even in a subsidence area, the upper limit of a marine Holocene formation is 1 m high (Sakaguchi 1968). Hence, considering the influence of land subsidence, the sea level was estimated to be a few meters higher than at present. Thus, it can be said that the sea level height in the Climatic Optimum ca. 6,000 years ago was slightly more (2–3 m) than today. Therefore, Shepard' or Curray's curves which indicate that the present sea level is the highest than ever before are not applicable to Japan.

Sequence of sea level change after the Numa Transgression

A question is asked: How has the sequence of the sea level been in these 6,000 years? For example, through the vertical cyclic change of facies of marine Holocene formation fluctuations of the sea level were estimated to be two or three times (Omoto & Ouchi 1978) or only once (Furukawa 1972). Especially, a small-scale regression in 2,000–4,000 years BP. was commonly recognized in the Japanese Islands through the following facts (Iseki 1977): buried channels which were observed in the coastal areas of the fluvial plain in about 3 m deep; black sand layers in coastal sand dunes with archaeological relics of the later Jōmon to Yayoi or Kofun period; buried forests which proved a cooler climate; peat beds below the sea level, and so on.

The period of this regression is thought to coincide with that of a cooler climate of 1.5°C less than today, which was estimated by the analysis of palynological data. Thus, this regression was possibly derived from the sea level lowering in close connection with the climatic change. The sea level in those days is assumed to have been 2–3 m lower than at the present.

At the same time, these fluctuations in the sea level might reflect intermittent tectonic uplifts. The Numa Terrace of a tectonically unstable area is often subdivided into a few steps. For example, in the Bōsō Peninsula, the terrace is classified into four steps, which are dated at ca. 5,500–6,000, 3,600, 2,800, and 300 years BP., respectively by the ^{14}C method. These steps are interpreted as a result of coseismic uplifts by huge earthquakes (Matsuda et al. 1978). In Kikai-jima Island, which is located in the northern part of Ryūkyū Islands, Holocene-raised coral reefs are also subdivided into four steps, whose ages are ca. 6,000–6,500, 4,000–5,000, 3,000 and 2,000 years BP. in the descending order (Ota et al. 1978). Such a step-making process may reflect seismic uplifts just like the terraces in the Bōsō Peninsula. However, the possibility of the sea level change is also expected in the formation of some of the steps, because, for example, the second step is wider and well-developed coral reefs are widely distributed, suggesting a sea level rise at the time of this step formation. Fluctuations in the relative sea level of about the same age as in Kikai-jima Island is assumed in the San-in area through the analysis of submerged coastal topographies and archaeological relics below the sea level (Toyoshima 1978).

The similarity in the number of steps and their age in different areas suggests a possible relation between the subdivision of the Holocene terrace and eustatic change in the sea level, although some of the steps in tectonically active areas must be attributed to coseismic uplifts. It is noteworthy that the regression prior to the formation of the second step such as in Kikai-jima Island about 5,000 years ago, which was not found in the past geological or stratigraphic works, is now alleged by the geomorphological method.

Fig. 223 shows a few examples of the sea level curves in the last 10,000 years. As shown in this figure, it is obvious that the sea level of Climatic Optimum of about 6,000 years ago (Numa Transgression) was slightly higher than the present level (possibly less than 5 m), and reached the present level after a few fluctuations. It must be added that the sea level curves in Fig. 223 are 'relative' ones which are resulted from the eustatic sea level change plus the regional tectonic uplift. A further study is proposed to find the absolute sea level change by eliminating the influence of the tectonic movement, especially of epeirogenetic ones.

The sea level during the late Würmian stage

Low stand of the sea level in the latest Würmian time

The amount of the drop of the sea level towards the Würm maximum is estimated at −100 to −140 m from either the depth of the mouth of the buried valley beneath the late Quaternary marine deposits (the Yurakucho Formation and its correlatives) (Kagami & Nasu 1964) or the depth of the abrasion surface at the edge of the continental shelf (Mogi & Sato 1975). Iseki (1975) illustrated it by diagrams showing the relation between the depth of a buried valley flat and its distance from −100 m and −140 m isobaths around Japan. As for the period of the lowest stand of the sea level, Kaizuka and Moriyama (1969) correlated the Mochi Terrace surface in the River Sagami Basin with the lowest stand of the sea level, examining the projected profile along the River Sagami. The radiometric age of the Mochi Terrace was estimated at about 17,000 years BP.

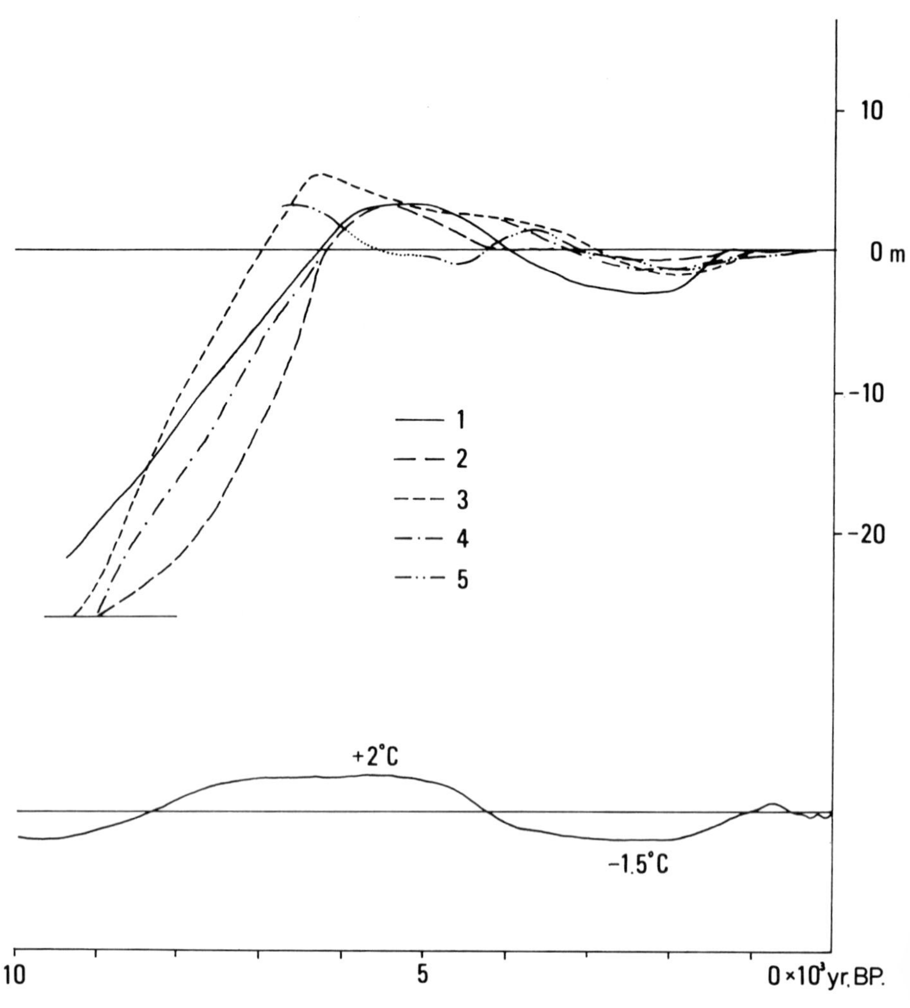

Figure 223. Changes in the sea level and the temperature in the Holocene.
Upper: sea level curve 1. Iseki 1977; 2. Maeda 1976; 3. Fujii and Fuji 1967; 4. Kaizuka et al. 1977; 5. Ota et al. 1978.
Lower: temperature curve, Fuji 1975a.

(Machida et al. 1974). On the other hand, the radiometric age of the lower part of the late Quaternary marine formation (the Nōbi Formation) in the north of Ise Bay is dated at 18,000 ± 500 years BP. (Furukawa 1972). These data indicate that the age of the lowest stand of the sea level in the latest Pleistocene falls between about 20,000 and 17,000 years BP.

A land connection between the Japanese Islands and the Asiatic continent at this time is assumed, on the basis of the submarine topography of the Tsugaru Strait and the Tsushima Strait. This is supported by the fact that *Alces* and

Myopus, both members of the Mammonth fauna, are reported from Honshu in the late Würmian age (Suzuki & Kamei 1969).

Problems on the Interstadial sea level of ca. 30,000 years BP.

As the fluvial Tachikawa Terrace surface (Tc–1 surface, ca. 30,000 years BP.) is buried beneath the present alluvial plain in the lower reach of the river around Tokyo Bay, the sea level of ca. 30,000 years BP. (Farmdale Interstadial) is supposed to have stood lower (Hatori et al. 1962). Kaizuka and others estimate the sea level of Tc–1 time at −40 m in South Kanto.

On the other hand, the terrace surfaces composed of marine deposits of ca. 30,000 years BP. have been reported from more than 10 localities of Japan, such as the Itami Terrace in Kinki (+31.6 m, 32,700 ± 3,300–2,500 years BP., Huzita & Maeda 1971), the Kawaminamibaru Terrace in Kyūshū (+60 m, 26,100 ± 900 years BP., Hoshino 1971), the Jinya Terrace in Hokkaido (+4–20 m, 32,100 ± 940 years BP., Omori 1975), etc. Therefore, it has become evident that a marine transgression occurred in some parts of Japan at about 30,000 years BP. In the Tokyo Bay area, however, the sea level prior to Tc–1 time is not obvious; so, it is uncertain that the sea level rose to −40 m in the Tc–1 time from the previous lower stand.

The sea level during the early Würmian stage

An elevated abrasion platform at the height of 20 to 40 m, named the Misaki Surface, is distributed in the southern tip of the Miura Peninsula. The radiometric age of the Misaki Surface is estimated at about 60,000 years BP., having been inferred from the fission-track age of the tephra layers covering the platform (Machida & Suzuki 1971). Abrasion platforms correlated to the Misaki Surface are distributed sporadically in other parts of the Japanese coast, but their radiometric ages are unknown except for Kikai-jima Island (Younger limestone member of the Ryūkyū limestone, 50,000–64,000 years ^{231}Pa BP., Konishi et al 1974).

Moreover, the marine terrace called the Obaradai Terrace, is well developed around Tokyo Bay at the height of 35 to 65 m, partly accompanying the buried valley beneath the terrace formation. Plant remains of temperate climate such as *Trochodendron aralioides* are contained in terrace deposits (Kokawa 1966).

The age of the Obaradai Surface is estimated at about 80,000 years BP., having been inferred from the fission-track age of tephra layers lying on the surface (Machida & Suzuki 1971). Marine surfaces correlated with the Obaradai are developed along many parts of the Japanese coast. The above-mentioned two marine surfaces of 60,000 years BP. and 80,000 years BP. may have been formed during the early Würmian Interstadial such as Brörup, Amersfoort, etc.

From a biogeographical point of view, *Mammonteus primigenius* reported from the terrace deposits of Hokkaido suggests the land connection between the Japanese Islands and the Asiatic continent in this stage. This is further supported by the fact that the Manchurian elements of land mammal such as *Macaca fuscata, Cervus nippon, Nyctereutes procyoniodes viverrinus, Sinomegaceros yabei,* etc. immigrated into Honshu through the land bridge in the Würm 1 (Kamei 1962).

A part of the Japan Alps was glaciated in the early stade of the Ikenotaira glacial stage which is characterized by the largest expanse of mountain glaciers in the area (Kobayashi & Shimizu 1966). The Porosiri Stadial of glaciation in the Hidaka range is also considered to belong to this stage (Ono & Hirakawa 1975).

The high sea level during the last interglacial

The Shimosueyoshi stage

The marine terrace surface named the *S* surface, which is most widely developed among the terraces is distributed at the height of 20 to 200 m along many parts of the Japanese coast (Ota 1975). It is correlated with the Shimosueyoshi Surface in the environs of Yokohama. In the lowland area such as Kanto Plain, Miyazaki Plain, etc., the S surface composes an elevated coastal plain (upland surface) in the rear of the present one. The surface of the Shimosueyoshi Upland is thickly covered by the Kanto Loam (volcanic ash formation), and the depositional top of the Shimosueyoshi Formation is nearly contemporaneous everywhere, because the lowest horizon of the Kanto Loam on it is nearly contemporaneous. Therefore, the emergence of the Shimosueyoshi Surface is supposed to have proceeded rapidly (Toma 1974).

The age of the Shimosueyoshi Surface is estimated at 120,000 to 130,000 years BP., having been inferred from the fission-track age of pumice layers on it (KlP-6, 128,000 ± 11,000 F.T. years BP. Machida & Suzuki 1971). This agrees with the fact that the Blake event (ca. 110,000 years BP.) in the Brunhes Normal Polarity Epoch is detected in the *S* surface making deposits (the Tsukahara Formation) along the Jōban coast (Manabe 1974).

The 120,000 to 130,000 years BP. surface is also reported from Kikai-jima Island (older limestone members of the Ryūkyū limestone, 99,000 to 124,000 ^{230}Th years BP., Konishi et al. 1974). The marine terrace surface showing the same radiometric age has been known in many parts of the world (Oka 1970).

In the Miura Peninsula one more marine terrace (named the Hikihashi Terrace) is distributed between the Shimosueyoshi and the Obaradai; its age is estimated at about 100,000 years BP. on the basis of the radiometric age of the overlying pumice layer (KmP–1, 98,000 ± 12,000 F.T. years BP., Machida 1975).

Shimosueyoshi Transgression

The marine formation which constitutes the Shimosueyoshi Terrace is named the Shimosueyoshi Formation and the marine transgression which brought about the Shimosueyoshi Formation is called the Shimosueyoshi Transgression. Marine deposits correlated with the Shimosueyoshi are usually less than 10 m thick on the rocky coast, and sometimes accompany buried valleys at their base. The Shimosueyoshi Marine Transgression began with a rapid rise of the sea level developing drowned valleys in its front and formed abrasion platforms together with gravel veneers overlying on it during the period of the highest stand of the sea level. A subsequent regression proceeded rapidly and a coastal plain emerged widely.

Keeping pace with the course of the transgression, there occurred paleoclimatic changes from the warm condition, as indicated by molluscs and plant remains of warm temperature like the present, to the cooler one, as suggested by seeds of *Menyanthes trifoliata* contained in the uppermost horizon of the Shimosueyoshi Formation. These aspects closely resemble those of the postglacial transgression, so that the Shimosueyoshi transgression is supposed to have been caused by a glacio-eustatic change in the sea level during the Last Interglacial. Therefore, it may be correlated with the Eem Interglacial in Europe.

The rate of eustatic sea level change

As the rate of eustatic rise of the sea level exceeds that of uplift even in the tectonically active coast of Japan, the sea invaded into the upheaving land surface, thereby forming drowned valleys. For example, in the southern tip of the Bōsō Peninsula where the seismic uplift during the Holocene time is as large as 3 mm/y (Naruse 1968), the marine transgression caused by the postglacial rise of the sea level formed the Numa coral bed, as stated before. Furthermore, in the vicinity of the Muroto Peninsula on the northeastern coast of Tosa Bay, though the uplift of 2 mm/y is believed to have continued since the late Pleistocene time, the Shimosueyoshi Marine Transgression was noticeable, forming a broad abrasion platform (the Murotozaki Surface 1) more than 1 km wide together with some buried valleys (Yoshikawa et al. 1964).

Against the rapid rise of the sea level, lowering from the highest stand seems to have proceeded slowly in the case of the postglacial transgression. Consequently, the date of the emergence of the marine terrace surface is slightly different at different places under the influence of the tectonic movement. For example, the marine surface due to the Numa Transgression in South Kanto where the tectonic movement was intense since the Pleistocene is dated about 6,000 to 7,000 years BP., while in the central part of the Kanto Plain which has been more or less stable since the late Pleistocene the same marine surface is dated at 4,120

± 100 years BP. (Sakaguchi 1968). This may be also true for the marine *S* surface, though it is not yet confirmed.

The sea level change during the middle to early Pleistocene

The aspects of the sea level change during the middle to early Pleistocene is not evident owing to either the poorly preserved marine surfaces, the tectonically deformed or dislocated terrace surfaces, or the deficiency of radiometric age determination. Recently, several cycles of sedimentation from transgression to regression have been distinguished in the middle Pleistocene Sagami Group in the Yokohama City area, each of which begins with marine deposits lying on the erosion surface and ends with aeolian volcanic ash layers called the Tama Loam (Kanto Quaternary Research Group 1974). Several marine surfaces composed of them are also reported (Endo & Uesugi 1972). Machida and others (1974) recognized the high stand of the sea level at 190,000, 220,000–230,000, 270,000–280,000, 350,000, 400,000 years BP. on the basis of the ages of tephra layers on the marine surfaces. On the other hand, the upper part of the middle to lower Pleistocene Osaka Group in the Kinki District is characterized by the alternation of marine clay beds (Ma0–Ma10) and freshwater beds (Itihara 1961). According to the results of the pollen analysis of the Osaka Group carried out by Tai (1973), the marine beds correspond to the warm periods, while the freshwater beds to the cooler climate. Thus, the frequent alternation of transgression and regression and the corresponding cyclic changes in the climatic condition recognized in the Osaka Group are supposed to have been caused by the glacio-eustatic changes in the sea level, and this may be true in the case of the Sagami Group. Recently, from the analysis of planktonic diatoms, it is elucidated that a marked fluctuation of the surface water temperature in the North Pacific begun between the Jaramillo event (ca. 900,000 years BP.) and the Brunhes/Matuyama boundary (690,000 years BP.) (Koizumi 1975). The fluctuation of terrestrial paleotemperature also became conspicuous through this period, and the cold climatic condition which had never been seen in the previous time of the Pleistocene appeared (Zagwijn 1975). As the beginning of the Günz Glacial age is supposed to be at 800,000 to 900,000 years BP. (Kukla 1975), so the eustatic fluctuation of the sea level may have become more noticeable since this time. The fact that the age of the lowest marine bed Ma0 and Ma1 of the Osaka Group is estimated at about 1 million years BP. may agree with the above argument.

Relation of the sea level change to the climatic change in Lake Biwa

Climatic change and sea level change

A pollen analysis of the 200 m core samples at 5 m intervals obtained from the bottom of Lake Biwa in 1971 was carried out by Fuji, and a depth-climate diagram since about 560,000 years BP. was drawn (Fuji & Horie 1972). Afterwards, they revised the diagram and illustrated the age-climate relationship as shown in Fig. 224 (Fuji & Horie 1977), based upon the fission-track ages of volcanic ash layers intercalated in the core (Nishimura & Yokoyama 1975). This diagram has been correlated with the paleoclimatic variations inferred from either the vertical change of $\delta^{13}C$ of organic carbon (Nakai & Shirai 1977) or that of the Σ organic-carbon/total organic carbon ratio (Handa 1975).

In the following section, the writers will compare the climatic change in Lake Biwa since 500,000 years BP. with the attitude of the sea level changes in Japan shown in the previous sections together with the $\delta^{18}O$ variation in the deep-sea core V28–238 from the western equatorial Pacific (Shackleton & Opdyke 1973). But a few problems have to be solved in combining the sea level change with the climatic change.

1) As already stated, 'eustatic' sea level changes reported from various districts of the world reflect both the voluminal change in ocean water and tectonic movement.

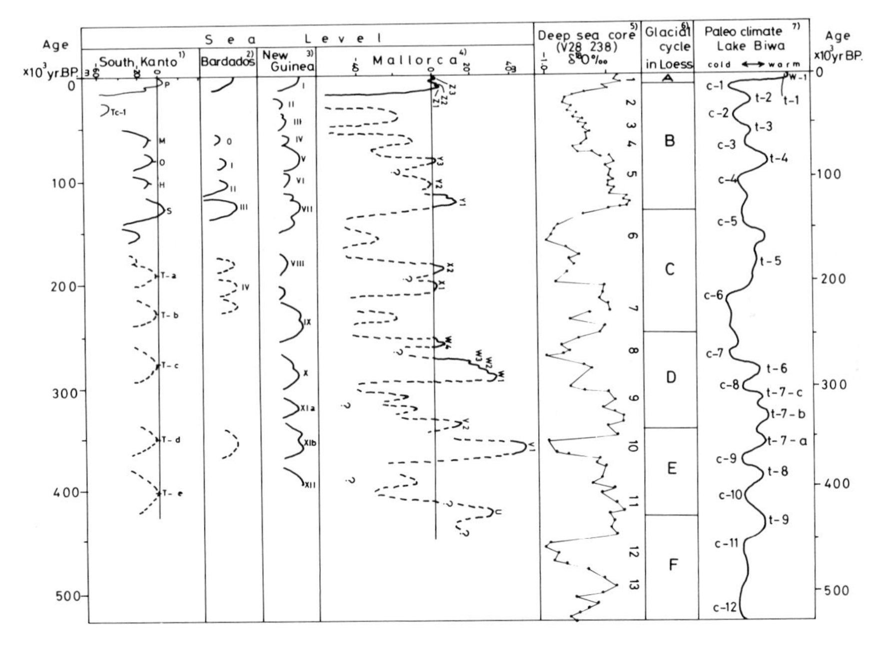

Figure 224. Changes in the sea level and paleoclimate since 500,000 years BP.
1. Machida (1975), P, Tc–1, M, O, H, S, T–a–T–e: marine surface
2. Mesolella et al. (1969), James et al. (1971), Bender et al. (1973), O, I–IV: marine terrace
3. Chappell (1974), I–XII: marine terrace
4. Butzer (1975), U–Z: sedimentary cycle (marine hemicycle)
5. Shackleton and Opdyke (1973), van Donk (1976), 1–13: stage
6. Kukla (1975), A–F: glacial cycle
7. Fuji and Horie (1977), w–1: warm climatic stage, t–1–t–9: temperate climatic stage, c–1–c–12: cold climatic stage.

2) The voluminal change in ocean water is caused by the wax and wane of continental ice sheets, but the correlation between the quantity of ice and the global paleoclimate has not been well established yet.

3) The relationship between the paleotemperature fluctuation in Lake Biwa and major climatic changes of the world has not been elucidated either.

Relation of the paleoclimate of Lake Biwa to the sea level change in Japan

By comparing the above-mentioned paleoclimate curve with the sea level curve of South Kanto by Machida (1975) in Fig. 224, the following facts can be pointed out:

1) The relative post-glacial high sea level (P) coincides with the warm period w-1.

2) The relative high stand of sea level about 30,000 years BP. (Tc–1) can be correlated with the temperate period of t–2.

3) The high sea levels inferred from the marine Misaki (M) and the Obaradai (O) Surfaces coincide with t–3 and t–4 respectively.

4) The marine Hikihashi Surface (H) is correlated with no temperate period but with a cold period of c–4.

5) The climatic amelioration during the formation of the Shimosueyoshi Surface (S) is not noticeable.

6) As for the marine Tama surfaces, the high sea levels deduced from the marine T–a, T–c, T–d surfaces may be correlated with the temperate periods. The temperate period expected by the T–e surface is possibly t–8 or t–9.

On the other hand, as for the high sea level periods inferred from the marine clay layers intercalated in the Osaka Group in Kinki, Sasajima (1977) correlated them with the ocean isotope records, estimating the ages of the clay layers on the basis of the magnetostratigraphic study of the Osaka Group (Maenaka et al. 1977). The writers give the following correlation using the estimated ages of the marine clay layers after Ishida (1970): Ma10 ≒ t–5, Ma9 ≒ t–6, Ma8 ≒ t–7. The Ma7 may presumably be correlated with t–8 or t–9.

In short, the high sea level periods in Japan can be correlated with the warm, temperate period of Lake Biwa during the last 500,000 years. However, a more reliable conclusion will be arrived at when more dates are obtained for the 200 m core samples of Lake Biwa.

Relation to the $\delta^{18}O$ variation in the deep-sea core V28–238

According to Shackleton and Opdyke (1973), the variation of ^{18}O record in deep-sea core samples rather reflects the quantity of ice stored on the continents than the temperature variation in ocean water. If so, the $\delta^{18}O$ diagram represents the eustatic changes in sea level. On the basis of the magnetostratigraphic and chronostratigraphic studies of the core (Shackleton & Opdyke 1976), it is proved that the V28–238 core has a constant accumulation rate except for about 130,000 years BP. during the last 900,000 years. Therefore, the age-$\delta^{18}O$ relationship in the V28–238 is reliable. Sasajima (1977) compared Fuji's paleoclimatic curve (1976b) with Emiliani's isotope temperature record depicted on the modified timescale (van Donk 1976), and found a rather significant correlation between them, except for the interval of Emiliani's stage number 4–5 or Fuji's t–4–c–5.

By comparing the $\delta^{18}O$ record of V28–238 with Fuji and Horie's paleoclimate curve (1977), the following correlation is possible: 6 max. ≒ c–5, 8 max. ≒ c–7, 10 ≒ c–9, 12 ≒ c–11. Though the interval of Emiliani's stage 5–7 is inconsistent with that of Fuji and Horie's c–4–c–6, a noticeable similarity exists between major trends of the two curves. In other words, the high sea level corresponds to the temperate period of Lake Biwa, and the low stand of sea level to the cold one.

The glacial-interglacial cycle detected in the loess stratigraphy in Central Europe is correlated with the oceanic $\delta^{18}O$ record of V28–238 (Kukla 1975) as shown in Fig. 224. It may be said that there is a significant relationship between the climatic changes in Lake Biwa and the glacial-interglacial cycle on a global scale.

Fossil mammals

Lake Biwa and fossil mammals

Faunal changes since the Pliocene time

Tadao Kamei

Introduction

Lake Biwa, the biggest lake in Japan, has been considered to be the best field for studying terrestrial environmental changes during the Late Cenozoic time. Above all, studies of the sediments in and around that lake are being carried out recently from various points of view, and many essential pieces of evidence are provided in approaching that subject.

In fact, hitherto, quite a few occurrences of mammalian fossils have been reported from Plio-Pleistocene sediments (Kobiwako (Paleo-Biwa) Group), terrace deposits and Neolithic settlements, which are distributed around Lake Biwa. But the importance of those materials in the interpretation of environmental changes is now being appreciated as the recent investigation has progressed. For example, the precise biostratigraphy in this area is going to be established, and the faunal succession since the Pliocene time is being clarified in relation to environmental changes. In these circumstances, as aspect of mammalian evolution of this area will be given in this article.

Fossil mammals around Lake Biwa

In 1804, a farmer who lived at Minamishō on the west coast of Lake Biwa found some fossil teeth and bones of an unknown animal when he was plowing his land on a hill. In the contemporary opinion at that time, those fossils were regarded as a 'dragon's bone', and the lord of this district received those fossils as a treasure. In those days, it was natural that the farmer who presented the treasure to the lord gained an honor as the discoverer of a supernatural being. Those fossils materials were painted by an artist as 'dragon's bone', and this picture remains even today (Fig. 225). From that fantastic picture we can easily recognize that people of that time imagined the tusk and jaw bone of the fossil elephant to be the dragon's horn and head respectively. Those pictures and the records of the discovery are now kept in the house of that farmer's descendant,

Horie, S. Lake Biwa

ISBN 90 6193 095 2. Printed in The Netherlands

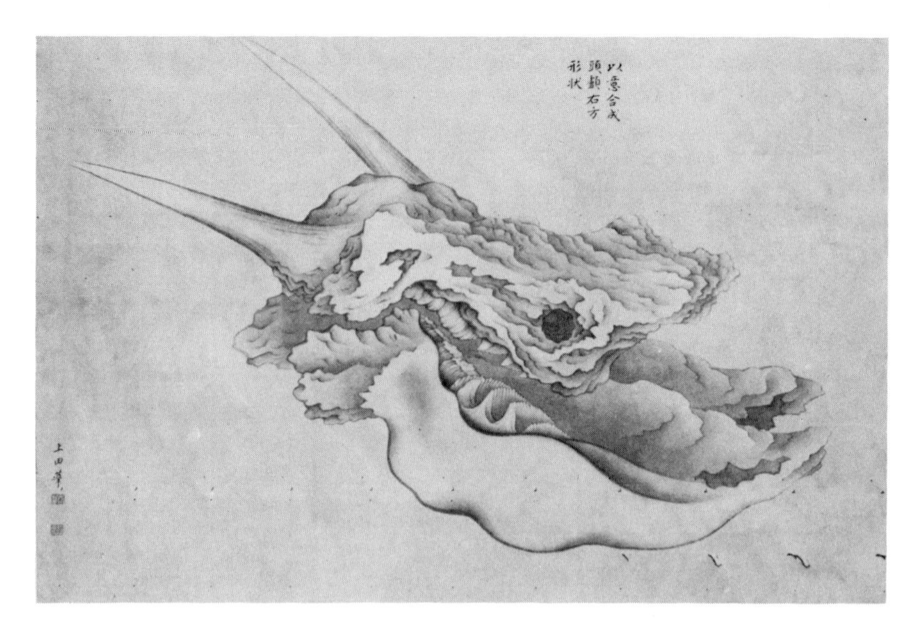

Fig. 225. 'Dragon's bone' (by courtesy of Mr. Ryū).

and the materials are deposited in the National Science Museum in Tokyo.

After seventy seven years, Naumann (1882) described those materials in 'Palaeontographica' together with the description of other Japanese fossil elephants (Fig. 226). He was a foreign teacher at the University of Tokyo, and was the first palaeontologist who conducted scientific studies on Japanese fossil elephants. But Naumann made a mistake in giving a specific name to those materials. He assigned the dragon's bone to *Stegodon insignis* Falconer et Cautley, a fossil elephant of the Siwalik hill of India, but Matsumoto (1924) corrected its name as *Stegodon orientalis* Owen. *S orientalis* is one of the representative fossil elephants of the Osaka Group (Plio-Pleistocene) in the Kyoto and Osaka Basins which are adjacent to Lake Biwa on the west. This fossil elephant is abundantly found also in the sea-bottom of Seto Inland Sea.

On the other hand, from the west coast of Lake Biwa where the farmer found the dragon's bone, fossil bone and teeth of another elephant have been reported. In 1957, Matsumoto & Ozaki described a specimen of a fossil elephant's molar yielded at Ono in the town of Shiga, and they named it as *'Archidiskodon' paramammonteus shigensis.* The type specimen of this species was known from the Kazusa Group (Plio-Pleistocene) of the South Kanto District, and the systematic position was in controversy. The present writer (1966) made a reexamination of this specimen of Lake Biwa, and included it in *'Elephas shigensis'* together with other specimens from five localities on the west coast of Lake Biwa and with specimens from the Osaka Group in the Osaka Basin. In his paper the writer stated that *'Elephas shigensis'* seems to be not a single

Fig. 226. Real figures of Dragon's bone (after Naumann, 1882).

species, but is a group which is transitional from Mammuthus meridionalis to *M. trogontherii.* Nowadays, this group is separable into two species, *Mammuthus paramammonteus shigensis* (Matsumoto et Ozaki) and *M. armeniacus proximus* (Matsumoto). These two species are widely discovered from the Early Pleistocene of the Japanese Islands.

The stratigraphical position of *M. paramammonteus shigensis* is lower than that of *Stegodon orientalis* without exception. The former is definitely a fossil elephant of the Early Pleistocene, while the latter is a representative of the early Middle Pleistocene not only in the Japanese Islands but also in East and Southeast Asia. Specimens of the former have been known from the west coast as well as from the east coast of Lake Biwa. In 1924, Makiyama described a fragmental molar of a fossil elephant found at the river bed of River Seri in the town of Taga as *'Parelephas trogontherii* Pohlig', but later he revised that specific name to *Elephas namadicus naumanni* (= *Palaeoloxodon naumanni*) (Makiyama 1938). Nevertheless, after a reexamination of that specimen it has become possible to treat it as *M. paramammonteus shigensis* as stated above. The characteristics of this species are beyond doubt quite different from those of *Palaeoloxodon.*

From the west coast and the east coast of Lake Biwa, molars and bones of other fossil elephants, *Stegodon insignis sugiyamai* (Tokunaga) and *S. shodoensis akashiensis* (Takai) have been discovered. These fossil elephants are commonly known from the lower part of the Osaka Group to the west, and are also found from the upper part of the Tōkai Group to the east of Lake Biwa which is stratigraphically correlative with the Kobiwako Group. Those elephants range from the latest Pliocene to the earliest Pleistocene.

As to the Pliocene fossil elephant, it should be mentioned that molars of a fossil elephant *Stegodon* cf. *elephantoides* Clift were recently found at Midoro, east of the city of Ueno, which is the southern extremity of the Kobiwako Group distribution. The occurrence of large tusks of a fossil elephant had also been reported from the Kobiwako Group, at Kosugi in the town of Iga, very near the place of the present find. Probably, those materials also belong to the same species. Adjacent to those areas, bones and teeth of this elephant have been found abundantly from the lower part of the Tōkai Group, and the stratigraphical position of those fossil materials is considered to be of the Middle Pliocene. It is very interesting that this elephant is closely related with *S. elephantoides* of the Upper Irrawaddi of Burma and of the Siwalik hill of India. This species is surely a member of the Indo-Malayan fauna.

In contrast to them, the occurrences of Middle to Late Pleistocene fossil elephant, Naumann's elephant *Palaeoloxodon naumanni* (Makiyama), are quite different. Detached and incomplete molar teeth of that elephant are found to have been rolled down on the river bed at Kyūtoku in the town of Taga on the east coast of Lake Biwa, and at River Seta, Nangō in the city of Ōtsu to the south of the lake. They might have been derived from terrace deposits or cave deposits nearby.

From the above-mentioned, it is clear that fossil elephants around Lake Biwa may play a role in the analysis of the course of Pliocene to Pleistocene mammalian succession in the Japanese Islands. Their course may be closely linked with the mammalian evolution in the northern Eurasia continents, and their history may reflect environmental turnover from non-glacial to glacial conditions. Contemporary with those fossil elephants, some kinds of artiodactyls like water buffalos, deers and bovines inhabited near and around Lake Biwa. Their teeth and bones were yielded in a fossil state from the Kobiwako Group, but unfortunately their stratigraphical positions and localities are left so vague in statements. Besides those fossil mammals, some pharyngeal teeth of carp and crocodile teeth were also found. A gigantic fossil crocodile of Malay Gavial *Tomistoma* was reported from the upper part of the Osaka Group which is adjacent to the west of Lake Biwa, and other occurrences of fossil crocodiles were known from several different horizons in that group. Therefore, it is possible to assume that crocodiles inhabited indigenously throughout the Osaka Basin and the Lake Biwa Basin in their preferable environment like that of today's Southeast Asia, during the period from the Early to Middle Pleistocene. All of those fossils from the Kobiwako Group have been usually found from muddy sediments in association with fresh water shells *Cristaria, Unio, Anodonta, Viviparus* etc. and occasionally impregnated by minerals of vivianite of blue color. This suggests that those remains were probably transported from their habitats nearby and buried under the lake floor with the condition not so far different from the present lake state.

The occurrence of a fossil horse mandible was reported from the bottom of Lake Biwa, off Kōtsuhama in the town of Shinasahi on the west coast of the lake. This horse was a smaller type of the Japanese Pleistocene equine, and from its pattern of teeth occlusion it may belong to Late Pleistocene wild horse *Equus hemionus nipponicus* Shikama et Onuki which is dominant in the last glacial sediments of Northeast and Central Japan. This may suggest the appearance of open land around Lake Biwa in the last glacial time.

There are some famous Neolithic settlements near the shore of Lake Biwa: namely, Ishiyama (Early Jōmon age) on the south coast and Shigasato (Middle Jōmon age). From those archaeological sites, numerous remains of mammals, reptiles, birds and amphibians have been obtained. Among the mammalian remains, the most dominant both in quantity and in quality are as follows: wild boar *Sus scrofa*, Japanese deer *Cervus (Sika) nippon*, Japanese serow *Capricornis crispus*, horse *Equus caballus*, Japanese monkey *Macaca fuscata*, Japanese black bear *Ursus (Solenarctos) thibetanus japonicus*, racoon dog *Nyctereutes procyonoides*, and dog *Canis familiaris palustris*. It readily follows that such assemblage was under the influence of artificial selection, because some domesticated species are included. In fact, the association of those animals does not represent a true natural assemblage, but it is possible to assume the mode of faunal turnover in post-glacial environmental changes. Large mammals like elephant had already disappeared in the Neolithic mammalian fauna. The mammalian association of the Neolithic was little more different from the present association around Lake Biwa.

In any event, many mammalian fossils and remains of the Pliocene and Pleistocene have been yielded from various beds and various localities around Lake Biwa. Those localities and their contents are listed in Table 63 and their sites of localities are shown in Fig. 227.

Faunal successions

In the surrounding areas of Lake Biwa, Plio-Pleistocene deposits of the Kobiwako Group are widely distributed and yield many fossil mammals here

Table 63. List of mammalian fossils around Lake Biwa.

No.	Locality	Mammalian Fossil	Horizon	Geological Age
1.	off Kōtsuhama Shin-Asahi town	*Equus nipponicus* (jaw bone)	lake sediments	probably Latest Pleistocene
2.	Ono, Shiga town	*Mammathus paramammonteus shigensis* (molar)	Katata Formation of Kobiwako Group	Early Pleistocene
3.	Otani, Shimoryuge Shiga town	*Buffelus* sp. (jaw bone)	ibid.	Middle Pleistocene
4.	Wani, Shiga town	*Mammuthus paramammonteus shigensis* (molar)	ibid	Early Pleistocene
5.	Wani, Shiga town	ibid (humerus)	ibid	Early Pleistocene
6.	Sayama, Mano town Ōtsu City	ibid (jaw bone)	ibid	Early Pleistocene
7.	Ieda, Mano town Ōtsu City	Cervid (vertebrate)	ibid	Early Pleistocene
8.	Mano town Ōtsu City	*Mammuthus paramammonteus shigensis* (molar)	ibid	Early Pleistocene
9.	Miyagidani, Ōgi Ōtsu City	ibid (molar)	ibid	Early Pleistocene
10.	Shimosakamoto Ōtsu City	ibid (molar)	ibid	Early Pleistocene
11.	Sakamotohoncho Ōtsu City	Bovid. (jaw bone)	ibid	Early or Middle Pleistocene
12.	Minamishō, Katata Ōtsu City	*Stegodon orientalis*, Cervid	ibid	Middle Pleistocene
13.	Shigasato Ōtsu City	*Cervus, Sus* etc	Neolithic settlement	Holocene (Middle Jōmon)
14.	Shishitobi, Nango Ōtsu City	*Palaeoloxodon naumanni* (molar)	unknown (river bed)	Late Pleistocene
15.	Kagamiyama; Yasu town	*Stegodon shodoensis akashiensis* (molar)	unknown	Earliest Pleistocene
16.	Nishi-sakuradani Hino town	ibid (molar, tibia)	Gamō Formation of Kobiwako Group	Earliest Pleistocene
17.	Rengeji Hino town	ibid (molar)	ibid	Earliest Pleistocene
18.	Hino town	Cervid (Femur)	ibid	Earliest Pleistocene

19.	Serigawa Taga town	*Mammuthus armeniacus proximus* (molar)	ibid	Early Pleistocene
20.	Serigawa Taga town	*Palaeoloxodon naumanni* (molars)	unknown (river bed)	Middle to Late Pleistocene
21.	Same Cave Taga town	*Nyctereutes, Sus, Ursus Capricornis* etc.	cave deposits	Holocene (Late Jōmon)
22.	Kosugi Iga town	*Stegodon* cf. *elephantoides* (tusks)	Aburahi Formation of Kobiwako Group	Middle Pliocene
23.	Midoro Ōyamada town	ibid (molars)	ibid	Middle Pliocene

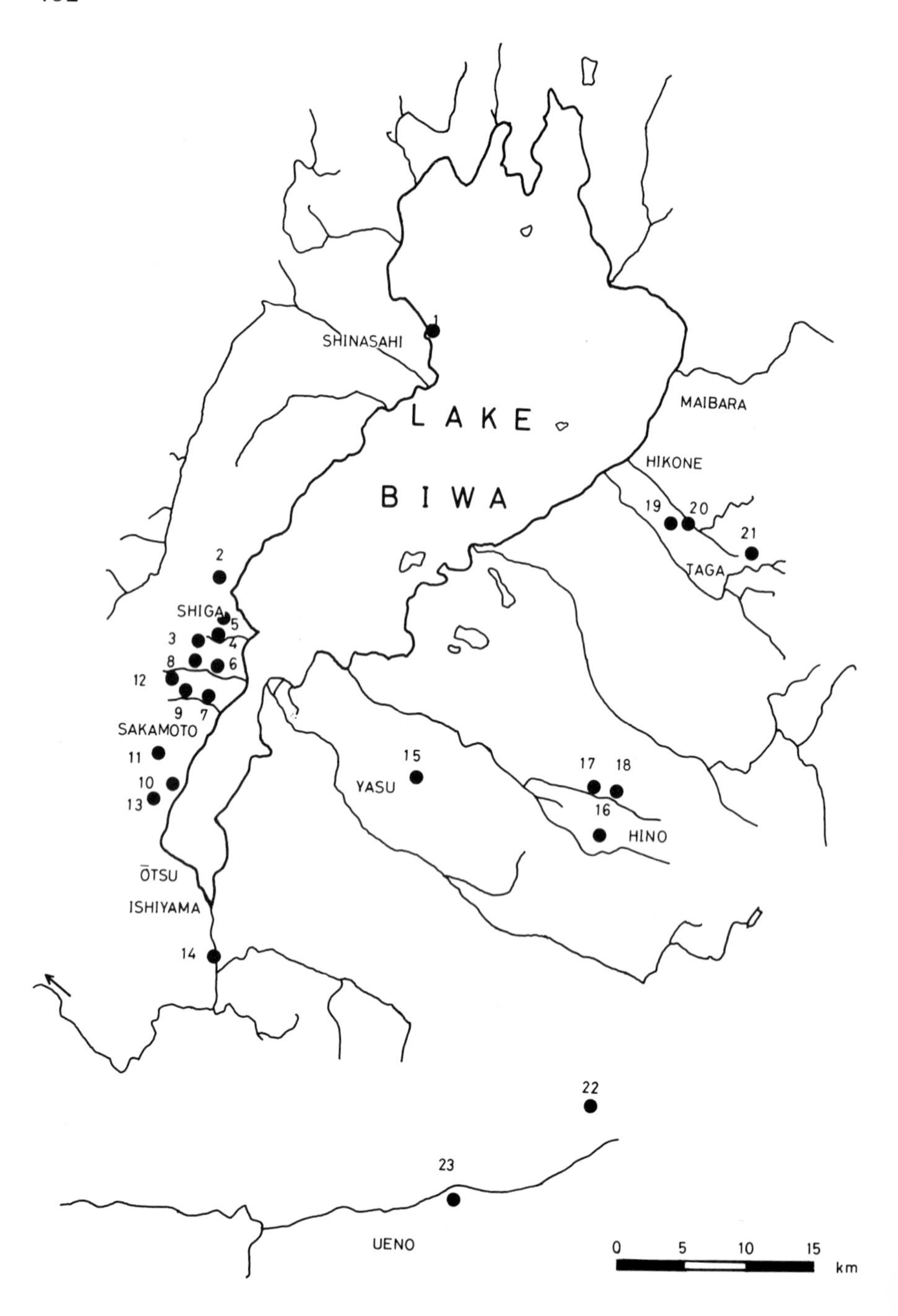

Fig. 227. Locality map of fossil mammals around Lake Biwa.

and there as mentioned in the preceding section. Those deposits consist mainly of alternative lacustrine and fluvial sediments, but none of brackish or marine sediments are intercalated at all. From the results obtained by a geological survey, it is confirmed that the initial basin depression occurred in the area to the southeast of the present lake position, and successively the sedimentation center was forced to shift continually toward northwest until its configuration came to coincide with the present Lake Biwa. Therefore, the name of Kobiwako is given generically to those ancient sediments which cover a large area. Recently, a great number of stratigraphical and biostratigraphical studies on the Kobiwako Group have been conducted by many researchers, and their results are summarized in Ikebe & Yokoyama (1976) and Maenaka et al. (1977). Moreover, the precise analysis from various view-points are being carried on for the deep core sediments which were obtained from the near shore and bottom of Lake Biwa, and the results are useful in evaluating those stratigraphical studies.

Both the Osaka Group on the west and the Tōkai Group on the east are rather contemporaneous with and equivalent to the Kobiwako Group around Lake Biwa, and those three sedimentary sequences cover a vast area from Osaka Bay to Ise Bay. Though each group has its own isolated distribution from basin to basin, common fossil mammals have been found, and the correlation of their stratigraphical positions among those deposits is possible by means of tephrochronology. In addition, magnetostratigraphy and fission-track age determination contribute to the consideration of world-wide correlation of those mammalian fossils. Kamei & Setoguchi (1970) proposed a tentative biostratigraphical zonation throughout those three sedimentary sequences based upon fossil mammals, and the validity of zonation is recognized to be appropriate to Plio-Pleistocene deposits of other regions in the Japanese Islands. In this section however, that will be discussed in a somewhat revised form.

The mammalian zones of the Osaka, Kobiwako and Tōkai Groups are summarized in an ascending order as follows (Fig. 228). The characteristics of each zone are explained below.

1. **Se:**
   Stegodon cf. *elephantoides* Zone — Middle Pliocene
2. **Ss:**
   Stegodon insignis sugiyamai Zone — Late Pliocene
3. **Sa:**
   Stegodon shodoensis akashiensis Zone — Earliest Pleistocene
4. **Ms – Mpx:**
   Mammuthus paramammonteus shigensis – M. armeniacus proximus Zone — Early Pleistocene
5. **So:**
   Stegodon orientalis Zone — early Middle Pleistocene
6. **Pn:**
   Palaeoloxodon naumanni Zone — late Middle to early Late Pleistocene

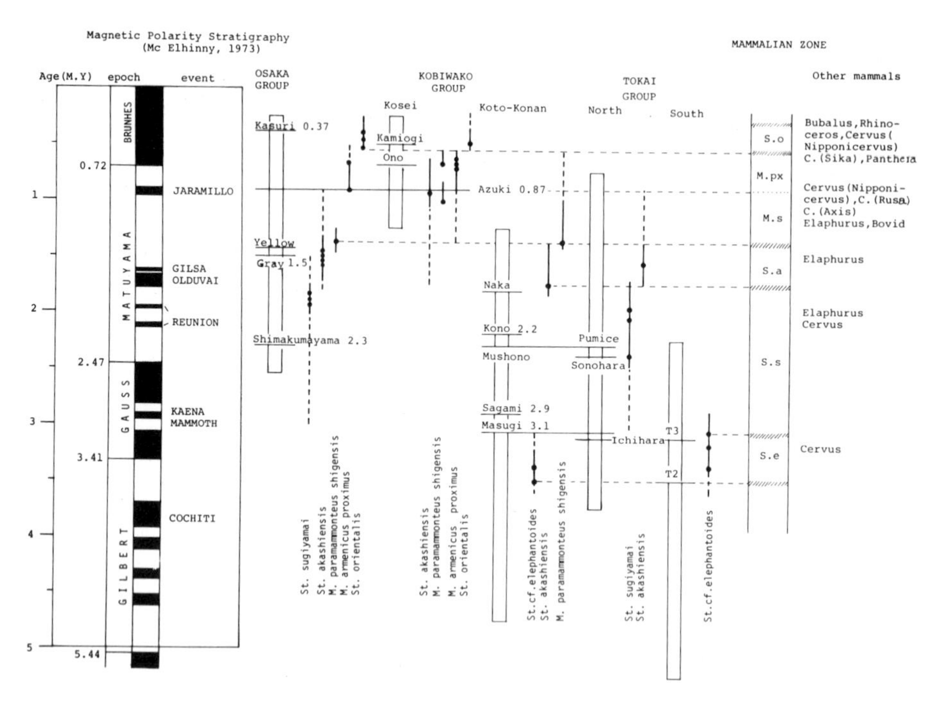

Fig. 228. Correlation chart of Proboscidean fossils in the Osaka, Kobiwako and Tōkai Groups (Kamei & Otsuka 1981).

7. **Mp:**
   Mammuthus primigenius Zone — Latest Pleistocene
8. **S–C:**
   Sus scrofa and *Cervus* (*Sika*) *nippon* Zone — Holocene

Zone of Stegodon cf. elephantoides

The guide species of this zone is *S.* cf. *elephantoides* Clift which is closely related to the fossil elephant abundantly found in the Upper Irrawaddy series of Burma, and in the Dhok Pathan zone of the Siwaliks of India. This species has been known from the lower part of the Tōkai Group which is distributed south-east from Lake Biwa, and is recently found in the basal part of the Kobiwako Group to the south of the lake. Apart from this fossil elephant, no other remarkable mammalian remains have been known from this zone, but some occurrences of unnamed fossil deer teeth and bones have been reported.

Until now, the representatives of the Lower Pliocene mammals like *Gomphotherium*, *Stegolophodon* and *Hipparion* are not observed in this zone. The time span of this zone is estimated by fission-track age determination to

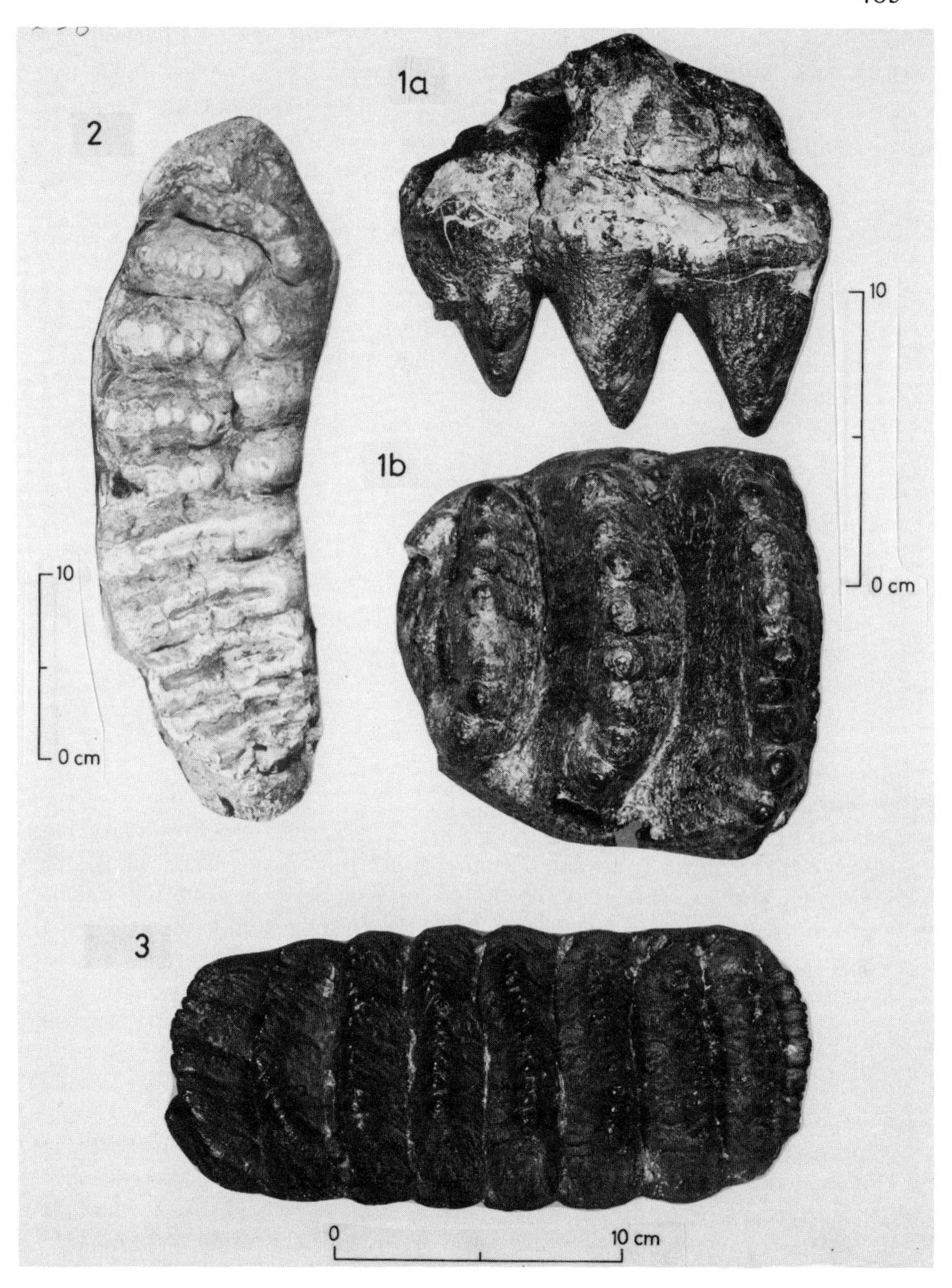

Plate 26.

1. Stegodon cf. *elephantoides* Clift
Posterior half of upper third molar, locality 23 (Mie Prefectural Museum). a: palatal view, b: occlusal view

2. *Stegodon* cf. *elephantoides* Clift
Lower left third molar, from the lower part of the Tōkai Group (Mie Prefectural Museum), occlusal view.

3. *Stegodon orientalis* Owen
Upper right second molar, from the upper of the Osaka Group (plast copy in Kyoto University, Department of Geology and Mineralogy), occlusal view.

have a range from 3.5 million years BP. to 3.0 million years BP. By the magnetostratigraphy, this zone is correlated with the Gilbert/Gauss boundary. The plant remains of *Pinus trifolia* flora (Miki, 1969) indicate that the environment of this early stegodon was warm temperate forests.

Zone of Stegodon insignis sugiyamai

In the Osaka Group to the west of Lake Biwa, *S. insignis sugiyamai* (Tokunaga) has been reported from sediments between the lower of the Shimakumayama tephra horizon (2.3 ± 0.2 million years) and the upper of the Senriyama tephra horizon which is lower than the Gray tephra horizon (1.5 ± 0.2 million years). In magnetostratigraphy, this zone records the Réunion and Olduvai events of the Matuyama Reversal. Therefore, the time interval of this zone is estimated roughly between the time of 3.0 million years BP. and 1.8 million years BP. The faunal lacuna is present in the lower part of this zone, but it is supplemented by paleobotanical evidence. That is to say, the faunal vacancy is substituted by the stage of the transitional flora, implying the extinguishing of some pre-existing warm temperate floral components like *Liquidambar*, *Pseudolarix* and *Keteleeria*. From pollen stratigraphy, the first cold phase is recognized to be near the base of this zone.

Zone of Stegodon shodoensis akashiensis

A molar of *Stegodon shodoensis akashiensis* (Takai) was found from the Naka tephra horizon of the Kobiwako Group at Hino Hill, on the east coast of Lake Biwa. Some tusks and bones of a fossil elephant have been known from the same horizon of this area, and those materials seem to belong to this species. On the west coast of Lake Biwa, an occurrence of this elephant is confirmed, but its exact horizon and locality are not so clear.

This fossil elephant is familiar to the Japanese palaeontologists, and is called the Akashi elephant. The specimens of the Akashi elephant have been frequently reported from the Osaka Group and the Tōkai Group, and those localities are adjacent to Lake Biwa. Therefore, it is certain that the Kobiwako Group comprises the same stratigraphical horizon as the Osaka Group and the Tōkai Group. In the Osaka group, this fossil elephant concentrates in deposits between the Senriyama tephra horizon of the lower and the Yellow tephra horizon of the upper. From this evidence, the time lapse of this zone is considered to be from 1.8 million years BP. to 1.5 million years BP. By magnetostratigraphy this zone is correlated to the Olduvai and Gilsa events. On the other hand, it may be natural that the range of this elephant extends to the upper horizon above the zone boundary. In fact, the material from the west coast of Lake Biwa seems to support this view.

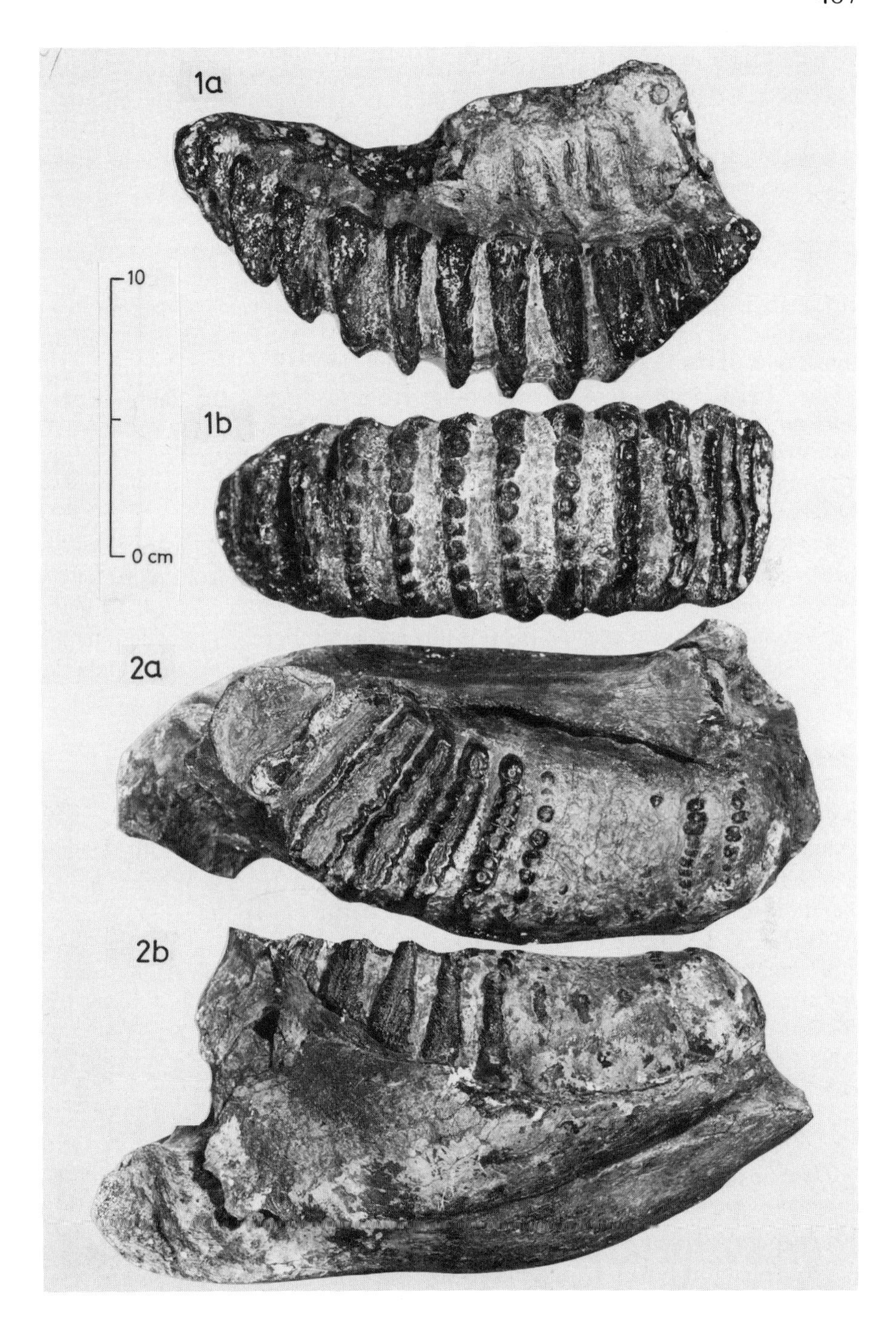

Plate 27

1. *Stegodon shodoensis akashiensis* (Takai)
   Upper right third molar, from the lower part of the Osaka (INOUE collection), a: buccal view, b: occlusal view.
2. *Stegodon insignis sugiyamai* (Tokunaga)
   Right mandible with third molar, from the lower part of the Osaka Group, (Kyoto University, Department of Geology and Mineralogy), a: occlusal view, b: lingual view (Makiyama 1938).

This zone is also characterized by the presence of an early form of Péré David's deer *Elaphurus* which is one of the important members of the Nihowan mammals of North China. From the paleobotanical viewpoint, a decline of the warm temperate flora and a rise of the cool temperate flora are characteristic in this zone (Age of extinction of the *Metasequoia* flora) (Itihara 1960).

Zone of Mammuthus paramammonteus shigensis and M. armeniacus proximus

In the areas around Lake Biwa, *M. paramammonteus shigensis* (Matsumoto et Ozaki) and *M. armeniacus proximum* (Matsumoto) have been known from the upper part of the Kobiwako Group in some hills of both sides of the lake. Also in the Osaka Group, those species range from just below the Yellow tephra horizon to the Hacchoike tephra horizon, and this time-interval is estimated to be from 1.5 million years BP. to 0.6 million years BP. by fission-track age determination. This zone records the Jaramillo event of the Matuyama Reversal in magnetostratigraphy.

It is conspicuous that the *Metasequoia* flora of warm temperate forests became completely extinct in the middle of this zone at the cold phase of the Azuki tephra horizon (0.87 million years) in the Osaka Group. Climatic oscillation of cold and warm is assumed by pollen stratigraphy to be initiated in this zone. This zone is also characterized by an increase in the species number of the Nihowan deers like *Elaphurus*, *Cervus* (*Nipponicervus*), *Rusa* and *Axis*.

Zone of Stegodon orientalis

As noted earlier, the 'dragon's bone' from the west coast of Lake Biwa was actually fossil bones and teeth of *Stegodon orientalis* Owen. The stratigraphical position of this fossil elephant is in the Kamioogi tephra horizon of the Kobiwako Group, and this horizon is correlated with the Sakura tephra horizon of Ma.* 6 in the Osaka Group (Hayashi 1974).

The range of this elephant in the Osaka Group is from Ma. 6 horizon to Ma. 8 horizon, and is estimated from 0.6 million years BP. to 0.3 million years BP. Fission-track age of the Kasuri tephra immediately below Ma. 8 is 0.38 $\pm$0.03 m.y. From the Kobiwako Group, water buffalo *Buffelus* sp., bovine and deers were found in association with this elephant.

Zone of Palaeoloxodon naumanni

Naumann's elephant *Palaeoloxodon naumanni* (Makiyama) is the most popular fossil elephant in Japan. Numerous fossil bones and teeth of this elephant have been reported from the Japanese Islands including Hokkaido and Kyushu as

*Ma. is the abbreviation for marine clay. The number is counted in an ascending order from 0 to 12.

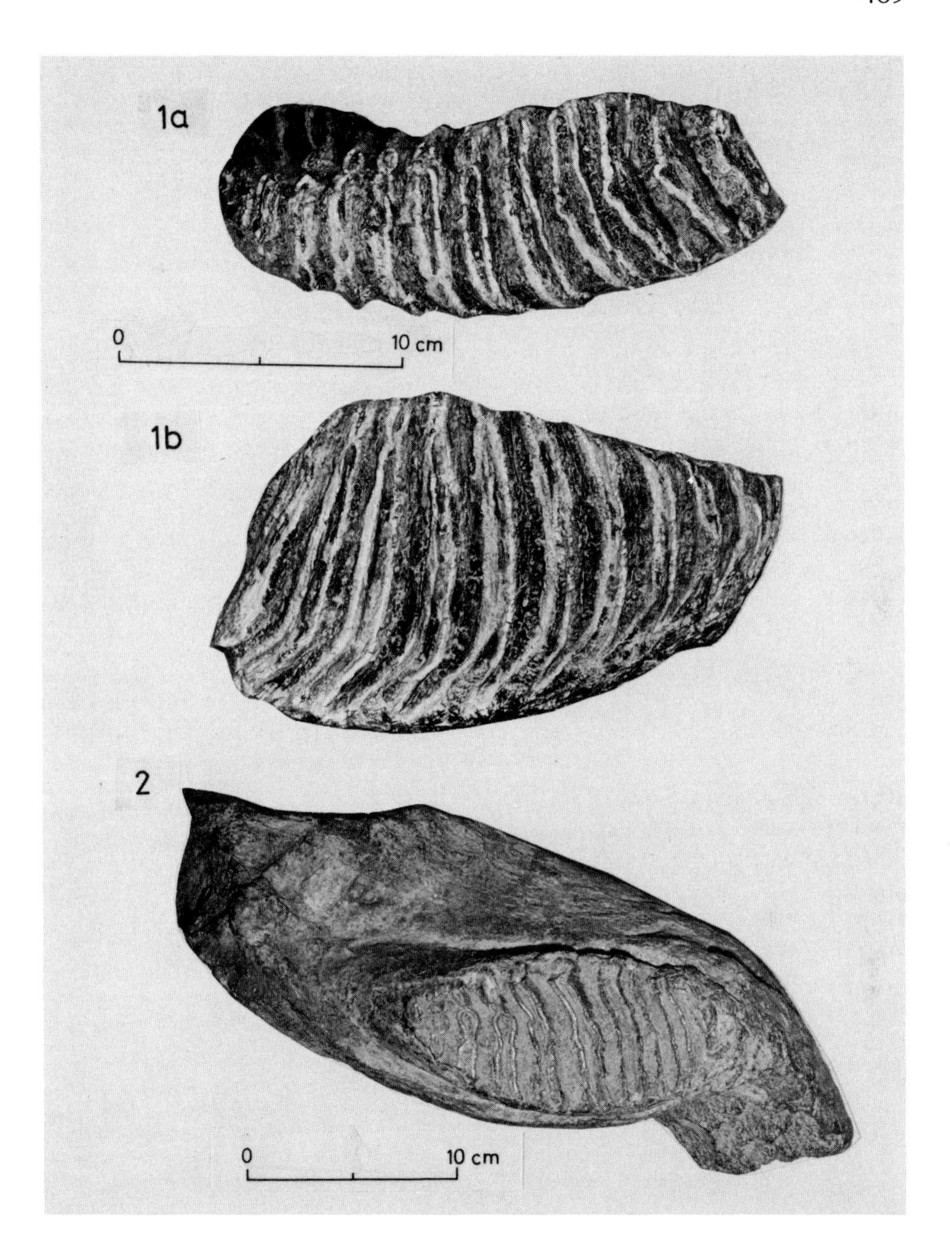

Plate 28.

1. *Mammuthus paramammonteus shigensis* (Matsumoto et Ozaki)
   Lower left second molar, locality 2 (plast copy in Kyoto University, Department of Geology and Mineralogy. a: occlusal view, b: lingual view (Kamei 1966).
2. *Mammuthus paramammonteus shigensis* (Matsumoto et Ozaki)
   Left mandible with third molar, locality 6. (Mano Primary school). Occlusal view (Kamei 1966).

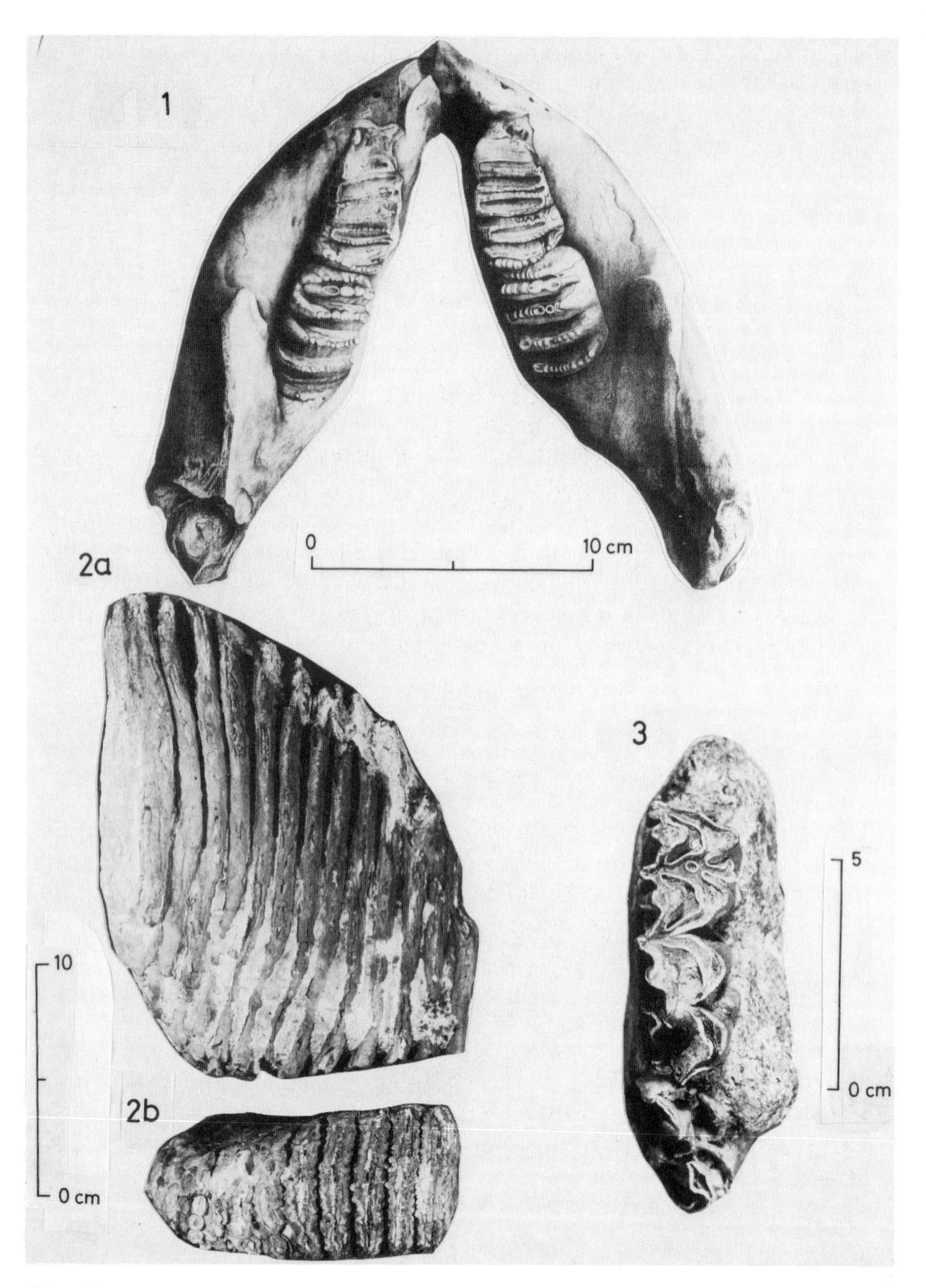

Plate 29.
1. *Stegodon orientalis* Owen
   Mandible with molars, locality 12, from the plate III in the paper of Naumann (1882).
2. *Palaeoloxodon naumanni* (Makiyama).
   Upper left third molar, locality 20, (Taga Middle School) a: palatal view, b: occlusal view.
3. *Buffelus* sp.
   A part of upper jaw bone with molars, locality 3, (Kyoto University, Faculty of Engineering), occlusal view (Hiki 1915).

well as from the sea bottom around those islands, namely, Seto Inland Sea, Japan Sea, Yellow Sea, and East China Sea. Outside of the Japanese Islands, this fossil elephant was reported from the Northeast Asia continent. In those areas, the range of this elephant is considered to be from the Middle to Late Pleistocene, but the oldest one is assumed to be about 0.3 million years and the last race of this elephant is known from sediments of 17,000 years BP. in the Japanese Islands. Therefore, it seems that this fossil elephant had a long range of survival, but it is also possible to divide faunal assemblages associated with this elephant into two groups with large and micro mammal association. One of them is a group without glacial mammals before the end of the Last Inter-glacial, and the other is a group with glacial mammals like moose and gigantic deer etc. which appeared after the end of the Last Inter-glacial. In this way, this zone is defined by the association of the former group.

In the areas south and southeast of Lake Biwa, several molar teeth of *P. naumanni* have been reported. Nevertheless, all of them are materials rolled down on the river bed, and their exact horizons are unknown. In Japan, bones and teeth of this fossil elephant have been generally found from terrace deposits and cave deposits. Accordingly, materials around Lake Biwa are supposed to have been derived from those terrace deposits or cave deposits. Hitherto, none of the trace of that elephant can be detected from the Kobiwako Group of the order.

Zone of Mammuthus primigenius

There are no fossil bones and teeth of wooly mammoth *Mammuthus primigenius* (Blumenbach) around Lake Biwa. This famous animal is known only in Hokkaido in the Japanese Islands. In that area molars of this elephant were discovered from terrace deposits of about 40,000 years BP. at Cape Erimo, the southern extremity of Hokkaido. It is currently believed that the Tsugaru Strait between Hokkaido and Honshu was the barrier to southward migration of wooly mammoth. But, as regards the Latest Pleistocene mammals of Honshu, it is possible to find out that some boreal mammals like bison, wild horse, moose, gigantic deer, wolf, brown bear and others which coexisted with wooly mammoth in Northeast Asia were associated with Naumann's elephant in situ. This suggests that the migration of wooly mammoth fauna was possible at the time of the Last glaciation.

Zone of Sus scrofa and Cervus (Sika) nippon

Holocene sediments are characterized by the presence of mammalian remains which are the same species as those living today and by the absence of large Pleistocene mammals like Naumann's elephant. The mammalian remains excavated at the Neolithic settlements around Lake Biwa are changed little in composition from the present faunal association of this area, if the quantitative ratio of each species is disregarded. Therefore, the Holocene fauna of Lake Biwa is represented by the association of wild boar *Sus scrofa* Linne and Japanese deer *Cervus* (*Sika*) *nippon* Temminck.

Environmental changes around Lake Biwa

Middle Pliocene

The most ancient mammal known in the area around Lake Biwa is reported from the southeastern distribution of the Kobiwako Group. In this place the basal part of this sedimentary sequence is developed, but until now only the fossil elephant *Stegodon* cf. *elephantoides* has furnished information about its faunal character. This elephant, an early form of *Stegodon*, was a large animal with a long straight tusk (2 m long), and was a close relative to the elephants of Burma and India. Undoubtedly, the Indo-Malayan fauna was spread homogeneously and widely from South Asia to North Asia. Migration routes for land animals and land connections through those areas might have been complete in those days.

The areas where fossil bones and teeth of *Stegodon* cf. *elephantoides* were found are estimated to be the place where the depression of the Lake Biwa basin was initiated. Those areas are distributed to the south and southeast of Lake Biwa, and most of the sediments in which the fossils are preserved are fluvial and thick around the basement hills. From the sedimentological viewpoint, it is assumed that a wide alluvial plain occupied a vast area and the topographical relief was very low. Thick forests of evergreen broad leaf trees like *Carya*, *Nissa*, *Liquidambar* covered densely this vast alluvial plain. Those forests were the continuum of the Mio-Pliocene subtropical forest.

Late Pliocene and Earliest Pleistocene

In this age, the large elephant of *Stegodon* was replaced by a dwarf form of *Stegodon*. Those elephants like *S. insignis sugiyamai* and *S. shodoensis akashiensis*, are generally known as progressive forms of *Stegodon* with hypsodont molars (high crown). Morphologically they are similar to the relatives in India and Java, eg. *S. insignis* and *S. punjabicus* of India, and *S. airawana*, *S. trigonocephalus* and *S. bondlensis* of Java, but they are distinguished from those relatives of South-Asia by their dwarf forms. The Japanese species are undoubtedly descendants of the Indo-Malayan fauna of tropical and subtropical lands, and they are considered as an isolated and endemic form.

It is known that the subsidence of the Lake Biwa Basin increased and the depression center was continuously shifting toward northwest. In accordance with this crustal movement, the alternation of fluvial and lacustrine sediments was developed in the area southeast of the present Lake Biwa. Such environment as lake shore and riverside may have provided a preferable habitat for those *Stegodon* which were dwellers in the forested area.

The forest vegetation of this area was transitional from evergreen forest of the warm temperate to deciduous forest of the cool temperate. Some warm-temperate forms of *Liquidambar*, *Pseudolarix* and *Keteleeria* were extinguished at the beginning of this time. Associated with this, the increment of cool-temperate mammals is detected by the increase in number of Nihowan deers like *Elaphurus*, *Cervus* (*Nipponocervus*) and *Rusa*. This may suggest that the migration route to the southern tropical and subtropical regions was interrupted by this age, but the land connection to the adjacent continent was still maintained.

Early Pleistocene

In this period, the position of the ancient lake and its surrounding geomorphology became nearly the same as the present Lake Biwa. But the extent of the lake may have been much wider than the present lake shape. Around the lake of those days, a preferable habitat for early mammoth, *Mammuthus paramammonteus shigensis* and *M. armeniacus proximus* may have been prepared. Those early mammoths were immigrants from the continent of Northeast Asia and replaced the niche of southern forest elephant of *Stegodon* successively. *M. paramammonteus shigensis* corresponds to *M. meridionalis* of Europe and *M. armeniacus proximus* is closely related to *M. armeniacus* (= *trogontherii*) or *M. columbi* of North America.

Remarkable climatic fluctuation is characteristic of this age. The warm-temperate Arcto-tertiary flora which had survived by this age diminished rapidly in number and variety. The full extinction of the *Metasequoia* flora which was the last of the Arcto-tertiary is detectable just below the Azuki tephra horizon in the Osaka basin (Itihara 1960; Itihara et al. 1973, 1977), and this conspicuous floral and climatic change is dated at 0.87 million years BP. and is recorded at the time of the Jaramillo event in magnetostratigraphy. As stated above, it is noteworthy that this climatic deterioration was accompanied by the advent of the early mammoth to this place. In fact, the worldwide expansion of the early mammoth was contemporaneous with this age, and a number of remains of Nihowan deers in the Japanese Islands suggests immigration from the continent at that time. Low relief topography around the ancient Lake Biwa covered by woodland forests may have been a suitable biotope for those immigrants.

Early Middle Pleistocene

It is well known that *Stegodon orientalis* is one of the representative mammals of the Sino-Malayan or Wanxian fauna which was widely distributed in Southeast Asia and Middle China in the Middle Pleistocene. It extended its territory to the Japanese Islands up to 39°N. latitude. Therefore, it is certain that the land connection to those southern regions was maintained in this time, and the migration route was revived again.

Though it is known that the cold phase intervened between Ma. 6 and Ma. 7 of the Osaka Group, a vigorous climatic recovery at the Ma. 8 horizon is indicated by the occurrence of warm-temperate fossil flora represented by *Syzygium*, *Lagerstroemia* etc.

There is no doubt that the warm-temperate forest expanded its territory further north than the present, and immigrants from the south, *Stegodon*, *Buffelus*, deers, perhaps rhinos and tiger lived around the ancient lake of Lake Biwa after the retreat of the early mammoth.

Middle Pleistocene to Late Pleistocene

In the beginning of this period, the crustal movement was much accelerated throughout the Japanese Islands. Due to rapid upheaval rate of mountains, topographical contrast between mountains and lowland came to be accented in and around the Lake Biwa area. Consequently, the configuration of the present Lake Biwa was settled at that time, and the piedmont and fan sediments built up terrace topography in and around the lake.

At the same time with this topographical change, the cool-temperate woodland mammals of the Zhoukoudian (Choukoutian) fauna of the continent immigrated into the Japanese Islands and replaced some of the preexisted warm-temperate forest fauna. In this way, warm-temperate forest elephant *Stegodon* completely disappeared, and cool-temperate woodland and forest elephant *Palaeoloxodon naumanni* (Naumann's elephant) came to be dominant among the Japanese mammals.

In association with this Naumann's elephant, many kinds of extinct and extant mammals of the Zhoukoudian fauna immigrated gradually into the Japanese Islands. Representative of them are Young's tiger *Panthera yongii*, Chinese rhinocerus *Rhinoceros sinensis*, wild boar *Sus lydekkeri*, black bear *Ursus* (*Solenarctos*) *thibetanus*, and probably Japanese monkey, Japanese serow, and gigantic deer. Unfortunately, they are barren as fossils in terrace deposits around Lake Biwa, but it is certain that they also inhabited there and wandered from place to place.

It is true that the Japanese Pleistocene mammals were influenced by the glacial eustasy. The sea-level change during the Last Interglacial played an important role in the formation of the present geography of the Japanese

Islands. Though it is considered that the Tokara strait between Honshu and Ryukyu was established earlier than this time, other main straits like the Korean on the west and the Tsugaru on the north were related to the upheaving sea-level during that time. It is significant that all land connections to the continental regions were completely broken at this high sea-level stage.

It is uncertain whether the ancestral forms of living mammals in the Japanese Islands were immigrants which came before that high sea-level stage or not, and it is yet in controversy whether the land mammals could or could not have immigrated from the continent to the Japanese Islands during the Last glacial time. Nevertheless, the Latest Pleistocene mammals of the Japanese Islands are so peculiar that they contain some new elements of the boreal fauna. Naumann's elephant survived to the Latest Pleistocene up to 17,000 years BP., and was often accompanied by gigantic deer *Megaloceros* and moose *Alces alces* which occur with woolly mammoth fauna in the continental provinces.

Though no reliable information about the Latest Pleistocene mammals is yet obtained from sediments around Lake Biwa, the molar teeth of Naumann's elephant from the river bed and the jaw bone of a wild horse from the lake bottom may indicate the Latest Pleistocene mammal association of the surrounding areas of the lake. At that time, coniferous forests came down to lowland area and grass-land vegetation invaded into bushy fields near the lake shore (Fuji, 1976a).

Latest Pleistocene to Holocene

The climate of the Latest Pleistocene around Lake Biwa was cooler and drier as compared with the present climatic condition, but the evidence from archaeological sites points to the recovery of temperate forest vegetation in the Neolithic time. Plant remains from those archaeological settlements reveal the presence of mixed forests of evergreen and deciduous broad leaf trees which covered densely areas around Lake Biwa. As to mammals, large mammals like elephant and gigantic deer already disappeared, but medium to small sized mammals like Japanese deer and boar were more in number and variety rather than the present faunal association.

For the Neolithic people, those animals may have been important game animals for their hunting, and nuts and chestnuts in those forests were available as essential foodstuffs. From the Neolithic settlements, in addition, numerous remains of soft-shelled turtle, freshwater fish and freshwater shells have been found in the state of baking and cooking. Therefore, Lake Biwa and its surrounding areas may have been fertile lands for those ancient people.

After the Neolithic time, those forests have been gradually destroyed by the increase of human activities, and secondary forests have appeared around Lake Biwa. Subsequently, the development of farming and fishing by human beings have modified the nature of Lake Biwa, and the mammals which inhabited there have been deprived of their habitat. Japanese deer, Japanese serow and Japanese monkey are observed only in the mountain regions far from Lake Biwa.

Pollen analysis

Norio Fuji

Introduction

The vegetational and climatic changes during the last 560,000 years in Lake Biwa from the viewpoint of pollen analysis have been determined from a 200 m core drilled in Lake Biwa.

The existence and character of several Quaternary glaciations are documented by studies of glacial geology, topography, paleoclimatology, and paleopedology in many countries. In the Japanese Islands, however, crustal movements, volcanic activity, and erosion have dominated. Most of the islands were not glaciated. It is therefore difficult to depend on glacial geology and topography or paleopedology to study the history of glaciation in this area. The detailed chronology of stadials and interstadials during the last glacial age is complex. So are correlations with earlier glaciations in Europe and North America. Therefore, rather than to develop a terminology based on inaccurate correlations, it seems desirable to establish a stratigraphic nomenclature that is not primarily based on meager and inaccurate evidence from mountainous regions, but rather on sedimentary sequences in basins outside the glaciated regions. Extensive deposits from glacial and interglacial times can be studied in deep bore holes, although not in surface exposures. Pollen analysis of the stratigraphic sequences in minute details during the last two million years can reveal a continuous record of the change of vegetation and climate in the non-glaciated area. Because of changes in the sea level during the glacial and stadial times, as explained by glacial eustatic theory, it is difficult to collect continuous samples covering the Quaternary period, which is characterized by alternations of warm and cold climatic times.

Although a complete sequence of sediments of both the cold and warm periods is found in the deep-sea cores, we cannot evaluate changes in climate over short time periods because of the extremely slow sedimentation rate. Thus, the following conditions have to be satisfied if climatic changes can be investigated by means of pollen, geochemical, and paleomagnetic analyses:
1. the rate of sedimentation must be 'large',

Horie, S. Lake Biwa

ISBN 90 6193 095 2. Printed in The Netherlands

2. the sediments must be entirely 'fine-grained materials',
3. sedimentation must be 'continuous' since the early Pleistocene.

In unglaciated areas, the bottom sediments of ancient lakes may cover the entire Pleistocene. Lake Biwa in Japan is such a lake. Lake Biwa is important for other reasons:
1. a strikingly negative gravity anomaly exists in Lake Biwa, amounting to -50 milligals,
2. a great number of endemic species of animals and plants occur in the lake,
3. judging from geological studies around the lake, the Pliocene-Pleistocene boundary and the Pleistocene-Holocene boundary exist in the lake sediments.

The present vegetation around Lake Biwa belongs to the flora of the central part of the Temperate zone. In this region, the snowline and the tree line are recognized at the height of about 4,300 m and 2,300–2,600 m respectively.

Location of the boring and core samples

A 200 m core was collected from the bottom of 65 meters below the surface of this lake water, 20 meters above the present sea level, at a location between the Oki-no-shima Island in the lake and Ōmi-maiko in Shiga Prefecture, Central Japan, during the autumn of 1971 (Fig. 229). Samples were taken for primary pollen analyses at intervals of about 5 m throughout this core, and for secondary analyses at intervals of 25 cm. The core samples are composed mainly of soft homogeneous clay and include at least 30 thin ash layers. Samples taken from the same horizon have been analyzed for paleomagnetism, geochemical and organic chemical components, mechanical analyses of grains, diatoms and animal microfossils. The dating of the 200 m core was made by the C^{14} and fission-track methods. Extrapolation of the radiometric dates suggests that the age of the 200 meter horizon is about 560,000 years BP. As we have found no palynological studies covering this long period of time, the pollen diagram from Lake Biwa is not correlated with any diagrams from other localities.

Preparation for pollen analyses

Pollen grains in the sediments were concentrated in the laboratory by slight modifications of the process described by Faegri and Iversen (1964). Sediments containing coarse organic materials as moss, small leaves, and small shells were strained through a fine screen. As most sediments were predominantly silt and clay, they were left for two hours in hydrofluoric acid to remove the silicate minerals. They were then treated by standard acetolyses. Pollen grains from most samples were stained with safranin, and the preparation was mounted in

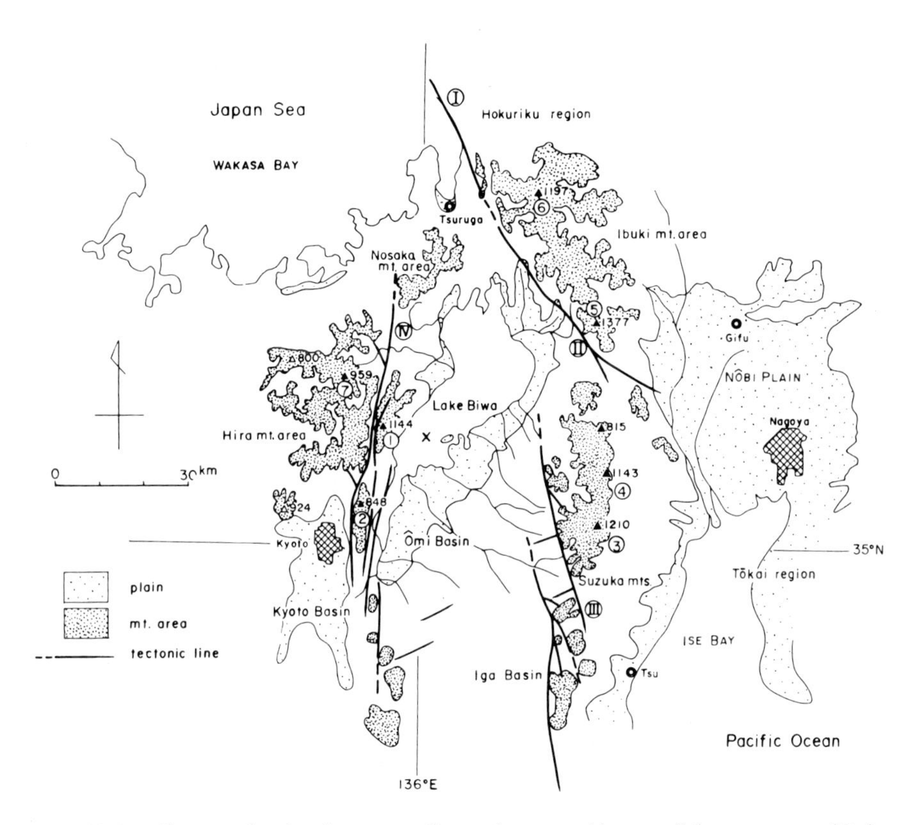

Fig. 229. Locality map showing the area studies, and topographic map of the areas around Lake Biwa (after Fuji 1976b).

1: Mt. Hira
2: Mt. Hiei
3: Mt. Gozaisho
4: Mt. Fujihara-dake
5: Mt. Ibuki
6: Mt. Mikuni-ga-dake
7: Mt. Mikuni-dake

X: Boring location
I: Kaburaki tectonic line
II: Yanagase tectonic line
III: Suzuka tectonic line
IV: Hira tectonic line

glycerine jelly. In order to obtain a pollen sum of about 500, one slide (size of covering glass area: 24 × 24 mm) was generally enough.

Method for interpretation of paleovegetation and climatic history

The writer depends upon pollen analyses for reconstructing a vegetation during a geological age. For the interpretation of pollen spectra obtained from the 200 m core sample the writer used two methods, namely, (1) pollen spectra of the modern samples taken in and around Lake Biwa and from various localities of some climatic zones throughout the Japanese Islands, and (2) the warmth index (month-degrees) (Fig. 230). According to the law of total effective temperature, the distribution of a plant is influenced more with the total temperature than with the mean annual temperature. The total temperature is given by the formula

$\Sigma(e_m - p)$, where e_m: mean month temperature,

p: physiological zero point, which is shown generally to be 5°C in the Japanese Islands.

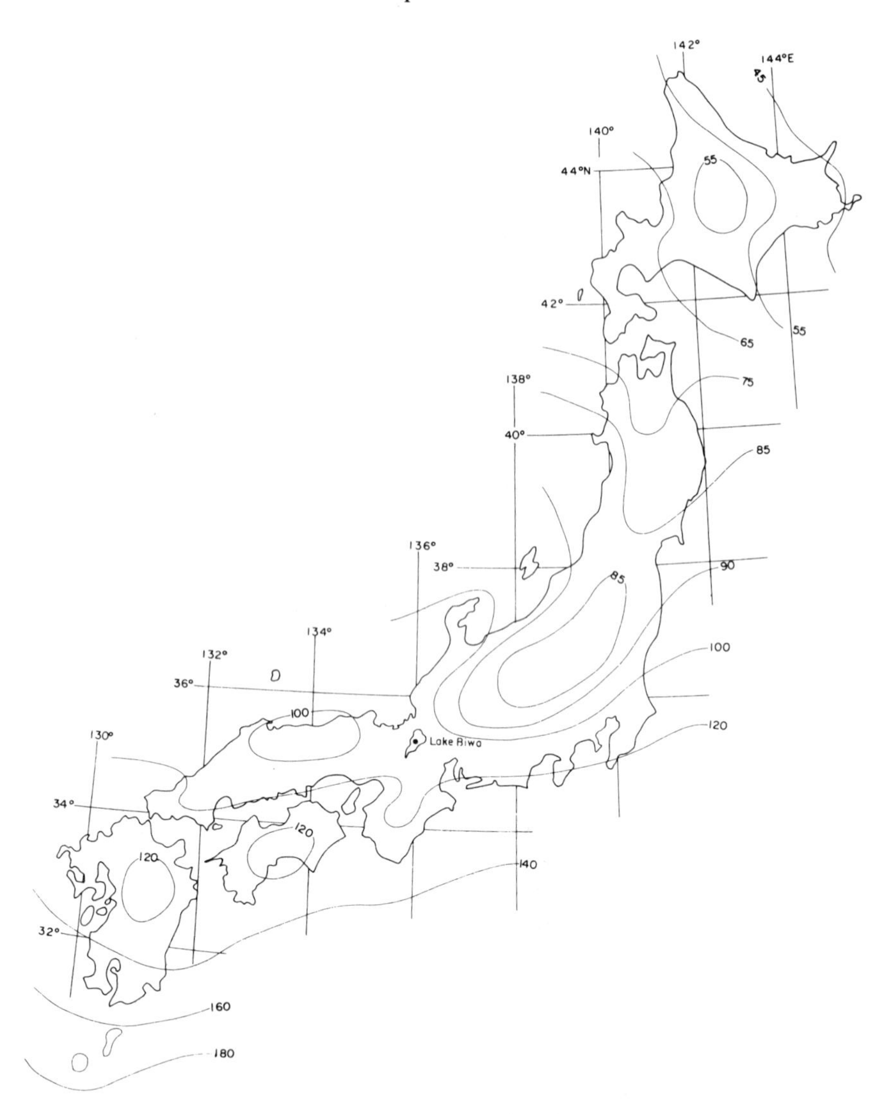

Fig. 230. Iso-warmth index (month-degrees) map of the Japanese Islands (Fuji 1976b).

When $e_m < p$, the $(e_m - p)$ value is omitted from this formula.

The total temperature is generally called the 'warmth index' (Kira & Iida 1958). The relationship among values of the warmth index (month-degrees), major plants growing in the Japanese Islands, temperature, climatic-zones, and latitude is summarized in Fig. 231. On the basis of the distribution of the warmth index of every plant, all of the arboreal genera found from the samples are distributed into one of the five groups, which are the Subpolar of Subalpine zone (warmth index month-degrees: 15° – 55°), the Cool Temperate zone (45° – 90°), the Cool Temperate zone-Temperate zone (55° – 140°), the middle part of the Cool Temperate zone – Warm Temperate zone (70° – 140°), and the southern part of the Temperate zone – the Subtropical zone (100° – 180°).

Construction of diagrams

Shown from the left to the right in pollen diagrams (Fig. 232) are depth of the core, C^{14} and fission-track dates, paleomagnetic stratigraphy, estimated absolute year, summary diagram, pollen profiles of each taxon and assemblage zones. The pollen sum used on the pollen types concerned depends on the local situation and on the kind of deposit. The pollen diagram (Fig. 233) shows all non-arboreal pollen (herbs), aquatic and marsh plants, and ferns, excluded from the pollen sum but calculated as a percentage over the pollen sum.

Percentages in both diagrams are shown on two scales; the scale with 10 × exageration permits the accurate plotting of minor curves and minor fluctuations. The summary diagram is drawn to facilitate the discussion and interpretation of paleovegetation and climatic history. This summary diagram is composed of changes of the ratio between total AP and total NAP, and of percentages of Subpolar plants, Cool Temperate plants, Cool Temperate–Temperate plants, plants of the middle area of the Cool Temperate zone and of the Warm Temperate zone, and plants of the southern area of Temperate zone and of the Subtropical zone, calculated on the basis of the warmth index (month-degrees) as will be described in a later chapter. There follow curves for boreal conifers.

Remarks of identification of pollen grains

Most of pollen grains from the core samples are found in the present Japanese Islands, and their descriptions can be found in the published references. Some of the critical genera are discussed below.

Abies

In the Japanese Islands, six species belong to *Abies*. *A. mariesii*, *A. veitchii*, *A. sachalinensis*, and *A. sachalinensis* var. *mayriana* grow in the Subpolar or

Plant Name
Abies Mariesii
Abies Veitchii
Picea jezoensis var. *hondoensis*
Picea jezoensis
Abies sachaliensis
Betula Ermani
Tsuga diversifolia
Larix leptolepis
Betula platyphylla
Abies homolepis
Fagus crenata
Quercus crispula
Quercus dentata
Fagus japonica
Carpinus cordata
Carpinus japonica
Carpinus laxiflora
Carpinus Tschonoskii
Tsuga Sieboldii
Castanea crenata
Quercus serrata
Zelkova serrata
Abies firma
Thujopsis dolabrata
Cryptomeria japonica
Chamaecyparis pisifera
Chamaecyparis obtusa
Torreya nucifera
Quercus variabilis
Quercus acutissima
Quercus myrsinaefolia
Quercus stenophylla
Ilex integra
Celtis sinensis
Pinus densiflora
Pinus Thunbergii
Camellia
Quercus acuta
Quercus paucidentata
Machilus Thunbergii
Quercus glauca
Quercus gilva
Shiia
Podocarpus macrophylla
Puerarla lobata
Podocarpus Nagi
Ficus Wightiana
Cycas revoluta

Warmth Index (month-degrees): 0 10 20 30 40 50 60 70 80 90 100 110 120 130 140 150 160 170 180

Warmth Index (month-degrees)	0 10 20 30 40 50 60 70 80 90 100 110 120 130 140 150 160 170 180
Annual mean temperature (° C)	2 5 7 10 12 14 16 18 21
Mean temperature on Jan. (°C)	-6 -4 -2 1 4 6 8 14
Mean temperature on July (°C)	16 20 22 24 24 25 27 28
Climatic Zones	Polar Alpine; Subpolar zone Subalpine; Cool Temperate; Temperate; Warm Temperate; Subtropical
Latitude in Japan — Pacific side	43 42.5 39 36 35 32 31 [illegible]
Latitude in Japan — Japan Sea side	44 44 37

Fig.231. Relationship between the warmth index (month degrees), plants, temperature, climatic zones and latitude (after Fuji 1976b).

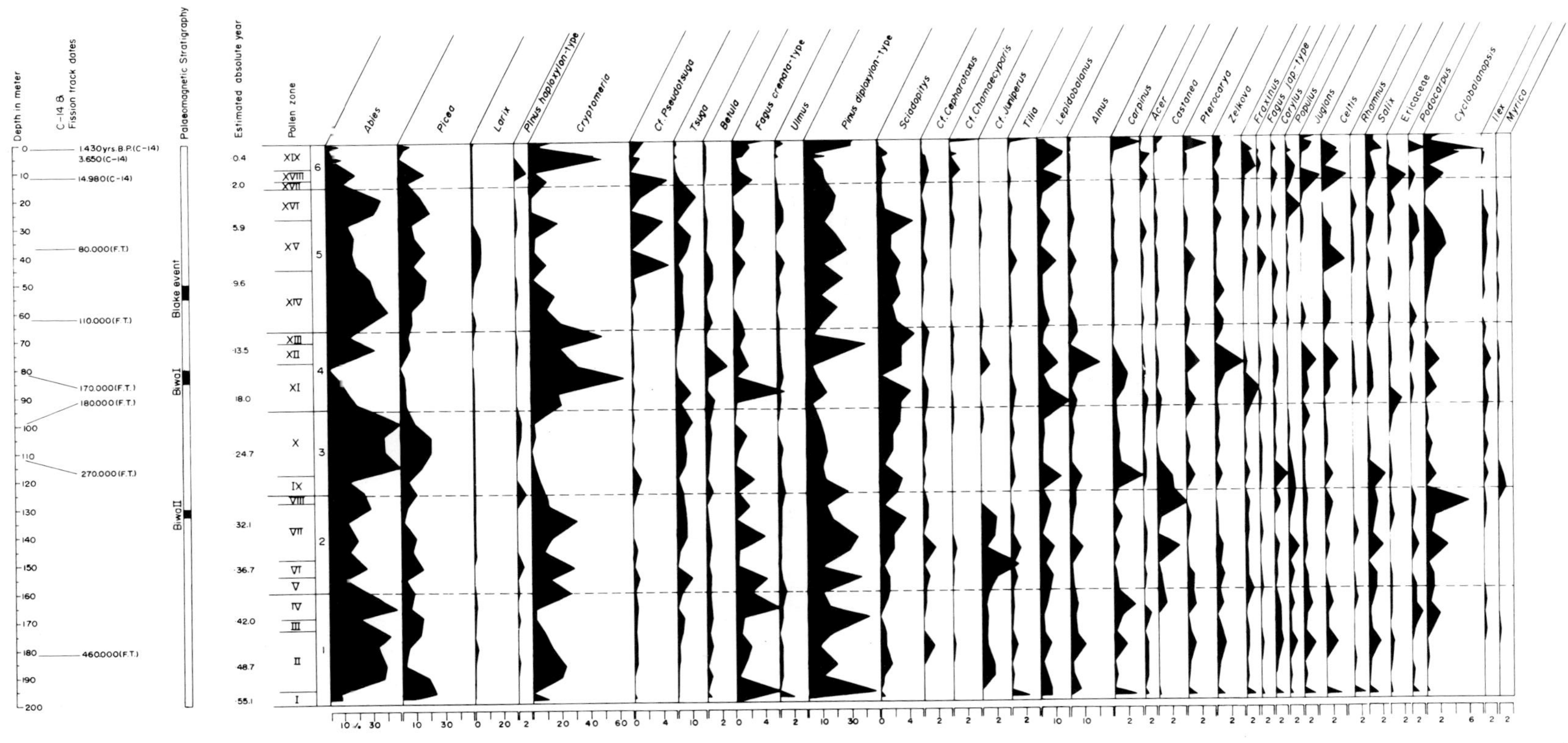

Fig. 232. Arboreal pollen diagram on the basis of samples at 5 meter interval from the 200 meter core (after Fuji 1976b).

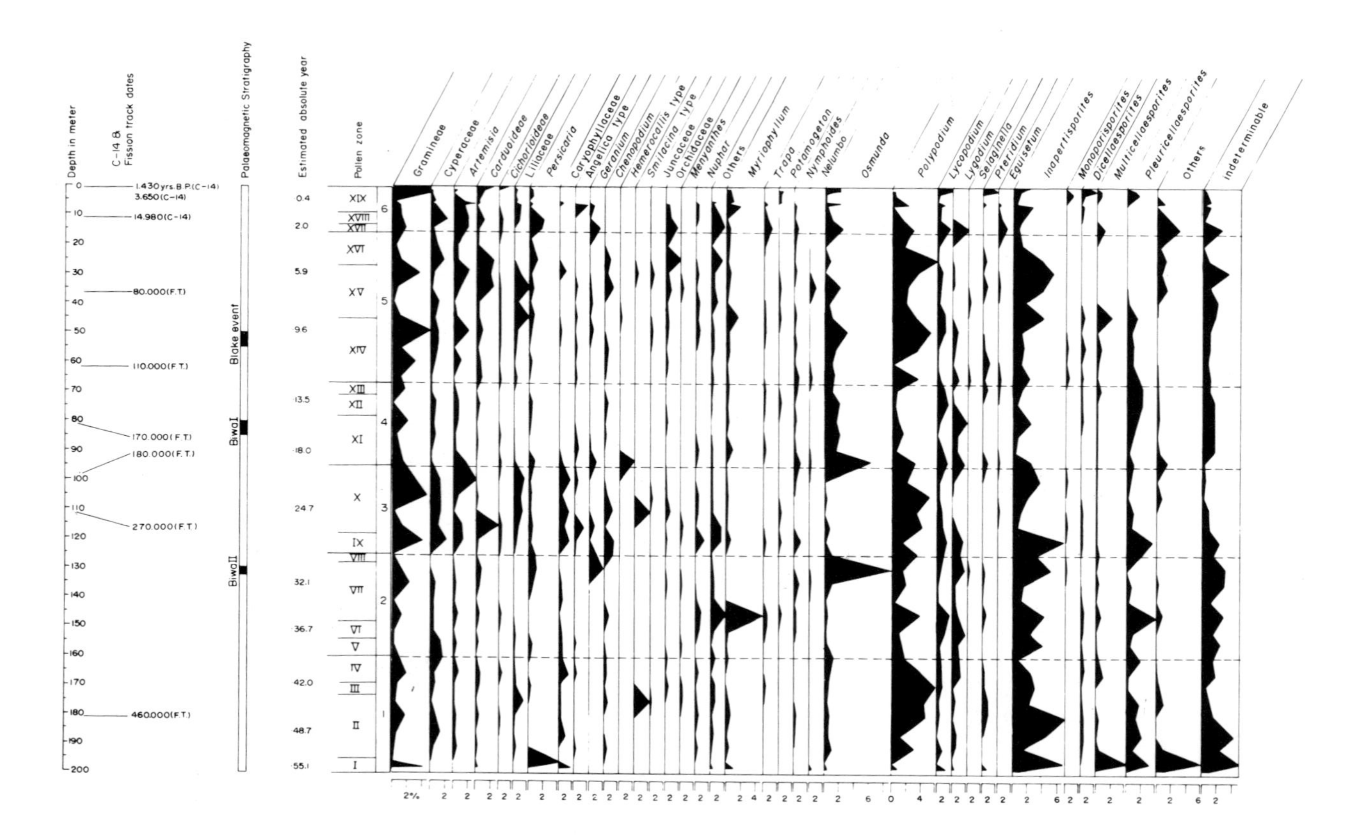

Fig. 233. Non-arboreal pollen diagram on the basis of samples at 5 meter interval from the 200 meter core (after Fuji 1976b).

Subalpine zone. *Abies homolepis* and *A. firma* grow in the Cool Temperate and Temperate zones. Pollen grains of these two species, however, are not distinguished from other species of *Abies* from the viewpoint of the shape of pollen grains.

Picea

Six species of *Picea* are found in the Islands. Although *Picea* is not classified into species on the basis of pollen morphology, *P. jezoensis* and *P. jezoensis* var. *hondoensis*, *P. glehnii*, *P. bicolor* var. *reflex*, and *P. bicolor* grow in the Subpolar or Subalpine zone, and *Picea polita* is found in the Cool Temperate zone.

Pinus

Genus *Pinus* includes about six species. *Pinus pumila* and *P. pentaphylla* are found in the Subpolar or Subalpine zone, *P. koraiensis* grows in the Cool Temperate and Subpolar, or Subalpine zones, and *P. densiflora*, *P. himekomatsu*, and *P. thunbergii* grow in the Temperate and Warm Temperate zones. *Pinus*, however, is divided easily into two types by the pollen morphology: they are *P. haploxylon*-type and *P. diploxylon*-type. Species with five leaves belong to the former; they grow in the Subpolar or Subalpine zone at present. The latter includes species with two leaves, such as *P. densiflora*, *P. himekomatsu*, and *P. thunbergii*.

Taxodiaceae

For genera belonging to Taxodiaceae, *Cryptomeria* and *Taxodium* grow in the Japanese Islands after the early Pleistocene. However, in the surroundings of Lake Biwa, the Plio-Pleistocene Kobiwako (Paleo-Biwa) Group is distributed widely and yields pollen grains of *Metasequoia* and *Glyptostrobus*. Accordingly, these pollen grains are found sometimes as secondary (reworked) fossils in younger deposits. Pollen grains of *Metasequoia* and *Cryptomeria* are discriminated from other genera by grain size and shape of papilla. According to the measurement of modern species (Fuji 1973a), *Metasequoia*: 20 × 23 microns in width and length; *Cryptomeria japonica*: 30–33 × 34–38 microns; *Sequoia sempervirens*: 36 – 42 × 34 – 38 microns; *Glyptostrobus pencilis*: 26 – 28 × 30 – 33 microns; *Taxodium distichum*: 22 – 25 × 25 – 29 microns. The secondary pollen grains are not shown on the pollen diagram.

Tsuga

In the present Japanese Islands, *Tsuga diversifolia* is found in the Subalpine or Subpolar zone and *T. sieboldii* in the Cool Temperate zone and in the northern part of the Temperate zone. They are not differentiated from other species.

Betula

Species such as *Betula ermani* is found in the Subpolar or Subalpine zone, and *B. platyphylla* in the Cool Temperate zone. Their pollen grains are not differentiated.

Fagus

In the present Japanese Islands, *Fagus* has two species, *F. crenata* and *F. japonica*. According to the writer's observations, *Fagus crenata* measures 38–40 × 45–48 microns and can thus be distinguished from *F. japonica* which measures 29–32 × 33 microns. *F. crenata* grows in the Cool Temperate zone, and *F. japonica* is found in the middle part of the Cool Temperate and Temperate zones.

Quercus

Quercus is divided into two subgenera, *Lepidobalanus* and *Cyclobalanopsis* by pollen grain-size and morphology. An evergreen *Quercus* belongs to the latter, and deciduous *Quercus* to the former. Pollen grains of *Cyclobalanopsis* can be distinguished from those of the *Lepidobalanus* by a combination of features (van Campo & Elhai 1956; van der Spoel-Walvius 1963). As shown below, grains of *Cyclobalanopsis* are in general smaller than those of *Lepidobalanus*. The thickness of the pollen grain wall of both subgenera varies from 0.8 to 1.8 microns. The *Lepidobalanus* usually shows a somewhat thicker wall than *Cyclobalanopsis*. The wall in *Cyclobalanus*, however, is relatively thicker because of the small size of the grain. The costae in *Cyclobalanopsis* are generally heavier than those of *Lepidobalanus*. The shape of costae and colpi in *Cyclobalanopsis* differ from that of *Lepidobalanus*. The colpi in *Cyclobalanopsis* often show a poroid area or a constriction (Bottema 1975).

Cyclobalanopsis (evergreen *Quercus*)
Quercus acuta : 21–23 × 26–27
Q. glauca : 19–20 × 21–23
Q. phillyraeoides : 20–22 × 23–26

Lepidobalanus (deciduous *Quercus*)
Quercus serrata : 22–23 × 24–28
Q. dentata : 32–34 × 36–38
Q. variabilis : 29–30 × 31–33
Q. acutissima : 28–30 × 37–38

Cyclobalanopsis occurs in the Temperate and Warm Temperate zones, and

Lepidobalanus in the Cool Temperate zone and in the northern part of the Warm Temperate zone.

The use of the suffix '-type' after the name of a species or a genus implies that the fossil may belong to the named species or genus or to others of closely similar or identical morphology which have not been distinguished by the writer. Thus, *Pinus haploxylon*-type includes pollen grains of *Pinus* with five leaves such as *Pinus pumila*, *P. pentaphylla*, and *P. koraiensis*. This type is not divided into the three above-mentioned species on the basis of pollen morphology. The use of 'cf.' as in Cf. *Cephalotaxus* indicates that most grains could be referred to this genus.

Zoning of pollen assemblages, interpretation of vegetation and climatic history based upon the samples selected at 5-m intervals

In order to facilitate descriptions and discussions of the pollen assemblages, the pollen diagrams are divided into nineteen pollen zones based on conspicuous changes in pollen percentages. Changes in the ratio of total AP to total NAP can be established, but changes in the values of one or two pollen types may also lead to the establishment of pollen zones which have similar assemblages of pollen grains and spores. The writer has attempted to indicate these under the same letter code in order to facilitate comparisons. (Figs. 232, 233)

Zone I, depth 195–198 m: The pollen assemblage of this zone is defined by *Pinus diploxylon*-type (46%), *Picea* (24%), *Fagus crenata*-type (21%), and *Cryptomeria* (10%), and is characterized by a very high pollen percentage of plants thriving in the middle part of the Cool Temperate–Temperate zone. The mean annual temperature of the summit area and lowland at the time of this zone may have been higher than 5°C and 13°C respectively and the climate was probably more or less arid.

Zone II, depth 173–195 m: This zone is characterized by large amounts of boreal conifers, especially *Abies* and *Picea*, and Cool Temperate–Temperate plants. The climate in the lakeside at that time may be compared with that in the middle part of the Cool Temperate zone. In the lowland areas, *Cryptomeria*, *Pinus diploxylon*-type, *Alnus*, *Fagus crenata*-type, *Carpinus*, *Zelkova*, *Juglans*, and *Lepidobalanus* grew, and, at the mountain sides, *Abies*, *Tsuga*, and *Betula* grew. *Pinus pumila* and *Abies* might have prevailed in the summit area during the later part of this period.

Zone III, depth 168–173 m: Although the pollen percentages of boreal conifers are almost the same as in Zone II, the pollen values of broadleaved deciduous trees decrease, and plants of the Warm Temperate–Subtropical zones such as broadleaved evergreen trees and *Podocarpus* attain relatively higher values. The climate at this time may have been as warm as that of the present day and relatively arid.

Zone IV, depth 160–168 m: This zone is characterized by an increase in plants of the Cool Temperate zone and of the Cool Temperate–Temperate zones, and by a decrease in plants of the middle part of the Cool Temperate–Temperate zones and of the southern Temperate–Subtropical zones. Judging from the pollen assemblages, the climate at this time may have been as cool as that in the southern part of the Cool Temperate zone and was probably wet as indicated by the large value of *Cryptomeria.*

Zone V, depth 154–160 m: This pollen zone is characterized by decreases in Subpolar or Subalpine plants and in Cool Temperate–Temperate plants, and by sharp increases in middle Cool Temperate–Temperate plants. In the mountainous area, broadleaved deciduous trees, *Abies* (perhaps *A. firma*), *Picea* (perhaps *P. polita*), and at the lakeside, *Pinus densiflora* prevailed. There were also scattered *Cryptomeria*, *Zelkova*, *Juglans*, *Salix*, and *Alnus*. The climate may have been similar to that of the present day.

Zone VI, depth 148–154 m: Relatively high percentages of Subpolar or Subalpine plants and Cool Temperate–Temperate plants, and decreases in the percentage of middle Cool Temperate–Temperate plants and southern Temperate–Subtropical plants are the characteristic features of this zone. The climate of the lowland at the time represented by this zone may have been similar to that of the middle part of the Cool Temperate zone.

Zone VII, depth 128–148 m: The main elements of this zone are *Pinus diploxylon*-type (28–34%), *Cryptomeria* (12–30%), and *Abies* (13–29%). This zone is characterized by boreal conifers which lasted for a time with only long small fluctuations in abundance. The climatic condition at this time was similar to that of the present day with temporarily cool periods.

Zone VIII, depth 125–128 m: The pollen assemblage of this zone contains *Pinus diploxylon*-type (28%), *Abies* (25%), *Picea* (11%), and *Cryptomeria* (9%). The zone is notable for drastic increases in Subpolar or Subalpine plants and in boreal conifers and for decreases in Cool Temperate–Temperate and southern Temperate–Subtropical plants. The climate during this time may have been similar to that of the middle part of the Cool Temperate zone, and was probably relatively dry with the mean annual temperature of about 10°C at the lakeside.

Zone IX, depth 118–125 m: This zone is similar to Zones V and VII. The climatic condition at this time may have been somewhat milder than that in the northern part of the Temperate zone. *Abies* (perhaps *A. homolepis*) and *Picea* prevailed with a little *Betula* (perhaps *B. platyphylla*) and *Fagus crenata*-type in the higher part of mountainous areas. In the lowland area, broadleaved trees, especially *Lepidobalanus*, and *Pinus densiflora* probably prevailed.

Zone X, depth 95–118 m: This zone is characterized by a high concentration of boreal conifers, especially *Abies* (45%) and *Picea* (18%). The climate in the lowland at this time may have been as cold as that of the northern part of the Cool Temperate zone and dry, with a temporarily cooler interval during the middle of the period. On the other hand, in the mountainous areas, boreal conifers prevailed.

Zone XI, depth 78–95 m: This zone is characterized by an increase in plants other than boreal conifers. The climate in the lakeside area may have corresponded to that of the northern part of the Temperate zone or of the southern part of the Cool Temperate zone; wetter conditions probably prevailed.

Zone XII, depth 71–78 m: This zone is characterized by increases in boreal conifers and plants of the middle part of the Cool Temperate and Cool Temperate–Temperate zones. The climate in the lowland area around Lake Biwa may have been as cool as that of Zone VI and Zone VIII but, on the other hand, *Pinus haploxylon*-type (here perhaps *P. koraiensis*), *Abies* (perhaps *A. homolepis*), and *Picea* (perhaps *P. polita*) were present in the mountainous area.

Zone XIII, depth 67–71 m: This zone is characterized by a markedly drastic increase in plants growing in the Cool Temperate–Temperate zones, particularly *Cryptomeria*. The climate was probably similar to that of the southern part of the Cool Temperate zone, and was clearly wetter.

Zone XIV, depth 45–67 m: The main elements of the pollen assemblage of this zone are *Abies* (26–42%), *Pinus diploxylon*-type (21–26%), *Larix* (24%), *Picea* (17%), and *Cryptomeria* (11–17%). The zone is marked by a drastic decline in plants thriving in the Cool Temperate zone, and by a strong increase in boreal conifers. The climate appears to have deteriorated slightly during this time from cool conditions at the beginning of the period to relatively cold dry conditions in the middle and at the end of the period.

Zone XV, depth 27–45 m: The climate during the early part of this period may have been similar to that of the northern part of the Temperate zone. Later, however, it became cooler and resembled that in the southern part of the Cool Temperate zone.

Zone XVI, depth 16–27 m: During this period the climate deteriorated and, in the lakeside areas boreal conifers and broadleaved deciduous trees probably prevailed.

Zone XVII, depth 13–16 m: This zone is characterized by a drastic decline in boreal conifers. Instead of boreal conifers, plants of the Cool Temperate–Temperate zones, particularly *Cryptomeria* (12%) and *Lepidobalanus* (16%) were relatively abundant. The climate at that time may have been correlated to that of the northern part of the Temperate zone or of the southern part of the Cool Temperate zone.

Zone XVIII, depth 9–13 m: The climate at the time represented by this zone may have been the coldest throughout the last 560,000 years, and was almost comparable to that of Zone X. In the summit and mountainous areas, *Pinus pumila*, *Abies* (perhaps *A. mariesii* or *A. veitchii*), and *Picea* prevailed. Boreal conifers such as *Abies homolepis*, *A. firma*, *Picea* (? *P. polita*), *Pinus koraiensis*, *Betula* (perhaps *B. platyphylla*) and other broadleaved trees were distributed in the lowlands around Lake Biwa.

Zone XIX, depth 0–9 m: During the time represented by this zone, boreal conifers and Cool Temperate plants decreased in abundance and were replaced by plants growing near the middle of the Cool Temperate zone and in the southern part of the Temperate–Subtropical zone. At this time, *Lepidobalanus* (perhaps *Quercus crispula* and *Q. serrata*), *Acer*, *Cryptomeria japonica*, and *Chamaecyparis* prevailed in the mountainous area instead of *Abies*, *Larix*, and *Ulmus* as in the period represented by Zone XVIII. On the other hand, *Lepidobalanus* as *Quercus acutissima* and *Cyclobalanopsis* as *Quercus glauca* were distributed widely with *Salix* and *Alnus* along the sides of the lake and rivers. The climate of this time may have been 1°–2°C warmer than and more humid than the present.

Conclusion of pollen analyses at intervals of 5-m

1. Judging from the pollen diagrams, the core samples are divided into nineteen pollen zones: Zones I, II, III, ... XVIII, and XIX (Fig. 234).

2. The climatic curve and ages from Lake Biwa display similarity to the palaeotemperature curve (oxygen-isotope ratio determination) from the Caribbean Sea (Emiliani & Shackleton 1974). In spite of differences in details and in interpretation of curves which have no absolute ages, similarity exists in the major trends in the inferred climatic fluctuations of North America and Europe.

3. During the glacial stages or stadials, the typical vegetation thriving today in the Subpolar or Subalpine zone of Japan prevailed at the summit area and/or the montane area around Lake Biwa and, in the lowland around the lake, plants growing today in the Cool Temperate zone were distributed. In the interglacials and interstadials, the vegetation in the higher area was characterized mainly by plants of the Cool Temperate zone and of the present Temperate zone and, in the lowland, the vegetation was composed mainly of broadleaved deciduous and evergreen trees growing in the Warm Temperate zone.

4. Correlation between the writer's paleoclimatic curve from Lake Biwa and Emiliani's palaeotemperature curve from the Caribbean Sea is shown in Fig. 235.

5. The remarkable human influence that is represented by clearing forest for rice-cultivation and by cutting woods began about 3,000 years BP. (the Neolitic age). This inference is supported by the archaeological evidence around the lake.

Zoning of pollen assemblages, interpretation of vegetation and climatic history based upon the samples selected at 25 cm intervals

The pollen diagrams are divided into six zones: Z–1, Z–2, Z–3, Z–4, Z–5, and Z–6. (Figs. 236, 237)

Zone Z–1, depth 42–55 m: This zone is characterized by large amplitudes and long durations of every maximum (up to 85%) or minimum (down to ca. 15%) period in fluctuations of pollen values of boreal conifers which are *Pinus haploxylon*-type, *Larix*, *Abies*. *Picea*, and *Tsuga*. This is shown especially in the cases of *Abies* which ranges from 10 to 50% and of *Picea* (3 to 15%). In contrast

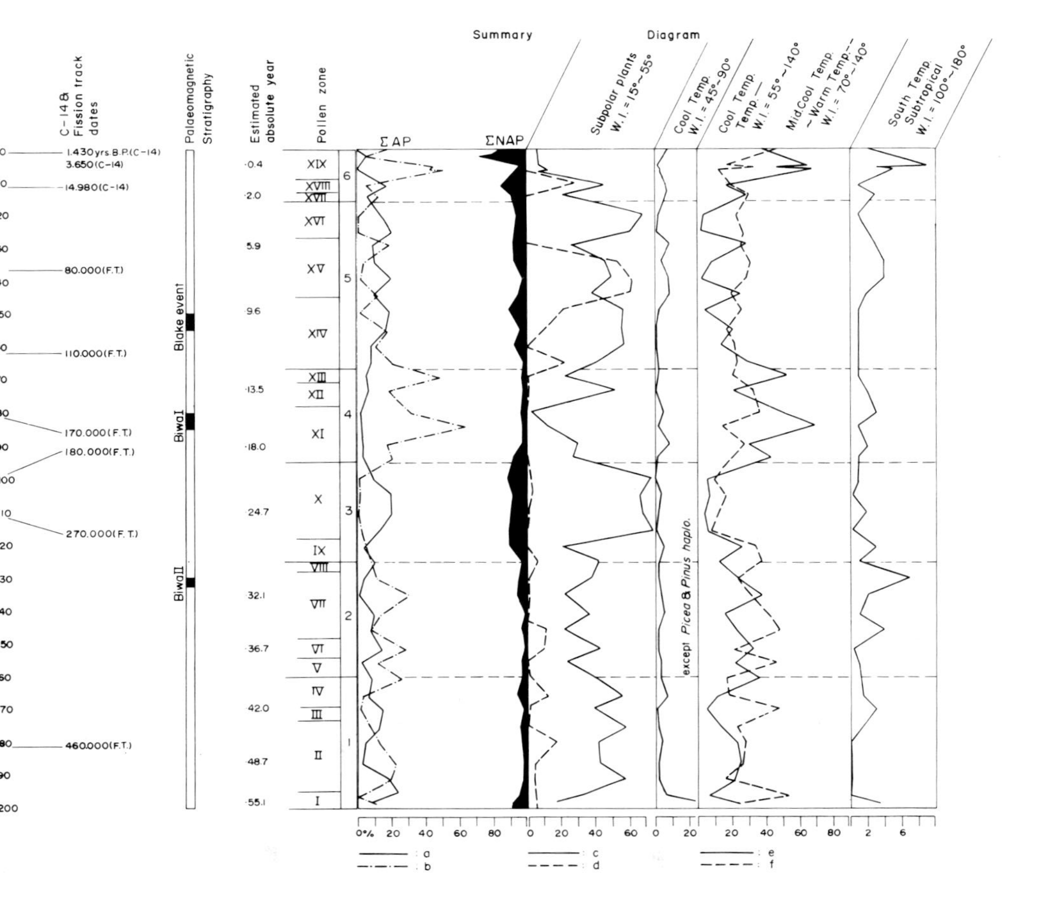

Fig. 234. Summary diagram of palynological analysis samples at 5 meter interval from the 200 meter core (after Fuji 1976b).

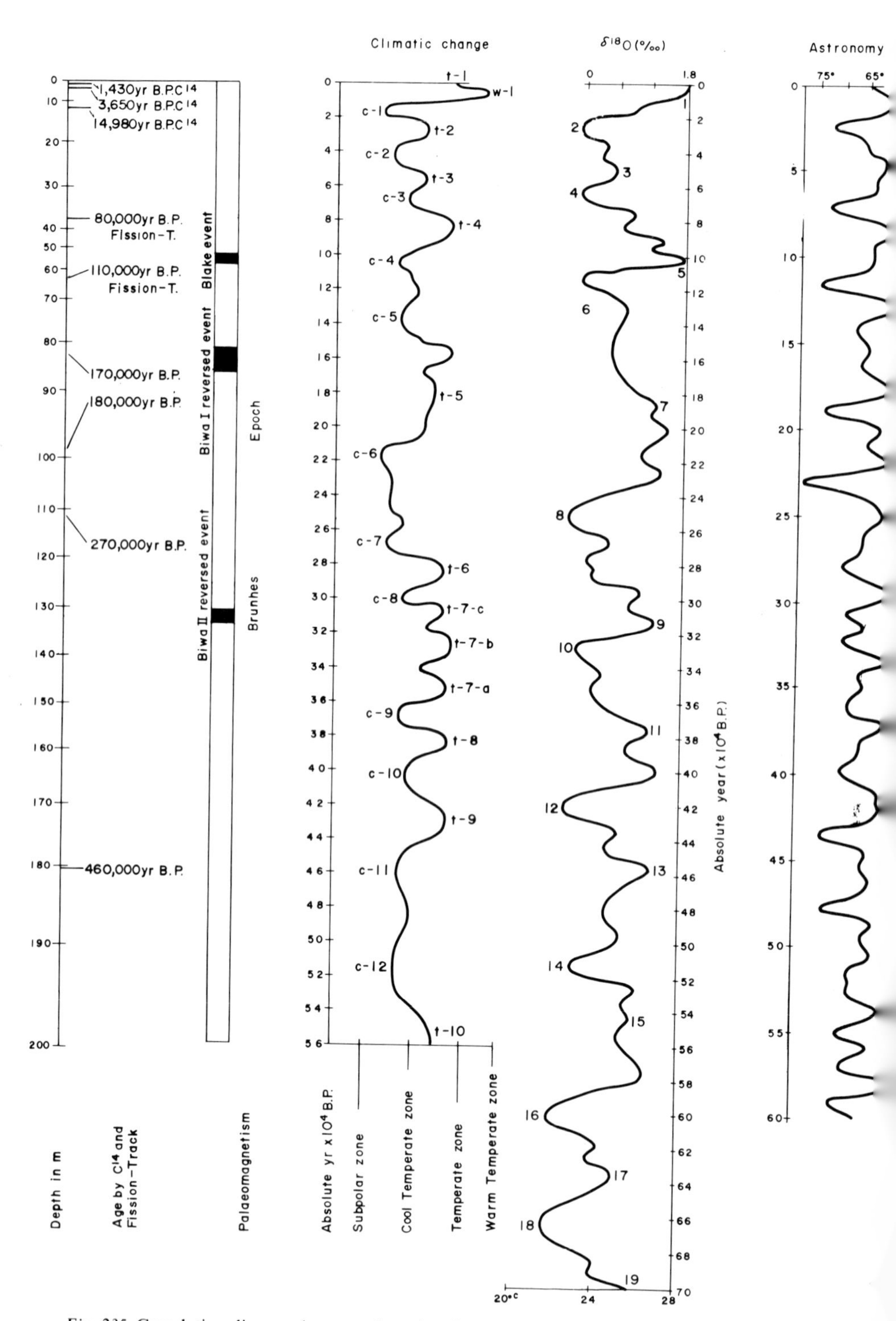

Fig. 235. Correlation diagram between the paleoclimatic change around Lake Biwa and paleotemperature curve based on the oxygen-isotope ratio determination from the Caribbean Sea by Emiliani & Shackleton (1974) (after Fuji 1976b).

to them. *Pinus diploxylon*-type, which includes the temperate *Pinus thunbergii* and *P. densiflora*, shows a low percentage (average 10%) in a period when boreal conifers are abundant, and an increase of the former is matched by a decrease in the values of the latter. In this zone, trees such as *Ulmus*, *Cryptomeria*, *Lepidobalanus*, and *Cyclobalanopsis* which grow in the Cool Temperate and Temperate zones have generally lower values than in Zone Z–2. *Juniperus* and *Chamaecyparis* show lower values than in Zone Z–2–Zone Z–6. Roughly speaking, although this zone is characterized by relatively low amounts of NAP (1–5%), *Artemisia* and Liliaceae each reach to 1.5%, and are very rare or lacking in Zone Z–2. On the basis of fluctuations in pollen values of boreal conifers and minor changes of the other types, this zone is divided into the following six subzones:

Subzone Z–1–a: depth 53.5–55 m.
In this subzone, *Picea* and *Pinus haploxylon*-type show lower values than the above.

Subzone Z–1–b: depth 50–53.5 m.
Boreal conifers reach to 80% near the top of this subzone, and *Ulmus* exceeds 5%. *Pinus diploxylon*-type is as low as 6%.

Subzone Z–1–c: depth 48.1–50 m.
Boreal conifers have values lower than 40%, with *Abies* as low as 15%. *Pinus diploxylon*-type exceeds 60%. Besides these conifer trees, *Chamaecyparis* is represented by a relatively high value (up to 3%).

Subzone Z–1–d: depth 45.9–48.1 m.
Boreal conifers have values up to 65–80% in this subzone, with *Abies* pollen averaging at 50%. *Pinus diploxylon*-type lowers to 10–15%.

Subzone Z–1–e, depth 45.2 – 45.9 m.
This subzone is characterized by relatively high values of Subpolar or Subalpine plants. *Abies*, *Pinus diploxylon*-type and *Ulmus* show 45–55%, 10–15% and about 5% respectively.

Subzone Z–1–f: depth 42–45.2 m.
In this subzone, boreal conifers have lower values (12%) again. *Pinus diploxylon*-type rises to more than 5%. Plants growing in the Warm Temperate–Subtropical zones, such as *Podocarpus* (about 3%) and *Cyclobalanopsis* (Evergreen *Quercus*, maximum 8%), show increased values.

Zone Z–2: depth 34–42 m.
Zone Z–2 shows steady curves for boreal coniferous trees, especially conspicuously in the cases of *Abies* (amplitude 20–40%) and *Picea* (amplitude 2–7%). Increases of *Tsuga* (to 27%), *Cryptomeria* (to 24%), and *Chamaecyparis* are notable. *Pinus diploxylon*-type shows generally lower values than Zone Z–1, and *Castanea* is essentially absent. This zone is divided into three subzones on the basis of pollen fluctuations of boreal conifer and broadleaved deciduous trees.

Subzone Z–2–a: depth 39.4–42 m.

Fig. 236 (overleaf). Arboreal pollen diagram from the upper 55 m of the 200 meter core (after Fuji 1976a).

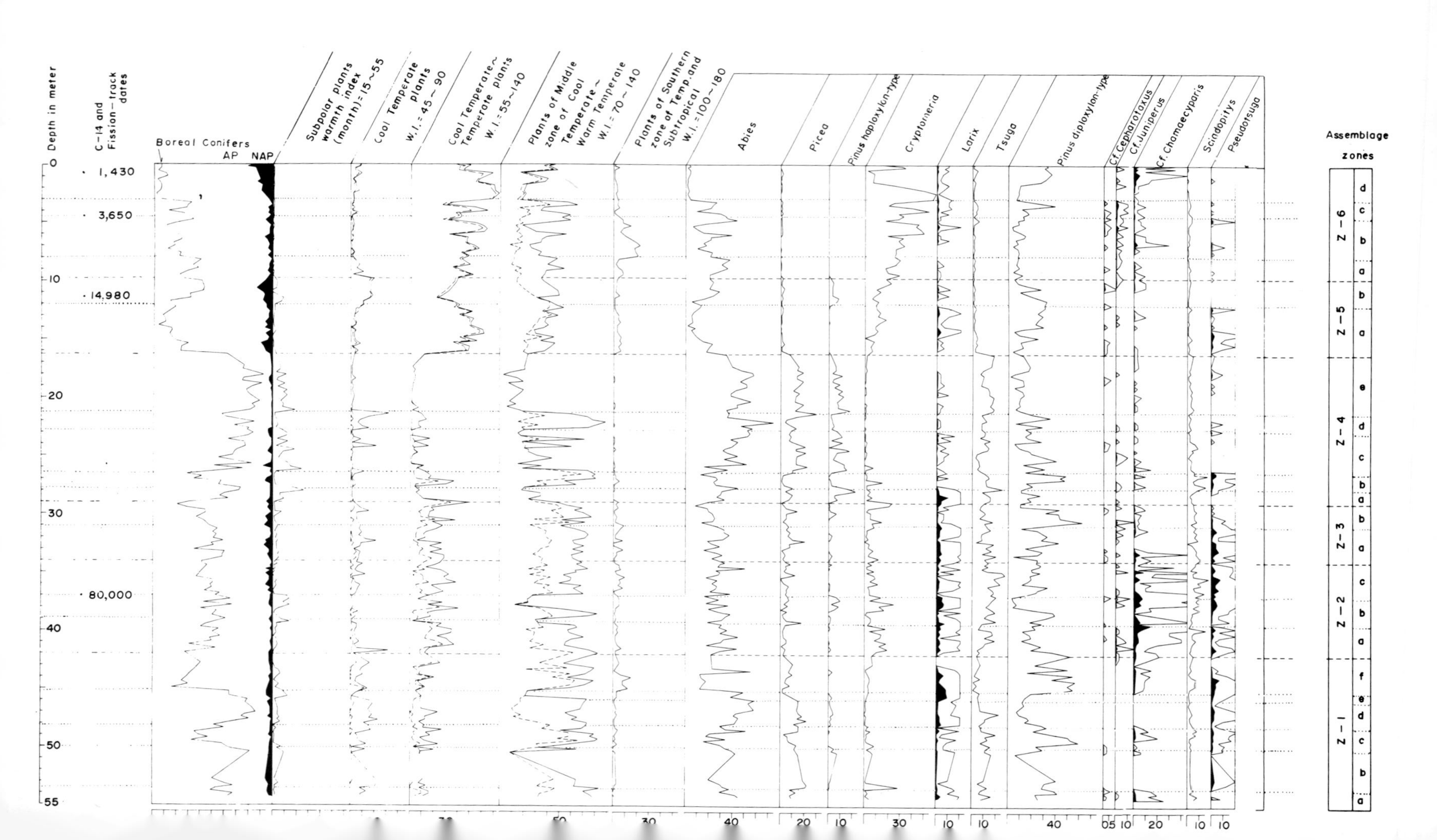
Depth in meter
C-14 and Fission-track dates
1,430
3,650
14,980
80,000
Boreal Conifers
AP
NAP
Subpolar plants warmth index (month)=15~55
Cool Temperate plants W.I.=45~90
Cool Temperate~ Temperate plants W.I.=55~140
Plants of Middle zone of Cool Temperate~ Warm Temperate W.I.=70~140
Plants of Southern zone of Temp. and Subtropical W.I.=100~180
Abies
Picea
Pinus haploxylon-type
Cryptomeria
Larix
Tsuga
Pinus diploxylon-type
Cf. Cephalotaxus
Cf. Juniperus
Cf. Chamaecyparis
Sciadopitys
Pseudotsuga
Assemblage zones
Z-6
Z-5
Z-4
Z-3
Z-2
Z-1

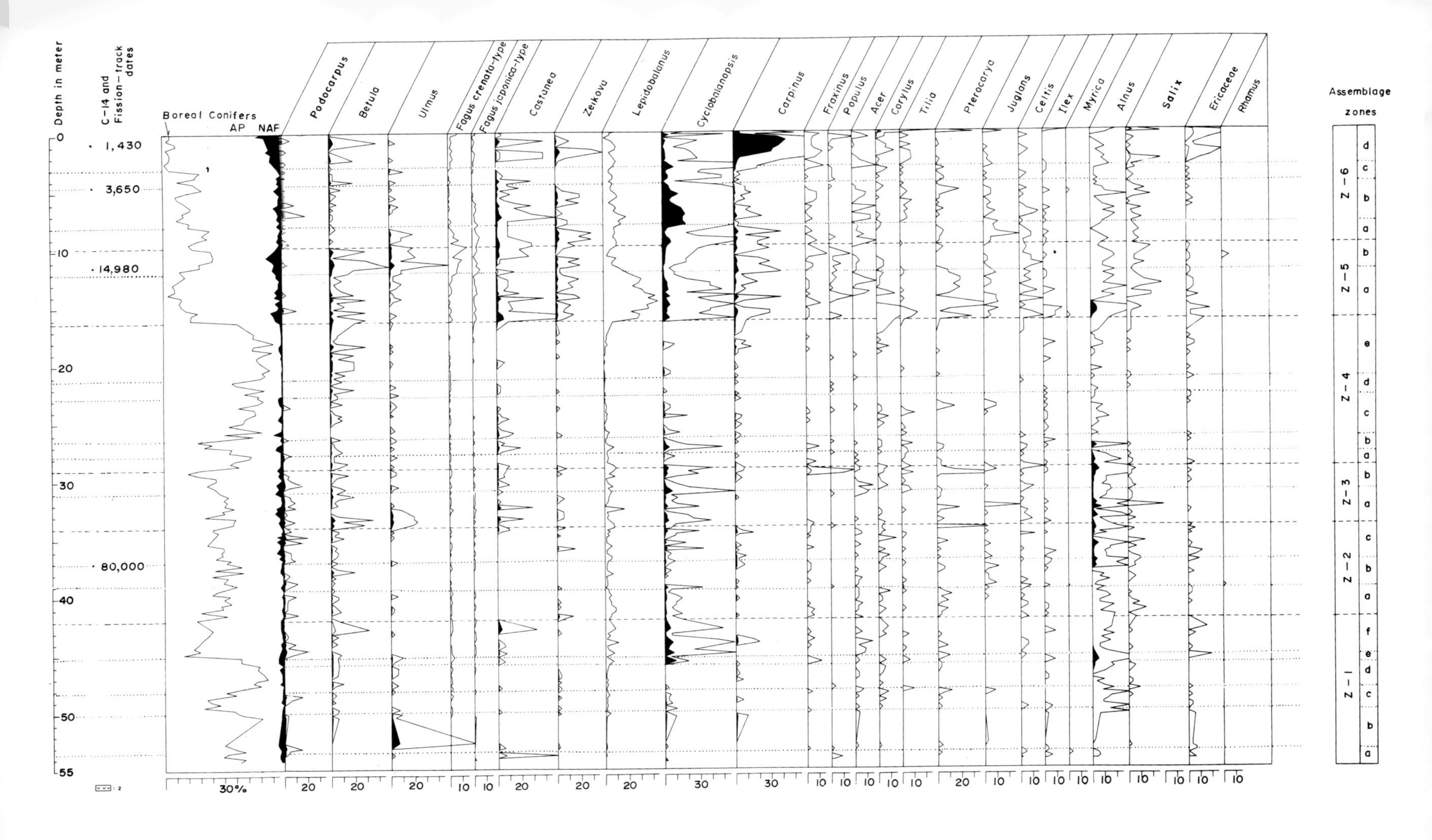

Depth in meter
C-14 and Fission-track dates
1,430
3,650
14,980
80,000
Boreal Conifers
AP
NAF
Podocarpus
Betula
Ulmus
Fagus crenata-type
Fagus japonica-type
Castanea
Zelkova
Lepidobalanus
Cyclobalanopsis
Carpinus
Fraxinus
Populus
Acer
Corylus
Tilia
Pterocarya
Juglans
Celtis
Ilex
Myrica
Alnus
Salix
Ericaceae
Rhamus
Assemblage zones
Z-6
Z-5
Z-4
Z-3
Z-2
Z-1
0
10
20
30
40
50
55
30%

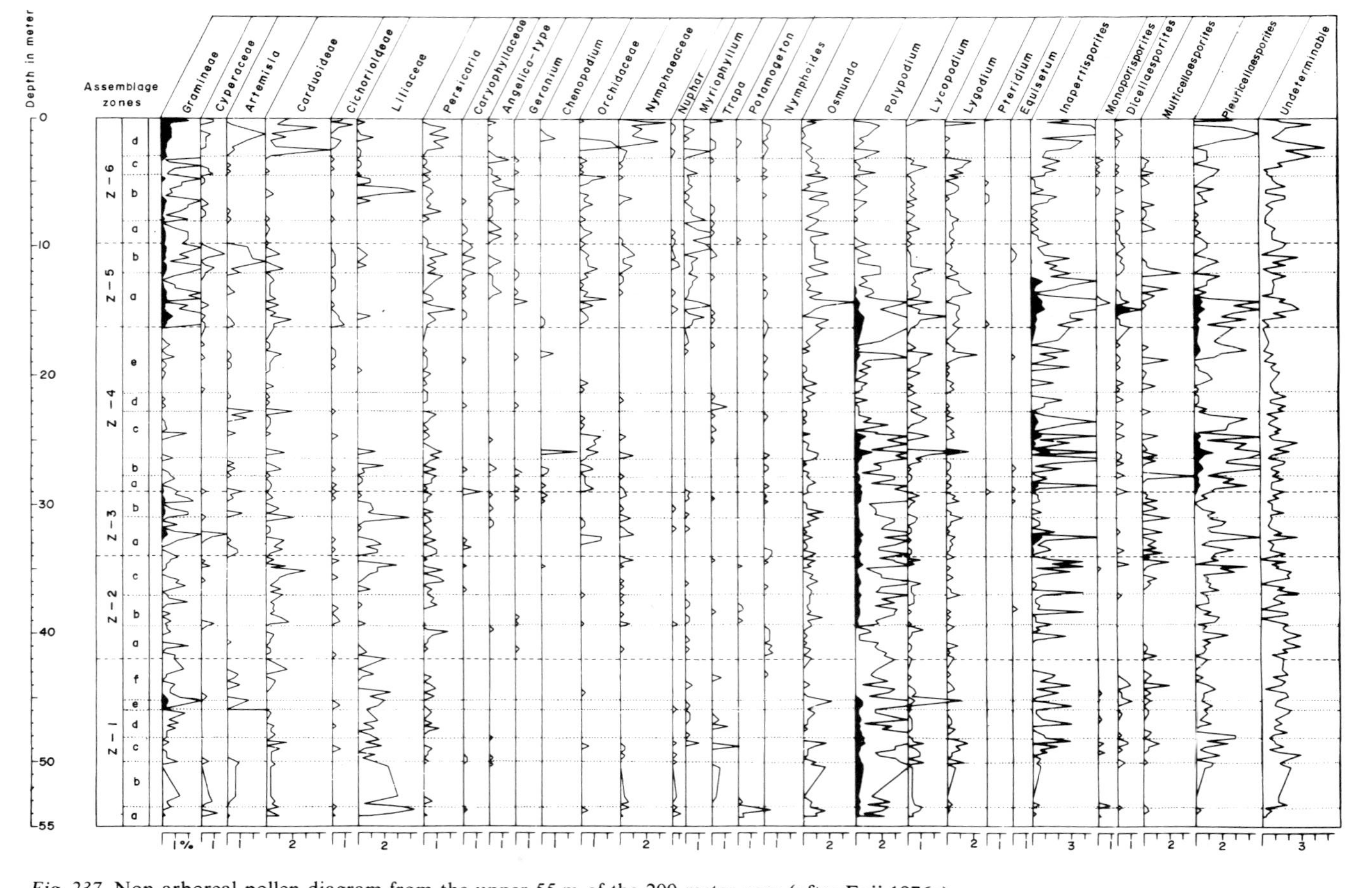

Fig. 237. Non-arboreal pollen diagram from the upper 55 m of the 200 meter core (after Fuji 1976a).

Cryptomeria and *Tsuga* increase in this subzone, as does *Juniperus*. Values for *Pinus haploxylon*-type and *Larix* rise gradually to the average of 3% and, at the same subzone some plants growing in the Cool Temperate zone increase. Decreases of some plants thriving in the Cool Temperate zone and the southern part of the Temperate–Subtropical zones are synchronized with increases of the *Pinus haploxylon*-type and *Larix* pollen values. In comparing Z–1 and Z–2, one sees that relatively high values of Cool Temperate–Temperate plants pollen from Zone Z–2 are derived mainly from an increase of *Cryptomeria*. On the contrary, some broadleaved evergreen trees, e.g., *Cyclobalanopsis*, the value is decreased to the average of 1% in the Subzone Z–2–a and absent in the Subzone Z–2–b. Judging from the phenomena stated above, the Subzone Z–2–a is distinguished from the Subzone Z–2–b.

Subzone Z–2–b: depth 37–39.4 m.
The most important character of Subzone Z–2–b is relatively high values of boreal conifers. An increase of boreal conifers, especially *Pinus haploxylon*-type and *Larix*, is matched by a decrease in the values of almost all broadleaved evergreen trees. Although *Cryptomeria* presents an average of 10% in Subzone Z–2–a and Subzone Z–2–b, it lowers to 5% in the Subzone Z–2–c.

Subzone Z–2–c: depth 34–37 m.
This subzone is more or less similar in nature to Subzone Z–2–a. In addition, an increase of plants belonging to the warm Temperate–Subtropical zones is characteristic of subzone Z–2–c. Boreal conifers and *Cryptomeria* decrease in quantity in this subzone. In contrast, plants of the Cool Temperate zone such as *Betula*, *Ulmus* and *Fagus crenata*-type, and *Cyclobalanopsis* and *Podocarpus* thriving in the Warm Temperate–Subtropical zones increase more or less as shown in the pollen diagram.

Zone Z–3: depth 29–34 m.
The total amounts of *Pinus haploxylon*-type and *Larix*, although only 3% on average and uniform in Zone Z–3, increase up to about 14% on average and 26% at maximum in Zone Z–4 where the amounts reach to the highest percentage among all the analyzed samples. Zone Z–3 is characterized by a very small amplitude of the fluctuation curve of the total amounts of the above mentioned two. *Cryptomeria* pollen values remain relatively high (to 12%) in Zone Z–3, except at the middle horizon of this zone. The *Chamaecyparis* value decreases distinctly in Zone Z–3. On the other hand, of broadleaved trees from this zone, the amount of *Quercus* is similar to or more (to 1%) than that from Zone Z–2, and is lowered markedly in Zone Z–4. The pollen assemblages from this zone show a transitional nature from Zone Z–2 to Zone Z–4. The zone is divided into Subzone Z–3–a and Z–3–b.

Subzone Z–3–a: depth 31–34 m.
At the beginning of this subzone, genera characteristic of the Cool Temperate zone such as *Ulmus* and *Castanea* increase markedly though temporarily. After this phase, their values decrease in Subzone Z–3–b. Boreal conifers remain abundant (36% in maximum).

Subzone Z–3–b: depth 29–31 m.
Cryptomeria and *Pinus diploxylon*-type increase, although *Abies* decreases.

Zone Z–4: depth 16.3–29 m.
This zone is characterized by a pronounced increase of boreal coniferous trees to the highest level in the diagram and correspondingly by a decrease of plants of the Cool Temperate–Temperate zones and/or the Warm Temperate–Subtropical zones. This zone is divided into five subzones on the basis of fluctuations of the pollen values of boreal coniferous trees, especially *Pinus haploxylon*-type, and some plants of the Cool Temperate–Temperate zones.

Subzone Z–4–a: depth 27.8–29 m.
The values for boreal conifers increase to 73% in this subzone as *Abies* and *Pinus haploxylon*-type increases. Cool Temperate plants, Cool Temperate–Temperate plants, and Warm Temperate–Subtropical plants all decrease.

Subzone Z–4–b: depth 26.4–27.8 m.
This subzone has much lower values for boreal coniferous pollen than the subzones below and above it, and higher percentages of plants growing in the Cool Temperate–Temperate zones.

Subzone Z–4–c, depth 22.8 – 26.4 m.
The boreal conifer values rise again to the average of 75% as *Abies*, *Picea*, and *Pinus haploxylon*-type increase. On the other hand, *Podocarpus* and *Pseudotsuga* drop out, and *Lepidobalanus* and *Cyclobalanopsis* decrease to very low values.

Subzone Z–4–d: depth 21.3–22.8 m.
This subzone shows another minor fluctuation below Z–4–b. A sharp drop of boreal conifers, especially *Pinus haploxylon*-type. *Abies* remains high, however.

Subzone Z–4–e: depth 16.3–21.3 m.
Boreal conifers reach to their maximum in the diagram (92%). *Podocarpus*, *Castanea*, *Fraxinus*, *Acer*, *Tilia*, *Pterocarya*, *Juglans*, *Celtis*, *Ilex* all have minimal values in the diagram.

Zone Z–5: depth 9.8–16.3 m.
Zone Z–5 is characterized by a very abrupt decline of boreal conifers (from 90% to 50%) and by a rise of plants belonging to the Cool Temperate–Temperate zones. Boreal coniferous trees, especially *Pinus haploxylon*-type, drop to a small value or nought, and *Picea*, *Tsuga*, and *Abies* decrease sharply. On the other hand, broadleaved trees of the Warm Temperate–Subtropical zones rise to 3% at the beginning of Zone Z–5 and increase to 7% at the top. Pollen of broadleaved evergreen trees increase, especially *Podocarpus*, *Cyclobalanopsis*, and *Cryptomeria*, as well as that of *Pseudotsuga*, *Betula*, *Ulmus*, *Fagus crenata*-type, *Fagus japonica*-type, *Castanea*, *Zelkova*, *Lepidobalanus*, and *Carpinus*. Among shrubs, the amounts of *Alnus* and *Salix* show also an increase in this zone. The NAP increases to 7%. Two subzones are recognized in Zone Z–5.

Subzone Z–5–a: depth 12.1––16.3 m.
This subzone is characterized by the lowest values of boreal coniferous trees, and by increases of *Cryptomeria*, *Pinus diploxylon*-type, *Pseudotsuga*, *Castanea*, *Lepidobalanus*, and *Cyclobalanopsis*, which thrive in the Cool Temperate–Temperate or Warm Temperate zones.

Subzone Z–5–b: depth 9.8–12.1 m.
In this subzone, plants growing in the Cool Temperate–Temperate zones and the Warm Temperate–Subtropical zones show smaller frequencies than in Subzone Z–6.

Subzone Z–6: depth 0–9.8 m.
This subzone is characterized by several fluctuations of boreal conifers, especially *Abies*, by gradual increases of some plants of the Cool Temperate–Temperate zones, and by a strong rise of pollen values of Warm Temperate–Subtropical plants. Of other boreal conifers, *Pinus haploxylon*-type goes to zero, and *Picea* and *Tsuga* remain low. *Cryptomeria, Chamaecyparis, Podocarpus*, and *Cyclobalanopsis* increase. Zone Z-6 is subdivided into four subzones.

Subzone Z–6–a: depth 8.0–9.8 m.
This subzone is basically transitional from Zone Z–5. At this subzone, the plants thriving in the Cool Temperate zone and the Cool Temperate–Temperate zones are abundant.

Subzone Z–6–b: depth 4.5–8.0 m.
This subzone is characterized by a sharp increase of plants such as *Cyclobalanopsis* belonging to the Warm Temperate–Subtropical zones, boreal trees *Abies* and *Larix*. *Tilia* also increases in this subzone.

Subzone Z–6–d: depth 0–4.5 m.
This subzone is characterized by a sharp decline of boreal conifers and *Cryptomeria* and by an abrupt increase in *Pinus diploxylon*-type and NAP. It is fairly clear that large areas of *Abies* and *Cryptomeria* woodland disappeared quite rapidly from the surroundings of Lake Biwa. However, the value of *Cryptomeria* rises very abruptly again at the uppermost horizon. In contrast to changes of *Abies* and *Cryptomeria*, *Pinus diploxylon*-type, *Chamaecyparis*, shrubs of open ground such as Ericaceae (up to 5%), *Alnus*, *Salix*, and *Corylus*, and NAP rise rapidly at this subzone. Human demand for timber during the last 3,650 years (Subzone Z–6–d), which is estimated by the data from C^{14} dating and the sedimentation rate, as well as man's grazing animals and cultivation, must have had its effect and contributed substantially toward those changes called 'destruction' or reconstruction of the natural vegetation around Lake Biwa.

Judging from a study of the recent vegetation around and in Lake Biwa (Shiga-ken 1974), *Pinus densiflora* and *Chamaecyparis* forests are one of the substitute communities which appeared in place of the natural communities of *Abies* and *Cryptomeria* destroyed by man over a long time. *Fagus crenata*, *Castanea*, and *Sasa* or *Miscanthus* assemblages are perhaps examples of the substitute communities in the *Querco-Fagetea crenatae* region. Also, Ericaceae, *Salix, Alnus,* and *Carpinus* assemblages, and *Phragmites* and *Bamboo, Sasa* communities are examples of the substitute communities appearing after the destruction of some natural vegetation in the *Camellietes japonicae* region. The most important and characteristic nature of this subzone is shown by plants of substitute communities that replaced some natural vegetation after its destruction.

Discussion on the interpretation of vegetation and climatic history based upon the samples selected at 25-cm intervals

Zone Z–1 is characterized by long and large fluctuations in pollen values of boreal coniferous trees. The maximum and minimum values in this amplitude reach to 85% and 15% respectively. The maximum value for *Pinus haploxylon*-type, which is one of representative plants of the Subpolar or Subalpine zone in the Japanese Islands, reach to 7% in Zone Z–1–b. Yet, pollen of broadleaved evergreen trees from the southern area of the Temperate–Subtropical zones is consistently present. Judging from the pollen assemblages and their frequencies from Subzone Z–1–b and Z–1–d, it is difficult to infer that all of those plants grew at almost the same elevation in or near the Ōmi Basin surrounding Lake Biwa at the same time (Z–1–b and Z–1–d), or that pollen grains of some boreal conifers would have been transported to this basin. *Pinus pumila,* which has *Pinus haploxylon* type pollen, grows now at the summit area of Mt. Haku-san, 2,300–2,700 meters above the present sea level, about 130 km north of Lake Biwa. The pollen grains of *Pinus haploxylon*-type were not detected in 18 modern samples collected in and around Lake Biwa. Accordingly, the present writer believes that the reconstruction of palaeovegetation must be considered from the viewpoint of palaeotopography.

At present, the uplift rate of the ground around Lake Biwa is within about 1 mm/yr. If it were uniform during the last 100,000 years, the total reaches to about 100 meters. If so, the topography around Lake Biwa must have been similar to the present one during the last 100,000 years (Fuji 1976b). According to this conclusion, the mountainous areas (about 1,200 meters in elevation) would have been in the western area of Lake Biwa, and the difference between the mountainous areas and lakeside area (about 100 meters in elevation) would have been perhaps about 1,100 meters. If this inference is correct, the writer can reconstruct the distribution of vegetation at that time as follows. 'The difference of elevation was 1,100 meters between the summit area and the lateside area. The difference in annual mean temperature is estimated to have been 5–5.5°C (–0.5°C/100 meters, because this area is a humid area).' Thus, if Subalpine plants grew in the summit area, the annual mean temperature there was 2–5°C. Therefore, the annual mean temperature at the lakeside area at that time was 7–10.5°C. Some thermophilous broadleaved deciduous trees and *Cyclobalanopsis* (evergreen *Quercus*) could grow at this area with this temperature.

During the time of Subzone Z–1–a and the lower horizon of Subzone Z–1–c, *Pinus haploxylon*-type (perhaps *P. pumila*) and *Abies* (*A. Mariesii*) could grow in the summit area, and *Betula* (e.g., *B. ermani* and *B. platyphylla*), *Tsuga* (e.g., *T. diversifolia*), and *Abies* (perhaps, *Abies homolepis*) grow on the mountain sides. At lower elevations, *Fagus crenata, Lepidobalanus* (*Quercus crispula* and/or *Q. dentata*), *Fagus japonica, Zelkova, Cryptomeria*, etc. could grow. In the lowland area around the lake, other deciduous *Quercus*, *Pinus diploxylon*-type (e.g., *P. densiflora*) and *Cyclobalanopsis* (evergreen *Quercus*) could be found. The annual mean temperature of that time might have been 10–11°C at the lakeside area. As the annual mean temperature of the present day is ca. 14°C, the temperature of that time was 3–4°C lower than that of the present day. *Podocarpus*, however, is thriving in the area where the annual mean temperature is 13–20°C. Therefore, if *Podocarpus* grew at the lowland area, *Pinus pumila* would not grow at the mountainous area at the same time. In this case, pollen grains of *Pinus haploxylon*-type therefore may not represent *Pinus pumila*, but might be derived from *Pinus koraiensis*. Thus, the phenomenon that *Pinus pumila* or *Podocarpus* could grow in the vegetation of the Ōmi Basin must have depended largely on the degree of climatic changes.

During the time of Subzone Z–1–f and the upper horizon of Subzone Z–1–c, the climatic condition might have been similar to that of the Hypsithermal phase of the Holocene epoch, because at the time the same broadleaved evergreen trees dominated and *Podocarpus* shows a relatively high frequency.

On the other hand, during the time of Subzone Z–1–b and Subzone Z–1–d, the amount of boreal coniferous trees increases to ca. 85%, and also *Ulmus* and *Abies* show high percentages. Therefore, in the mountainous areas of that time *Pinus haploxylon*-type (here perhaps *Pinus pumila*), *Abies* (apparently not *A. firma*), and *Picea* would grow, and in the lower areas *Abies* (here perhaps *A. firma*), *Ulmus*, *Betula*, *Fagus crenata*, and *Pinus diploxylon*-type (perhaps *P. densiflora*) would be distributed. Accordingly, the climatic condition of that time might have been milder than that of Subzone Z–4–e, which may be correlated to one, the ice advance phases in the entire Wisconsin stage. The age of Zone Z-1 covers an interval with climatic conditions more favorable for the growth of boreal coniferous trees than today in the Ōmi Basin, though this interval would have perhaps been short.

At the time of Subzone Z–2–a, boreal conifers and some plants of the Cool Temperate–Temperate zones increase, and some plants belonging to the Cool temperate zone and the Warm Temperate–Subtropical zones decrease. At the summit area, boreal conifers, especially *Abies*, *Picea* and *Pinus haploxylon*-type (not *P. pumila*, but here *P. pentaphylla* or *P. koraiensis*) predominated. On mountainous sites, on the other hand, *Abies* (perhaps *A. firma*) and *Tsuga* (perhaps *T. sieboldii*), and *Cryptomeria* and some shrubs grew near the lakeside areas. The change from rich broadleaved deciduous and evergreen forests (the end of Zone Z–1) to a dominant conifer forest of an apparently boreal character and *Cryptomeria*, *Chamaecyparis* forest (Subzone Z–2–a) suggests a climatic change from warm to cooler humid conditions.

During the time of Subzone Z–2–b, the expansion of boreal conifers, especially *Pinus haploxylon*-type and *Picea*, occurs on the high mountainous area and, at the foot of mountains and the lakeside, *Lepidobalanus*, *Cryptomeria*, *Chamaecyparis*, or *Tsuga* (perhaps *T. sieboldii*) dominate. Therefore, the climatic condition of this time might have become colder and wetter than that at the previous age (Z–2–a).

At the beginning of Subzone Z–2–c, although boreal conifers had decreased, the expansion of *Tsuga* (*T. sieboldii*) and *Picea* (here perhaps *P. polita*) occurs on the mountainous areas. On the other hand, in the lower phases, *Pinus diploxylon*-type (distinctly *P. densiflora*) and *Abies* occur instead of *Cryptomeria* and *Chamaecyparis*. Though pollen values of *Abies* from the previous time (Z–2–b) were as much as those of Subzone Z–2–c, species involved might have been *Abies homolepis* or *A. firma* of the Cool Temperate zone. Judging from the proof of the present distribution of *Abies firma*, the climate at the time of this subzone might have perhaps been as warm as the present.

Boreal conifers, e.g., *Abies*, *Picea* and *Larix*, occur with a relatively high percentage throughout the time of Zone Z–3. Cool Temperate plants e.g., *Betula* (here perhaps representing *Betula platyphylla*) and *Ulmus* (including many species) might have been represented more distinctly at the beginning of this time than those of Zone Z–2 were. In contrast to the plants described just above, some plants belonging to the southern area of the Temperate–Subtropical zones and the Cool Temperate–Temperate zones would decrease. Thus at the beginning of Zone Z–3, the expansion of *Abies*, *Picea*, *Larix*, etc. occurs at the higher part of the mountainous area, and *Betula platyphylla* (?) and *Ulmus* prevail on the mountain side. Around the lake, *Pinus densiflora*, *Castanea*, *Zelkova*, *Quercus* and shrubs such as *Alnus* and *Salix* grow with grasses. Judging from the evidence of pollen spectra, the climate in the early times of Zone Z–3 changed to a cooler condition than that of the end of Zone Z–2, being as humid as Zone Z–2. However, as suggested in the pollen diagrams, the climatic condition would be temporarily so mild that boreal conifers decreased more or less and, instead, *Podocarpus* and *Cyclobalanopsis* dominated somewhat.

It may be concluded that this age (Z–3) was a transitional phase from the warm condition of Z–2 to the cold condition of Z–4.

Boreal conifers, e.g., *Pinus haploxylon*-type, *Abies*, *Picea*, and *Tsuga* show a marked expansion immediately after the beginning of Zone Z–4. In contrast, *Cryptomeria*, *Pinus diploxylon*-type, *Pseudotsuga* and *Chamaecyparis* decreased noticeably, and *Podocarpus* is absent throughout this zone except at a few short intervals (the early times) when it is represented only sporadically. Besides conifers, broadleaved trees, though *Betula* decreased temporarily in the early half of the age (Z–4–b–c), were represented pronouncedly with a high percentage in the late half of the age (Z–4–d–e). Judging from a phenomenon that the expansion of Betula in the late age synchronized with that of boreal conifers, the species of *Betula* might have been *Betula ermani*, and the species from the early age perhaps *Betula platyphylla*. The reduction of *Betula* (*B. platyphylla*) early in Zone Z–4 would have been due to a severe climatic condition, and the later expansion of *Betula* would have resulted from the thriving of the other species (*Betula ermani*) under the severe climatic condition. Early in Zone Z–4, *Picea*, *Abies* (perhaps *A. Mariessi*), *Pinus haploxylon*-type (perhaps *P. pumila*) and *Betula* (*B. ermani*) occur in the higher part of the Ōmi Basin and spread gradually to lower elevations with a change in climatic conditions. In the lakeside area, forests by *Chamaecyparis*, *Quercus*, *Zelkova* and *Juglans* etc. were reduced. Instead of them, *Abies* (perhaps *A. homolepis*), *Tsuga*, and *Pinus diploxylon*-type (here perhaps *P. densiflora*) prospered.

Late in Zone Z–4, besides deciduous trees and conifers as described above, *Pinus diploxylon*-type was reduced markedly. Instead of them, boreal conifers prevailed not only in the higher mountainous areas but also at the mountain side and lower areas. Around the lake, relatively simple vegetation was perhaps distributed with the *Betula* community. Such palaeovegetation may be the same as the present vegetation of the middle Hokkaido district, for example, Obihiro. This inference is supported by the pollen spectra of contemporary samples from the Hokkaido district. The annual mean temperature at the coldest condition of this time may have been about 7°C (the northern part of the Cool Temperate zone), which is about 7°C lower than that of the present day in the Ōmi Basin. If so, the snow line at this time (Z–4–e) is estimated to have been 2,400–2,500 meters, and the forest line to be at about 1,000 meters above the present sea level. The height of the snow line corresponds to the height calculated by Kobayashi (1958) on the basis of the distribution of cirques and moraines in the Japan Alps east of Lake Biwa. This horizon may be correlated with the Würm II (in Europe) and the Tottabetsu I Subglacial epoch (Minato 1972); at that time the sea level was about 100 meters below the present sea level in the Baltic Sea and in the Japanese Islands (Minato 1972).

At the beginning (Z–5–a) of Zone Z–5, boreal conifers decreased markedly. An expansion of *Betula* (here perhaps *B. platyphylla), Ulmus, Fagus crenata*-type, *Fagus japonica*-type, *Castanea, Zelkova* (probably *Z.serrata*) and *Cyclobalanopsis* gradually replaced boreal conifers in the lowland and the mountain side. *Lepidobalanus* increased remarkably in the lower mountain areas. Judging from the pollen spectra, *Pinus haploxylon*-type (mainly *P. pumila*) and *Abies* (perhaps *A. Mariesii*) were limited only in the narrow summit area, perhaps above about 1,000 meters. *Abies* (probably *A. homolepis*), *Picea*, *Betula* (distinctly *B. platyphylla*), *Pinus haploxylon*-type (here perhaps *P. koraiensis*) and *Larix* (probably *L. leptolepis*) were distributed in the higher part of the mountainous area, and in the lower part *Pinus densiflora*, *Cryptomeria japonica* and broadleaved deciduous trees mentioned above prevailed. The expansion of Ericaceae, *Alnus, Salix* and *Gramineae* began actively around the lake. In addition, at the mountain side (400–1,000 meters in height), the change from a conifer forest of an apparently boreal character (Subzone Z–4–e) to a rich deciduous forest (Subzone Z–5–a) suggests a climatic change from a cold condition like the Subpolar zone to a cool condition as the middle-southern part of the Cool Temperate zone. The climatic condition of that time, although cooler than that of the present (14°C annual mean temperature), might have been milder than that of Subzone Z–4–e (about 7°C annual mean temperature). Also, judging from the existence of *Cryptomeria* and *Chamaecyparis*, it might have been more or less humid. However, the re-expansion of boreal conifers was represented temporarily in the beginning of the late time of Zone Z–5 and influenced the distribution of *Cryptomeria* and *Pinus densiflora*. The *Betula* pollen value is higher at this subzone than at the immediately preceding subzone (Z–5–a). This phenomenon perhaps resulted from the fact that *Betula* (here perhaps *B. ermani*) thriving in the cold conditions prevailed in the higher area and another species (perhaps *B. platyphylla*) that was adapted to the cool condition migrated to the lower area from the mountainous area at the previous age (Z–5–a).

Judging from this vegetation, the climate of this time (Z–5–b) must have changed to a severer climatic condition corresponding to that of Subzone Z–4–e. However, it cannot be concluded that the same severe climatic condition continued for the entire time of Subzone Z–5–b. According to the pollen diagrams, at the time of Subzone Z–5–b–β, though very short, plants belonging to the Subpolar or Subalpine zone decreased. In contrast to these plants, *Cyclobalanopsis* increased by a small amount. Therefore, the time when the climatic condition was milder than that of Subzone Z–5–b–α or γ was intercalated in the cold age of Z-5-b (Fig. 239).

In conclusion, the climatic condition shows the transitional character from the arid and cold condition correlated with the Wisconsin Stage to the humid and milder condition of the Holocene epoch.

In the beginning (Z-6-a) of Zone Z-6, Cool Temperate plants and boreal conifers decreased markedly. In contrast to the above-mentioned plants, the re-expansion of *Castanes*, *Zelkova*, *Cryptomeria*, *Chamaecyparis*, and *Cyclobalanopsis* was represented relatively pronouncedly at the beginning of Zone Z–6. That is, in the higher part of the mountainous area, *Pinus pumila* and *Betula* were not found, though *Pinus koraiensis* might have grown perhaps very little, and *Abies* (perhaps *A. homolepis*), *Cryptomeria japonica*, *Lepidobalanus* (here perhaps mainly *Quercus crispula*) and *Fagus crenata* prevailed in the lower part. At the lakeside area *Castanes, Zelkova serrata, Cryptomeria japonica,* and *Chamaecyparis obtusa* dominated instead of the *Pinus diploxylon*-type community. If so, the climatic condition at this time, though milder than that of Zone Z–5, might have been cooler than the present. However, in the middle stage (Z-6-b) of Zone Z-6, increase of *Cyclobalanopsis, Lepidobalanus, Acer, Alnus,* and *Salix* are matched by decrease of boreal conifers, especially *Abies.* At the mountainous area *Lepidobalanus* (perhaps *Quercus crispula* and *Q. serrata* mainly), *Acer*, *Cryptomeria japonica*, and *Chamaecyparis* belonging to the *Querco-Fagetea* region prevailed instead of *Abies*, *Larix*, and *Ulmus*. On the other hand, in the lower area, *Lepidobalanus* as *Quercus acutissima* and *Cyclobalanopsis* as *Quercus glauca* belonging to the *Camellietea japonicae* region were distributed widely with *Salix* and *Alnus* growing at the lakeside and riverside. Accordingly, the climate of this time might have been 1°–2°C warmer and more humid than the present, judging from a comparison of pollen spectra from this subzone (Z–6–b) and from the contemporary sample of the Shikoku district belonging to the Warm Temperate zone (Nakamura 1973).

At the time of Subzone Z–6–c, *Cryptomeria japonica*, *Larix leptolepis*, *Tsuga sieboldii*, *Fagus crenata*, and *Betula* (?), etc. increased, and *Cyclobalanopsis* and

Podocarpus decreased as shown in a pollen diagram. The climatic condition of this time might have been more or less cooler (1–2°C) than the present and humid.

Cryptomeria and *Abies* decreased more drastically. In contrast to these coniferous trees, *Pinus densiflora*, *Chamaecyparis*, NAP, *Zelkova*, *Tilia*, *Carpinus*, and shrubs as Ericaceae, *Alnus*, *Salix*, and *Corylus* increased markedly at the time of Subzone Z–6–d. Of these, the expansion of NAP, e.g., Gramineae, *Carduoideae, Artemisia,* and *Persicaria,* was abrupt in the lowland around the lake. Such a change of vegetation during this time occurred because of some special trees and grasses. If these special types disappeared by a natural cause, plants growing with these trees must have disappeared also. However, such disappearance cannot be found in the pollen spectra. Some special trees that are useful for man which were grown in the lowland around the lake, disappeared. The strong increase of NAP since about 2,750 years BP. (2.25 m below the surface) corresponds to the start of rice cultivation in the Ōmi Basin. According to the archaeological evidence, the evidence of rice cultivation was found from the latest Jomonian Shigasato Site (ca. 3,000 years BP.) at the western lakeside of Lake Biwa. In addition, *Cryptomeria* shows suddenly a high value in the uppermost 0–5 m of the 200–meter core. This drastic change in the *Cryptomeria* pollen curve may mean a *Cryptomeria japonica* plantation, which was dated at about 600 years BP. (34 cm in depth) on the basis of C^{14} dating.

Human demand for timber and cultivation must have had its effect and influenced substantially the change called *destruction* of natural vegetation.

In conclusion, our interpretation of the vegetational and climatic conditions for the last 100,000 years on the basis of the results of the writer's palynological investigation of the 200 meter core from Lake Biwa may be summarized as follows.

The vegetational history during the last 100,000 years is reconstructed by the parallel representation of vegetational zones found at the present time. The typical vegetation distributed in the present Subpolar or Subalpine zone in the Japanese Islands prevailed at the summit area and/or the montane area around the lake, and in the lowland around the lake some plants growing in the Cool Temperate zone were distributed. A cold and arid condition prevailed at that time in this area.

During the interglacial and interstadial intervals, the vegetation in the higher area was characterized mainly by the forest that is distributed today in the Cool Temperate zone, and with plants thriving in the present montane or Temperate zone. In the lowland the vegetation was composed mainly of broadleaved deciduous and evergreen trees growing in the present Temperate–Warm Temperate zones.

Conclusions of pollen analyses at intervals of 25 cm

1. Judging from the pollen diagrams, the upper 50 m core samples are divided into six pollen assemblage zones: Z–1, Z–2, Z–3, Z–4, Z–5, and Z–6; and

comprise the late Sangamon Interglacial, the Wisconsin Glacial stages and the Holocene.

2. The climatic curve and ages from Lake Biwa display a clear similarity to the palaeotemperature curve (oxygen-isotope ratio determination) from Greenland (Dansgaard et al., 1971) and to the curve of the eustatic movements during the Wisconsin Stage in the Japanese Islands (Minato 1974) (Figs. 238, 239). In spite of differences in details and in interpretation of the curves which have no absolute ages, similarity exists among the major trends in the inferred climatic fluctuations of North America, Europe, Greenland and the Japanese Islands.

3. The vegetational history during the last 100,000 years is reconstructed by the parallel representation of the vegetational zone found in the present Japanese Islands. During the glacial stage or stadial, the typical vegetation thriving in the Subpolar or Subalpine zone in the present Islands prevailed at the summit area and/or the montane area around Lake Biwa; at the lowland around the lake, plants growing in the Cool Temperate zone were distributed. In the interglacial and interstadial, the vegetation in the higher area was characterized mainly by plants of the Cool Temperate zone and the present Temperate zone; in the lowland the vegetation was composed mainly of broadleaved deciduous and evergreen trees growing in the Warm Temperate zone.

4. The boundary between Zone Z-2 and Zone-3 corresponds to the boundary between the Sangamon and Wisconsin Stages, and is inferred to be about 70,000 years BP. (about 34 m below the bottom of the lake) from the viewpoints of the chronology and climatic change as shown in Figs. 238 and 239.

5. Zones Z-1 and Z-2 are marked by the vegetation roughly similar to that of the present lakeside area, but during the age of Zone Z-2 the climate was as mild and humid as that of the present or somewhat cooler than that of the present, and as humid as the present. These zones may be correlated with the Sangamon Interglacial Stage.

6. The climate during the period of Zone Z-3 was cool, although it includes the time of a temporary climatic deterioration. Judging from the vegetation at the period of the cool condition, the annual mean temperature may have been 10 – 11°C at the lakeside area.

During the age of Zone-4 the palaeoclimate was generally stable. Zone Z-5 comprises the cold condition and is characterized further by greatly varying conditions. According to the pollen spectra from Lake Biwa, the coldest climatic condition correlated with the Woodfordian Stadials might have been a relatively short period around the lake. The annual mean temperature at this period is estimated at about 7°C.

Zone Z-3, Z-4 and Z-5 may be correlated with the Wisconsin Glacial Stage.

7. Zone Z-6 may be correlated with the Holocene epoch from the viewpoints

---

Fig. 238. Correlation diagram between the temperature curve based on the oxygen-isotope ratio determination from the Greenland (Dansgaard et al. 1971) and paleoclimatic change around Lake Biwa (Fuji 1976a).

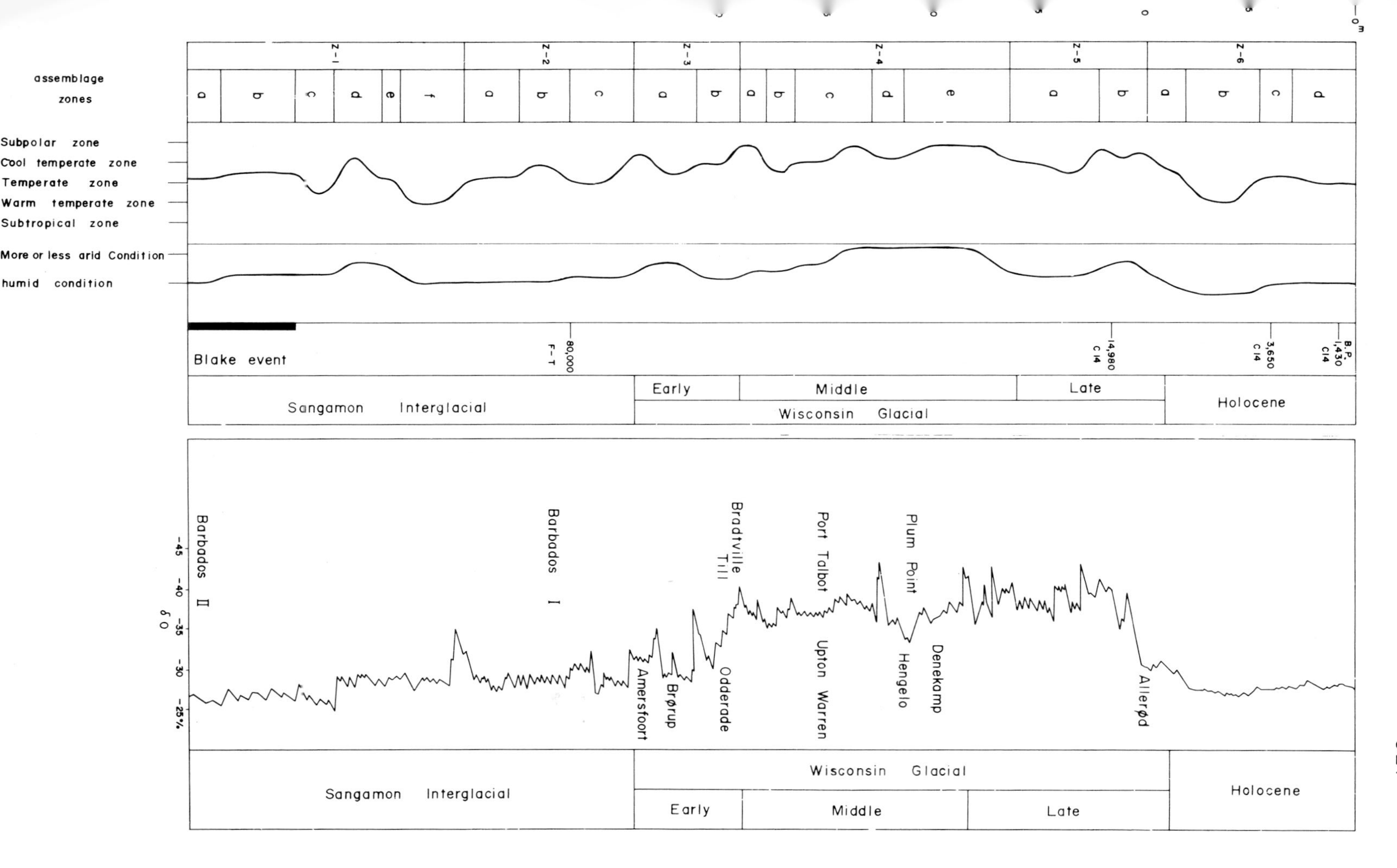

0m
assemblage zones
Z-1
Z-2
Z-3
Z-4
Z-5
Z-6
a
b
c
d
e
f
Subpolar zone
Cool temperate zone
Temperate zone
Warm temperate zone
Subtropical zone
More or less arid Condition
humid condition
Blake event
80,000 F-T
14,980 C14
3,650 C14
B.P. 1,430 C14
Early
Middle
Late
Sangamon Interglacial
Wisconsin Glacial
Holocene
Barbados II
Barbados I
Bradtville Till
Port Talbot
Plum Point
Amersfoort
Brørup
Odderade
Upton Warren
Hengelo
Denekamp
Allerød
-45
-40
-35
-30
-25%
δ O

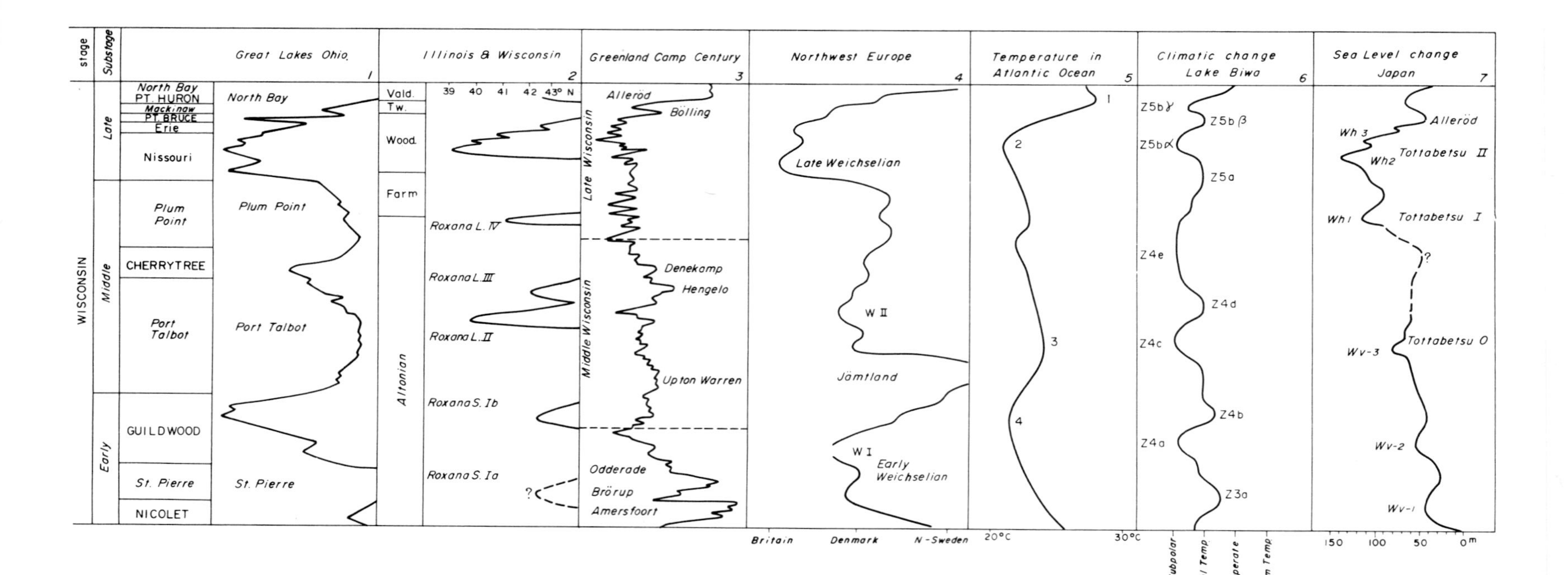

Fig. 239. Correlation diagram of the changes of the ice sheet development (1, 2 and 4), climate (6), temperature (3 and 5), and sea level fluctuation in Japan (7). 1: Dreimanis & Karrow 1972; 2: Willman & Frye 1970; 3: Dansgaad et al. 1971; 4: Malakovsky et al. 1969; 5: Emiliani & Shackleton 1974; 6: Fuji 1976a; 7: Minato 1972 (Fuji 1976a).

of the chronology and climate. Although the beginning of the warmest age during the Holocene epoch in the Japanese Islands corresponds to that of Sweden, the end does not correspond.

8. Man's noticeable influence, represented by clearing forests for rice-cultivation and by cutting woods, began in the latest Jōmonian age on the Neolithic period. This inference is supported by the archaeological evidence around the lake.

9. The history of the *Cryptomeria japonica* plantation which flourished as a response to human action during the late Holocene epoch is clearly recorded in the pollen sequence.

Diatom analysis

Shinobu Mori and Shoji Horie

Historical review

During the last one century, diatom investigators have been engaged in many researches. Among them, the most remarkable was B. W. Skvortzow who carried out his study not only in Lake Biwa but in other lakes such as Kizaki and Ikeda. In the case of Lake Biwa, he mentioned as follows (Skvortzow 1936):

'Several years ago Prof. Dr. Tamiji Kawamura, of Kyoto, sent me a tube of diatom clay from Biwa Lake, Nippon. Biwa Lake, one of the largest in Nippon, is north of Osaka, Honshu Island, in 35° 23′ north latitude. Its altitude is 86.3 meters, its area 644.8 square kilometers, and its maximum depth 95 meters.

A careful examination of the diatom sample yielded more than two hundred forms of siliceous algae. From systematic and geographic points of view the diatoms from Biwa Lake are of great interest. Some of these diatoms are essentially tropical, others are characteristic of alpine and arctic regions. Among the species found in Biwa Lake, the following seem to inhabit warmer climates:

Melosira solida	*Neidium obliquestriatum*
Melosira americana	*Navicula Lambda*
Melosira undulata	*Navicula Pusio*
Amphipleura pellucida v. *recta*	*Amphora delphinea* var. *minor*
Cymbella tumidula	*Gomphonema Berggrenii*

The northern elements are widely represented in Biwa Lake by many large species of *Stauroneis*, *Navicula*, *Pinnularia*, *Gomphonema*, and *Cymbella*. *Didymosphenia geminata*, a common diatom in the northern part of Asia and of Europe, was also found. It was peculiar to find in Biwa Lake some species of American origin. *Melosira solida*, known from Arizona, was very abundant; *Stephanodiscus carconensis*, reported from Klamath Lake, Oregon, was represented by thousands of specimens. A distinct species, *Melosira americana*, known from tropical America, was also common in Biwa Lake. About eighty

Our thanks are offered to Mr. K. Hashizume who gave us valuable suggestions.

Horie, S. Lake Biwa

ISBN 90 6193 095 2. Printed in The Netherlands

different diatoms known from Kizaki Lake were recovered in Biwa Lake. Several forms, of frequent occurrence in Kizaki Lake, were not found in the Kawamura gathering, which can scarcely be because my sample is not sufficiently large. Over seventy new species and varieties of algae are described from Biwa Lake, and some are very distinct and peculiar.

From the ecological point of view the following diatoms from Biwa Lake are plankton species:

Melosira granulata and var
Melosira solida
Cyclotella comta and var
Stephanodiscus carconensis
Coscinodiscus lacustris var
Attheya Zachariasi
Chaetoceros sp
Asterionella gracillima
Asterionella formosa

The other diatoms belong to a bottom formation and include large forms; such as, *Melosira undulata, Opephora Martyi, Synedra Ulna* and var., and various species of *Eunotia, Cocconeis, Achnanthes, Navicula, Pinnularia, Cymbella, Gomphonema,* and *Surirella.* The last genus was very richly represented in the lake.

All of the diatoms listed in this note are fresh-water species, and only a few forms can be referred to brackish-water species; they are *Navicula crucicula* var., *Nitzschia tryblionella, N. Lorenziana,* and *N. Clausii.* This note is illustrated with drawings by the author, and they may be useful in future investigations.'

Based on descriptions given by Skvortzow, Negoro (1960) emphasized them again as follows (Table 64):

Table 64. Diatom composition in Lake Biwa (Negoro 1960).

Abundant
1. *Melosira solida* Eulenstein
2. *Stephanodiscus carconensis* Grun.
3. *Attheya Zachariasi* Brun.

Very common
1. *Navicula hasta* Pant.
2. *Surirella nipponica* Skv.

Common
1. *Melosira solida* Eulenstein var. *nipponica* Skv.
2. *Stephanodiscus carconensis* Grun. var. *pusilla* Grun.
3. *Opephora Martyi* Herib.
4. *Synedra ulna* (Nitzsch) Ehr. var. *amphirhynchus* (Ehr.) Grun.
5. *Synedra parasitica* W. Sm.
6. *Achnanthes minutissima* Kütz.
7. *Achnanthes Clevei* Grun.
8. *Achnanthes lanceolata* Bréb.
9. *Achnanthes lanceolata* Bréb. var. *nipponica* Skv.
10. *Caloneis nipponica* Skv.
11. *Neidium dubium* (Ehr.) Cleve fo. *constricta* Hustedt
12. *Neidium Hitchcockii* Ehr.
13. *Neidium obliquestriatum* A.S. var. *nipponica* Skv.
14. *Neidium obliquestriatum* A.S. var. *elongata* Skv.
15. *Diploneis ovalis* (Hilse) Cleve var. *oblongella* (Naegeli) Cleve

Table 64. continued

16. *Diploneis ovalis* (Hilse) Cleve var. *bipunctata* Skv.
17. *Diploneis puella* (Schum.) Cleve
18. *Stauroneis phoenicenteron* Ehr.
19. *Navicula costulata* Grun. var. *nipponica* Skv.
20. *Navicula menisculus* Schum.
21. *Navicula lanceolata* (Agardh) Kütz. var. *cymbula* (Donk.) Cleve
22. *Navicula Pusio* Cleve
23. *Pinnularia molaris* Grun.
24. *Pinnularia interrupta* W. Smith
25. *Pinnularia borealis* Ehr.
26. *Pinnularia major* (Kütz.) Cleve var. *linealis* Cleve
27. *Pinnularia cucumis* Skv.
28. *Pinnularia Lacus Biwa* Skv.
29. *Pinnularia nipponica* Skv.
30. *Cymbella cuspidata* Kütz.
31. *Cymbella hybrida* Grun.
32. *Cymbella turgidula* Grun. var. *nipponica* Skv.
33. *Gomphonema intricatum* Kütz. var. *pumila* Grun.
34. *Gomphonema lanceolata* Ehr. var. *insignis* (Gregory) Cleve
35. *Epithemia zebra* (Ehr.) Kütz. var. *saxonica* (Kütz.) Grun.
36. *Epithemia turgida* (Ehr.) Kütz.
37. *Epithemia Hyndmanii* W. Smith
38. *Rhopalodia parallela* (Grun.) O. Müll.
39. *Rhopalodia gibba* (Ehr.) O. Müll. var. *ventricosa* (Ehr.) Grun.
40. *Nitzschia triblionella* Hantzsch var. *victoride* Grun.
41. *Nitzschia interrupta* (Reich.) Hust.
42. *Nitzschia acicularis* W. Smith var. *nipponica* Skv.
43. *Surirella biseriata* Bréb.
44. *Surirella robusta* Ehr. var. *splendida* (Ehr.) Van Heurck
45. *Surirella elegans* Ehr.
46. *Surirella elegans* Ehr. *norvegica* (Eulenst.) Brun. fo. *obtusa* A. Mayer
47. *Surirella linealis* W. Smith var. *constricta* (Ehr.) Grun.
48. *Surirella ovata* Kütz.
49. *Surirella ovata* Kütz. var. *pinnata* (W. Smith)

Fossil analysis of Lake Yogo

On these phenomena of the inhabiting diatoms listed above, some paleo-limnological work has been carried out. Horie has obtained a core at Lake Yogo, north of Lake Biwa, and verified the symptoms of eutrophication that occurred many years ago. That event at the 6-m level might have occurred because of the drop of the lake level in this closed lake (Fig. 240). A more interesting fact is the beautiful parallelism of nitrogen and carbon contents in the core sample (Fig. 241). It strongly suggests that eutrophication caused by the drop in the lake level was accompanied by the supply of allochthonous organic material. Eutrophication in this tiny lake was brought about by the appearance of a warm, dry climate, which caused an increase in both nitrogen and carbon. More organic materials were produced as the climate warmed, supplying more carbon of allochthonous origin to the lake. At the same time, the drier climate brought about a drop in the level of the closed lake, causing a decrease in redox potential and therefore increasing metabolic activity and increasing the nitrogen

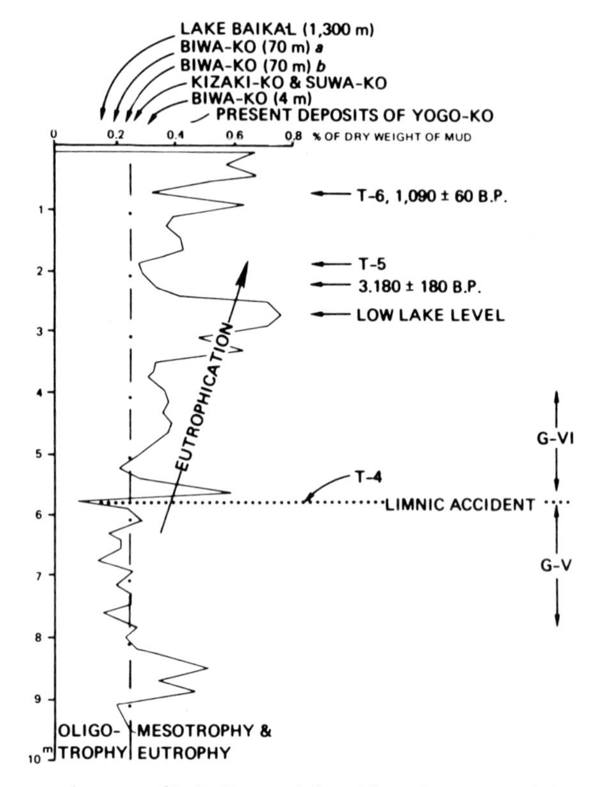

Fig. 240. Developmental process of Lake Yogo as inferred from the amount of nitrogen. (Horie 1966)

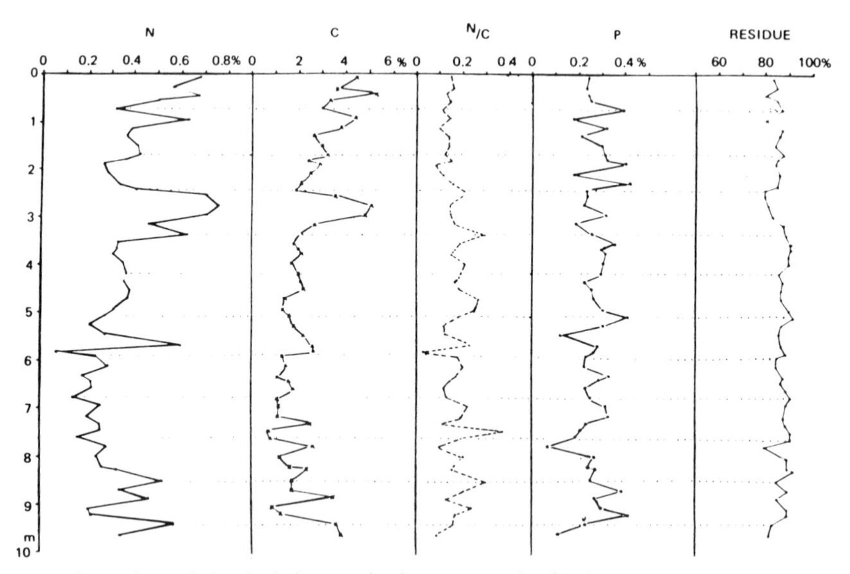

Fig. 241. Fluctuations of chemical elements in the core sample of Lake Yogo. (Horie 1967a)

Table 65. Diatom fossils in the core sample of Lake Yogo (after Negoro 1968).

Depth (cm)	Amount	Dominant species
7	*	*Melosira italica, Melosira granulata*
27	*	
47	**	
55	****	
75	****	
95	***	
115	****	
127	****	*Stephanodiscus carconensis*
147	***	
167	****	
177	**	
186	*	
206	*	
226	****	
236	**	
254	***	*Melosira italica, Melosira granulata*
274	***	*Melosira italica, Melosira granulata*
294	***	*Melosira italica, Melosira granulata*
311	*	*Cyclotella comta*
331	**	*Melosira italica, Melosira granulata*
351	*	*Cyclotella comta*
361	**	*Stephanodiscus carconensis*
373	**	*Cyclotella comta*
393	**	*Melosira italica, Melosira granulata*
413	**	*Cyclotella comta*
433	*	
447	*	
467	*	
487	*	
501	*	*Cyclotella comta*
521	*	
541	0	
561	*	*Stephanodiscus carconensis*
576	0	
588	0	
608	0	
628	0	
638	**	*Stephanodiscus carconensis*
651	*	*Stephanodiscus carconensis*
671	0	
691	*	*Stephanodiscus carconensis*
711	*	
725	**	*Stephanodiscus carconensis*
736	*	
756	*	*Cyclotella comta*
776	*	
796	*	
816	**	*Melosira islandica?*
825	**	
845	**	
865	**	
885	**	*Cyclotella comta*
905	*	*Stephanodiscus carconensis*
918	*	
938	*	*Melosira islandica?*
958	**	

****: very abundant, ***: abundant, **: common, *: rare.

content of the water. Negoro had worked on diatom fossils in the core denoting correlation between chemical data and microfossil data. It indicates that lake trophy tends to alternate between oligotrophy and eutrophy (Table 65).
Thus, climatic change probably controls trophy in closed lake basins in which everything, be it may of both autochthonous and allochthonous origin, must accumulate for many years, but it is difficult to find similar evidence in a core sample of an open lake, even in a shallow lake like Lake Yogo; organic detritus and nutrient salts are frequently lost by discharge. Moreover, the phosphorus content in this core is considered to be of inorganic origin.

Fossil analysis of Lake Biwa

On the other hand, in the core of Lake Biwa, a core more than 1.2 meter long was not obtained for a technical reason until 1965 when Horie succeeded in obtaining a core more than 6 m long. Therefore, before 1965, Negoro had worked only with a short core obtained in the pelagic area and his initial step was to clarify the present diatom distribution (Table 66).

This table shows that diatom valves, found in the lake bottom deposits off both the eastern and western shores, consist mainly of *Stephanodiscus carconensis* (var. *pusilla* is included) and *Melosira solida* throughout the columnar

Table 66. Analytical result on diatom fossil in the core of Lake Biwa (Negoro 1960).

Depth (cm)	I. *Stephanodiscus carconensis* (%)	I'. *St. carconensis* var. *pusilla* (%)	II. *Melosira solida* (%)	III. Others (%)
3	46.3	39.9	5.9	7.9
10	24.0	50.7	18.7	6.6
16	35.8	40.9	16.0	7.3
23	28.0	47.7	15.9	8.4
30	27.9	51.4	13.5	7.2
36	23.7	55.5	14.1	6.7
43	24.6	52.7	13.6	9.1
46	32.8	50.8	9.4	7.0
53	35.9	49.0	9.6	5.5
59	31.2	52.0	11.6	5.2
64	38.1	48.0	9.9	4.0
70	34.4	51.5	9.8	4.3
75	29.2	57.7	7.1	6.0
81	29.2	55.5	8.3	6.3
87	25.4	53.5	14.1	7.0
93	21.3	55.1	15.5	8.1
100	17.4	65.2	7.6	9.8
106	25.0	50.0	16.1	8.9
112	44.0	38.9	13.9	2.8
Average	(30.3)	(50.8)	(12.1)	(6.8)

section, with others in extremely small quantities. Although the frequency varies in some degree throughout the core, the stability as a whole is recognized all the layers over. The mean cell number of *Stephanodiscus carconensis* (var. *pusilla* is included) and *Melosira solida* denotes 81.1% and 12.1% of the whole, respectively. The remainder accounts for only 6.8% on average. In other words, this fact proves that diatom fossils in lake sediments of the pelagic area of the main basin of Lake Biwa were mostly derived from the offshore plankton; and the diatom which is distributed in the coastal area or is carried into it by inflowing rivers seldom mixed in that part of the lake.

In 1967, Negoro published a paper dealing with Horie's new core more than 6 meters long. His result is shown in Table 67 and Fig. 242.

In the core *Stephanodiscus carconensis* (var. *pusilla* is included) and *Melosira solida* are found to be apparently dominant ones. Furthermore, Negoro paid attention to the total cell number. The total cell number of diatoms contained in a given quantity of dry materials varies more or less in each horizon and decreases as a whole gradually from the older times towards the present. As Horie (1969) had stated two years later, the age of this core covers the middle and late Holocene. In other words, the Hypsithermal Interval found in the chemical analyses coincides well with the number of total cells of diatoms in Fig. 242. So far as these features are concerned, warm climate might give rise to a more productive state in the limnetic condition though the biogeochemical feature does not necessarily coincide with it.

Table 67. Diatom fossils found in 6 meters core (cell/1mg of dry samples) (Negoro 1967).

Depth (m)	*Stephanodiscus carconensis*	*St. carc.* var. *pusilla*	*Melosira solida*	Others	Total
0.06	4699	2880	4017	303	11899
0.57	6214	4396	10307	455	21372
1.07	5760	3183	15081	0	24024
1.23	4850	3638	13263	303	22054
1.73	2880	3335	18871	152	25238
2.23	4244	4699	13793	0	22736
2.43	3789	2577	13263	152	19781
2.90	5002	3183	14399	303	22887
3.03	6214	4244	17886	0	28344
3.57	4850	4699	13187	152	22888
4.11	5457	8639	17431	152	31679
4.26	7275	8033	12656	303	28267
4.80	3941	6669	24251	303	35164
5.33	3638	4396	17961	152	26147
5.47	3638	8639	13793	909	26979
5.90	7730	7124	14172	303	29329
6.34	5911	9397	16521	303	32132
Average	5064	5278	14756	250	25348
%	19.98	20.82	58.21	0.99	100.00

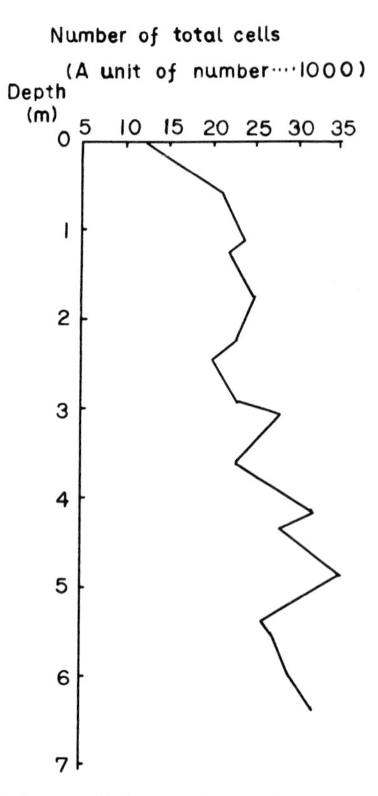

Fig. 242. Vertical variation of diatom cells in 1 mg dry sample of 6 meters core in Lake Biwa (after Negoro 1967).

Fossil analysis of the 200 meters core of Lake Biwa

The most striking feature of the succession of diatom flora was verified by Mori's work on the 200 m core obtained in the center of Lake Biwa in 1971.

Mori studied diatoms in the 200 meter sample and reported some interesting facts (Morish. 1974, 1975; Mori and Horie 1975).

Fig. 243 shows the occurrence spectrum of some main planktonic diatoms in the core sample. This figure clearly shows us that the quality and the quantity of planktonic diatoms in Lake Biwa have changed remarkably throughout the past approximately 0.55 million years. Each of the planktonic diatoms found in the core sample changed as follows (Pl. 30, 31).

Melosira solida Eulenstein (Pl. 30, a–d)

This diatom was found commonly at the layers of 3.3–75.0 m, 129.9–140.6 m and 183.5–197.2 m in depth, and occurred abundantly especially at the layers of 3.3–11.0 m, 58.6–75.0 m and 129.9–140.6 m in depth. It may be assumed that it was a temperate — warm climate in the period when *M. solida* occurred

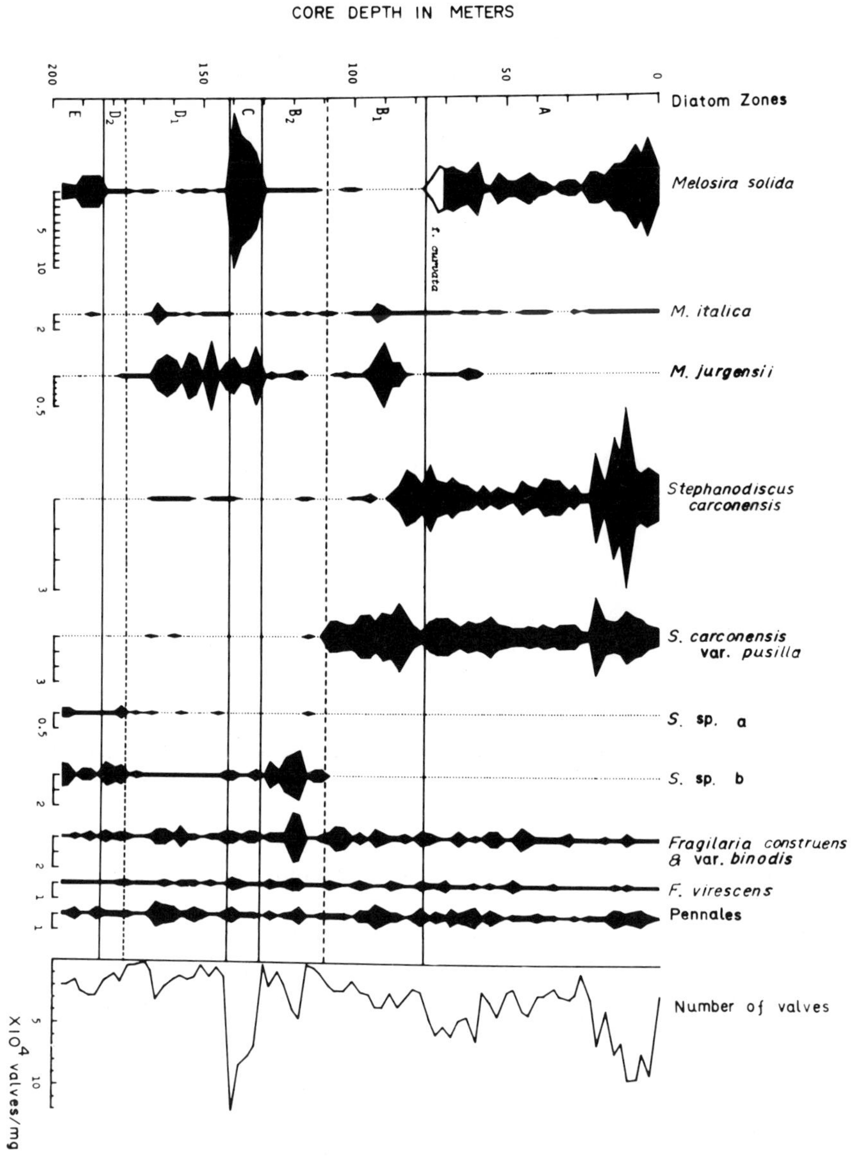

Fig. 243. Succession of Diatom Composition in 200 meters core (after Morish. 1974).

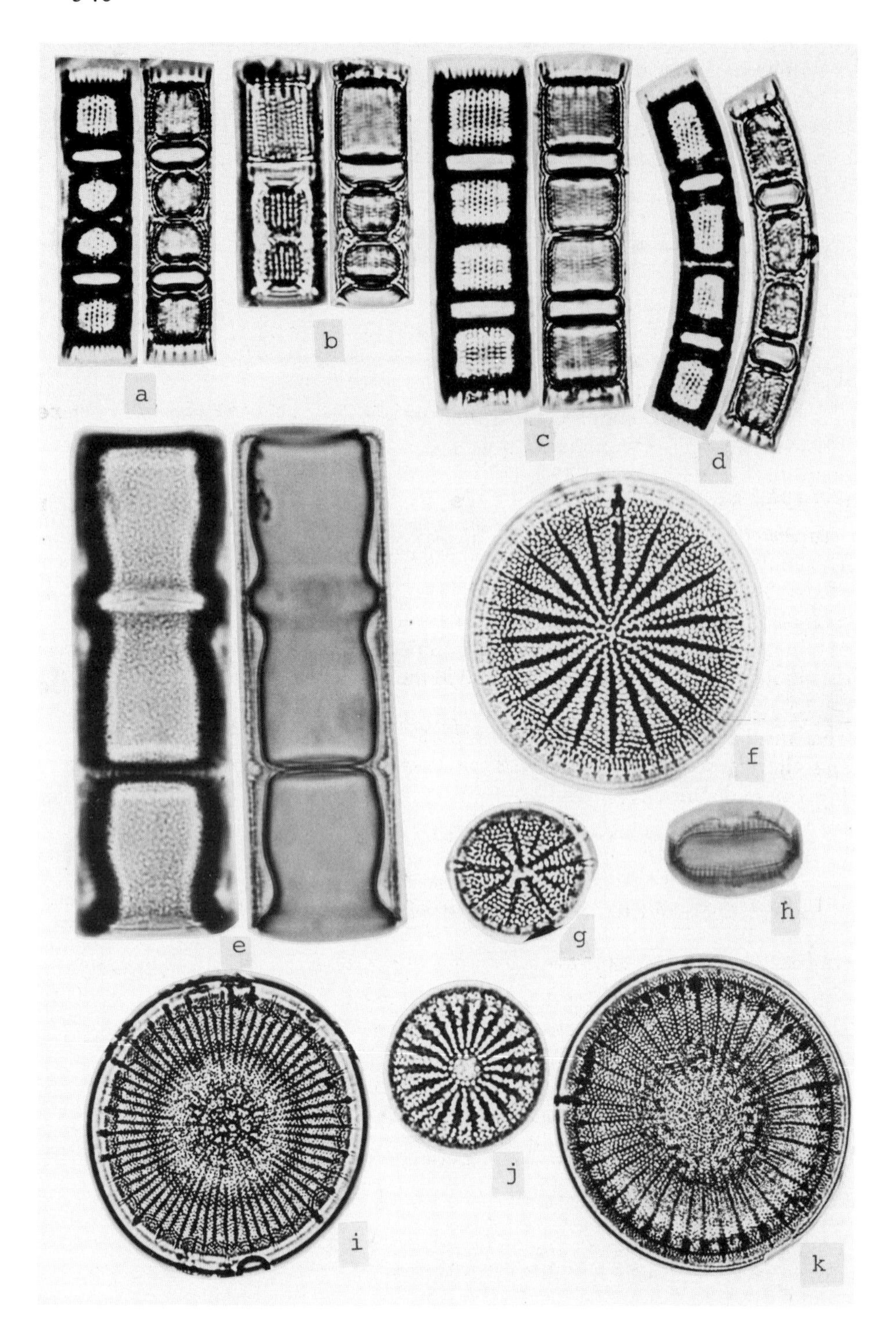

Plate 30.

abundantly. As seen in a–d of Plate 30, this species has morphologically varied according to the stratified formation.

Melosira jurgensii C. A. Agardh (Pl. 30, e)*

This diatom occurred at the layers of 59.0–177.6 m in depth in varying quantities. At present, it does not inhabit in Lake Biwa and was found only in the core sample. In case the reader is interested in pondering over paleolimnology of Lake Biwa, it is important to speculate how the diatom in the brackish water had lived in Lake Biwa.

Stephanodiscus carconensis Grunow (Pl. 30, f)

This indicator diatom in the water system of Lake Biwa was found commonly at the layers of 3.3–85.3 m in depth, and occurred abundantly especially at the layers of 9.3–15.1 m in depth. It was found merely sporadically at layers lower than 85.3 m in depth.

S. carconensis var. *pusilla* Grunow (Pl. 30, g, h)

This variety, occurred commonly at the layers of 3.3–107.1 m in depth in varying quantities. At the layers of 85.3–107.1 m it did not coexist with *S. carconensis*, but it was found commonly.

Stephanodiscus sp. a (Pl. 30, i)

This diatom, regarded as a new species, was found commonly at the layers of 175.5–197.2 m in depth. At present, it does not inhabit Lake Biwa, but was found abundantly in the above lowest layer of the core sample. It was also obtained in and around the Kamiōgi volcanic ash layer of the Kobiwako Group distributed over the Shiga Hills located in the southwestern part of Lake Biwa.

Stephanodiscus sp. b (Pl. 30, j)

This tiny diatom was found commonly at the layers of 107.1–197.2 m in depth, and alternated with *S. carconensis* var. *pusilla* at the upper layer more than 107.1 m in depth.

* Crawford (1978) revised this species name to *M. lineata* (Dillw.) C. Ag.

Plate 30. Diatom fossils in 200 m core sample (after Morish. 1974).

a–d: *Melosira solida* Eulenstein
- a: Normal type. From 10.7 m in core depth. Diameter 8 μ.
- b: Chain of normal and thin valves. From 17.7 m in core depth. Diameter 10 μ.
- c: Thin valves type. From 14.7 m in core depth. Diameter 12 μ.
- d: Curved valves type. From 72.3 m in core depth. Diameter 8 μ.

e: *Melosira jurgensii* C. A. Agardh. From 152.5 m in core depth . Diameter 21 μ.
f: *Stephanodiscus carconensis* Grunow. From 25.4 m in core depth. Diameter 44 μ.
g: *S. carconensis* var. *pusilla* Grunow. From 25.4 m in core depth. Diameter 19 μ.
h: *S. carconensis* var. *pusilla* Grunow. Girdle view. From 3.3 m in core depth. Diameter 19 μ.
i: *Stephanodiscus* sp. a
From 177.5 m in core depth. Diameter 54 μ.
j: *Stephanodiscus* sp. b
From 177.5 m in core depth. Diameter 23 μ.
k: *Stephanodiscus carconensis* Grunow. From the Kosaji clay bed of Kobiwako Group. Diameter 57 μ.

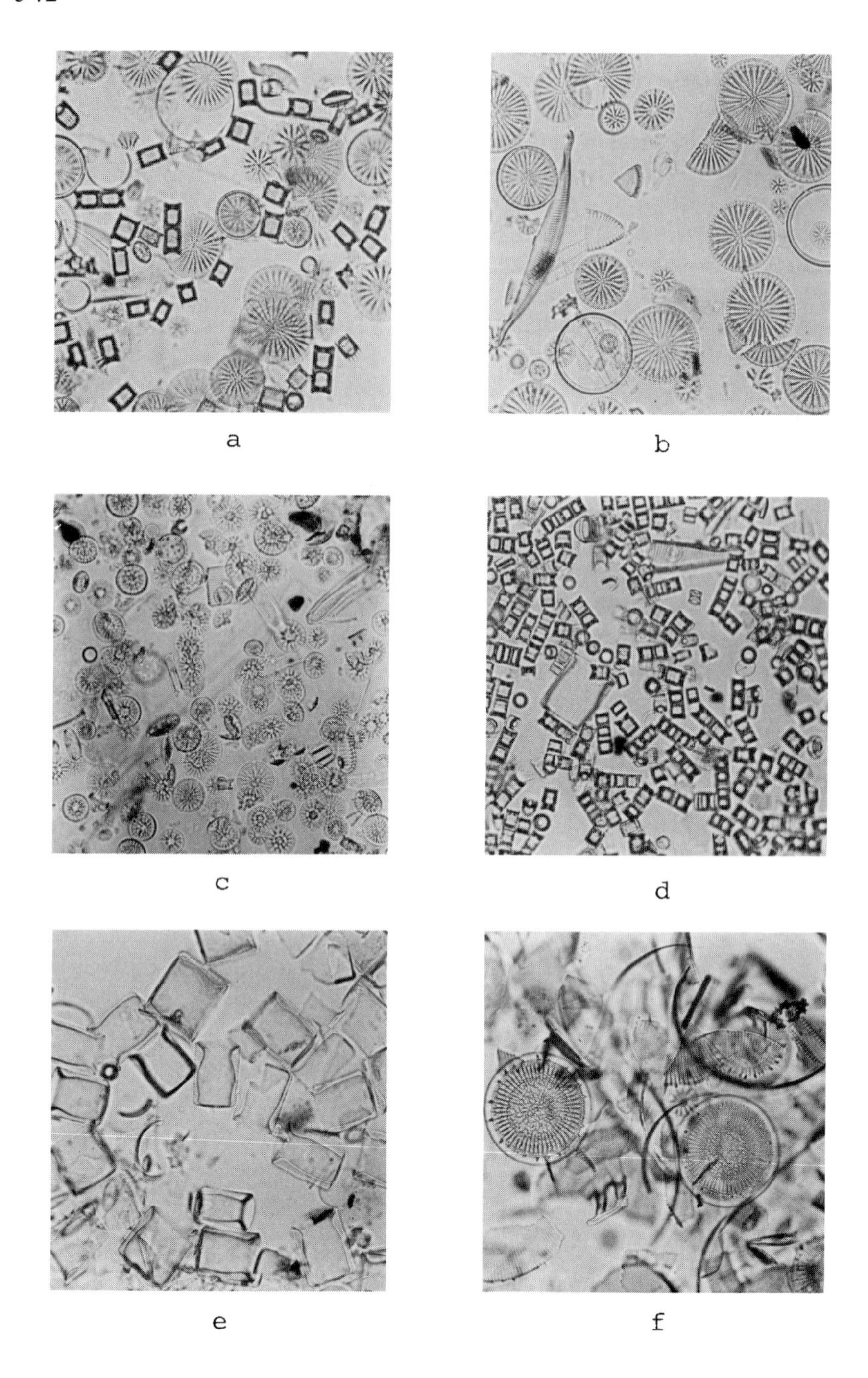

Plate 31

According to the above-mentioned occurrence spectrum of *Melosira solida* and specific variations of *Stephanodiscus*, Mori classified planktonic diatoms into five divisions (A, B_1, B_2, C, D_1, D_2, E) in 1975. E flora of planktonic diatom were gradually changed toward A flora. It is considered that the layer of 75.0 – 183.5 m in depth of these divisions corresponds to the formation in the age of Rokko Movement.

It is a very noticeable fact that some horizons were found in the core at which the variation of diatom abundance and species was correlated with the event of oscillating geomagnetic field, studied by Kawai et al. (1972). The correlations are shown below. Among the correlations one at the Biwa event II was the most conspicuous.

B (Blake event) ....Extinct horizon of *M.* cf. *jurgensii*
C (Biwa event I) ....Decrease horizon of *M.* cf. *jurgensii*
C-D (Weak magnetism) ...B1 diatom subzone
D (Biwa event II) ....Decrease horizon of *M. solida* & *M.* cf. *jurgensii*

It can be considered that an oscillating geomagnetic field had some influence on the diatoms existence, though it might be indirectly.

Plate 31. Diatom assemblages in 200 m core sample.

a: Diatom assemblage dominated *Melosira solida* and *Stephanodiscus carconensis*. From 12.3 m in core depth.
b: Diatom assemblage dominated *Stephanodiscus carconensis* and var. *pusilla*. From 77.5 m in core depth.
c: Diatom assemblage dominated *Stephanodiscus* sp. b. From 107.3 m in core depth.
d: Diatom assemblage dominated *Melosira solida*. From 137.6 m in core depth.
e: Diatom assemblage dominated *Melosira jurgensii*. From 147.3 m in core depth.
f: Diatom assemblage dominated *Stephanodiscus* sp. a. From 182.6 m in core depth.

Animal microfossils

Sadami Kadota

Introduction

Lake Biwa has a very long history. Besides this, its large area and considerable depth provide diverse habitats for organisms. More than 700 species of plants and animals are known from the lake, of which about 50 that have been found only in Lake Biwa are considered endemic. Roughly 200 species of plankton organisms are known, roughly half of which are zooplankton, comprising 79 species of Rotifera, 19 Protozoa, 13 Cladocera, and 3 Copepoda. The 70-some species of aquatic insects belong primarily to the orders Ephemeroptera, Plecoptera, Odonata, Trichoptera, and Diptera. Most important of these is the dipteran family Chironomidae, numbering about 30 species. There are also about 40 extant species of mollusks (Gastropoda and Pelecypoda) and about 50 species of fishes. In addition there are species of Porifera, Hirudinea, Bryozoa, Oligochaeta, decapod crustaceans (Macrura), and Aves.

Of the many species of animals in the lake, most are completely decomposed after death and hence do not contribute any recognizable morphological fragments to the sediments. Those that have resistant skeletons, principally of chitin, calcite/aragonite, or silica, can be preserved in the sediments. From the species represented and the stratigraphic changes in their abundance, interpretations can be made of changing conditions in the lake and its watershed over time. Data on sediment chemistry, palynology, algae, macrophytes, and other components of the sediments are all necessary for an integrated interpretation of the paleolimnology and paleoclimatology.

Most studies of microfossils to date, exemplified by Lundqvist's (1927) work on lakes of Sweden, have been based on fairly small lakes with comparatively shallow deposits. Jenkin et al. (1941) have worked on somewhat larger lakes, but the 200 meter core from Lake Biwa obtained in the autumn of 1971 is, to the writer's knowledge, the first core of this length raised from a large and deep lake. The assignment of the writer was to analyze the animal microfossils in the core.

The writer wishes to express his deep gratitude to Dr. D.G. Frey for his help.

Horie, S. Lake Biwa

ISBN 90 6193 095 2. Printed in The Netherlands

Main animal microfossils and their distribution in the sediments of Lake Biwa

The following discussion of the various groups of animals already recovered from the sediments of Lake Biwa or likely to be recovered is referenced to the literature review by Frey (1964). At least some species in each group of animals can leave skeletal remains in sediments, although sometimes only under very specialized conditions. In offshore sediments remains composed of chitin or silica are preserved best. In sediments deposited in shallower water or in deepwater with a low oxygen demand in the sediments and overlying water, calcareous remains (as of mollusks, ostracods, and fishes) can also be preserved. Remains accumulated in offshore sediments are mainly those of microscopic animals. Remains of larger animals, such as Isopoda, Amphipoda, Mysidacea, Decapoda, Insects (excluding the Diptera), mites, spiders, mollusks, and vertebrates are not yet represented in the writer's studies.

Protozoa

The only two protozoans recovered from the sediments of Lake Biwa are *Trachelomonas* and *Codonella*. *Trachelomonas* is a flagellate with a bottle-shaped test of chitin, whereas the ciliate *Codonella* has a bell-shaped test. Both species are rare. Maximum concentrations of *Trachelomonas* are only 6/cc and of *Codonella* 11/cc. Both have been found only in very shallow sediments, *Trachelomonas* in two horizons near 4.5 m and *Codonella* at 3.5, 5, and 9.5 m. According to Frey (1964) the most commonly encountered protozoans in lake sediments are rhizopods, but flagellates (particularly dinoflagellates and Chrysophyceae) can also be abundant.

Porifera

The siliceous spicules of freshwater sponges commonly are well preserved and abundant in lake sediments. These are mostly flesh spicules, which are not diagnostic of the species. Only when birotule spicules are present can the species be identified. *Spongilla* occurs in Lake Biwa today, but its spicules have not yet been recovered from the sediments.

Coelenterata

Wesenberg-Lund (1896) predicted that the distinctive resting eggs of *Hydra* should occur in lake sediments, but at least up to the time of Frey's (1964) literature review they had not been reported. As *Hydra* does occur in Lake Biwa, its resting eggs should be looked for.

Turbellaria

The eggs capsules or cocoons, particularly of the neorhabdocoeles, are often abundant in lake sediments. However, very few of them can be identified to species, because taxonomists have largely neglected these structures in their studies. Harmsworth (1968) found about 40 different morphotypes of eggs capsules in the sediments of Blelham Tarn, England, but could put provisional names on only two of them. In Lake Biwa turbellarian egg capsules reach a maximum abundance of 523/cc at 10.5 m. These egg capsules are present at all samples in the layers shallower than 55 m deep, but their concentrations in the sediments gradually tend to decrease at greater depths.

Rotifera

The resting eggs of rotifers occur in lake sediments and can be abundant. Nipkow (1961) determined how long the eggs of 16 species remained viable in the sediments of Lake Zürich. Regretfully the resting eggs of very few species have been described, and hence it is not even possible to decide which of the many small egg-like structures in sediments are rotifer resting eggs, much less what species are represented. Müller (1970) was the first to study stratigraphic changes in abundance of resting eggs of several species of rotifers. These microfossils undoubtedly occur in the sediments of Biwa, but up to now they have not been recognized.

Bryozoa

The very distinctive statoblasts of the Phylactolaemata are preserved in lake sediments, and their differences in morphology are sufficiently great to permit recognition at least to genus, frequently to species. Wesenberg-Lund (according to Frey, 1964) described and illustrated fossil statoblasts of *Plumatella fungosa* or *repens*, *P. fruticosa*, and *Cristatella mucedo*. Species of the genus *Plumatella* are generally the most common bryozoans in lakes, and their statoblasts can be abundant enough for construction of close-interval stratigraphies. Although Bryozoa live in Lake Biwa, their statoblasts have not yet been found in the sediments.

Crustacea

The class Crustacea includes not only microscopic forms, such as copepods, cladocerans, and ostracods, but also larger forms, such as decapods. Like all arthropods, these organisms grow stepwise by shedding their exoskeleton periodically. The cast exoskeletons (=exuviae) and the exoskeletons of those animals that died without molting are subject to normal breakdown processes, which tend to separate the remains into their various components — mandibles, antennules, heads, shells, postabdomens, caudal spines, etc. — and then to attack them biochemically to the extent permitted by the structure of their chitin. As a result of this differential destruction, the fossil record of the crustaceans, like fossil records in general, is very patchy: some groups and species preserve well, perhaps quantitatively, whereas others preserve scarcely at all.

Among the Anostraca only antennae and mandibles have been reported. In the Notostraca, Bennie (1894) reported finding many body segments, tail segments, appendages, and mandibles of *Lepidurus*. Of the modern species of Conchostraca, only *Limnadia* has been recovered from lake sediments. Ostracods are abundantly represented by their shells, which, being largely calcareous, tend to occur chiefly in shallow-water deposits.

The Cladocera constitute the major group of crustacean microfossils in freshwater sediments. Frey (1964) listed 90 species of Cladocera that had been recovered from lake sediments. Of these, the Holopedidae (2 species), Sididae (3 species), Daphniidae (17 species), Macrothricidae (3 species), Leptodoridae (1 species), and Polyphemidae (2 species) are represented chiefly by postabdomens, postabdominal claws, caudal spines, mandibles, antennal segments, and ephippia, whereas the Bosminidae (3 species) and Chydoridae (59 species) are represented not only by these components but also by headshields and shells. The macrothricids and chydorids occur chiefly in the littoral zone. Species of the other families are predominantly planktonic, even though some prefer the littoral zone to the limnetic.

The 17 species of Cladocera recovered to date from the Lake Biwa sediments are given in Table 68. In addition to these species, *Diaphanosoma brachyurum*, *Moina macrocopa*, *Bosminopsis deitersi*, *Chydorus sphaericus*, and others presently occur in the lake but have not been recovered from the sediments. For purposes of describing and interpreting the stratigraphy, the writer has grouped the Cladocera into *Sida*, *Daphnia*, *Bosmina*, *Leptodora*, *Monospilus*, and other Cladocera (entirely chydorids: Table 68).

Sida. This species, which is associated with macrophytes in the littoral zone, is rare and discontinuous in its occurrence in the offshore sediments of Lake Biwa. Its maximum abundance was only 17/cc.

Daphnia. Attempts to recover remains of *Daphnia* were made only between 35.5 and 55.5 m. Both species listed in Table 68 were found. As they occur continuously to a depth of 46 m but only discontinuously thereafter at 47 and 52.2 m, it is very likely that they also occur in shallower sediments. The greatest concentrations of remains was 132/cc at 45 m (Kadota 1976).

The postabdomens and postabdominal claws of *Daphnia* occur commonly in lake sediments and are important in interpretations of paleoliminology and paleoecology (Frey 1964). Mandibles also occur but can be identified to species with even less success than the postabdomens and postabdominal claws. Mandibles, possibly of *Daphnia*, were found in the Biwa sediments. In addition, the ephippia of *Daphnia*, which occur in the geological record at least as far back as mid-Tertiary (Frey 1964), can occur, but as their numbers are not closely related to the size of the producing population they cannot be used in studies of past productivity.

Bosmina. This genus of Cladocera is widely distributed over the world. its exoskeletal fragments are highly resistant to biological degradation, with the result that nearly all components are abundant in sediments. Most common are headshields with their rigidly attached antennules and shells. Particularly the headshields and the much less common postabdominal claws are useful in determining the species present. In Lake Biwa the remains of *Bosmina* are the most abundant cladoceran microfossils, with a maximum of 7,656/cc at 5 m. In the uppermost 55 m the taxon occurs continuously from the surface to 23 m, with a few discontinuities, most notably between 24.5 and 26 m, below here. Besides the peak at 5 m, lesser peaks occur at 11.0, 22.5, 28.5, 40.5, and 53.5–55.0 m, although the taxon in general tends to be much less common at depths greater than 38 m.

Leptodora. This largest cladoceran is a predator on other zooplankters. Its mandibles were present in all samples to a depth of 25 m, with a maximum of 172/cc at 12.5 m. From 27.5 to 37.5 m there are gaps in continuity, and below 41 m the species is rarely present.

Monospilus. For reasons unknown, *Monospilus* is particularly well preserved in the sediments of Lake Biwa and is present in almost all samples that have been examined. Interestingly, it is a heavy-bodied benthic chydorid that does not occur in the plankton at all. It is abundant from 16.5 m to the surface, with a maximum of 507/cc at 10.5 m. At greater depths it tends to be less abundant, as likewise are most of the other animal remains, but with a minor increase at 27.5–29.5 m.

Other cladocera. Included here are the 11 species of chydorids other than *Monospilus* listed in Table 68. Although occurring in all samples studied, they are most abundant in the uppermost 17 m, with a maximum of 2,453/cc at 10.5 m. Lesser peaks of abundance occur at 26.5–31.5, 40.5, and 48 m.

Ostracoda. Very few species of ostracods are planktonic, nearly all of them being heavy-bodied animals living among macrophytes in the littoral zone or in the sediments. Those with calcareous shells are generally not found in organic offshore sediments, although those with chitinous shells occasionally are recovered. No remains have thus far been recovered from the Biwa sediments.

Table 68. Species of Cladocera identified from the uppermost 200 meters of Lake Biwa sediments and the particular kinds of remains recovered.

Species	Shell	Head-shield	Post-abdomen	Claw	Antennule	Mandibles
Sididae						
Sida crystallina				X		
Daphniidae						
Daphnia pulex type				X		
Daphnia longispina type				X		
Bosminidae						
Bosmina longirostris sens. lat	X	X			X	X
Leptodoridae						
Leptodora kindtii						X
Chydoridae						
Kurzia latissima		X?				
Oxyurella tenuicaudis		X	X	X		
Camptocercus rectirostris		X	X	X		
Graptoleberis testudinaria	X	X				
Alona affinis		X	X	X		
Alona quadrangularis		X	X	X		
Alona costata			X	X		
Alona guttata		X	X	X		
Alona rectangula			X	X		
Alona intermedia			X	X		
Monospilus dispar	X	X	X	X		
Alonella excisa		X				

Copepoda. Although one of the dominant groups of zooplankton all over the world, copepods regretfully are seldom preserved in freshwater sediments. Their chitin seems particularly susceptible to biodegradation, with the result that except for the spermatophores of harpacticoids, particularly *Canthocamptus,* most other records from lakes in humid environments are strictly from surficial sediments (Frey 1964). No remains of copepods have been identified from Lake Biwa.

Insects

Freshwater sediments contain abundant exoskeletal fragments of aquatic insects, particularly the immature stages, as well as lesser numbers of terrestrial insects that are blown in or washed in. Insects are particularly important in paleolimnology and paleoecology because of their great diversity and their often highly restricted ecological requirements for food and habitat. Most important in paleolimnology are the aquatic Diptera of the families Chironomidae and

Helidae and *Chaoborus* of the family Culicidae. The latter is a benthic/planktonic predator, whereas the others occur as substrate organisms in the sediments or in association with macrophytes in shallow water. The species of the family Chironomidae, which generally are the most common insect remains, are particularly useful for interpreting past conditions in the hypolimnion. The exoskeletal fragments of the larvae most commonly encountered are intact headcapsules and isolated labia and mandibles in the Chironomidae, head capsules in the Heleidae, and mandibles, basal antennal segments, and occasionally premandibular scales in *Chaoborus*.

In the top 55 m of Biwa sediments, the remains of aquatic insects (chiefly *Chironomus*) as well as terrestrial insects were recovered in all samples (Kadota 1973, 1974, 1975, 1976). Maximum abundance was 1,029/cc at 5 m, with lesser peak at 15, 22.5, 27.5, and 39 m. As with other animal microfossils, the concentrations of insect remains generally declined with depth, perhaps reflecting the long-term consequences of a slow, continuous degradation of chitin.

In summary, remains of various groups of animals occur in the sediments of lakes, of which the Cladocera, Chironomidae, Turbellaria, and Protozoa, being richest in number of species and abundance of remains, are most important. For *Codonella* and *Trachelomonas* in the Protozoa, *Leptodora, Sida, Bosmina,* and *Monospilus*, the remains not only vary in concentration from one level to another but also are completely absent in some samples. One of the most important problems for further study is to identify the conditions that control these changes in occurrence and abundance.

Vertical distribution of animal microfossils in relation to climatic change*

In the 200 meter core the writer (1973) analyzed the kinds and abundance of microfossils at 5-m intervals and grouped them into the following categories: *Trachelomonas* and *Codonella* in the Protozoa, the egg capsules or cocoons of Turbellaria, *Leptodora, Sida, Bosmina, Monospilus,* other Cladocera, *Chironomus*, and other insects. In addition *Pediastrum* and *Staurastrum* in the green algae were analyzed. Of the total animal remains recovered, Cladocera comprised 63%, Turbellaria 25.1%, Insects 10.9%, and Protozoa 1.0%.

The core can be divided into the following zones based on differences in abundance of the animal microfossils and major shifts in their relative abundance.

10 – 100 m: Abundance of total animal remains tends to increase steadily except for a decline at 25 m. Turbellaria and Cladocera are prominent, chironomids strictly subsidiary. Protozoans appear in the upper part.

*The numbers reported in the 2nd paragraph under this section, are not the mean number/cc in the 200 m core, and the author is not certain that the percentage composition of the four groups given above is based on all the samples in the core. Mean frequencies *must* include zero values as well as those in which the particular organism is represented.

105—175 m: Microfossils less abundant than above. Relative abundance of insects decreases toward the top of the zone as Cladocera and Turbellaria become more abundant.

180 197.2 m: Quantity of microfossils extremely small (<28/cc), consisting entirely of insects.

The stratigraphic distribution of the green algae is the converse of the animals in that they are most abundant in the deeper sediments and decrease steadily toward the surface.

An attempt was made to relate the stratigraphic changes in abundance of the various groups of microfossils to estimations of climatic change attempted by Fuji (1973b, 1974), Fuji and Horie (1977), Nakai (1973), Nakai and Shirai (1978), and Yamamoto (1974a). On the basis of the palynological study, Fuji and Horie (1977) recognized 12 stages of cold climate, 10 of temperate climate, and 1 of warm climate. Nakai and Shirai (1978) concluded that their curves for organic carbon in the sediments and the composition of stable isotopes of carbon agreed with the climatic changes proposed by Fuji. Nakai found peaks of warm climate at 90, 140, and 190 m and of cold climate at 50, 135, 145, and 170 m. No clear relationship exists between the peaks in abundance of animal microfossils and the warm periods (Fig. 244). However, the green algae seem to peak in abundance during the warm periods and to be absent during the cold periods estimated by Fuji and Horie (1977). The abundance of animal microfossils in these samples is low.

In Fig. 244C are given the results of Yamamoto (1974a) on grain-size analysis in the top 40 m of core, and in Figs. 244A and 244B, respectively, the density of animal microfossils and green algae at 25-cm intervals as analyzed by the writer (1974, 1975, 1976). The animal microfossils exhibit strong peaks in abundance at roughly 5, 10, 15 m, which are in good agreement with peaks of coarse material in the sediments shown in Fig. 244C. Yamamoto suggests that the peaks of grain size coarser than 44μm represent pluvial periods with reduced temperatures, but the temperature curve proposed by Fuji & Horie (1977) in Fig. 245 indicates that two of the three microfossil peaks may be related to warm periods rather than cold. Hence, the interpretations of the data are not yet in complete agreement. Regardless of the temperature involved, however, the peaks in the coarse fraction of sediments suggest greater precipitation and inflow to the lake, which could have supplied larger quantities of organic and inorganic nutrients to the lake, thereby secondarily increasing the production of animals.

Fuji (1973b) reported that evergreen trees growing in the warm temperate zone such as *Quercus* and *Ilex* are abundant between 4.5 and 7.0 m, an interval corresponding to the so-called Flandrian transgression which would be a warm period. The writer's analysis of the abundance of green algae here (Fig. 244B) also shows a marked increase at these levels. Furthermore, the abundance of animal microfossils (Fig. 244A) increases markedly beginning about 16 m. Below this level green algae are in low abundance and discontinuous in the record. All the higher frequencies of algae in this part of the record occur

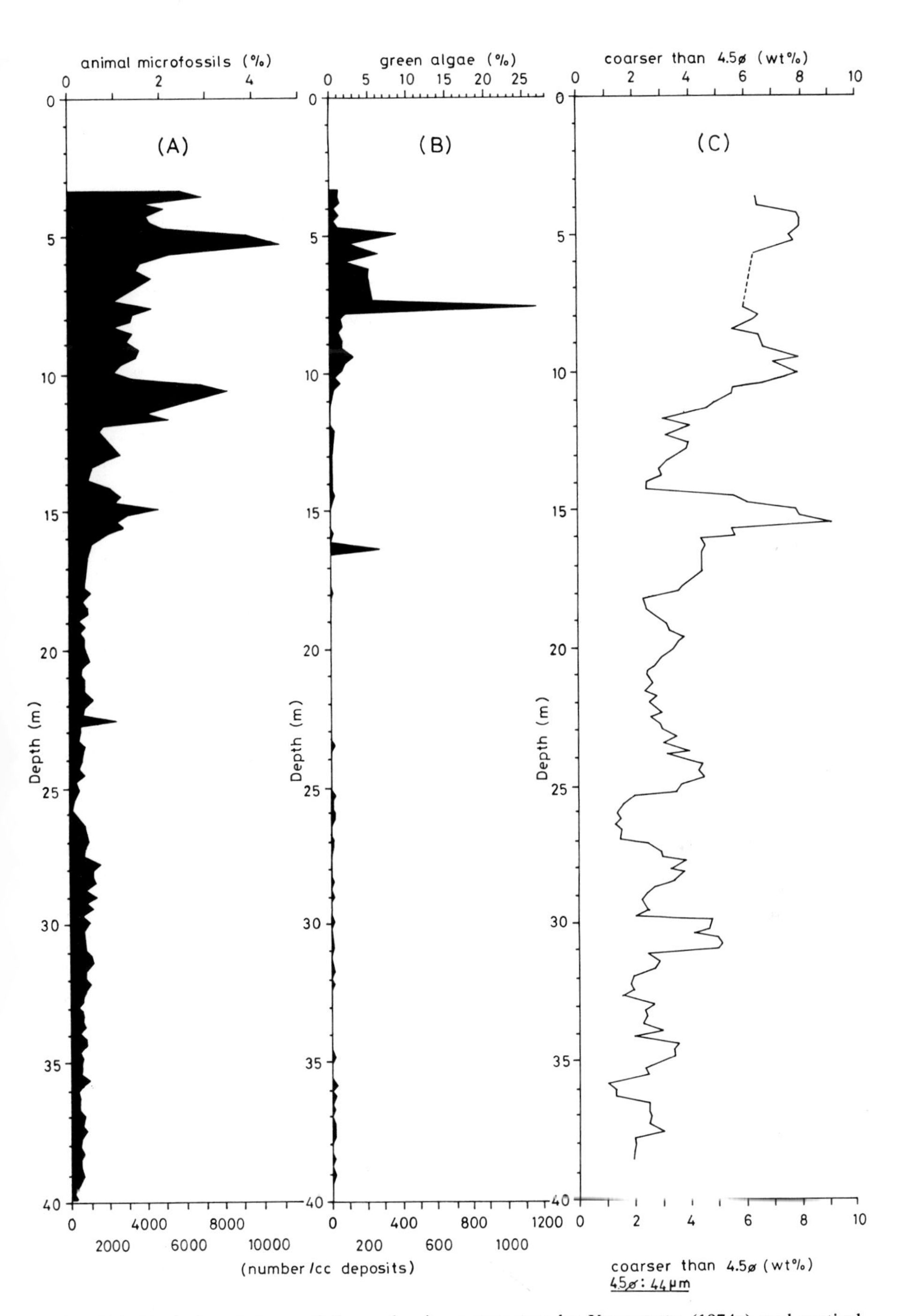

Fig. 244. Vertical variations of the grain size parameters by Yamamoto (1974a) and vertical distributions of microfossils by Kadota.

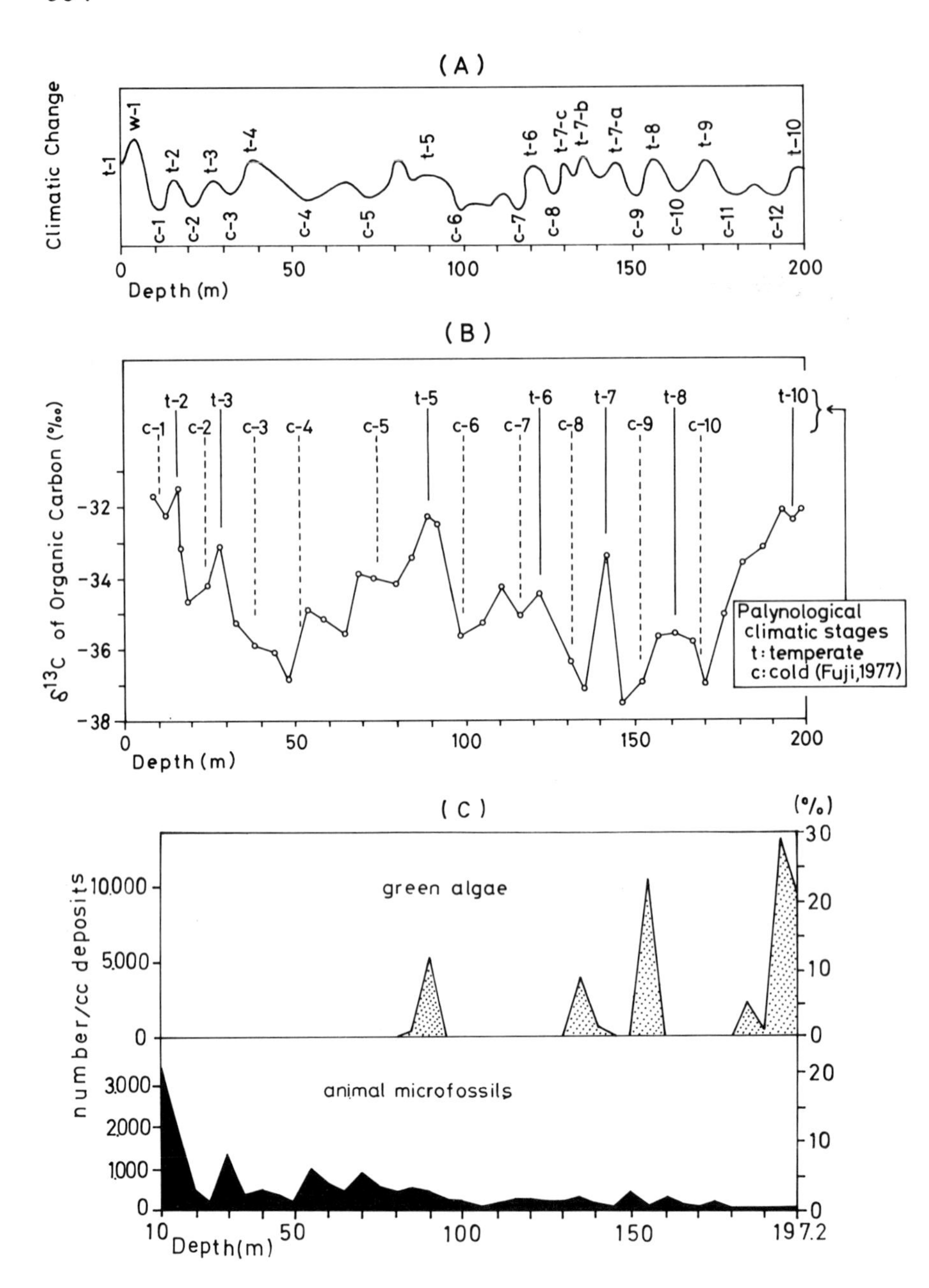

Fig. 245. Comparison of paleoclimate curve, $\delta^{13}C$ variation of organic materials and microfossils in the core sample from Lake Biwa.

A: paleoclimate curve estimated by Fuji and Horie (1977).

B: variation of $\delta^{13}C$ values determined by Nakai and Shirai (1978).

C: vertical distribution of animal microfossils and green algae (*Pediastrum, Staurastrum*) (after Kadota 1973.)

between 17 m and the surface. According to Fuji's (1974) interpretation, a preceding long, cold climatic stage transformed into a warm climatic stage at about this level. By this interpretation both the algae and the animals responded positively to the warmer temperature or to other related conditions.

Thus, it appears that the stratigraphic changes in occurrence and abundance of microfossils as found by the writer are clearly related to the climatic changes interpreted by Fuji (1973b, 1974, 1975b), Nakai (1973, 1975), and Yamamoto (1974a, 1975, 1976). Further studies are needed, however, to determine the actual conditions controlling the increases and decreases of the organisms in the lake over time.

Organic geochemistry

Ryoshi Ishiwatari and Kazuko Ogura

Introduction

Organic compounds found in a sediment core generally originate from organisms in the lake water and the surface sediment, and from surroundings of the lake at the time of the sediment formation. And so, they would provide information on the paleoenvironment of the lake as well as physico-chemical conditions of the sediment where they have existed. Although many investigations have been conducted on various types of sediment, data are still insufficient to establish the correlation between specific organic compounds and organisms, and also to establish directions of long-term organic reactions in sediments.

Organo-geochemical studies have been conducted mainly by the Nagoya University and the Tokyo Metropolitan University groups for a long 200 meter core sediment sample of Lake Biwa.

Organic compounds of the Lake Biwa sediments

Hydrocarbons

Aliphatic hydrocarbons. Normal alkanes have been studied in lake sediments in attempts to relate their composition to the biological residues within the sediments (Cranwell 1973, 1976; Brooks et al. 1976). Since alkanes are biologically and chemically more stable than oxygenated compounds, their distribution pattern may be well preserved in sediments for a long period. Most algae contribute normal heptadecane (n-C_{17} alkane), whereas alkanes with a long carbon chain ($>C_{21}$) may originate from higher plant materials.

Fig. 246 shows the distribution of normal alkanes at various depths of Lake Biwa sediment (Ishiwatari 1976a). The carbon number of the n-alkanes found in the sediment ranges from 17 to 35, showing a maximum in the C_{27}–C_{31} region. The ratio of odd-numbered homologues to even-numbered ones in the sediment

Horie, S. Lake Biwa

ISBN 90 6193 095 2. Printed in The Netherlands

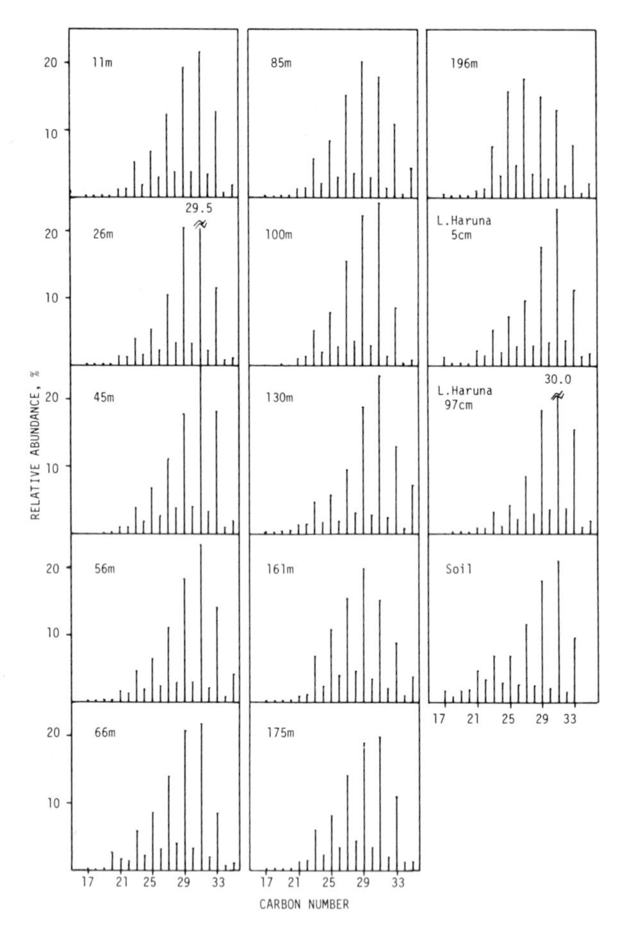

Fig. 246. Carbon-number distribution of normal alkanes from Lake Biwa sediments and its comparison with those from Lake Haruna sediments and a soil (Ishiwatari 1976a).

samples are high, indicating that the alkanes are contributed mainly by lipids of higher plants. The n-alkane distribution observed in Lake Biwa is similar to those reported by other writers for most lacustrine sediments (Cranwell 1973; Brooks et al. 1976). The fact that n-alkane composition is similar among all the core sediment samples indicates that the trophic level of Lake Biwa has not varied much during its history. Moreover, as far as n-alkanes are concerned, their main source may be out of the lake, since the relative contribution of n-C_{17} which is characteristic of algae is extremely low. A similar n-alkane distribution was observed for sediment samples from Lake Haruna and from soil surrounding Lake Haruna, as shown in Fig. 246.

In most layers, the C_{31} alkane is predominant. However, the C_{29}, C_{29} and C_{27} alkanes are the most abundant for the layers of 85, 161 and 196 m, respectively.

This suggests that the kind of higher plant as a source material for n-alkanes in those layers is different from that in most layers. It is of interest to note that all of those layers show high pheopigment content, as shown in Fig. 247.

The total n-alkane content of the sediment samples is below 9 μg/g-dry sediment, as shown in Fig. 247. The n-alkane content in layers shallower than 100 m is about twice higher than that in deeper layers, suggesting that organic matter supply into the lake was relatively low for the latter layers.

Fig. 248 gives the relation between the n-alkane and perylene content of sediment samples from Lake Biwa. There seems to be a positive relation between them, suggesting that their origins are similar.

If we assume that the n-alkanes in the Lake Biwa sediment come mainly from sources other than the lake, the pheopigments/n-alkanes ratio would indicate the relative contribution of autochthonous and allochthonous organic matter. Fig. 247 (right) shows the vertical variation of pheopigments/n-alkanes. Although the amount of data is quite limited, there seems to be a relation between the variation of pheopigments/n-alkanes and that of the paleoclimate inferred by Fuji and Horie (1972) based on palynology. In a colder climate, seemingly, the pheopigments/n-alkanes is low, whereas the ratio is high in a warmer climate.

Aromatic hydrocarbons. Many kinds of aromatic hydrocarbon have been reported to be present in soils and sediments. In particular, perylene has been well investigated and believed to be a good indicator of soil-derived organic matter (Blumer 1961; Orr & Grady 1967; Taguchi & Sasaki 1971; Aizenshtat

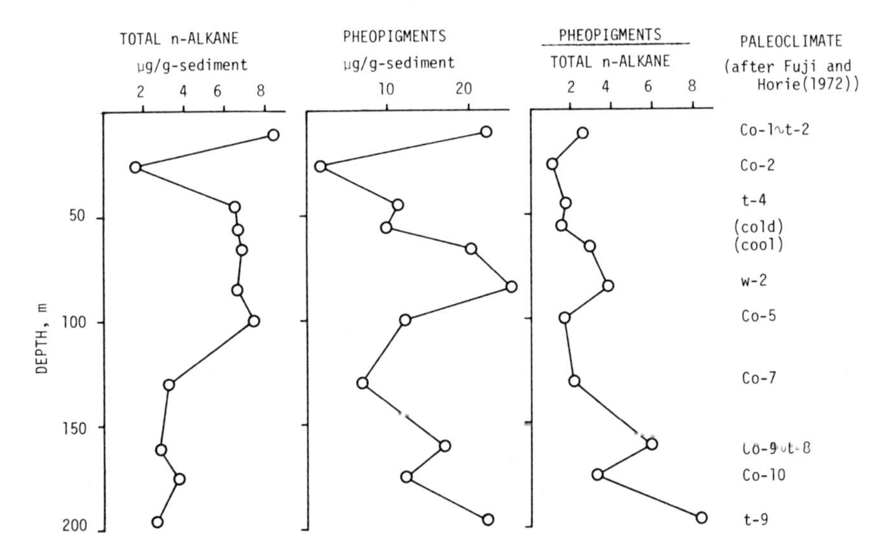

Fig. 247. Vertical distribution of total n-alkane and pheopigments in Lake Biwa sediment (Ishiwatari 1967a).

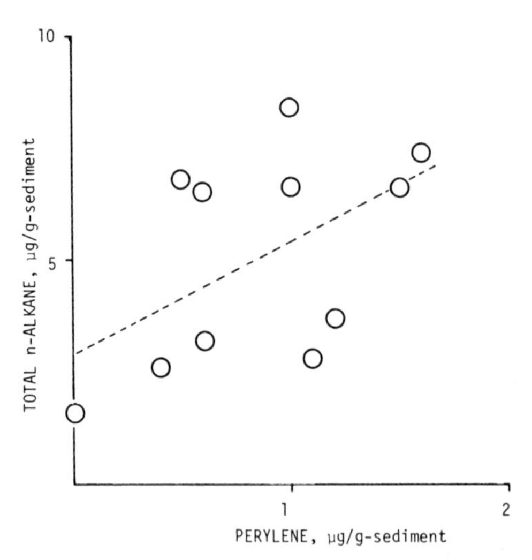

Fig. 248. Total n-alkane *vs.* perylene for the sediment samples from Lake Biwa (Ishiwatari 1967a).

1973; Ishiwatari & Hanya 1975; Ishiwatari et al. 1977, 1980; Wakeham 1977; Laflamme & Hites 1978).

Aromatic hydrocarbons in Lake Biwa sediments were determined by gas chromatography-mass spectrometry (GC/MS; Ishiwatari 1975). They include biphenyl ($C_{12}H_{10}$), fluorene ($C_{13}H_{10}$), phenanthrene or anthracene ($C_{14}H_{10}$), fluoranthene ($C_{16}H_{12}$), chrysene ($C_{18}H_{12}$), pyrene ($C_{16}H_{12}$), benzpyrene ($C_{20}H_{12}$) and perylene ($C_{20}H_{12}$). Fig. 249 gives a representative gas chromatogram (mass fragmentogram) of those hydrocarbons from Lake Biwa. Table 69 shows their quantitative data.

Perylene is the most abundant of all aromatic hydrocarbons in the sediment samples (0.004 – 1.6μg/g-dry sediment; $0.01-2.0\times10^{-2}\%$ of total organic matter (TOM)), and its amount is generally higher than those for another lake sediment ($0.1–0.7\times10^{-2}\%$ of TOM: Ishiwatari et al. 1977) and for a nearshore marine sediment ($0.1–0.9\times10^{-2}\%$ of TOM: Aizenshtat 1973).

The origin of perylene is considered to be different from that of the other aromatic hydrocarbons. The former is believed to be a reduction product of the corresponding quinone which arises predominantly from land organisms (Aizenshtat 1973), and the latter might come from natural combustion (forest fires), as claimed by Youngblood & Blumer (1975). A relatively high concentration of phenanthrene/anthracene and an unidentified aromatic hydrocarbon (molecular weight = 274) in layers deeper than 85 m was observed, suggesting that they have formed as products of chemical reaction after burial.

The unknown compound having mass spectral peaks at m/e 218 and m/e 274 (Fig. 249) was recently identified to be trimethyltetrahydrochrysene ($C_{21}H_{22}$) (Ishiwatari 1979). This compound is considered to be derived from pentacyclic triterpenes such as β-amyrin by loss of ring A and dehydrogenation in the early stage of sedimentation (Spyckerelle et al. 1977).

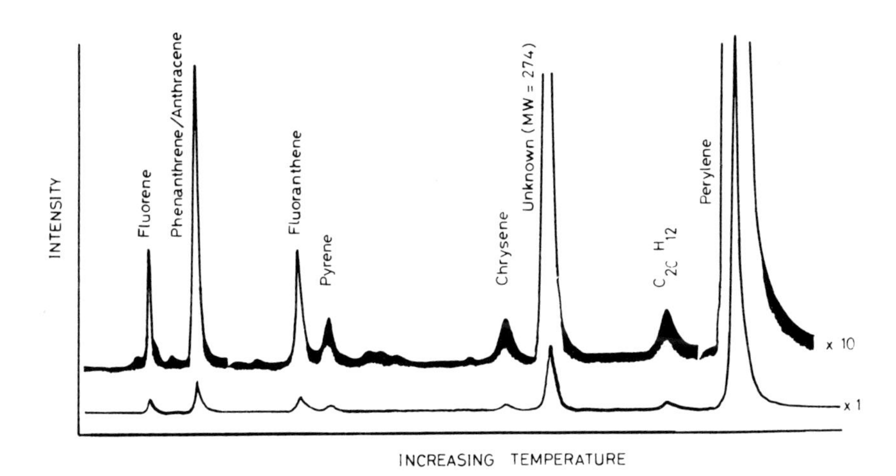

Fig. 249. Gas chromatogram (mass fragmentogram) of polynuclear aromatic hydrocarbons extracted from 45 m in depth in Lake Biwa sediment.
Analysis was made on a Shimadzu-LKB 9000 GC/MS with a 2 m by 3 mm glass column packed with Chromosorb W coated with 1% OV 17. Column temperature was programmed from 150° to 280°C at a rate of 5°C/min. Mass fragmentography was run at a mass number (m/e) corresponding to the molecular ion for each compound. (After Ishiwatari, 1975)

Aliphatic carboxylic acids

Fatty acids (aliphatic monocarboxylic acids) are the major lipid components of all living organisms. It is generally known that even carbon numbered normal fatty acids C_4 to C_{26} are found in natural fats; those C_{26} to C_{38} are principally contained in waxes of insect and plants (Kvenvolden 1967). Bacterial lipids are noted for being rich in branched-chain fatty acids (Kaneda 1967). The fatty acid pattern of a sediment depends on the nature of the acids supplied by organisms and the relative survivability, both chemical and biological, of the different acids (Leo & Parker 1966). Therefore, such fatty acids are not only a potential indicator of the contribution by organisms to the sedimentary organic matter but also that of biological and chemical environments.

Table 70 gives the analytical results of organic compounds including mono- and dicarboxylic acids in various depths of Lake Biwa sediment (Ishiwatari & Hanya 1975).

Monocarboxylic acids. The normal C_{12}–C_{32} monocarboxylic acids, branched (iso and anteiso) C_{15} and C_{17} monocarboxylic acids were found to be present in the sediment samples of Lake Biwa (Ishiwatari & Hanya 1973, 1975). They show a marked even carbon number predominance common to most biosynthetic lipids. As shown in Fig. 250, there are two maxima, C_{16} acid in the region C_{14}–C_{20} and C_{24}–C_{28} acids in the region C_{22}–C_{32}, in the distribution of fatty acids. The fatty acids ranging from C_{14} to C_{20} may be derived from phytoplankton lipids (autochthonous), while the C_{26}–C_{28} acids may come from the

Table 69. Vertical distribution of polynuclear aromatic hydrocarbons in Lake Biwa sediment (ng/g-dry sediment) (Ishiwatari 1975).

Molecular weight	Molecular formula	Identification	Depth (m)										
			11	26	45	56	66	85	100	130	161	175	196
154	$C_{12}H_{10}$	Biphenyl	–	–	–	–	–	2	?	2	8	2	–
168	$C_{13}H_{12}$	unknown	–	–	–	–	–	2	?	6	63	4	6
166	$C_{13}H_{10}$	Fluorene	–	–	2	–	–	–	–	1	–	1	2
178	$C_{14}H_{10}$	Phenanthrene/ Anthracene	32	2	9	8	7	11	10	12	39	17	89
202	$C_{16}H_{12}$	Fluoranthene	26	2	5	5	4	5	8	5	13	6	19
202	$C_{16}H_{12}$	Pyrene	14	1	3	3	3	3	3	2	7	4	9
228	$C_{18}H_{12}$	Chrysene	18	5	3	5	5	6	5	5	6	5	11
274		unknown	?	2	37	+	46	34	34	21	79	28	78
252	$C_{20}H_{12}$	unknown	13	7	4	5	6	6	2	5	7	6	8
252	$C_{20}H_{12}$	Benzpyrene	–	6	–	–	–	–	–	2	4	3	4
252	$C_{20}H_{12}$	Perylene	960	4	630	1000	490	1470	1590	560	1080	1240	350

* –: not detected; +: detected; ?: not measured.

Table 70. Analytical results of organic compounds in Lake Biwa sediment (Ishiwatari & Hanya 1975).

Depth m	Total organic matter (TOM)* mg/g	Pheopigment		Monocarboxylic acid		Dicarboxylic acid	
		μg/g	% of TOM	μg/g	% of TOM	μg/g	% of TOM (×10²)
11	18.2	22.4	0.12	48.8	0.27	2.3	1.3
26	5.8	1.7	0.03	12.5	0.22	3.7	6.4
45	11.5	11.2	0.10	7.4	0.07	0.9	0.8
56	12.4	9.7	0.08	47.3	0.38	6.3	5.1
66	12.1	20.1	0.17	35.2	0.29	5.3	4.4
85	12.9	25.0	0.19	53.4	0.41	4.8	3.7
100	7.9	11.9	0.15	33.7	0.43	1.0	1.3
130	7.4	6.7	0.09	17.0	0.23	0.6	0.8
161	10.1	16.8	0.17	36.2	0.36	0.3	0.3
175	6.6	12.1	0.18	25.1	0.38	0.4	0.6
196	9.0	21.8	0.24	30.4	0.34	0.4	0.4

* After Handa (1972); TOM = (Organic carbon) × 1.8.

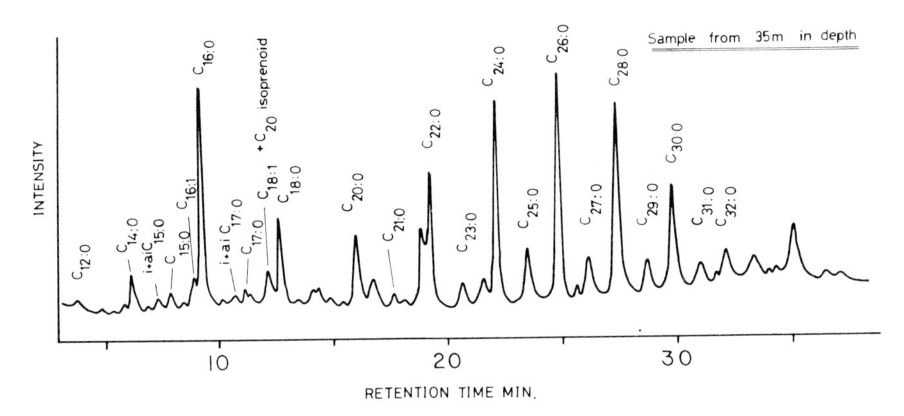

Fig. 250. Typical gas chromatogram (total ion monitor trace) of aliphatic monocarboxylic acid methyl esters derived from Lake Biwa sediment.
Analysis was made on a Shimadzu-LKB 9000 GC/MS with a glass column (2 m × 3 mm) of 1% OV 1 Chromosorb W. column temperature was programmed from 100° to 280°C at a rate of 5°C/min.
*Symbols: i = iso; ai = anteiso. $C_{n:m}$: n = carbon chain length; m = number of unsuration (after Ishiwatari & Hanya 1973).

surrounding soils (allochthonous). Fig. 251 gives the vertical variation in the normal monocarboxylic acid distribution in the sediment of Lake Biwa. A striking feature of the distribution is the relatively high abundance of C_{20}–C_{30} acids as compared with those having shorter carbon chains. Thus, the ratio of abundance of n-C_{16} and n-C_{18} acids to that of n-C_{20}–n-C_{30} acids (Lower Molecular Weight Fatty Acids/Higher Molecular Weight Fatty Acids: LFA/HFA), which is a possible index of contribution of autochthonous organic matter

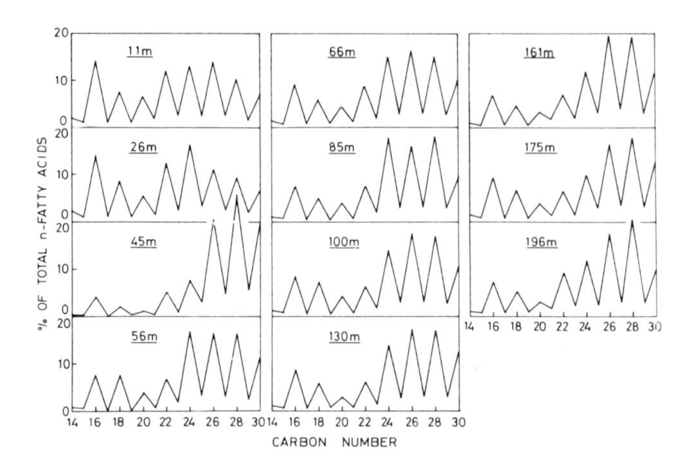

Fig. 251. Vertical variation of normal monocarboxylic acid distribution in Lake Biwa sediment (Ishiwatari 1975, Ishiwatari & Hanya 1975).

relative to allochthonous organic matter in sediments (Cranwell 1974; Ishiwatari & Hanya 1973, 1975), is significantly low for the sediment layers (0.18 – 0.37 for the 11 – 196 m layers).

In order to see whether such a low LFA/HFA observed for the 200-m core sample is due to their post-depositional change or not, changes of fatty acid composition were examined in a 5-m core sample taken from Lake Biwa (Ishiwatari & Kawamura 1978). The results indicated that the LFA/HFA ratio at the surface layer is rather high and shows essentially a decreasing tendency with depth although some minima and maxima are seen at various depths (Table 71). They concluded that the decrease of the LFA/HFA ratio with depth is due to chemical reactions involving formation, decomposition and polymerization of fatty acids, and presented possible courses of their formation and transformation in the sediment, as shown in Fig. 252. In their study, a possible formation of $C_{26}-C_{30}$ fatty acids by oxidation of the corresponding fatty alcohols during early diagenesis was stated.

Mono- and polyunsaturated fatty acids, $C_{16:1}$, $C_{16:2}$, $C_{18:1}$, $C_{18:2}$, $C_{18:3\alpha}$, were found in the sediment samples, in particular, in the near surface layers (Kawamura et al. 1978, 1980). The vertical distribution of polyunsaturated fatty acids, as shown in Fig. 253, seems to indicate that there were several times favorable for the production of a larger amount of polyunsaturated fatty acids during the past 6,000 years. This fact might be interpreted in terms of changes of paleoclimate. A colder climate is supposed to be favorable for the preferential production of polyunsaturated fatty acids in aquatic organisms.

Dicarboxylic acids. The normal C_{14}–C_{30} α,ω-dicarboxylic acids were found in the sediment samples (Ishiwatari & Hanya 1975). The total amount of the dicarboxylic acids in the sediments is 0.003–0.06% of TOM and 1/3 to 1/20 of

Table 71. Analytical results of fatty acids for a 5-meter core sample from Lake Biwa (Ishiwatari & Kawamura 1978).

Depth cm	Total carbon mg/g[b]	Total fatty acids[a]		LFA/HFA[d]
		μg/g[b]	% of TOM[c]	
0–5	6.4	86.4	0.80	1.60
10–15	4.2	38.9	0.56	0.91
20–25	4.5	44.9	0.60	0.79
30–35	4.6	35.9	0.47	0.73
40–45	3.6	35.7	0.59	0.77
50–55	4.9	48.7	0.60	1.34
99–104	8.9	79.1	0.53	0.43
150–153	9.2	43.9	0.29	0.90
199–204	9.9	71.5	0.43	0.63
249–254	10.9	77.4	0.43	0.74
300–305	12.6	59.9	0.28	0.43
348–354	12.3	80.4	0.39	0.61
395–403	13.4	60.2	0.27	0.45
448–453	12.8	79.4	0.37	0.54
468–475	13.5	76.1	0.34	0.47

[a]Total saturated mono-carboxylic acids.
[b]dry sediment.
[c]TOM = Total carbon × 1.67.
[d]$(C_{16}+C_{18})/\Sigma C_{20}-C_{30}$ (see text).

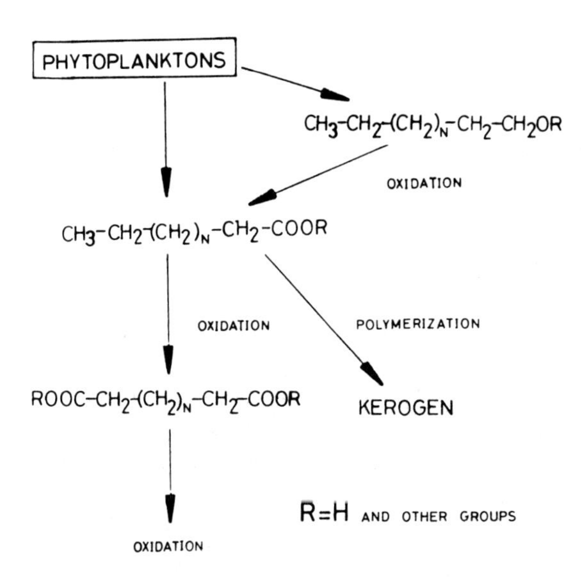

Fig. 252. Possible pathways of formation and transformation of fatty acids in Lake Biwa sediment (after Ishiwateri & Kawamura 1977).

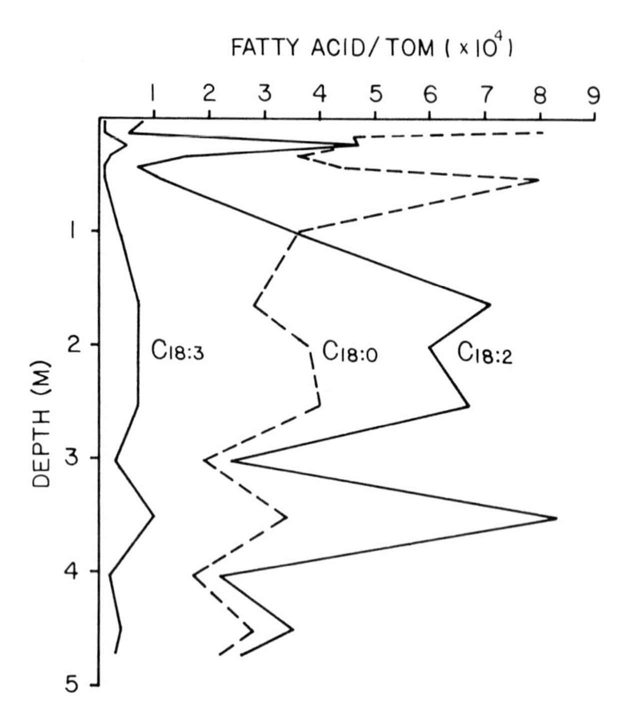

Fig. 253. Vertical distribution of C_{18} saturated and polyunsaturated fatty acids in Lake Biwa sediment. TOM content is the same as in Table 71 (Kawamura & Ishiwatari 1981).

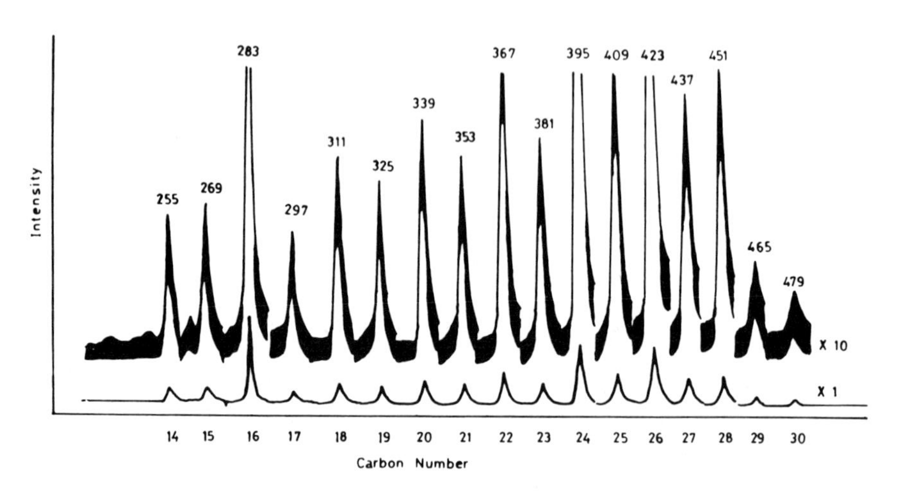

Fig. 254. Gas chromatogram (Mass fragmentogram) of dicarboxylic acid methyl esters from Lake Biwa sediment (100 m in depth) (Ishiwatari 1975).
Conditions the same as in Fig. 250. Number indicates the mass number (m/e) scanned, which corresponds to M-31 of each ester.

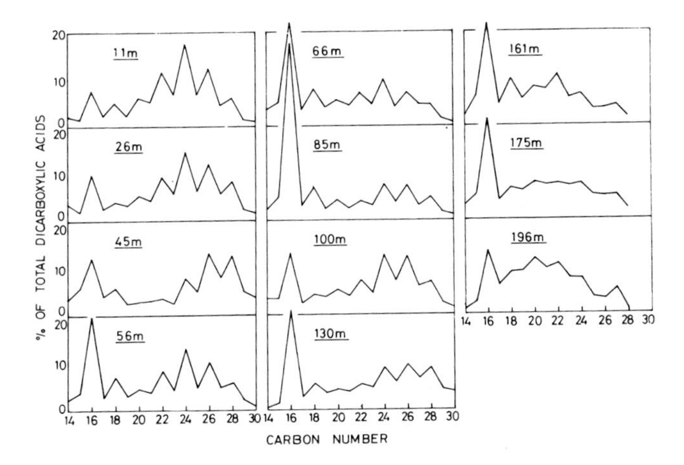

Fig. 255. Vertical variation of α,ω-dicarboxylic acid distribution in Lake Biwa sediment (Ishiwatari 1975, Ishiwatari & Hanya 1975).

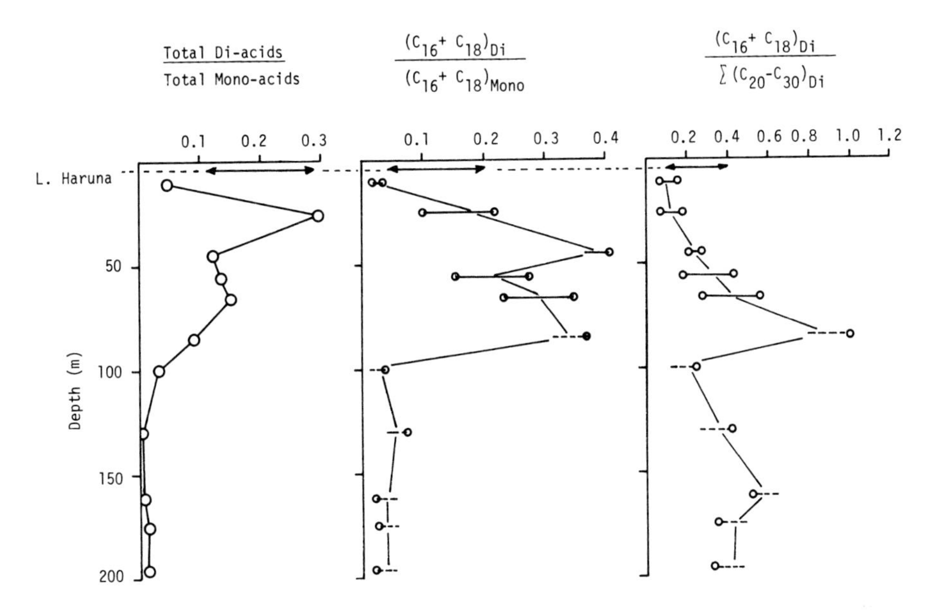

Fig. 256. Variation of various parameters with depth in Lake Biwa sediment (Ishiwatari 1976b).

the monocarboxylic acid fraction, as given in Table 70. Figs. 254 and 255 show a typical gas chromatogram of dicarboxylic acids and the vertical variation of their relative abundance in Lake Biwa sediment, respectively. The distribution patterns have two maxima at C_{16} and C_{24}–C_{28} regions and are quite similar to those of monocarboxylic acids, indicating a close relation in origin between mono- and dicarboxylic acids. A possible course of the formation of α,ω-dicarboxylic acids is omega oxidation of the corresponding monocarboxylic acids, as shown in Fig. 252. The occurrence of this mechanism in Lake Biwa sediments seems to be supported by the fact that the ratio of total dicarboxylic acids to the total monocarboxylic acids as well as the ratio of n-C_{16} and n-C_{18} dicarboxylic acids to n-C_{16} and n-C_{18} monocarboxylic acids increase with depth in upper layers and become low in deeper layers, as shown in Fig. 256 (Ishiwatari 1976b).

Sterols

Since Schwendinger & Erdman (1964) identified sterols in recent sediments, sterols have been found in various sediments (Attaway & Parker 1970; Henderson et al. 1971; Ogura & Hanya 1973; Gaskell & Eglinton 1974).

As they are widely distributed in both oceanic and terrestrial biota, sterols in the sediments have been thought to be the marker of origins in the environments, so called chemical fossils.

In the sediments of Lake Biwa, both of surface and core sediments, representative sterols found in other sediments such as cholesterol, brassicasterol, campesterol, stigmasterol, β-sitosterol were determined. In addition to them, large amounts of stanols (saturated sterols) such as 5α-cholestan-3β-ol (cholestanol), 24-methylcholestan-3β-ol (campestanol), 24-ethylcholestan-3β-ol (stigmastanol), 5β-cholestan-3β-ol (coprostanol) and 5β-ethylcholestan-3β-ol were also detected. The latter two stanols, 5β-cholestan-3β-ol and 5β-ethylcholestan-3β-ol, are usually found only in areas highly polluted by feces of man and domestic animals. In the 200-m core sample of Lake Biwa, whether the 5β-stanols are the evidence of mammals or only the transformation products of cholestanol and stigmastanol is not yet determined.

The ratios of cholestanol to the summation of cholesterol and cholestanol rapidly increased with the depth of core sediments and reached to 80% or more at the depth of 105 m, as shown in Fig. 257. Steel & Henderson (1972) found

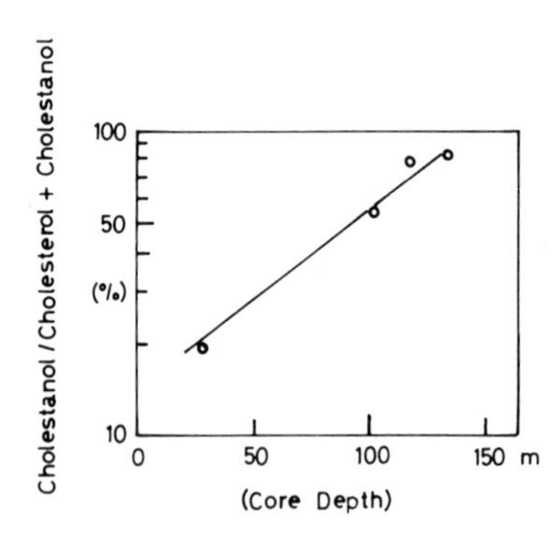

Fig. 257. Variation of the ratios of cholestanol to the sum of cholesterol and cholestanol with depth (Ogura & Hanya, 1973).

only saturated sterol, both of 5α- and 5β-stanols in Green River Shale. The mechanisms of saturation of these sterols in such old sediments cannot be interpreted by biological degradation or higher resistivity of saturated sterols, which are transformed from unsaturated sterols by organisms at early stages or derived directly from aquatic biota, than unsaturated sterols. It must be the results of hydrogenation by deep layer diagenesis.

The formation rate of cholestanol from cholesterol was calculated as follows on the assumption that the age of the 200 m core sample is 0.5 m.y.:

$$X = X_0 \cdot e^{2.8t}$$

where X is the percentage of cholestanol in the summation of cholesterol and cholestanol at several depths, X_0 is that of the depth of 14 m, and t is the age of the sediments in the unit of million years.

The degradation rate of sterols in the core sample, though the fluctuation of the contents which depend on environments is as shown in Fig. 258, was calculated also as follows:

$$Y = Y_0 \cdot e^{-2t}$$

where Y is the concentration of sterols at several depths, Y_0 is the initial concentration of the line, in which sterols are assumed to decrease according to the first-order kinetics, and t is the age of the sediments in the unit of million years.

Cholesterol has been widely found in aquatic animals and plants, but is dominant in animals: for example, the concentration of cholesterol in zooplankton (major: *Bosmina*) was 12.4 mg/g in dry weight and the percentage of cholesterol in total sterols was 92.6%. On the other hand, brassicasterol is

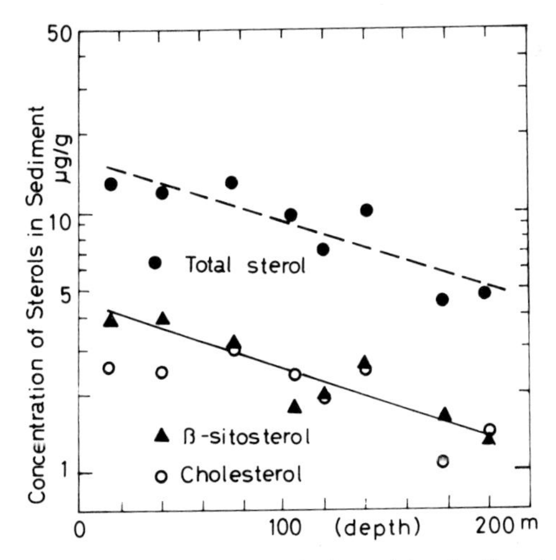

Fig. 258. Variation of the concentrations of sterols in dry weight of sediments with depth (original data were given by Ogura; in Horie 1972).

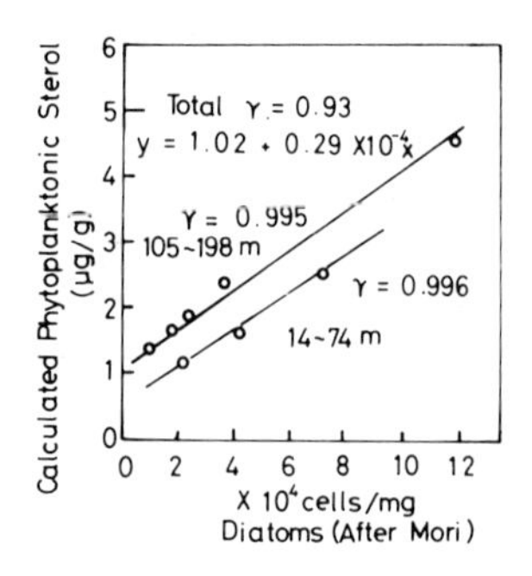

Fig. 259. Correlation between diatoms and calculated phytoplanktonic sterol (Ogura 1977).

dominant in diatoms: for example, in *Asterionella*, the concentration of brassicasterol was 374 μg/g in dry weight and the percentage was 50.3% in total sterols. β-sitosterol is mainly derived from terrestrial plants and is a major component of soil sterols. From relative abundances of each of the sterols in zooplankton, diatoms and soils, their contributions in the core sample were calculated (Ogura 1976, 1977). There was a good correlation between the numbers of fossil diatoms (Mori, Sh. 1974) and the contents of diatom's sterols as shown in Fig. 259. In the figure, two regression lines were obtained. One is constructed with the sediments from 14 m to 74 m in depth and the other corresponds to deeper sediments. The two lines, obtained from the calculation, seem to differ mainly due to the differences of species of diatoms in the sediments. The species of fossil diatoms in the core sample of the upper sediments (from 3.3 m to 74.5 m) is dominated by *Melosira solida, Stephanodiscus carconensis* and var. *pusilla* and the assemblage is correlated to that of Lake Biwa in recent times (Mori, Sh. 1974). In Fig. 259, the depth of the abrupt changes of the species observed by sterols and fossil diatoms and the good correlation between the numbers of fossil diatoms and the contents of diatoms' sterols prove that organic compounds such as sterols in old sediments should be chemical fossils which represent the aquatic environments of the ages.

Lipids

Lipids in sediments refer to substances soluble in benzene, chloroform, petroleum ether, diethyl ether, or similar organic solvents, and include hydrocarbons, chlorinoid pigments, fatty acids and esters, carotenoid pigments, resin alkaloids, heterocyclic compounds and others. Lipids in lake sediments are derived from not only aquatic biota (autochthonous) but also wax of higher plants (allochthonous) in the drainage area in that age. Lipids in surface sediments are more abundant in an eutrophic lake than in an oligotrophic lake and show the productivity of the lake. But the high productivity in the lake may not always coincide with the high temperature. Furthermore, they are subject to degradations at early age by microorganisms and, sometimes, to reproductions arising from other substances such as kerogens in deeper sediments. In the 200 m core sample, the latter case cannot be observed at the present time.

Swain (1970a) has listed the lipid materials in non-marine source organisms expected to occur in sediments, and also shown the benzene-methanol (8:2) extractable matter in peat deposits, oligotrophic lakes and eutrophic lakes of the glaciated region (6.5%, 0.30% and 0.83% respectively in weight). Smith (1954) has reported on the organic matter extracted from the fresh water sediments of Stony Lake and Lake Wapalann NJ, (11,200 and 2,230 ppm in weight, respectively). As the carbon contents of the organic materials mentioned above are estimated to be 50%, the lipids of the sediments are about 5,600 and 1,100 μg/g in carbon. In Japan, the lipids materials in the sediments of Lake Haruna are about 3,000–1,000 μg-C/g and about 5.0–2.0% of total organic carbons.

Lipids concentrations were determined in the sediments of the 200 m-long core sample of Lake Biwa, which had been extracted with chloroform-methanol (2:1) at intervals of 25 cm and partly at intervals of 50 cm. In this study, to estimate the paleoclimate, the relation between the lipids and total numbers of fossils, such as diatoms and pollens, carbon isotopic ratios and other organic compounds, is to be discussed. The details of the analytical method used on lipids were previously described (Ogura & Ishiwatari 1975).

Fig. 260 shows the vertical changes of total lipids concentrations at every one meter observed in the 200 m long core sediments and total numbers of diatoms (Mori, Sh. 1974) and pollens (personal communication with Dr. N. Fuji).

The relation of diatoms and total lipids are shown in Fig. 261a. The values of lipids are chosen from the one meter average concentrations nearest to the depth in which fossils, carbon isotopic compositions and other organic compounds were analysed. In the upper part of the core sample from 10 m to 79 m, an approximately linear relationship exists between them (closed circle in the figure), but that relationship cannot be observed below the depth of 83 m. The relationship between the percentages of total lipids to total organic cargon (L-C/TOC) and the numbers of diatoms is plotted in Fig. 261b. No marked correlation can be observed as a whole between them. However, it seems that two groups can be distinguished in the figure. One group shows relatively low and constant L-C/TOC as against the change of numbers of diatoms. Another group shows a negative correlation between L-C/TOC and numbers of diatoms.

The relation between numbers of pollens and total lipids is also shown in Fig. 262a. Lipids concentrations were not noticeably affected by large amounts of pollens. The same pattern is observed in the relation of L-C/TOC with pollens as in the case of diatoms. These two kinds of fossils may be part of the sources of lipids, but this result indicates that the contributions of these fossils on the lipids must be considered as to the species, sizes, lipids contents and the differences of rates of degradation between fossils and lipids in addition to the numbers.

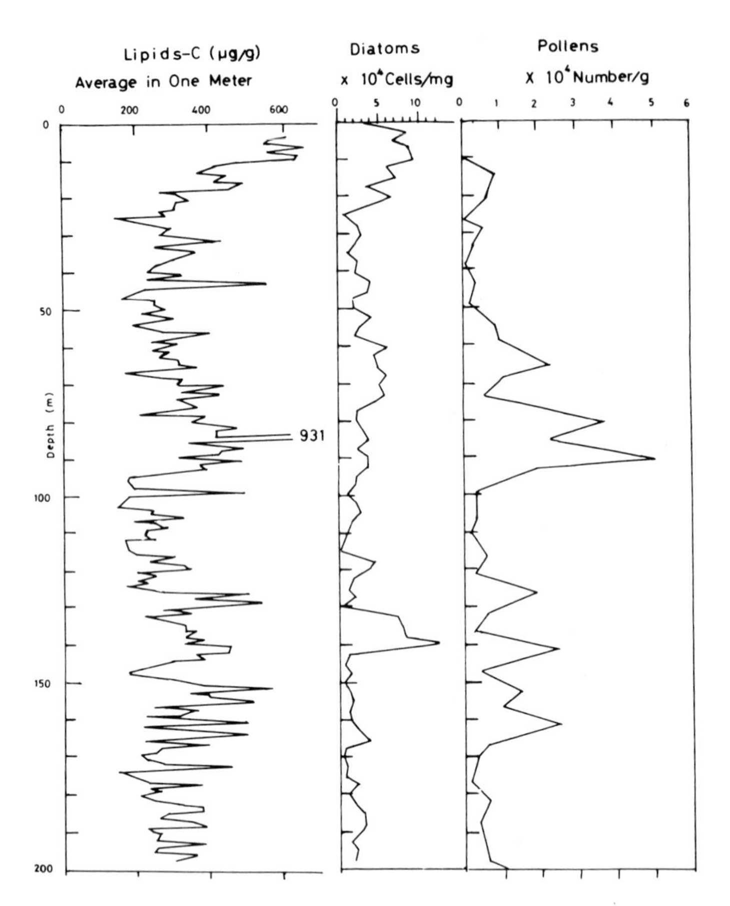

Fig. 260. The variation of average lipids carbon in one meter, cells of diatoms (After Mori 1974) and absolute number of pollens (After Fuji) with depth in the 200 m core sample of Lake Biwa (Ogura 1978).

The relationship between L-C/TOC and carbon isotopic composition of organic carbon $\delta^{13}C$ is shown in Fig. 263 (Nakai 1972). The low value of $\delta^{13}C$ must be derived from materials which contain lipids components. If the carbon isotopic composition shows the paleotemperature as described by Nakai (Nakai 1973), the L-C/TOC may play an important role in the estimation of the paleotemperature.

In his review of the lipids in marine water, Jeffrey (1970) reports that the chloroform extracts from the semi-tropical water contain 10–20% of total dissolved organic carbon. On the other hand, the chloroform extracts from the Antarctic oceanic water contain 40–50% of total dissolved organic carbon and 0.28–0.32 mg-C/liter as a lipid material. The oceanic lipid concentrations in the Gulf of Mexico are about half of those in the Antarctic where 0.12–0.16 mg-C/liter constitutes 15–20% of total dissolved organic carbon. Plankton samples

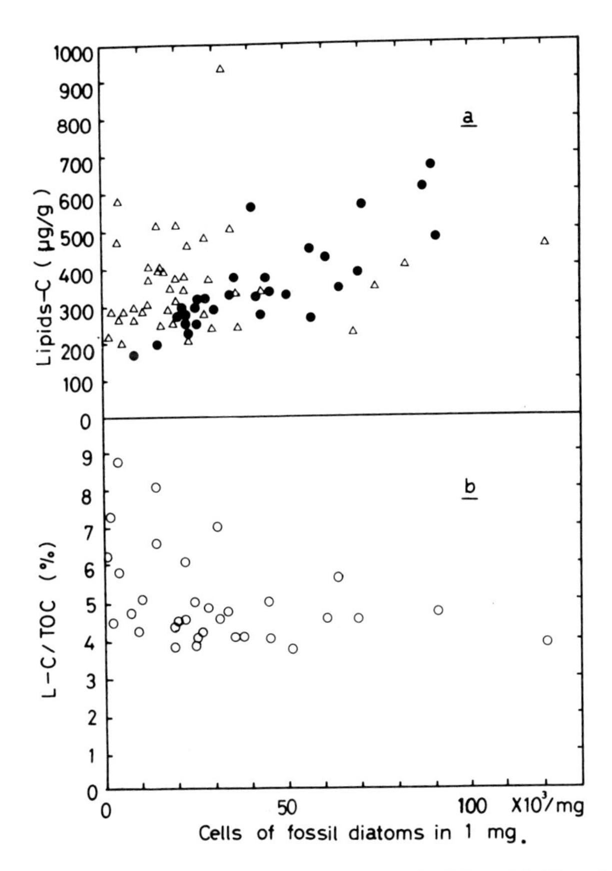

Fig. 261. The relationship between cells of diatoms (After Mori 1974) and lipids carbon in the upper figure a and the ratio of lipids carbon to total organic carbon (L-C/TOC) in the lower figure b (Ogura 1978).

from the Gulf of Mexico contain 10–20% lipids on the dry weight basis, while Antarctic plankton samples contain 20–40% lipids in dry weight. High concentrations of lipids in such cold waters are due not only to the total productivities but to the lipid percentages of biomass in the waters. In addition, the rate of degradation of lipids is considered to be low in such cold waters.

If the observation that the lipids contents in organic carbon are high in a cold region can be applied to the sediments of Lake Biwa and if the organic materials that remain in the sediments reflect the paleoenvironments which are unchanged in their relative abundance, L-C/TOC should show the paleoclimates. However, the total organic carbon in the lake sediments is composed of both autochthonos and allochthonous organic matter. For this reason, the summation of carbohydrates and proteins carbon, which are the chemical fossils of biological production, is a factor more suitable than total organic carbon. The ratios of lipids carbon to the summations of carbohydrates and proteins carbon are

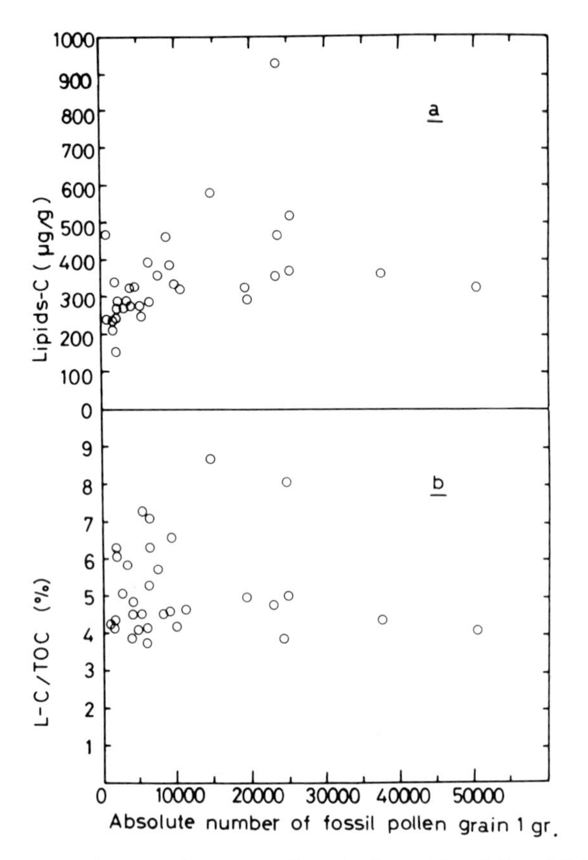

Fig. 262. The relationship between absolute number of pollens (After Fuji) and lipids carbon in the upper figure a and the ratio of lipids carbon to total organic carbon (L-C/TOC) in the lower figure b (Ogura 1978).

calculated with the data of carbohydrates and proteins carbon being used in accordance with the results of Handa (Handa 1972, 1973). The vertical profiles of the ratios of lipids carbon to the summations of carbohydrates and proteins carbon L-C/CP and L-C/TOC with depth are shown in Fig. 264. The ratios of Σ organic carbon, which includes lipids, carbohydrates and proteins, to total organic carbon in the 200 m-core sample have been discussed by Handa previously (Handa 1973), but the present writer proposes the ratio of lipids/carbohydrates and proteins as a more reasonable factor of the paleotemperature. The value of L-C/CP increases with increasing depth because of the differences of rates of degradation between lipids and carbohydrates and proteins. The relatively high values of L-C/CP, which shows the cold climates, in the 200 m-long core sample, are obtained from the depth of 19 m, 49 m, 85 m, 104 m, 110 m, 130 m, 151 m, 160 m, 186 m and 197 m. A comparison of this data with carbon isotopic composition and the data of granulometry (Yamamoto 1974a)

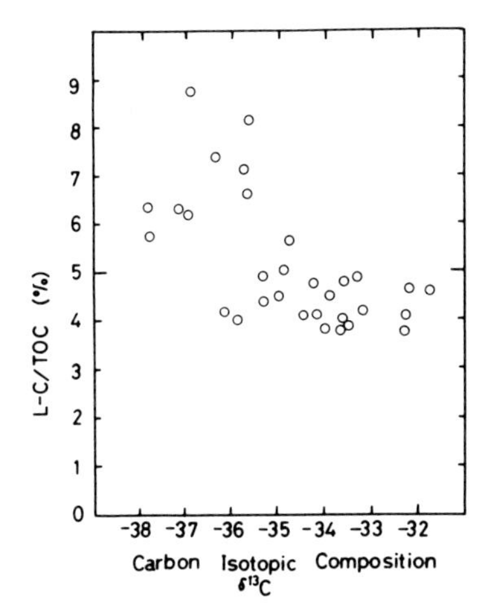

Fig. 263. The relationship between carbon isotopic composition of total organic carbon (After Nakai 1972) and the ratio of lipids carbon to total organic carbon (L-C/TOC) in the 200 m core sample of Lake Biwa (Ogura 1978).

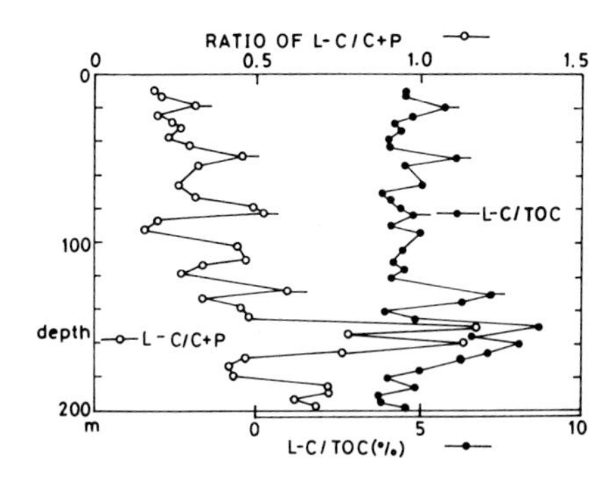

Fig. 264. The vertical variation of the ratio of lipids carbon to the summation of carbohydrate and protein carbon (L-C/C+P) (After Handa 1972 and 1973) compared with the variation of the ratio of lipids carbon to total organic carbon (L-C/TOC) in the 200 m core sample (Ogura 1978).

produces a good agreement between them, but some ambiguities are found below the depth of 151 m. It is expected that, in the near future, the analysis of the sediments at the same depth will eliminate this discrepancy.

Humic compounds and kerogen

In a 200 meter core sample of Lake Biwa, 40 – 80% of the total organic matter has not been determined as specific compounds (Handa 1972, 1973). They belong to so-called humic compounds and kerogen, and their percentage in the total organic matter in the sediment increases gradually with depth, as shown in

Fig. 265. Kerogen and humic acid are formed possibly through polymerization of biolipids, proteins and carbohydrates by unknown mechanisms. They are more resistant to bacterial and chemical decomposition than other organic compounds. Bccause of these characteristics, the distribution and properties of kerogen and humic acid are believed to preserve information of paleoenvironments.

Kerogen and humic acid were isolated from some sediment samples of Lake Biwa and characterized (Ishiwatari 1977). Table 72 shows the amount of humic acid and kerogen isolated from various layers. Small amounts of humic acid are present in the 11 m layer. In the layers deeper than 45 m, humic acid can no longer be isolated. Their percentage of kerogen actually isolated to the residual organic matter (ROM: Total organic carbon minus the amounts of lipid-, protein- and carbohydrate-carbon) increases with depth. For the 130 m layer, 64% of total organic matter (TOM) was actually isolated as kerogen, corresponding to 80% of ROM. These facts indicate that polymerization of organic materials in the sediment has been gradually occurring with time.

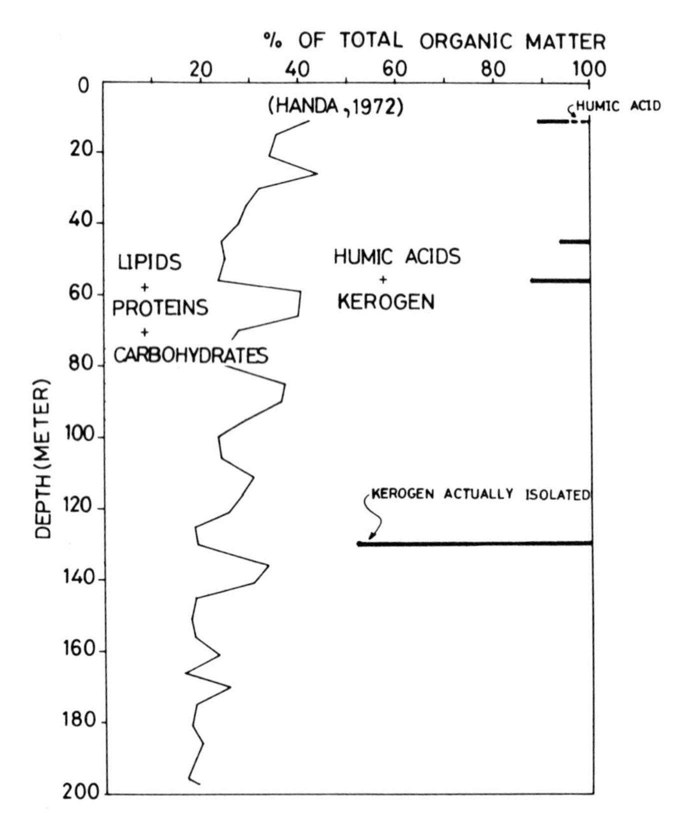

Fig. 265. Vertical changes of organic composition and the amount of kerogen isolated in Lake Biwa sediment (afte Ishiwatari 1976c).

Table 72. The amount of humic acid and kerogen isolated from sediment samples of Lake Biwa (Ishiwatari 1977).

Depth m	Total organic matter[a] mg/g	Kerogen		Humic acid	
		mg/g (as free basis)	% of TOM	mg/g	% of TOM
11	16.2	1.00	6.2	0.70[b]	4.3
45	10.2	0.59	5.8	0.0	0.0
56	11.0	1.31	11.9	0.0	0.0
130	6.6	4.2	64.0	0.0	0.0

[a]TOM = (Total organic carbon determined by Handa (1972)) × 1.6.
[b]Ash content was not determined.

Table 73. Elemental composition and $\delta^{13}C$ of the kerogens from Lake Biwa (Ishiwatari 1977).

Depth	C %	H %	Atomic H/C ratio	$\delta^{13}C$ of kerogen (A) ‰	$\delta^{13}C$ of total organic matter (Nakai 1972) (B) ‰	Difference (A)–(B) ‰
11	61.54	6.57	1.27	−24.6	−31.8	7.2
45	64.06	6.92	1.29	−27.5	−36.2	8.7
56	64.60	6.56	1.21	−25.2	−33.9	8.7
130	–	–	–	−25.5	−36.4	10.9

Table 73 gives results of elemental and $\delta^{13}C$ analyses of the kerogens. Carbon content of the lacustrine kerogens is 62–65%, which is much higher than those for kerogens from marine sediments (52–56%; Ishiwatari et al. 1978) and humic acids from other lacustrine surface sediments (46 – 57%; Ishiwatari 1967). The atomic H/C ratio of the Lake Biwa kerogens is 1.2 – 1.3. This value is slightly lower than those for a marine kerogen less contaminated by terrestrial materials (1.4), but higher than that for kerogen from a bay sediment (1.1: Ishiwatari et al. 1978).

The kerogens isolated show a $\delta^{13}C$ range of −25 to −28‰. These values fall in the range previously reported for TOM in lake and river sediments, and also in lacustrine plants and animals (Nakai 1973). However, $\delta^{13}C$ of the kerogens are 7 – 11‰ heavier than those reported by Nakai (1973) for TOM of the corresponding sediment layers. So far, the reason for the difference is unknown.

Infrared spectroscopy revealed the presence of the following functional groups in the kerogens: OH/NH (3,400 cm^{-1}), aliphatic CH_2 and CH_3 (2,940–2,840 and 1460 – 1380 cm^{-1}), C=O groups of ketone, quinone groups and/or aromatic C=C groups (1,700 – 1,600 cm^{-1}), peptide bonds (1,540 cm^{-1}) and C–O of acid, ester or alcohol and C–O of carbohydrates (1,200 – 1,100 cm^{-1}). The 1,540 cm^{-1} band is characteristic of sedimentary humic compounds and kerogens mainly contributed by autochthonous materials. The above results on the kerogens indicate that they are largely autochthonous.

Paleogeochemistry of mercury

Keiichiro Fuwa

Mercury is one of the unusual metals which has a liquid form at room temperature with a very low melting point of −38.89°C and a rather high specific gravity of 13.5. Consequently, mercury behaves uniquely in nature. The atomic number of 80 shows that this element was evolved later in the process of chemical evolution and, therefore, its abundance in the universe is necessarily limited. The Clarke number on the earth's crust, 0.2 ppm, reflects this condition.

In the human history, mercury has been known from the ancient age in connection with primitive religions both in the West and East, and actually used for metallurgy since 2,300 years ago.

Some of the geochemical concentrations, such as the accumulation in the air or soil near or above the noble metal mine areas, have been observed and reported, as well as those as one of the environmental pollutants in the very recent history. But there is not much of analytical results of mercury for us to discuss its paleogeochemistry, partially because of the difficulty of analyzing this element before the atomic absorption method was developed in recent years. The results of the mercury contents shown here for the 1,000 m core of Lake Biwa sediments indicate roughly that, over the past 1,000,000 years, there has been no drastic change of mercury content in the environment near this area, except for the pollution in modern civilization (Table 74).

This may have resulted from the volatile and mobile nature of this element.

The development of a more precise and more sensitive analytical method will reveal more accurate knowledge of the behavior of mercury in ancient times.

Horie, S. Lake Biwa

ISBN 90 6193 095 2. Printed in The Netherlands

Table 74. Mercury contents in 1,000-meter core samples from Lake Biwa. (Takahashi et al., 1977, 1979)

Sample No. (Depth: m)	(ppm*)	No. (Depth: m)	(ppm*)	No. (Depth: m)	(ppm*)	No. (Depth: m)	(ppm*)
33-1 (252.0)	0.034	85-2 (312.3)	0.085	113-1 (340.9)	0.075	139-5 (362.5)	0.035
33-2 (252.6)	0.042	85-3 (312.6)	0.088	113-2 (341.2)	0.049	141-1 (363.3)	0.086
37-2 (261.3)	0.058	85-9	0.057	115-2 (345.3)	0.036	143-3 (363.9)	0.068
39-2 (262.2)	0.077	85-13	0.048	117-2 (346.6)	0.055	143-6 (364.2)	0.033
41-1 (266.2)	0.114	85-16	0.035	117-5 (346.0)	0.073	145-2 (364.5)	0.054
41-2 (266.5)	0.075	87-1 (313.2)	0.045	117-7 (346.3)	0.089	147-3 (365.5)	0.082
43-1 (268.5)	0.042	87-2 (313.6)	0.086	121-1 (347.4)	0.083	149-4 (366.4)	0.037
49-2 (281.3)	0.060	89-1 (318.4)	0.106	121-2 (347.6)	0.054	149-6 (366.6)	0.058
51-5 (283.2)	0.068	91-0 (318.9)	0.070	123-2 (347.8)	0.044	149-8 (366.9)	0.028
53-2 (284.1)	0.043	93-1 (321.4)	0.059	123-4 (348.1)	0.048	151-3 (367.2)	0.059
59-2 (297.2)	0.075	95-2 (322.1)	0.080	123-6 (348.4)	0.040	151-6 (367.5)	0.043
61-2 (299.9)	0.029	95-6 (322.5)	0.095	125-5 (349.1)	0.065	151-8 (367.7)	0.046
63-3 (300.3)	0.052	95-10 (322.8)	0.068	127-2 (356.1)	0.033	153-2 (369.2)	0.113
69-2 (301.8)	0.065	97-1 (326.6)	0.044	127-4 (356.4)	0.038	153-4 (369.5)	0.065
75-1 (305.3)	0.061	99-1 (326.9)	0.046	129-3 (357.1)	0.062	153-6 (369.8)	0.043
79-2 (307.5)	0.087	99-5 (327.2)	0.042	129-5 (357.7)	0.060	155-4 (370.2)	0.041
79-4 (307.9)	0.060	99-8 (327.8)	0.078	131-6 (359.1)	0.061	159-2 (375.2)	0.044
81-2 (310.4)	0.086	101-3 (328.2)	0.090	131-8 (359.3)	0.047	159-5 (375.6)	0.039
81-5 (310.9)	0.066	101-7 (328.5)	0.044	133-2 (359.5)	0.049	159-8 (375.9)	0.070
83-1 (311.3)	0.079	103-2 (329.0)	0.042	133-4 (359.7)	0.037	161-1 (377.8)	0.047
83-2 (311.4)	0.055	105-2 (329.3)	0.054	133-6 (359.9)	0.049	161-3 (378.1)	0.043
83-3 (311.6)	0.046	105-5 (329.7)	0.055	135-4 (360.8)	0.030	161-5 (378.5)	0.066
83-4 (311.7)	0.060	107-2 (330.3)	0.040	137-2 (361.3)	0.019	161-7 (379.0)	0.036
85-1 (312.3)	0.071	107-4 (330.5)	0.019	137-4 (361.4)	0.052	163-2 (379.2)	0.017
85-1 (312.2)	0.122	111-1 (340.4)	0.050	139-3 (362.2)	0.054	165-2 (379.8)	0.026

Numerals in the parentheses denote depth in the core.
* $\mu g/g$ wet sample.

Table 74. continued

Sample No. (Depth: m)	(ppm*)	No. (Depth: m)	(ppm*)	No. (Depth: m)	(ppm*)	No. (Depth: m)	(ppm*)
165-6 (380.2)	0.158	189-2 (402.3)	0.069	213-1 (421.4)	0.105	231-1 (438.5)	0.071
167-2 (380.4)	0.034	189-4 (402.5)	0.056	213-3 (421.5)	0.106	233-1 (439.9)	0.080
167-4 (380.6)	0.037	189-6 (402.7)	0.057	213-5 (421.8)	0.043	233-2 (441.0)	0.089
167-6 (380.8)	0.081	193-2 (404.4)	0.052	213-7 (422.1)	0.083	233-3 (441.1)	0.038
167-8 (381.0)	0.061	195-2 (405.3)	0.038	215-1 (422.2)	0.033	233-4 (441.2)	0.082
167-10 (381.2)	0.022	195-4 (405.5)	0.026	215-3 (422.5)	0.035	233-6 (441.4)	0.054
169-1 (381.9)	0.065	195-6 (405.7)	0.044	215-5 (422.7)	0.041	235-1 (441.7)	0.041
171-1 (382.3)	0.067	197-1 (405.8)	0.058	217-1 (423.7)	0.053	235-2 (441.8)	0.061
171-2 (382.4)	0.042	197-6 (406.9)	0.067	217-4 (424.1)	0.036	235-4 (442.0)	0.042
179-3 (391.3)	0.041	199-1 (407.2)	0.076	217-6 (424.7)	0.043	235-5 (442.1)	0.066
179-6 (391.6)	0.048	199-3 (407.4)	0.060	219-1 (428.7)	0.066	235-6 (442.3)	0.054
179-8 (391.8)	0.037	199-5 (407.6)	0.089	219-2 (428.8)	0.037	237-1 (443.7)	0.051
181-6 (392.2)	0.026	199-7 (407.8)	0.036	219-4 (429.0)	0.047	237-3 (443.9)	0.051
183-1 (393.8)	0.068	199-8 (408.1)	0.058	219-5 (429.1)	0.017	239-1 (444.0)	0.035
183-2 (394.0)	0.052	203-2 (410.4)	0.071	221-2 (429.4)	0.075	**239-3 (444.2)**	0.093
183-4 (394.2)	0.039	203-3 (410.5)	0.047	221-4 (429.6)	0.052	239-5 (444.3)	0.047
183-5 (394.3)	0.050	203-5 (410.7)	0.088	221-6 (429.8)	0.017	239-7 (444.5)	0.044
183-7 (394.6)	0.056	203-6 (410.8)	0.038	221-8 (429.9)	0.043	239-9 (444.7)	0.030
185-1 (396.7)	0.055	203-8 (410.9)	0.023	223-1 (431.1)	0.043	239-11 (445.0)	0.022
185-3 (396.9)	0.066	205-1 (411.1)	0.061	225-1 (431.9)	0.033	241-1 (445.2)	0.125
187-1 (397.2)	0.085	205-3 (411.2)	0.053	225-3 (432.2)	0.041	241-3 (445.3)	0.072
187-2 (397.3)	0.086	205-5 (411.5)	0.026	227-1 (434.2)	0.043	241-5 (445.5)	0.030
187-3 (397.4)	0.049	209-2 (415.8)	0.049	227-3 (434.4)	0.045	241-7 (445.8)	0.039
187-5 (397.6)	0.051	211-2 (421.1)	0.030	227-5 (434.5)	0.043	241-9 (446.0)	0.024
187-7 (397.8)	0.108	211-3 (421.2)	0.066	227-8 (435.0)	0.081	243-1 (446.8)	0.043

Table 74. continued

Sample No. (depth: m)	(ppm*)	No. (Depth: m)	(ppm*)	No. (Depth: m)	(ppm*)	No. (Depth: m)	(ppm*)
243-2 (446.9)	0.038	259-3 (460.7)	0.060	271-5 (466.1)	0.017	589-2 (784.6)	0.070
243-4 (447.1)	0.114	259-7 (461.1)	0.062	271-8 (466.4)	0.034	593-3 (786.2)	0.054
243-6 (447.3)	0.035	259-8 (461.2)	0.095	∼∼∼∼∼∼∼∼		597-3 (787.9)	0.050
245-1 (447.4)	0.062	261-1 (461.3)	0.026			603-2 (790.3)	0.084
245-3 (447.6)	0.045	261-3 (461.5)	0.019	353-7 (546.6)	0.068	607-3 (792.0)	0.053
235-5 (447.8)	0.030	261-5 (462.0)	0.062	501-3 (706.2)	0.072	611-4 (794.6)	0.040
247-1 (449.6)	0.057	263-2 (462.3)	0.066	505-3 (712.8)	0.104	617-4 (796.9)	0.126
247-3 (449.8)	0.048	263-4 (462.5)	0.052	513-2 (724.4)	0.050	621-2 (798.6)	0.061
247-6 (450.0)	0.055	263-5 (462.6)	0.081	519-2 (729.6)	0.223	625-2 (799.4)	0.026
249-2 (450.3)	0.042	263-7 (462.8)	0.045	523-3 (736.3)	0.069	629-2 (801.3)	0.114
249-3 (450.4)	0.040	263-10 (463.1)	0.067	527-5 (738.1)	0.055	635-3 (803.7)	0.035
249-6 (450.6)	0.024	265-2 (463.3)	0.028	531-3 (740.0)	0.044	639-1 (804.7)	0.031
249-8 (450.8)	0.028	265-4 (463.5)	0.019	535-3 (741.9)	0.055	641-2 (805.8)	0.041
251-1 (453.5)	0.082	265-6 (463.6)	0.054	539-2 (743.5)	0.068	645-1 (807.4)	0.015
251-3 (453.8)	0.054	265-8 (463.9)	0.070	543-4 (749.0)	0.056	649-2 (809.8)	0.026
253-1 (454.0)	0.098	267-2 (464.1)	0.074	547-3 (750.5)	0.044	653-2 (811.0)	0.094
253-3 (454.2)	0.068	267-3 (464.2)	0.053	551-1 (753.4)	0.035	657-3 (812.7)	0.050
253-5 (454.3)	0.072	267-5 (464.5)	0.050	555-3 (755.2)	0.039	661-2 (814.5)	0.113
253-7 (454.5)	0.070	267-7 (464.7)	0.028	559-3 (757.4)	0.047	665-4 (817.5)	0.062
253-9 (454.7)	0.043	269-1 (465.0)	0.053	565-4 (763.0)	0.062	669-1 (818.4)	0.025
253-11 (454.9)	0.042	269-3 (465.2)	0.024	569-3 (765.9)	0.064	673-1 (819.8)	0.034
257-2 (458.0)	0.026	269-4 (465.3)	0.061	573-3 (768.3)	0.026	679-1 (822.0)	0.081
257-4 (458.2)	0.040	269-6 (465.5)	0.075	577-4 (774.8)	0.046	683-2 (823.7)	0.026
257-6 (458.4)	0.063	271-1 (465.7)	0.045	581-7 (776.7)	0.293	689-1 (826.2)	0.019
257-9 (458.6)	0.057	271-3 (465.9)	0.028	585-4 (783.8)	0.087		

Table 74. continued

Sample No. (Depth: m)	(ppm*)	No. (Depth: m)	(ppm*)	No. (Depth: m)	(ppm*)	No. (Depth: m)	(ppm*)
695-3 (831.5)	0.088	713-3 (840.6)	0.046	729-2 (850.5)	0.017	745-3 (859.7)	0.080
701-3 (834.7)	0.064	717-3 (843.1)	0.048	733-1 (852.5)	0.037	753-3 (865.4)	0.019
705-3 (836.2)	0.051	721-0 (845.4)	0.016	737-2 (854.4)	0.027	759-1 (868.0)	0.033
709-2 (838.9)	0.087	725-4 (848.8)	0.062	741-1 (856.7)	0.035	767-1 (872.3)	0.017

Inorganic chemical aspects

Mutsuo Koyama, Munetsugu Kawashima and
Takejiro Takamatsu

Chemical composition of 200 meter and 1,000 meter core samples

General profile

More than one hundred samples of both the 200 meter and 1,000 meter cores have been analyzed by means of neutron activation and X-ray fluorescence spectrometry. Since numerical data have already been published elsewhere, the writers present the data in a summary form below. In Table 75, the averages and ranges of concentrations of elements determined are shown for comparison. As is obvious, the average contents of K, Cr, La, Th, Hg, and Zn are almost the same in both of the cores. On the other hand, those of Fe, Mn, As, Cs, Co, Sc and Cu are greater in the 200 meter core samples than in the 1,000 meter ones. Especially noticeable are Mn and As contents; that is, in the 200 meter core more than three times as much Mn and As are found as in the 1,000 meter core. As has been discussed in the foregoing section, Mn, As and P are particularly concentrated in the bottom surface of the pelagic part. Furthermore, the concentrations of those elements beneath the bottom surface layer are still kept higher in the pelagic part than in the littoral, although these ions have a tendency to escape from the deeper layers of sediments to the bottom surface.

If those observations apply to the longer cores, namely, the 200 meter and 1,000 meter cores, the paleolimnological situation could be imagined as follows. For example, sediments located at 29.6, 31.5, 36.1, 48.5, 59.8, 174.9, and 191.1 meters in the 200 meter core contain less manganese than 1,000 ppm.

The fact means that these layers are composed of sediments deposited at the time of shallow water depth or that something happened which caused the dilution of manganese in the sediments, possibly floods.

In the following figures, bottom surface samples are marked with open circles and sampling stations must be referred to Kobayashi et al. 1975.

Horie, S. Lake Biwa

ISBN 90 6193 095 2. Printed in The Netherlands

Table 75. Average, minimum and maximum contents of representative elements in the 200 and 1,000 meter core samples from Lake Biwa (Koyama, M. et al. 1976).

Element	200 meter core					1,000 meter core					
	Method	No.	Max.	Min.	Mean	Mean	Min.	Max.	No.	Method	Unit
Na	NAA	158	1.79	.570	.784	.967	.188	1.40	77	NAA	%
K	NAA	149	3.77	1.49	2.77	2.67	1.92	4.46	77	NAA	%
Fe	NAA	143	6.14	4.12	5.01	3.29	1.30	8.93	66	NAA	%
Mn	NAA	158	.463	.072	.166	.050	.016	.245	77	NAA	%
Cs	NAA	136	15.4	9.00	11.7	8.48	1.40	15.4	66	NAA	ppm
Co	NAA	141	42.3	14.8	17.2	11.9	2.20	21.2	66	NAA	ppm
Cr	NAA	140	81.3	48.0	63.4	53.1	26.4	70.0	66	NAA	ppm
La	NAA	121	46.8	32.9	38.3	41.2	28.8	60.8	66	NAA	ppm
Sc	NAA	140	17.6	11.7	15.3	13.2	4.5	16.3	66	NAA	ppm
Th	NAA	140	20.0	12.1	16.6	16.0	4.6	23.7	66	NAA	ppm
As	NAA	83	97	6	35.9	10	5	31	53	XFS	ppm
Cu	AA*	37	66	34	52	26	10	42	53	XFS	ppm
Zn	AA*	37	160	98	133	141	93	201	53	XFS	ppm
Pb	AA*	37	43	18	27	30	10	47	53	XFS	ppm
Hg	AA**	37	220	10	60	56	17	158	251	ZAA	ppb

NAA = Neutron Activation Analysis.
XFS = X-ray fluorescence spectrometry.
ZAA = Zeeman effect Atomic Absorption spectrophotometry.
AA = ordinary Atomic Absorption spectrophotometry.
* = Kobayashi et al. 1975.
** = Itasaka et al. 1973.
No. refers to the number of data treated.

Correlation of distribution of elements in the 200 meter core and in the bottom surface samples

Linear correlation coefficients among elements determined on the bottom surface and the 200 meter core samples have been examined.

In Table 76, correlation coefficients calculated for the bottom surface samples are shown in the area enclosed with a dotted line and those of the core samples in the area enclosed with a solid line, respectively. In the table, coefficients higher than 0.6 in the absolute value are underlined either with a dashed line or with a solid line.

Alkali elements. In the bottom surface samples, no linear correlation among alkali elements can be observed. However, it can be seen that most of the coastal samples which are rich in sandy particles contain less potassium than sodium. In the core samples, the correlation coefficient between sodium and potassium is relatively high and those of sodium with other alkali elements decreases as the atomic number increases. The opposite trend can be noted in the correlations of cesium with other alkali elements, that is, the lighter the alkali element, the smaller the coefficients with cesium.

Alkali elements and others. In the bottom surface samples, sodium is inversely correlated with such elements as manganese, chromium, iron, cobalt and scandium. On the other hand, different profiles can be noticed in the core samples. Namely, positive correlations of sodium with manganese and of potassium with manganese become noticeable. Furthermore, correlations of rubidium and of cesium with such elements as manganese, iron, cobalt and scandium become considerable.

Iron and cobalt. The correlation coefficients in the bottom surface and the core were 0.72 and 0.81 respectively. A close examination of correlation plots presented in Fig. 266 reveals that the coastal samples of the bottom surface are located fairly close to the line drawn by the least squares method, whereas the pelagic ones shift to the Fe-rich side.

Iron, cobalt, chromium and scandium. Both in the bottom surface and in the core samples, correlation coefficients between cobalt and scandium were more than 0.90, an excellent result in view of the difference in chemical nature between the two. It should be pointed out that the linear fits in the bottom surface and in the core samples have different slopes in Fig. 267. It is obvious that the core samples are depleted of cobalt or enriched in scandium. A similar correlation applies to the pair of iron and scandium, as shown in Fig. 268.

Distribution profiles of chromium and scandium between the core and the bottom surface samples are almost the same as shown in Fig. 269. That means

Table 76. Correlation coefficients among contents of elements both in the bottom surface and the core samples (Koyama, M. 1975).

	Na	K	Mn	Fe	Rb	Cs	Sc	La	Ce	Eu	Th	Cr	Co	Sb	
Na	1.00	−0.28	−0.49	−0.78		−0.39	−0.59	0.03	−0.38	−0.51	0.19	−0.72	−0.63	0.00	Na
K	0.76	1.00	−0.42	0.12		−0.21	−0.08	−0.16	−0.07	−0.31	0.03	−0.04	−0.16	−0.19	K
Mn	0.69	0.76	1.00	0.50		0.61	0.27	−0.09	0.02	0.08	−0.03	0.23	0.16	−0.12	Mn
Fe	−0.18	0.03	−0.12	1.00		0.22	0.79	−0.05	0.24	0.38	−0.45	0.64	0.72	−0.11	Fe
Rb	0.60	0.84	0.68	0.55	1.00										Rb
Cs	0.24	0.42	0.18	0.80	0.74	1.00	−0.01	−0.03	0.39	−0.16	0.66	0.04	−0.17	−0.45	Cs
Sc	−0.04	0.23	−0.10	0.91	0.62	0.85	1.00	−0.09	0.25	0.41	−0.55	0.62	0.91	0.18	Sc
La	0.03	0.24	0.23	0.37	0.26	0.34	0.37	1.00	−0.12	−0.49	−0.13	0.02	0.10	−0.10	La
Ce	0.35	0.47	0.18	0.63	0.74	0.77	0.73	0.55	1.00	0.70	0.33	0.75	0.31	0.01	Ce
Eu	−0.06	0.13	−0.10	0.32	0.10	0.21	0.41	−0.04	0.28	1.00	0.28	0.10	0.06	0.26	Eu
Th	0.29	0.46	0.18	0.65	0.75	0.78	0.75	0.66	0.97	0.25	1.00	−0.45	−0.67	−0.59	Th
Cr	−0.12	0.10	−0.10	0.77	0.36	0.76	0.72	0.15	0.42	0.21	0.40	1.00	0.68	0.11	Cr
Co	−0.08	0.18	−0.08	0.81	0.60	0.80	0.90	0.02	0.56	0.33	0.56	0.80	1.00	0.27	Co
Sb	0.02	−0.12	−0.12	−0.11	−0.26	−0.38	−0.18	−0.08	0.13	0.23	0.22	0.11	0.24	1.00	Sb
	Na	K	Mn	Fe	Rb	Cs	Sc	La	Ce	Eu	Th	Cr	Co	Sb	

remark ———: |correlation coefficient| > 0.8
- - - - - -: 0.8 > |correlation coefficient| > 0.6

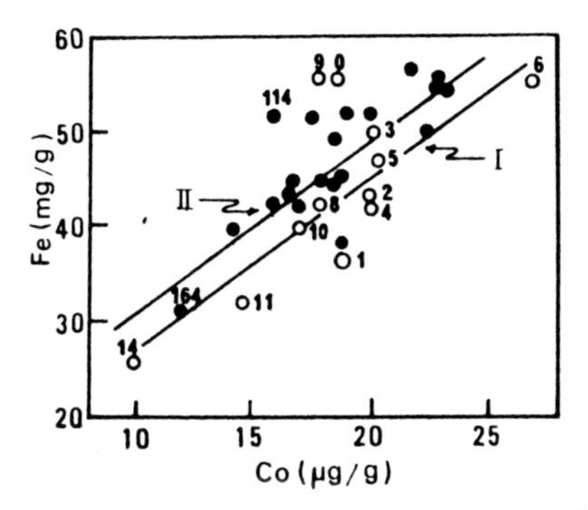

Fig. 266. Correlation between Co and Fe contents. I: for bottom surface samples; II: for core samples (Koyama, M. 1975).

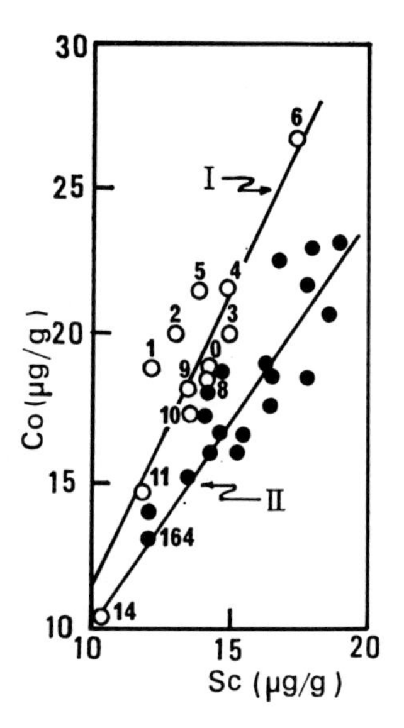

Fig. 267. Correlation between Sc and Co contents. I: for bottom surface samples; II: for core samples. (Koyama, M. 1975).

both the elements behave almost similarly once they have been incorporated in the bottom surface sediment. Accordingly, the relative contents in the core samples as compared to those in the bottom surface samples are in the following order:

$$Sc \gtrsim Cr > Fe > Co.$$

Thorium and scandium. In the core sample, a fairly good correlation coefficient of 0.75 was calculated. In the bottom surface samples, the correlation is not straightforward.

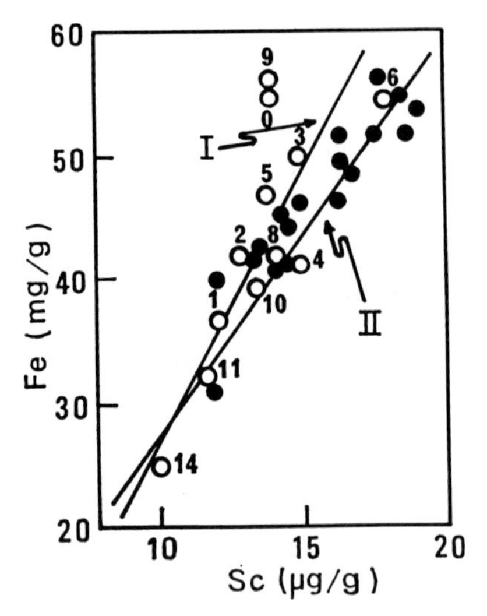

Fig. 268. Correlation between Sc and Fe contents. I: for bottom surface samples; II: for core samples (Koyama, M. 1975).

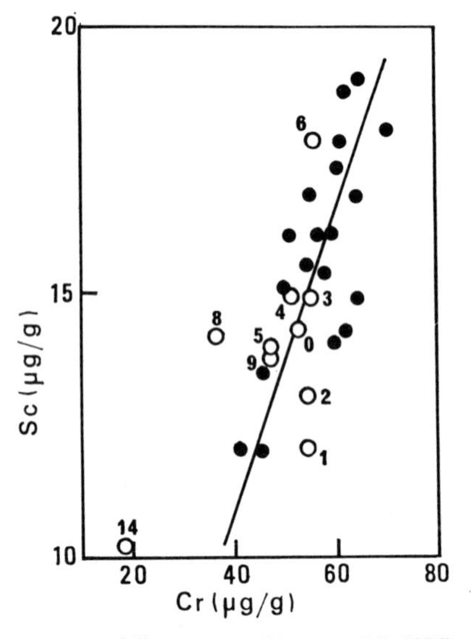

Fig. 269. Correlation between Cr and Sc contents (Koyama, M. 1975).

However, it can be pointed out that a fairly good correlation appears within the part surrounded by a solid line in Fig. 271, in which the bottom surface samples from the northern and eastern coasts and the pelagic part are included.

Those from the southern and western coasts of the lake tend to be located in the thorium-rich side. Since thorium and scandium are not chemically labile as manganese or iron are in the sedimentary environments, the distribution of those elements might indicate the main origin of the materials of sediments.

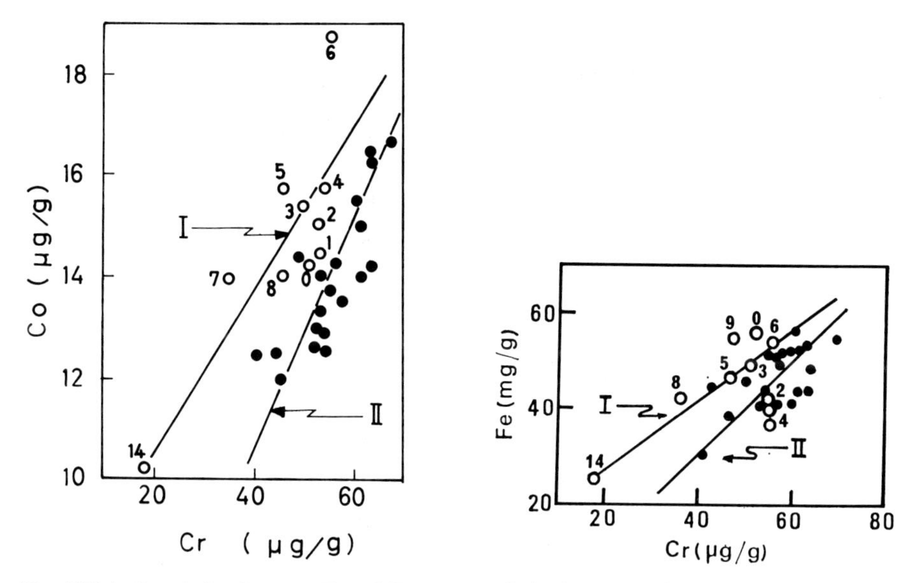

Fig. 270(a). Correlation between Cr and Co contents. I: for bottom surface samples; II: for core samples (Koyama, M. 1975).

Fig. 270(b) Correlation between Cr and Fe contents. I: for bottom surface samples; II: for core samples (Koyama, M. 1975).

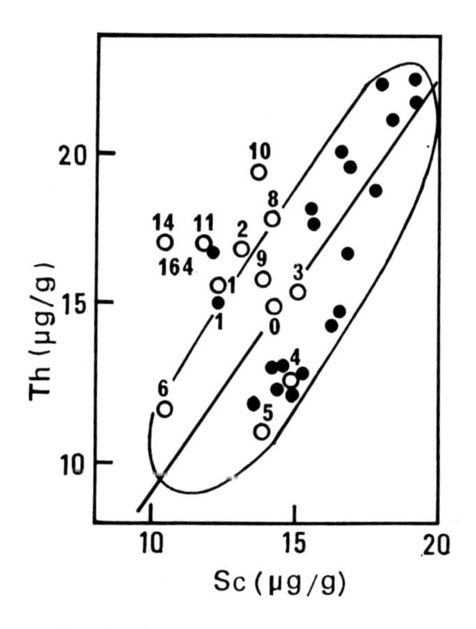

Fig. 271. Correlation between Th and Sc contents (Koyama, M. 1975).

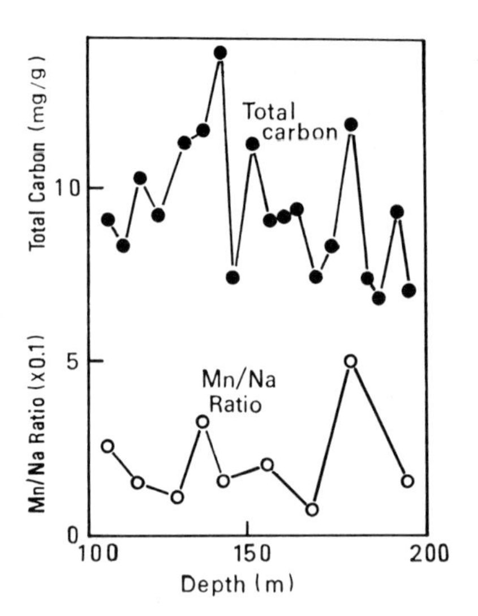

Fig. 272. Comparison of distribution pattern of total carbon and that of Mn/Na ratio. Total carbon data are those obtained by T. Koyama (1972) (Koyama, M. 1975).

Manganese and total carbon. The samples with the same serial numbers which had been previously analyzed for total carbon, C^{13}, particle size, etc. were analyzed on manganese. The samples are those of a part of the core at the depth of 100 meters below the bottom surface. As shown in Fig. 272, the distribution patterns of both components are alike: the higher the total carbon content is, the higher the Mn/Na ratio is.

Summary on chemical properties of the elements and the sedimentary environments

What can be found by analyzing lake sediments is an essential question in this sort of work.

Generally speaking, inorganic compositions of lake sediments are correlated with the following factors as have been mentioned previously.

1. Geological environments: natures of soils and rocks in and around the lake, the shape and depth of the lake, etc.

2. Physical environments: climate, water flow into and within the lake, temperature, etc.

3. Chemical environments: quality of the lake water which includes redox potential, pH salt content, etc.

4. Biological environments: biological productivity, species which concentrate special elements, etc.

What we really observe in the fresh sediments or core samples is a complex combination of direct and indirect effects of these factors. However, by focusing our attention on chemical properties of the elements in relation to their behavior in sediments, we can find a clue for estimating what kind of elements are likely to stay in sediments during the fixation. The supposition that such an estimation is possible offers us a key to the clarification of what is recorded and not recorded in the inorganic compositions of sediments.

According to the idea mentioned above, the elements treated here can be classified into the following categories.

1. Alkali elements have no chemical activity with respect to redox and hydroxide formation reactions. Among them, sodium tends to escape to the hydrosphere from the sediments. Other heavier alkaline ions among which cesium is the extreme case, tends to be incorporated into the matrices of aluminosilicate minerals. Potassium has a character between rubidium and sodium.

2. Manganese, chromium, iron and cobalt are susceptible to redox potential and the pH of the environments: $Mn^{++} \rightleftharpoons MnO_2$, $Cr^{+++} \rightleftharpoons CrO_4^{--} \rightleftharpoons Cr_2O_7^{--}$, $Fe^{++} \leftrightarrows Fe^{+++}$, $Co^{++} \leftrightarrows Co^{+++}$. Except for chromium, these ions in higher valency states, once deposited on the bottom surface of the lake, are likely to be reduced and become soluble when the environment turns into the reducing atmosphere.

3. Scandium, cerium and thorium are considered to be chemically inert in the natural environments and will stay in mineral precipitates. By taking this classification and the results of data treatments into account, the following ordering of survivability of elements in the core samples can be induced.

$$Th \approx Ce \gtrsim Sc \gtrsim Cr > Fe > Co > Mn.$$

In the above discussion, the role of sulfide formation is not considered. In view of the fact that the concentration of reactive sulfide ions in the bottom surface sediment reaches to one percent of the dry sample, a few tens percent of ferrous ions will be combined with sulfide ions at most.

Manganous ions which are expected to be the most mobile in the above ordering are fixed in the oxidizing layers by being trapped with manganese dioxide.

$$MnO_2\downarrow + Mn^{++} \rightarrow Mn(II) \cdot MnO_2\downarrow.$$

This reaction occurs readily in laboratory experiments. The above precipitation reaction and dissolution in the reducing environments clearly explain the surface concentration of manganese in the lake bottom. In this connection, it should be mentioned that manganese has a correlation with the total organic carbon. The fact suggests that the rate of accumulation on the fresh sediments and that of disappearance from deeper sediments are parallel for both manganese and total carbon.

As for anions, arsenic and phosphorous are important species. The absorption of ferric hydroxide and manganese dioxide may explain the sharp concentration gradient near the bottom surface. In the case of arsenic, bacterial actions may produce arsine derivatives which diffuse out to the bottom surface from the deeper sedimental layers, thus reducing the arsenic concentration in the core samples.

Biochemical organic compounds

Nobuhiko Handa

Introduction

The metal porphyrins which were first isolated from many crude oils and shales in 1934 (Treibs 1934) led organic chemists to claim two major findings – first, that the oils were certainly originated from biological organisms and, second, that the conditions of the oil formation could not have involved high temperature. Since then, many works have been conducted to analyze total organic carbon and nitrogen, amino acids and proteins, carbohydrates, lipid materials, hydrocarbons, carotenoids, porphyrins and heterocyclic and carbocyclic compounds in the lake and marine sediments. The purpose of these works is to determine the source materials and their significance in the interpretation of the sedimentary environment.

In the present work, sedimentary samples from Lake Biwa have been analyzed for several organic compounds. It has been found that they are of value in interpreting the climate and the paleoecological condition in which the sediment was formed.

Carbohydrate

Carbohydrate determination of geochemical samples including recent and ancient sediments, rocks and fossils has been conducted by several workers. A comprehensive review of geochemical studies of carbohydrates is provided by Vallentyne (1963) for those up to 1962 and by Swain (1970b) for later works up to 1969.

Paracas (1970) first identified some of the monosaccharides from Paleozoic and from pre-Cretaceous Mesozoic rocks by the aid of the paper-chromatographic technique. Later, some additional monosaccharides were identified by Rogers (1970) in Devonian and other rocks; he employed both the chromatographic and the enzymatic methods.

Numerous aldonic and biuronic acids with several organic acids were found

Horie, S. Lake Biwa

ISBN 90 6193 095 2. Printed in The Netherlands

in Baltic Sea and Black Sea sediments and characterized by the combined gas chromatography and mass spectrometry (GC-MS) (Mopper & Larsson 1978). The low abundance of uronic acids characteristic of terrigenous plants and the similarity of the biuronic composition to that of marine algae indicate a marine algal source of the acids in the sediments.

Polysaccharides, such as cellulose, starch, laminarin and chitin have been found in geological samples by Abderhalden and Heyns (1933) and by Swain and his co-workers (1967). The determination of their chemical nature, however, is still incomplete.

In making carbohydrate analyses of sedimentary samples from Lake Biwa, our aims are primarily to ascertain the total amounts of carbohydrate materials present and their chemical nature, and secondarily to discuss the behavior of carbohydrates in any form during the course of sedimentation and diagenesis on the basis of the results obtained.

Sedimentary samples were taken from a core sediment of 200 m in length which had been obtained from the bottom of Lake Biwa in 1971 and analyzed for carbohydrate by both the phenol-sulfuric acid method and the anthrone method (Handa 1972). Total carbohydrate of the sedimentary samples was determined in the range from 0.37 mgC/g dry sediment to 1.87 mgC/g dry sediment by the phenol-sulfuric acid method and from 0.12 mgC/g dry sediment to 0.86 mgC/g dry sediment by the anthrone method, respectively. It is obvious that the carbohydrate values obtained by the phenol-sulfuric acid method are about three times as much as those obtained by the anthrone method throughout the samples as shown in Fig. 273. The values differ between the two methods for the following reason: As it has previously been reported by Handa (1966), the sugar chromatophores in phenol and anthrone reactions in concentrated sulfuric acid are different in relative absorbances depending on the monosaccharide species. More intensive differences among monosaccharides were observed in the anthrone method than those in the phenol-sulfuric acid method. Extreme cases were observed in both arabinose and xylose. The chromatophores of these monosaccharides in the anthrone reaction give only 3% as much as that of glucose. Thus, this seems to be one of the main reasons for the differences between the carbohydrate content of the sediment as determined by the phenol-sulfuric acid method and the anthrone method.

A few grams of fresh sedimentary samples were placed in a test tube and soaked in 2 ml of 72% sulfuric acid at room temperature for 2 hrs. After dilution with distilled water, the reaction mixture was kept at 100°C overnight to hydrolyze the polysaccharides to monosaccharides. The monosaccharides were converted to corresponding sugar alcohol peracetates, which were separated and identified or elucidated by GC and GC-MS respectively.

An acid hydrolysate of the sedimentary samples from Lake Biwa gave fifteen peaks as sugars consisting of the sedimentary carbohydrate in the gas chromatogram as shown in Fig. 274. A mixture of the authentic specimens of the sugar alcohol acetates was also run in the identical condition of gas chromatography

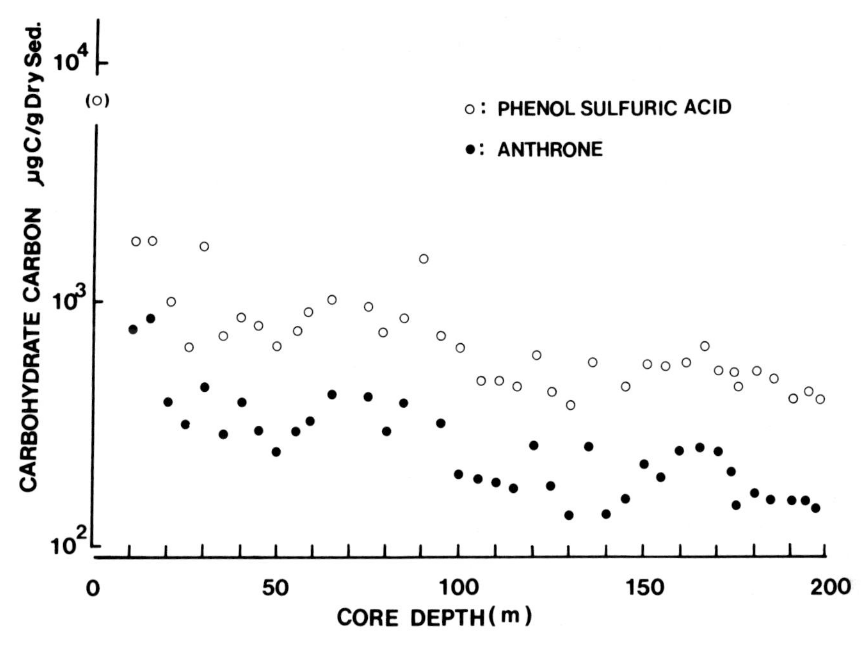

Fig. 273. Vertical profile of carbohydrate carbon in the 200 meter core sample from Lake Biwa. Carbohydrate carbon determined by phenol sulfuric and method was referred from Handa (1974).

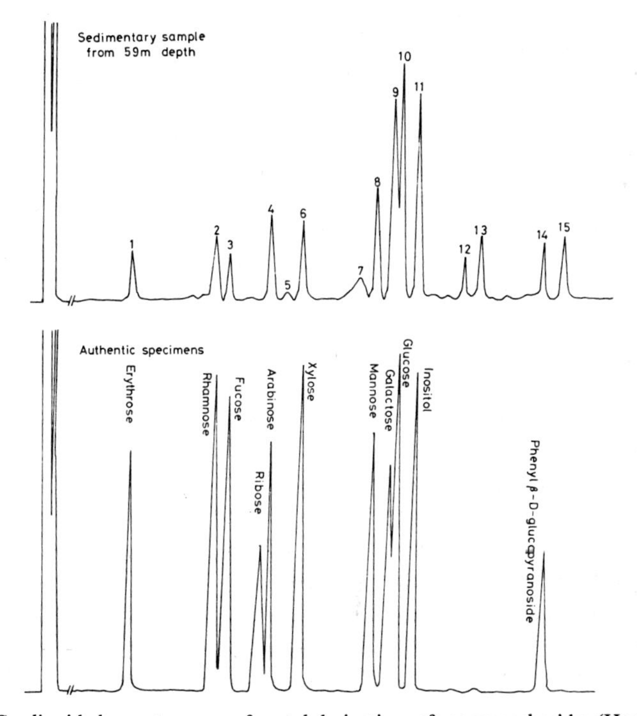

Fig. 274. Gas liquid chromatograms of acetyl derivatives of monosaccharides (Handa 1977).

(GC). Retention time of the authentic specimens indicates that monosaccharides that appeared in the peaks ♯1, 2, 3, 4, 6, 8, 9, 10 and 11 in the sedimentary sample correspond to sugar alcohol peracetates derived from erythrose, rhamnose, fucose, ribose, arabinose, xylose, mannose, galactose, glucose and inositol (internal standard) respectively. Mass spectral analyses of each of the monosaccharide derivatives were conducted by the combined gas chromatography-mass spectrometer (Handa 1976, 1977) and mass spectra of these derivatives agreed entirely with those of the derivatives of the authentic monosaccharides.

For monosaccharides that appeared at the peaks ♯12 and 13, the ion greatest in mass was m/e 433 which was formed as a result of the splitting out of CH_3 radical (m/e 15) from the molecular ion. This fact indicates that the molecular ions (M^+) of these monosaccharides must be m/e 448. As regards the fragmentation pattern of these monosaccharide derivatives, it can be concluded that these monosaccharides are methyl hexoses although steric configuration of H and OH groups of these sugars cannot be determined by the mass spectroscopic method.

For monosaccharides that appeared at the peaks ♯14 and 15, mass spectroscopic analyses were applied to elucidate their chemical structures. The ion greatest in mass was recorded at m/e 360 which was formed as a result of splitting out of two of $CH_2OCOCH_3^+$ (m/e 73) from the molecular ion (M^+). Thus, M^+ must be m/e 506. As regards the fragmentation pattern of these monosaccharide derivatives, it can be concluded that these monosaccharides are heptoses although further steric configuration of these compounds is still obscure.

The vertical profile of monosaccharide composition of the sedimentary samples from various depths are shown in Table 77. Glucose, galactose and mannose are major fractions of the monosaccharides comprising the sedimentary carbohydrate, while tetrose, pentoses, methyl pentoses, methyl hexoses and heptoses are minor fractions throughout the sedimentary samples. However, the concentration of mannose was found to be quite variable due to the sedimentary samples. Low values of the mannose were measured at the depths of 50–56 m and 79–85 m where low values of Σ organic carbon/total organic carbon as shown in Fig. 275, and of the concentration of chlorophyll derivatives and carotenoids were found (Handa 1975). Mori, Sh. (1974) also found low values of the content of fossil diatom in the sedimentary samples from these depths stated above. It can be concluded from the data, that a sedimentary sample in which the number of fossil diatom is measured to be small gives low concentration of the sedimentary mannose. Considering the fact that glucuronomannan is a sole component of the cell wall polysaccharide of the diatom (Ford & Percival 1965), it is most likely that there exists a close relationship between the content of fossil diatoms and mannose concentration in the sedimentary samples from Lake Biwa. Sources of the methyl hexoses and heptoses found in the sedimentary samples from this lake are still uncertain; however, land plant may be the most probable source of these sugars (Smith & Montgomery 1959).

Table 77. Monosaccharide composition of sedimentary samples at various depths of 200-meter core sample from Lake Biwa (Handa 1977).

		Sedimentary samples (depth, m)							
	Peak ♯	11	15	50	56	59	79	85	95
Erythrose	1	15.6%	13.1%	28.0%	25.4%	6.7%	2.9%	7.0%	1.4%
Rhamnose	2	5.9	6.8	2.1	2.6	6.6	6.4	5.3	9.4
Fucose	3	8.0	9.3	1.9	1.9	4.9	4.7	3.5	8.8
Arabinose	4	9.0	10.4	8.5	6.3	8.0	5.8	7.8	7.3
Unknown	5	1.0	1.8	3.1	1.5	1.2	2.1	1.9	1.2
Xylose	6	2.7	3.5	8.3	5.2	5.6	3.7	6.5	5.8
Unknown	7	1.7	2.0	5.0	8.0	1.7	10.0	10.2	0.3
Mannose	8	9.5	8.7	3.3	4.8	11.5	2.2	3.5	11.0
Galactose	9	14.6	12.4	7.7	8.2	16.2	16.0	13.5	14.1
Glucose	10	23.4	22.1	15.0	19.2	25.5	29.0	26.3	26.0
Methylhexose-I	12	2.5	2.5	3.8	4.5	3.2	4.1	3.9	4.3
Methylhexose-II	13	3.3	3.0	7.0	6.9	4.5	7.1	4.9	5.2
Heptose-I	14	1.2	1.5	2.7	2.5	1.5	2.8	2.9	1.8
Heptose-II	15	1.7	2.0	4.0	3.5	2.5	4.0	3.0	3.1
Total concentration	(A)*	0.85	1.19	0.29	0.39	0.40	0.34	0.46	0.40
(mgC/g dry sed.)	(B)**	0.76	0.86	0.24	0.30	0.32	0.29	0.37	0.31

*Determined by the anthrone method.

**Determined by gas liquid chromatography by use of inositol as the internal standard.

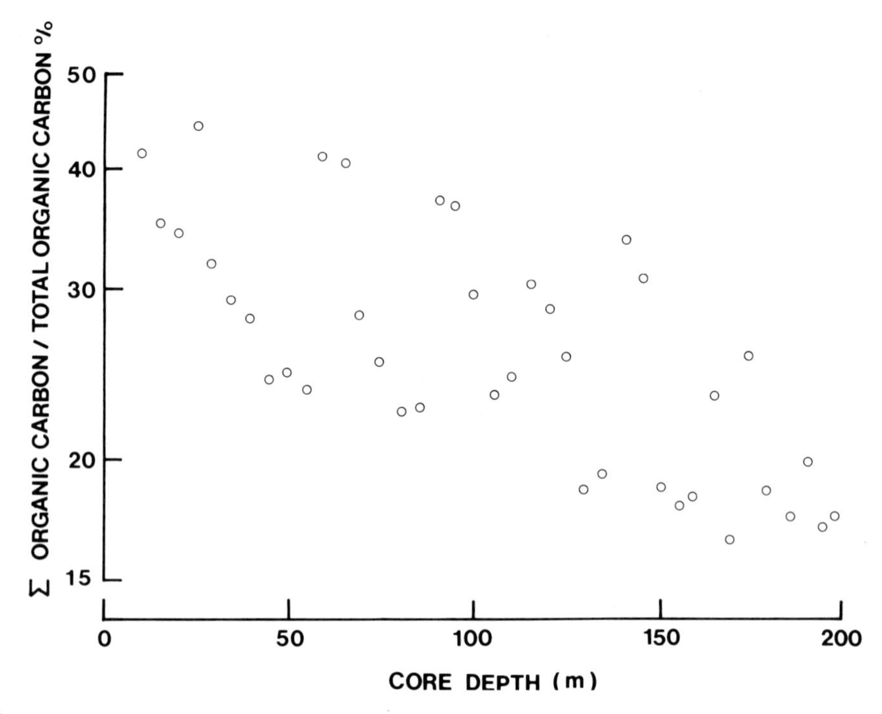

Fig. 275. Vertical profile of Σ organic carbon/total organic carbon in the 200 meter core sample from Lake Biwa.

Protein and amino acids

A sizable fraction of the bulk weight in many organisms, often over 50 percent on the dry weight basis, consists of protein, which accounts for most of the nitrogen compounds in living organisms.

Once entering into the sediment, the protein undergoes biological degradation followed by thermal alteration. Thus, the protein has been found to be preserved as the residues in many different deposits of the Holocene sediment and less than one percent of the total amino acid residues are free, while less than 40 percent of the amino acid residues are in combined forms as protein in the Pleistocene samples. None of the amino acids have been detected in peptide linkage in the Miocene shell.

In a study of the early diagenesis of the combined amino acids in a series of radiocarbon-dated *Mytilus* shells (Hare 1963), it was found that the amino acid composition of the combined amino acids changed progressively with time. An especially marked change in the ratio of glycine to alanine concentrations was observed in the Recent through Pleistocene samples. Apart from this, Degens and Love (1965) reported a remarkable similarity in the amino acid ratios of the fossil specimens compared to a Recent *Planabis.*

Studies on the amino acids in marine (Rittenberg et al. 1963) and lake (Jones & Vallentyne 1960) sediment cores show generally decreasing total amounts with depth but highly irregular distributions with respect to many individual amino acids. Some amino acids not normally present in proteins have been found in the sedimentary samples as well as in a series of fossil shell of *Mercenaria* (Hare & Mitterer 1965–66). These nonprotein amino appear to be the degradation and racemization products of the original amino acid constituents of the shell proteins. A series of D-isomers of the amino acids, D-allo-isoleucine, γ-aminobutyric acid, and β-alanine etc. appear most abundantly in shells, bones and sedimentary samples.

Each of the amino acids has its characteristic activation energy, which can be determined by the kinetic study of the decomposition at a series of the temperatures. Time and temperature are the main variables for this reaction. These two unknowns can be determined by analyzing more than two of the amino acid reactions. Hare (1969) suggests that decomposition studies at several temperatures between 88°C and 225°C may be useful in predicting reaction rates at temperature between 0°C and 30°C and are thus of potential use in both geochronologic and geothermometric studies.

In making protein and amino acids analyses of the sedimentary samples from Lake Biwa, our aims are primarily to know the total amount of amino acids (in combined and free forms) and secondarily to discuss the behaviors of the amino acids during the course of the sedimentation and the diagenesis.

The total amount of protein carbon in the sedimentary sample from Lake Biwa was determined in a range from 0.22 mgC/g dry sediment to 1.80 mgC/g dry sediment, as is shown in Fig. 276. The values tend to decrease with the depth of the core sample, with some variation.

The total amount of lipid carbon in the sedimentary sample was also determined. The values range from 0.07 mgC/g dry sediment to 0.66 mgC/g dry sediment. The values tend to decrease with depth.

The vertical profile of the ratio of protein carbon to the total organic carbon is shown in Fig. 277. The ratio tend to decrease in value as the depth of the sample increases. Relatively high values were observed, with some intervals, at the depth of 11 m, 26 m, 59–66 m, 90–100 m 116–121 m, 140–145 m and 175 m. This vertical profile of the values is basically identical with that of carbohydrate carbon.

The ratio of lipid carbon to the total organic carbon tends to decrease with the depth of the sedimentary sample, as shown in Fig. 277. The vertical profile of the values is similar to those of carbohydrate and protein. Some additional maxima were observed at the depths of 11 m and 161 m.

Carbohydrate, protein including amino acids, and lipid are major constituents of living organisms. The total of these organic materials in the sedimentary samples tends to decrease as the depth of the sedimentary sample increases. The ratio of the sum of the organic carbon of these materials (Σ organic carbon) to the total organic carbon of the sedimentary sample was

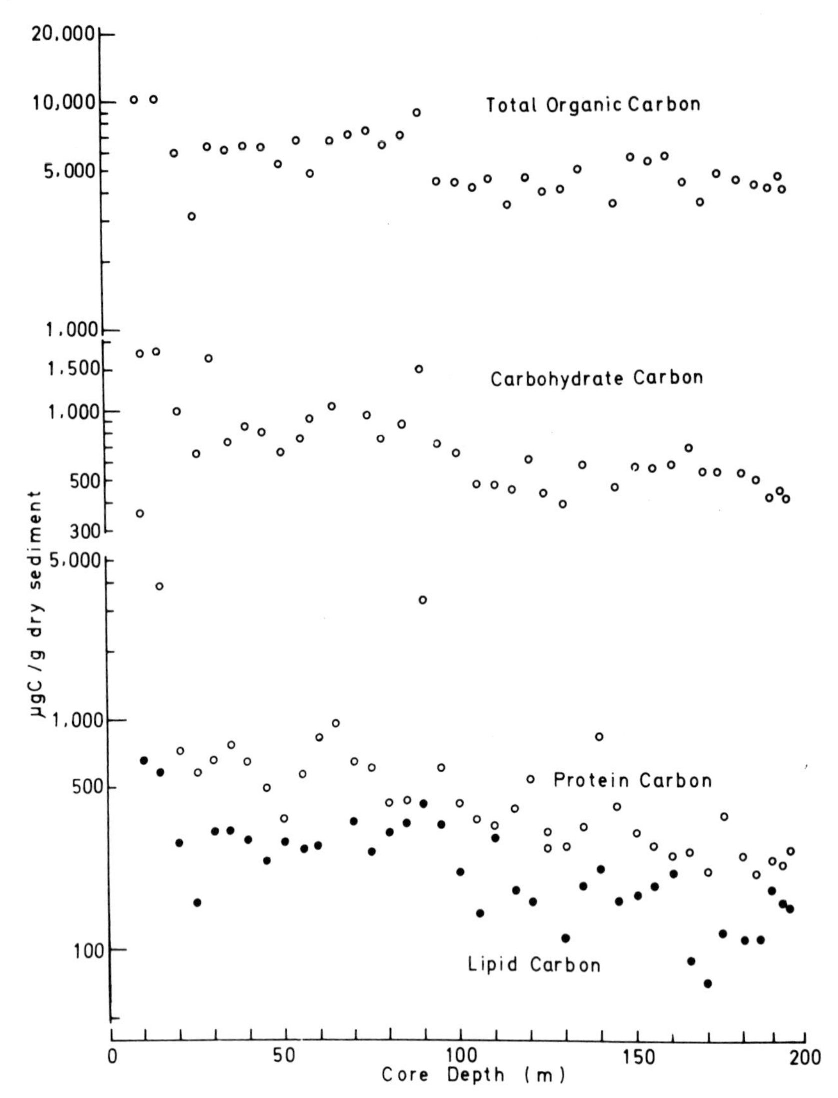

Fig. 276. Vertical profile of total organic carbon, carbohydrate carbon, protein carbon and lipid carbon. (Handa 1976.)

calculated. The vertical profile of the ratio is shown in Fig. 277. The value tends to decrease as the depth of the sample increases. Relatively high values were observed, with some intervals, at the depths of 11 m, 26 m, 59–66 m, 90–100 m, 116–121 m, 140–145 m, 166 m, and 175 m.

Fuji and Horie (1972) established the paleotemperature curve of the 200 m core sample on the basis of a palynological examination. A comparison of their data with the vertical change in the ratio of the Σ organic carbon shows that

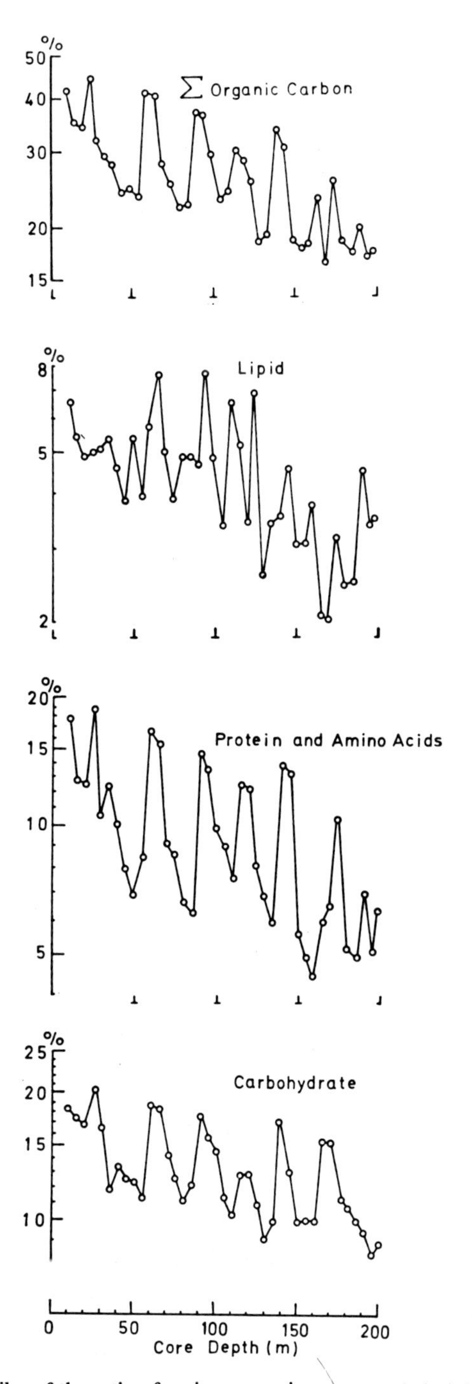

Fig. 277. Vertical profiles of the ratio of various organic components to total organic carbon in a 200 meter core sample from Lake Biwa (after Handa; Koyama, T. et al. 1973).

high values of the ratio correspond to the cold-climate stages, while the low values correspond to the warm- or temperate-climate stages as shown in Fig. 278 although there occur a few exceptions. The facts clearly indicate that the content of carbohydrate, protein and lipid in the lake sediment has a close relation to the climatological condition of the period when the sediment was formed.

It, however, is difficult to determine exactly how the past climatological changes affected the content of these organic materials in the sedimentary samples. Judging from the data concerning the behavior of organic materials during the sedimentation processes in an aquatic environment at the present time, we believe that a more detailed chemical characterization of these organic materials may give important clues to solving this problem.

The linear combination of protein carbon was calculated by the method of least squares. The relationship between the protein carbon ($N_{\text{protein carbon}}$ mgC/g dry sediment) and the depth of the sample (Zm) is given by the following equation:

$$N_{\text{protein carbon}} = 0.98\ e^{-0.0014Z} \qquad [1]$$

If it takes 2.5×10^5, 5.0×10^5 and 7.5×10^5 years for the formation of a sediment 175 m thick, the decomposition rate constant of the sedimentary protein can be calculated as shown in Table 78. According to Nishimura and Yokoyama (1973), the age of the volcanic ash found in the sedimentary sample of 175 m depth was determined to be 0.47×10^6 years BP. Thus, the rate of the

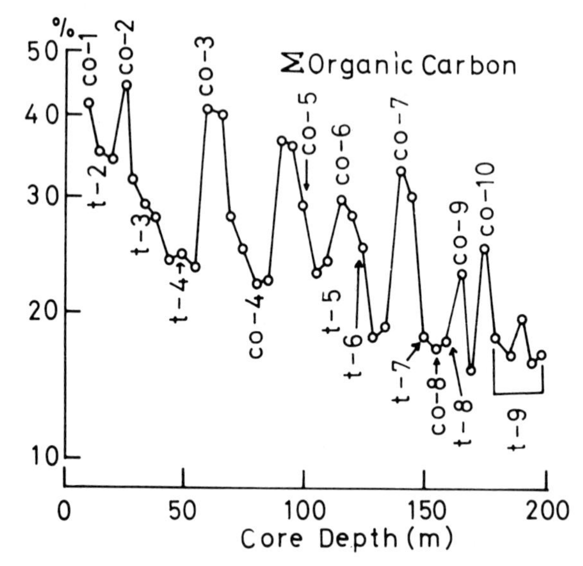

Fig. 278. Vertical profile of the ratio of Σ organic carbon to the total organic carbon in a 200-m core sample from Lake Biwa. co – 1 – co – 10, t – 2 – t – 9 in the figure are the cold, temperate, and warm climate stages, as estimated by Fuji and Horie (1972) (after Handa 1973).

degradation of protein including amino acids is assumed to be 0.48×10^{-6} years^{-1}. This figure is far less than that of microbiological degradation of organic matter. Thus, this fact suggests that the thermal degradation process plays an important role in causing vertical decreases in the amount of protein in the sediment.

The linear combination of carbohydrate and lipids carbons were also calculated by applying the method of least squares and the following equations were obtained:

$$N_{\text{carbohydrate carbon}} = 1.18\, e^{-0.0056Z} \qquad [2]$$

$$N_{\text{lipid carbon}} = 0.45\, e^{-0.0012Z} \qquad [3]$$

The decay rates of organic matter are shown in Table 78. Judging from the decomposition rate constants of these organic materials, the results obtained indicate that carbohydrate is the most stable compound in the thermal degradation process, followed by lipid and protein in that order.

Amino acids and amino sugars were released when the sedimentary samples were treated with hydrochloric acid (6N). The resulting amino acids and amino sugars were separated by the liquid chromatography and the data shown in Table 79 were obtained by Terashima and Mizuno (1974). The most abundant amino acids are glycine, alanine, alutamic acid and aspartic acid, followed by lysine including ornithine, threonine, proline, valine and leucine in the sedimentary sample from 10 m depth. This pattern of the amino acid composition is similar to those of aquatic higher plants reported by Swain et al. (1964) and soils (Stevenson 1956), while it is quite different from those obtained in the hydrolysate of the bulk protein of several unicellular algae in that the latter is more abundant in arginine than alanine. These facts may indicate that soils and terrestrial higher plants must be the most probable sources of the sedimentary amino acids rather than unicellular algae. Aquatic higher plants may not contribute to a greater extent, because of the lack of the data concerning the occurrence of the aquatic higher plant community.

It is obvious that there is a downward decrease in most of the protein amino acids from as little as one-tenth in phenylalanine to as much as one-fiftieth in

Table 78. Decay rate of organic matter in a 200 m core sample (Handa 1972, 1973).

	Rate constant		
	Time required for the formation of a sediment 175 m thick		
	7.5×10^5y	5.0×10^5y	2.5×10^5y
	10^{-6}y^{-1}	10^{-6}y^{-1}	10^{-6}y^{-1}
Protein carbon	0.32	0.48	0.96
Lipid carbon	0.28	0.42	0.84
Organic carbon	0.12	0.18	0.36
Carbohydrate carbon	0.24	0.36	0.72

Table 79. Contents of amino acids and amino sugars in the core sample (in μg/g dry sample) (after Terashima & Mizuno 1974).

Depth of sample (m)	10	15	20	25	35	45	75	95	110	125	130	150	170	195
Amino acids														
Lysine + Orinithine	199	39	50	31	73	66	61	41	28	28	16	25	19	15
Histidine	36	34	22	6	14	14	11	5	6	5	5	2	3	2
Arginine	75	106	10	14	35	30	14	16	5	3	5	5	3	2
Aspartic acid	325	325	214	69	157	109	80	57	32	19	20	16	7	–
Glutamic acid	244	209	122	41	115	78	50	40	21	12	13	13	13	–
Threonine	168	158	76	23	57	49	48	30	14	10	10	8	5	6
Serine	138	128	60	18	45	38	38	24	11	7	7	8	7	6
Proline	165	140	62	22	58	32	33	22	13	7	7	8	3	7
Glycine	297	260	104	52	116	82	64	57	23	20	17	21	16	15
Alanine	316	290	127	46	123	96	86	40	28	20	19	20	14	15
Cystine	+	+	+	–	–	–	++	+	+	–	+	–	–	–
Valine	176	166	76	23	66	61	59	30	16	13	12	13	9	12
Methionine	10	13	4	3	3	1	1	9	1	1	1	6	–	–
Iso Leucine	100	94	38	16	37	35	33	20	9	8	7	7	5	7
Leucine	150	144	63	17	55	51	42	29	14	10	10	10	7	9
Tyrosine	53	43	16	11	16	13	9	7	4	2	4	11	2	4
Phenylalanine	109	104	36	15	35	31	28	18	7	5	5	5	3	12
β-Alanine	41	36	16	15	26	21	8	26	9	13	14	12	6	10
γ-Aminobutyric acid	22	22	14	8	28	21	27	15	14	13	11	11	8	12
Total	2624	2311	1110	430	1059	828	692	486	255	196	183	201	130	134
Amino sugars														
Glucosamine	–	–	7	–	–	+	–	7	13	2	13	14	16	4
Galactosamine	–	–	–	–	–	–	–	–	2	–	4	2	4	–
Total	–	–	7	–	–	+	–	7	15	2	17	16	20	4

aspartic acid, while no consistent vertical trend is observed in cystine and methionine. Only a trace of eystine was found in the hydrolysate of the sedimentary samples. This is due to its conversion to cysteic acid as observed in the treatment of protein with hydrochloric acid in general. Methionine, too, was found only in low concentration, perhaps due to the decomposition of the amino acid to alanine, glycine and aminobutyric acid (Vallentyne 1965).

Non-protein amino acids, β-alanine and γ-aminobutyric acid were measured in concentration of certain magnitudes throughout the sedimentary samples, but no vertical trends were observed in their concentration. However, the ratio of β-alanine and γ-aminobutyric acid to the total amino acids tends to increase with depth as shown in Fig. 279. These facts clearly indicate that only thermal degradation processes give rise to the vertical decrease in concentration of the protein-amino acids.

Chlorophyll derivatives and carotenoids

In recent years, much attention has been given by many workers to the sedimentary plant pigments because several investigations in the past decades have shown that plant pigments preserved in the sediments may follow lake productivity in the past.

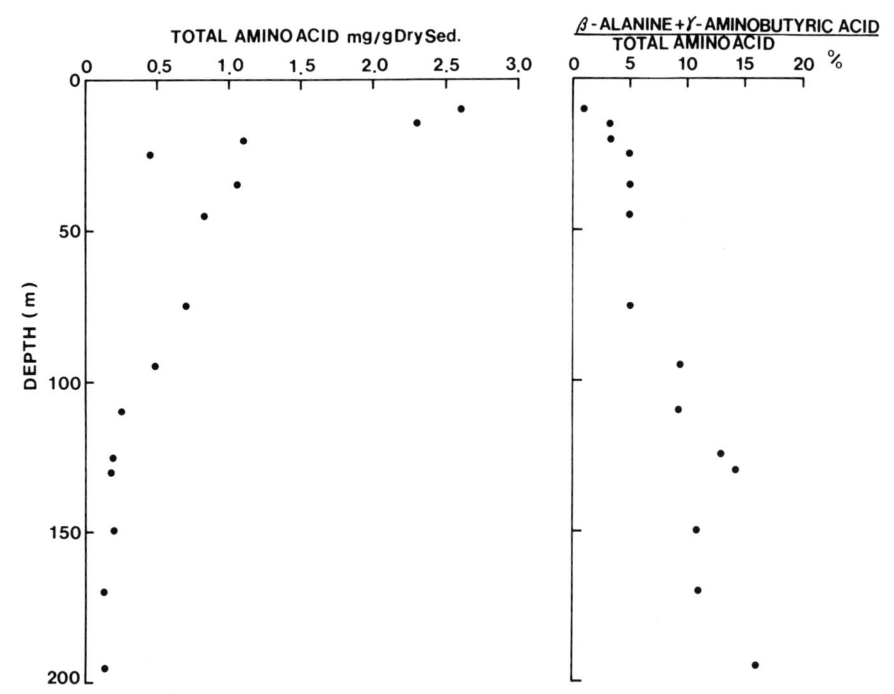

Fig. 279. Vertical distribution of amino acids in the core sample (after Terashima & Mizuno; Koyama, T. et al. 1973).

In his study of the organic matter of the surface sediment from various lakes in the Midwest area of the United States, Gorham (1960) suggested that the concentration of chlorophyll derivatives might be a sensitive indicator of lake fertility. Later, Gorham (1961) reported that the pigments in the sediment provided important clues in estimating the magnitude of the primary production in the postglacial age. Subsequently, Gorham and Sanger (1967) reported that highly productive lakes show a concentration of sedimentary pigments greater than those of the upland litter and humus layer and that most of the sedimentary organic matter in such lakes must have been derived from autochthonous phytoplankton population of the lake.

Quantitative determination of chlorophyll derivatives in various sedimentary samples has been attempted by Vallentyne (1955) and Wetzel (1970). These workers provided the sedimentary chlorophyll degradation product (SCDP) unit which is an arbitrary relative unit based on the absorbance of light at the wavelength of maximum absorption in the red end of the spectrum when the pigments are dissolved in 90% aqueous acetone.

Applying this method to the lacustrine sediment, Wetzel (1970) found good correlation between the quantities of pigment degradation products and organic matter and other paleolimnological indicators of productivity. The writer also reported that there was a major peak of the pigment products per gram of organic matter following the Boreal period and a subsequent regression of productivity with several lesser peaks.

In the 200 meter core sample from Lake Biwa, chlorophyll derivatives and carotenoids were determined at successive levels to ascertain their total amounts. Vertical profiles of the plant pigments were also discussed in relation to the data of organic analyses and traditional microfossil studies (Handa 1975).

Chlorophyll derivatives in the core sample from Lake Biwa were determined at the range from 22 to 347 units/g dry sediment. Fluctuation of the values was found to occur obviously throughout the depth of the core sample as shown in Fig. 280. Relatively higher values of chlorophyll derivatives were observed at the depth of 11–25 m, 59–70 m, 90–99 m and 140–145 m where relatively higher values of Σ organic carbon/total organic carbon were observed. Σ organic carbon is defined as the sum of carbohydrate, amino acids and protein and lipid carbons which are major constituents of autochthonous plankton cells rather than allochthonous leaf litter and soil humic materials. Gorham and Sanger (1972) concluded through extensive analyses of chlorophyll derivatives of woodland soils and swamp, pond and lake sediments that allochthonous contribution of chlorophyll derivatives to a lake sediment is usually minor while autochthonous sources are dominant. Wetzel (1970) also reported that the direct ingress of autumnal decidous leaf fall into a nearly closed basin of the lake would be minor, even when the lake was completely surrounded by forests. It can be concluded from these facts that the vertical change in the concentration of chlorophyll derivatives in Lake Biwa is mainly due to the change in the phytoplankton population in the past. The vertical change in Σ organic

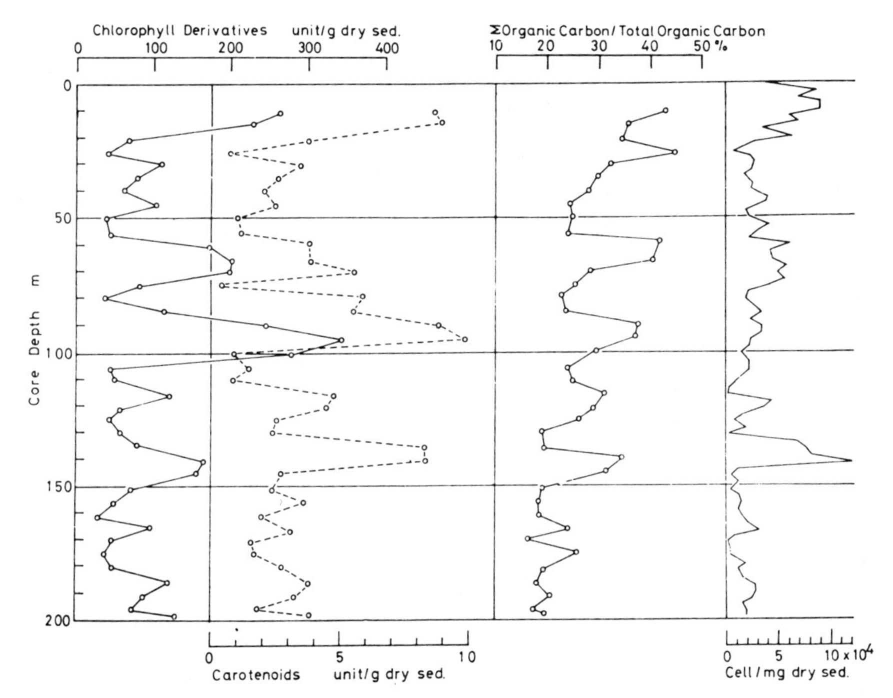

Fig. 280. Vertical profiles of chlorophyll derivatives, carotenoids, Σ organic carbon/total organic carbon and fossil diatoms in the 200 meter core (after Mori, Sh. 1974; Handa 1975).

carbon/total organic carbon can be consistently interpreted to be a result of the change in the phytoplankton population in the past.

Mori, Sh. (1974) reported that a dramatic change in species composition of fossil diatoms occurred in the 200 meter core sample; we have analyzed their organic materials. A relatively larger number of fossil diatom cells is counted at the depths of 11–25 m, 59–70 m, 115 m and 135–145 m, where higher values of chlorophyll derivatives and Σ organic carbon/total organic carbon were found. This fact indicates that the diatom population has been one of the main producers of organic materials in this lake, although it is difficult to make a quantitative statement on the relationship between the amounts of chlorophyll derivatives and the fossil diatom population. Algal population other than diatoms may also have an important role for the organic production in the lake, especially at the level of 90–99 m depth where higher values were obtained in chlorophyll derivatives and Σ organic carbon/total organic carbon, but not in diatom population.

Carotenoids have been found in all of the green plants as integral constituents of photosynthetic tissues and also in non-photosynthetic bacteria and fungi. These organisms are able to produce their own carotenoids, which can be converted into animal carotenoids after the organisms of the higher trophic levels feed these microorganisms.

The principal carotenoids in the chloroplast of higher plants are β-carotene, lutein, violaxanthin and neoxanthin, while only β-carotene has been demonstrated to occur in algal classes as the major component of the carotenoids. The diversity of nature has been shown in xanthophylls of the algae, while fucoxanthine is the major xanthophyll of diatoms which make up approximately 85 percent by weight of the plant biomass of the oceans and lakes (Harvey 1957). Since algal organic matter must be a source of the sedimentary organic materials, much attention has been paid to the sedimentary carotenoids to deduce the algal biomass of the aquatic environment in the past.

It has been shown that the algal decay in lakes favors the preservation of carotenoids, but these materials are more susceptible than chlorophyll derivatives to microbiological degradation during the processes of decay of autumn leaves in soil (Gorham & Sanger 1967). Sanger and Gorham (1970) reported that the amount of carotenoids in lake sediments and its diversity on thin-layer chromatogram might be useful indicators of past trophic levels of the lake. Thus, carotenoids were determined in the 200 meter core sample. The vertical profile of carotenoids was found to be almost the same as that of chlorophyll derivatives as shown in Fig. 280. The results obtained suggest that fluctuation of the trophic levels has occurred in Lake Biwa. A more detailed analysis of carotenoids will give us more precise information on this point.

References

Abderhalden, E. and Heyns, K., 1933. Nachweis von chitin in Flügelresten von Coleopteren des Oberen Mitteleoecäns (Fundstelle Geiseltal). Biochem. Zeitschr. 259: 320-321.

Aizenshtat, Z., 1973. Perylene and its geochemical significance. Geochim. Cosmochim. Acta 37: 559-567.

Attaway, D. and Parker, P. L., 1970. Sterols in recent marine sediments. Science 169: 674-676.

Bender, M. L., Taylor, F. T. and Matthews, R. K., 1973. Helium-uranium dating of corals from middle Pleistocene Barbados reef tracts. Quaternary Res. 3: 142-146.

Bennie, J., 1894. Arctic plants in the old lake deposits of Scotland. Ann. Scot. Nat. Hist. 46-52.

Blumer, M., 1961. Benzpyrenes in soil. Science 134: 474-475.

Bottema, S., 1975. Reconstruction of the Late Quaternary vegetation of northwestern Greece. In: Problems of Balkan flora and vegetation, Sofia, Pub. House Bulgarian Acad. Sci., 58-63.

Brooks, P. W., Eglinton, G., Gaskell, S. J., McHugh, D. J., Maxwell, J. R. and Philp, R. P., 1976. Lipids of recent sediments. Part I: Straight-chain hydrocarbons and carboxylic acids of some temperate lacustrine and sub-tropical lagoonal/tidal flat sediments. Chem. Geol. 18: 21-38.

Bucha, V., 1971. Archaeomagnetic dating. In: Dating techniques for the archaeologist, ed. H. N. Michael and E. K. Ralph, Cambridge and London, MIT Press, 57–117.

Budyko, M. I., 1974. Klimat i Zhizn. (Climate and Life). Translated D. H. Miller. New York, London, Academic Press, 108 pp.

Butzer, K. W., 1975. Pleistocene littoral-sedimentary cycles of the Mediterranean basin. A Mallorquin view. In: After the Australopithecines, Stratigraphy, Ecology and culture change in the middle Pleistocene, ed. K.W. Butzer and B. Isaac, The Hague, Paris, Mouton Pub., 25-71.

Chappell, J., 1974. Geology of coral terraces, Huon Peninsula, New Guinea. A study of Quaternary tectonic movements and sea level changes. Geol. Soc. Amer., Bull. 85: 553-570.

Cox, A., 1969. Geomagnetic reversals. Science 163: 237-245.

Craig, H., 1953. The geochemistry of the stable carbon isotopes. Geochim. Cosmochim. Acta 3: 53-92.

Craig, H., 1957. Isotopic standards for carbon and oxygen and correction factors for mass-spectrometric analysis of carbon dioxide. Geochim. Cosmochim. Acta 12: 133-149.

Craig, H., 1961. Standard for reporting concentrations of deuterium and oxygen-18 in natural waters. Science 133: 1833-1834.

Cranwell, P. A., 1973. Chain-length distribution of n-alkanes from lake sediments in relation to post-glacial environmental change. Freshwater Biol. 3: 259-265.

Cranwell, P. A., 1974. Monocarboxylic acids in lake sediments: indicators, derived from terrestrial and aquatic biota, of paleoenvironmental trophic levels. Chem. Geol. 14: 1-14.

Cranwell, P. A., 1976 Decomposition of aquatic biota and sediment formation: lipid components of two blue-green algal species and of detritus resulting from microbial attack. Freshwater Biol. 6: 481-488.

Crawford, R. M., 1978. The taxonomy and classification of the diatom genus *Melosira* C. A. Agardh. III. *Melosira lineata* (Dillw.) C. A. Ag. and *M. varians* C. A. Ag. Phycol. 17: 237-250.

Creer, K. M., Thompson, R., Molyneux, L. and Mackereth, F. J. H., 1972. Geomagnetic secular variation recorded in the stable magnetic remanence of recent sediments. Earth Planet. Sci. Lett. 14: 115-127.

Dansgaard, W., 1964. Stable isotopes in precipitation. Tellus 16: 436-468.

Dansgaard, W., Johnsen, S. J., Clausen, H. B. and Langway, C. C., Jr., 1969. Ice cores and paleoclimatology. In: Nobel Symposium 12, Radiocarbon Variations and Absolute Chronology, ed. I. U. Olsson, Stockholm, Almqvist and Wiksell, 337-351.

Dansgaard, W., Johnsen, S. J., Clausen, H. B. and Langway, C. C., Jr., 1971. Climatic record revealed by the Camp Century ice core. In: The Late Cenozoic Glacial Ages, ed. K.K. Turekian, New Haven, Yale Univ. Press, 37-56.

Degens, E. T. and Love, S., 1965. Comparative studies of amino-acids in shell structures of *Gyraulus trochiformis*, Stahl, from the Tertiary of Steinheim, Germany. Nature 205: 876-878.

Degens, E. T., Guillard, R. R. L., Sackett, W. M. and Hellebust, J. A., 1968. Metabolic fractionation of carbon isotopes in marine plankton – I. Temperature and respiration experiments. Deep-Sea Res. 15: 1-9.

Degens, E. T., 1969. Biogeochemistry of stable carbon isotopes. In: Organic Geochemistry; methods and results, ed. G.Eglinton and M. T. J. Murphy, Berlin-Heidelberg-New York, Springer Verlag, 304-329.

Denham, C. R. and Cox, A., 1971. Evidence that the Laschamp Polarity Event did not occur 13300–30400 years ago. Earth Planet. Sci. Lett. 13: 181-190.

Deuser, W. G. and Degens, E. T., 1967. Carbon isotope fractionation in the system CO_2(gas) –CO_2(aqueous) –HCO_3^-(aqueous). Nature 215: 1033-1035.

Dreimanis, A. and Karrow, P. F., 1972. Glacial history of the Great Lakes – St. Lawrence Region, the classification of the Wisconsin Stage, and its correlatives. Internat. Geol. Congr. 1972, 24th Sess., Sec. 12, Quaternary Geology, ed. J. E. Gill, 5-15.

Emiliani, C., 1966a. Isotopic paleotemperatures. Science 154: 851-857.

Emiliani, C., 1966b. Paleotemperature analysis of Caribbean cores P6304-8 and P6304-9 and a generalized temperature curve for the past 425,000 years. Jour. Geol. 74: 109-126.

Emiliani, C., 1972. Quaternary paleotemperatures and the duration of the high-temperature intervals. Science 178: 398-401.

Emiliani, C. and Shackleton, N. J., 1974. The Brunhes Epoch: Isotopic paleotemperatures and geochronology. Science 183: 511-514.

Endo, K. and Uesugi, Y., 1972. On the older tephra of Oiso and Yokohama areas. Quaternary Res. (Daiyonki Kenkyu) 11: 15-28.

Epstein, S., Buchsbaum, R., Lowenstam, H. A. and Urey, H. C., 1953. Revised carbonate-water isotopic temperature scale. Geol. Soc. Amer., Bull. 64: 1315-1326.

Faegri, K. and Iversen, J., 1964. Textbook of Pollen Analysis, Copenhagen, Munksgaard, 237 pp.

Fairbridge, R. W., 1961. Eustatic changes in sea level. In: Physics and Chemistry of the Earth, Progress Series, 4, ed. L. H. Ahrens, F. Press, K. Rankama and S. K. Runcorn, London, Oxford, New York, Paris, Pergamon Press, 99-185.

Fleischer, R. L. and Price, P. B., 1964. Techniques for geological dating of minerals by chemical etching of fission fragment tracks. Geochim. Cosmochim. Acta 28: 1705-1714.

Ford, C. W. and Percival, E., 1965. Carbohydrates of *Phaeodactylum tricornutum*, Part II. A sulphated glucuronomannan. Jour. Chem. Soc.: 7042-7046.

Frey, D. G., 1964. Remains of animals in Quaternary lake and bog sediments and their interpretation. Arch. Hydrobiol. Beih. Ergebn. Limnol. 2: 1-114.

Fuji, N., 1966. Climatic changes of Postglacial age in Japan. Quaternary Res. (Daiyonki Kenkyu) 5: 149-156.

Fuji, N. and Horie, S., 1972. Palynological study on 200 meters core sample of Lake Biwa in Japan. Proc. Japan Acad. 48: 500-504.

Fuji, N., 1973a. Change of climate during the Wisconsin Stage based on palynological study of 200 meters core sample of Lake Biwa in Japan. Proc. Japan Acad. 49: 737-741.

Fuji, N., 1973b. Palynological study on core samples from Lake Biwa-ko. Jap. Jour. Limnol. 34: 97-102.

Fuji, N., 1974. Palynological investigations on 12-meter and 200-meter core samples of Lake Biwa in Central Japan. Paleolimnology of Lake Biwa and the Japanese Pleistocene, 2, ed. S. Horie, 227-235.

Fuji, N., 1975a. The coastal sand dunes of Hokuriku district, central Japan. Quaternary Res. (Daiyonki Kenkyu) 14: 195-220.

Fuji, N., 1975b. Climatic change of the Wisconsin Stage in the Japanese Islands from the viewpoint of palynological research. Paleolimnology of Lake Biwa and the Japanese Pleistocene, 3, ed. S. Horie, 321-353.

Fuji, N., 1976a. Palaeoclimatic and palaeovegetational changes around Lake Biwa, Central Japan during the past 100,000 years. Paleolimnology of Lake Biwa and the Japanese Pleistocene, 4, ed. S. Horie, 316-356.

Fuji, N., 1976b. Palynological investigation on a 200-meter core samples from Lake Biwa in Central Japan. Paleolimnology of Lake Biwa and the Japanese Pleistocene, 4, ed. S. Horie, 357-421.

Fuji, N. and Horie, S., 1977. Palynological study of a 200-meter core sample from Lake Biwa, Central Japan. I. Palaeoclimate during the last 600,000 years. Proc. Japan Acad. 53: 139-142.

Fujii, S. and Fuji, N., 1967. Postglacial sea level in the Japanese Islands. Jour. Geosci., Osaka City Univ. 10: 43-51.

Furukawa, H., 1972. Alluvial deposits of the Nohbi Plain, Central Japan. Mem. Geol. Soc. Japan, No. 7, 39-59.

Gaskell, S. J. and Eglinton, G., 1974. Short-term diagenesis of sterols. In: Advances in Organic Geochemistry, Proceedings of the 6th Internat. meeting on Organic Geochemistry Sept. 18-21, 1973, Rueil-Malmaison, France, ed. B. Tissot and F. Bienner, 1973, Paris, Editions Technip, 963-976.

Gorham, E., 1960. Chlorophyll derivatives in surface muds from the English lakes. Limnol. Oceanogr. 5: 29-33.

Gorham, E., 1961. Chlorophyll derivatives, sulphur, and carbon in sediment cores from two English lakes. Can. Jour. Bot. 39: 333-338.

Gorham, E. and Sanger, J., 1967. Plant pigments in woodland soils. Ecol. 48: 306-308.

Gorham, E. and Sanger, J. E., 1972. Fossil pigments in the surface sediments of a meromictic lake. Limnol. Oceanogr. 17: 618-622.

Griffiths, D. H., King, R. F., Rees, A. I. and Wright, A. E., 1960. The remanent magnetism of some recent varved sediments. Proc. Roy. Soc., London, A256: 359-383.

Handa, N., 1966. Examination on the applicability of the phenol sulfuric acid method to the determination of dissolved carbohydrate in sea water. Jour. Oceanogr. Soc. Japan 22: 79-86.

Handa, N., 1972. Organogeochemical studies of a 200 meters core sample from Lake Biwa. The determination of carbohydrate and organic carbon. Proc. Japan Acad. 48: 510-515.

Handa, N., 1973. Organogeochemical studies of a 200-meter core sample from Lake Biwa. II. The determination of protein and lipid. Proc. Japan Acad. 49: 265-270.

Handa, N., 1974. Geochemical studies on organic materials in a 200-meter core from Lake Biwa. Paleolimnology of Lake Biwa and the Japanese Pleistocene, 2, ed. S. Horie, 184-193.

Handa, N., 1975. Organogeochemical studies of a 200-meter core sample from Lake Biwa. III. The determination of chlorophyll derivatives and carotenoids. Proc. Japan Acad. 51: 442-446.

Handa, N., 1976. Identification of monosaccharides of the sedimentary carbohydrate and the vertical change in monosaccharide composition. Paleolimnology of Lake Biwa and the Japanese Pleistocene, 4, ed. S. Horie, 214-232.

Handa, N., 1977. Organogeochemical studies of a 200-meter core sample from Lake Biwa. IV. - Identification of monosaccharides comprising the sedimentary carbohydrate and vertical change in monosaccharide composition. - Proc. Japan Acad., 53, Ser. B, 51-55.

Hare, P. E., 1963. Amino acids in the proteins from aragonite and calcite in the shells of *Mytilus californianus*. Science 139: 216-217.

Hare, P. E. and Mitterer, R. M., 1966. Nonprotein amino acids in fossil shells. Carnegie Inst. Wash. Year Book 65: 362-364.

Hare, P. E., 1969. Geochemistry of proteins, peptides, and amino acids. In: Organic Geochemistry, – Methods and results –, ed. G. Eglinton and M. T. J. Murphy, Berlin, Heidelberg, New York, Springer Verlag, 438-463.

Harvey, H. W., 1957. The chemistry and fertility of sea waters, 2nd ed., Cambridge, Cambridge University Press, 240 pp.

Harmsworth, R. V., 1968. The developmental history of Blelham Tarn (England) as shown by animal microfossils, with special reference to the cladocera. Ecol. Monogr. 38: 223-241.

Harrison, C. G. A., 1966. The paleomagnetism of deep sea sediments. Jour. Geophys. Res. 71: 3033-3043.

Hasegawa, Y., 1975. Significance of diatom thanatocoenoses in the Neolithic sea-level change problem. Pacific Geol. 10: 47-78.

Hatori, K., Inokuchi, M., Kaizuka, S., Naruse, Y., Sugimura, A. and Toya, H., 1962. Latest Quaternary features of Tokyo Bay and its environs. Quaternary Res. (Daiyonki Kenkyu) 2: 69-90.

Hayashi, T., 1974. The Kobiwako Group in the Katata Hills, Shiga Prefecture, Japan. Jour. Geol. Soc. Japan 80: 261-276.

Henderson, W., Reed, W. E., Steel, G. and Calvin, M., 1971. Isolation and identification of sterols from a Pleistocene sediment. Nature, 231, 308-310.

Hendy, C. H. and Wilson, A. T., 1968. Palaeoclimatic data from speleothems. Nature 219: 48-51.

Hiki, T., 1915. Bubalina remains from the Province of Omi. Mem. Coll. Eng., Kyoto Univ. 1: 245-247.

Hirooka, K., 1971. Archaeomagnetic study for the past 2,000 years in southwest Japan. Mem. Fac. Sci., Kyoto Univ., Ser. Geol. Mineral. 38: 167-207.

Horie, S., 1966. Paleolimnological study on ancient lake sediments in Japan. Verh. Internat. Verein. Limnol. 16: 274-281.

Horie, S., 1967a. On the problem of the crustal deformation in lake basin. Disaster Prev. Res. Inst. Ann., Kyoto Univ., No. 10, A, 599-606.

Horie, S., 1969. Late Pleistocene limnetic history of Japanese ancient lakes Biwa, Yogo, Suwa, and Kizaki. Mitt. Internat. Verein. Limnol., No. 17, 436-445.

Horie, S., Mitamura, O., Kanari, S., Miyake, H., Yamamoto, A. and Fuji, N., 1971. Paleolimnological study on lacustrine sediments of Lake Biwa-ko. Disaster Prev. Res. Inst. Ann., Kyoto Univ., No. 14, B, 745-762.

Horie, S., (ed. 1972). Paleolimnology of Lake Biwa and the Japanese Pleistocene, 1, 93 pp.

Horie, S. and Yamamoto, A., 1977. Climatic change. In: Nihon no Daishiki-kenkyu, ed. Japan Assoc. Quaternary Res., Tokyo, Tokyo Univ. Press, 177-188.

Horie, S., Yaskawa, K., Yamamoto, A., Yokoyama, T. and Hyodo, M., 1980. Paleolimnology of Lake Kizaki. Arch. Hydrobiol. 89: 407-415.

Hoshino, Y., 1971. Some problems on the geomorphic development in the Miyazaki Plain, Southeastern Kyūshū. Quaternary Res. (Daiyonki Kenkyū) 10: 99-109.

Huzita, K. and Maeda, Y., 1971. Itami terrace with special reference to the late Pleistocene transgression in Japan. Jour. Geosci., Osaka City Univ. 14: 53-70.

Ikebe, N. and Yokoyama, T., 1976. General explanation of the Kobiwako Group – Ancient lake deposits of Lake Biwa –. Paleolimnology of Lake Biwa and the Japanese Pleistocene, 4, ed. S. Horie, 31-51.

Inman, D. L., 1952. Measures for describing the size distribution of sediments. Jour. Sed. Petrol. 22: 125-145.

Iseki, H., 1975. On the basal gravel beds of the recent deposits. Jour. Geogr., 84: 247-264.

Iseki, H., 1977. Holocene sea level changes. In: Nihon no Daishiki-kenkyu, ed. Japan Assoc. Quaternary Res., Tokyo, Tokyo Univ. Press, 89-97.

Ishida, S., 1970. The Osaka Group – The cyclic sediments of lacustrine and bay in Plio-Pleistocene, Japan. Quaternary Res. (Daiyonki Kenkyu) 9: 101-112.

Ishiwatari, R., 1967. Elementary composition of humic acids from lake sediments. Chikyu Kagaku, (Geochemistry) No. 1, 7-14.

Ishiwatari, R. and Hanya, T., 1973. Organic geochemistry of a 200-meter core sample from Lake Biwa. I. Identification of fatty acids by combined gas chromatography-mass spectrometry. Proc. Japan Acad. 49: 731-736.

Ishiwatari, R., 1975. Vertical distribution of mono- and di-carboxylic acids and polynuclear aromatic hydrocarbons in a 200-meter core sample from Lake Biwa. Paleolimnology of Lake Biwa and the Japanese Pleistocene, 3, ed. S. Horie, 285-299.

Ishiwatari, R. and Hanya, T., 1975. Organic geochemistry of a 200-meter core sample from Lake Biwa. II. Vertical distribution of mono- and di-carboxylic acids and polynuclear aromatic hydrocarbons. Proc. Japan. Acad. 51: 436-441.

Ishiwatari, R., 1976a. Normal alkane distribution in a 200-meter core sample from Lake Biwa. ---A preliminary report. Paleolimnology of Lake Biwa and the Japanese Pleistocene, 4, ed. S. Horie, 240-245.

Ishiwatari, R., 1976b. Fatty acids and perylene in a 200-meter core sample from Lake Biwa--- A discussion. Paleolimnology of Lake Biwa and the Japanese Pleistocene, 4, ed. S. Horie, 246-252.

Ishiwatari, R., 1976c. Isolation and characterization of kerogen and humic acid from a 200-meter core sample of Lake Biwa. Paleolimnology of Lake Biwa and the Japanese Pleistocene, 4, ed. S. Horie, 233-239.

Ishiwatari, R., 1977. Organic geochemistry of a 200-meter core sample from Lake Biwa. III. Isolation and characterization of kerogen and humic acid. Proc. Japan Acad., 53(B): 47-50.

Ishiwatari, R. and Kawamura, K., 1977. Variation of fatty acid composition in a sediment of Lake Biwa. Paleolimnology of Lake Biwa and the Japanese Pleistocene, 5, ed. S. Horie, 181-194.

Ishiwatari, R., Ogura, K. and Horie, S., 1977. Characterization of organic compounds in a sediment of Lake Haruna --- I. Fatty acids. Paleolimnology of Lake Biwa and the Japanese Pleistocene, 5, ed. S. Horie, 221-233.

Ishiwatari, R. and Kawamura, K., 1978. Organic geochemistry of a 200 meter core sample from Lake Biwa. IV. Variation of fatty acid composition in the upper 5 meter layers. Proc. Japan Acad. 54, Ser. B, 75-80.

Ishiwatari, R., Rohrback, B. G. and Kaplan, I. R., 1978. Hydrocarbon generation by thermal alteration of kerogen from different sediments. Amer. Assoc. Petrol. Geol. Bull. 62: 687-692.

Ishiwatari, R., 1979. Trimethyltetrahydrochrysene and 17βH-bishomohopanoic acid in Lake Biwa and Lake Haruna sediments: A research note. Paleolimnology of Lake Biwa and the Japanese Pleistocene, 7, ed. S. Horie, 290-301.

Ishiwatari, R., Ogura, K. and Horie, S., 1980. Organic geochemistry of a lacustrine sediment (Lake Haruna, Japan). Chem. Geol. 29: 261-280.

Itasaka, O., Ohyama, Y., Ohnishi, T and Hori, T., 1973. On the water qualities of Lake Biwa-ko and the heavy metal contents of bottom matters of the lake. Mem. Fac. Educ., Shiga Univ., Nat. Sci., No. 23, 25-36.

Itihara, M., 1960. Some problems of the Quaternary sedimentaries, Osaka and Akahi Areas. Earth Sci. (Chikyu Kagaku), No. 49, 15-25.

Itihara, M., 1961. Some problems of the Quaternary sedimentaries in the Osaka and Akasi Areas, Japan. Jour. Inst. Polytech., Osaka City Univ., Ser. G., 5, 13-30.

Itihara, M., Kamei, T., Mitsunashi, T., Suzuki, K. and Kuwano, Y., 1973. The basis of the Plio-Pleistocene boundary in Japan. Jour. Geosci., Osaka City Univ. 16: 25-49.

Itihara, M., Kamei, T., Mistunashi, T., Suzuki, K. and Kuwano, Y., 1977. The second report on the Plio-Pleistocene boundary in Japan. Neogene-Quaternary Boundary Proc., II Symposium, Bologna 1975, Gionale di Geologia (2) XLI, fasc. I-II, 265-273.

James, N. P., Mountjoy, E. W. and Omura, A., 1971. An early Wisconsin reef terrace at Barbados, West Indies, and its climatic implications. Geol. Soc. Amer. Bull. 82: 2011-2018.

Jeffrey, L. M., 1970. Lipids of marine waters. In: Organic Matter in Natural Waters, ed. D. W. Hood, Institute of Marine Science Univ. Alaska Occasional Pub., No. 1, 55-76.

Jenkin, B. M., Mortimer, C. H. and Pennington, W., 1941. The study of lake deposits. Nature 147: 496-500.

Johnsen, S. J., Dansgaard, W., Clausen, H. B. and Langway, C. C., Jr., 1972. Oxygen isotope profiles through the Antarctic and Greenland ice sheets. Nature 235: 429-434.

Johnson, E. A., Murphy, T. and Torreson, O. W., 1948. Pre-history of the Earth's magnetic field. Terr. Mag. Atomospheric Electr. 53: 349-372.

Jones, J. D. and Vallentyne, J. R., 1960. Biogeochemistry of organic matter. - I. Polypeptides and amino acids in fossils and sediments in relation to geothermometry. Geochim. Cosmochim. Acta 21: 1-34.

Kadota, S., 1973. A quantitative study of microfossils in the core sample from Lake Biwa-ko. Jap. Jour. Limnol. 34: 103-110.

Kadota, S., 1974. A quantitative study of the microfossils in a 200-meter core sample from Lake Biwa. Paleolimnology of Lake Biwa and the Japanese Pleistocene, 2, ed. S. Horie, 236-245.

Kadota, S., 1975. A quantitative study of the microfossils in a 200-meter-long core sample from Lake Biwa. Paleolimnology of Lake Biwa and the Japanese Pleistocene, 3, ed. S. Horie, 354-367.

Kadota, S., 1976. A quantitative study of the microfossils in a 200-meter-long core sample from Lake Biwa. Paleolimnology of Lake Biwa and the Japanese Pleistocene, 4, ed. S. Horie, 297-307.

Kagami, H. and Nasu, N., 1964. The Paleo-Kuji River – A sample of burried valley on the shelf due to the sea level rise during the post glacial age.- In: Studies on Oceanogr., 538-549.

Kaizuka, S., 1955. On the age of submarine shelves of southern Kanto. Geogr. Rev. Japan 28: 15-26.

Kaizuka, S. and Moriyama, A., 1969. Geomorphology and subsurface geology of the alluvial plain of the lower Sagami River, central Japan. Geogr. Rev. Japan 42: 85-105.

Kaizuka, S., Naruse, Y. and Matsuda, I., 1977. Recent formations and their basal topography in and around Tokyo Bay, Central Japan. Quaternary Res. 8: 32-50.

Kajiyama, H. and Itihara, M., 1972. The developmental history of the Osaka Plain with reference to the radio-carbon dates. Mem. Geol. Soc. Japan, No. 7 101-112.

Kamei, T., 1962. Some problems on the succession of the Quaternary mammalian faunas in Japan. Earth Sci. (Chikyu Kagaku), No. 60-61, 23-34.

Kamei, T., 1966. Notes on *Elephas shigensis* (MATSUMOTO and OZAKI) from the Osaka Group and the Paleo-Biwa group. Mem. Coll. Sci., Univ. Kyoto, Ser. B, 32: 381-399.

Kamei, T. and Setoguchi, T., 1970. Some remarks on mammalian faunas in the early Pleistocene. Quaternary Res. (Daiyonki Kenkyu) 9: 158-163.

Kamei, T. and Otsuka, H., 1981. The Plio-Pleistocene stratigraphy of Japan in relation to Proboscidean evolution. Proc. Neogene Quaternary boundary field Conf., India, 1979, 83-88.

Kanari, S. and Takenoya, Y., 1975. An estimation of the deposition rate of Lake Biwa during the past 560,000 years. Paleolimnology of Lake Biwa and the Japanese Pleistocene, 3, ed. S. Horie, 235-240.

Kaneda, T., 1967. Fatty acids in the genus *Bacillus*. I. Iso- and anteiso-fatty acids as characteristic constituents of lipids in 10 species. Jour. Bacteriol. 93: 894-903.

Kanto Loam Research Group, 1960. Some problems of the Kanto Loam (III). Earth Sci. (Chikyu Kagaku), No. 46, 1-18.

Kanto Quaternary Research Group, 1974. Problems on the Quaternary system in Yokohama City area, south Kanto (1), (2). Earth Sci. (Chikyu Kagaku), 28: 155-171, 218-235.

Kaseno, Y., Miura, S. and Fujii, S., 1972. On the formation processes of the coastal plain of Hokuriku Region, Japan. Mem. Geol. Soc. Japan, No. 7, 91-100.

Kawai, N., 1951. Magnetic polarization of Tertiary rocks in Japan. Jour. Geophys. Res. 56: 73-79.

Kawai, N., Yaskawa, K., Nakajima, T., Torii, M. and Horie, S., 1972. Oscillating geomagnetic field with a recurring reversal discovered from Lake Biwa. Proc. Japan Acad. 48: 186-190.

Kawai, N., 1974. Restrained photosynthesis during Brunhes field transition. Paleolimnology of Lake Biwa and the Japanese Pleistocene, 2, ed. S. Horie, 59-64.

Kawai, N., Yaskawa, K., Nakajima, T., Torii, M. and Natsuhara, N., 1975. Voice of geomagnetism from Lake Biwa. Paleolimnology of Lake Biwa and the Japanese Pleistocene, 3, ed. S. Horie, 143-160.

Kawai, N., Otofuji, Y. and Kobayashi, K., 1976. Paleomagnetic study of deep-sea sediments using thin sections. Jour. Geomag. Geoelectr. 28: 395-412.

Kawai, N., Sato, T. and Sueishi, T., 1977. The study of palaeomagnetism using the Pacific ocean sediments. Paleolimnology of Lake Biwa and the Japanese Pleistocene, 5, ed. S. Horie, 264-275.

Kawai, N., Sato, T., Sueishi, T. and Kobayashi, K., 1977. Paleomagnetic study of deep-sea sediments from the Melanesian Basin. Jour. Geomag. Geoelectr. 29: 211-223.

Kawamura, K., Ishiwatari, R. and Yamazaki, M., 1978. Polyunsaturated fatty acids in surface lacustrine sediments: A preliminary report. Paleolimnology of Lake Biwa and the Japanese Pleistocene, 6, ed. S. Horie, 206-213.

Kawamura, K., Ishiwatari, R. and Yamazaki, M., 1980. Identification of polyunsaturated fatty acids in surface lacustrine sediments. Chem. Geol. 28: 31-39.

Kawamura, K. and Ishiwatari, R., 1981. Polyunsaturated fatty acids in a lacustrine sediment as a possible indicator of paleoclimate. Geochim. Cosmochim. Acta. 45: 149-155.

Kent, D. V., 1973. Post-depositional remanent magnetisation in deep-sea sediment. Nature 246: 32-34.

Kira, T. and Iida, K., 1958. Ecological system in the mountainous areas. In: Seibutsu to Kankyō, ed. H. Tamiya et al., Gendai Seibutsugaku Kōza, 5, Tokyo, Kyoritsu-shuppan, K.K., 231-269.

Kobayashi, J., Morii, F., Muramoto, S., Nakashima, S., Teraoka, H. and Horie S., 1975. Distribution of arsenic, cadmium, lead, zinc, copper, and manganese contained in the bottom sediment of Lake Biwa. Jap. Jour. Limnol. 36: 6-15.

Kobayashi, K., 1958. Quaternary glaciation of the Japan Alps. Jour. Fac. Liberal Arts Sci., Shinshu Univ. No. 8 13-67.

Kobayashi, K. and Shimizu, H., 1966. Significance of the Ikenotaira Interstadial indicated by moraines on Mt. Kumazawa of the Kiso Mountain Range, central Japan. Jour. Fac. Sci., Shinshu Univ. 1: 97-113.

Koizumi, I., 1975. Diatom events in Late Cenozoic deep-sea sequences in the north Pacific. Jour. Geol. Soc. Japan 81: 567-578.;

Kokawa, S., 1966. Late Cenozoic floras of the Boso Peninsula, Japan, I. – Upper Pleistocene floral change. Jour. Biol., Osaka City Univ. 17: 105-149.

Konishi, K., Omura, A. and Nakamichi, O., 1974 Radiometric coral ages and sea level records from the late Quaternary reef complexes of the Ryukyu Islands. Proc. Second Internat. Coral Reef Symp. 2: 595-613.

Koyama, M., 1975. Distribution of inorganic elements in 200-meter-long core samples from Lake Biwa-ko by neutron activation analysis (II). Correlation between distribution patterns of elements and sedimentary environments. Paleolimnology of Lake Biwa and the Japanese Pleistocene, 3, ed. S. Horie, 306-315.

Koyama, T., 1972. Geochemical studies of a 200 meters core sample from Lake Biwa. The determination of carbon and nitrogen. Proc. Japan Acad. 48: 505-509.

Koyama, T., Handa, N., Ishiwatari, R., Ogura, K., Mizuno, A., Terashima, M., Koyama, M., Hori, T., Okuda, J., Fujinaga, T., Takeuchi, T. and Horie, S., 1973. Chemical studies of a 200 meter core sample from Lake Biwa-ko. Jap. Jour. Limnol. 34: 75-88.

Koyama, M., Matsushita, R., Kawashima, M., Takamatsu, T., Okamoto, K., Fuwa, K., Horie, S., Hori, T. and Fujinaga, T., 1976. Inorganic compositions of 200 meter and 1000 meter core samples from Lake Biwa. Paleolimnology of Lake Biwa and the Japanese Pleistocene, 4, ed. S. Horie, 276-296.

Koyama, T., 1976. Geochemical studies on a 200-meter core sample from Lake Biwa -. The vertical distribution of carbonate - and stable organic-carbon and acid soluble phosphate-phosphorus -. Paleolimnology of Lake Biwa and the Japanese Pleistocene, 4, ed. S. Horie, 205-213.

Krishnaswami, S., Lal, D., Prabhu, N. and Macdougall, D., 1974. Characteristics of fission tracks in zircon: Applications to geochronology and cosmology. Earth Planet. Sci. Lett. 22: 51-59.

Kukla, G. J., 1975. Loess stratigraphy of central Europe. In: After the Australopithecines. Stratigraphy, Ecology and culture change in the middle Pleistocene, ed. K. W. Butzer and Isaac, G. LL., Hague, Paris, Mouton Pub. 99-188.

Kvenvolden, K. A., 1967. Normal fatty acids in sediments. Jour. Amer. Oil Chem. Soc. 44: 628-636.

Laflamme, R. E. and Hites, R. A., 1978. The global distribution of polycyclic aromatic hydrocarbons in recent sediments. Geochim. Cosmochim. Acta 42: 289-303.

Lal, D. and Somayajulu, B. L. K., 1975. On the importance of studying magnetic susceptibility stratigraphy and geochronology of Lake Biwa sediments. Paleolimnology of Lake Biwa and the Japanese Pleistocene, 3, ed. S. Horie, 530-535.

Lamb, H. H., 1971. Volcanic activity and climate. Palaeogeography, Palaeoclimatology, Palaeoecology 10: 203-230.

Leo, R. F. and Parker, P. L., 1966. Branched - chain fatty acids in sediments. Science 152: 649-650.

Lundqvist, G., 1927. Bodenablagerungen und Entwicklungstypen der Seen. Binnengewässer 2: 124 s.s.

Lindqvist, J., 1974. Outlines of the Weichsel Glacial in Sweden. Geol. Fören. I Stockholm Förh. 96, 327-339.

Machida, H. and Suzuki, M., 1971. Absolute age of volcanic ash and chronology of the late Quaternary by fission track method. Kagaku 41: 263-270.

Machida, H., Arai, F., Murata, A., and Hakamata, K., 1974. Correlation and chronology of the middle Pleistocene tephra layers in south Kanto. Jour. Geogr. 83: 302-338.

Machida, H., 1975. Pleistocene sea level of south Kanto, Japan, analysed by tephrochronology. Roy. Soc. New Zealand Bull. (Quaternary Studies), 13, 215-222.

Maeda, Y., 1976. The sea-level changes of Osaka Bay from 12,000 B.P. to 6,000 B.P. Kokogaku to Shizen-kagaku No. 9 31-41.

Maenaka, K., Yokoyama, T. and Ishida, S., 1977. Paleomagnetic stratigraphy and biostratigraphy of the Plio-Pleistocene in the Kinki District, Japan. Quaternary Res. 7: 341-362.

Makiyama, J., 1924. The occurrence of *Elephas trogontherii* in Japan. Jap. Jour. Geol. Geogr. 3: 55-57.

Makiyama, J., 1938. Japonic Proboscidea. Mem. Coll. Sci., Kyoto Imp. Univ., Ser. B, 14: 1-59.

Manabe, K., 1974. Geomagnetic reversals recorded in the Late Pleistocene sediments in Odaka, Fukushima Prefecture. Quaternary Res. (Daiyonki Kenkyu) 13: 21-25.

Matsuda, T., Ota, Y., Ando, M. and Yonekura, N., 1978. Fault mechanism and recurrence time of major earthquakes in the southern Kanto district, Japan, as deduced from coastal terrace data. Amer. Geol. Soc., Bull. 89: 1610-1618.

Matsumoto, E., 1975. Sedimentation rates in Lake Biwa by Pb-210 method. Paleolimnology of Lake Biwa and the Japanese Pleistocene, 3, ed. S. Horie, 228-234.
Matsumoto, H., 1924. Preliminary note on fossil elephants in Japan. Jour. Geol. Soc., Tokyo 31: 255-272.
Matsumoto, H. and Ozaki, H., 1957. On a new geological subspecies of *Archidiskodon paramammonteus* MATSUMOTO discovered at Ono, Shiga town, Province of Omi. Bull. National Sci. Museum, Tokyo 4: 355-357.
Matsushima, Y. and Ohshima, K., 1974. Littoral molluscan fauna of the Holocene Climatic Optimum (5,000-6,000 y. B.P.) in Japan. Quaternary Res. (Daiyonki Kenkyu) 13: 135-159.
Matsushima, Y., 1976. The alluvial deposits in the southern part of the Miura Peninsula, Kanagawa Prefecture. Bull. Kanagawa Pref. Museum, No. 9 87-162.
Mesolella, K., Matthews, R. K., Broecker, W. S. and Thurber, D. L., 1969. The astronomical theory of climatic change: Barbados data. Jour. Geol. 77: 250-274.
Miki, S., 1969. *Protosequoia* (n.g.) in Taxodiaceae from *Pinus trifolia* beds in central Honshu, Japan. Proc. Japan Acad.. 45: 727-732.
Milankovitch, M., 1930. Mathematische Klimalehre und Astronomische Theorie der Klimaschwankungen. In: Handbuch der Klimatologie. 1: Tl. A, herausg. W. Köppen und R. Geiger, Berlin, Gebrüder Borntraeger, 176 ss.
Minato, M., 1966. The final stage of land bridges in the Japanse Islands. Earth Sci. (Chikyu Kagaku), No. 85-86, 2-11.
Minato, M., 1972. Late Quaternary geology in northern Japan. In: Quaternary Geology, ed. J. G. Fyles et al., 24th Internat. Geol. Congr. in 1972 (Montreal, Canada), Sec. 12, 63-71.
Minato, M., 1974. Nihon no Daiyonkei. Tokyo, Tsukiji-shokan, 167 pp.
Mogi, A. and Sato, T., 1975. Submarine geomorphology of the continental margin around Japan, I, II. Kagaku 45: 551-559, 622-629.
Mopper, K. and Larsson, K., 1978. Uronic and other organic acids in Baltic Sea and Black Sea sediments. Geochim. Cosmochim. Acta 42: 153-163.
Mori, Sh., 1974. Diatom succession in a core from Lake Biwa. Paleolimnology of Lake Biwa and the Japanese Pleistocene, 2, ed. S. Horie, 247-254.
Mori, Sh., 1975. Vertical distribution of diatoms in core samples from Lake Biwa. Paleolimnology of Lake Biwa and the Japanese Pleistocene, 3, ed. S. Horie, 368-391.
Mori, Sh. and Horie, S., 1975. Diatoms in a 197.2 meters core sample from Lake Biwa-ko. Proc. Japan Acad. 51: 675-679.
Müller, H., 1970. Ökologische Veränderungen im Otterstedter See im Laufe der Nacheiszeit. Ber. Naturhist. Ges. 114: 33-47.
Nagata, T., Hirao, K. and Yoshikawa, H., 1949. Remanent magnetization of 'Pleistocene' deposits - Palaeomagnetism of Japan. - Jour. Geomag. Geoelectr. 1: 52-58.
Nakai, N., 1960. Carbon isotope fractionation of natural gas in Japan. Jour. Earth Sci., Nagoya Univ. 8: 174-180.
Nakai, N., 1972. Carbon isotopic variation and the paleoclimate of sediments from Lake Biwa. Proc. Japan Acad. 48: 516-521.
Nakai, N., 1973. Carbon isotopic composition of Lake Biwa-ko sediments and the environmental change in the depositional history. Jap. Jour. Limnol. 34: 89-96.
Nakai, N., 1975. The climatic variation during late-Pleistocene based on carbon isotope study of 200-meter-long core samples from Lake Biwa, Japan. Paleolimnology of Lake Biwa and the Japanese Pleistocene, 3, ed. S. Horie, 250-256.
Nakai, N. and Shirai, Y., 1977. Paleotemperature variations in central Japan based on the stable isotope studies of Lake Biwa sediments and speleothems. Paleolimnology of Lake Biwa and the Japanese Pleistocene, 5, ed. S. Horie, 149-164.
Nakai, N. and Shirai, Y., 1978. Paleoclimatic features in central Japan based on the stable isotopic compositions of Lake Biwa sediments and speleothems. Proc. Japan Acad. 54, Ser. B: 81-86.

Nakajima, T., Yaskawa, K., Natsuhara, N., Kawai, N. and Horie, S., 1973. Very short period geomagnetic excursion 18,000 yr BP. Nature Phys. Sci. 244: 8-10.

Nakajima, T. and Kawai, N., 1974. Secular geomagnetic variation excavated from Lake Biwa Basin. Paleolimnology of Lake Biwa and the Japanese Pleistocene, 2, ed. S. Horie, 65-76.

Nakamura, J., 1973. Palynological aspects of the Late Pleistocene in Japan. Quaternary Res. (Daiyonki Kenkyu) 12: 29-37.

Nakata, T., Takahashi, T. and Koba, M., 1978. Holocene-emerged coral reefs and sea-level changes in the Ryukyu Islands. Geogr. Rev. Japan 51: 87-108.

Naruse, Y., 1968. Quaternary crustal movements in the Kanto region. Mem. Geol. Soc. Japan, No. 2, 29-32.

Naumann, E., 1882. Ueber Japanische Elephanten der Vorzeit. Palaeontographica 28: 1-40.

Negoro, K., 1960. Studies on the diatom-vegetation of Lake Biwa-ko (First report). Jap. Jour. Limnol. 21: 200-220.

Negoro, K., 1967. An analytical study of diatom shells in the bottom deposits of Lake Biwa-ko, based on a new core-sample. Jap. Jour. Limnol. 28: 132-135.

Negoro, K., 1968. An analytical study of diatom shells in the bottom deposits of Lake Yogo-ko. Mem. Fac. Sci., Kyoto Univ., Ser. Biol. 1: 121-124.

Nier, A. O., 1950. A redetermination of the relative abundances of the isotopes of carbon, nitrogen, oxygen, argon and potassium. Phys. Rev. 77: 789-793.

Nipkow, F., 1961. Die Rädertiere im Plankton des Zürichsees und ihre Entwicklungsphasen. Schweiz. Zeitschr. Hydrol. 23: 398-461.

Nishimura, S. and Yokoyama, T., 1973. Fission-track ages of volcanic ashes in 200 m core sample of Lake Biwa, Japan. Proc. Japan Acad. 49: 615-618.

Nishimura, S. and Yokoyama, T., 1974. Fission-track ages of volcanic ashes of core samples of Lake Biwa and Kobiwako Group. Paleolimnology of Lake Biwa and the Japanese Pleistocene, 2, ed. S. Horie, 38-46.

Nishimura, S., 1975. On the value of the decay constant for spontaneous fission of uranium-238. Mem. Fac. Sci., Kyoto Univ., Ser. Geol. Mineral. 41: 15-19.

Nishimura, S. and Yokoyama, T., 1975. Fission-track ages of volcanic ashes of core samples of Lake Biwa and the Kobiwako Group (2). Paleolimnology of Lake Biwa and the Japanese Pleistocene, 3, ed. S. Horie, 138-142.

Ogura, K. and Hanya, T., 1973. The cholestanol-cholesterol ratio in a 200-meter core sample of Lake Biwa. Proc. Japan Acad. 49: 201-204.

Ogura, K. and Ishiwatari, R., 1975. Total lipids in the sediments of 200 M-long core samples. Paleolimnology of Lake Biwa and the Japanese Pleistocene, 3, ed. S. Horie, 277-284.

Ogura, K., 1976. Estimation of paleoenvironments from the characterization of the origin of sterols in the sediments. Paleolimnology of Lake Biwa and the Japanese Pleistocene, 4, ed. S. Horie, 259-266.

Ogura, K. and Ishiwatari, R., 1976. Total lipids in the sediments of 200 M-long core samples. II. Observed lipid concentrations. Paleolimnology of Lake Biwa and the Japanese Pleistocene, 4, ed. S. Horie, 267-275.

Ogura, K., 1977. Relationship between the calculated amounts of sterols with respect to origins and the carbon isotopic ratio ($\delta^{13}C$). Paleolimnology of Lake Biwa and the Japanese Pleistocene, 5, ed. S. Horie, 195-205.

Ogura, K., 1978. Concentrations of total lipids and their significance as an index of paleoclimates in the 200 m long core sample of Lake Biwa. Proc. Japan Acad. 54, Ser. B: 145-150.

Oka, Y., 1970. Some problems on late Pleistocene sea-level. Geogr. Sci. (Chiri-Kagaku) No. 14 11-21.

Omori, H., 1975. Coastal terraces in the Esashi district of the Oshima Peninsula in Hokkaido. Quaternary Res. (Daiyonki Kenkyu) 14: 63-76.

Omoto, K. and Ouchi, Y., 1978. Some data on the Holocene sea-level changes in the Sendai Coastal Plain, northern Japan. Geogr. Rev. Japan 51: 158-175.

Ono, Y. and Hirakawa, K., 1975. Glacial and periglacial morphogenetic environments around the Hidaka Range in the Würm glacial age. Geogr. Rev. Japan 48: 1-26.

O'Neil, J. R., Clayton, R. N. and Mayeda, T. K., 1969. Oxygen isotope fractionation in divalent metal carbonates. Jour. Chem. Phys. 51: 5547-5558.

Opdyke, N. D., 1972. Paleomagnetism of deep-sea cores. Rev. Geophys. Space Phys. 10: 213-249.

Orr, W. L. and Grady, J. R., 1967. Perylene in basin sediments off southern California. Geochim. Cosmochim. Acta 31: 1201-1209.

Ota, Y., 1975. Late Quaternary vertical movement in Japan estimated from deformed shorelines. Roy. Soc. New Zealand Bull. (Quaternary Studies), 13: 231-239.

Ota, Y., Matsuda, T. and Naganuma, K., 1976. Tilted marine terraces of the Ogi Peninsula, Sado Island, central Japan, related to the Ogi earthquake of 1802. Zisin (Jour. Seismol. Soc. Japan), Second Ser. 29: 55-70.

Ota, Y., Machida, H., Hori, N., Konishi, K. and Omura, A., 1978. Holocene raised coral reefs of Kikai-jima (Ryukyu - Islands). An approach to Holocene sea level study. - Geogr. Rev. Japan 51: 109-130.

Otuka, Y., 1948. Nihontō no Oitachi. Tokyo, Kyoto, Ōyashima Shuppan K.K., 133 pp.

Paracas, J. G., 1970. Geochemistry of carbohydrate. University of Minnesota Doctoral Dissertation (1959), referred to Non-marine organic geochemistry, ed. F. M. Swain, Cambridge, Cambridge University Press, 211-245.

Paterson, W. S. B., Koerner, R. M., Fisher, D., Johnsen, S. J., Clausen, H. B., Dansgaard, W., Bucher, P. and Oeschger, H., 1977. An oxygen-isotope climatic record from the Devon Island ice cap, Arctic Canada. Nature 266: 508-511.

Pearman, G. I., Francey, R. J. and Fraser, P. J. B., 1976. Climatic implications of stable carbon isotopes in tree rings. Nature 260: 771-773.

Rittenberg, S. C., Emery, K. O., Hulsemann, J., Degens, E. T., Fay, R. C., Reuter, J. H., Grady, J. R., Richardson, S. H. and Bary E. E., 1963. Biogeochemistry of sediments in experimental Mohole. Jour. Sediment. Petrol. 33: 140-172.

Rogers, M. A., 1970. Organic geochemistry of some Devonian black shales from eastern North America. - Carbohydrate. University of Minnesota Doctoral Dissertation (1965), referred to Non-marine organic geochemistry, ed. F. M. Swain, Cambridge, Cambridge University Press, 211-245.

Sackett, W. M., Eckelmann, W. R., Bender, M. L. and Be Allan, W. H., 1965. Temperature dependence of carbon isotope composition in marine plankton and sediments. Science 148: 235-237.

Sakaguchi, Y., 1963. On the postglacial sea level changes in Japan. Quaternary Res. (Daiyonki Kenkyu) 2: 211-219.

Sakaguchi, Y., 1968. Age of emergence of Central Kanto Plain during the Holocene. Quaternary Res. (Daiyonki Kenkyu) 7: 57-58.

Sanger, J. E. and Gorham, E., 1970. The diversity of pigments in lake sediments and its ecological significance. Limnol. Oceanogr. 15: 59-69.

Sasajima, S., 1977. Paleoenvironmental changes deduced from a 200 M core sample in Lake Biwa, central Japan: A review. Paleolimnology of Lake Biwa and the Japanese Pleistocene, 5, ed. S. Horie, 19-35.

Schwendinger, R. B. and Erdman, J. G., 1964. Sterols in recent aquatic sediments. Science 144: 1575-1576.

Shackleton, N. J. and Opdyke, N. D., 1973. Oxygen isotope and palaeomagnetic stratigraphy of equatorial Pacific core V28-238; Oxygen isotope temperatures and ice volumes on a 10^5 year and 10^6 year scale. Quaternary Res. 3: 39-55.

Shackleton, N. J. and Opdyke, N. D., 1976. Oxygen-isotope and paleomagnetic stratigraphy of Pacific core V28-239, late Pliocene to latest Pleistocene. Geol. Soc. Amer. Mem. 145: 449-464.

Shiga-ken, 1974. Shiga-ken no Genzon Shokusei to Kichō Shizen. 32 sheets, Otsu.

Shigesawa, K., 1957. Studies on the Shigaraki Clay from Shiga Prefecture. Mem. Fac. Liberal Arts Educ., Shiga Univ., No. 7, 63-68.

Shigesawa, K. and Nishikawa, K., 1963. On the tuffaceous sediments from KOBIWAKO-SO in Hino-cho district, Shiga Prefecture. Mem. Fac. Liberal Arts Educ., Shiga Univ., No. 13, 27-30.

Skvortzow, B. W., 1936. Diatoms from Biwa Lake, Honshu Island, Nippon. Philippine Jour. Sci. 61: 253-296.

Smith, F. and Montgomery, R., 1959. The Chemistry of Plant Gums and Mucilages and some related polysaccharides. New York, Reinhold Pub. Corp., 627 pp.

Smith, J. D. and Foster, J. H., 1969. Geomagnetic reversal in Brunhes Normal Polarity Epoch. Science 163: 565-567.

Smith, P. V., Jr., 1954. Studies on origin of petroleum: Occurrence of hydrocarbons in recent sediments. Bull. Amer. Assoc. Petrol. Geol. 38: 377-404.

Spyckerelle, C., Greiner, A. Ch., Albrecht, P. and Ourisson, G., 1977. Aromatic hydrocarbons from geological sources. III. A tetrahydrochrysene derived from triterpenes in recent and old sediments: 3,3,7-trimethyl-1,2,3,4-tetrahydrochrysene. Jour. Chem. Res. (M): 3746-3777.

Steel, G. and Henderson, W., 1972. Isolation and characterization of stanols from the Green River Shale. Nature 238: 148-150.

Stevenson, F. J., 1956. Isolation and identification of some amino compounds in soils. Soil Sci. Soc. Amer. Proc. 20: 201-204.

Sueishi, T., Kawai, N. and Kobayashi, K., 1977. Magnetic stability of deep-sea sediments. Rock Magnetism and Paleogeophycs 4: 29-35.

Sugimura, A., 1950. On the submarine terraces along the coast of Kanto Region and the others. Geogr. Rev. Japan 23: 10-16.

Sugimura, A. and Naruse, Y., 1954, 1955. Changes in sea level, seismic upheavals, and coastal terraces in the southern Kanto region, Japan. (I) (II). Jap. Jour. Geol. Geogr. 24: 101-113, 26: 165-176.

Suzuki, K. and Kamei, T., 1969. Changes in flora and migration of mammalian faunas during the Quaternary. Kagaku 39: 19-27.

Swain, F. M., Venteris, G. and Ting, F., 1964. Relative abundance and order of stability of amino acids in some aquatic plants and associated freshwater sediments. Jour. Sediment. Petrol 34: 25-45.

Swain, F. M., Bratt, J. M. and Kirkwood, S., 1967. Carbohydrate components of some Paleozoic plant fossils. Jour. Paleontol. 41: 1549-1554.

Swain, F. M., 1970a. Bitumens of non-marine sediments and sedimentary rocks. In: Non-marine Organic Geochemistry, Cambridge, Cambridge University Press, 113-173.

Swain, F. M., 1970b. Carbohydrates of non-marine deposits. In: Non-marine Organic Geochemistry. Cambridge, Cambridge University Press, 211-245.

Taguchi, K. and Sasaki, K., 1971. Organo-sedimentological study of the 'Uetsu geosynclinal sediments', with special reference to the significance of polycyclic aromatic compounds in the sediments. Mem. Geol. Soc. Japan, No. 6, 89-96.

Tai, A., 1973. A study on the pollen stratigraphy of the Ōsaka Group, Pliocene-Pleistocene deposits in the Ōsaka basin. Mem. Fac. Sci., Kyoto Univ., Ser. Geol. Mineral. 39: 123-165.

Takahashi, J., Okamoto, K. and Fuwa, K., 1977. Determination of mercury in the 1,000 meter core sample. Paleolimnology of Lake Biwa and the Japanese Pleistocene, 5, ed. S. Horie, 165-169.

Takahashi, J., Tao, H., Okamoto, K., Haraguchi, H. and Fuwa, K., 1979., Determination of marcury in the 1,000 meter core samples of Lake Biwa (Part II). Paleolimnology of Lake Biwa and the Japanese Pleistocene, 7, ed. S. Horie, 251-256.

Terashima, M. and Mizuno, A., 1974. Preliminary result of amino acid and amino-sugar determination on a 200-meter core sample from Lake Biwa. Paleolimnology of Lake Biwa and the Japanese Pleistocene, 2, ed. S. Horie, 219-224.

Toma, T., 1974. Tephrochronological considerations on emergence of the Shimosueyoshi surface in Yokohama and its surroundings, Japan. Quaternary Res. (Daiyonki Kenkyu) 13: 199-215.
Toyoshima, Y., 1978. Postglacial sealevel change along San'in district, Japan. Geogr. Rev. Japan 51: 147-157.
Treibs, A., 1934. Chlorophyll- und Häminderivate in bituminösen Gesteinen, Erdölen, Erdwachsen und Asphalten. Ann. Cem. 510: 42-62.
Umitsu, M., 1976. Geomorphic development of the Tsugaru Plain in the Holocene Period. Geogr. Rev. Japan 49: 714-735.
Vallentyne, J. R., 1955. Sedimentary chlorophyll determination as a paleobotanical method. Can. Jour. Bot. 33: 304-313.
Vallentyne, J. R., 1963. Geochemistry of carbohydrates. In: Organic Geochemistry, ed. I. A. Breger, Oxford, London, Paris, New York, Pergamon Press, 456-502.
Vallentyne, J. R., 1965. Two aspects of the geochemistry of amino acids. In: The Origin of Prebiological System and their Molecular Matrices, ed. S. W. Fox, New York, London, Academic Press, 105-125.
Van Campo, M. and Elhai, H., 1956. Étude comparative du pollens de quelques chênes. Application à une tourbière normande. Bull. Soc. Bot. France. 103: 254-260.
van der Spoel-Walvius, M. R., 1963. Les caractèristiques de l'exine chez quelques especes de *Quercus*. Acta Bot. Neerlandica 12: 525-532.
Van Donk, J., 1976. O^{18} record of the Atlantic ocean for the entire Pleistocene epoch. Geol. Soc. Amer. Mem. 145: 147-163.
Wakeham, S. G., 1977. Synchronous fluorescence spectroscopy and its application to indigenous and petroleum-derived hydrocarbons in lacustrine sediments. Environment-Sci. Technol. 11: 272-276.
Weertman, J., 1976. Milankovitch solar radiation variations and ice age ice sheet sizes. Nature 261: 17-20.
Wesenberg-Lund, C., 1896. Om Ferskvandsfaunaens Kitin- og Kisellevninger i Trvelagene. Medd. Dansk Geol. Foren. No. 3 51-84.
Wetzel, R. G., 1970. Recent and postglacial production rates of a marl lake. Limnol. Oceanogr. 15: 491-503.
Wickman, F. E., 1952. Variation in the relative abundance of the carbon isotopes in plants. Geochim. Cosmochim. Acta 2: 243-254.
Willman, H. B. and Frye, J. C., 1970. Pleistocene stratigraphy of Illinois. Illinois St. Geol. Surv., Bull. No. 94, 204 pp.
Wilson, A. T. and Grinsted, M. J., 1977. $^{12}C/^{13}C$ in cellulose and lignin as paleothermometers. Nature 265: 133-135.
Wollin, G., Ericson, D. B., Ryan, W. B. F. and Foster, J.H., 1971. Magnetism of the Earth and climatic changes. Earth Planet. Sci. Lett. 12: 175-183.
Yabe, H., 1929. The latest land connection of the Japanese Islands to the Asiatic continent. Proc. Imp. Acad. 5: 167-169.
Yamamoto, A., 1973. Pluvials in the post-glacial age. Pre-print, Jap. Soc. Limnol. symp. II-9.
Yamamoto, A., 1974a. Grain size analysis of bottom sediments and its application to the paleoclimatology in Lake Biwa. Paleolimnology of Lake Biwa and the Japanese Pleistocene, 2, ed. S. Horie, 113-134.
Yamamoto, A., 1974b. An estimation of the paleoclimatic changes in Quaternary Age, based on the grain size analysis of bottom sediments in Lake Biwa. Bull. Osaka Electro-Communication Univ., Nat. Sci. No. 10 127-150.
Yamamoto, A., 1974c. Granulometric data for the 200 meters core sample from Lake Biwa (I) Bull. Osaka Electro-Communication Univ., Nat. Sci. No. 10 151-185.
Yamamoto, A., Kanari, S., Fukuo, Y. and Horie, S., 1974. Consolidation and dating of the sediments in core samples from Lake Biwa. Paleolimnology of Lake Biwa and the Japanese Pleistocene, 2, ed. S. Horie, 135-144.

Yamamoto, A., 1975. Grain sizes of the core sediments and variations of paleoprecipitation in Lake Biwa during the last 3 hundred thousand years. Paleolimnology of Lake Biwa and the Japanese Pleistocene, 3, ed. S. Horie, 209-225.
Yamamoto, A. and Higashihara, M., 1975. Granulometric data for the 200 meters core sample from Lake Biwa. (II). Bull. Osaka Electro-Communication Univ., Nat. Sci. No. 11 257-268.
Yamamoto, A., 1976. Palcoprecipitational change estimated from the grain size variations in the 200 M-long core from Lake Biwa. Paleolimnology of Lake Biwa and the Japanese Pleistocene, 4, ed. S. Horie, 179-203.
Yamamoto, A. and Higashihara, M., 1976. Granulometric data for the 200 meters core sample from Lake Biwa (III). Bull. Osaka Electro-Communication Univ., Nat. Sci. No. 12 205-215.
Yamamoto, A., 1977a. On the grain size and density of bottom sediments from Osaka Bay. Pre-print, Oceanogr. Soc. Japan., 189.
Yamamoto, A., 1977b. The structure of density and grain size variations seen near the 130 M-layer in the 200 M-long core sample from Lake Biwa. Paleolimnology of Lake Biwa and the Japanese Pleistocene, 5, ed. S. Horie, 125-136.
Yamamoto, A. and Higashihara, M., 1978. Granulometric data for the 200 meters core sample from Lake Biwa (IV). Bull. Osaka Electro-Communication Univ., Nat. Sci. No. 14 117-130.
Yaskawa, K., 1973. Significance of paleomagnetic investigation of lake sediments. Jap. Jour. Limnol. 34: 55-62.
Yaskawa, K., Nakajima, T., Kawai, N., Torii, M., Natsuhara, N. and Horie, S., 1973. Palaeomagnetism of a core from Lake Biwa (1). Jour. Geomag. Geoelectr. 25: 447-474.
Yonekura, N., 1975. Quaternary tectonic movements in the outer arc of southwest Japan with special reference to seismic crustal deformations. Bull. Dept. Geogr., Univ. Tokyo, No. 7, 19-71.
Youngblood, W. W. and Blumer, M., 1975. Polycyclic aromatic hydrocarbons in the environment: homologous series in soils and recent marine sediments. Geochim. Cosmochim. Acta 39: 1303-1314.
Yoshikawa, T., Kaizuka, S. and Ota, Y., 1964. Mode of crustal movement in the late Quaternary on the southeast coast of Shikoku, southwestern Japan. Geogr. Rev. Japan 37: 627-648.
Zagwijn, W. H., 1975. Variations in climate as shown by pollen analysis, especially in the lower Pleistocene of Europe. In: Ice ages: Ancient and modern, ed. A. E. Wright and F. Moseley, Liverpool, Seel House Press, 137-152.

Biogeography of Lake Biwa

Masuzo Uéno

The characteristics of the fauna and flora

Pisces

The fish inhabiting the present Lake Biwa belong to at least sixty-three species, including subspecies and those introduced from foreign countries. In this respect, the lake differs markedly from other Japanese lakes, in none of which the constituent species of their fish fauna exceed twenty. Sixty-three species stated contain a number of fishes peculiar to Lake Biwa, that is, the 'endemic species' which never occur in other fresh waters in Japan. Such forms are: a cyprinid *Gnathopogon elongatus caerulescens* (Sauvage), two subspecies of silver carp *Carassius auratus grandoculis* Temminck et Schlegel and *C. auratus cuvieri* Temm. et Schl., two species of catfish *Parasilurus lithophilus* Tomoda and *P. biwaensis* Tomoda, as well as a goby *Chaenogobius isaza* Tanaka. In this connection, four cyprinids, *Squalidus biwae* (Jordan et Snyder), *S. japonicus* Sauvage, *Ischikauia steenackeri* (Sauvage) and *Opsariichthys uncirostris* (Temm. et Schl.) are also noticed, though they are not strictly endemic because of their occurrence also somewhere in other districts especially around the lake. A rare loach *Leptobotia curta* (Temm. et Schl.) belongs to this category, too. It is presumed, however, that the center of their distribution is undoubtedly Lake Biwa, so that it will be safe to say without any fear of objection that they belong to the species either endemic or semi-endemic to that lake.

Mollusca

The mollusks inhabiting Lake Biwa can be referred to forty-four species including subspecies, the number of which is seventy percent of all molluscan species recorded from Japanese fresh waters. It is conspicuous that more than the half or eighteen species of the molluscan fauna of the present Lake Biwa are endemic. The Unionidae is especially striking in having the following six

Horie, S. Lake Biwa

ISBN 90 6193 095 2. Printed in The Netherlands

endemic forms out of eleven. Those are: *Lanceolaria oxyrhyncha* (von Martens), *Unio biwae* Kobelt, *Inversidens reiniana* (Kobelt), *Hyriopsis schlegeli* (v. Martens), *Cristaria plicata clessini* (Kobelt) and *Anodonta calipygos* Kobelt. Among such bivalves, the group of *Inversidens, Hyriopsis schlegeli, Anodonta calipygos* as well as *Corbicula sandai* Reinhardt of the Corbiculidae are found in so large amounts as to have economic values to the fisheries of Lake Biwa. Of the gastropods, pond snail *Heterogen longispira* (Smith) of the Viviparidae is peculiar to the lake, where it is found on the bottom mud as deep as 30 m.

Planktonic animals and plants

Among the microscopic life, the animals consist of at least one hundred species, while the plants are known to be about ninety. Out of the two hundred animals and plants in all, fifty-two are true pelagic species, that is, twenty-five species of nannoplankton, sixty-four percent of which belongs to green algae.

Most of freshwater plankton organisms are generally regarded as cosmopolitan in their geographical distribution, having only a few forms of limited distribution. Lake Biwa, however, has several prominent peculiar species in regard to its plankton. Two diatoms, *Melosira solida* Eulenstein and *Stephanodiscus carconensis* Grunow, are restricted only to this lake in Japan, though they are recorded in certain limited districts in North America. A green alga *Pediastrum biwae* Negoro is a similar example. *Ceratium hirundinella* (O. F. Müller) is a dinoflagellate of world-wide distribution, but in Lake Biwa it occurs as a peculiar type (subspecies). *Eodiaptomus japonicus* (Burckhardt), though this copepod is found also in some other bodies of fresh water in Japan, is originally an inhabitant of Lake Biwa, as pointed out previously in regard to *Ischikauia steenackeri*, *Opsariichthys uncirostris* and two species of *Squalidus*.

It has been believed that plankton animals became evolved from littoral forms and gradually adapted to pelagic life. A pelagic species *Daphnia biwaensis* Uéno peculiar to Lake Biwa is such an example and seems to be derived from a shallow water inhabitant *Daphnia pulex* Leydig.

The diversity of habitats in Lake Biwa furnished a wide variety of ecological niches for organisms, some of which changed their habitat, as has been known in *Anisogammarus annandalei* (Tattersall). This amphipod, 10–15 mm long, is of originally benthonic species, but floats up as a plankter at midnight in the upper layer below the depth of about 20 m. A similar habit is also observed in the case of *Chaenogobius isaza*. This cold stenothermal goby of originally benthonic species migrates into the upper layer as a nekton at night during spring and early summer. Its large individuals feed on *Anisogammarus annandalei*. This goby belongs in its habit to 'bentho-nektonic' designated by the investigators of Lake Baikal, where such a phenomenon is especially conspicuous (Kozhov 1963). In that lake there live four bentho-nektonic species of the cottid fishes, that is, *Comephorus baicalensis* Pallas, *C. dybowskii* Korotnev, *Cottocomephorus grewingki* Dybowsky and *Ct. comephoroides* (Berg). Similarly, a benthoplanktonic amphipod *Macrohectopus branickii* Dybowsky occurs also in that lake and is eaten by the four cottids as food. It has a habit well resembling that of *Chaenogobius isaza* in Lake Biwa in its relation to *Anisogammarus annandalei.*

In Lake Biwa the bentho-planktonic forms are also found in the algae. Those are *Melosira solida* and *Stephanodiscus carconensis*, stated already, both of which are dominant constituents of the diatom flora of the bottom mud in summer. In a way similar to an amphipod, these algae also float up as plankters during winter and early spring. Including such microscopic plants, benthonic fishes and amphipod which have become evolved into either benthos or planktonic or nektonic habit are important indicators of the peculiarity of Lake Biwa, as has been known from Lake Baikal.

As illustrated above, the richness of the fauna and flora of the present Lake Biwa may be attributed to the presence of rich and diversified conditions which might satisfy the ecological niche requirements by organisms. These conditions are probably caused by factors, such as, the great extent and morphological features of the lake basin, the diversity of environments for life, and the lapsed time long enough to receive various kinds of organisms into the lake.

The origin and evolution of the fauna

For the purpose of discussing the origin and evolution of the fauna of Lake Biwa, it will be necessary to analyze what kinds of biogeographical elements have constituted the life in the lake, and whether or not endemism took place as to intra-lacustrine speciation or subspeciation.

In the districts like North and Middle Europe where the standards are well established to determine the glacial and interglacial periods, historical observations of the fauna and flora under consideration are comparatively easy. The case is different with Japan. As to the last period of the Quaternary in Japan, where there was glaciation on a small scale only at the summit of its high mountains, there is a void in every respect of glaciation. Japan is said to be a district of the 'Eiszeitliche Refugien' or 'Refugialgebiet' called by the German zoologist Reinig (1937). The successful results of 200 meter long core-boring below the bottom at the central part of Lake Biwa in 1971 have led us to the expectation of understanding various conditions of the Quaternary period in this 'refuge' district (Horie 1973).

The greater part of the fauna of Lake Biwa is of boreal origin and is believed to be derived from the stocks of the Eurasian freshwater fauna, together with some brackish water or marine ones. The endemic species in this lake indicated as relics of the old freshwater fauna which were once distributed widely in Eurasia before the ice age may be determined by comparative studies with their closely related forms distributed on the continent. *Opsariichthys uncirostris* (Temm. et Schl.) and *Ischikauia steenackeri* Sauvage are good examples of this kind. The close relative of the former fish is *O. uncirostris bidens* Günther (= *O. bidens* Günther) distributed in China. The latter is a fish closely allied to the Chinese genus *Xenocypris* Günther, which consists of *X. davidi* Bleeker, *X. argentea* Günther, *X. fangi* Tchung and others. Similarly to those fishes, a large bivalve *Hyriopsis cumingii* (Lea) of the Yangtsekiang is a closely allied species to *H. schlegeli* (v. Martens) of Lake Biwa.

Similarly to the fishes, the greater part of the freshwater mollusks of this lake, too, were distributed already either in the preglacial time or in the early Pleistocene at the latest. This suggestion is verifiable by the fossil mollusks found in the clay or sand of the 'Kobiwako-sō' (Paleo-Biwa Group) distributed around the present lake, because the fossil species are identical with those still surviving in the present Lake Biwa.

In most cases, endemism occurs prominently in benthonic inhabitants. This is especially conspicuous in ancient lakes like Lakes Baikal, Tanganyika and Biwa as well, where speciation took place within the lakes themselves. Some geographical barriers serving as an isolating factor, may cause the speciation of organism groups in lakes. Different environments, littoral, sublittoral or profound substrata, as well as in regard to open water (pelagic), all making geographical barriers, may have had the possibility to bring about the isolation of the population of some species into a few separate smaller populations from which geographical subspeciation or speciation took place again within the lake during the lapse of time.

In connection with intra-lacustrine speciation or subspeciation, the confirmation of breeding isolation within the lake is noticeable among the populations of an amphipod *Anisogammarus annandalei*. Narita (1976) found two groups of ecological forms of that species, inshore and offshore, in Lake Biwa. The inshore group stays persistently near shore throughout the year, while the offshore group is always restricted to the deep bottom of cold water in summer and is dispersed over every depth in winter. The inshore group breeds in spring and grows in autumn, but the offshore one breeds in autumn and is followed by growth during winter and spring.

More than a half of the forty-four molluscan species of Lake Biwa belong to the endemic species or subspecies as can be seen from the following figures.

Family	Endemic species or subspecies	Family	Endemic species or subspecies
Viviparidae	1	Succineidae	0
Valvatidae	0	Unionidae	6
Bithyniidae	0	Corbiculidae	1
Thiaridae	5	Sphaeriidae	2
Lymnaeidae	1		Total = 18
Physidae	0		
Planorbiidae	2		
Ancylidae	0		

One of the three species of the unionid bivalve genus *Inversidens* is peculiar to the lake. Out of the eight species and subspecies of the genus *Semisulcospira*, five are endemic. Those pleurocoelids are: *Semisulcospira decipiens decipiens* Westerlund), *S. d. multigranosa* Boettger, *S. d. reticulata* Kajiyama et Habe, *S. nakasekoae* Kuroda and *S. niponica* (Smith). The occurrence of two species and three subspecies belonging to the same genus (*Semisulcospira*) and inhabiting the same lake together is not an usual situation. *Radix* (*Biwakoia*) *onychia* (Westerlund), as well as *Gyraulus biwaensis* (Preston) and *G. amplificatus* (Mori) of the Planorbidae are all recognized to be endemic to the lake.

The diversity and change of their habitats within the lake existed not only on the substratum but also in the water mass. In the course of evolution the lake experienced the climatic change, during which times, corresponding to the glacial and interglacial ages, the dimictic circulation of lake water probably took place occasionally. In the present Lake Biwa, the circulation of water is warm monomictic.

Some boreal elements which had dispersed into the lake in its early days migrated into the deep cold water during the interglacial times when the lake water became warmer. They constituted the deep bottom fauna of the lake which is represented by oligochaetes and a minute snail *Valvata*. As the deep bottom inhabitants, there occur two more rare species of worms, *Ancyrobdella biwae* Oka and *Bdellocephala annandalei* Ijima et Kaburaki. The former is a strange glossiphonid leech, 17 mm long, found at a depth of 80 m. The latter is a large dendrocoelid turbellarian, 40–50 mm long, living in the bottom mud deeper than 40 m.

The significance of the pattern of dispersal from marine or brackish waters will be a special subject for discussion. From the geological and topographical points of view, Takaya (1963) has suggested that the dispersal of marine animals into Lake Biwa was confined only to the earliest Pleistocene. One of the examples of such immigrants is a stickleback *Gasterosteus aculeatus* (Linnaeus) which is a widely distributed small fish in the Holarctic Region. In a number of cold springs (rheocrenes) distributed close to the east coast of Lake Biwa, there lives a land-locked form, *G. aculeatus microcephalus* (Girard) (sometimes designated as *G. aculeatus* f. *gymnurus*) whose area of distribution extends to the bodies of water of the surrounding districts of the lake in Shiga, Gifu and Mié Prefectures. The original species *G. aculeatus* which is known as one of the preglacial elements was distributed in Japan already in the early Pleistocene, during which time it differentiated into a peculiar freshwater form without its catadromous habit. It is also notable to find the occurrence of a small lacustrine race of an osmerid *Plecoglossus altivelis* Temm. et Schl., which came to the lake in the early Pleistocene and lost its catadromous habit (*vide* Kawanabe 1975). A similar case is observed with the lacustrine race of a trout *Oncorhynchus rhodurus* Jordan et McGregor surviving in Lake Biwa, though it has been transplanted into a few lakes in Japan.

There is one more noticeable marine element in the lake. That is *Kamaka biwae* Uéno of boreal origin (Fig. 281). This small and strange-shaped amphipod belonging to the marine family Corophiidae was obtained from the shallow bottom of Lake Biwa in 1942 (Uéno 1943). Its nearest relative is *Kamaka kuthae* Dershavin which is distributed in the bodies of brackish or fresh water on the coast around the Okhotsk Sea (Kamchatka, the Kuriles, Sakhalin and Hokkaido). The occurrence in Lake Biwa of such a glacial-marine relic is of great zoogeographical interest.

A similar view will also be applied to *Gasterosteus*, *Anisogammarus* and *Corbicula sandai*, all of which came to be inhabitants of Lake Biwa in the early Pleistocene. The last-named dioecious bivalve which had been a brackish water inhabitant became adapted to the freshwater habitat of Lake Biwa during the ice age, and came to change drastically even its reproductive process in omitting its plankton larval stage. Its fossils were found after the period corresponding to the Günz-Mindel interglacial age designated in the European Alps (cit. Kawanabe 1975).

Biogeography of microfossils

The materials useful for elucidating the lacustrine biogeography are the microfossils of autochthonous origin. The 200 m long core recently obtained by

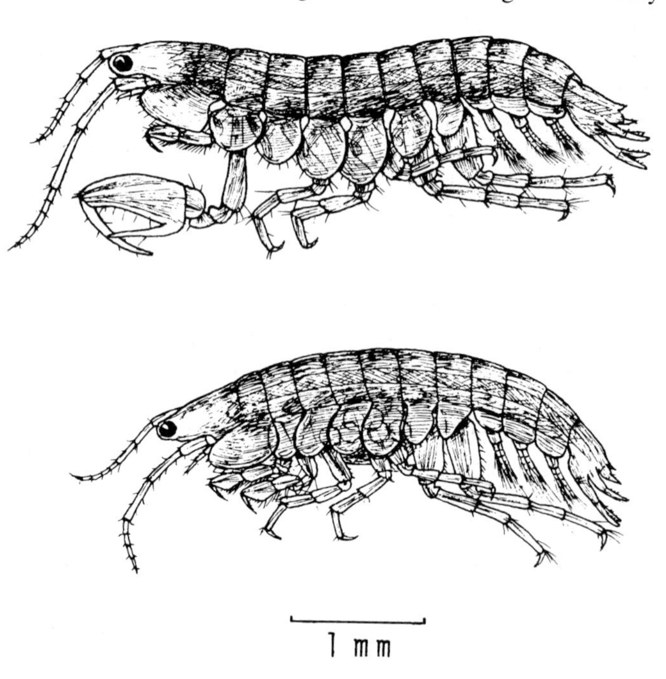

Fig. 281. Kamaka biwae (Uéno 1943). Upper: ♂ Lower: ♀

boring from the deep bottom of Lake Biwa has supplied important data which verify the chronological history of the fauna and flora of Lake Biwa over the past half million years.

The most helpful microfossils for the purpose mentioned above are the remains of Cladocera shells and fragments of chitinous body parts of chironomid larvae. Cladocera are roughly classified ecologically into two kinds, littoral and planktonic inhabitants. Kadota (1973, 1974) who made microfossil analysis of the core found that shells of *Bosmina* and *Leptodora* as planktonic and chydorids as littoral. *Bosmina longirostris* (O. F. Müller) and *Leptodora kindtii* (Focke) are the main components of the zooplankton in the present Lake Biwa during the warm seasons.

In the core stated *Bosmina* occurred from the depth of 70 m upward, while *Leptodora* was first determined in a much shallower part (35 m deep) of the core. The fossil Chydoridae first occurs in the part of the core, 175 m deep, and intermittently continuing upward to 10 m deep. At least seven genera, that is, *Alona*, *Alonella*, *Camptocercus*, *Graptoleberis*, *Monospilus* and *Oxyurella*, found in the core were identified with those living in the present lake. Among those chydorids, *Monospilus* is the leading genus found throughout the 200 m core. Since the shells and post-abdominal claws were possibly carried away by a water current from the littoral to the pelagic region, the occurrence of the remains of the Chydoridae in the core taken at the center of the lake will not be unnatural. Kadota (1976) later found the post-abdominal claws of *Daphnia* from the parts of the core shallower than 53 m deep. He distinguished two kinds of claws of that genus, the *longispina*-type without large basal pecten and the *pulex*-type with distinct basal pecten. In the present Lake Biwa, there are found three species of the genus *Daphnia*, that is, *D. hyalina* (Leydig) and *D. galeata* G. O. Sars of the former type of claws, and *D. biwaensis* Uéno of the latter type. Of these three species, *D. hyalina* and *D. biwaensis* are true pelagic inhabitants, but *D. galeata* occurs more abundantly in the shallow littoral region during the cold seasons. *Sida*, the other littoral form, was found in the upper parts of the core shallower than 85 m deep. The results obtained by Kadota suggest that the Chydoridae established their inhabitation in the lake already in the glacial time, and that *Bosmina* and *Daphnia*, too, joined the faunal components of the lake sometime during the Pleistocene.

Kadota (1976) pointed out also that there were found Turbellaria represented by its egg capsules at every layer throughout the core, but he has given no taxonomic distinction.

It is difficult to understand that none of the silicic spicules of the freshwater sponges was present at all throughout the 200 m long core. Kadota (1974, 1976) quite failed to determine even fragmentary spicules of any sponge species. In the present Lake Biwa there are found four species of sponges, that is, *Eunapius fragilis* (Leidy), *Spongilla lacustris* (Linnaeus), *Radiospongilla cerebellata* (Bowerbank) and *Stratospongilla clementis* (Annandale). They are prominent components of the littoral fauna of Lake Biwa, occurring not only widely but also in abundant quantities in various habitats of shallow water.

It is worthy of notice that chironomid larvae occurred even in the deepest part of the core (197.2 m deep) and were found continuously as upward as nearly the uppermost sediment. If the taxa of those chironomids could be determined, it may be possible to appreciate the trophic conditions of this lake at any time in the past, in the same way as modern limnologists set up the classification of lakes into oligotrophic or eutrophic types or systems. Either the cladoceran shells or the chitinous body parts of chironomids are difficult to decompose and remain permanently in the bottom sediments, as undecomposed silicic shells of diatoms, which will be dealt with below.

Diatoms are important fossils like Cladocera, because the group is not only considerably rich in their taxa, but also remarkable in the varieties of their ecological requirements for habitats. By his study of the 200 m long core, Shinobu Mori (1974) found two hundreds taxa of diatoms, among which *Melosira solida* Eulenstein and *Stephanodiscus* sp. were dominant in the parts of the core (197.2 m to 177.5 m deep). These species reminds us of the diatom flora in the sediments of the Paleo-Biwa Group. The former species, *M. solida* occurs discontinuously upward in the core and it is found even in the uppermost part, where it still survives as a representative of the benthonic diatom flora of the present Lake Biwa, as stated already. In the part of the core (100 m deep) *Stephanodiscus carconensis* was found to be dominant, its upward sequence reaching the present deposit in which it still survives.

So far as our observations are concerned, it is suspected that the life in Lake Biwa seems to keep a balance between the number of species and the population density of the respective species. Such a stability of life in the lake has gradually been established and probably kept for very many years. The 200 m long core recently obtained by boring from the deep bottom of the lake has supplied important data which verify the stability or constancy of the fauna and flora of this lake over the past half million years. The remains of autochthonous organisms obtained in the core are indispensable to the reconstruction of a lake history together with the analysis of allochthonous remains, especially the pollen derived from the districts around the past lake (Fuji 1976a, 1976b).

References

Fuji, N., 1976a. Palaeoclimatic and palaeovegetational changes around Lake Biwa, Central Japan during the past 100,000 years. Paleolimnology of Lake Biwa and the Japanese Pleistocene, 4: ed. S. Horie, 316-356.

Fuji, N., 1976b. Palynological investigation on a 200-meter core sample from Lake Biwa in Central Japan. Paleolimnology of Lake Biwa and the Japanese Pleistocene, 4, ed. S. Horie, 357-421.

Horie, S., 1973. An outline of the paleolimnological works of Lake Biwa-ko, Japan. Jap. Jour. Limnol. 34: 49-54.

Kadota, S., 1974. A quantitative study of microfossils in a 200-meter core sample from Lake Biwa. Paleolimnology of Lake Biwa and the Japanese Pleistocene, 2, ed. S. Horie, 236-245.

Kadota, S., 1976. A quantitative study of the microfossils in a 200-meter-long core sample from Lake Biwa. Paleolimnology of Lake Biwa and the Japanese Pleistocene, 4, ed. S. Horie, 236-245.

Kawanabe, H., 1975. On the origin of Aya-Fish (Pisces, Osmeridae) in Lake Biwa (Preliminary notes). Paleolimnology of Lake Biwa and the Japanese Pleistocene, 3, ed. S. Horie, 317-320.

Kozhov, M., 1963, Lake Baikal and its Life. The Hague, Dr. W. Junk, 344 pp.

Mori, Sh., 1974. Diatom succession in a core from Lake Biwa. Paleolimnology of Lake Biwa and the Japanese Pleistocene, 2, ed. S. Horie, 247-254.

Narita, T., 1976. Occurrence of two ecological forms of *Anisogammarus annandalei* (Tattersall) (Crustacea: Amphipoda) in Lake Biwa. Physiol. Ecol. 17: 551-556.

Reinig, W. F., 1937. Die Holarktis. – Ein Beitrag zur diluvialen und alluvialen Geschichte der zirkumpolaren Faunen- und Florengebiete. Jena, Gustav Fischer, 124 SS.

Takaya, Y., 1963. Stratigraphy of the paleo-Biwa group and the paleogeography of Lake Biwa with special reference to the origin of the endemic species in Lake Biwa. Mem. Coll. Sci., Univ. Kyoto, Ser. B, 30: 81-119.

Uéno, M., 1943. *Kamaka biwae*, a new amphipod of marine derivative found in Lake Biwa. Bull. Biogeogr. Soc. Japan 13: 139-143.

The influence of human activities on Lake Biwa

– The influence of the great public works –

Setsuo Okuda

Introduction

The history of human activities around Lake Biwa is very old, and the remains show that men had lived on the shore since the early part of the Jōmon era (B.C. 4,000).

As for agriculture, the remains show that rice-field cultivation had already been performed on the shore in the Yayoi era, and the marks of Jōri-sei which can be observed in many places on the shore also show that arable land had been adjusted in the age when the ancient Ritsu-Ryō-sei was enforced (about 645 A.D.). Because of good rice crops under the blessed climate and the abundant water supply, the farm land around Lake Biwa went on expanding. Particularly after the drop in the water level of the lake due to the dredging work of the River Seta to be noted later, the farm land was expanded as far as the zone close to the lake, and a great deal of land was reclaimed.

As for fishery, various kinds of lacustrine products were important near Kyoto where marine products were hard to obtain, so that many types of fishing methods have been developed since older times and many fishing ports have been built. Especially in recent years, the transportation of young Ayu (sweet fish) to other parts of Japan and the culture of fresh-water pearls are the staples of fishery.

Water transportation on the lake had played an important part as a section of main traffic facilities that link Hokuriku (the northern region of Lake Biwa) with Kyoto. After the opening of the railway, however, it declined, and is now mainly used as a sight-seeing line.

As for industry, the textile industry introduced in the Meiji era (1868–1912), declined during World War II. After the War, with the development of traffic facilities on land, the inner-land type of industry has been advanced centering in the south of the lake.

Recent rapid development in commerce and industry and the extension of the commuting range to Kyoto and Osaka led to a rapid increase in population and the conversion of farm land to housing sites in the south of the lake.

Horie, S. Lake Biwa

ISBN 90 6193 095 2. Printed in The Netherlands

Thus, for long time, various types of human activities have been actively pursued around Lake Biwa. Particularly of late, the increase of activities that are deeply concerned with the water use of Lake Biwa have resulted in the notable influence of human activities upon the lake. Of these influences, those on water quality and on the ecological system were already mentioned in other sections. In this section, the influences of modern large-scale public works, namely, the dredging works of the River Seta and the reclaiming or filling-up projects of the shore, on the physical circumstances of the lake will be noted.

Historical progress of the dredging works in the River Seta and their influence on the water level of Lake Biwa

Although many rivers flow into Lake Biwa, natural outlet is limited only to the River Seta. As a result, the water level rises after a heavy rain or a thaw and it does not fall for a long period of time, causing severe damages on the shore.

Inhabitants on the shore desired to have the River Seta dredged and to increase the amount of water outflow in order to reduce the damage, but Tokugawa Shogunate (the feudal government) did not permit the dredging works for the following reasons: First, for military purposes, the government wanted to secure a wading point in order to let the army into Kyoto from the eastern part of Japan and did not want to reduce the water level for the sake of defending the Hikone castle and the Zeze castle. Secondly, economically, the government was afraid of a sand movement towards downstream because of its wishes to protect Osaka, where merchant princes were powerful, from a flood disaster and to maintain the water transportation of the lower part of the River Yodo. Actually, the technique of those days could not keep the dredging effective for a long time, and it was impossible to obtain a united cooperation of more than fifty possessions on the shore. Hence, in spite of frequent flood disasters, transient small-scale dredging works were carried out only six times in the Tokugawa Era (1603–1867).

In the Meiji era, the improvement plans of the River Yodo were supervised by foreign engineers. But as a result of the great floods in 1885 and in 1889, inhabitants on the shore strongly cried for the dredging work, so that the dredging was permitted under such a condition that it had no bad influence on the lower part, and the plan was undertaken on the basis of a careful hydrological investigation that was novel in those days.

As a result of the works from 1900 to 1908, the Nangō-Arai Dam (with running water 3.64 m wide and with 32 gates – also as a control gate of the amount of the flowing water — was finished, dredging the River Seta by 4.4×10^4 m^3, and scraping Mt. Dainichi by 5.5×10^4 m^3. Fig. 282 is a plane figure that shows the river and the plan before the works. Plate 32 is an aerial photograph of the same area in 1975.

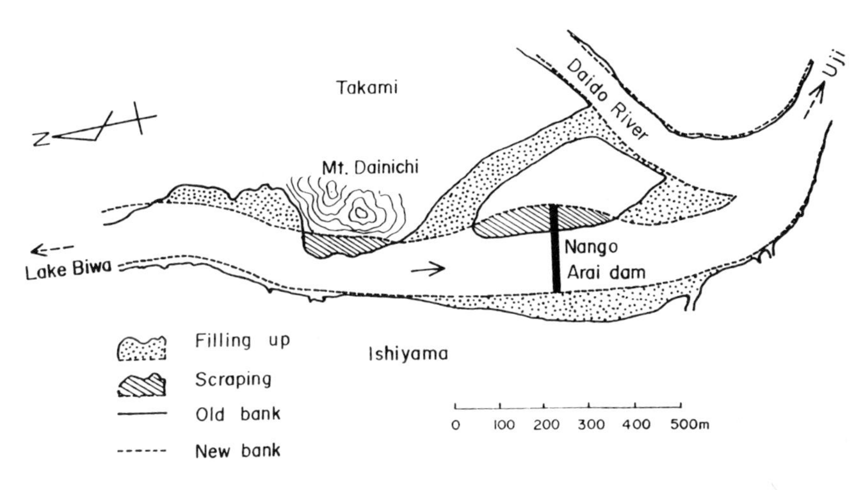

Fig. 282. Old and new river channels before and after dredging at River Seta (after Kensetsu-shō, Kinki Chihō Kensetsu-Kyoku 1974).

Plate 32. Aerial photograph of the area of Fig. 282 taken in 1975, offered by Asia Air Survey.

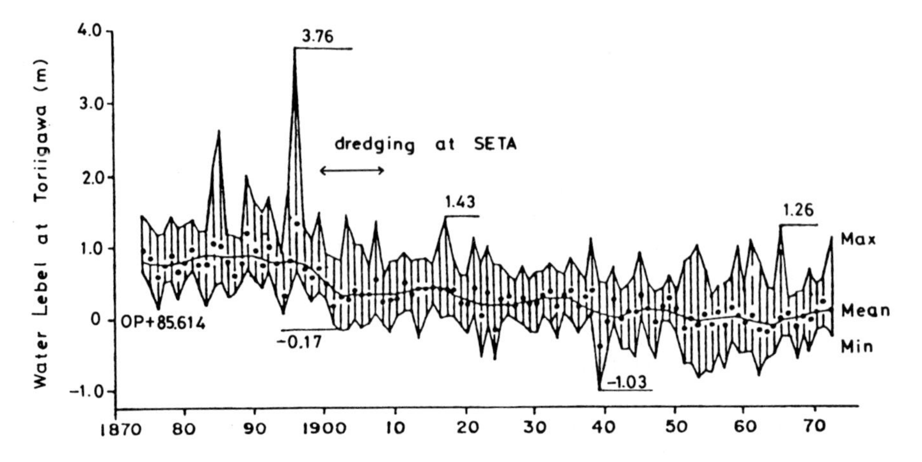

Fig. 283. Long-term change in the water level of Lake Biwa (after Kensetsu-shō, Kinki Chihō Kensetsu-Kyoku 1974).

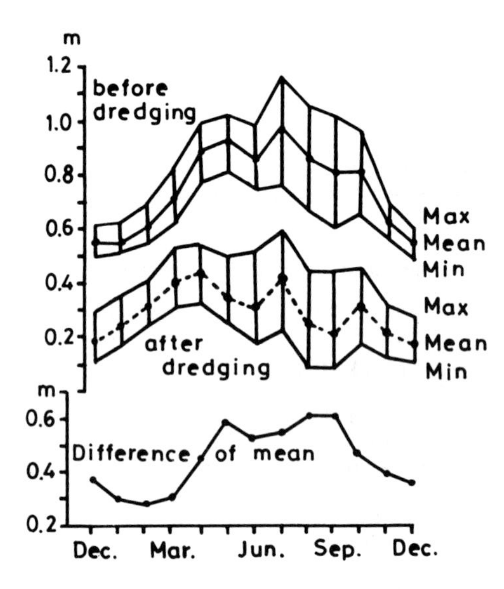

Fig. 284. Seasonal variations in the water level of Lake Biwa (after Kensetsu-shō, Kinki Chihō Kensetsu-Kyoku 1974).

The dredging lowered the hydraulic height of the river bed by about 1 meter so that the average water level of Lake Biwa dropped by about 0.5 m as shown in Fig. 283. As for the seasonal effects, it dropped about 0.6 m in summer (the rainy season) and about 0.3 m in winter (the dry season) as shown in Fig. 284.

The water level of the lake, of course, depends on the natural inflow of water and the artificial discharge. Still more, the improvement of the River Seta is proceeded and the water demand of the lower part increases, so that the average level continues to fall now as shown in Fig. 283.

Although the level dropped, the flooded area of the arable land on the shore is not reduced so much because the arable land was extended as far as the lower area on the shore as the level was stabilized. On the other hand, though the Arai Dam enabled control of the amount of the water discharge, interests often clash between the upper and lower sides in case of a flood or of a drought concerning the water level of Lake Biwa (upper side) and the amount of the flowing water of the River Yodo (lower side). Conflicts on the rules of the dam control still remain and an all-round solution containing another kind of a counter plan is desirable in the future.

Reclaiming and filling-up in Lake Biwa and in its surroundings

As the average water level is lowered by the dredging of the River Seta as mentioned above and the danger of the flood is lessened, the arable land has been extended as far as the lower shore. Particularly after World War II, when an increased yield of food was strongly demanded, the nation and the prefecture of Shiga promoted the reclaiming works of the shallow part of the shore and the surrounding inland lake in order to create rice fields on a large scale. Moreover in the 1950s, during rapid economic growth, the prefecture of Shiga and the city of Ōtsu filled up the shore to extend the city area and to create a recreation area. Main reclaiming or filling-up projects are listed in chronological order in Table 80. Their locations are shown in Fig. 285.

Table 80. Polder and reclaimed lands around Lake Biwa (limited to those larger than 10 ha)

Location No*		Area (ha)	Constructed in
1	Sho Naka-no-ko Pold.	342.2	1947
2	Irie Pold.	305.2	1947
3	Suikei Naiko Pold.	201.3	1947
4	Matsubara Pold.	73.7	1947
5	Nodanuma Pold.	43.8	1951
6	Yotsugawa Naiko Pold.	19.7	1951
7	Shiozu Naiko Pold.	16.9	1951
8	Nukekawa Naiko Pold.	12.5	1951
9	Setaura Recl.	31.9	1963
10	Shiozu Shaba Naiko Pold.	13.6	1963
11	Konohama Recl.	124.3	1966
12	Dai Naka-no-ko Pold.	1127.7	1968
13	Sone-numa Pold.	87.0	1968
14	Ōtsu Kogan Recl.	41.9	1968
15	Tsuda Naiko Pold.	121.5	1971
16	Hayasaki Pold.	91.9	1971
17	Off Yabase Recl.	73.0	1977

*The number corresponds to the location given in Fig. 285.

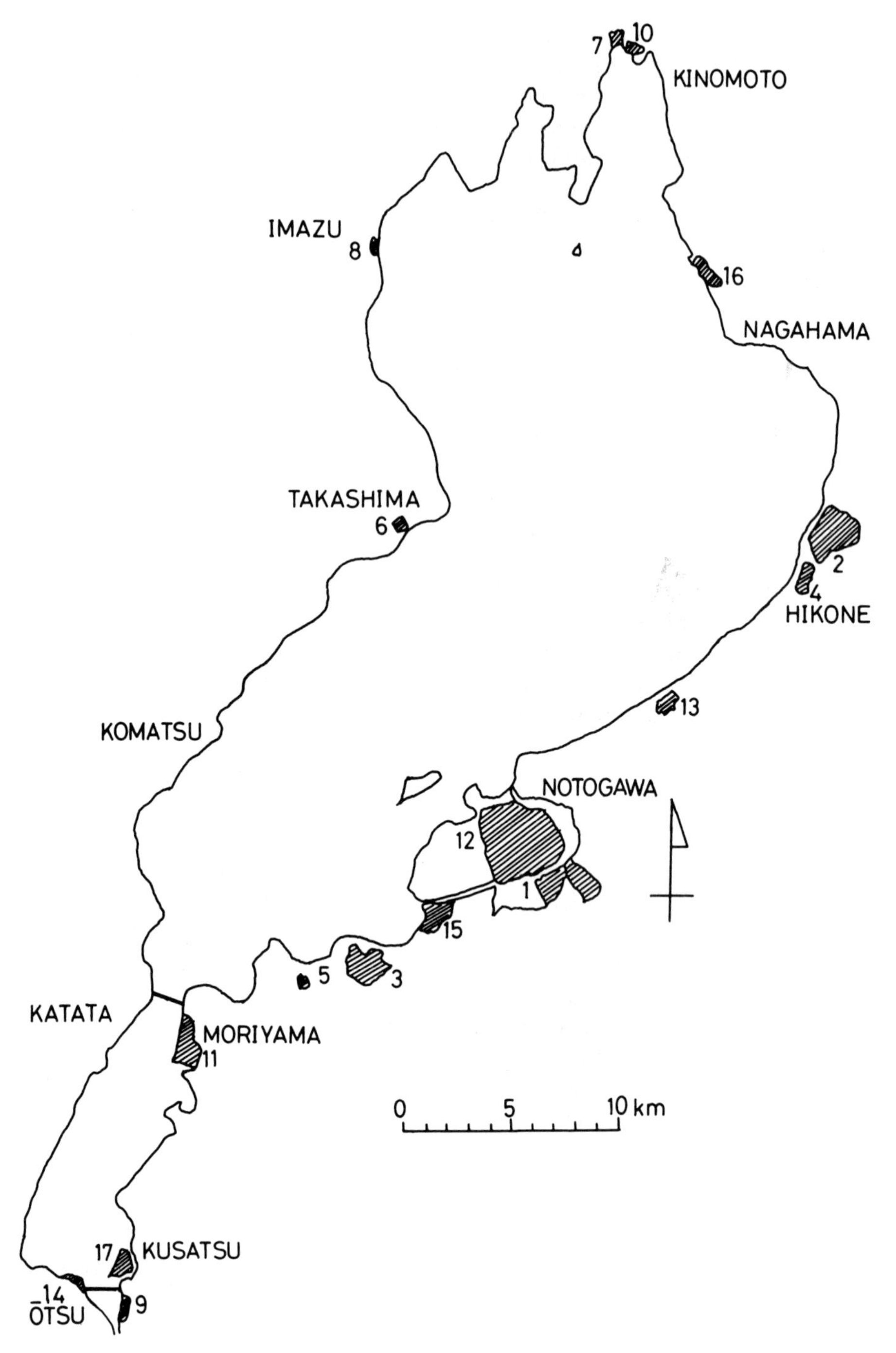

Fig. 285. Location of polder and reclaimed lands around Lake Biwa (numbers in the figure correspond with those in Table 80).

No less than 27 km^2 of the water area — nearly 4 per cent of Lake Biwa — was filled up in a very short time. This has exerted various influences on the environment of the lake. First of all, an investigation of water quality shows that polluted water and suspended soil flow out a long way carried on the circulatory current or the density stratified flow in the lake. Hydrologically, the reduction of the total water area by 4 per cent seems to have no appreciable influence on the change of the water level. But as the shallow water area which was filled up has high biological productivity and powerful self-purifying action, chemists and biologists assert that the reduction of the shallow water area would accelerate the deterioration of water quality and inflict a great loss on fishery.

In this section, the influence of the latest large-scale public works on the lake has been noted. Needless to say, the development of industry and the increase in population around the lake inevitably have a deleterious influence on the environment of the lake, especially on the deterioration of water quality.

Now, various human activities have come to impinge upon every circumstance of Lake Biwa. To maintain the excellent natural environment of Lake Biwa, to recover the quality in the polluted water, to promote the use of water resources, and to improve the welfare of the inhabitants concerned, the government has planned an all-round program of development of the lake and is making an effort to adopt it as a state enterprise. Nevertheless, with regard to individual specific problems, many points remain to be resolved in co-operation with the scientists and the inhabitants. Scientific investigations and the establishment of rational measures based thereon are strongly desired.

Addenda

*the Jōmon era — 6,000–2,000 years ago, the primitive times in Japan, when men had used the Jōmon-type earthenware.

*the Yayoi era — 2,000–1,500 years ago, the primitive times following the Jōmon era, when men had used the Yayoi-type earthenware.

*Jōri-sei — the ancient way of land division, with streets laid out in checkerboard squares.

*Ritsu-Ryō-sei — a system based on Ritsu-Ryō. Ritsu is a kind of the criminal law, Ryō the civil law.

Reference

Kensetsu-shō, Kinki Chihō Kensetsu-Kyoku, 1974. Yodogawa Hyakunen-shi. Osaka, 1822 pp.

Index

A

B

C

L

M

N

O

Q

R

S

T

Z